More information about this series at http://www.springer.com/series/7409

Lecture Notes in Computer Science 10710

Commenced Publication in 1973
Founding and Former Series Editors:
Gerhard Goos, Juris Hartmanis, and Jan van Leeuwen

Giuseppe Nicosia · Panos Pardalos
Giovanni Giuffrida · Renato Umeton (Eds.)

Machine Learning, Optimization, and Big Data

Third International Conference, MOD 2017
Volterra, Italy, September 14–17, 2017
Revised Selected Papers

 Springer

Editors
Giuseppe Nicosia
University of Catania
Catania
Italy

Panos Pardalos
University of Florida
Gainesville, FL
USA

Giovanni Giuffrida
University of Catania
Catania
Italy

Renato Umeton
Harvard University
Cambridge, MA
USA

ISSN 0302-9743   ISSN 1611-3349   (electronic)
Lecture Notes in Computer Science
ISBN 978-3-319-72925-1   ISBN 978-3-319-72926-8   (eBook)
https://doi.org/10.1007/978-3-319-72926-8

Library of Congress Control Number: 2017962876

LNCS Sublibrary: SL3 – Information Systems and Applications, incl. Internet/Web, and HCI

Printed on acid-free paper

This Springer imprint is published by Springer Nature
The registered company is Springer International Publishing AG
The registered company address is: Gewerbestrasse 11, 6330 Cham, Switzerland

Preface

MOD is an international conference embracing the fields of machine learning, optimization, and data science. The third edition, MOD 2017, was organized during September 14–17, 2017 in Volterra (Pisa, Italy), a stunning medieval town dominating the picturesque countryside of Tuscany.

The key role of machine learning, reinforcement learning, artificial intelligence, large-scale optimization, and big data for developing solutions to some of the greatest challenges we are facing is undeniable. MOD 2017 attracted leading experts from the academic world and industry with the aim of strengthening the connection between these institutions. The 2017 edition of MOD represented a great opportunity for professors, scientists, industry experts, and postgraduate students to learn about recent developments in their own research areas and to learn about research in contiguous research areas, with the aim of creating an environment to share ideas and trigger new collaborations.

As chairs, it was an honor to organize a premiere conference in these areas and to have received a large variety of innovative and original scientific contributions.

During this edition, six plenary lectures were presented:

Yi-Ke Guo, Department of Computing, Faculty of Engineering, Imperial College London, UK. Founding Director of Data Science Institute

Panos Pardalos, Department of Systems Engineering, University of Florida, USA. Director of the Center for Applied Optimization

Ruslan Salakhutdinov, Machine Learning Department, School of Computer Science at Carnegie Mellon University, USA. Director of AI Research at Apple

My Thai, Department of Computer and Information Science and Engineering, University of Florida, USA

Jun Pei, Hefei University of Technology, China

Vincenzo Sciacca, Cloud and Cognitive Division – IBM Rome, Italy

There were also two tutorial speakers:

Domenico Talia, Dipartimento di Ingegneria Informatica, Modellistica, Elettronica e Sistemistica Università della Calabria, Italy

Xin–She Yang, School of Science and Technology – Middlesex University London, UK

Moreover, the conference hosted the second edition of the industrial session on "Machine Learning, Optimization and Data Science for Real-World Applications":

Luca Maria Aiello, Nokia Bell Labs, UK

Pierpaolo Basile, University of Bari, Italy

Carlos Castillo, Universitat Pompeu Fabra in Barcelona, Spain

Moderator: Aris Anagnostopoulos, Sapienza University of Rome, Italy

We received 126 submissions from 46 countries and five continents; each manuscript was independently reviewed by a committee formed by at least five members through a blind review process. These proceedings contain 49 research articles written by leading scientists in the fields of machine learning, artificial intelligence, reinforcement learning, computational optimization, and data science presenting a substantial array of ideas, technologies, algorithms, methods, and applications.

For MOD 2017, Springer generously sponsored the MOD Best Paper Award. This year, the paper by Khaled Sayed, Cheryl Telmer, Adam Butchy, and Natasa Miskov-Zivanov titled "Recipes for Translating Big Data Machine Reading to Executable Cellular Signaling Models" received the MOD Best Paper Award.

This conference could not have been organized without the contributions of these researchers, and so we thank them all for participating. A sincere thank you also goes to all the Program Committee, formed by more than 300 scientists from academia and industry, for their valuable work of selecting the scientific contributions.

Finally, we would like to express our appreciation to the keynote speakers, tutorial speakers, and the industrial panel who accepted our invitation, and to all the authors who submitted their research papers to MOD 2017.

September 2017

Giuseppe Nicosia
Panos Pardalos
Giovanni Giuffrida
Renato Umeton

Organization

General Chair

Renato Umeton   Harvard University, USA

Conference and Technical Program Committee Co-chairs

Giuseppe Nicosia   University of Catania, Italy and University of Reading, UK
Panos Pardalos   University of Florida, USA
Giovanni Giuffrida   University of Catania, Italy

Tutorial Chair

Giuseppe Narzisi   New York University Tandon School of Engineering, USA

Industrial Session Chairs

Ilaria Bordino   UniCredit R&D, Italy
Marco Firrincieli   UniCredit R&D, Italy
Fabio Fumarola   UniCredit R&D, Italy
Francesco Gullo   UniCredit R&D, Italy

Organizing Committee

Piero Conca   CNR, Italy
Jole Costanza   Italian Institute of Technology, Milan, Italy
Giorgio Jansen   University of Catania, Italy
Giuseppe Narzisi   New York University Tandon School of Engineering, USA
Andrea Patane'   University of Oxford, UK
Andrea Santoro   Queen Mary University London, UK
Renato Umeton   Harvard University, USA

Technical Program Committee

Agostinho Agra   Universidade de Aveiro, Portugal
Kerem Akartunali   University of Strathclyde, UK
Richard Allmendinger   The University of Manchester, UK
Aris Anagnostopoulos   Università di Roma La Sapienza, Italy
Davide Anguita   University of Genoa, Italy

Takaya Arita	Nagoya University, Japan
Jason Atkin	The University of Nottingham, UK
Chloe-Agathe Azencott	Institut Curie Research Centre, Paris, France
Jaume Bacardit	Newcastle University, UK
James Bailey	University of Melbourne, Australia
Baski Balasundaram	Oklahoma State University, USA
Elena Baralis	Politecnico di Torino, Italy
Xabier E. Barandiaran	University of the Basque Country, Spain
Cristobal Barba-Gonzalez	University of Malaga, Spain
Helio J. C. Barbosa	Laboratório Nacional de Computacao Cientifica, Brazil
Roberto Battiti	University of Trento, Italy
Lucia Beccai	Istituto Italiano di Tecnologia, Italy
Aurelien Bellet	Inria Lille, France
Gerardo Beni	University of California at Riverside, USA
Khaled Benkrid	The University of Edinburgh, UK
Peter Bentley	University College London, UK
Katie Bentley	Harvard Medical School, USA
Heder Bernardino	Universidade Federal de Juiz de Fora, Brazil
Daniel Berrar	Tokyo Institute of Technology, Japan
Adam Berry	CSIRO, Australia
Luc Berthouze	University of Sussex, UK
Martin Berzins	SCI Institute, University of Utah, USA
Mauro Birattari	IRIDIA, Université Libre de Bruxelles, Belgium
Leonidas Bleris	University of Texas at Dallas, USA
Christian Blum	Spanish National Research Council, Spain
Paul Bourgine	École Polytechnique Paris, France
Anthony Brabazon	University College Dublin, Ireland
Paulo Branco	Instituto Superior Tecnico, Portugal
Juergen Branke	University of Warwick, UK
Larry Bull	University of the West of England, UK
Tadeusz Burczynski	Polish Academy of Sciences, Poland
Robert Busa-Fekete	Yahoo! Research, NY, USA
Sergiy I Butenko	Texas A&M University, USA
Stefano Cagnoni	University of Parma, Italy
Yizhi Cai	University of Edinburgh, UK
Guido Caldarelli	IMT Lucca, Italy
Alexandre Campo	Université Libre de Bruxelles, Belgium
Angelo Cangelosi	University of Plymouth, UK
Salvador Eugenio Caoili	University of the Philippines Manila, Philippines
Timoteo Carletti	University of Namur, Belgium
Jonathan Carlson	Microsoft Research, USA
Celso Carneiro Ribeiro	Universidade Federal Fluminense, Brazil
Michelangelo Ceci	University of Bari, Italy
Adelaide Cerveira	Universidade de Tras-os-Montes e Alto Douro, Portugal
Uday Chakraborty	University of Missouri – St. Louis, USA

Xu Chang	University of Sydney, Australia
W. Art Chaovalitwongse	University of Washington, USA
Antonio Chella	Università di Palermo, Italy
Ying-Ping Chen	National Chiao Tung University, Taiwan
Haifeng Chen	NEC Labs, USA
Keke Chen	Wright State University, USA
Gregory Chirikjian	Johns Hopkins University, USA
Silvia Chiusano	Politecnico di Torino, Italy
Miroslav Chlebik	University of Sussex, UK
Sung-Bae Cho Yonsei	University, South Korea
Anders Christensen	Lisbon University Institute, Portugal
Dominique Chu	University of Kent, UK
Philippe Codognet	University Pierre and Marie Curie – Paris 6, France
Carlos Coello Coello	CINVESTAV-IPN, Mexico
George Coghill	University of Aberdeen, UK
Pietro Colombo	University of Insubria, Italy
David Cornforth	University of Newcastle, UK
Luís Correia	University of Lisbon, Portugal
Chiara Damiani	University of Milan-Bicocca, Italy
Thomas Dandekar	University of Würzburg, Germany
Ivan Luciano Danesi	Unicredit Bank, Italy
Christian Darabos	Dartmouth College, USA
Kalyanmoy Deb	Michigan State University, USA
Nicoletta Del Buono	University of Bari, Italy
Jordi Delgado	Universitat Politecnica de Catalunya, Spain
Ralf Der	MPG, Germany
Clarisse Dhaenens	Université Lille, France
Barbara Di Camillo	University of Padua, Italy
Gianni Di Caro	IDSIA, Switzerland
Luigi Di Caro	University of Turin, Italy
Luca Di Gaspero	University of Udine, Italy
Peter Dittrich	Friedrich Schiller University of Jena, Germany
Federico Divina	Pablo de Olavide University of Seville, Spain
Stephan Doerfel	Kassel University, Germany
Devdatt Dubhashi	Chalmers University, Sweden
George Dulikravich	Florida International University, USA
Juan J. Durillo	University of Innsbruck, Austria
Omer Dushek	University of Oxford, UK
Marc Ebner	Ernst-Moritz-Arndt-Universität Greifswald, Germany
Pascale Ehrenfreund	The George Washington University, USA
Gusz Eiben	VU Amsterdam, The Netherlands
Aniko Ekart	Aston University, UK
Talbi El-Ghazali	University of Lille, France
Michael Elberfeld	RWTH Aachen University, Germany
Michael T. M. Emmerich	Leiden University, The Netherlands
Andries Engelbrecht	University of Pretoria, South Africa

Geir Hasle	SINTEF ICT, Norway
Carlos Henggeler Antunes	University of Coimbra, Portugal
Francisco Herrera	University of Granada, Spain
Arjen Hommersom	Radboud University, The Netherlands
Vasant Honavar	Pennsylvania State University, USA
Fabrice Huet	University of Nice Sophia Antipolis, France
Hiroyuki Iizuka	Hokkaido University, Japan
Takashi Ikegami	University of Tokyo, Japan
Bordino Ilaria	Unicredit Bank, Italy
Hisao Ishibuchi	Osaka Prefecture University, Japan
Peter Jacko	Lancaster University Management School, UK
Christian Jacob	University of Calgary, Canada
Yaochu Jin	University of Surrey, UK
Colin Johnson	University of Kent, UK
Gareth Jones	Dublin City University, Ireland
Laetitia Jourdan	Inria/LIFL/CNRS, France
Narendra Jussien	Ecole des Mines de Nantes/LINA, France
Janusz Kacprzyk	Polish Academy of Sciences, Poland
Theodore Kalamboukis	Athens University of Economics and Business, Greece
George Kampis	Eotvos University, Hungary
Dervis Karaboga	Erciyes University, Turkey
George Karakostas	McMaster University, Canada
Istvan Karsai	ETSU, USA
Jozef Kelemen	Silesian University, Czech Republic
Graham Kendall	Nottingham University, UK
Didier Keymeulen	NASA – Jet Propulsion Laboratory, USA
Daeeun Kim	Yonsei University, South Korea
Zeynep Kiziltan	University of Bologna, Italy
Georg Krempl	University of Magdeburg, Germany
Erhun Kundakcioglu	Ozyegin University, Turkey
Renaud Lambiotte	University of Namur, Belgium
Doron Lancet	Weizmann Institute of Science, Israel
Pier Luca Lanzi	Politecnico di Milano, Italy
Sanja Lazarova-Molnar	University of Southern Denmark, Denmark
Doheon Lee	KAIST, South Korea
Jay Lee	Center for Intelligent Maintenance Systems – UC, USA
Eva K. Lee	Georgia Tech, USA
Tom Lenaerts	Université Libre de Bruxelles, Belgium
Rafael Leon	Universidad Politecnica de Madrid, Spain
Shuai Li	Cambridge University, UK
Lei Li	Florida International University, USA
Xiaodong Li	RMIT University, Australia
Joseph Lizier	The University of Sydney, Australia
Giosue' Lo Bosco	Università di Palermo, Italy
Daniel Lobo	University of Maryland Baltimore County, USA
Fernando Lobo	University of Algarve, Portugal

Daniele Loiacono	Politecnico di Milano, Italy
Jose A. Lozano	University of the Basque Country, Spain
Paul Lu	University of Alberta, Canada
Angelo Lucia	University of Rhode Island, USA
Dario Maggiorini	University of Milan, Italy
Gilvan Maia	Universidade Federal do Cear, Brazil
Donato Malerba	University of Bari, Italy
Lina Mallozzi	University of Naples Federico II, Italy
Jacek Mandziuk	Warsaw University of Technology, Poland
Vittorio Maniezzo	University of Bologna, Italy
Marco Maratea	University of Genoa, Italy
Elena Marchiori	Radboud University, The Netherlands
Tiziana Margaria	University of Limerick and Lero, Ireland
Omer Markovitch	University of Groningen, The Netherlands
Carlos Martin-Vide	Rovira i Virgili University, Spain
Dominique Martinez	LORIA, France
Matteo Matteucci	Politecnico di Milano, Italy
Giancarlo Mauri	University of Milan-Bicocca, Italy
Mirjana Mazuran	Politecnico di Milano, Italy
Suzanne McIntosh	NYU Courant Institute, and Cloudera Inc., USA
Peter Mcowan	Queen Mary University, UK
Gabor Melli	Sony Interactive Entertainment Inc., Japan
Jose Fernando Mendes	University of Aveiro, Portugal
David Merodio-Codinachs	ESA, France
Silja Meyer-Nieberg	Universität der Bundeswehr München, Germany
Martin Middendorf	University of Leipzig, Germany
Taneli Mielikainen	Nokia, Finland
Kaisa Miettinen	University of Jyvaskyla, Finland
Orazio Miglino	University of Naples "Federico II", Italy
Julian Miller	University of York, UK
Marco Mirolli	ISTC-CNR, Italy
Natasa Miskov-Zivanov	University of Pittsburgh, USA
Carmen Molina-Paris	University of Leeds, UK
Sara Montagna	Università di Bologna, Italy
Marco Montes de Oca	Clypd, Inc., USA
Sanaz Mostaghim	Otto von Guericke University Magdeburg, Germany
Mohamed Nadif	University of Paris Descartes, France
Hidemoto Nakada	NIAIST, Japan
Amir Nakib	Università Paris EST Creteil, Laboratoire LISSI, France
Mirco Nanni	CNR – ISTI, Italy
Sriraam Natarajan	Indiana University, USA
Chrystopher L. Nehaniv	University of Hertfordshire, UK
Michael Newell	Athens Consulting, LLC
Giuseppe Nicosia	University of Catania, Italy
Xia Ning	IUPUI, USA
Wieslaw Nowak	N. Copernicus University, Poland

Eirini Ntoutsi	Leibniz University of Hanover, Germany
Michal Or-Guil	Humboldt University of Berlin, Germany
Mathias Pacher	Goethe-Universität Frankfurt am Main, Germany
Ping-Feng Pai	National Chi Nan University, Taiwan
Wei Pang	University of Aberdeen, UK
George Papastefanatos	IMIS/RC Athena, Greece
Luis Paquete	University of Coimbra, Portugal
Panos Pardalos	University of Florida, USA
Andrew J. Parkes	Nottingham University, UK
Andrea Patane'	University of Oxford, UK
Joshua Payne	University of Zurich, Switzerland
Jun Pei	University of Florida, USA
Nikos Pelekis	University of Piraeus, Greece
Dimitri Perrin	Queensland University of Technology, Australia
Koumoutsakos Petros	ETH, Switzerland
Juan Peypouquet	Universidad Tecnica Federico Santa Maria, Chile
Andrew Philippides	University of Sussex, UK
Vincenzo Piuri	University of Milan, Italy
Alessio Plebe	University of Messina, Italy
Silvia Poles	Noesis Solutions NV
Philippe Preux	Inria, France
Mikhail Prokopenko	University of Sydney, Australia
Paolo Provero	University of Turin, Italy
Buyue Qian	IBM T. J. Watson, USA
Chao Qian	University of Science and Technology of China, China
Gunther Raidl	TU Wien, Austria
Helena R. Dias Lourenco	Pompeu Fabra University, Spain
Palaniappan Ramaswamy	University of Kent, UK
Jan Ramon	Inria, France
Vitorino Ramos	Technical University of Lisbon, Portugal
Shoba Ranganathan	Macquarie University, Australia
Cristina Requejo	Universidade de Aveiro, Portugal
John Rieffel	Union College, USA
Laura Anna Ripamonti	Università degli Studi di Milano, Italy
Eduardo Rodriguez-Tello	Cinvestav-Tamaulipas, Mexico
Andrea Roli	Università di Bologna, Italy
Vittorio Romano	University of Catania, Italy
Andre Rosendo	University of Cambridge, UK
Samuel Rota Bulo	Fondazione Bruno Kessler, Italy
Arnab Roy	Fujitsu Laboratories of America, USA
Alessandro Rozza	Parthenope University of Naples, Italy
Kepa Ruiz-Mirazo	University of the Basque Country, Spain
Florin Rusu	University of California Merced, USA
Jakub Rydzewski	N. Copernicus University, Poland
Nick Sahinidis	Carnegie Mellon University, USA
Lorenza Saitta	University of Piemonte Orientale, Italy

Francisco C. Santos	INESC-ID Instituto Superior Tecnico, Portugal
Claudio Sartori	University of Bologna, Italy
Frederic Saubion	Université d'Angers, France
Andrea Schaerf	University of Udine, Italy
Oliver Schuetze	CINVESTAV-IPN, Mexico
Luis Seabra Lopes	Universidade of Aveiro, Portugal
Roberto Serra	University of Modena and Reggio Emilia, Italy
Marc Sevaux	Lab-STICC, Université de Bretagne-Sud, France
Ruey-Lin Sheu	National Cheng Kung University, Taiwan
Hsu-Shih Shih	Tamkang University, Taiwan
Patrick Siarry	Université de Paris 12, France
Alkis Simitsis	HP Labs, USA
Johannes Sollner	Emergentec Biodevelopment GmbH, Germany
Ichoua Soumia	Embry-Riddle Aeronautical University, USA
Giandomenico Spezzano	CNR-ICAR, Italy
Antoine Spicher	LACL University of Paris Est Creteil, France
Pasquale Stano	University of Salento, Italy
Thomas Stibor	GSI Helmholtz Centre for Heavy Ion Research, Germany
Catalin Stoean	University of Craiova, Romania
Reiji Suzuki	Nagoya University, Japan
Domenico Talia	University of Calabria, Italy
Kay Chen Tan	National University of Singapore, Singapore
Letizia Tanca	Politecnico di Milano, Italy
Charles Taylor	UCLA, USA
Maguelonne Teisseire	Cemagref – UMR Tetis, France
Tzouramanis Theodoros	University of the Aegean, Greece
Jon Timmis	University of York, UK
Gianna Toffolo	University of Padua, UK
Joo Chuan Tong	Institute of HPC, Singapore
Nickolay Trendafilov	Open University, UK
Soichiro Tsuda	University of Glasgow, UK
Shigeyoshi Tsutsui	Hannan University, Japan
Aditya Tulsyan	MIT, USA
Ali Emre Turgut	IRIDIA-ULB, France
Karl Tuyls	University of Liverpool, UK
Jon Umerez	University of the Basque Country, Spain
Renato Umeton	Harvard University, USA
Ashish Umre	University of Sussex, UK
Olgierd Unold	Politechnika Wroclawska, Poland
Giorgio Valentini	Università degli Studi di Milano, Italy
Edgar Vallejo	ITESM Campus Estado de Mexico, Mexico
Sergi Valverde	Pompeu Fabra University, Spain
Werner Van Geit	EPFL, Switzerland
Pascal Van Hentenryck	University of Michigan, USA
Ana Lucia Varbanescu	University of Amsterdam, The Netherlands

Best Paper Awards

MOD 2017 Best Paper Award

"Recipes for Translating Big Data Machine Reading to Executable Cellular Signaling Models"
Khaled Sayed*, Cheryl Telmer**, Adam Butchy*, and Natasa Miskov-Zivanov*
*University of Pittsburgh, USA
**Carnegie Mellon University, USA
Springer sponsored the MOD 2017 Best Paper Award with a cash prize of EUR 1,000.

MOD 2016 Best Paper Award

"Machine Learning: Multi-site Evidence-Based Best Practice Discovery"
Eva Lee, Yuanbo Wang and Matthew Hagen
Eva K. Lee, Professor Director, Center for Operations Research in Medicine and HealthCare H. Milton Stewart School of Industrial and Systems Engineering, Georgia Institute of Technology, Atlanta, GA, USA

MOD 2015 Best Paper Award

"Learning with Discrete Least Squares on Multivariate Polynomial Spaces Using Evaluations at Random or Low-Discrepancy Point Sets"
Giovanni Migliorati
Ecole Polytechnique Federale de Lausanne – EPFL, Lausanne, Switzerland

Contents

Recipes for Translating Big Data Machine Reading to Executable Cellular Signaling Models

Khaled Sayed[1], Cheryl A. Telmer[2], Adam A. Butchy[3],
and Natasa Miskov-Zivanov[1,3,4(✉)]

[1] Department of Electrical and Computer Engineering,
University of Pittsburgh, Pittsburgh, PA, USA
{k.sayed, nmzivanov}@pitt.edu
[2] Department of Biological Sciences, Carnegie Mellon University,
Pittsburgh, PA, USA
ctelmer@cmu.edu
[3] Department of Bioengineering, University of Pittsburgh,
Pittsburgh, PA, USA
aabl33@pitt.edu
[4] Department of Computational and Systems Biology,
University of Pittsburgh, Pittsburgh, PA, USA

Abstract. Biological literature is rich in mechanistic information that can be utilized to construct executable models of complex systems to increase our understanding of health and disease. However, the literature is vast and fragmented, and therefore, automation of information extraction from papers and of model assembly from the extracted information is necessary. We describe here our approach for translating machine reading outputs, obtained by reading biological signaling literature, to discrete models of cellular networks. We use outputs from three different reading engines, and demonstrate the translation of different features using examples from cancer literature. We also outline several issues that still arise when assembling cellular network models from state-of-the-art reading engines. Finally, we illustrate the details of our approach with a case study in pancreatic cancer.

Keywords: Machine reading · Big data in literature · Text mining
Cell signaling networks · Automated model generation

1 Introduction

Biological knowledge is voluminous and fragmented; it is nearly impossible to read all scientific papers on a single topic such as cancer. When building a model of a particular biological system, one example being cancer microenvironment, researchers usually start by searching for existing relevant models and by looking for information about system components and their interactions in published literature.

Although there have been attempts to automate the process of model building [1, 2], most often modelers conduct these steps manually, with multiple iterations

© Springer International Publishing AG 2018
G. Nicosia et al. (Eds.): MOD 2017, LNCS 10710, pp. 1–15, 2018.
https://doi.org/10.1007/978-3-319-72926-8_1

between (i) information extraction, (ii) model assembly, (iii) model analysis, and (iv) model validation through comparison with most recently published results. To allow for rapid modeling of complex diseases like cancer, and for efficiently using ever-increasing amount of information in published work, we need representation standards and interfaces such that these tasks can be automated. This, in turn, will allow researchers to ask informed, interesting questions that can improve our understanding of health and disease.

The systems biology community has designed and proposed a standardized format for representing biological models called the *systems biology markup language* (SBML). This language allows for using different software tools, without the need for recreating models specific for each tool, as well as for sharing the built models between different research groups [3]. However, the SBML standard is not easily understood by biologists who create mechanistic models, and thus requires an interface that allows biologists to focus on modeling tasks while hiding the details of the SBML language [4–7].

To this end, the contributions of the work presented in this paper include:

- A **representation format** that is straightforward to use by both machines and humans, and allows for efficient synthesis of models from big data in literature.
- An **approach to effectively use** state-of-the-art machine reading output to create executable discrete models of cellular signaling.
- A **proposal for directions** to further improve automation of assembly of models from big data in literature.

In Sect. 2, we briefly describe cellular networks, our modeling approach, and our framework that integrates machine reading, model assembly and model analysis. In Sect. 3, we present details of our model representation format, while Sect. 4 outlines our approach to translate reading output to the model representation format. Section 5 discusses other issues that need to be taken into account when building interface between big data reading and model assembly in biology. Section 6 describes a case study that uses our translation methodology. Section 7 concludes the paper.

2 Background

2.1 Cellular Networks

Intra-cellular networks include signal transduction, gene regulation, and metabolic networks [8]. Signaling networks are characterized by protein phosphorylation and binding events, which transduce extracellular signals across the plasma membrane and through the cytoplasm [9]. Gene regulatory networks involve translocation of signaling proteins from the cytoplasm to the nucleus, where the integration of these protein signals act on the genome, resulting in changes in gene expression and cellular processes [10]. The regulation of metabolic networks incorporates phosphorylation and binding, as do signaling networks, and also integrates allosteric regulation, other protein modifications, and subcellular compartmentalization [11].

Inter-cellular networks assume interactions between cells of the same or different types. These interactions occur via signaling molecules such as growth factors and cytokines, synthesized and secreted by one cell, and bound to itself or other cells in its surroundings, or via a cell-cell contact.

At all levels of signaling, there are feedforward and feedback loops and crosstalk between signaling pathways to either maintain homeostasis or amplify changes initiated by extracellular signals [12].

2.2   Modeling Approach

When generating executable models, we use a discrete modeling approach previously described in [13]. As illustrated in the example in Fig. 1, we represent system components as model *elements* (A, B, and C in the example), where each element is defined as having a discrete number of levels of activity. Each element has a list of regulators called *influence set*. In our example, A is a positive regulator of C, B and C are positive regulators of A, and C activates itself while B inhibits itself. Additionally, each element has a corresponding *update rule*, a discrete function of its regulators. In our example, A is a conjunction of B and C, while C is a disjunction of A and C. Although the model structure is fixed, the simulator that we use [14] is stochastic, and thus, allows for closely recapitulating the behavior of biological pathways and networks.

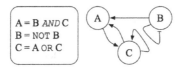

Fig. 1. Toy example illustrating our modeling approach.

2.3   Framework Overview

To automatically incorporate new reading outputs into models, we have developed a reading-modeling-explanation framework, called DySE (Dynamic System Explanation), outlined in Fig. 2. This framework allows for (*i*) expansion of existing models or assembly of new models from machine reading output, (*ii*) analysis and explanation of models, and (*iii*) generation of machine-readable feedback to reading engines. We focus here on *the front end of the framework, the translation from reading outputs to the list of elements and their influence sets, with context information, where available.*

3   Model Representation Format

To enable comprehensive translation from reading engine outputs to executable models, the models are first represented in tabular format. It is important to note here that the tabular representation does not include final update rules, that is, the tabular version of the model is further translated into an *executable model* that can be

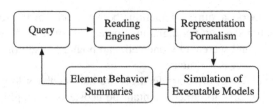

Fig. 2. DySE framework.

simulated. Each row in the model table corresponds to one specific model element (i.e., modeled system component), and the columns are organized in several groups: (*i*) information about the modeled system component, (*ii*) information about the component's regulators, and (*iii*) information about knowledge sources. This format enables straightforward model extension to represent both additional system components as new rows in the table, and additional component-related features by including new columns in the table. The addition of new columns occurs with improvements in machine reading.

The first group of fields in our representation format includes **system component-related** information. This information is either used by the executable model, or kept as background information to provide specific details about the system component when creating a hypothesis or explaining outcomes of wet lab experiments.

A. Name – full name of element, e.g., "Epidermal growth factor receptor".
B. Nomenclature ID – name commonly used in the field for cellular components, e.g., "EGFR" is used for "Epidermal growth factor receptor".
C. Type – these are types of entities used by reading engines as listed in Table 1.
D. Unique ID – we use identifiers corresponding to elements that are listed in databases, according to Table 1.
E. Location – we include subcellular locations and the extracellular space, as listed in Table 2.
F. Location identifier – we use location identifiers as listed in Table 2.
G. Cell line – obtained from reading output.
H. Cell type – obtained from reading outputs.

Table 1. Element type and ID database.

Element type	Database name
Protein	UniProt [16]
Protein family	Pfam [17], InterPro [18]
Protein complex	Bioentities [19]
Chemical	PubChem [20]
Gene	HGNC [21]
Biological process	GO [15], MeSH [22]

Table 2. The list of cellular locations and their IDs from the Gene Ontology [15] database.

Location name	Location ID
Cytoplasm	GO:0005737
Cytosol	GO:0005829
Plasma membrane	GO:0005886
Nucleus	GO:0005634
Mitochondria	GO:0005739
Extracellular	GO:0005576
Endoplasmic reticulum	GO:0005783

I. `Tissue type` – obtained from reading output.

J. `Organism` – obtained from reading output.

K. `Executable model variable` – variable names currently include above described fields B, C, E, and H.

The second group of fields in our representation includes **component regulators-related** information that is mainly used by executable models, with a few fields used for bookkeeping, similar to the first group of fields.

L. `Positive regulator nomenclature IDs` – list of positive regulators of the element.

M. `Negative regulator nomenclature IDs` – list of negative regulators of the element.

N. `Interaction type` – for each listed regulator, in case it is known whether interaction is direct or indirect.

O. `Interaction mechanism` – for each known direct interaction, if the mechanism of interaction is known. Mechanisms that can be obtained from reading engines are listed in Table 3.

P. `Interaction score` – for each interaction, a confidence score obtained from reading.

The third group of fields in our representation includes **interaction-related provenance** information.

Q. `Reference paper IDs` – for each interaction, we list IDs of published papers that mention the interaction. This information is obtained directly from reading output.

R. `Sentences` – for each interaction, we list sentences describing the interaction. This information is obtained directly from reading output.

It is worth mentioning that this representation format can be converted into the SBML format to be used by different software tools and shared between different working groups. Additionally, the tabular format provides an interface that can be easily created or read by biologists, and generated or parsed by a machine.

4   From Reading to Model

We obtain outputs from three types of reading engines, namely REACH [2], RUBICON [24], and Leidos table reading (LTR) [25]. These reading engines provide output files with similar but not exactly the same format. In Table 3, we list the interaction mechanisms that can be obtained from these three reading engines, and in the following sub-sections we outline their differences and the advantages of each reading engine.

Table 3. Intracellular interactions (mechanisms) recognized by the three reading engines.

Reading engine	Recognized mechanisms
REACH [23]	Activation, Inhibition, Binding, Phosphorylation, Dephosphorylation, Ubiquitination, Acetylation, Methylation, Increase or Decrease Amount, Transcription, Translocation
RUBICON [24]	Activation, Inhibition, Promotes, Signaling, Reduce, Induce, Supports, Attenuates, Stimulate, Antagonize, Synergize, Increase and Decrease Amount, Abrogates
LTR [25]	Binding, Phosphorylation, Dephosphorylation, Isomerizations

4.1   Simple Interaction Translation

The first type of reading engine, REACH [2], can extract both direct and indirect interactions, as well as interaction mechanisms, where available. The simplest and most common reading outputs are those that include only a regulated element and a single regulator, each of them having one of the entity types listed in Table 1, with the interaction mechanism being one of the mechanisms described in Table 3. Such interactions have straightforward translation to our representation format, that is, they are translated into a single table row with some or all of the fields described in Sect. 3. Given that our modeling formalism accounts for positive and negative regulators, while reading engines can also output specific mechanisms where available in text, we assume in the translation that Phosphorylation, Acetylation, Increase Amount, and Methylation represent positive regulations, and Dephosphorylation, Ubiquitination, Decrease Amount, and Demethylation represent negative regulations. Additionally, we treat Transcription events as positive regulation.

4.2   Translation of Translocation Interaction

We translate translocation events (moving components from one cellular location to another) using the formalism described in [26]. This formalism requires including two

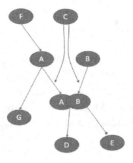

Fig. 3. Schematic representation of a situation common to many biological signaling pathways where the regulation of complex formation, A binding to B, is regulated by a third protein, C, so that the A/B complex can activate D and inhibit E. F can regulate A that is able to regulate G without forming a complex.

separate model elements for the translocated component, one at the original and one at the new location. Additionally, in the translocation type of interaction, translocation regulators can be listed.

4.3   Translation of Complexes

Binding interaction mechanism represents formation of protein complexes in most cases. However, in order to include both individual proteins and complexes in which they participate within a single model, we defined rules for incorporating complexes listed in reading outputs into our model representation format.

A generic example is shown in Fig. 3. If an element in the reading output file is a complex, we incorporate that output into our model representation format by creating a separate table row for each component of the protein complex, and change the regulation set as described in the example outlined in Fig. 3. If the formation of complex AB is regulated by C, then we create two rows; one for element A, which is also positively regulated by F, and one for element B. The positive regulation rule for element A becomes (C AND B) OR F, while the positive regulation rule for element B becomes (C AND A). Additionally, if an element is regulated by a complex, we list all components of that complex as positive regulators for the element. In the example in Fig. 3, the positive regulation rule for element D is (A AND B) because D is regulated by the complex AB. An example of how complexes are translated from reading output into our representation format is shown in Table 4.

Table 4. Converting REACH output for complexes into our modeling representation format.

Column name		Element			Positive regulator		Mech. type	Paper ID	Evidence
		Name	Type	ID	Name	ID			
REACH output		{FAK, PTP-PEST}	{Protein, Protein}	{Q05397, Q05209}	PIN1	Q13526	Binding	PMC 3272802	PIN1 stimulates the binding of FAK to PTP-PEST
DySE format	Comp. 1	FAK	Protein	Q05397	PIN1 *AND* PTP-PEST	(Q13526, Q05209)		PMC 3272802	
	Comp. 2	PTP-PEST	Protein	Q05209	PIN1 *AND* FAK	(Q13526, Q05397)		PMC 3272802	

4.4   Translation of Nested Interactions

REACH reading engine can also detect nested interactions, where some of the participants are interactions themselves. The following sub-sections show several examples of these interactions.

Positive Regulation of Activation. As shown in Fig. 4(a), REACH can find and output interactions where element A is activating element B, while element C is positively regulating the interaction between A and B. We also include in this and the following examples element D. In this case, we assume that D is a negative regulator of B. This means that C will activate B only when A is active. If A is inactive, only D will inhibit B, while C will not have any effect on B. The following is an example of the aforementioned situation that can occur in text, and is extracted by REACH as described above: "*In fact, RANKL induced phosphorylation of Akt was enhanced by the addition of TNF-alpha*". Here, RANKL is a positive regulator of Akt, and this activation is further regulated by TNF-alpha.

Positive Regulation of Inhibition. Figure 4(b) illustrates an example of a nested interaction where A inhibits B, and C positively regulates this inhibition, which means that C will increase the inhibition of B by A, when A is active/high. Here, we also assume that element D is a positive regulator of B. If A is inactive/low, only D will activate B, and C will not have any effect on B. The following text represents an example sentence for such situation: "*This conclusion was supported by the finding that nilotinib also induced dephosphorylation of the BCR-ABL1 target CrkL*". Here, the inhibition of CrkL by BCR-ABL1 is enhanced with nilotinib.

Negative Regulation of Activation. The example in Fig. 4(c) shows that C negatively regulates the activation of B by A. So, if A is inactive/low, only D will activate B, and C will not have any effect on B. An example text for this situation is "*These data provide evidence that PDK1 negatively regulates TGF-β signaling through modulation of the direct interaction between the TGF-β receptor and Smad3 and -7*".

Negative Regulation of Inhibition. Figure 4(d) shows that C negatively regulates the inhibition of B by A. Therefore, if A is inactive/low, only D will activate B, and C will not have any effect on B.

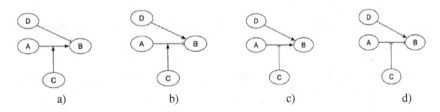

a)   b)   c)   d)

Fig. 4. Examples of nested interactions. (a) Positive regulation of Activation interaction, (b) Positive regulation of Inhibition interaction, (c) Negative regulation of Activation interaction, (d) Negative regulation of Inhibition interaction

4.5   Translation of Direct and Indirect Interactions

RUBICON [24] provides two reading outputs, one for *direct interactions* and one for *indirect interactions*. For the indirect interactions, it creates a chain of elements that starts with the regulator and ends with the regulated element, and includes the intermediate elements, also found in the read paper, forming a path from the regulator to the regulated elements.

The RUBICON reader output file with direct interactions, has two special fields, different from REACH: Confidence and Tags. The Confidence column indicates how *confident* the reading engine is about the extracted interaction, and the values in this column can be LOW, MODERATE, and HIGH. The Tags column includes *epistemic tags* such as 'implication', 'method', 'hypothesis', 'result', 'goal', or 'fact'. Table 5 shows reading output examples from RUBICON for the direct and chain interactions. Due to space constraints, and given that RUBICON does not provide information for all the columns, Table 5 includes a subset of columns from our representation.

The second reading output file from RUBICON contains indirect interactions that form a path from the regulator to the regulated element. This output file also includes a column called "Connection" and in this column, it lists intermediate elements on a path, followed by their IDs. For example, if there is a path of the form A → B → C, element B will be included in the connection column.

Table 5. RUBICON output examples for both Direct and Chain.

Column name	Element		Positive regulator		Mech. type	Connection	Paper ID	Evidence	Confidence	Tags
	Name	ID	Name	ID						
Direct	TNF alpha	P01375	IL-2	P60568	Induced	*NA*	PMC 149405	In addition, cytokines including TNFalpha, TNFbeta and flt3 ligand were induced by IL-2 as detected by the arrays	Low	Results
Chain	Apoptosis	GO: 0006915	imatinib	5291	Enhances, induced	TRAIL, ID: P50591	PMC 4896164	Treatment with imatinib enhances TRAIL induced apoptosis	–	Goal

4.6   Translation from Table Reading Output

The third reading engine, LTR, performs table reading and generates reading output in the tabular format with some or all of the fields described in Sect. 3. The LTR output also contains information about Cell Line and Binding sites. Additionally, this output includes much more specific, connected information than those offered by RUBICON or REACH. Where RUBICON or REACH look at all the interactions listed in a paper, the nature of their search returns information on many different experiments and contexts. LTR is able to focus on one table at a time. As tables tend to describe a highly specific experiment about interacting components, such output can provide detailed information about parts of the network, which can be valuable in finding answers to specific questions. An example of an LTR output is shown in Table 6.

Table 6. LEIDOS output example illustrating the effects of the negative regulator (TiO_2) on two different molecules. As both sites affected by the negative regulator are serine residues, this provides additional context that the negative regulator might be a serine-specific.

Element			Negative regulator		Cell line	Organism	Paper ID	Evidence
Name	ID	Site	Name	ID				
AKT1	P31749	S124	TiO_2	CHEBI: 32234	HeLa	Human	PMC 3251015	Resource3.xls. table.serial.txt
Gab2	Q9UQC2	S264	TiO_2	CHEBI: 32234	HeLa	Human	PMC 3251015	Resource4.xls. table.serial.txt

5 Matching Reading and Modeling

Due to the writing style in biology, reading engines often encounter texts that are hard to interpret even by human readers. In the following, we outline several situations where it is critical to correctly interpret interactions listed in reading outputs to enable accurate model expansion. When there are contradictions among reading outputs, or between reading output and an existing model, a feedback to reading can be generated in the form of new queries to guide further literature search and reading. Queries are designed using AND, OR and NOT to define more precisely the search space and also to remove papers that would describe information that is not relevant (e.g., focusing on different cell type).

5.1 Protein Families

Reading engines often come across entities that represent protein families instead of specific proteins. In such cases, there is no unique protein ID, instead either all IDs of proteins from that family need to be listed, or a unique protein family ID should be used. Since our goal is to automate the assembly of models from machine reading output, we need to be able to accurately treat such protein family entities in the reading output. There are several issues that can arise when protein families are outputs as interaction entities in reading output, described in the following example.

Example 1: Let us assume that either an existing model or previous reading output include an interaction that describes positive regulation of ERK1 by MEK1 (MEK1 $\rightarrow$ ERK1), where both MEK1 and ERK1 are specific proteins that have unique IDs in protein databases. We list below other similar interactions that may be recognized by reading, and propose methods to resolve such situations.

a. Reading output MEK $\rightarrow$ ERK, where both MEK and ERK are listed as protein families. In order to incorporate both the original interaction and the new one within the same model, we can treat the new interaction as generalization. Furthermore, this is also an example of a situation where a feedback to reading engines can be created, to obtain more information about the interaction. For example, queries that could result from the scenario described here are:

   • Search for other (non-MEK1) MEK family members and their interactions with ERK1;
   • Search for other (non-ERK1) ERK family members and their interactions with MEK1;
   • Search for other MEK (non-MEK1) and ERK (non-ERK1) family members, and their mutual interactions.

b. Reading output MEK1 $\rightarrow$ ERK, where MEK1 is a protein and ERK is a protein family. In this case, the feedback to reading could be:

   • Search for other ERK family members and their interactions with MEK1.

c. Reading output: MEK $\rightarrow$ ERK1, where MEK is a protein family and ERK1 is a protein. In this case, the feedback to reading could be:

- Search for other MEK family members and their interaction with ERK1.

d. Reading output: MEK → p38, MEK protein family activating protein p38. This case requires additional knowledge that would either already exist in the model or other reading outputs, or would need to be curated by a human expert. MEK3, and not MEK1, therefore, adding the original interaction (MEK1 → ERK1) to the model, and then incorporating connection between MEK1 (as a member of MEK family) and p38 in the model would make it incorrect. The feedback to reading in this case could be:

- Search for interaction between MEK1 and p38 to confirm or disconfirm the interaction MEK → p38.

5.2 Cell Type

Often, the modeling goal is to include multiple cell types, for example, model of cancer microenvironment could include cancer cell and several types of immune cells. In such cases, it is important to know to which cell type to assign the interaction that is extracted from text by machine reading. When cell type is taken into account, depending on the information that exists in the reading output, the relationship between similar reading outputs, or between reading outputs and an existing model, can be interpreted in several ways and the following example illustrates one such case.

Example 2: Let us assume that the machine reading output lists interaction A → B (A regulates B), but no information is given about cell type to which this interaction belongs. The model assembly step needs to decide to which cell to add this interaction, and therefore, different scenarios are possible, some of them described here:

- A is already listed in interactions in more than one cell type in the model;
- B is already listed in interactions in more than one cell type in the model;
- Neither A nor B is listed in other interactions;
- Both A and B are listed in interactions in exactly one cell type in the model (same or different).

The model assembly step, which adds new reading output to existing model, needs to either take into account previously defined assumptions (e.g., always add interactions to one predetermined cell type, or add interactions to all cell types, or skip the interaction that does not indicate cell type, etc.). Another approach is a feedback to reading engines that requests additional search for evidence of cell type in the paper.

5.3 Cellular Location

In some cases, it is important to know the location of elements participating in interactions. For example, translocation of element from one cellular location to another may take time, or it may be known that a particular element can affect another element only in a specific location. In order to accurately model such location-dependent interactions, the machine reading output should include the information about subcellular locations or extracellular space, the effect of location on interactions and on timing of cellular events (e.g., translocation). The following examples illustrate two such case.

Example 3: Let us assume that new reading output includes interaction A → B (A regulates B), but the interaction location is different from the one that exists in the current model. This can either be interpreted as a contradiction, or a feedback to reading engines can be generated in the form of a query to initiate literature search for further evidence of new interaction location. Additionally, the confidence obtained from reading can be compared with the confidence for the interaction in the model, to decide how to treat the reading output.

Example 4: Let us assume that an existing model includes interaction A → B (A positively regulates B) at a specific location, and reading output includes interaction A-|B (A negatively regulates B), but without location information. This can either be interpreted as a contradiction, or, as in previous examples, a feedback to reading engines can be formed to search for further evidence of new interaction location. It is possible that the new interaction is observed at a different location, thus, the opposite regulation sign will not be interpreted as contradiction.

5.4 Contradicting Interaction Type

In the case of contradiction among individual reading outputs, or between new reading output and an existing model, a feedback to reading engines can be created to initiate new literature search. The following example illustrates one such case.

Example 5: Let us assume that an existing model includes interaction A → B (A positively regulates B), while in reading output A-|B (A negatively regulates B). Assuming that the location information matches, there are several ways to handle this situation. The difference between reading outputs and model can be interpreted as a contradiction, or the new interaction may be interpreted as indirect, forming a negative feed-forward loop with the one existing in the model. In this case, a feedback to reading engines can request search for further evidence for elements on a path between A and B.

5.5 Negative Information

When it is well known that some interactions do not exist, such information is not stored in models. However, the reading output may include such interactions and the following example shows how these situations can be resolved.

Example 6: Let us assume that the previous reading output or an existing model includes interactions MEK1 → ERK1 and MEK3 → p38. There are several other reading outcomes that could occur:

a. New reading output includes interaction NOT (MEK3 → ERK), where MEK3 is interpreted as a protein, and ERK is interpreted as a protein family. This is in agreement with the model, however, reading output that indicates that an interaction does not exist is not used to extend the model.

b. New reading output includes interaction NOT (MEK $\rightarrow$ ERK1), where MEK is interpreted as a protein family and ERK1 is interpreted as a protein. This new reading output would contradict the model or other reading output, assuming that an interaction MEK1 $\rightarrow$ ERK1 (from Example 1) already exists in the model or in other reading output. However, when taking into account the fact that MEK3 does not indeed regulate ERK1, such reading output could also be interpreted as corroboration. To resolve this, a search for further evidence in the paper that confirms that the MEK from the reading output is <u>not</u> MEK1 could be conducted.

6 Case Study

To illustrate the utility of the translation from output of automated reading to the model representation format, we show an example of two queries, followed by a summary of reading results that we obtained from the three reading engines. The summary includes numbers of unique extensions that were identified by our interaction classifier tool, which compares reading outputs with baseline model.

The first query that we used is related to molecule GAB2. The original model does not contain GAB2 and we were interested in extending the model to incorporate GAB2. The query that we used is:

```
GAB2 AND (phosphatidylinositol OR proliferation OR SHC1 OR PI-3
kinase OR PI3K OR PIK3 OR GRB2 OR PTPN11 OR 14-3-3 OR SFN OR
YWHAH OR HCK OR AKT OR beta-catenin OR Calcineurin OR SERPINE1)
NOT (Fc-epsilon receptor OR osteoclast OR mast cell)
```

Note that GAB2 was identified in 1998 so the protein and gene have the same name and this results in a confusion in the literature search. In Tables 7 and 8, we show the number of papers returned by REACH and RUBICON reading engines using the GAB2 and Beta-catenin queries respectively, the events extracted from all of the papers analyzed, and the unique extensionsthat were found by comparison to two existing models, Normal and Cancer.

Table 7. Results from GAB2 query.

	REACH	RUBICON
Number of papers	249	249
Extracted events	4800	5082
Unique extensions	4618	3317

Table 8. Results from β-catenin query.

	REACH	RUBICON
Number of papers	351	351
Extracted events	2809	2024
Unique extensions	2532	1906

The second query that we used is related to molecule β-catenin. The original model does not contain β-catenin and we were interested in extending the model to incorporate this molecule. The query that we used is:

```
(beta-catenin OR B-catenin OR β-catenin OR catenin beta-1 OR
CTNNB1) AND (Wnt OR AXIN1 OR AXIN2 OR AXIN OR APC OR CSNK1A1 OR
GSK3B OR TCF OR LEF OR TCF/LEF OR CDK2 OR PTPN6 OR CCEACAM1 OR
insulin OR PML OR RANBP2 OR YAP1 OR GSK3 OR HSPB8 OR SERPINE1 OR
AKT OR PTPN13 OR ACAP1 OR MST1R) NOT (neuroblasts OR neurogenesis
OR anoikis OR cardiac OR EMT OR breast OR embryonic OR osteoblast
OR synapse OR muscle OR renal)
```

In this case, the β-catenin protein was identified in 1989 and the human gene in 1996 so the protein and gene have different names. However, using Greek letters in the name requires using various related terms in the query to increase the chance of capturing the right molecule in papers.

These two examples of search terms and the corresponding reading results emphasize the fact that a careful construction of search terms is critical – with proper selection of search terms, we can tailor the reading output for relevant context.

7   Conclusion

This paper describes a representation format that we created for the purpose of automating assembly of models from machine reading outputs. The proposed representation format allows for capturing biological interactions at the molecular level, and it can be easily used by both human experts and machines. The tabular formatting described in this paper allows for the transit of files through the pipeline from reading of scientific literature (text written by scientists), to executable model (computer readable mathematical model that can be simulated). The format is critical to have all of the tools communicate with each other and also retain readability for biologists to evaluate the work of the machines. Manual reading and annotation of thousands of papers would take many weeks instead of hours.

By using this format, our automated framework rapidly assembles and validates executable models from big data in literature, with the runtimes and comprehensiveness not previously possible. Such formalized representation of research findings for the purpose of creating dynamic models will significantly speed up the process of collecting data from literature, and it will facilitate the reusability of existing scientific results, increase our knowledge and improve our understanding of biological systems. This, in turn, should lead to rapidly designing new disease treatments and effectively guiding future studies.

References

1. Miskov-Zivanov, N.: Automation of biological model learning, design and analysis. In: Proceedings of the 25th Edition on Great Lakes Symposium on VLSI. ACM (2015)
2. Valenzuela-Escárcega, M.A., et al.: A domain-independent rule-based framework for event extraction. In: ACL-IJCNLP 2015, p. 127 (2015)
3. Hucka, M., et al.: The systems biology markup language (SBML): a medium for representation and exchange of biochemical network models. Bioinformatics **19**(4), 524–531 (2003)
4. Droste, P., et al.: Visualizing multi-omics data in metabolic networks with the software Omix—a case study. Biosystems **105**(2), 154–161 (2011)
5. Büchel, F., et al.: Qualitative translation of relations from BioPAX to SBML qual. Bioinformatics **28**(20), 2648–2653 (2012)
6. Faeder, J.R., Blinov, M.L., Hlavacek, W.S.: Rule-based modeling of biochemical systems with BioNetGen. In: Systems Biology, pp. 113–167 (2009)
7. Hedengren, J.D., et al.: Nonlinear modeling, estimation and predictive control in APMonitor. Comput. Chem. Eng. **70**, 133–148 (2014)
8. Albert, R.: Scale-free networks in cell biology. J. Cell Sci. **118**(21), 4947–4957 (2005)
9. Pawson, T., Scott, J.D.: Protein phosphorylation in signaling–50 years and counting. Trends Biochem. Sci. **30**(6), 286–290 (2005)
10. Erwin, D.H., Davidson, E.H.: The evolution of hierarchical gene regulatory networks. Nat. Rev. Genet. **10**(2), 141–148 (2009)
11. Schuster, S., Fell, D.A., Dandekar, T.: A general definition of metabolic pathways useful for systematic organization and analysis of complex metabolic networks. Nat. Biotechnol. **18**(3), 326–332 (2000)
12. Schmitz, M.L., et al.: Signal integration, crosstalk mechanisms and networks in the function of inflammatory cytokines. Biochimica et Biophysica Acta (BBA)-Molecular Cell Research **1813**(12), 2165–2175 (2011)
13. Miskov-Zivanov, N., Marculescu, D., Faeder, J.R.: Dynamic behavior of cell signaling networks: model design and analysis automation. In: Proceedings of the 50th Annual Design Automation Conference. ACM (2013)
14. Sayed, K., et al.: DiSH simulator: capturing dynamics of cellular signaling with heterogeneous knowledge (2017). arXiv preprint arXiv:1705.02660
15. GO. Gene Ontology Database. http://geneontology.org/page/go-database
16. UniProt. UniProt Database. http://www.uniprot.org/
17. Pfam. Pfam Database. http://pfam.xfam.org/
18. InterPro. InterPro Database. https://www.ebi.ac.uk/interpro/
19. Bioentities. Bioentities Database. https://github.com/sorgerlab/bioentities
20. PubChem. PubChem Database. https://pubchem.ncbi.nlm.nih.gov/
21. HGNC. Database of Human Gene Names. http://www.genenames.org/
22. MeSH. MeSH Database. https://www.ncbi.nlm.nih.gov/mesh
23. REACH. Reading and Assembling Contextual and Holistic Mechanisms from Text (2016). http://agathon.sista.arizona.edu:8080/odinweb/
24. Burns, G.A., et al.: Automated detection of discourse segment and experimental types from the text of cancer pathway results sections. In: Database 2016, p. baw122 (2016)
25. Sloate, S., et al.: Extracting protein-reaction information from tables of unpredictable format and content in the molecular biology literature. In: Bioinformatics and Artificial Intelligence (BAI), New York (2016)
26. Sayed, K., Telmer, C.A., Miskov-Zivanov, N.: Motif modeling for cell signaling networks. In: 2016 8th Cairo International Biomedical Engineering Conference (CIBEC). IEEE (2016)

Improving Support Vector Machines Performance Using Local Search

S. Consoli$^{(\boxtimes)}$, J. Kustra, P. Vos, M. Hendriks, and D. Mavroeidis

Philips Research, High Tech Campus 34, 5656 AE Eindhoven, The Netherlands
sergio.consoli@philips.com

Abstract. In this paper, we propose a method for optimization of the parameters of a Support Vector Machine which is more accurate than the usually applied grid search method. The method is based on Iterated Local Search, a classic metaheuristic that performs multiple local searches in different parts of the space domain. When the local search arrives at a local optimum, a perturbation step is performed to calculate the starting point of a new local search based on the previously found local optimum. In this way, exploration of the space domain is balanced against wasting time in areas that are not giving good results. We show a preliminary evaluation of our method on a radial-basis kernel and some sample data, showing that it is more accurate than an application of grid search on the same problem. The method is applicable to other kernels and future work should demonstrate to what extent our Iterated Local Search based method outperforms the standard grid search method over other heterogeneous datasets from different domains.

1 Introduction

Support Vector Machine (SVM) is a popular supervised learning technique to analyze data with respect to classification and regression analysis [29]. SVM models have been successfully applied in numerous applications, such as character recognition [9], text categorization [14], image classification [25] and have recently entered the healthcare domain to solve classification problems such as protein recognition [24], genomics [3] and cancer classification [10,30].

The performance of a SVM is dependent on the parameters setting of the underlying model. The parameters are usually set by training the SVM on a specific dataset and are then fixed when applied to a certain application. Finding the optimal setting of those parameters is an art by itself and as such many publications on the topic exist [6,12,18,28,31][1]. Of the techniques used, grid search (or parameter sweep) is one of the most common methods to determine optimal parameter values [5]. Grid search involves an exhaustive searching through a manually specified subset of the hyperparameter space of a learning algorithm,

---

[1] Note that automatic configuration for algorithms is the same problem faced when doing hyper-parameter tuning in machine learning; it is just another wording.

© Springer International Publishing AG 2018
G. Nicosia et al. (Eds.): MOD 2017, LNCS 10710, pp. 16–28, 2018.
https://doi.org/10.1007/978-3-319-72926-8_2

guided by some performance metric (e.g. cross-validation). This traditional app-roach, however, has several limitations. Firstly, this approach is vulnerable to local optimum. Although a multi-resolution grid search may overcome this lim-itation, it does not provide an absolute guarantee that it will find the absolute minimum. Secondly, setting an appropriate search interval is an ad-hoc app-roach which, likewise, does not guarantee the absolute minimum. Moreover, it is a computationally expensive approach when intervals are set to capture wide ranges.

If the parameters to be set are constrained to assume only a fixed set of values, it has been shown in the literature that a classic random walk performs better than grid search [4]; but this only applies for fixed grids to explore, which is not the case when tuning a SVM where the parameters vary in a continuous search space. As an alternative to grid search approaches and its limitations, gradient descent has been proposed in literature for SVM parameter tuning [16]. Gradient descent, or steepest descent optimization finds the local minimum by taking the gradient (or the approximate gradient) at each parameter step as a directional indication instead of exploring all possible directions. Although this approach is able to get better solutions than the grid search, it has however the disadvantage to be sensitive to initial settings of the parameters. That is, when the provided initial parameter setting produces a starting solution that is excessively far from the optimal solution within the search domain, the algorithm then may converge to a local optimum instead of the optimal minimum.

In this paper, we describe a method to tackle the parameters setting prob-lem in SVMs using an intelligent optimization procedure based on Iterated Local Search (ILS) [21]. This is a popular metaheuristic which has been shown to be a promising approach for several real world optimization problems due to its strong global search capability [26]. ILS has been previously used with success to address the problem of automatically configuring the parameters of com-plex, heuristic algorithms in order to optimize performance on a given set of benchmark instances [13,19]. In this paper we describes a further application of parameter tuning via ILS specifically to SVMs. The goal is to exploit the maximum generalization capability of SVMs by selecting an optimal setting of kernel parameters.

2 Support Vector Machines

SVMs were developed in 1995 by Cortes and Vapnik [9] with the specific aim of binary classification. Given the input parameters $x \in X$ and their corresponding output parameters $y \in Y = \{-1, 1\}$, the separation between classes is achieved by fitting the hyperplane $f(x)$ that has the optimal distance to the nearest data point used for training of any class.

$$f(x) = \sum_{i=1}^{n} \alpha_i y_i < x_i, x > +b, \tag{1}$$

where n is the total number of parameters. The goal is to find the hyperplane which maximizes the minimum distances of the samples on each side of the plane. However, the solution for the above problem is not always possible, since fitting a plane could result in samples being on the *wrong* side of the plane. To account for this, a penalty is associated with the instances which are misclassified and added to the minimization function. This is done via the parameter C in the minimization formula:

$$\underset{f(x)=\omega^T x+b}{\arg\min} \frac{1}{2}\|\omega\|^2 + C\sum_i^n c(f, x_i, y_i). \qquad (2)$$

By varying C, a trade-off between the accuracy and stability of the function is defined. Larger values of C result in a smaller margin, leading to potentially more accurate classifications, however overfitting can occur. The above approach only allows for the separation of linear data. In most real world problems, this is not the case. To overcome this issue, a mapping of the data into a richer feature space, including non-linear features is applied prior to the hyperplane fitting. For the purpose of this mapping, kernel functions $k(x, x')$ are used. Several kernel functions have been proposed, such as polynomial, hyperbolic or Gaussian radial-basis functions. We focus this paper on the latter:

$$K(x_i, x') = exp(-\gamma\|x_i - x'\|^2), \gamma > 0. \qquad (3)$$

When a Gaussian radial-basis (RBF) function is used as the kernel of the SVM function, γ defines the variance of the RBF, practically defining the shape of the kernel function peaks: lower γ values set the bias to low and corresponding high γ to high bias.

3   Iterated Local Search

Iterated Local Search (ILS) [21] is a popular explorative local search method for solving discrete optimization problems. It belongs to the class of trajectory optimization methods, i.e. at each iteration of the algorithm the search process designs a trajectory in the search space, starting from an initial state and dynamically adding a new better solution to the curve in each discrete time-step. Thus this process can be seen as the evolution in time of a discrete dynamical system in the state space. The generated trajectory is useful because it provides information about the behavior of the algorithm and its dynamics.

Iterated Local Search mainly consists of two steps, the first to reach local optima performing a walk in the search space, while the second to efficiently escape from local optima [20]. The aim of this strategy is to prevent getting stuck in local optima of the objective function. Iterated Local Search is probably the most general scheme among explorative optimization strategies. It is often used as framework for other metaheuristics or can be easily incorporated as a subcomponent in some of them to build effective hybrids.

The algorithm initializes the search by selecting an initial candidate solution. The construction of the initial solution should be both computationally not expensive and, possibly, a good starting point for local search. The fastest way is to generate randomly the initial solution, although this does not guarantee to have a good-quality starting point. However constructive heuristics may also be adopted in order to quickly find high-quality starting points, like, e.g. GRASP [26]. Afterwards, a locally optimal solution is achieved by applying a local search, whose characteristics have a considerable influence on the performance of the entire algorithm. The core of the algorithm mainly consists of the following three phases:

(1) A *perturbation* applied to the current candidate solution, say s;
(2) A *local search* performed to the modified solution in order to find a local optimum, say s';
(3) The application of an *acceptance criterion* to decide which of the two local optima, s or s', has to be chosen to continue the search process.

The specific steps have to be properly designed and set to find a good trade-off between diversification (exploration) and intensification (exploitation) of the optimization process, which is an essential task for a heuristic in order to quickly identify regions in the search space with high quality solutions, without wasting too much time in regions with a low quality.

In ILS, both the perturbation and the acceptance criterion mechanisms can use aspects of the search history (long- or short-term memory). For example, stronger perturbation should be applied when the same local optima are repeatedly encountered. The role of the perturbation (usually probabilistic to avoid cycling) is to modify the current candidate solution to help the search process to effectively escape from local minima, in order to eventually find different better points. Typically, the strength of perturbation has a strong influence on the length of the subsequent local search phase. It can be either fixed (independently of the problem size) or variable. However, the latter one is in general more effective because the bigger the problem size is, the larger should be the strength. A more sophisticated adaptive strength scheme is also possible in which the perturbation strength is increased when more diversification is needed, and decreased when intensification seems preferable (Variable Neighbourhood Search and its variants [11] belong to this category). The acceptance criterion has also a strong influence on the behavior and performances of ILS. The two extremes are:

- Accepting the new local optimum only in case of improvement (strong intensification: iterative improvement mechanism);
- Always accepting the new solution (high diversification: random walk in the search space). In this case a number of the old solutions can be also kept in memory.

Between these extremes, there are several intermediate choices. In Simulated Annealing [1], for example, it is possible to adopt a so-called "cooling schedule": accepting all the improving candidate solutions and also the non-improving ones with a probability that is a function of a temperature parameter, T, and of the difference of objective function values [26]. In Simulated Annealing, the cooling schedule for the temperature T can be either monotonic (non-increasing in time) or non-monotonic (adapted to tune the balance between diversification and intensification). The non-monotonic schedule is particularly effective if it exploits the history of the search: instead of constantly decreasing the temperature, it is increased when more diversification seems to be required.

4   Our ILS Method for SVM Parameters Tuning

The accuracy of a SVM model is largely dependent on the selection of its model parameters used in training and predicting [8]. The most common kernel functions are: linear, polynomial, radial-basis, and sigmoid. In our case we choose a radial-basis kernel function since it is particularly suited to model non-linear effects and therefore is indicated for many classification and regression problems. Although our study has been tailored to a radial-basis kernel function, our routine is generalizable to the parameters setting of any other kernel function choice.

A radial-basis kernel functions requires two parameters to be properly tuned: the penalty parameter (C), and the gamma of the kernel function (γ). The C parameter is a general penalizing parameter for classification, determining the trade-off between the fitting error minimization and model complexity. The γ parameter of the kernel function defines the nonlinear mapping from the input space to some high dimensional feature space. It is critical, as in any regularization scheme, that proper values are chosen for both parameters C and γ.

The C parameter affects the trade-off between complexity and proportion of non-separable samples and it must be set by the user [7]. If C is chosen too large, we have a high penalty for non-separable points and we may store many support vectors and overfit. If it is too small, we may have under-fitting [2]. The parameter C controls indeed the trade-off between errors of the SVM on training data and margin maximization; $C = \infty$ leads to hard margin a SVM [27]. In some SVMs implementations C has been selected equal to the range of output values [22] to get a reasonable and quick setting of this parameter; however this choice is quite questionable since it does not consider possible outliers effects in the training data. In other situations, tuning of parameter C is performed in practice by means of trial-and-error, i.e. by trying to vary C through a wide range of values and assessing the optimal performance obtained, either within a separate validation set, or by cross-validation using only training data [27].

For a SVM the value of the γ parameter of the kernel function should also be selected. This parameter has an effect on the smoothness of the SVMs response and it affects the number of support vectors, so both the complexity and the generalization capability of the network depend on its value [15]. The value of

γ strongly affects therefore the level of accuracy of the approximated function. There is also some connection between observation noise in the training data set and the value of γ. Fixing this parameter can be useful if the desired accuracy of the approximation can be specified in advance. If γ is larger than the range of the target values we cannot expect a good result. If instead γ assumes low values, tending to zero, we can expect overfitting since the kernel function would always tends to output one, and the classifier would be able to perform only majority prediction over the labels. Therefore γ must be chosen to reflect the data in some way. Choosing γ to be a certain accuracy does of course only guarantee that accuracy on the training set. An optimal setting of γ requires the knowledge of noise level. As shown in [7], the noise variance can be estimated directly from training data, i.e. by fitting very flexible (high-variance) estimator to the data. Alternatively, one can first apply least-modulus regression to the data, in order to estimate noise level. In [17] the authors proposed asymptotically optimal γ values proportional to noise variance, in agreement with [7]. Similarly, in [22] the authors propose to choose the γ value so that the percentage of support vectors in the SVM regression model is around 50% of the number of samples. However, one can easily show examples when optimal generalization performance is achieved with the number of support vectors larger or smaller than 50%. A robust compromise can be to impose the condition that the percentage of support vectors be equal to 50%. A larger value of γ can be utilized (especially for very large and/or noisy training sets) [22]. The main practical drawback of such proposals is that they do not reflect sample size. Intuitively, the value of γ should be smaller for larger sample size than for small sample size (with same noise level).

Hence, summarizing, both C and γ values affect model complexity (but in a different way). Grid search [5] is one of the most common methods used to set C and γ parameters. However, since this approach is vulnerable to local optimum, we employ an Iterated Local Search procedure to produce optimal, or near-optimal, kernel parameters in order to get an effective SVM having the maximum generalization capability over the considered dataset.

The implemented ILS uses the grid search as an inner local search routine, which is iterated in order to make it fine-grained by using an automated optimization strategy, which at the end produces the best parameters C and γ found to date and improves the accuracy and robustness of the subsequent classifier. The algorithm pseudo-code is shown in Algorithm 1. Given training dataset D and a Support Vector Machine model Θ, the procedure first generates an initial solution. To get an initial solution, we use an initial solution produced by the grid search. The grid search exhaustively generates candidates from a grid of the parameter values, C and γ, specified in the arrays $range_\gamma \in \Re^+$ and $range_C \in \Re^+$. We choose arrays containing five different values for each parameter, so that the grid search method will look to 25 different parameters combinations. The range values are taken as different powers of 10 from -2 to 2, i.e.: $[10^{-2}, 10^{-1}, 1, 10^1, 10^2]$, in order to provide an initial solution belonging to a promising region of the search space. Solution quality is evaluated as the

---

Algorithm 1. ILS procedure for the generation of the SVM parameters

---

Input: A training dataset D and a Support Vector Machine model Θ;
Output: The best set of parameters γ, C of Θ;
- Let $range_\gamma^{1\times5} \in \Re^+$ and $range_C^{1\times5} \in \Re^+$ be the arrays (1×5) containing the range values for, respectively, the parameters γ and C where to perform the grid search;

begin
    · Generate the initial solution:
    $[\gamma, C] \leftarrow grid\text{-}search(D, \Theta, range_\gamma, range_C)$ //Initialization;
    · Accuracy evaluation: $Acc \leftarrow 10\text{-}fold\text{-}CV(\gamma, C)$;
    while *termination conditions not met* **do**
        · $[range_\gamma, range_C] \leftarrow new\text{-}ranges(range_\gamma, \gamma, range_C, C)$ //Perturbation;
        · $[\gamma', C'] \leftarrow grid\text{-}search(D, \Theta, range_\gamma, range_C)$ //Local search;
        · Accuracy evaluation: $Acc' \leftarrow 10\text{-}fold\text{-}CV(\gamma', C')$;
        //Acceptance criterion:
        if *(Acc' > Acc)* **then**
           | · Set $\gamma \leftarrow \gamma'$, $C \leftarrow C'$, $Acc \leftarrow Acc'$;
        end
    end
4   ⇒ The best set of parameters γ, C of Θ.
end

---

accuracy of the SVM by means of 10-fold cross validation. In general, in a k-fold cross validation, the original sample is randomly partitioned into k equal sized subsamples. Of the k subsamples, a single subsample is retained as the validation data for testing the model, and the remaining $k-1$ subsamples are used as training data. The cross-validation process is then repeated k times, with each of the k subsamples used exactly once as the validation data. The k results from the subsamples can then be averaged to produce a single estimation. The advantage of this method over repeated random sub-sampling is that all observations are used for both training and validation, and each observation is used for validation exactly once. Although in general k remains an unfixed parameter, we used a 10-fold cross-validation since $k = 10$ is the most common used setting [23]. The evaluated accuracy is stored in the variable Acc.

Afterwards, the *perturbation phase*, which represents the core idea of ILS, is applied to the incumbent solution. The goal is to provide a good starting point (i.e. parameters ranges) for the next *local search phase* of ILS (i.e. the grid search in our case), based on the previous search experience of the algorithm, so that to obtain a better balance between exploration of the search space against wasting time in areas that are not giving good results. Ranges are set as:

$$range_\gamma = [\gamma * 10^{-2}, \gamma * 10^{-1}, \gamma, \gamma * 10, \gamma * 10^2]$$
$$\equiv [\gamma_{inf-down}, \gamma_{inf-up}, \gamma, \gamma_{sup-down}, \gamma_{sup-up}], \qquad (4)$$

$$range_C = [C * 10^{-2}, C * 10^{-1}, C, C * 10, C * 10^2]$$
$$\equiv [C_{inf-down}, C_{inf-up}, C, C_{sup-down}, C_{sup-up}]. \qquad (5)$$

Imagine that the grid search gets the set of parameters γ', C' as a new incumbent solution, whose evaluated accuracy is Acc' by means of 10-fold cross-validation.

Then the *acceptance criterion* of this new solution is that it produces a better quality, that is an increased accuracy, than the best solution to date. If it does not happen, the new incumbent solution is rejected and the ranges are updated automatically with the following values:

$$\gamma_{inf-down} = rand(\gamma_{inf-down} * 10^{-1}, \gamma_{inf-down}) \text{ and } C_{inf-down}$$
$$= rand(C_{inf-down} * 10^{-1}, C_{inf-down}) \tag{6}$$

$$\gamma_{inf-up} = rand(\frac{\gamma - \gamma_{inf-up}}{2}, \gamma) \text{ and } C_{inf-up} = rand(\frac{C - C_{inf-up}}{2}, C) \tag{7}$$

$$\gamma_{sup-down} = rand(\frac{\gamma_{sup-down} - \gamma}{2}, \gamma) \text{ and } C_{sup-down} = rand(\frac{C_{sup-down} - C}{2}, C) \tag{8}$$

$$\gamma_{sup-up} = rand(\gamma_{sup-up} * 10) \text{ and } C_{sup-up} = rand(C_{sup-up} * 10). \tag{9}$$

That is, indifferently for γ and C, the values of the *inf-down* and *sup-up* components are random values always taken farther the current parameter (γ or C), in order to increase the diversification capability of the metaheuristic; while the values of the *inf-up* and *sup-down* components are random values always taken closer the current parameter, in order to increase the intensification strength around the current parameter. This perturbation setting allows a good balance among the intensification and diversification factors. Otherwise, if in the acceptance criterion the new incumbent solution, γ' and C', is better than the current one, γ and C, i.e. $Acc' > Acc$, then this new solution becomes the best solution to date ($\gamma \leftarrow \gamma'$, $C \leftarrow C'$), and $range_\gamma$ and $range_C$ are updated by following Eqs. 1 and 2. This procedure continues iteratively until the termination conditions imposed by the user, such as maximum allowed CPU time, maximum number of iterations reached, or the maximum number of iterations between two successive improvements, are satisfied and, at the end of the algorithm, the best combination of γ and C parameters is produced as output of the procedure.

5   Experimental Analysis

To test the performance of the proposed algorithm we run some preliminary experiments on real clinical data which was available to us. We took a random sample of 1,500 instances from this dataset related to patients who had a surgery for prostate cancer, with attributes like pre-surgical information on laboratory data, histology data, time to progression free survival, etc. The aim is to predict if these patients progress to biochemical recurrence after surgery given the values taken from the other attributes using a SVM model. Instances with missing values were discarded from selection. Scaling in $[-1, 1]$ was employed to avoid feature values in greater numerical ranges dominating those in smaller numerical ranges, as well as to avoid the numerical difficulties during the calculation.

The proposed ILS used to tune the SVM model was implemented using the R platform. For the SVM implementation and the grid search we used the open source package *e1071* available in R^2. We implemented the ILS algorithm from scratch. For the grid search, the searching space for the parameters γ and C were set to $[10^{-5}, 10^{-4}, ..., 1, ..., 10^4, 10^5]$ for both parameters, for an overall of $11 * 11 = 121$ combinations of (γ, C) parameters tested each time. In order to ensure a fair computational comparison among the two methods, we store the overall computational time required by the grid search and we use it as stopping condition for the ILS algorithm. Our computational experiments were conducted on an Intel Quad-Core i5 5300U CPU (2.3 GHz) with 16 GB of RAM.

In order to guarantee valid results, a k-fold cross validation, with $k = 10$, was used to evaluate the classification accuracy [23]. That is, data was divided into ten subsets and, each time, one of the ten subsets was used as the test set and the other nine subsets formed the training set. Then the average error across all ten trials was computed. To test the performance of the two SVM models, we evaluated sensitivity, specificity, and the area under the Receiver Operating Characteristic curve (AUC), which are the typical statistical measures used for the performance of a binary classification test. Sensitivity, also called the true positive rate, or recall in some cases, measures the proportion of positives that are correctly identified as such. Specificity, also referred to as true negative rate, measures the proportion of negatives that are correctly identified as such. The receiver operating characteristic (ROC) curve is a graphical display that gives the measure of the predictive accuracy of a logistic model. The curve displays the true positive rate and false positive rate. AUC is the area under the ROC curve, which is one of the best methods for comparing classifiers in two-class problems.

Table 1. Computational results for the 10-fold cross validation.

n	Sensitivity		Specificity		AUC	
	ILS	Grid	ILS	Grid	ILS	Grid
1	0.63	0.63	0.73	0.64	0.72	0.68
2	0.74	0.74	0.69	0.65	0.77	0.66
3	0.75	0.65	0.69	0.69	0.75	0.70
4	0.68	0.68	0.80	0.63	0.80	0.68
5	0.75	0.67	0.80	0.68	0.84	0.80
6	0.52	0.48	0.82	0.79	0.73	0.68
7	0.78	0.78	0.67	0.65	0.78	0.74
8	0.58	0.58	0.75	0.67	0.69	0.64
9	0.75	0.71	0.74	0.67	0.78	0.76
10	0.64	0.59	0.63	0.62	0.70	0.66
Avg:	0.68	0.65	0.73	0.67	0.76	0.70

---

[2] https://cran.r-project.org/web/packages/e1071/index.html.

Our results are shown in Table 1. The table shows sensitivity, specificity and AUC values over the 10-fold cross validation obtained by the two SVM models tuned with ILS and grid search, respectively. It can be observed that the average accuracy achieved by the developed SVM tuned with the ILS is always higher than that of the SVM tuned with the grid search. Indeed, sensitivity, specificity and AUC values obtained by the first method are always larger, or equal, to those obtained by the SVM with the grid search method. When ILS obtains a same value of the grid search, is because it obtained better performance on the other measures, like for example it happens in the first subsample, where they get the same sensitivity, but specificity and AUC are considerably larger in the ILS method. On the average, the ILS and grid search methods obtain, respectively, the following measures: sensitivity, 0.68 against 0.65; specificity, 0.73 against 0.67, AUC, 0.76 against 0.70. The superiority of the Iterated Local Search tuning method against the grid search is further evidenced by Fig. 1, where the bar plots of the AUC for the 10-fold cross validation for the two methods are shown (ILS, in green, and grid search, in blue).

It is evident that the proposed ILS for tuning the SVM obtains more appropriate parameters, performing better than the grid search method over the considered sample dataset. The better performance of the proposed method can be attributed to its adaptive control behaviour during the tuning phase of the parameters, while the grid search method, being a local search method in nature, can be stuck on local optima during the search process, wasting time in less promising areas of the search space.

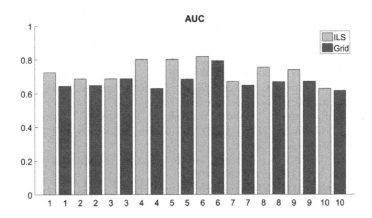

Fig. 1. Bar plot of the AUC obtained by the two approaches. (Color figure online)

6   Conclusions and Future Research

We have defined an implementation of Iterated Local Search to optimize the C and γ values for a radial-basis kernel. On our preliminary computational experiments over same real sample clinical data available to us, we have shown that our method outperforms a standard grid search for the task of setting C and γ parameters on this data set. By exploiting promising regions of the search domain, derived from the search history of previously found local optima, the Iterated Local Search method provides additional exploration of the search space over grid search while taking care not to waste too much time in areas where the results are not good. While the method was applied to a radial-basis kernel in this paper, it translates too other kernels as well. Future research should investigate to what extent this method can outperform grid search on other kernels and on other heterogeneous datasets, coming also from different domains. Furthermore, the efficiency of the method on different sizes of data sets and feature sets should be investigated. Extending the method to include a penalty for model complexity in terms of the number of included features could provide a nice encompassing method for feature selection. In future work we plan also to compare the performance of the proposed algorithm against other common metaheuristics for tuning SVM hyperparameters, in particular evolutionary algorithms. In addition we aim at a better statistical analysis of the results, showing also how the grid search results improve using a denser grid, and how this affects the computational time in comparison to the ILS.

References

1. Aarts, E., Korst, J., Michiels, W.: Simulated annealing. Search Methodologies: Introductory Tutorials in Optimization and Decision Support Techniques, pp. 187–210 (2005)
2. Alpaydin, E.: Introduction to Machine Learning (Adaptive Computation and Machine Learning). The MIT Press, Cambridge (2009)
3. Anaissi, A., Goyal, M., Catchpoole, D.R., Braytee, A., Kennedy, P.J.: Ensemble feature learning of genomic data using support vector machine. PLoS One 11(6), 1 June 2016, Article Number e0157330 (2016)
4. Balaprakash, P., Birattari, M., Stützle, T.: Improvement strategies for the F-race algorithm: sampling design and iterative refinement. In: Bartz-Beielstein, T., Blesa Aguilera, M.J., Blum, C., Naujoks, B., Roli, A., Rudolph, G., Sampels, M. (eds.) HM 2007. LNCS, vol. 4771, pp. 108–122. Springer, Heidelberg (2007). https://doi.org/10.1007/978-3-540-75514-2_9
5. Bergstra, J., Bengio, Y.: Random search for hyper-parameter optimization. J. Mach. Learn. Res. **13**(1), 281–305 (2012)
6. Ceylan, O., Taşkn, G.: SVM parameter selection based on harmony search with an application to hyperspectral image classification. In: 24th Signal Processing and Communication Application Conference (SIU), pp. 657–660 (2016)
7. Cherkassky, V., Ma, Y.: Practical selection of SVM parameters and noise estimation for SVM regression. Neural Networks **17**(1), 113–126 (2004)
8. Conca, P., Stracquadanio, G., Nicosia, G.: Automatic tuning of algorithms through sensitivity minimization. In: Pardalos, P., Pavone, M., Farinella, G.M., Cutello, V. (eds.) MOD 2015. LNCS, vol. 9432, pp. 14–25. Springer, Cham (2015). https://doi.org/10.1007/978-3-319-27926-8_2
9. Cortes, C., Vapnik, V.N.: Support-vector networks. Mach. Learn. **20**(3), 273–297 (1995)

10. Gatos, I., Tsantis, S., Spiliopoulos, S., Karnabatidis, D., Theotokas, I., Zoumpoulis, P., Loupas, T., Hazle, J.D., Kagadis, G.C.: A new computer aided diagnosis system for evaluation of chronic liver disease with ultrasound shear wave elastography imaging. Med. Phys. **43**(3), 1428–1436 (2016)
11. Hansen, P., Mladenović, N., Moreno-Pérez, J.A.: Variable neighbourhood search: methods and applications. Ann. Oper. Res. **175**(1), 367–407 (2010)
12. Hutter, F., Hoos, H.H., Leyton-Brown, K.: Sequential model-based optimization for general algorithm configuration. In: Coello, C.A.C. (ed.) LION 2011. LNCS, vol. 6683, pp. 507–523. Springer, Heidelberg (2011). https://doi.org/10.1007/978-3-642-25566-3_40
13. Hutter, F., Stützle, T., Leyton-Brown, K., Hoos, H.H.: ParamILS: an automatic algorithm configuration framework. J. Artif. Intell. Res. **36**(1), 267–306 (2009)
14. Joachims, T.: Text categorization with support vector machines: learning with many relevant features. In: Nédellec, C., Rouveirol, C. (eds.) ECML 1998. LNCS, vol. 1398, pp. 137–142. Springer, Heidelberg (1998). https://doi.org/10.1007/BFb0026683
15. Kecman, V.: Learning and Soft Computing. The MIT Press, Cambridge (2001)
16. Keerthi, S.: Efficient tuning of SVM hyperparameters using radius/margin bound and iterative algorithms. IEEE Trans. Neural Networks **13**(5), 1225–1229 (2002)
17. Kwok, J.T., Tsang, I.W.: Linear dependency between ϵ and the input noise in ϵ-support vector regression. IEEE Trans. Neural Networks **14**(3), 544–553 (2003)
18. Lameski, P., Zdravevski, E., Mingov, R., Kulakov, A.: SVM parameter tuning with grid search and its impact on reduction of model over-fitting. In: Yao, Y., Hu, Q., Yu, H., Grzymala-Busse, J.W. (eds.) RSFDGrC 2015. LNCS (LNAI), vol. 9437, pp. 464–474. Springer, Cham (2015). https://doi.org/10.1007/978-3-319-25783-9_41
19. López-Ibáñez, M., Dubois-Lacoste, J., Pérez-Cáceres, L., Birattari, M., Stützle, T.: The irace package: Iterated racing for automatic algorithm configuration. Operat. Res. Perspect. **3**, 43–58 (2016)
20. Lourenço, H.R.: Job-shop scheduling: computational study of local search and large-step optimization methods. Eur. J. Oper. Res. **83**(2), 347–364 (1995)
21. Lourenço, H.R., Martin, O.C., Stützle, T.: Iterated local search: framework and applications. In: Gendreau, M., Potvin, J.Y. (eds.) Handbook of Metaheuristics. International Series in Operations Research & Management Science, vol. 146, pp. 363–397. Springer, Boston (2010). https://doi.org/10.1007/978-1-4419-1665-5_12
22. Mattera, D., Haykin, S.: Support vector machines for dynamic reconstruction of a chaotic system. In: Schölkopf, B., Burges, C.J.C., Smola, A.J. (eds.) Advances in Kernel Methods, pp. 211–241. MIT Press, Cambridge (1999)
23. McLachlan, G.J., Do, K.-A., Ambroise, C.: Analyzing Microarray Gene Expression Data. Wiley, New York (2004)
24. Melvin, I., Ie, E., Kuang, R., Weston, J., Stafford, W.N.N., Leslie, C.: SVM-Fold: a tool for discriminative multi-class protein fold and superfamily recognition. BMC Bioinform. **8**(Suppl. 4), S2 (2007)
25. Osuna, E., Freund, R., Girosit, F.: Training support vector machines: an application to face detection. In: Proceedings of the 1997 Conference on Computer Vision and Pattern Recognition (CVPR 1997), pp. 130–137. IEEE Computer Society (1997)
26. Pardalos, P.M., Resende, M.G.C.: Handbook of Applied Optimization. Oxford University Press, Oxford (2002)
27. Shawe-Taylor, J., Cristianini, N.: Kernel Methods for Pattern Analysis. Cambridge University Press, New York (2004)

28. Sherin, B.M., Supriya, M.H.: Selection and parameter optimization of SVM kernel function for underwater target classification. In: 2015 IEEE Underwater Technology (UT), pp. 1–5 (2015)
29. Vapnik, V.N.: The Nature of Statistical Learning Theory. Springer, New York (2000)
30. Vos, P.C., Hambrock, T., Hulsbergen van de Kaa, C.A., Futterer, J.J., Barentsz, J.O., Huisman, H.J.: Computerized analysis of prostate lesions in the peripheral zone using dynamic contrast enhanced MRI. Med. Phys. **35**(3), 888–899 (2008)
31. Yang, C., Ding, L., Liao, S.: Parameter tuning via kernel matrix approximation for support vector machine. J. Comput. **7**(8), 2047–2054 (2012)

Projective Approximation Based Quasi-Newton Methods

Alexander Senov[✉]

Control of Complex Systems Laboratory, Institute of Problems of Mechanical
Engineering, V.O., Bolshoj pr., 61, St. Petersburg 199178, Russia
alexander.senov@gmail.com
http://math.spbu.ru/eng

Abstract. We consider a problem of optimizing convex function of vector parameter. Many quasi-Newton optimization methods require to construct and store an approximation of Hessian matrix or its inverse to take function curvature into account, thus imposing high computational and memory requirements. We propose four quasi-Newton methods based on consecutive projective approximation. The idea of these methods is to approximate the product of the function Hessian inverse and function gradient in a low-dimensional space using appropriate projection and then reconstruct it back to original space as a new direction for the next estimate search. By exploiting Hessian rank deficiency in a special way it does not require to store Hessian matrix neither its inverse thus reducing memory requirements. We give a theoretical motivation for the proposed algorithms and prove several properties of corresponding estimates. Finally, we provide a comparison of the proposed methods with several existing ones on modelled data. Despite the fact that the proposed algorithms turned out to be inferior to the limited memory Broyden-Fletcher-Goldfarb-Shanno (L-BFGS) one, they have important advantage of being easy to extent and improve. Moreover, two of them do not require the function gradient knowledge.

Keywords: Least-squares · Function approximation
Convex optimization · Iterative methods · Quadratic programming
Quasi-Newton methods · Projective methods

1 Introduction

Mathematical optimization is a very popular and widely used tool in multiple science and engineering problems that are to maximising or minimising some quantity. It is a core technique for solving many machine learning problems, particularly in big data related areas (e.g., see [2,11]).

Assuming function differentiability one may choose among many zero- or first-order iterative optimization algorithms (for their list, see [3,12,15,17]). However, most of them treat function as linear in particular point neighbourhood, ignoring its quadratic constituent.

© Springer International Publishing AG 2018
G. Nicosia et al. (Eds.): MOD 2017, LNCS 10710, pp. 29–40, 2018.
https://doi.org/10.1007/978-3-319-72926-8_3

If the function were twice differentiable and its Hessian was known, then one may use Newton method thus achieving quadratic convergence rate. Unfortunately, second derivative is often unknown in machine learning problems, or its calculation become infeasible in high-dimensional case especially frequently arising in big data areas. Quasi-Newton methods that iteratively approximate Hessian matrix on the basis of the parameter changes and the function gradient changes seem to be a reasonable trade-off between Newton method and first-order optimization methods, combining high convergence speed of Newton method and weak function requirements (e.g. knowledge of Hessian).

Most quasi-Newton algorithms work in iterative fashion updating current Hessian matrix approximation consequently performing rank one or rank two update based on gradient and point values at each iteration: Davidon-Fletcher-Powell update [6,8], Symmetric Rank 1 update [5] and possibly the most widely used, Broyden-Fltecher-Goldfarb-Shanno update [4,8]. This approach has one major drawback: it requires storing entire Hessian matrix or its inverse in memory. To overcome this obstacle, the limited-memory BFGS (L-BFGS) method was proposed in [16]. Despite the fact that quasi-Newton methods mentioned above do use history of points and gradient vectors to approximate Hessian matrix, they use these vectors in iterative fashion and do not attempt to reconstruct the objective function itself.

In contrast, another type of optimization algorithms—surrogate optimization methods—do use entire or truncated history of estimates and function values at those points. Surrogate optimization methods iteratively approximate the objective function by another function called *surrogate* on the basis of a set of points and function values at these points and take the optimum estimate based on obtained surrogate. Despite many advantages, most surrogate models share common drawbacks: they are memory and time consuming and, what is most important, their quality depends on the chosen surrogate model adequacy with respect to original function [10].

Additionally, so-called *multi-step* optimization methods do use history too [17]. Most of them use only fixed amount of history, e.g. *two-step Heavy-ball method* uses information from two past steps only. Others do use parametrized amount of history, like *multi-step quasi-Newton method* [9] as well as L-BFGS [16], but in slightly different way (e.g., not using the projection trick, or without explicit quadratic approximation).

There are a number of previous work regarding the projection idea. E.g. in [14] authors proposed method which use a variant of L-BFGS algorithm but storing gradient and estimates steps vectors not in original space but project it into a low-dimensional space spanned on last m gradient and estimate vectors. Compressed sensing is a powerful technique for signal recovery solely based on projection idea [7]. In fact, L-BFGS and many quasi-Newton methods use a low-rank Hessian matrix approximation that can be interpreted as implicit projection.

In this paper we propose several optimization methods that implicitly approximate Hessian matrix using *projective approximation*. By projective approximation we understand the following procedure which we denote by acronym *PAR* performed on each iteration.

1. **P**rojection step: rectangular matrix with orthonormal rows is constructed and then used to project parameters values history to a low-dimensional space.
2. **A**pproximation step: projected parameters together with corresponding function values are been approximated with quadratic polynomial.
3. **R**econstruction step: finally, Hessian matrix is calculated using estimated quadratic polynomial coefficients and rectangular matrix from step 1.

This procedure is described in detail in Sect. 3. Worth noting that in paper [18] author utilized the similar procedure for gradient descent method acceleration.

The paper is organized as follows. In Sect. 2 we give some preliminary information for further discussion. In Sect. 3 we propose four quasi-Newton algorithms based on projective approximation idea together with some explanation. In Sect. 4 we provide theoretical motivation behind the proposed algorithms and prove some of their properties. Further, in Sect. 5 we perform a comparative analysis of these algorithms together with gradient descent (GD), BFGS and L-BFGS methods on modelled data. Finally, Sect. 6 brings the conclusion.

2   Preliminaries

2.1   Notation Remarks

We use small light symbols x for scalars and indexes (mainly), small bold symbols $\mathbf{x}$ for vectors, $\mathbf{x}^{(i)}$ for i-th vector element, capital light symbols X for constants and sets (except parameter matrix Θ), capital bold symbols $\mathbf{X}$ for matrices, $\mathbf{X}_{i,\cdot}$ for i-th row and $\mathbf{X}_{\cdot,j}$ for j-th column, $\mathrm{diag}^{-1} : \mathbb{R}^{d \times d} \to \mathbb{R}^d$ is an operator which transform a matrix to a vector consisting of the matrix diagonal elements: $\mathrm{diag}^{-1}(\mathbf{X})^{(i)} = (\mathbf{X})_{i,i}$. Specifically we denote t as iteration index, T as total number of iterations, d as dimension of original space, q as dimension of projective space, $\mathbf{P} \in \mathbb{R}^{q \times d}$ as matrix with orthonormal rows used for projection, $\mathbf{I}$ as identity matrix (its size follows from the context), $^\top$ as transpose sign, $\hat{}$ as estimate sign, f as function and $\nabla_{\mathbf{x}} f$ or ∇f as function gradient with respect to parameter vector $\mathbf{x}$.

2.2   Quadratic Response Surface Methodology

Quadratic response surface methodology (QRSM) is a surrogate optimization method where 2nd order polynomial constructed via polynomial regression used as a surrogate. It uses a set of points and corresponding objective function measurements to interpolate them in $\mathbb{R}^{d+1}$ using quadratic least squares technique [1]. I.e., having a set of points $\{\mathbf{x}_i\}_{i=1}^n \subset \mathbb{R}^d$ and corresponding function values $\{y_i\}_{i=1}^n \subset \mathbb{R}$, quadratic least squares method estimates the 2nd order polynomial coefficients by solving the following optimization problem analytically (since polynomial is linear in all coefficients):

$$\hat{\mathbf{A}}, \hat{\mathbf{b}}, \hat{c} = \underset{\mathbf{A} \in \mathbb{R}^{d \times d}, \, \mathbf{A}^{\top} = \mathbf{A}, \, \mathbf{b} \in \mathbb{R}^{d}, \, c \in \mathbb{R}}{\operatorname{argmin}} \sum_{i=1}^{n} \left(\frac{1}{2} \mathbf{x}_i^{\top} \mathbf{A} \mathbf{x}_i + \mathbf{b}^{\top} \mathbf{x}_i + c - y_i \right)^2 \quad (1)$$

argmin of obtained 2nd order polynomial which is equal to $-\hat{\mathbf{A}}^{-1}\hat{\mathbf{b}}$ is then used as a next objective function optimum estimate.

One of the polynomial least squared and particularly of quadratic response surface methodology drawbacks is that they are inapplicable in high-dimensional problems [1]: straightforward 2nd order polynomial reconstruction requires $\mathcal{O}\left(Nd^2 + d^4\right)$ in memory and $\mathcal{O}\left(d^6 + d^4 N\right)$ in time. Clearly this is inappropriate in case of big d.

2.3   Quasi-Newton Optimization Methods

Quasi-Newton algorithms iteratively improve optimum point estimate $\mathbf{x}_t$ by moving it in the direction to $\widehat{\mathbf{H}}_t^{-1}\mathbf{g}_t$, where matrix $\widehat{\mathbf{H}}$ is the Hessian estimate and $\mathbf{g}$ is either the function gradient in $\mathbf{x}_t$ or its estimate. Some quasi-Newton algorithms maintain a low-rank Hessian approximation and exploit its rank deficiency to directly approximate resulted vector $\left(\nabla^2 f(\mathbf{x}_t)\right)^{-1} \nabla f(\mathbf{x}_t)$ without explicit Hessian matrix construction and inversion (L-BFGS is one of examples [16]). A general optimization procedure including both this approaches presented in Algorithm 1 where all the logic is in the CalcDirection procedure.

Algorithm 1. QuasiNewtonNoHessian($f, \mathbf{x}_0,$ CalcDirection, lineSearch, T)

1: **for** $t \leftarrow 1$ to T **do**
2:     $\triangle\mathbf{x}_t \leftarrow$ CalcDirection($f, \mathbf{x}_{t-1}$)
3:     $\lambda_t \leftarrow \operatorname{argmin}_\lambda f(\mathbf{x}_{t-1} + \lambda \triangle\mathbf{x}_t)$
4:     $\mathbf{x}_t \leftarrow \mathbf{x}_{t-1} + \lambda_t \triangle\mathbf{x}_t$
5: **return** $\mathbf{x}_T$

---

3   Algorithm Descriptions

In this section we describe four quasi-Newton algorithms all exploiting the projective-approximation-reconstruction idea briefly described above. First, we present an algorithms pseudocode and then give some comments and explanations.

Algorithm 2 contains general PAR-based implementation of *CalcDirection* procedure from Algorithm 1. Unfortunately we do not provide any study of the parameters q and m influence on the algorithm quality due to paper size restrictions. However, to knowledge there is usually no need to select $q > 1$ and optimal value for parameter m is highly dependant on the particular function.

As one can see, entire projection-approximation-reconstruction pipeline is encapsulated in the PAR($\mathbf{y}, \mathbf{X}, \mathbf{G}$) procedure. Four proposed Algorithms 3, 4, 5 and 6) describe particular implementations of the PAR procedure. In turn all of

Algorithm 2. GeneralProjectiveApproximationCalcDirection$(f, \mathbf{x})$

Parameters:
1: q — projective space dimension
2: m — number of estimates, gradients and function values stored
   Initialize:
3: $\mathbf{G} \leftarrow \mathbf{0} \in \mathbb{R}^{m \times d}$
4: $\mathbf{X} \leftarrow \mathbf{0} \in \mathbb{R}^{m \times d}$
5: $\mathbf{y} \leftarrow \mathbf{0} \in \mathbb{R}^m$
6: $t \leftarrow 0$
   Evaluate:
7: $t \leftarrow t + 1$
8: **if** $t \leq m$ **then**
9:      $y_t \leftarrow f(\mathbf{x})$
10:     $\mathbf{X}_{t,\cdot} \leftarrow \mathbf{x}^\top$
11:     $\mathbf{G}_{t,\cdot} \leftarrow \nabla f(\mathbf{x})^\top$
12:     **return** $\nabla f(\mathbf{x})$
13: **else**
14:     $\mathbf{y} \leftarrow (y_2, \ldots, y_m, f(\mathbf{x}))^\top$
15:     $\mathbf{X} \leftarrow [\mathbf{X}_{2,\cdot}^\top, \ldots, \mathbf{X}_{m,\cdot}^\top, \mathbf{x}]^\top$
16:     $\mathbf{G} \leftarrow [\mathbf{G}_{2,\cdot}^\top, \ldots, \mathbf{G}_{m,\cdot}^\top, \nabla f(\mathbf{x})]^\top$
17:     $\triangle\mathbf{x} \leftarrow \text{PAR}(\mathbf{y}, \mathbf{X}, \mathbf{G})$
18:     **return** $\triangle\mathbf{x}$

Algorithm 3. PAR1$(y, \mathbf{X}, \mathbf{G})$

1: $\mathbf{P} \leftarrow \text{Gram-Schmidt}(\mathbf{G}_{m,\cdot}, \ldots, \mathbf{G}_{m-q+1,\cdot})$
2: $\mathbf{Z} \leftarrow \mathbf{X}\mathbf{P}^\top$ ▷ Projection
3: $\widehat{\mathbf{Q}} \leftarrow \text{QuadraticLeastSquares}(\mathbf{Z}, \text{diag}^{-1}(\mathbf{G}\mathbf{X}^\top) - \mathbf{y})$ ▷ Approximation
4: $\triangle\mathbf{x} \leftarrow -\mathbf{P}^\top\widehat{\mathbf{Q}}^{-1}\mathbf{P}\mathbf{G}_{m,\cdot} + (\mathbf{I} - \mathbf{P}^\top\mathbf{P})\sum_{i=1}^{m}\mathbf{X}_{i,\cdot}$ ▷ Reconstruction
5: **return** $\triangle\mathbf{x}$

Algorithm 4. PAR2$(y, \mathbf{X}, \mathbf{G})$

1: $\mathbf{P} \leftarrow \text{Gram-Schmidt}(\xi_1, \ldots, \xi_q)$ ▷ $\xi_i \sim \mathcal{N}(0, \mathbf{I}_d)$
2: $\mathbf{Z} \leftarrow \mathbf{X}\mathbf{P}^\top$ ▷ Projection
3: $\widehat{\mathbf{Q}} \leftarrow \text{QuadraticLeastSquares}(\mathbf{Z}, \text{diag}^{-1}(\mathbf{G}\mathbf{X}^\top) - \mathbf{y})$ ▷ Approximation
4: $\triangle\mathbf{x} \leftarrow -\mathbf{P}^\top\widehat{\mathbf{Q}}^{-1}\mathbf{P}\mathbf{G}_{m,\cdot} + (\mathbf{I} - \mathbf{P}^\top\mathbf{P})\sum_{i=1}^{m}\mathbf{X}_{i,\cdot}$ ▷ Reconstruction
5: **return** $\triangle\mathbf{x}$

Algorithm 5. PAR3$(y, \mathbf{X}, \mathbf{G})$

1: $\mathbf{P} \leftarrow \text{Gram-Schmidt}(\mathbf{G}_{m,\cdot}, \ldots, \mathbf{G}_{m-q+1,\cdot})$
2: $\mathbf{Z} \leftarrow \mathbf{X}\mathbf{P}^\top$ ▷ Projection
3: $\widehat{\mathbf{Q}}, \widehat{\mathbf{p}} \leftarrow \text{QuadraticLeastSquares}(\mathbf{Z}, \mathbf{y})$ ▷ Approximation
4: $\triangle\mathbf{x} \leftarrow \mathbf{X}_{m,\cdot} - \mathbf{P}^\top\widehat{\mathbf{Q}}^{-1}\widehat{\mathbf{p}} + (\mathbf{I} - \mathbf{P}^\top\mathbf{P})\sum_{i=1}^{m}\mathbf{X}_{i,\cdot}$ ▷ Reconstruction
5: **return** $\triangle\mathbf{x}$

Algorithm 6. PAR4$(y, \mathbf{X}, \mathbf{G})$

1: $\mathbf{P} \leftarrow$ Gram-Schmidt$(\xi_1, \ldots, \xi_q)$ ▷ $\xi_i \sim \mathcal{N}(0, \mathbf{I}_d)$
2: $\mathbf{Z} \leftarrow \mathbf{X}\mathbf{P}^\top$ ▷ Projection
3: $\widehat{\mathbf{Q}}, \widehat{\mathbf{p}} \leftarrow$ QuadraticLeastSquares$(\mathbf{Z}, \mathbf{y})$ ▷ Approximation
4: $\triangle \mathbf{x} \leftarrow \mathbf{X}_{m,\cdot} - \mathbf{P}^\top \widehat{\mathbf{Q}}^{-1} \widehat{\mathbf{p}} + (\mathbf{I} - \mathbf{P}^\top \mathbf{P}) \sum_{i=1}^m \mathbf{X}_{i,\cdot}$ ▷ Reconstruction
5: **return** $\triangle \mathbf{x}$

---

them a largely based on QuadraticLeastSquares and Gram-Schmidt procedures. QuadraticLeastSquares procedure solves the (1) problem and Gram-Schmidt method perform consecutive vectors orthonormalization using modified (stabilized) Gram-Schmidt orthogonalization (MGS) method (see, e.g. [13]). Random gaussian vectors $\xi_i \sim \mathcal{N}(0, \mathbf{I}_d)$ generated by consequently generating their elements $\xi_i^{(j)} \sim \mathcal{N}(0, 1)$.

Algorithms 3 and 5 construct matrix $\mathbf{P}$ using orthonormalization of last q gradient vectors as rows, while Algorithms 4 and 6 use q random orthonormal vectors. Algorithms 3 and 4 use quadratic least squares method to estimate quadratic dependency between projected vectors $\mathbf{Z}$ and $\text{diag}^{-1}(\mathbf{G}\mathbf{X}^T) - \mathbf{y}$ instead of raw function values y since it removes linear term (see Remark 2). Thus, there is no need to estimate linear coefficients of the approximating polynomial. These algorithms use special form of "reconstruction" of the estimated quadratic coefficients from the low-dimensional to the original space: in addition to $\mathbf{P}^\top \widehat{\mathbf{Q}}^{-1} \mathbf{P} \mathbf{G}_{m,\cdot}$ term which approximates $(\nabla^2 f(\mathbf{x}))^{-1} \nabla f(\mathbf{x})$ (see Proposition 3) special *projection correction* term $(\mathbf{I} - \mathbf{P}^\top \mathbf{P}) \sum_{i=1}^m \mathbf{X}_{i,\cdot}$ is used to reduce an effect of the projection (see Proposition 1). In contrast, Algorithms 5 and 6 reconstruct linear term with quadratic least squares and use it in reconstruction step, where $(\nabla^2 f(\mathbf{x}))^{-1} \nabla f(\mathbf{x})$ is approximated by $\mathbf{X}_{m,\cdot} - \mathbf{P}^\top \widehat{\mathbf{Q}}^{-1} \widehat{\mathbf{p}}$ (see Proposition 2).

Table 1. PAR algorithms classification by the way projection, approximation and reconstruction steps are performed.

Algorithm	Projection	Approximation	Reconstruction
PAR1	Orthonormalized gradients	Quadratic least squares fixed linear coefficients	Use Hessian approximation from Proposition 3
PAR2	Random projection	Quadratic least squares fixed linear coefficients	Use Hessian approximation from Proposition 3
PAR3	Orthonormalized gradients	Quadratic least squares	Direct minimum reconstruction from Proposition 2
PAR4	Random projection	Quadratic least squares	Direct minimum reconstruction from Proposition 2

Thus, Algorithms 3, 4, 5 and 6 can be categorized by the way how projection, approximation and reconstruction steps are performed. This categorization is presented in Table 1. Finally, we have

Remark 1. General Algorithm 2 utilize $\mathcal{O}\left(md\right)$ in memory and $\mathcal{O}\left(d\right)$ in operations at each iteration. In addition, Algorithms 3, 4, 5 and 6 utilize $\mathcal{O}\left(q^2m + q^4\right)$ in memory and $\mathcal{O}(qd + q^{4.76} + mq^4 + md)$ in number of operations.

4   Theoretical Ground

In this section we present several propositions that motivates the proposed algorithms and estimate some of their properties. At first, we bring the context and then describe the propositions in this context. Proofs are given in the Appendix A.

Here is the context. Consider function $f : \mathbb{R}^d \to \mathbb{R}$: $f(x) = \frac{1}{2}\mathbf{x}^\top\mathbf{H}\mathbf{x}+\mathbf{b}^\top\mathbf{x}+c$, where $\mathbf{H} \in \mathbb{R}^{d\times d}$, $\mathbf{H} \succ 0$, $\mathbf{b} \in \mathbb{R}^d$, $c \in \mathbb{R}$, $\mathbf{P} \in \mathbb{R}^{q\times d}$, $\mathbf{PP}^\top = \mathbf{I}_q$, $q < d$. Moreover, consider sequence of in some way related points $\{\mathbf{x}_t\}_{t=1}^K$ (e.g. consecutive estimates from optimization algorithm). In Proposition 3 we additionally assume that function f has additive error term with centralized normal distribution. We use quadratic function as a approximation of convex function in some point neighbourhood. The noise term includes both errors/uncertainties and model inadequacy.

Remark 2. Note, that difference between gradient and function itself removes linear term in quadratic function: $\nabla f(\mathbf{x}) - f(x) = \frac{1}{2}\mathbf{x}^\top\mathbf{H}\mathbf{x} - c$.

Now assume that we obtain $\widehat{\mathbf{z}}$—minimum with respect to the projection. If matrix $\mathbf{P}$ were square invertible matrix, we would simply set $\widehat{\mathbf{x}} = \mathbf{P}^{-1}\widehat{\mathbf{z}}$. Unfortunately, it obviously is not: any point $\mathbf{z} \in \mathbb{R}^q$ corresponds the entire set $\{\mathbf{x} \in \mathbb{R}^d : \mathbf{Px} = \mathbf{z}\}$. Hence we need to impose additional restrictions to pick specific point from this set. Since we are extending original sequence $\{\mathbf{x}_t\}_1^K$ picking the closest point to the original sequence in terms of Euclidean distance sounds reasonable. The following proposition gives an explicit expression for $\widehat{\mathbf{x}}$ that minimizes this distance.

Proposition 1. *Consider* $\{\mathbf{x}_1, \ldots, \mathbf{x}_K\} \subset \mathbb{R}^d$. *Then*

$$\widehat{\mathbf{x}} = \underset{\{\mathbf{x}\in\mathbb{R}^d : \mathbf{Px}=\widehat{\mathbf{z}}\}}{\operatorname{argmin}} \sum_{t=1}^K \|\mathbf{x}_t - \mathbf{x}\|_2^2 = (\mathbf{I} - \mathbf{P}^\top\mathbf{P})\frac{1}{K}\sum_1^K \mathbf{x}_t + \mathbf{P}^\top\widehat{\mathbf{z}}.$$

Having $\widehat{\mathbf{z}}$, the following proposition gives an explicit estimate of difference between corresponding $\widehat{\mathbf{x}}$ and $\operatorname{argmin} f$. It demonstrates that $\widehat{\mathbf{x}}$ is the best estimate in terms of the projection $\mathbf{P}$ and it obviously benefits from closeness of the preceding $\mathbf{x}_i$-th to the function minimum point.

Proposition 2. *Consider* $\{\mathbf{x}_t\}_1^K \subset \mathbb{R}^d$ *and corresponding function values* $\{y_t\}_1^K$, $y_t = f(\mathbf{x}_t)$. *Denote* $\bar{\mathbf{x}} = \sum_1^K \mathbf{x}_i/K$ *and* $\hat{\mathbf{x}} = -\mathbf{P}^\top\mathbf{P}\mathbf{H}^{-1}\mathbf{b} + (\mathbf{I} - \mathbf{P}^\top\mathbf{P})\bar{\mathbf{x}}$. *Then*

$$\|\operatorname{argmin} f - \hat{\mathbf{x}}\|_2^2 = \left\|(\mathbf{I} - \mathbf{P}^\top\mathbf{P})(\mathbf{H}^{-1}\mathbf{b} - \bar{\mathbf{x}})\right\|_2^2.$$

Finally, the following proposition consider a problem of Hessian estimation with projective approximation approach. It gives an upper bound for Hessian matrix estimation error in terms of Frobenius norm. Worth noting, that it consider not deterministic but stochastic variant of function f with additional noise term ε. Thus, it gives an upper bound as a random variable.

Proposition 3. *Consider function* $f(\mathbf{x}) = \frac{1}{2}\mathbf{x}^\top\mathbf{H}\mathbf{x} + \mathbf{b}^\top\mathbf{x} + c + \varepsilon$, *where* $\varepsilon \sim \mathcal{N}(0, \sigma_\varepsilon^2)$ *and* $\{\mathbf{x}_t\}_1^K \subset \mathbb{R}^d$ *with corresponding projections* $\{\mathbf{z}_t = \mathbf{P}\mathbf{x}_t\}_1^K \subset \mathbb{R}^q$ *and function values* $\{y_t = f(\mathbf{x}_t)\}_1^K \subset \mathbb{R}$. *Let* $\widehat{\mathbf{Q}}$ *be an estimation of Hessian matrix obtained via quadratic least squares applied to a set of points* $\{\mathbf{z}_t, y_t\}_1^K$. *Denote* $\widehat{\mathbf{H}} = \mathbf{P}^\top\widehat{\mathbf{Q}}\mathbf{P}$. *Then*

$$\left\|\mathbf{H} - \widehat{\mathbf{H}}\right\|_F^2 \le \left\|(\mathbf{I} - \mathbf{P}^\top\mathbf{P})\mathbf{H}(\mathbf{I} - \mathbf{P}^\top\mathbf{P})\right\|_F^2 + q^2\xi,$$

where $\xi \sim q^2 C(\mathbf{X}\mathbf{P}^\top)\chi^2(1)$ *and* C *is a positive scalar-valued function.*

Remark 3. One may note, that $\left\|(\mathbf{I} - \mathbf{P}^\top\mathbf{P})\mathbf{H}(\mathbf{I} - \mathbf{P}^\top\mathbf{P})\right\|_F^2$ achieve its minimum with respect to $\mathbf{P}$ if subspace spanned on rows of $\mathbf{P}$ coincides with subspace spanned on eigenvectors of $\mathbf{H}$ corresponding to q largest eigenvalues.

5   Modelling

In this section we evaluate four proposed algorithms together with gradient descent, BFGS and L-BFGS algorithms on simple modelling example. We use function $f(\mathbf{x}) = \frac{1}{2}\mathbf{x}^\top\mathbf{H}\mathbf{x} + \mathbf{b}^\top\mathbf{x} + c + \varepsilon$, where $\varepsilon \sim \mathcal{N}(0, \sigma_\varepsilon^2)$, $\mathbf{x} \in \mathbb{R}^8$, matrix $\mathbf{H} = \mathbf{L}\mathbf{L}^\top$ where $\mathbf{L}_{i,j} \sim \mathcal{N}(0,1)$ and $\mathbf{b}^{(i)} \sim \mathcal{N}(0,1)$.

Each algorithm was evaluated as follows: we run algorithm one hundred times on randomly generated f till absolute Euclidean norm between estimate and argmin f became smaller than 0.1 or till number of iterations exceeds 1000. We have used the following setup for each algorithm:

- GD: initial step rate equals to 0.01 with momentum equal to 0.9;
- L-BFGS: m (history size) equals to 3;
- PAR1, PAR2, PAR3, PAR4: q equals to 1, m equals to 3.

We consider number of iterations as an informative measure for methods comparison since on each iteration of gradient descent, L-BFGS, PAR1, PAR2, PAR3, PAR4 and BFGS algorithms function and its gradient are evaluated exactly once due to caching.

Table 2 contains 25%, 50% and 75%-percentiles of number of iterations for each algorithm. In addition, we present number of iterations as boxplot for each algorithm on Fig. 1 for clarity. As one can see, all proposed algorithms demonstrate much better convergence then gradient descent algorithm, especially Algorithms 3 and 5. However, L-BFGS algorithm surpass all of them: even PAR1 and PAR3 algorithms are at least twice as slow as L-BFGS. However, few interesting conclusions can be made. First, projection matrix constructed as orthonormalized gradients in few past points perform much better than random one. Second, explicit argmin estimation which is done in Algorithms 5 and 6 seems to work a little bit better than reconstruction approach used in Algorithms 3 and 4.

Table 2. Number of iterations percentiles

Algorithm	25%	50%	75%
GD	832	1000	1000
BFGS	34	56	84
L-BFGS	71	96	163
PAR1	96	178	391
PAR2	122	513	1000
PAR3	83	152	238
PAR4	126	592	1000

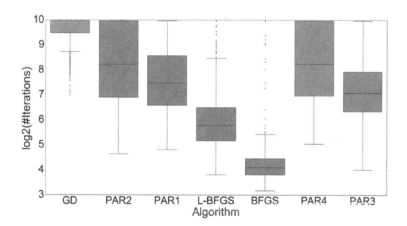

Fig. 1. Boxplots visualizing iterations number distribution for each algorithm. The bottom and top of each box are the first and third quartiles, and the band inside each boxes is the median, whiskers correspond to 0.05- and 0.95- quantiles.

6   Conclusion

We propose four quasi-Newton algorithms all sharing the same *projective approximation* idea based on implicit approximation of product between inverse Hessian and gradient in the low-dimensional space obtained through specifically constructed projection. We motivate these algorithms by several propositions and prove some properties of corresponding estimates, howbeit, no convergence proof is given. We perform a comparative analysis of the proposed algorithms with gradient descent, BFGS and L-BFGS algorithms. Despite the fact that the algorithms proposed turned to be less effective in number of iterations than L-BFGS, they have few important properties. First one is that projection-approximation-reconstruction steps which are implicit in L-BFGS algorithm are explicitly expressed in the proposed algorithms. The second and most important one is that since projective-approximation-reconstruction steps are expressed explicitly they can be easily improved or modified. In example, one can replace quadratic least squares by another approximation algorithm (e.g. by adding regularization, by using more complex and even non-convex surrogates), or modify the projection procedure. Thus, it opens up ample opportunities for further research. With regard to future research direction, we plan to apply the proposed algorithms for estimating parameters of deep convolutional neural networks, since in that case the problem of the parameter space high dimensionality is especially relevant.

Acknowledgments. This work was supported by Russian Science Foundation (project 16-19-00057).

A   Proofs

Proof (of Proposition 1)

$$\partial_{\mathbf{x}} \underset{\{\mathbf{x}\in\mathbb{R}^d : \mathbf{Px}=\widehat{\mathbf{z}}\}}{\operatorname{argmin}} \sum_{t=1}^{K} \|\mathbf{x}_t - \mathbf{x}\|_2^2$$

$$= \partial_{\mathbf{x}} \sum_{t=1}^{K} \left\|\mathbf{P}^{\top}\mathbf{P}\mathbf{x}_t + \left(\mathbf{I} - \mathbf{P}^{\top}\mathbf{P}\right)\mathbf{x}_t - \left(\mathbf{P}^{\top}\widehat{\mathbf{z}} + \left(\mathbf{I} - \mathbf{P}^{\top}\mathbf{P}\right)\mathbf{x}\right)\right\|_2^2$$

$$= \partial_{\mathbf{x}} \sum_{t=1}^{K} \left\|\mathbf{P}^{\top}\mathbf{P}\mathbf{x}_t - \mathbf{P}^{\top}\widehat{\mathbf{z}}\right\|_2^2 + \partial_{\mathbf{x}} \sum_{t=1}^{K} \left\|\left(\mathbf{I} - \mathbf{P}^{\top}\mathbf{P}\right)\left(\mathbf{x}_t - \mathbf{x}\right)\right\|_2^2$$

$$= \sum_{t=1}^{K} 2(\mathbf{I} - \mathbf{P}^{\top}\mathbf{P}) \left(\left(\mathbf{I} - \mathbf{P}^{\top}\mathbf{P}\right)\mathbf{x}_t - \left(\mathbf{I} - \mathbf{P}^{\top}\mathbf{P}\right)\mathbf{x}\right)$$

$$= 2(\mathbf{I} - \mathbf{P}^{\top}\mathbf{P}) \sum_{t=1}^{K} \mathbf{x}_t - 2K\left(\mathbf{I} - \mathbf{P}^{\top}\mathbf{P}\right)\mathbf{x} = 0.$$

Since $(\mathbf{I} - \mathbf{P}^{\top}\mathbf{P})$ is not invertible the above equation has infinite number of solutions. Hence, we are free to choose any one of them, e.g. $\mathbf{x} = \frac{1}{K}\sum_1^K \mathbf{x}_t$.   □

Proof (of Proposition 2). From Proposition 1 and the fact that $\operatorname{argmin}_{\mathbf{z} \in \mathbb{R}^q} f(\mathbf{P}^\top \mathbf{z} + \mathbf{v}) = -\mathbf{P}\mathbf{H}^{-1}\mathbf{b}$ it follows that $\widehat{\mathbf{x}} = (\mathbf{I} - \mathbf{P}^\top \mathbf{P}) \overline{\mathbf{x}} - \mathbf{P}^\top \mathbf{P}\mathbf{H}^{-1}\mathbf{b}$. Hence

$$
\begin{aligned}
\|\operatorname{argmin} f - \widehat{\mathbf{x}}\|_2^2 &= \left\|-\mathbf{H}^{-1}\mathbf{b} - \widehat{\mathbf{x}}\right\|_2^2 \\
&= \left\|-(\mathbf{I} - \mathbf{P}^\top \mathbf{P})\overline{\mathbf{x}} + \mathbf{P}^\top \mathbf{P}\mathbf{H}^{-1}\mathbf{b} - \mathbf{H}^{-1}\mathbf{b}\right\|_2^2 \\
&= \left\|(\mathbf{I} - \mathbf{P}^\top \mathbf{P})\left(\mathbf{H}^{-1}\mathbf{b} - \overline{\mathbf{x}}\right)\right\|_2^2.
\end{aligned}
$$

$\square$

Proof (of Proposition 3). First,

$$
\|\mathbf{H} - \widehat{\mathbf{H}}\|_F^2 = \|(\mathbf{I} - \mathbf{P}^\top \mathbf{P})\mathbf{H}(\mathbf{I} - \mathbf{P}^\top \mathbf{P})\|_F^2 + \|\mathbf{P}^\top(\mathbf{P}\mathbf{H}\mathbf{P}^\top - \widehat{\mathbf{Q}})\mathbf{P}\|_F^2
$$

One can note that $\widehat{\mathbf{Q}}_{i,j}$ are normally distributed variables s.t. $\mathrm{E}\left[\widehat{\mathbf{Q}}\right] = \mathbf{P}\mathbf{H}\mathbf{P}^\top$ (for example, see [1]). Moreover, consider a vectorization of the matrix $\widehat{\mathbf{Q}}$ upper triangle $\widehat{\theta}$:

$$
\widehat{\theta} = \left(\widehat{\mathbf{Q}}_{1,1}, \widehat{\mathbf{Q}}_{1,2}, \ldots, \widehat{\mathbf{Q}}_{q-1,q}, \widehat{\mathbf{Q}}_{q,q}\right),
$$

— its covariance matrix is equal to $\boldsymbol{\Sigma}_\theta = \frac{\sigma_\varepsilon}{m}\ddot{\mathbf{Z}}\ddot{\mathbf{Z}}^\top$, where $\ddot{\mathbf{Z}}_{i,\cdot}$ consist of quadratic elements of $\mathbf{z}_i = \mathbf{P}\mathbf{x}_i$:

$$
\ddot{\mathbf{Z}}_{i,\cdot} = \left(\mathbf{z}_i^{(1)}\mathbf{z}_i^{(1)}, \mathbf{z}_i^{(1)}\mathbf{z}_i^{(2)}, \ldots, \mathbf{z}_i^{(q-1)}\mathbf{z}_i^{(q)}, \mathbf{z}_i^{(q)}\mathbf{z}_i^{(q)}\right)^\top.
$$

Next, denote θ as a vectorization of the $\mathbf{P}\mathbf{H}\mathbf{P}^\top$ matrix upper triangle and consider eigendecomposition of $\boldsymbol{\Sigma}_\theta = \mathbf{U}\boldsymbol{\Lambda}\mathbf{U}^\top$. Then, vector $\widehat{\beta} = \mathbf{U}\widehat{\theta}$ would have gaussian distribution with covariance matrix $\boldsymbol{\Lambda}$, and

$$
\|\mathbf{P}^\top(\mathbf{P}\mathbf{H}\mathbf{P}^\top - \widehat{\mathbf{Q}})\mathbf{P}\|_F^2 \le \|\mathbf{P}^\top\|_F^2\|\mathbf{P}\mathbf{H}\mathbf{P}^\top - \widehat{\mathbf{Q}}\|_F^2\|\mathbf{P}\|_F^2 = q^2\|\mathbf{P}\mathbf{H}\mathbf{P}^\top - \widehat{\mathbf{Q}}\|_F^2.
$$

$$
\|\mathbf{P}\mathbf{H}\mathbf{P}^\top - \widehat{\mathbf{Q}}\|_F^2 = \|\widehat{\theta} - \theta\|_2^2 = \left(\mathbf{U}(\widehat{\theta} - \theta)\right)^\top \left(\mathbf{U}(\widehat{\theta} - \theta)\right) = \sum_{i=1}^{q^2} \xi_2 \sim \sum_{i=1}^{q^2} \lambda_i^2 \chi^2 \ (1).
$$

Thus, $C(\mathbf{X}\mathbf{P}^\top) = \max_i \lambda_i^2$.

$\square$

References

1. Box, G.E., Draper, N.R., et al.: Empirical Model-Building and Response Surfaces. Wiley, New York (1987)
2. Boyd, S.: Global optimization in control system analysis and design. In: Control and Dynamic Systems V53: High Performance Systems Techniques and Applications: Advances in Theory and Applications, vol. 53, p. 1 (2012)

3. Boyd, S., Vandenberghe, L.: Convex Optimization. Cambridge University Press, Cambridge (2004)
4. Broyden, C.G.: The convergence of a class of double-rank minimization algorithms 1. General considerations. IMA J. Appl. Math. **6**(1), 76–90 (1970)
5. Conn, A.R., Gould, N.I., Toint, P.L.: Convergence of quasi-Newton matrices generated by the symmetric rank one update. Math. Program. **50**(1–3), 177–195 (1991)
6. Davidon, W.C.: Variable metric method for minimization. SIAM J. Optim. **1**(1), 1–17 (1991)
7. Donoho, D.L.: Compressed sensing. IEEE Trans. Inf. Theor. **52**(4), 1289–1306 (2006)
8. Fletcher, R.: Practical Methods of Optimization, 2nd edn. Wiley, New York (1987)
9. Ford, J., Moghrabi, I.: Multi-step quasi-Newton methods for optimization. J. Comput. Appl. Math. **50**(1–3), 305–323 (1994)
10. Forrester, A., Keane, A.: Recent advances in surrogate-based optimization. Prog. Aerosp. Sci. **45**(1), 50–79 (2009)
11. Granichin, O., Volkovich, Z.V., Toledano-Kitai, D.: Randomized Algorithms in Automatic Control and Data Mining, vol. 67. Springer, Heidelberg (2015). https://doi.org/10.1007/978-3-642-54786-7
12. Granichin, O.N.: Stochastic approximation search algorithms with randomization at the input. Autom. Remote Control **76**(5), 762–775 (2015)
13. Hoffmann, W.: Iterative algorithms for Gram-Schmidt orthogonalization. Computing **41**(4), 335–348 (1989)
14. Krause, A.: SFO: a toolbox for submodular function optimization. J. Mach. Learn. Res. **11**, 1141–1144 (2010)
15. Nesterov, Y.: Introductory Lectures on Convex Programming Volume I: Basic Course. Citeseer (1998)
16. Nocedal, J.: Updating quasi-Newton matrices with limited storage. Math. Comput. **35**(151), 773–782 (1980)
17. Polyak, B.T.: Introduction to optimization. Translations series in mathematics and engineering. Optimization Software Inc., Publications Division, New York (1987)
18. Senov, A.: Accelerating gradient descent with projective response surface methodology. In: Battiti, R., Kvasov, D.E., Sergeyev, Y.D. (eds.) LION 2017. LNCS, vol. 10556, pp. 376–382. Springer, Cham (2017). https://doi.org/10.1007/978-3-319-69404-7_34

Intra-feature Random Forest Clustering

Michael Cohen[✉]

Galvanize, San Francisco, CA, USA
u6357432@anu.edu.au

Abstract. Clustering algorithms are commonly used to find structure in data without explicitly being told what they are looking for. One key desideratum of a clustering algorithm is that the clusters it identifies given some set of features will generalize well to features that have not been measured. Yeung et al. (2001) introduce a Figure of Merit closely aligned to this desideratum, which they use to evaluate clustering algorithms. Broadly, the Figure of Merit measures the within-cluster variance of features of the data that were not available to the clustering algorithm. Using this metric, Yeung et al. found no clustering algorithms that reliably outperformed k-means on a suite of real world datasets (Yeung et al. 2001). This paper presents a novel clustering algorithm, intra-feature random forest clustering (IRFC), that does outperform k-means on a variety of real world datasets per this metric. IRFC begins by training an ensemble of decision trees of limited depth to predict randomly selected features given the remaining features. It then aggregates the partitions that are implied by these trees, and outputs however many clusters are specified by an input parameter.

Keywords: Cluster analysis · Random forest · Unsupervised learning Ensemble · Figure of Merit

1 Introduction

One of the central challenges for unsupervised learning has been the lack of a universally accepted validation metric (Giancarlo et al. 2008). Giancarlo et al. review several possible validation techniques, but they only evaluate those validation metrics by their ability to identify the number of clusters that a dataset should be partitioned into given some ground truth. Furthermore, many of those validation measures are only coincident with the key desiderata of a clustering algorithm. Singh and Kim (1988) articulate the aim of clustering as follows: "The purpose of cluster analysis is to place objects into groups suggested by the data such that objects in a given group have tendency to be similar to each other, and objects in different clusters tend to be dissimilar." Of the seven metrics that Giancarlo et al. (2008) review, three of them only consider the stability of cluster assignments under modified conditions: Clest (Dudoit and Fridlyand 2002) considers stability of clusters after effectively subsetting the data, Consensus Clustering (Monti et al. 2003), the stability after bootstrapping, ME (Model Explorer) (Ben-Hur et al. 2001), the stability after adding random noise. While stability of assignments is certainly a nice feature, it is obviously not the primary aim of a clustering algorithm, or else we could create a perfect clustering algorithm that outputs the same assignments

© Springer International Publishing AG 2018
G. Nicosia et al. (Eds.): MOD 2017, LNCS 10710, pp. 41–49, 2018.
https://doi.org/10.1007/978-3-319-72926-8_4

every time, completely arbitrarily. Three other metrics considered by Giancarlo, et al. only validate cluster assignments with data that the clustering algorithm had access to: WCSS (Within Cluster Sum-of-Squares) (Kaufman and Rousseeuw 2009), the Gap Statistic (Tibshirani et al. 2001), and the KL (Krzanowski-Lai) Index (Krzanowski and Lai 1988). These approaches have analogous risks to validating on the training set in supervised learning, as opposed to withholding a test set. Yeung's Figure of Merit (Yeung et al. 2001), the remaining metric considered by Giancarlo et al., directly measures the suitability of a partition according to Singh's articulation of the purpose of clustering. To show this, we first review how the Figure of Merit is calculated, then we present a wide variety of examples from the literature of applications of clustering algorithms, and demonstrate that in all these applications, a clustering algorithm is useful insofar as it scores well according to Yeung's Figure of Merit. Naturally, that is strong support for the utility of this metric.

To compute the Figure of Merit, features are withheld one by one, analogously to k-fold cross validation. Then, the clustering algorithm is applied to the remaining features, and the within-cluster variance of the withheld feature is divided by the total variance of that feature. In other words, how much does the variance of the withheld feature go down when it is separated into clusters? Finally, that quantity is averaged over all features, again analogously to k-fold cross validation. An important distinction, however is that while k-fold cross validation withholds data points one at a time, this method withholds features. A lower Figure of Merit indicates "that objects in a given group have tendency to be similar to each other, and objects in different clusters tend to be dissimilar," (Singh 1998) even with respect to features the clustering algorithm did not have access to.

Consider the following applications of clustering. Hilas and Mastorocostas (2008) use clustering to detect telecommunication fraud. The clustering algorithm is therefore useful insofar as the clusters identified will have relatively low variance with respect to a feature the algorithm did not have access to: the Boolean value of whether the event was fraudulent. Masulli and Schenone (1999) use clustering of medical images to support diagnosis. The algorithm is useful insofar as members of a given group have lower variance with respect to their diagnoses than the entire sample does. The true diagnosis, again, is a feature the algorithm did not have access to. Iliadis (2005) use clustering to identify forest types to assist in fire risk estimation. The algorithm is useful insofar as clusters are created in such a way that the variance of a new feature (likelihood fire, in this case) is minimized within clusters. The utility of Li et al.'s (2009) transcriptomic clusters is their ability to discriminate glioma subtypes, a feature not available to the clustering algorithm. Harrigan's (1985) use of clustering to identify "strategic groups" among competitors in an industry is relevant insofar as companies in the same clusters deserve strategic treatment that is more similar than that of companies in different clusters. Companies in the same cluster are expected to respond similarly to a broad range of treatments, and to a greater degree than two companies selected at random, so however "expected company behavior" is measured, there should be lower within-cluster variance than total variance. Becker et al. (2011) identify clusters in people flow using cellular data, and use categories of movement patterns to evaluate the comparative utility of different urban developments to members of different clusters. If the variance of individuals' utilities from Project X is no less for members of a particular cluster than it is for the population as a whole, the cluster assignments are unhelpful. When Chicco et al. (2003) cluster electricity customers, they expect members of any given cluster to respond similarly to service regulations, but differently from members of other clusters. Again, this feature is not included in the original clustering, and the clustering algorithm is useful insofar as the clusters identified minimize within-cluster variance with respect to that new feature. Wang (2010) describes the utility of unsupervised market segmentation to the service industry. The utility arises from the expectation that customers in different market segments will respond uniquely to different sorts of targeting. Therefore, "customer response to Campaign X" needs to have a within-cluster-variance that is lower than the total variance in order for the clustering to be useful. Pham (1998) demonstrates the utility of clustering radar signals for the identification of aircrafts, aircraft identity being feature the clustering algorithm was not provided. Pavlidis et al. (2003) use clustering in financial forecasting; their clustering algorithm obviously does not have access to future financial data, but it is tasked with making partitions that identify data points with similar future-behavior. Park's (2002) forecasting task is to predict freeway traffic with the assistance of unsupervised methods, and the case is analogous to financial forecasting. This is a miniscule sample of the applications of clustering, but they begin to support the following generalization: cluster assignments are likely to be useful when and only when novel features tend to have low within-cluster-variance compared to their total variance.

An extraordinary variety of clustering algorithms have been proposed (Xu and Wunsch 2005), as well as many cluster ensembling methods for cluster analysis (Vega-Pons and Ruiz-Shulcloper 2011). In this paper, we present Intra-Feature Random Forest Clustering (IRFC), which represents a single clustering algorithm that implements the driving thesis of cluster ensembling: an ensemble of partitions, benefitting from a wisdom-of-crows effect, will generally outperform a single partition (Vega-Pons and Ruiz-Shulcloper 2011). IRFC, unsurprisingly given the name, also borrows extensively from the supervised learning algorithm Random Forest, in that both make use of an ensemble of decision tree regressors (Breiman 2001). The strong performance of IRFC with respect to Yeung's Figure of Merit is quite analogous to the strong performance of a random forest regressor with respect to the root-mean-square error of its predictions.

This paper first lays out the algorithm for IRFC, then describes its performance according to Yeung's Figure of Merit on a suite of real world data. Ultimately, it's strong performance with respect to that metric justifies the algorithm's utility in applied settings.

2   The Algorithm

IRFC consists of two stages: train many limited-depth decision trees to derive an ensemble of partitions, then aggregate the partitions into a single one. For the first stage, the following parameters are used. nTrees is the number of trees to use. MaxDepth is the maximum depth for each decision tree. PredictionFraction is the fraction of features that are to be trained on for each tree. SetOfPoints is a matrix where each row is a data point and each column is a feature. RandomForestTransform implements stage 1.

```
RandomForestTransform (SetOfPoints, nTrees,
MaxDepth, PredictionFraction)

Output := empty matrix;
nFeatures := PredictionFraction * (number of fea-
tures in SetOfPoints);
FOR i in nTrees DO
  TempY := ChooseRandomFeatures(nFeatures, SetOf-
Points);
  // ChooseRandomFeatures returns nFeatures columns,
  selected randomly from SetOfPoints
  TempX := remaining columns from SetOfPoints;
  Assignments := DecisionTreeRegressor(MaxDepth,
TempX, TempY);
  // DecisionTreeRegressor uses a decision tree pre-
  dict TempY from TempX, and returns the leaf_id of
  each data point. Since MaxDepth is restricted,
  each leaf with contain many data points.
  // DecisionTreeRegressor trains the decision tree
  as in a random forest: bootstrapping data, and
  randomly restricting which features can be split
  on at any given node.
  column i of Output := Assignments;
END FOR
RETURN Output
```

RandomForestTransform thus transforms each data point to a vector of length nTrees, where each entry in the vector represents the leaf_id that the data point was assigned by the nth decision tree. For the second stage of the algorithm, k-medoids is employed on the transformed data (optionally Minkowski weighted), using the Jaccard distance between the rows. In this circumstance, the Jaccard distance between two rows represents the fraction of trees for which the points are assigned to *different* leaves. The cluster assignments generated by k-medoids represent the output of IRFC.

As one increases the parameter nTrees, the first stage of IRFC slows, and performance improves, but of course those improvements fall off asymptotically. For different applications, the question of what the optimal value of nTrees is (such that gains to performance no longer justify the additional computation) is hard to predict in advance. By way of example, for the performance evaluation below, nTrees was set to 100. The parameter MaxDepth is similarly situational. The relatively small effect of attempting to tune it did not yield significant insight to optimal values, but values between 2 and 6 were successful for the datasets tested.

The computational complexity of the algorithm is $O(n^2k)$, where n is the number of data points, and k the number of features. n^2k is the complexity of creating a Jaccard distance matrix between the transformed data points (for use by k-medoids), and this is the slowest step.

A sample implementation of IRFC in Python can be accessed at https://github.com/mkc1000/random_forest_cluster. The first stage of IRFC can be found in rf_transform.py, the second in kmedoids.py, and the two are synthesized in rf_cluster.py.

At this point we can briefly call out the difference between IRFC and an algorithm that is superficially similar: Breiman's (2003) unsupervised random forest method. This involves training a random forest classifier to distinguish real data from "fake data" that is generated by a poor generative model. (In the generative model, a new vector of fake data is generated by sampling independently from each feature.) While their similarities are rather minimal, both unsupervised random forests and IRFC clearly employ random forests to do unsupervised learning, so it is worth disambiguating the two.

3 Performance Evaluation

The metric chosen for evaluation was Yeung's Figure of Merit, for the reasons discussed above. This metric was chosen before the algorithm was designed, to ensure that a positive result would not merely reflect the abundance of evaluation metrics for clustering algorithms. The greater the number of clusters that an algorithm outputs, the easier it is to have a small Figure of Merit. Therefore, when comparing algorithms, one must hold the number of clusters constant, and repeat over many values of the number of clusters.

Among existing clustering methods, k-means is perhaps best suited theoretically to perform well on this metric. K-means explicitly attempts to minimize the within-cluster variance of the features that it is trained on, which is plausibly an optimal heuristic for minimizing the within-cluster variance of the features it is not trained on. Indeed, Yeung et al. (2001) found experimentally that no other algorithms they tested reliably outperformed k-means on the real world data that was included in their analysis. Albaum et al. (2011) confirmed this finding on different datasets.

Therefore, IRFC was chiefly compared with k-means across four datasets and for many different values of k. Two other algorithms, DBSCAN (Ester et al. 1996) and agglomerative clustering, were tested alongside IRFC and k-means to further illustrate their inferiority to k-means on the datasets used here, as Yeung and Albaum found on a different suite of datasets. For the iris dataset, DBSCAN performed poorly, and the other algorithms performed equivalently. Across all datasets, IRFC, k-means, and agglomerative clustering performed approximately equivalently for k equal to 2. In all other cases, IRFC outperformed the other algorithms (Fig. 1). (A more detailed explanation of the construction of Fig. 1 is provided below.) That IRFC outperforms DBSCAN and agglomerative clustering is not particularly remarkable, since there is no reason to expect those two to be paragons by this metric. Outperforming k-means, however, preliminarily indicates that it outperforms all existing clustering algorithms according

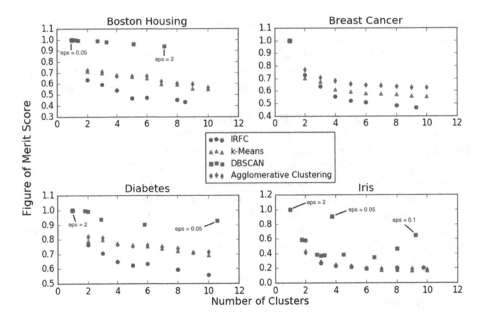

Fig. 1. Figures of Merit for cluster assignments given the number of clusters. IRFC and k-means are compared across four standard datasets

to the FOM, since k-means had previously been found dominant. For a description of the datasets, see Table 1.

Table 1. Descriptions of the datasets used. All features were scaled to a mean of 0 and a standard deviation of 1 prior to clustering

	Dimensions	Example features
Boston housing	506 × 13	Per capita crime rate (by town), average number of bedrooms per house, nitrogen oxides concentration, full-value property-tax rate per 10 k
Breast cancer	569 × 30	Mean radius (of tumors), mean area, mean concave points, worst radius, worst area, worst concave points
Diabetes	442 × 10	Age, body mass index, blood pressure
Iris	150 × 4	Petal length, petal width, sepal length, sepal width

Each data point in Fig. 1 represents an average of n runs of an algorithm with a given parameter setting on a given dataset, where n is the number of features in the dataset. Each of the n runs excludes one feature and uses this for the FOM score. Thus, the y-axis represents the average score from these runs, and the x-axis represents the average number of clusters generated with this parameter setting, which varies from run to run for some algorithms. For DBSCAN the parameter varied is epsilon, although for certain data (the breast cancer data, for example), this is not sufficient to result in any nontrivial partitions at all. For the remaining algorithms, the number of clusters is a parameter to the algorithm, and this is varied directly. A bit of marginalia is required to explain how

IRFC can yield a different number of clusters on different runs while holding constant a parameter that ostensibly specifies that quantity. In the k medoids step, if two clusters are assigned the same data point as a centroid, one of the two clusters evaporates, so the "number of clusters" parameter in IRFC could more precisely be described as the maximum number of clusters returned.

The roughly equivalent performance of k-means and IRFC on the iris dataset may simply reflect that the clusters in the iris dataset are very easily learned. The particularly small number of features in the iris data should also hamper IRFC. Since one feature at a time was withheld to validate the cluster assignments, both algorithms only had access to three features while clustering. In the other datasets, every decision tree in IRFC was trained on a unique subset of features, but for the iris data, there was significant redundancy, weakening the ability of IRFC to make use of the wisdom-of-crowds effect.

To acquire a single quantitative measure of the performance of an algorithm on a dataset, one can evaluate the area under the Figure of Merit curve, and divide by a normalizing factor. The resulting Integrated Figure of Merit simply compresses the information from Fig. 1 (Table 2).

Table 2. Integrated Figure of Merit for IRFC, k-Means, and Agglomerative Clustering across four standard datasets

	IRFC	K-Means	Aggl. clust.
Boston	0.531	0.658	0.673
Breast cancer	0.570	0.627	0.687
Diabetes	0.660	0.766	0.775
Iris	0.301	0.274	0.277

The Integrated Figure of Merit (IFOM) uses the algorithmic parameter for the number of clusters, so DBSCAN does not fit neatly into this analysis, lacking such a parameter. Since its poor performance is nevertheless evident above, and since k-means is most relevant com-parison to IRFC, DSBCAN's incompatibility with the IFOM is not particularly problematic.

4   Conclusions

For k greater than 2, IRFC was found to generally outperform k-means according to Yeung et al.'s Figure of Merit metric. While unsupervised learning may be used for other purposes, if one's goal is to predict which data points will have similar values for an unmeasured feature, IRFC is likely to be optimally effective for this task. Future research may consider the following extensions. This algorithm could be modified to output a hierarchical clustering model. Other methods besides Jaccard k-medoids could be used to aggregate the clusters. Vega-Pons and Ruiz-Shulcloper (2011) review several worthy candidates for effective cluster aggregation. Other supervised learning models besides decision trees could be used in the first stage, in such a way that an implied data compression could be extracted. For example, if a neural network with a hidden layer of minimal width were used to predict one subset of features from another, the activations in the hidden layer would represent a continuous rather than discrete compression of the data. Hopefully, Intra-Feature Random Forest Clustering can inspire a family of clustering algorithms that make similar use of supervised learning methods.

References

Albaum, S.P., Hahne, H., Otto, A., Haußmann, U., Becher, D., Poetsch, A., Goesmann, A., Nattkemper, T.W.: A guide through the computational analysis of isotope-labeled mass spectrometry-based quantitative proteomics data: an application study. Proteome sci. **9**(1), 1 (2011)

Becker, R.A., Caceres, R., Hanson, K., Loh, J.M., Urbanek, S., Varshavsky, A., Volinsky, C.: A tale of one city: using cellular network data for urban planning. IEEE Pervasive Comput. **10**(4), 18–26 (2011)

Ben-Hur, A., Elisseeff, A., Guyon, I.: A stability based method for discovering structure in clustered data. In: Pacific Symposium on Biocomputing, vol. 7, pp. 6–17, December 2001

Breiman, L.: Random forests. Mach. Learn. **45**(1), 5–32 (2001)

Breiman, L.: Random Forests Manual v4.0. Technical report, UC Berkeley (2003). ftp:// ftp.stat.berkeley.edu/pub/users/breiman/Using_random_forestsv4.0.pdf

Chicco, G., Napoli, R., Piglione, F.: Application of clustering algorithms and self organising maps to classify electricity customers. In: Power Tech Conference Proceedings, 2003 IEEE Bologna, vol. 1, 7 pp. IEEE, June 2003

Dudoit, S., Fridlyand, J.: A prediction-based resampling method for estimating the number of clusters in a dataset. Genome Biol. **3**(7), 1 (2002)

Ester, M., Kriegel, H.P., Sander, J., Xu, X.: A density-based algorithm for discovering clusters in large spatial databases with noise. In: Kdd, vol. 96, no. 34, pp. 226–231, August 1996

Harrigan, K.R.: An application of clustering for strategic group analysis. Strateg. Manag. J. **6**(1), 55–73 (1985)

Hilas, C.S., Mastorocostas, P.A.: An application of supervised and unsupervised learning approaches to telecommunications fraud detection. Knowl. Based Syst. **21**(7), 721–726 (2008)

Iliadis, L.S.: A decision support system applying an integrated fuzzy model for long-term forest fire risk estimation. Environ. Model Softw. **20**(5), 613–621 (2005)

Kaufman, L., Rousseeuw, P.J.: Finding Groups in Data: An Introduction to Cluster Analysis, vol. 344. Wiley, Hoboken (2009)

Krzanowski, W.J., Lai, Y.T.: A criterion for determining the number of groups in a data set using sum-of-squares clustering. Biometrics **44**, 23–34 (1988)

Li, A., Walling, J., Ahn, S., Kotliarov, Y., Su, Q., Quezado, M., Oberholtzer, J.C., Park, J., Zenklusen, J.C., Fine, H.A.: Unsupervised analysis of transcriptomic profiles reveals six glioma subtypes. Cancer Res. **69**(5), 2091–2099 (2009)

Masulli, F., Schenone, A.: A fuzzy clustering based segmentation system as support to diagnosis in medical imaging. Artif. Intell. Med. **16**(2), 129–147 (1999)

Monti, S., Tamayo, P., Mesirov, J., Golub, T.: Consensus clustering: a resampling-based method for class discovery and visualization of gene expression microarray data. Mach. Learn. **52**(1–2), 91–118 (2003)

Park, B.: Hybrid neuro-fuzzy application in short-term freeway traffic volume forecasting. Transp. Res. Rec. J. Transp. Res. Board **1802**, 190–196 (2002)

Pavlidis, N.G., Tasoulis, D.K., Vrahatis, M.N.: Financial forecasting through unsupervised clustering and evolutionary trained neural networks. In: The 2003 Congress on Evolutionary Computation (CEC 2003), vol. 4, pp. 2314–2321. IEEE, December 2003

Pham, D.T.: Applications of unsupervised clustering algorithms to aircraft identification using high range resolution radar. In: Proceedings of the IEEE 1998 National Aerospace and Electronics Conference (NAECON 1998), pp. 228–235. IEEE, July 1998

Singh, C., Kim, Y.: An efficient technique for reliability analysis of power systems including time dependent sources. IEEE Trans. Power Syst. **3**(3), 1090–1096 (1988)

Tibshirani, R., Walther, G., Hastie, T.: Estimating the number of clusters in a data set via the gap statistic. J. R. Stat. Soc. Ser. B (Stat. Methodol.) **63**(2), 411–423 (2001)

Vega-Pons, S., Ruiz-Shulcloper, J.: A survey of clustering ensemble algorithms. Int. J. Pattern Recognit. Artif. Intell. **25**(03), 337–372 (2011)

Wang, C.H.: Apply robust segmentation to the service industry using kernel induced fuzzy clustering techniques. Expert Syst. Appl. **37**(12), 8395–8400 (2010)

Xu, R., Wunsch, D.: Survey of clustering algorithms. IEEE Trans. Neural Netw. **16**(3), 645–678 (2005)

Yeung, K.Y., Haynor, D.R., Ruzzo, W.L.: Validating clustering for gene expression data. Bioinformatics **17**(4), 309–318 (2001)

Dolphin Pod Optimization
A Nature-Inspired Deterministic Algorithm for Simulation-Based Design

Andrea Serani[✉] and Matteo Diez

CNR-INSEAN, National Research Council–Marine Technology Research Institute,
Rome, Italy
andrea.serani@insean.cnr.it, matteo.diez@cnr.it

Abstract. A novel nature-inspired, deterministic, global, and derivative-free optimization method, namely the dolphin pod optimization (DPO), is presented for solving simulation-based design optimization problems with costly objective functions. DPO is formulated for unconstrained single-objective minimization and based on a simplified social model of a dolphin pod in search for food. A parametric analysis is conducted to identify the most promising DPO setup, using 100 analytical benchmark functions and three performance criteria, varying pod size and initialization, coefficient set, and box-constraint method, resulting in more than 140,000 optimization runs. The most promising setup is compared with deterministic particle swarm optimization, central force optimization, and DIviding RECTangles and finally applied to the optimization of a destroyer hull form for reduced resistance and improved seakeeping.

Keywords: Dolphin pod optimization · Deterministic optimization
Global optimization · Derivative-free optimization

1 Introduction

Simulation-based design (SBD) methods integrate numerical simulations, design modification tools, and optimization algorithms. SBD has been widely applied in many engineering fields, including aerospace, automotive, and naval applications, where the overall computation cost of the optimization process is determined by the simulation tool, the design space dimensionality, and the efficiency of the optimization algorithm. The numerical simulations are typically affected by the presence of residuals and, for this reason, the objective function is likely noisy. Furthermore, most simulation tools (such as commercial software) do not directly provide derivatives and, generally, the existence of local minima cannot be excluded a priori. For these reasons, global derivative-free optimization algorithms have been developed and applied to SBD, providing global approximate solutions to the design problem. Although complex SBD applications are often solved by metamodels, their development and assessment require benchmark solutions, with simulations directly connected to the optimization algorithm. These solutions may be achieved only if affordable and effective optimization algorithms are available. When global techniques are used with CPU-time

G. Nicosia et al. (Eds.): MOD 2017, LNCS 10710, pp. 50–62, 2018.
https://doi.org/10.1007/978-3-319-72926-8_5

expensive solvers, the optimization process is computationally expensive and its effectiveness and efficiency remain an algorithmic and technological challenge.

Metaheuristic algorithms for global optimization have been developed and extensively applied in the last decade, such as particle swarm optimization (PSO) [1] and ant colony optimization (ACO) [2]. Several new algorithms appeared recently, such as firefly algorithm (FA) [3], cuckoo search (CS) [4], and bat algorithm (BA) [5]. These metaheuristics are usually stochastic, implying that statistically significant results can be obtained only through extensive numerical campaigns. Such an approach could be too expensive in SBD optimization for industrial applications, when CPU-time expensive computer simulations are used directly as analysis tool. For this reason, deterministic global derivative-free algorithms have been developed and applied in the context of SBD, such as deterministic PSO (DPSO, [6]) and central force optimization (CFO, [7]). Other deterministic global methods have been developed and applied to SBD based on non-heuristic approaches, such as DIviding RECTangles (DIRECT, [8]).

The objective of the present work is to introduce and assess a novel nature-inspired deterministic global derivative-free method based on a simplified social model of a dolphin pod in search for food: the dolphin pod optimization (DPO). The present DPO belongs to the class of deterministic swarm-intelligence methods. To the authors' knowledge, this is a little explored field, where the global search ability of swarm-intelligence systems is exploited in a deterministic way. DPO is formulated for unconstrained single-objective minimization and intended for SBD optimization problems with costly objective functions. The novelty stems from formulating the global search by defining the pod dynamics as a spring-mass system subject to internal and external forces. Specifically, DPO is formulated considering the essential elements of the cetacean intelligence: *congregation*, *self-awareness*, *communication*, and *memory*. Würsig writes on cooperative foraging [9]:

> "Individual feeding may be enhanced by the presence of the group due to rapid and efficient information transfer concerning where, for example, the major concentration of prey is and what the extent of the prey school may be."

Accordingly, the general rules of the algorithm are that each dolphin:

- wants to stay in group since hunting is more efficient (congregation)
- quantifies the concentration of preys he locates (self-awareness)
- communicates to the pod the location and concentration of preys (communication)
- is able to reconstruct the food distribution known to the pod (memory)
- is willing to modify his speed and course based on how concentrated and far preys are (intelligence)
- follows his rational deterministic will.

Based on these rules, one may assume that each dolphin is subject to a pod attraction force and a force related to the food distribution or fitness (representing the objective function). Based on these forces, the pod may be modelled as a spring-mass system (Fig. 1). The integration of the system's dynamics provides the mathematical formulation of DPO and allows to study the system stability with beneficial effects on the search capability and efficiency.

DPO shares some common features with other well-known swarm-intelligence methods, such as PSO, CFO, and gravitational search algorithm (GSA, [10]), where the particle (or agent) position and speed are described by the particle system dynamics. Nevertheless, compared to PSO, CFO, and GSA, the DPO formulation presents significant differences as it is the only algorithm that at the same time is deterministic, based on agent position and absolute fitness, memory based, fully informed,

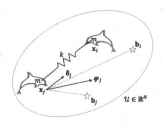

Fig. 1. Pod model

and finally formulated by a rigorous integration of the agent dynamics. Table 1 compares the current DPO formulation with PSO, CFO, and GSA. The DIRECT algorithm is also included in the table as a good example of deterministic global derivative-free method based on a rigorous mathematical framework (non-heuristic). Moreover, DPO differentiates also from other somehow-related meta-heuristics: differently from FA, CS, and BA, the current method is based on a deterministic concept and the system state depends not only on the agent positions and comparative fitness, but also directly on the objective function values (absolute fitness), through the food attraction force. Dolphin behaviour characteristics [11,12] also inspired other metaheuristic methods, such as the dolphin echolocation algorithm [13,14].

The effectiveness and efficiency of DPO are influenced by the choice of four main parameters: the number of dolphins interacting during the optimization,

Table 1. Comparison of DPO formulation with PSO, CFO, GSA, and DIRECT

	DPO	PSO (DPSO)	CFO	GSA	DIRECT
Metaheuristics metaphor or search method	Dolphin behaviour	Bird/bee behaviour	Gravitational law	Gravitational law	Lipschitzian optimization
Deterministic	✓	DPSO only	✓		✓
Agent based	✓	✓	✓	✓	
System state depends on agent position	✓	✓	✓	✓	
System state depends directly on agent absolute fitness	✓		✓	✓	
Memory based	✓	✓			
Fully informed	✓		✓	✓	
Rigorous solution of agent dynamics	✓				

the initialization of the pod in terms of position and velocity, the set of coefficients controlling the pod dynamics, and finally the method used to handle the box constraints. The analysis includes a parametric study of 100 analytical benchmark functions [15], with dimensionality from 2 to 50, with a full-factorial combination of: number of dolphins, initialization scheme, coefficient set, and box-constraint method. Three metrics are used to evaluate the algorithm performance and the most significant parameters for DPO are identified. DPO results are compared with deterministic global derivative-free algorithms, namely DPSO, CFO, and DIRECT. Finally, DPO is applied to two SBD problems, pertaining to the hull-form optimization of a USS Arleigh Burke-class destroyer, namely the DTMB 5415 model (an early version of the DDG-51) in calm water and waves, respectively. The SBD optimization results obtained by DPO are finally compared to DPSO, CFO, and DIRECT. A preliminary version of this work was presented in [16].

2   Dolphin Pod Optimization

Consider an optimization problem of the type

$$\text{Minimize } f(\mathbf{x})$$
$$\text{subject to } \mathbf{l} \leq \mathbf{x} \leq \mathbf{u} \tag{1}$$

where $f(\mathbf{x})$ is the objective function, $\mathbf{x} \in \mathbb{R}^N$ is the variable vector with $N \in \mathbb{N}^+$ the number of variables, and $\mathbf{l}$ and $\mathbf{u}$ are the lower and the upper bounds for $\mathbf{x}$, respectively.

Now consider a foraging pod of dolphins located at $\mathbf{x}_j$, exploring the variable space with the aim of finding an approximate solution for problem 1. The pod is modelled as a dynamical system where the dynamics of the j-th individual depends on a pod attraction force $\boldsymbol{\delta}_j$, a food attraction force $\boldsymbol{\varphi}_j$, as well as the drag, proportional to $\dot{\mathbf{x}}_j$ (Fig. 1)

$$\ddot{\mathbf{x}}_j + \xi \dot{\mathbf{x}}_j + k \boldsymbol{\delta}_j = h \boldsymbol{\varphi}_j \tag{2}$$

where

$$\boldsymbol{\delta}_j = \sum_{i=1}^{N_d} (\mathbf{x}_j - \mathbf{x}_i) \quad \text{and} \quad \boldsymbol{\varphi}_j = \sum_{i=1}^{N_d} \frac{2\hat{f}(\mathbf{x}_j, \mathbf{b}_i)}{1 + \|\mathbf{x}_j - \mathbf{b}_i\|^\alpha} \mathbf{e}(\mathbf{b}_i, \mathbf{x}_j) \tag{3}$$

with

$$\hat{f}(\mathbf{x}_j, \mathbf{b}_i) = \frac{f(\mathbf{x}_j) - f(\mathbf{b}_i)}{\rho} \quad \text{and} \quad \mathbf{e} = \frac{\mathbf{b}_i - \mathbf{x}_j}{\|\mathbf{b}_i - \mathbf{x}_j\|} \tag{4}$$

In the above equations, ξ, k, and $h \in \mathbb{R}^+$ define the pod dynamics; $N_d \in \mathbb{N}^+$ is the pod size; $\alpha \in \mathbb{R}^+$ tunes the food attraction force; $\mathbf{x}_j \in \mathbb{R}^N$ is the vector-valued position of the j-th individual; $f(\mathbf{x}) \in \mathbb{R}$ is the objective function (representing the food distribution); $\mathbf{b}_i$ is the best position ever visited by the i-th individual; $\rho = f(\mathbf{w}) - f(\mathbf{b})$ is a dynamic normalization term for

f, where $\mathbf{b} = \text{argmin}\{f(\mathbf{b}_j)\}$ is the best position ever visited by the pod and $\mathbf{w} = \text{argmax}\{f(\mathbf{x}_j)\}$ the worst position occupied by the pod individuals at the current time instance; $\mathbf{b}_i$, $\mathbf{b}$, and $\mathbf{w}$ are defined in the variable space.

Using the explicit Euler integration scheme yields:

$$
\begin{cases}
\mathbf{v}_j^{n+1} = (1 - \xi\Delta t)\mathbf{v}_j^n + \Delta t(-k\boldsymbol{\delta}_j + h\boldsymbol{\varphi}_j) \\
\mathbf{x}_j^{n+1} = \mathbf{x}_j^n + \mathbf{v}_j^{n+1}\Delta t
\end{cases}
\tag{5}
$$

where $\mathbf{x}_j^n$ and $\mathbf{v}_j^n$ represent the j-th dolphin position and velocity at the n-th iteration, respectively. Equation 5 represents a fully informed formulation, where each individual knows the story of the whole pod.

The integration step Δt, in Eq. 5, must guarantee the stability of the explicit Euler scheme, at least for the free dynamics. To this aim, consider the free dynamics of the k-th component of $\mathbf{x}$ (k-th variable), say a. Consider the dynamics of a for the j-th dolphin

$$
\ddot{a}_j + \xi\dot{a}_j + k\delta_j = 0
\tag{6}
$$

and finally for the entire pod

$$
\begin{Bmatrix} \dot{\mathbf{a}} \\ \dot{\mathbf{c}} \end{Bmatrix} = \begin{bmatrix} \mathbf{0} & \mathbf{I} \\ -\mathbf{K} & -\mathbf{G} \end{bmatrix} \begin{Bmatrix} \mathbf{a} \\ \mathbf{c} \end{Bmatrix} = \mathbf{A} \begin{Bmatrix} \mathbf{a} \\ \mathbf{c} \end{Bmatrix}
\tag{7}
$$

where

$$
\mathbf{K} = -k \begin{bmatrix} N_d - 1 & -1 & \cdots & -1 \\ -1 & N_d - 1 & \cdots & -1 \\ \vdots & \vdots & \ddots & \vdots \\ -1 & \cdots & -1 & N_d - 1 \end{bmatrix}
\tag{8}
$$

and $\mathbf{G} = \xi\mathbf{I}$, with $\mathbf{I}$ the $[N_d \times N_d]$ identity matrix. The solution of Eq. 7 is stable if $Re(\lambda) \leq 0$ where $\lambda = -\gamma \pm i\omega$ are eigenvalues of $\mathbf{A}$. This yields

$$
\Delta t \leq \left. \frac{2\gamma}{\gamma^2 + \omega^2} \right|_{\min} = \Delta t_{\max}
\tag{9}
$$

The DPO pseudo-code is shown in Algorithm 1.

3   DPO Setting Parameters

The number of dolphins used (N_d) is defined as $N_d = 2^r N$, with $r \in \mathbb{N}[2,4]$ therefore ranging from $4N$ to $16N$. The initialization of dolphins' location and velocity is performed using a deterministic and homogeneous distribution, following the Hammersley sequence sampling (HSS) [17], applied

Table 2. Coefficient set

Coefficient ID	ξ	q	p	α
1		0.01	0.1	2	0.5
2		0.1	1	4	1
3		1	10	8	2

Algorithm 1. DPO pseudo-code

1: Normalize $\mathbf{x}$ into a unit hypercube $\mathcal{U}$
2: Initialize the pod of N_d dolphins (position and velocity)
3: Evaluate $\Delta t_{\max}$
4: **while** $n \leq$ max number of iterations **do**
5:    **for** $j = 1, N_d$ **do**
6:        Evaluate $f(\mathbf{x}_j)$
7:        Update $\mathbf{b}_j$ and $f(\mathbf{b}_j)$
8:    **end for**
9:    Update $\mathbf{b}$, $\mathbf{w}$, $f(\mathbf{b})$, and $f(\mathbf{w})$
10:    **for** $j = 1, N_d$ **do**
11:        Evaluate the attraction forces, $\boldsymbol{\delta}_j$ and $\boldsymbol{\varphi}_j$
12:        Update $\mathbf{v}_j$ and $\mathbf{x}_j$
13:    **end for**
14: **end while**
15: Output the best solution found, $\mathbf{b}$ and $f(\mathbf{b})$

---

to three different sub-domains, defined as: (A) domain, (B) domain boundary, and (C) domain and boundary. The HSS is implemented using a basis of 2,000 prime number and following Eq. 13 in [6]. A non-null initial velocity is used (see Eq. 15 in [6]). Provided that all the design variables are normalized such that the domain is confined in a unit hypercube $\mathcal{U}$ (i.e. $-0.5 \leq \mathbf{x} \leq 0.5$), the following positions are used for the coefficients controlling the pod dynamics: $k = h = q/N_d$; $\Delta t = \Delta t_{\max}/p$; $\xi \Delta t < 1$, where q defines the weight for the attraction forces ($\boldsymbol{\delta}_j$ and $\boldsymbol{\varphi}_j$) and p defines the integration time step. Table 2 summarizes the coefficient sets used in the current analysis. The dolphins are confined within $\mathcal{U}$ using an inelastic (IW) and an elastic (EW) wall-type approach [6]. Specifically, in the IW approach, if a dolphin is found to violate one of the bounds in the transition from two consecutive iterations, it is placed on that bound setting to zero the associated velocity component, whereas, in the EW approach, the associated velocity component is reversed. The full-factorial combination of parameters results in a total of 1458 different setups. Finally, the number of function evaluations or evaluation budget ($N_{\max}$) is assumed as $N_{\max} = 2^c N$, where $c \in \mathbb{N}\,[7, 12]$ and therefore ranges from $128N$ to $4096N$.

4   Performance Metrics

Three performance metrics are used to assess the algorithm performances and defined as follows [6]:

$$\varepsilon_x = \sqrt{\frac{1}{N} \sum_{k=1}^{N} \left(\frac{x_{k,\min} - x^{\star}_{k,\min}}{R_k} \right)^2}, \quad \varepsilon_f = \frac{f_{\min} - f^{\star}_{\min}}{f^{\star}_{\max} - f^{\star}_{\min}}, \quad \varepsilon_t = \sqrt{\frac{\varepsilon_x^2 + \varepsilon_f^2}{2}} \quad (10)$$

ε_x is a normalized Euclidean distance between the minimum position $\{x_{k,\min}\}_{k=1}^{N}$ found by the algorithm and the analytical minimum position $\{x^{\star}_{k,\min}\}_{k=1}^{N}$, where $R_k = |u_k - l_k|$ is the range of the k-th variable. ε_f is the associated normalized distance in the function space, $f_{\min}$ is the minimum found by the algorithm, $f^{\star}_{\min}$ is the analytical minimum, and $f^{\star}_{\max}$ is the analytical maximum of the function $f(\mathbf{x})$ in the search domain. ε_t is a combination of ε_x and ε_f and used for an overall assessment. Additionally, the relative variability $(\sigma_{r,s}^2)$ for the performance metrics [18] is used to assess the impact of each tuning parameter s on the algorithm performance.

5   Numerical Results

A preliminary study of analytical benchmark functions is used to identify the most promising setup for DPO. The analyses are conducted setting apart functions with less or more than ten variables and DPO is compared to DPSO, CFO, and DIRECT. Finally, the optimization of the DTMB 5415 is performed by DPO with the most promising setup, and a comparison with DPSO, CFO, and DIRECT is provided. DPSO and CFO are used as suggested in [6,7], respectively, fort both analytical benchmark functions and SBD problems.

5.1   Analytical Benchmark Functions

100 benchmark functions are used, including a wide variety of problems, such as continuous and discontinuous, differentiable and non-differentiable, separable and non-separable, scalable and non-scalable, unimodal and multimodal, with $2 \leq N \leq 50$ (see Tables A.11 and A.12 in [6]).

Figures 2a and b show the relative variability $\sigma_{r,s}^2$ for ε_t, associated to the DPO parameters for $N < 10$ and $N \geq 10$, respectively (each group is composed by 50 functions). The pod initialization is the most significant parameter overall. For $N < 10$, the coefficient p (used to define the time step) becomes more significant as the number of function evaluations increases. The coefficient q (defining the attraction force intensity) shows an opposite trend. For $N \geq 10$, the initialization and the coefficient ξ are the most significant. The coefficient α is the least important overall.

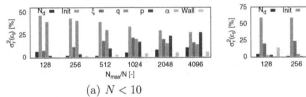
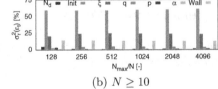

(a) $N < 10$ (b) $N \geq 10$

Fig. 2. Relative variability $\sigma_r^2\%$ of ε_t for DPO parameters

Table 3 shows the best performing parameters based on ε_t, varying the budget of function evaluations. Average values and standard deviation (STD) of performance among all setups are provided. Finally the best performing setup based on budget-averaged performance is shown. Budget-averaged values (bold

Table 3. Best performing setups for DPO based on ε_t, conditional to number of variables and budget of function evaluations

N	$N_{\max}/N$	N_d/N	Init	ξ, q, p, α	Wall	ε_t	Average	STD
<10	128	16	A	3, 2, 1, 2	IW	0.571E–01	1.568E–01	5.335E–02
	256	4	A	3, 2, 3, 1	EW	0.450E–01	1.317E–01	5.118E–02
	512	16	A	3, 2, 2, 2	EW	0.345E–01	1.145E–01	4.958E–02
	1024	4	C	2, 2, 3, 3	EW	0.333E–01	1.032E–01	4.827E–02
	2048	16	A	2, 1, 3, 1	EW	0.307E–01	0.962E–01	4.774E–02
	4096	8	C	2, 2, 3, 1	EW	0.276E–01	0.921E–01	4.782E–02
	Average	**4**	**A**	**3, 2, 3, 1**	**EW**	0.494E–01	1.157E–01	4.685E–02
≥ 10	128	4	B	2, 1, 2, 1	EW	0.107E+00	2.686E–01	1.064E–01
	256	4	B	2, 1, 3, 1	EW	0.998E–01	2.441E–01	9.740E–02
	512	4	B	2, 1, 3, 1	EW	0.943E–01	2.308E–01	9.220E–02
	1024	4	B	2, 1, 3, 1	EW	0.935E–01	2.230E–01	8.897E–02
	2048	4	B	2, 1, 3, 1	EW	0.930E–01	2.183E–01	8.640E–02
	4096	4	B	2, 1, 3, 1	EW	0.930E–01	2.150E–01	8.439E–02
	Average	**4**	**B**	**2, 1, 3, 1**	**EW**	0.100E+00	2.333E–01	9.143E–02

character in Table 3) are used to define a reasonable guideline for the use of DPO in SBD optimization. The suggested guideline corresponds to: $N_d = 4N$, pod initialization on domain (A), $\xi = 1$, $q = 1$, $p = 8$, $\alpha = 0.5$, and elastic wall-type for problem with $N < 10$; $N_d = 4N$, pod initialization on domain boundary (B), $\xi = 0.1$, $q = 0.1$, $p = 8$, $\alpha = 0.5$, and elastic wall-type for problem with $N \geq 10$.

Figures 3a and b show a comparison among the suggested DPO setups, DPSO, CFO, and DIRECT for $N < 10$ and $N \geq 10$, respectively. On average, DPO outperforms DPSO, CFO, and DIRECT, specially for problems with a large number of variables.

Finally three illustrative examples of the algorithms convergence are shown in Fig. 4 for (a) Schubert, (b) Levy n.15, and (c) Griewank functions with dimensionality equal to $N = 2$, 10, and 20, respectively. DPO shows a faster convergence than DPSO, CFO, and DIRECT.

5.2   Hull-Form SBD Optimization Problem

The optimization aims at improving separately (I) calm-water and (II) seakeeping performances. For problem I, the objective function f is a normalized ratio of the total resistance over the displacement in calm water at 18 kn, whereas for problem II f is a seakeeping merit factor based on the root mean square of the vertical acceleration of the bridge at 30 kn in head wave ($0°$) and on the roll angle at 18 kn in stern long-crested wave ($150°$). Modifications of the parent hull are performed using orthogonal functions, defined over surface-body

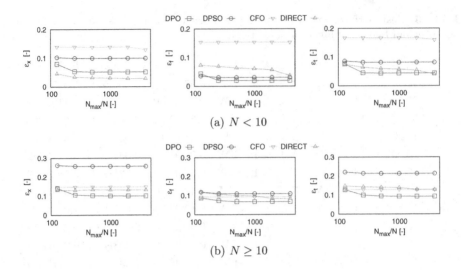

Fig. 3. Comparison of average performance of DPO, DPSO, CFO, and DIRECT

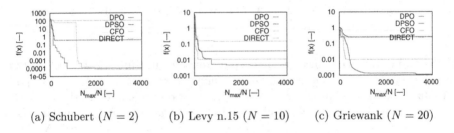

(a) Schubert ($N = 2$)   (b) Levy n.15 ($N = 10$)   (c) Griewank ($N = 20$)

Fig. 4. DPO, DPSO, CFO, and DIRECT performance for illustrative functions

patches, and 90%-confidence dimensionality reduction based on the Karhunen-Loève expansion, leading to a number of design variable $N = 6$ [19].

Problem I is solved using the code WARP, developed at CNR-INSEAN. Wave resistance computations are based on the double-model linear potential flow (PF) theory. The frictional resistance is estimated using a flat-plate approximation, based on the local Reynolds number. Problem II is solved using the code SMP, developed at the David Taylor Naval Ship Research and Development Center. SMP provides a potential flow solution based on linearized strip theory. Details of the computational domain for the free-surface, the hull grid, and the validation of PF analysis versus the experimental data can be found in [20]. For each problem, a budget of 4800 function evaluations is used.

Optimization results are summarized in Table 4. DPO achieves an objective function reduction close 13% and 32%, for problem I and II, respectively.

For problem I, DPO, DPSO, and DIRECT have a similar objective function reduction even if DPSO and DIRECT show a faster convergence to the global minimum, as shown in Fig. 5a. The final solutions found by DPO, DPSO, and

Table 4. SBD optimization results (non-normalized variables)

Prob.	Algorithm	x_1	x_2	x_3	x_4	x_5	x_6	$\Delta f\%$
I	DPO	0.999	0.030	0.116	0.353	0.221	0.581	-12.73
	DPSO	1.000	-0.044	0.069	0.370	0.229	0.579	-12.78
	CFO	0.652	0.204	0.220	0.240	0.180	0.229	-9.28
	DIRECT	0.996	0.025	0.033	0.370	0.247	0.551	-12.68
II	DPO	-0.022	0.417	0.127	-0.421	0.874	-0.997	-32.20
	DPSO	0.146	0.356	0.049	-0.545	0.884	-0.873	-31.31
	CFO	0.565	-0.852	0.440	0.224	0.289	-0.220	-12.91
	DIRECT	-0.173	-0.787	-0.140	-0.543	0.993	-0.993	-26.65

DIRECT are very close (see Fig. 5b) showing almost the same modification compared to the original design (see Fig. 6), whereas CFO reaches a less-significant objective improvement, probably corresponding to a local optimum.

Problem II shows a greater objective function reduction by DPO, compared with DPSO and especially CFO and DIRECT, probably trapped in local minima (see Fig. 7a). The optimal design variables found by DPO, DPSO, CFO, and DIRECT fall in different regions of the design space. In particular, the solution found by CFO and DIRECT is pretty far from those by DPO and DPSO. This is also reflected by the hull sections, shown in Fig. 8.

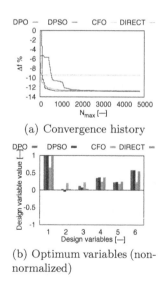

(a) Convergence history

(b) Optimum variables (non-normalized)

Fig. 5. Problem I, optimization results

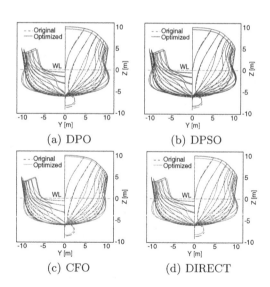

(a) DPO

(b) DPSO

(c) CFO

(d) DIRECT

Fig. 6. Problem I, optimized hull sections compared to the original

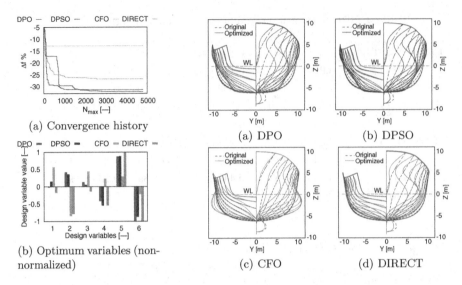

(a) Convergence history

(b) Optimum variables (non-normalized)

Fig. 7. Problem II, optimization results

(a) DPO

(b) DPSO

(c) CFO

(d) DIRECT

Fig. 8. Problem II, optimized hull sections compared to the original

6   Conclusions and Future Work

Dolphin pod optimization is a novel global derivative-free optimization algorithm based on a simplified social model of a dolphin pod in search for food. DPO uses a rigorous deterministic formulation, providing opportunity to define a set of coefficients that guarantee the pod convergence (at least for the free-dynamics).

A parametric analysis has been conducted using 100 analytical benchmark functions and three performance metrics, varying the number of dolphins, their initialization, the coefficient set, and the box-constraint method. For $N < 10$ and a low budget of function evaluations, coefficient q (force magnitude) and pod initialization are the most significant parameters. For a higher budget of function evaluations, coefficient p (time step) and pod initialization are the most significant parameters. Finally, for $N \geq 10$ the pod initialization is found the most significant parameter for all budgets.

The most promising setup has been identified: a number of dolphins N_d equal to 4 times the number of variables N; a pod initialization with a distribution over the whole design variables domain for $N < 10$ and only on the domain boundary for $N \geq 10$; a set of coefficients corresponding to: $\xi = 1$, $q = 1$, $p = 8$, and $\alpha = 0.5$ for $N < 10$, and to $\xi = 0.1$, $q = 0.1$, $p = 8$, and $\alpha = 0.5$ for $N \geq 10$; an elastic wall-type approach. DPO performance has been found on average better than DPSO, CFO, and DIRECT. DPO has been applied to two hull-form optimization problems for the improvement of (I) calm water and (II) seakeeping performances. These have shown comparable results of DPO, DPSO, and DIRECT. DPO has been found more effective for problem II.

Future work includes extensions to different formulations for the food attraction force φ, the possibility of using different coefficients for pod and food attraction forces, and comparison to other dolphin methods [13,14].

Acknowledgements. The work is supported by ONRG, NICOP grant N62909-15-1-2016, under the administration of Dr Woei-Min Lin, Dr. Salahuddin Ahmed, and Dr. Ki-Han Kim, and by the Italian Flagship Project RITMARE.

References

1. Kennedy, J., Eberhart, R.: Particle swarm optimization. In: Proceedings of the Fourth IEEE Conference on Neural Networks, Piscataway, pp. 1942–1948 (1995)
2. Dorigo, M., Maniezzo, V., Colorni, A.: Ant system: optimization by a colony of cooperating agents. IEEE Trans. Syst. Man Cybern. Part B **26**(1), 29–41 (1996)
3. Yang, X.S.: Nature-Inspired Metaheuristic Algorithms. Luniver Press (2008)
4. Yang, X.S., Deb, S.: Cuckoo search via levy flights. In: Proceedings of World Congress on Nature and Biologically Inspired Computing (NaBic 2009), India, pp. 210–214. IEEE Publications, USA (2009)
5. Yang, X.S.: A new metaheuristic bat-inspired algorithm. In: Gonzlez, J., Pelta, D., Cruz, C., Terrazas, G., Krasnogor, N. (eds.) NICSO 2010. SCI, vol. 284, pp. 65–74. Springer, Heidelberg (2010). https://doi.org/10.1007/978-3-642-12538-6_6
6. Serani, A., Leotardi, C., Iemma, U., Campana, E.F., Fasano, G., Diez, M.: Parameter selection in synchronous and asynchronous deterministic particle swarm optimization for ship hydrodynamics problems. Appl. Soft. Comput. **49**, 313–334 (2016)
7. Formato, R.A.: Central force optimization: a new metaheuristic with applications in applied electromagnetics. Prog. Electromagn. Res. **77**, 425–491 (2007)
8. Jones, D., Perttunen, C., Stuckman, B.: Lipschitzian optimization without the Lipschitz constant. J. Optim. Theor. Appl. **79**(1), 157–181 (1993)
9. Würsig, B.: Delphinid foraging strategies. In: Dolphin Cognition and Behavior: A Comparative Approach. Psychology Press (2013)
10. Rashedi, E., Nezamabadi-pour, H., Saryazdi, S.: GSA: a gravitational search algorithm. Inf. Sci. **179**(13), 2232–2248 (2009). Special Section on High Order Fuzzy Sets
11. Connor, R.C., Wells, R.S., Mann, J., Read, A.J.: The bottlenose dolphin: social relationships in a fission-fusion society. In: Cetacean Societies: Field Studies of Bottlenose Dolphins and Whales, pp. 97–126. University of Chicago Books, Chicago (2010)
12. Bruck, J.N.: Decades-long social memory in bottlenose dolphins. Proc. R. Soc. Lond. B Biol. Sci. **280**(1768) (2013)
13. Kaveh, A., Farhoudi, N.: A new optimization method: dolphin echolocation. Adv. Eng. Softw. **59**, 53–70 (2013)
14. Kaveh, A., Hosseini, P.: A simplified dolphin echolocation optimization method for optimum deisgn of trusses. Int. J. Optim. Civ. Eng. **4**(3), 381–397 (2014)
15. Jamil, M., Yang, X.S.: A literature survey of benchmark functions for global optimisation problems. Int. J. Math. Model. Numer. Optim. **4**, 150–194 (2013)
16. Serani, A., Diez, M.: Dolphin pod optimization. In: Tan, Y., Takagi, H., Shi, Y. (eds.) ICSI 2017. LNCS, vol. 10385, pp. 63–70. Springer, Cham (2017). https://doi.org/10.1007/978-3-319-61824-1_7

17. Wong, T.T., Luk, W.S., Heng, P.A.: Sampling with Hammersley and Halton points. J. Graph. Tools **2**(2), 9–24 (1997)
18. Campana, E.F., Diez, M., Iemma, U., Liuzzi, G., Lucidi, S., Rinaldi, F., Serani, A.: Derivative-free global ship design optimization using global/local hybridization of the DIRECT algorithm. Optim. Eng. **17**(1), 127–156 (2015)
19. Diez, M., Serani, A., Campana, E.F., Volpi, S., Stern, F.: Design space dimensionality reduction for single- and multi-disciplinary shape optimization. In: AIAA/ISSMO Multidisciplinary Analysis and Optimization (MA&O), AVIATION 2016, Washington D.C., 13–17 June 2016
20. Diez, M., Serani, A., Campana, E.F., Goren, O., Sarioz, K., Danisman, D.B., Grigoropoulos, G., Aloniati, E., Visonneau, M., Queutey, P., Stern, F.: Multi-objective hydrodynamic optimization of the DTMB 5415 for resistance and seakeeping. In: Proceedings of the 13th International Conference on Fast Sea Transportation, FAST 2015, Washington, D.C. (2015)

Contraction Clustering (RASTER)
A Big Data Algorithm for Density-Based Clustering in Constant Memory and Linear Time

Gregor Ulm$^{(\boxtimes)}$, Emil Gustavsson, and Mats Jirstrand

Fraunhofer-Chalmers Research Centre for Industrial Mathematics,
Chalmers Science Park, 412 88 Gothenburg, Sweden
{gregor.ulm,emil.gustavsson,mats.jirstrand}@fcc.chalmers.se
http://fcc.chalmers.se

Abstract. Clustering is an essential data mining tool for analyzing and grouping similar objects. In big data applications, however, many clustering methods are infeasible due to their memory requirements or runtime complexity. CONTRACTION CLUSTERING (RASTER) is a linear-time algorithm for identifying density-based clusters. Its coefficient is negligible as it depends neither on input size nor the number of clusters. Its memory requirements are constant. Consequently, RASTER is suitable for big data applications where the size of the data may be huge. It consists of two steps: (1) a contraction step which projects objects onto *tiles* and (2) an agglomeration step which groups tiles into clusters. Our algorithm is extremely fast. In single-threaded execution on a contemporary workstation, it clusters ten million points in less than 20 s—when using a slow interpreted programming language like Python. Furthermore, RASTER is easily parallelizable.

Keywords: Algorithms · Big data · Machine learning
Unsupervised learning · Clustering

1 Introduction

The goal of clustering is to aggregate similar objects into groups in which objects exhibit similar characteristics. When attempting to cluster very large amounts of data, i.e. data in excess of 10^{12} elements [10], two limitations of many well-known clustering algorithms become apparent. First, they operate under the premise that all available data fits into memory, which does not necessarily hold in a big data context. Second, their time complexity is unfavorable. For instance, one of the most used clustering algorithms is DBSCAN [8]. It runs in $\mathcal{O}(n \log n)$ at best, where n stands for the number of objects. There are standard implementations that use a distance matrix, but its space requirements of $\mathcal{O}(n^2)$ make big data applications infeasible. In addition, the logarithmic factor is problematic in a big data context. Linear-time clustering methods have been described, for instance

© Springer International Publishing AG 2018
G. Nicosia et al. (Eds.): MOD 2017, LNCS 10710, pp. 63–75, 2018.
https://doi.org/10.1007/978-3-319-72926-8_6

in surveys by Kumar et al. [12,13], but those have large coefficients, which makes them inapplicable to big data clustering.

In this paper, we introduce CONTRACTION CLUSTERING (RASTER).[1] In the taxonomy presented by Fahad et al. [9], it is a grid-based clustering algorithm. RASTER has been designed for clustering big data. It scales linearly in the size of its input. In addition, it is able to handle cases where the available data do not fit into memory as its memory requirements are constant. Its two distinctive phases are parallelizable; the most computationally intensive first one trivially so, while the second one can be expressed in the divide-and-conquer paradigm of algorithm design. In a review paper, Shirkhorshidi et al. [19] state that distributed clustering algorithms are needed for efficiently clustering big data. However, our algorithm is a counter example to that claim as it exhibits excellent performance metrics even in single-threaded execution on a single machine.

The novelty of RASTER is that it is a straight-forward, easy to implement, and extremely fast big data clustering algorithm. It requires only one pass through the input data and does not need to retain its input. In addition, key operations like projecting to tiles and neighborhood lookups are performed in constant time. Another benefit of our approach is that the parameters required are intuitive, and that there is no need to compute distances between points. Furthermore, it is easily parallelizable.

The remainder of this paper is organized as follows. Section 2 contains the problem description, followed by a detailed description of RASTER in Sect. 3. In Sect. 4 we discuss results. Related work is covered in Sect. 5. Lastly, we outline possible future work in Sect. 6.

2   Problem Description

In this section we give a brief overview of the clustering problem (Sect. 2.1), describe the motivating use case behind RASTER (Sect. 2.2), and highlight limitations of common clustering methods (Sect. 2.3).

2.1   The Clustering Problem

Clustering is a standard approach in machine learning for grouping similar items, with the goal of dividing a dataset into subsets that share certain features. It is an example of unsupervised learning, which implies that there are many valid ways of clustering data points. Elements belonging to a cluster are normally more similar to other elements in it than to elements in any other cluster. If an element does not belong to a cluster, it is classified as noise. An element normally only belongs to at most one cluster. However, fuzzy clustering methods [17,22] can identify non-disjoint clusters, meaning that elements may be part of multiple

---

[1] The chosen shorthand may not be immediately obvious: RASTER operates on an implied grid. Resulting clusters can be made to look similar to the dot matrix structure of a *raster graphics* image. Furthermore, the name *RASTER* is an agglomerated contraction of the words *contraction* and *clustering*.

overlapping clusters. Yet, this is not an area we are concerned with in this paper. Instead, we focus on problems that are in principle solvable by common clustering methods such as DBSCAN [8] or k-means clustering [16].

2.2 Motivating Use Case

Large-scale processing of telemetry data in the form of GPS coordinates is required for the analysis of vehicle transportation networks. A central problem in that domain is the identification of *hubs*, which are locations within a road network at which vehicles stop so that a particular action can be performed, e.g. warehouses, delivery points, or bus stops. Conceptually, a hub is a node in such a network, and a node is the center of a cluster.

GPS coordinates are provided as coordinate pairs. Their precision is defined by the number of place values after the decimal point. For instance, the coordinate pair (40.748441, −73.985664) specifies the location of the Empire State Building in New York City with a precision of 11.1 cm. A seventh decimal place value would specify a given location with a precision of 1.1 cm, while truncating to five decimal place values would lower precision to 1.1 m. High-precision GPS measurements require special equipment, while consumer-grade GPS is accurate to within about ten meters under open sky [7]. Thus, for the purpose of clustering, lower-precision GPS coordinates could be used, without losing a significant amount of information. This is the key insight that led to the discovery of RASTER.

2.3 Limitations of Common Clustering Methods

In this subsection we briefly describe why two standard clustering algorithms, DBSCAN and k-means clustering, are not suitable for our big data clustering use case.

DBSCAN identifies density-based clusters. Its time complexity is $\mathcal{O}(n \log n)$ in the best case. This depends on whether a query identifying the neighbors of a particular data point can be performed in $\mathcal{O}(\log n)$. In a pathological case, or in the absence of a fast lookup query, its time complexity is $\mathcal{O}(n^2)$. DBSCAN is comparatively fast. However, when working with many billions or even trillions of data points, clustering becomes infeasible due to the logarithmic factor, provided all data even fits into memory.[2]

In k-means clustering, the number of clusters k, has to be known in advance. The goal is to determine k partitions of the input data. Two aspects make k-means clustering less suitable for our use case. First, when dealing with big data, estimating a reasonable k is non-trivial. Second, its time complexity is unfavorable. An exact solution requires $\mathcal{O}(n^{dk+1})$, where d is the number of dimensions [11]. Lloyd's algorithm [15], which uses heuristics, is likewise not applicable as its time complexity is $\mathcal{O}(dnki)$, where i is the number of iterations

---

[2] On a contemporary workstation with 16 GB RAM, the `scikit-learn` implementation of DBSCAN cannot even handle one million data points.

until convergence is reached. There have been recent improvements [2,4], but their time complexity still seems unfavorable for huge data sets.[3]

3   RASTER

This section contains a thorough presentation of RASTER, consisting of a high-level description of the algorithm (Sect. 3.1), a discussion of the concept of *tiles* and their role in creating clusters (Sect. 3.2), a detailed description of the algorithm, including its time complexity (Sect. 3.3), an outline of one possible parallelization strategy (Sect. 3.4), brief remarks on using data of higher dimensionality (Sect. 3.5), and a potential weakness of our algorithm, including a workaround (Sect. 3.6).

3.1   High-Level Description

The goal of RASTER is to reduce a very large number n of 2-dimensional points to a more manageable number of points that specify the *approximate area* of clusters in the input data, without retaining its input data. Figure 1 provides a visualization. Our algorithm uses an implicit 2-dimensional grid of a coarser resolution than the input data. Each square of this grid is referred to as a *tile*; each point is mapped to exactly one tile. A tile containing at least a user-specified threshold number t of observations is labeled as a *significant tile*. Afterwards, *RASTER clusters* are constructed from adjacent significant tiles.

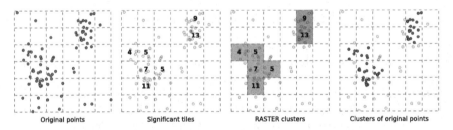

Original points	Significant tiles	RASTER clusters	Clusters of original points

Fig. 1. A visualization of RASTER

3.2   Tiles and RASTER Clusters

A key component of RASTER is the deliberate reduction of the precision of its input data. In general, this operation is a projection of points to tiles, which could, for instance, be achieved by truncating or rounding. The goal is the identification of RASTER clusters, which is attained via two distinct and consecutive steps: *contraction* and *agglomeration*. The contraction step first determines the number of observations per tile, and then discards all non-significant tiles.

---

[3] We have identified tens of thousands of clusters with RASTER in a huge real-world data set, which shows that k-means clustering would have been *highly* unsuitable.

The agglomeration step constructs RASTER clusters out of adjacent significant tiles. To illustrate the idea of a *tile*, consider a grid consisting of squares of a fixed side length. A square may contain several coordinate points. Reducing the precision by one decimal digit means removing the last digit of a fixed-precision coordinate. For instance, with a chosen decimal precision of 2, the coordinates $(1.005, 1.000)$, $(1.009, 1.002)$, and $(1.008, 1.006)$ are all truncated (*contracted*) to the tile identified by the corner point $(1.00, 1.00)$. Thus, $(1.00, 1.00)$ is a tile with three associated points. This tile would be classified as significant when using a threshold value of $t \geq 3$ and disregarded otherwise.

In the subsequent *agglomeration* step, RASTER clusters are constructed, which consist of significant tiles that are at most a Manhattan distance of δ steps apart. A value of $\delta = 1$ means that significant tiles need to be directly adjacent; with a larger δ, a cluster could contain significant tiles without direct neighbors. Finally, constructing a RASTER cluster is an iterative process that agglomerates significant tiles, taking δ into account.

3.3   The Algorithm

In this subsection we present some explanations that accompany the RASTER pseudocode in Fig. 2.[4] The algorithm consists of three sequential loops. The first two *for*-loops constitute the *contraction* step. The subsequent *while*-loop constitutes the *agglomeration* step.

Mapping to a tile consists of associating a data point p to a value representing a tile. The case of 2-dimensional values was illustrated in the previous subsection. RASTER does not exhaustively check every possible tile value, but instead only retains tiles that were encountered while processing data. Due to the efficiency of hash tables, the first *for*-loop runs in $\mathcal{O}(n)$, where n is the number of points. The mapping function is performed in $\mathcal{O}(1)$, which is also the time complexity of the various hash table operations we use. After the first *for*-loop all points are mapped to a tile. The second *for*-loop traverses all keys of the hash table *tiles*. Only significant tiles are retained. The intermediate result of this loop is a hash table of significant tiles and their respective number of observations. Deleting an entry is an $\mathcal{O}(1)$ operation. At most, and only in the pathological case where there is exactly one observation per tile, there are n tiles. In any case, it holds that $m \leq n$, where m is the number of tiles. Thus, this step of the algorithm is performed in $\mathcal{O}(m) \leq \mathcal{O}(n)$.

The subsequent *while*-loop performs the *agglomeration* step, which constructs clusters from significant tiles. The pseudocode does not specify the definition of clusters, but implies that a cluster is either a set of significant tiles, or a set of tuples, where each tuple consists of the coordinates of a significant tile and the total number of observations. There are at most n tiles. In order to determine the tiles a cluster consists of, take one tile from the set *tiles* and recursively determine all neighboring tiles in $\mathcal{O}(n)$ in a depth-first manner. This is conceptually similar

---

[4] Reference implementations in several programming languages are available at https://gitlab.com/fraunhofer_chalmers_centre/contraction_clustering_raster.

to the well-known flood fill algorithm. Neighborhood lookup with a hash table is in $\mathcal{O}(1)$, as the locations of all its neighboring tiles are known. For instance, when performing agglomerations with $\delta = 1$, the neighbors of any coordinate pair (x, y) are the squares to its left, right, top, and bottom in the grid. Thus, the third loop runs likewise in $\mathcal{O}(m) \leq \mathcal{O}(n)$. Each of the three loops runs in $\mathcal{O}(n)$ in the worst case, leading to a total time complexity of $\mathcal{O}(n)$.

The result is a set C of sets, where each $c \in C$ is a cluster of a finite amount of points that each uniquely identify a significant tile. An illustration of how such a cluster may look like is given in Fig. 1, which shows the count of observations per tile and, considering a threshold value of $t = 4$, highlights those that are classified as significant tiles and agglomerated into clusters.

According to our specification, the result is a set of sets, where each set is a cluster that is specified only by its constituent significant tiles. It is trivial to modify RASTER to retain either all points of a cluster, or the count of observations per tile. Indeed, one key element of our algorithm is that information regarding data points per tile is only maintained as an aggregate by keeping track of their sum. This is a useful approach when clustering very large amounts of data. However, if data fits into memory, one could as well retain all points per tile or all unique points per tile.

```
input: data points, threshold t,
       distance δ
output: collection of clusters clusters

                        ▷ initialization
initialize hash table tiles, set clusters
                        ▷ contraction
for point in points do
    map point to corresponding tile
    if tile ∉ tiles then
        add tile with value 1 to tiles
    else
        increment value for tile by 1
    end if
end for
for tile in tiles do
    if value for tile < t then
        remove tile from tiles
    end if
end for
                        ▷ agglomeration
while tiles ≠ ∅ do
    select arbitrary tile from tiles
    determine cluster c containing tile,
        considering δ
    remove tiles in c from tiles
    add c to clusters
end while
```

Fig. 2. RASTER

3.4 Parallel RASTER

RASTER is easily parallelizable. An obvious target is mapping to tiles, which is *embarrassingly parallel* in nature as no projection depends on the result of any other projection. Thus, this step could be part of a separate parallel *for*-loop. Updating the hash table *tiles* could be sped up with a concurrent hash table, which affects the first two loops of the algorithm. The agglomeration

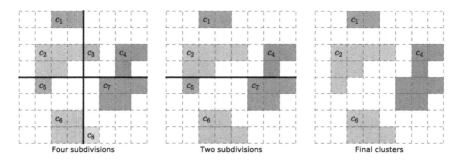

Fig. 3. The agglomeration step as a divide-and-conquer algorithm

step could be parallelized via the divide-and-conquer paradigm of algorithm design by subdividing the input into partitions of equal size and recursively processing them. Figure 3 shows how partitions can be processed in parallel. However, adjacent RASTER clusters c_1 and c_2 that cross the boundaries of a partition need to be joined. A cluster is considered complete once none of its neighbors is a significant tile and it does not touch any partition boundaries.

3.5 Generalizing to Higher Dimensions

While we have focused on 2-dimensional geospatial data, it is possible to generalize RASTER to d dimensions. Generalizing from $\mathbb{R}^2$, irrespective of dimension, a similar case can be constructed for $\mathbb{R}^n$. For the former, the reduction for each decimal value is 10^2 per tile. For the latter, it is 10^n. In the case of $\delta = 1$, the number of neighbors to consider per tile is $2d$.

3.6 Minimum Cluster Size in Disadvantageous Grid Layouts

We consider truncation of a fixed number of decimal digits to be the standard behavior of RASTER. As long as mapping to tiles is performed in a consistent manner, any mapping can be chosen. Yet, for any possible mapping a corner case can be found that illustrates that a significant tile may not be found. Thus, the identification of significant tiles may depend on the chosen grid.

Assume a threshold of $t = 4$ for a significant tile, and a tile with four points. If all points were located in the same tile of a grid, a significant tile would be detected. However, those four points could also be spread over adjacent tiles, as illustrated by Fig. 4. One could of course shift the grid by choosing a different projection, but an adversary could easily place all points on different tiles in the new grid. In order to alleviate this problem, a threshold $u < t$ for the number of observations in a tile needs to be picked. To be on the safe side, a value of $u = \frac{t}{4}$ is recommended. Alternatively, an additional step could be added to make RASTER

Fig. 4. Any RASTER grid is vulnerable

more precise. With a complexity of $\mathcal{O}(m)$, where m is the number of tiles, one could determine whether a group of four tiles contains at least t points.[5]

4   Results

In this section we present results of executing RASTER. A visualization of the use case this algorithm was designed for (Sect. 4.1) is followed by results with sample datasets for density-based clustering algorithms (Sect. 4.2). Afterwards, we discuss empirical runtime measurements, including a comparison with DBSCAN (Sect. 4.3).

The default implementation of RASTER does not retain its input data, but instead keeps track of the number of observations per tile. That number is dropped when agglomerating significant tiles to clusters. This leads to clusters of significant tiles, which would not have been ideal for visualizations. Therefore, the results in this section were achieved with a trivially modified version of our algorithm that retains all points per significant tile. For $\delta = 2$, we made a minor modification to RASTER to take tiles forming a square around any coordinate into account, instead of the Manhattan distance. Lastly, as an added post-processing step, we disregarded clusters containing less than certain threshold number of data points, which is specified in each case below.

In the figures in this section, the entirety of the input data is visualized as translucent gray points. Clusters in the colors red and blue were plotted on top of those gray points. Thus, all remaining non-red and non-blue points are noise.

4.1   Ideal Data

RASTER was designed for the identification of centers of groups of points which are placed increasingly tightly the closer they are to the center of a cluster. In Fig. 5, we use an artificial data set, where a roughly Gaussian distribution of points is spread around a center point. This is similar to patterns encountered in the motivating use case we described in Sect. 2.2. Points plotted in red constitute the resulting RASTER cluster. In order to generate a larger cluster, we chose a precision of 2 decimal values when mapping to tiles

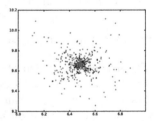

Fig. 5. RASTER applied to ideal data

and the parameters $t = 2$, $\delta = 2$. We disregarded clusters that did not contain at least 10 elements.

---

[5] In case it is not obvious why this is in linear time: For each row in a grid, take the current and next row into account. Start, for instance, with the tile in the top left corner and take its right neighbor as well as the two tiles adjacent in the next row into account.

4.2   Sample Datasets

Even though RASTER was specifically designed for the pattern shown in Fig. 5, it also performs well as a general-purpose density-based clustering algorithm. We illustrate this by showing the results of applying RASTER to two standard scikit-learn data sets: noisy_circles and noisy_moons, which both contain 1500 points. For the experiments below, we tried several parameters. We did not preprocess the input data, for instance by rescaling coordinates.

In the noisy_circles data set shown in Fig. 6, RASTER identifies two distinct clusters, but ignores part of the outer circle. The chosen parameters were $t = 3$ and $\delta = 1$. The chosen precision for mapping to tiles was one decimal value. Resulting clusters had to contain more than 25 data points. In the noisy_moons data set, density-based clusters are likewise reliably identified. Figure 7 shows output similar to noisy_circles, as the lower arc in the image is not fully included in the identified cluster. The chosen parameters were $t = 7$ and $\delta = 2$, as well as a precision of 1 for mapping to tiles. Clusters needed to have a size greater than 15. With minor parameter tweaking of t the shape of the lower arc could be clearly identified, as shown in Fig. 8, but at the cost of adding a minimal amount of noise to the cluster containing the upper arc.

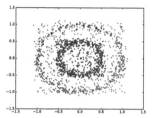

Fig. 6. RASTER applied to noisy circles

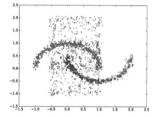

Fig. 7. RASTER applied to noisy moons with threshold $t = 7$

4.3   Empirical Runtime

RASTER is a fast clustering algorithm, running in linear time. In order to quantify this, we performed several measurements for two variations of our algorithm, i.e. one that retains all unique points and one that retains a count of the number of observations per significant tile. The experiments were run on an Oracle VirtualBox virtual machine, hosting Ubuntu Linux 16.04. We allocated 16 GB RAM. The CPU was an Intel i7-6700K clocked at 4.0 GHz. The implementation was executed by a Python 3.5 inter-

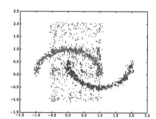

Fig. 8. RASTER applied to noisy moons with threshold $t = 6$

preter. We generated ideal data as described in Sect. 4.1. The chosen parameters were $t = 5$ and $\delta = 1$. No post-processing to filter out clusters below a certain minimum size was performed. The resulting clusters were rather dense. The reason behind that choice was to cause a higher workload for the agglomeration step. When processing real-world data with many small and dense clusters, agglomeration would merely add a negligible cost, due to the fact that clusters are very dense but also quite small.

The difference between both variations of RASTER is modest. To highlight one of the measurements illustrated by Fig. 9: ten million points are processed in 17.8 versus 23.1 s, while the slower runtime was achieved when retaining all data points. For an input size of 10^8 points, memory was insufficient for the version of RASTER that retains its input. Memory requirements when aggregating data are bounded by the number of tiles in a grid and the size of the used data types for storing the count of observations per tile, which are both finite with GPS data. Thus, memory requirements are constant. We also compared RASTER, in the variation that retains data points, with the DBSCAN implementation of scikit-learn 0.18.1, cf. Fig. 10. As parameters, we set $\epsilon = 0.3$ and the minimum sample size to 10. The DBSCAN experiment ended prematurely as it ran out of memory with one million points.

5 Related Work

An early approach to grid-based spatial data mining was STING [20]. A key difference of that algorithm is that it performs statistical queries, using distributions of attribute values.

WaveCluster [18] shares some similarities with RASTER. It can reduce the resolution of the input, which leads to output that is visually similar to RASTER clusters that are defined by its significant tiles. WaveCluster runs in linear time. However, because the computation of wavelets is costlier than the operations RASTER performs, its empirical runtime is presumably worse.

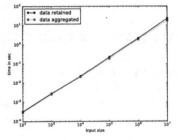

Fig. 9. Two variations of RASTER

There are also conceptual similarities between a sub-method of CLIQUE [1] and RASTER. Yet, differences are that the former is mainly concerned with subspace clustering of high-dimensional data. Further, RASTER is more flexible with regards to neighborhood lookup. It is also a faster operation with our algorithm. A minor difference is that CLIQUE uses a density-ratio threshold. Lastly, the empirical runtime of

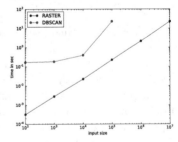

Fig. 10. RASTER vs DBSCAN

RASTER can be assumed to be lower, largely due to the very cheap cost of projection onto tiles and neighborhood lookup.

The idea of counting observations within a grid has been explored by Baker and Valleron [3]. They presented a solution to a problem in spatial epidemiology whose initial step seems similar to the contraction step performed by RASTER.

The approach taken by GRPDBSCAN, discovered by Darong and Peng [6], seems to have aspects in common with the RASTER variant that retains all input data. Their description is not fleshed out, however, so it is not clear how similar their algorithm is to ours.

Lastly, there is some similarity between our algorithm and blob detection in image analysis [5]. A direct application of that method to finding clusters would arguably require very dense cluster centers, which could be achieved by mapping points to fewer tiles. Also, blob detection is computationally more expensive than RASTER.

6   Future Work

There are several ways to build upon RASTER. One obvious direction would be arbitrary resolutions in the reduction step via rounding. Currently, reductions truncate by one decimal place value. Smaller reduction steps are easy to implement. However, RASTER is more general and could use any projection, so it may be worth exploring different areas of application.

Xiaoyun et al. introduced GMDBSCAN [21], a DBSCAN-variant that is able to detect clusters of different densities. The inability to detect clusters of different densities is a weakness of DBSCAN that is shared by RASTER. For more general purpose-applications, it may be worth investigating a similar approach for RASTER as well. A starting point is an adaptive distance parameter for significant tiles. While this paper only considered fixed values of 1 and 2 for δ, one could certainly consider arbitrary values instead.

RASTER does not distinguish between significant tiles. Yet, one could think of cases in which some of those tiles contain a very large number of observations, while others barely reach the specified threshold value. Thus, one could consider an adaptive approach to RASTER-clustering, for instance by subdividing such tiles into smaller segments, with the goal of determining more accurate cluster shapes. This idea is related to adaptive mesh refinement, suggested by Liao et al. [14]. A related idea is to change the behavior of RASTER when detecting a large number of adjacent tiles that have not been classified as significant. This may prompt a coarsening of the grid size for that part of the input space.

For practical use, it may be worthwhile to add a contextual relaxation value ϵ for the threshold value of significant tiles. For instance, in the vicinity of several significant tiles, a neighboring tile with $t - \epsilon$ observations may be considered part of the agglomeration, in particular if it has multiple significant tiles as neighbors.

Acknowledgments. This research was supported by the project *Fleet telematics big data analytics for vehicle usage modeling and analysis (FUMA)* in the funding program *FFI: Strategic Vehicle Research and Innovation (DNR 2016-02207)*, which is administered by VINNOVA, the Swedish Government Agency for Innovation Systems.

References

1. Agrawal, R., Gehrke, J., Gunopulos, D., Raghavan, P.: Automatic subspace clustering of high dimensional data for data mining applications. In: International Conference on Management of Data, vol. 27, pp. 94–105. ACM (1998)
2. Bachem, O., Lucic, M., Hassani, H., Krause, A.: Fast and provably good seedings for k-means. In: Advances in Neural Information Processing Systems, pp. 55–63 (2016)
3. Baker, D.M., Valleron, A.J.: An open source software for fast grid-based datamining in spatial epidemiology (FGBASE). Int. J. Health Geogr. **13**(1), 46 (2014)
4. Capó, M., Pérez, A., Lozano, J.A.: An efficient approximation to the k-means clustering for massive data. Knowl. Based Syst. **117**, 56–69 (2017)
5. Danker, A.J., Rosenfeld, A.: Blob detection by relaxation. IEEE Trans. Pattern Anal. Mach. Intell. **1**, 79–92 (1981)
6. Darong, H., Peng, W.: Grid-based DBSCAN algorithm with referential parameters. Phys. Procedia **24**, 1166–1170 (2012)
7. van Diggelen, F., Enge, P.: The world's first GPS MOOC and worldwide laboratory using smartphones. In: Proceedings of the 28th International Technical Meeting of The Satellite Division of the Institute of Navigation (ION GNSS+2015), pp. 361–369. ION (2015)
8. Ester, M., Kriegel, H.P., Sander, J., Xu, X., et al.: A density-based algorithm for discovering clusters in large spatial databases with noise. In: KDD, vol. 96, pp. 226–231 (1996)
9. Fahad, A., Alshatri, N., Tari, Z., Alamri, A., Khalil, I., Zomaya, A.Y., Foufou, S., Bouras, A.: A survey of clustering algorithms for big data: taxonomy and empirical analysis. IEEE Trans. Emerg. Top. Comput. **2**(3), 267–279 (2014)
10. Hathaway, R.J., Bezdek, J.C.: Extending fuzzy and probabilistic clustering to very large data sets. Comput. Stat. Data Anal. **51**(1), 215–234 (2006)
11. Inaba, M., Katoh, N., Imai, H.: Applications of weighted voronoi diagrams and randomization to variance-based k-clustering: (extended abstract). In: Proceedings of the Tenth Annual Symposium on Computational Geometry, SCG 1994, pp. 332–339. ACM, New York (1994)
12. Kumar, A., Sabharwal, Y., Sen, S.: Linear time algorithms for clustering problems in any dimensions. In: Caires, L., Italiano, G.F., Monteiro, L., Palamidessi, C., Yung, M. (eds.) ICALP 2005. LNCS, vol. 3580, pp. 1374–1385. Springer, Heidelberg (2005). https://doi.org/10.1007/11523468_111
13. Kumar, A., Sabharwal, Y., Sen, S.: Linear-time approximation schemes for clustering problems in any dimensions. J. ACM (JACM) **57**(2), 5 (2010)
14. Liao, W.k., Liu, Y., Choudhary, A.: A grid-based clustering algorithm using adaptive mesh refinement. In: 7th Workshop on Mining Scientific and Engineering Datasets of SIAM International Conference on Data Mining, pp. 61–69 (2004)
15. Lloyd, S.: Least squares quantization in PCM. IEEE Trans. Inf. Theory **28**(2), 129–137 (1982)
16. MacQueen, J., et al.: Some methods for classification and analysis of multivariate observations. In: Proceedings of the Fifth Berkeley Symposium on Mathematical Statistics and Probability, Oakland, CA, USA, vol. 1, pp. 281–297 (1967)
17. Pham, D.L.: Spatial models for fuzzy clustering. Comput. Vis. Image Underst. **84**(2), 285–297 (2001)
18. Sheikholeslami, G., Chatterjee, S., Zhang, A.: Wavecluster: a multi-resolution clustering approach for very large spatial databases. In: VLDB, vol. 98, pp. 428–439 (1998)

19. Shirkhorshidi, A.S., Aghabozorgi, S., Wah, T.Y., Herawan, T.: Big data clustering: a review. In: Murgante, B., et al. (eds.) ICCSA 2014. LNCS, vol. 8583, pp. 707–720. Springer, Cham (2014). https://doi.org/10.1007/978-3-319-09156-3_49

20. Wang, W., Yang, J., Muntz, R., et al.: Sting: a statistical information grid approach to spatial data mining. In: VLDB, vol. 97, pp. 186–195 (1997)

21. Xiaoyun, C., Yufang, M., Yan, Z., Ping, W.: GMDBSCAN: multi-density DBSCAN cluster based on grid. In: IEEE International Conference on e-Business Engineering, ICEBE 2008, pp. 780–783. IEEE (2008)

22. Yang, M.S.: A survey of fuzzy clustering. Math. Comput. Modell. 18(11), 1–16 (1993)

Deep Statistical Comparison Applied on Quality Indicators to Compare Multi-objective Stochastic Optimization Algorithms

Tome Eftimov[1,2], Peter Korošec[1,3(✉)], and Barbara Koroušić Seljak[1]

[1] Computer Systems Department, Jožef Stefan Institute,
Jamova cesta 39, 1000 Ljubljana, Slovenia
{tome.eftimov,peter.korosec,barbara.korousic}@ijs.si
[2] Jožef Stefan International Postgraduate School,
Jamova cesta 39, 1000 Ljubljana, Slovenia
[3] Faculty of Mathematics, Natural Science and Information Technologies,
Glagoljaška ulica 8, 6000 Koper, Slovenia
http://cs.ijs.si/

Abstract. In this paper, a study of how to compare the performance of multi-objective stochastic optimization algorithms using quality indicators and Deep Statistical Comparison (DSC) approach is presented. DSC is a recently proposed approach for statistical comparison of meta-heuristic stochastic optimization algorithms over single-objective problems. The main contribution of DSC is the ranking scheme that is based on the whole distribution, instead of using only one statistic such as average or median. Experimental results performed by using 6 multi-objective stochastic optimization algorithms on 16 test problems show that the DSC gives more robust results compared to some standard statistical approaches that are recommended for a comparison of multi-objective stochastic optimization algorithms according to some quality indicator.

Keywords: Multi-objective optimization · Quality indicators
Deep statistical comparison · Single problem analysis
Multiple problem analysis

1 Introduction

In real-world applications and systems, a lot of problems involve simultaneous optimization of several conflicting objective functions [3]. Finding an optimal solution for this kind of problems is really a challenging task because a single optimal solution does not exist and there is a set of alternative solutions. Each solution that belongs to the set of alternative solutions is optimal in a manner that no other solution from the search space is superior to it when all objective functions are considered. These solutions are known as *Pareto-optimal* solutions and the set is known as *Pareto-optimal* set. The representation of the *Pareto-optimal* set in the objective space is known as *Pareto-optimal* front.

© Springer International Publishing AG 2018
G. Nicosia et al. (Eds.): MOD 2017, LNCS 10710, pp. 76–87, 2018.
https://doi.org/10.1007/978-3-319-72926-8_7

Stochastic optimization algorithms [2] can be assumed as one efficient technique of finding a good approximation to the *Pareto optimal* front in multiobjective optimization. They usually do not guarantee to identify optimal trade-offs but try to find a good approximation, a set of solutions that are not far away from the optimal front. For this reason, over the last years, many multi-objective stochastic optmizaiton algorithms have been developed. Performance analysis of a new algorithm compared with the state-of-the-art is a crucial task.

In single-objective optimization, to analyze the performance of meta-heuristic stochastic optimization algorithms, the best obtained solution by an algorithm needs to be used. In the case of minimization problems, the solution with the lowest value is the best solution. However, in multi-objective stochastic optimization algorithms, the result is usually an approximation of the *Pareto-optimal* front, called an approximation set, so a vector of real numbers that reflects different aspects of the quality of the solution can be assigned to the approximation set. The quality is measured in terms of some criteria that are related to the convergence and diversity properties. A large number of quality indicators have been proposed to compare the performance of different stochastic optimization algorithms in multi-objective optimization. Some of them are: hypervolume [22], generational distance [21], inverse generational distance [21], epsilon [15], spread [3], generalized spread [3], etc.

In comparative studies, algorithms are used to solve a number of benchmark problems [6]. Meta-heuristics are non-deterministic techniques, meaning we do not have any guaranty that the result will be the same for every run. So to test the quality of the algorithm, it is not enough to perform just one run, but many of them, from which we can draw some conclusions. By calculating quality indicator for each approximation set that is obtained from multiple runs on a single problem, the high-dimensional data is transformed into one-dimensional data. Additionally, this data must be analyzed with some statistical tests to ensure the significance of the results, otherwise the conclusions may be wrong because the differences between the algorithms could have occurred by chance. Further, if algorithms need to be compared over multiple multi-objective problems, an average or a median of the quality indicator data for an algorithm on one problem needs to be calculated. This value is a representative value involved in the multiple-problem scenario for this algorithm on that problem.

The use of average or median can have a negative outcome to the relevancy of results of statistical test [7]. Averaging is sensitive to outliers that need to be considered especially because the algorithms could have poor runs. Even more, in the case when poor runs do not exist, averages can be in some ϵ-neighborhood, which is the set of all numbers whose distance from a number is less than some specified number ϵ, and the algorithms will obtain different rankings. Only in the case of ties, average rankings are assigned. To exceed the problem of sensitivity to outliers, medians are sometimes used because they are more robust to outliers. However, medians can be in some ϵ-neighborhood, and according to them the algorithms will obtain different rankings.

For these reasons, in our previous work [8], an approach was proposed, which removes the sensitivity of the simple statistics to the data and enables calculation of more robust statistics without fear of outliers influence or some errors inside ϵ-neighborhood. The approach is known as *Deep Statistical Comparison (DSC)* and is used to compare meta-heuristics stochastic optimization algorithms over single-objective problems.

In this paper, a study of the DSC in the domain of multi-objective optimization is presented. The data for comparisons consists of quality indicators, which needs to be analyzed in a proper way, in order to give correct conclusions. The rest of the paper is organized as follows. Section 2 gives an overview of the related works. In Sect. 3, the DSC is reintroduced. Section 4 presents the experimental study, followed by the discussion of the results. The conclusions of the paper are presented in Sect. 5.

2   Related Work

Over last years, there are many studies that address the problem of comparing approximation sets in a quantitate manner. Some of them include unary indicators [18]. An unary indicator is a real number assigned to each approximation set that reflects a certain quality aspect. Other studies are based on binary indicators [18]. A binary indicator is a real number that is assigned to pairs of approximation sets. Another approach is the attainment function, which consists of estimating the probability of attaining arbitrary goals in objective space from multiple approximation sets [10]. Riquelme et al. [18] presented a study of a large number of metrics that can be used to compare the performance of different algorithms in multi-objective approach. In the paper, they presented a review and analysis of 54 multi-objective optimization metrics with a discussion about the usage, tendency, and advantages/disadvantages of the most cited ones in order to give researchers enough information when choosing metrics is necessary. This review indicates that the hypervolume is the most used metric, followed by the generational distance, the epsilon indicator, and the inverted generational distance.

By comparing multi-objective stochastic optimization algorithms according to quality indicators, we transform the comparison problem from high-dimensional space into one-dimensional space. Then, the statistical methodology that needs to be applied is the same as the methodology used in the case of single-objective optimization. The difference is only in the content of the data. In the single-objective optimization the data consists of best solutions, while in the case of the multi-objective optimization the data consists of quality indicators of the approximation sets.

García et al. [11] presented a study on the use of nonparametric tests for analyzing the evolutionary algorithms' behaviour over single-objective optimization problems, following the study that has already been presented by Demšar for machine learning classifiers [5]. The study is conducted in two ways: single-problem analysis and multiple-problem analysis. We call this approach the common approach because it is the most used approach for statistical comparison of meta-heuristic stochastic optimization algorithms.

In the paper [8], we introduced a novel approach, known as *Deep Statistical Comparison* (DSC), for the statistical comparison of stochastic optimization algorithms over multiple single-objective problems. The DSC enables calculation of more robust statistics and the incorrect conclusions caused by the presence of outliers or ranking scheme used by some standard statistical tests can be avoided.

3 Deep Statistical Comparison

Deep Statistical Comparison (*DSC*) is a recently proposed approach for statistical comparison of meta-heuristc stochastic optimization algorithms over multiple single-objective problems [8]. The main contribution of DSC is the ranking scheme, which is based on the whole distribution, instead of using only one statistic to describe the distribution, such as average or median. The approach consists of two steps. The first step uses a newly proposed ranking scheme to obtain data that will be later used for statistical comparison. The ranking scheme is based on comparing distributions. By using some statistical test for comparing distributions, (e.g. the two-sample *Kolmogorov-Smirnov test* or the two-sample *Anderson-Darling* test [9]), all pairwise comparisons between the algorithms need to be made, and the obtained p-values are organized in a matrix. Further, because multiple pairwise comparisons are made, these p-values are corrected with *Bonferroni correction* [11] in order to control the FWER [16], which is the probability of making one or more false discoveries, or type I errors, among all hypotheses when performing multiple hypotheses tests. After correction, this matrix is checked for transitivity, and according to it the algorithms obtain their rankings. The second step is a standard omnibus statistical test, which uses the data obtained by the *DSC* ranking scheme as input data.

Contrary to common approach, the *DSC* gives more robust statistical results, which are not affected by outliers or misleading ranking scheme. The comparison between both approaches over single-objective problems together with the power analysis are presented in [8].

4 Results and Discussion

Two experiments are presented. In the first experiment, examples of multiple-problem analysis are presented according to different quality indicators, when the number of compared algorithms are 2 or 3, with an explanation on the level on single-problem analysis. In the second experiment, an example of multiple-problem analysis is presented in the case of multiple comparisons with a control algorithm.

4.1   Experimental Setup

The experimental data is the same as the data used in the paper [20]. Six algorithms are used for comparisons. Three of them are genetic algorithms NSGA-II, SPEA2 and IBEA and the other three are their differential evolution based variants $DEMO^{NS-II}$, $DEMO^{SP2}$, and $DEMO^{IB}$. Algorithms are compared on 16 test problems. The seven are DLTZ test problems [4] and the next nine are WFG test problems presented in [14]. Each of the 16 problems was used three times, each time with a different number of objectives (2, 3, and 4). More about the parameters of the test problems and the parameters of the algorithms can be found in [20]. All test problems assume minimization of all objectives. Each algorithm was run on each test problem 30 times. Before calculating the quality indicators, each approximated *Pareto* front was normalized.

Statistical comparisons were made separately according to different quality indicators, including: hypervolume, epsilon indicator, and r_2 indicator. All mentioned quality indicators are unary indicators. To calculate the hypervolume, the reference point $(1, \ldots, 1)$ was used. For the other quality indicators, the reference set consisted of non-dominated solutions acquired from all runs of each algorithm on a given problem.

The statistical test for comparing distributions used in the *DSC* ranking scheme is the two-sample *Anderson-Darling (AD)* test. The benefits of using it are presented in [9]. The significance level for it is set to 0.05. Further, the rankings are used for multiple-problem analysis. Because the required conditions for safe use of the parametric tests are not satisfied, an omnibus nonparametric statistical test needs to be selected. In our case, the *Wilcoxon signed-rank* test [17] is used for pairwise comparison over multiple problems, while if three or more algorithms are compared, the *Friedman* test [11] is used. The significance level for an omnibus statistical test is set to 0.05.

To see the benefit of using the DSC when multi-objective stochastic optimization algorithms are compared, the obtained quality indicator data for each problem over 30 runs is additionally analyzed by some standard statistical tests. If two algorithms are compared on a single problem, then the *Mann-Whitney rank sum* test [19] is used. If three or more than three algorithms are compared on a single problem, it is recommended to use the *Kruskal-Wallis* test [1]. The significance level for an omnibus statistical test is set to 0.05.

4.2   First Experiment

In this experiment, two examples are presented. In the first one, the statistical comparison is made between $DEMO^{NS-II}$ and NSGA-II, while in the second one between $DEMO^{SP2}$, $DEMO^{NS-II}$, and NSGA-II. In both examples, the algorithms are compared according to three quality indicators, hypervolume, epsilon indicator, and r_2 indicator, separately. The number of objective functions is set to 2 and 4, respectively.

Table 1 presents the DSC rankings of both algorithms, $DEMO^{NS-II}$ and NSGA-II. For each quality indicator, the first two columns correspond to the

DSC rankings and the third column corresponds to the obtained p-value from the *Mann-Whitney rank sum* test applied on the quality indicator data for a given problem.

Table 1. Statistical comparison of two algorithms, A_1=DEMO^{NS-II} and A_2=NSGA-II

F	Hypervolume			Epsilon			r_2		
	A_1	A_2	p_{value}	A_1	A_2	p_{value}	A_1	A_2	p_{value}
DTLZ1	1.00	2.00	*(.00)	1.00	2.00	*(.00)	2.00	1.00	*(.00)
DTLZ2	1.00	2.00	*(.00)	1.00	2.00	*(.00)	1.00	2.00	*(.00)
DTLZ3	1.00	2.00	*(.00)	1.00	2.00	*(.00)	1.00	2.00	*(.00)
DTLZ4	1.00	2.00	*(.00)	1.00	2.00	*(.00)	1.00	2.00	*(.00)
DTLZ5	1.00	2.00	*(.00)	1.00	2.00	*(.00)	1.00	2.00	*(.00)
DTLZ6	1.00	2.00	*(.00)	1.00	2.00	*(.00)	1.00	2.00	*(.00)
DTLZ7	1.00	2.00	*(.00)	1.00	2.00	*(.00)	1.00	2.00	*(.00)
WFG1	1.00	2.00	*(.00)	1.00	2.00	*(.00)	1.00	2.00	*(.00)
WFG2	1.00	2.00	*(.00)	1.00	2.00	*(.00)	1.00	2.00	*(.00)
WFG3	1.00	2.00	*(.00)	2.00	1.00	*(.02)	1.00	2.00	*(.00)
WFG4	1.00	2.00	*(.00)	2.00	1.00	*(.03)	1.50	1.50	(.70)
WFG5	1.00	2.00	*(.00)	2.00	1.00	*(.00)	2.00	1.00	*(.00)
WFG6	1.00	2.00	*(.00)	1.00	2.00	*(.00)	1.00	2.00	*(.00)
WFG7	1.00	2.00	*(.00)	1.00	2.00	*(.00)	1.00	2.00	*(.00)
WFG8	1.00	2.00	*(.01)	1.00	2.00	*(.00)	1.00	2.00	*(.00)
WFG9	1.00	2.00	*(.02)	2.00	1.00	*(.02)	1.50	1.50	*(.04)

* The null hypothesis is rejected, using $\alpha = 0.05$
p_{value} - the p-value of *Mann-Whitney ranks sum test*

Using this table, on a single problem level, there is a significant statistical difference between the performance of both algorithms according to the hypervolume and epsilon indicator. For example, the rankings according to the hypervolume on the DTLZ1 are 1.00 and 2.00. This means that there is a significant statistical difference between the performance of both algorithms on that problem, and the algorithm that is ranked with number 1, DEMO^{NS-II}, is significantly better than the algorithm that is ranked with number 2, NSGA-II. Then, the data is analyzed using the *Mann-Whitney rank sum* test, and the obtained p-value is 0.00, which is smaller than the used significance level, 0.05. To see which algorithm is better using the *Mann-Whitney rank sum* test, a one-sided test needs to be used, which is not needed for DSC since the ranking scheme has already presented this information. For this problem, the result from the *Mann-Whitney rank sum* test is the same as the result from the DSC.

According to the r_2 indicator, there are two problems, WFG4 and WFG9, for which the DSC rankings are 1.50 and 1.50, so there is no significant statistical difference between the performance of the algorithms. For the WFG4 problem, the p-value by the *Mann-Whitney rank sum* test is 0.70, which is greater than the significance level that is used, so the result is the same as in the case of the DSC. For the WFG9 problem, the p-value of the *Mann-Whitney rank sum* test is 0.04, which is lower than the used significance level, so the result is not the same as the result of the DSC and there is a significant statistical difference between the performance of the algorithms.

To see what happens on this problem, the result from the DSC is presented in detail. In Fig. 1a, the cumulative distribution functions (the step functions) and the mean values (the horizontal lines) of the r_2 quality indicator data for both algorithms on WFG9 are presented.

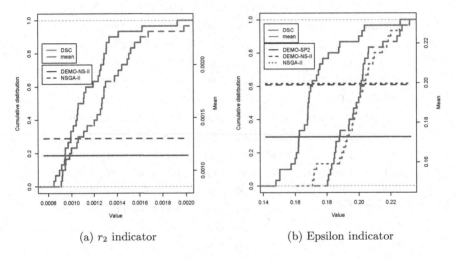

(a) r_2 indicator   (b) Epsilon indicator

Fig. 1. Cumulative distribution functions of a quality indicator on WFG9

From this figure, it is not clear if there is a difference between the cumulative distribution functions of r_2 quality indicator, so to check this the two-sample AD test is used. The p-value is 0.06, which is greater than the significance level, 0.05, so there is no significant statistical difference between the performance of the algorithms. When a standard statistical procedure is used, the result could be affected if outliers are presented, or if the differences between the obtained values for the r_2 quality indicator of both algorithms are in some small ϵ-neighbourhood and they influence on the ranking scheme. Even more, the question is how the result of the *Mann-Whitney rank sum* test is interpreted. If the two distributions have a different shape, the *Mann-Whitney rank sum* test is used to determine whether there are differences in the distributions of the two algorithms. However, if the two distributions have the same shape, the test is used to determine whether there are differences in the medians of the quality indicator between the two algorithms. So if the results are presented without checking the shape of the distributions, than this could lead to incorrect interpretation of the obtained results. The same explanation is also valid for the *Kruskal-Wallis* test.

From other side, the DSC ranking scheme use a statistical test for comparing distributions according to their shapes and parameters, so the problem of incorrect interpretation of the results is avoided.

After ranking the algorithms for each single problem, the rankings are further used for multiple problem analysis and compared using an omnibus statistical test. Because the required conditions for the safe use of parametric tests are not satisfied, the *Wilcoxon signed-rank* test is used in the case of pairwise comparison. The obtained p-values for the comparisons made according to each quality indicator, separately, are 0.00, so there is a significant statistical difference between the performance of both algorithms according to the three quality algorithms over multiple problems.

In the second example, the statistical comparison over multiple problems is made between three algorithms, $DEMO^{SP2}$, $DEMO^{NS-II}$, and NSGA-II. The DSC results are compared with the common approach. So to perform a multiple problem analysis by the common approach, an average of the quality indicator data of an algorithm over 30 runs on a given problem is set as a representative value for this algorithm on that problem. Table 2 presents the DSC rankings and the Friedman rankings. The Friedman ranking scheme is used when the data is analyzed following the common approach. The first six columns correspond to the rankings according to the hypervolume, while the next six columns correspond to the ranks according to the epsilon indicator.

By using this table, the rankings according to the hypervolume for both approaches differ in 4 problems, DLTZ3, DTLZ5, WFG8, and WFG9, while the rankings according to the epsilon indicator differ in 8 problems, DTLZ3, DTLZ5, WFG2, WFG3, WFG6, WFG7, WFG8, and WFG9. This happens because when the common approach is used, averaging is sensitive to outliers, so even when the difference between the averages are in some small ϵ-neighbourhood, the best one will receive ranking 1 and so on. In the case of DSC, the whole distribution is used, so if the quality indicator distributions of the algorithms are the same, they need to obtain the same ranking. So the rankings are obtained according to the whole distribution and not relying only on one statistic, which is the average in our example. To clarify this, in Fig. 1b, the cumulative distribution functions (the step functions) and the mean values (the horizontal lines) for the epsilon indicator of the algorithms on WFG9 are presented.

From this figure, one may assume that there is no difference between the cumulative distribution functions of epsilon indicator between the algorithms $DEMO^{NS-II}$ and NSGA-II, but both distributions differ from the cumulative distribution of $DEMO^{-SP2}$. To check this, the two-sample AD test is used. The p-values for each pairwise comparison are: 0.00 ($DEMO^{-SP2}$, $DEMO^{NS-II}$), 0.00 ($DEMO^{-SP2}$, NSGA-II), and 0.61 ($DEMO^{NS-II}$, NSGA-II). These value are further corrected by *Bonferroni correction*. In this example, the transitivity of the matrix used in the DSC ranking scheme is satisfied, so the set of all algorithms is split into two disjoint sets $\{DEMO^{-SP2}\}$ and $\{DEMO^{NS-II}$, NSGA-II$\}$ and the algorithms rankings are, 1.00, 2.50, 2.50. When a common approach is used, the rankings are 1.00, 2.00, 3.00, but it is obvious that averages of the epsilon

Table 2. Rankings of the algorithms, A_1=DEMO^{SP2}, A_2=DEMO^{NS-II}, and A_3= NSGA-II

F	Hypervolume						Epsilon					
	DSC			Friedman			DSC			Friedman		
	A_1	A_2	A_3	A_1	A_2	A_3	A_1	A_2	A_3	A_1	A_2	A_3
DTLZ1	2.00	1.00	3.00	2.00	1.00	3.00	1.00	2.00	3.00	1.00	2.00	3.00
DTLZ2	2.00	1.00	3.00	2.00	1.00	3.00	2.00	1.00	3.00	2.00	1.00	3.00
DTLZ3	1.50	1.50	3.00	1.00	2.00	3.00	1.50	1.50	3.00	1.00	2.00	3.00
DTLZ4	1.00	2.00	3.00	1.00	2.00	3.00	1.00	2.00	3.00	1.00	2.00	3.00
DTLZ5	2.50	2.50	1.00	3.00	2.00	1.00	2.00	2.00	2.00	1.00	2.00	3.00
DTLZ6	2.00	1.00	3.00	2.00	1.00	3.00	2.00	1.00	3.00	2.00	1.00	3.00
DTLZ7	2.00	1.00	3.00	2.00	1.00	3.00	2.00	1.00	3.00	2.00	1.00	3.00
WFG1	1.00	2.00	3.00	1.00	2.00	3.00	1.00	2.00	3.00	1.00	2.00	3.00
WFG2	1.00	2.00	3.00	1.00	2.00	3.00	1.00	2.50	2.50	1.00	2.00	3.00
WFG3	1.00	3.00	2.00	1.00	3.00	2.00	1.00	2.50	2.50	1.00	2.00	3.00
WFG4	1.00	2.00	3.00	1.00	2.00	3.00	2.00	1.00	3.00	2.00	1.00	3.00
WFG5	3.00	2.00	1.00	3.00	2.00	1.00	1.00	3.00	2.00	1.00	3.00	2.00
WFG6	1.00	2.00	3.00	1.00	2.00	3.00	1.00	2.50	2.50	1.00	2.00	3.00
WFG7	1.00	2.00	3.00	1.00	2.00	3.00	1.00	2.50	2.50	1.00	2.00	3.00
WFG8	1.00	2.50	2.50	1.00	3.00	2.00	1.00	2.50	2.50	1.00	3.00	2.00
WFG9	1.00	2.50	2.50	1.00	3.00	2.00	1.00	2.50	2.50	1.00	2.00	3.00

indicator for DEMO^{NS-II} and NSGA-II are in some small ϵ-neighbourhood, so this affects the ranking scheme, which could also influence the end result.

For multiple-problem analysis, the *Friedman* test is used with the rankings obtained on each single problem as an input data. The p-value for the DSC and the common approach according to the epsilon indicator is 0.00, so the result is that there is a significant difference between the performance of the algorithms over the multiple problems. When the comparison is made according to the hypervolume, the p-value using the DSC is 0.00, while the obtained p-value using the common approach is 0.01. The result is the same and there is a significant difference between the performance of the compared algorithms over the multiple problems that are used in the comparison. In this example, in the case of multiple-problem analysis, there is no difference between the result obtained by both approaches, however there is a difference on a single problem level. In general, the differences that exist on a single problem level can influenced the result for multiple-problem analysis.

4.3　Second Experiment

In typical comparison of multi-objective stochastic optimization algorithms usually more than 3 algorithms are used to compare against the proposed one.

For this reason, an example of multiple comparisons with a control algorithm is presented. More information about this scenario is presented in [8]. One way to do this is to obtain the rankings of the algorithms using the DSC ranking scheme, to used them as input data in an omnibus statistical test, and if there is a difference to continue with some post-hoc procedure [12,13]. When the number of algorithms increases, the correction of the p-values used in the DSC ranking scheme could influence the rankings. In order to avoid this, multiple *Wilcoxon* tests with the DSC rankings as input data can be used, one for each pairwise comparison. To calculate the true statistical difference for combining pairwise comparisons [11] in order to control the FWER the following equation is used:

$$p_{value} = 1 - \prod_{i=1}^{k-1} [1 - p_{value\,H_i}], \tag{1}$$

where k is the number of algorithms involved in the comparison.

In this example, the comparison is performed between $DEMO^{NS-II}$ as a control algorithm, and the other 5 algorithms: NSGA-II, $DEMO^{SP2}$, SPEA2, $DEMO^{IB}$, and IBEA, using multiple *Wilcoxon* tests. Table 3 presents the p-values for the pairwise comparisons according to the r_2 indicator.

Table 3. Multiple comparisons with a control algorithm ($DEMO^{NS-II}$) by using multiple *Wilcoxon* tests with the DSC ranking scheme

	1	2	3	4	5
$DEMO^{NS-II}$ vs.	NSGA-II	$DEMO^{SP2}$	SPEA2	$DEMO^{IB}$	IBEA
p_{value}	0.008	0.021	0.999	0.331	0.013

From this table, before the FWER is not controlled, the conclusion that the algorithm $DEMO^{NS-II}$ has a significant statistical different performance according to the r_2 indicator than the algorithms NSGA-II, $DEMO^{SP2}$, and IBEA, with a significance level $\alpha = 0.05$, can not be correct. The $DEMO^{NS-II}$ algorithm has a significant statistical different performance than each of these 3 algorithms since the p-values are smaller than $\alpha = 0.05$, considering pairwise comparisons. The true statistical difference for combining pairwise comparisons for these 5 hypotheses is calculated using Eq. 1. In our case the true p-value is 0.04, which is smaller than the significance level, $\alpha = 0.05$, and we can conclude that the $DEMO^{NS-II}$ has a significant statistical different performance than these 3 algorithms.

5   Conclusion

In this paper, to compare the performance of multi-objective stochastic optimization algorithms, a study for using quality indicators with deep statistical comparison approach (DSC) is presented. Deep Statistical Comparison approach is a recently proposed approach for the statistical comparison of meta-heuristic stochastic optimization algorithms over multiple single-objective problems. The main contribution of the DSC is that its ranking scheme is based on the whole distribution, instead of using only one statistic to describe the distribution, which can be average or median. By using the DSC, incorrect conclusions caused by the presence of outliers or misleading ranking scheme can be avoided.

The evaluation of the study is performed using the results for 6 multi-objective stochastic optimization algorithms tested on 16 test multi-objective problems. Two scenarios are tested in the experiments, a single-problem analysis, when the algorithms are compared according to some quality indicator over single multi-objective problem, and a multiple-problem analysis, when the algorithms are compared according to some quality indicator over multiple multi-objective problems. Experimental results show that the DSC give more robust result compared to some standard statistical approaches that are recommended to use in order to compare the performance of multi-objective stochastic optimization algorithms according to some quality indicator. For our future work, we are planning to find a way of how to combine different comparison results from different quality indicators, following the idea of ensemble learning.

Acknowledgments. This work is supported by the project ISO-FOOD, which received funding from the European Union's Seventh Framework Programme for research, technological development and demonstration under grant agreement No. 621329 (2014–2019) and the project that has received funding from the Slovenian Research Agency (research core funding No. L3-7538). We would like to thank Ph.D. Tea Tušar from the Department of Intelligent Systems at the Jožef Stefan Institute, for providing us the data involved in the experiments, which is also available on her website.

References

1. Breslow, N.: A generalized Kruskal-Wallis test for comparing K samples subject to unequal patterns of censorship. Biometrika **57**(3), 579–594 (1970)
2. Coello Coello, C.A., Lamont, G.B., Van Veldhuizen, D.A.: Evolutionary Algorithms for Solving Multi-objective Problems, vol. 5. Springer, Boston (2007). https://doi.org/10.1007/978-0-387-36797-2
3. Deb, K., Sindhya, K., Hakanen, J.: Multi-objective optimization. In: Sengupta, R.N., Gupta, A., Dutta, J. (eds.) Decision Sciences: Theory and Practice, pp. 145–184. CRC Press, Boca Raton (2016)
4. Deb, K., Thiele, L., Laumanns, M., Zitzler, E.: Scalable test problems for evolutionary multiobjective optimization. In: Abraham, A., Jain, L., Goldberg, R. (eds.) Evolutionary Multiobjective Optimization. Advanced Information and Knowledge Processing. Springer, London (2005). https://doi.org/10.1007/1-84628-137-7_6
5. Demšar, J.: Statistical comparisons of classifiers over multiple data sets. J. Mach. Learn. Res. **7**, 1–30 (2006)
6. Durillo, J.J., Nebro, A.J., Alba, E.: The jMetal framework for multi-objective optimization: design and architecture. In: 2010 IEEE Congress on Evolutionary Computation (CEC), pp. 1–8. IEEE (2010)

7. Eftimov, T., Korošec, P., Koroušić Seljak, B.: Disadvantages of statistical comparison of stochastic optimization algorithms. In: Proceedings of the Bioinspired Optimizaiton Methods and Their Applications, BIOMA 2016, pp. 105–118. JSI (2016)
8. Eftimov, T., Korošec, P., Koroušić Seljak, B.: A novel approach to statistical comparison of meta-heuristic stochastic optimization algorithms using deep statistics. Inf. Sci. **417**, 186–215 (2017)
9. Engmann, S., Cousineau, D.: Comparing distributions: the two-sample Anderson-Darling test as an alternative to the Kolmogorov-Smirnoff test. J. Appl. Quant. Methods **6**(3), 1–17 (2011)
10. da Fonseca, V.G., Fonseca, C.M., Hall, A.O.: Inferential performance assessment of stochastic optimisers and the attainment function. In: Zitzler, E., Thiele, L., Deb, K., Coello Coello, C.A., Corne, D. (eds.) EMO 2001. LNCS, vol. 1993, pp. 213–225. Springer, Heidelberg (2001). https://doi.org/10.1007/3-540-44719-9_15
11. García, S., Molina, D., Lozano, M., Herrera, F.: A study on the use of nonparametric tests for analyzing the evolutionary algorithms behaviour: a case study on the CEC'2005 special session on real parameter optimization. J. Heuristics **15**(6), 617–644 (2009)
12. Hochberg, Y.: A sharper Bonferroni procedure for multiple tests of significance. Biometrika **75**(4), 800–802 (1988)
13. Holm, S.: A simple sequentially rejective multiple test procedure. Scand. J. Stat. **6**(2), 65–70 (1979)
14. Huband, S., Barone, L., While, L., Hingston, P.: A scalable multi-objective test problem toolkit. In: Coello Coello, C.A., Hernández Aguirre, A., Zitzler, E. (eds.) EMO 2005. LNCS, vol. 3410, pp. 280–295. Springer, Heidelberg (2005). https://doi.org/10.1007/978-3-540-31880-4_20
15. Knowles, J., Thiele, L., Zitzler, E.: A tutorial on the performance assessment of stochastic multiobjective optimizers. Tik report no. 214, pp. 327–332 (2006)
16. van der Laan, M.J., Dudoit, S., Pollard, K.S.: Multiple testing. Part II. Step-down procedures for control of the family-wise error rate. Stat. Appl. Genet. Mol. Biol. **3**(1), 1–33 (2004)
17. Lam, F., Longnecker, M.: A modified Wilcoxon rank sum test for paired data. Biometrika **70**(2), 510–513 (1983)
18. Riquelme, N., Von Lücken, C., Baran, B.: Performance metrics in multi-objective optimization. In: 2015 Latin American Computing Conference (CLEI), pp. 1–11. IEEE (2015)
19. Ruxton, G.D.: The unequal variance t-test is an underused alternative to Student's t-test and the Mann-Whitney U test. Behav. Ecol. **17**(4), 688–690 (2006)
20. Tušar, T., Filipič, B.: Differential evolution versus genetic algorithms in multiobjective optimization. In: Obayashi, S., Deb, K., Poloni, C., Hiroyasu, T., Murata, T. (eds.) EMO 2007. LNCS, vol. 4403, pp. 257–271. Springer, Heidelberg (2007). https://doi.org/10.1007/978-3-540-70928-2_22
21. Van Veldhuizen, D.A., Lamont, G.B.: Multiobjective evolutionary algorithm research: a history and analysis. Technical report, Citeseer (1998)
22. Zitzler, E., Thiele, L.: Multiobjective evolutionary algorithms: a comparative case study and the strength pareto approach. IEEE Trans. Evol. Comput. **3**(4), 257–271 (1999)

On the Explicit Use of Enzyme-Substrate Reactions in Metabolic Pathway Analysis

Angelo Lucia[1]([✉]), Edward Thomas[1], and Peter A. DiMaggio[2]

[1] Department of Chemical Engineering,
University of Rhode Island, Kingston, RI 02881, USA
alucia@uri.edu
[2] Department of Chemical Engineering,
Imperial College London, London SW7 2AZ, UK

Abstract. Flux balance (or constraint-based) analysis has been the mainstay for understanding metabolic networks for many years. However, recently Lucia and DiMaggio [1] have argued that metabolic networks are more correctly modeled using game theory, specifically Nash Equilibrium, because it (1) captures the natural competition between enzymes, (2) includes rigorous chemical reaction equilibrium thermodynamics, (3) incorporates element mass balance constraints, and therefore charge balancing, in a natural way, and (4) allows regulatory constraints to be included as additional constraints.

The novel aspects of this work center on the explicit inclusion of enzyme-substrate reactions at the cellular length scale and molecular length scale protein docking information in metabolic network modeling. This multi-scale information offers the advantages of directly (1) computing cellular enzyme concentrations and activities, (2) incorporating genetic modification of enzymes, and (3) encoding the effects of age-related changes in enzymatic behavior (e.g., protein misfolding) within any pathway. Molecular length scale binding histograms are computed using protein-ligand docking and directly up-scaled to the cellular level. A small, proof-of-concept example from the Krebs cycle is presented to illustrate key ideas. Numerical results show that the proposed approach provides a wealth of quantitative enzyme information.

1 Introduction

While flux balance analysis (FBA) or constraint-based modelling (CBM) and its many variants have been used for metabolic pathway analysis for some time (see, for example, [2–12]), the Nash Equilibrium (NE) approach recently proposed by Lucia and DiMaggio [1] and extended in Lucia et al. [13] far outperforms all FBA and CBM methods because it is a first principles approach that incorporates rigorous chemical reaction equilibrium and elemental mass balances. As a result of its formulation, the NE approach has superior capabilities that naturally address (1) the competition among enzymes for resources in the metabolic pool, (2) substrate and co-factor charge balancing, and (3) regulatory controls.

© Springer International Publishing AG 2018
G. Nicosia et al. (Eds.): MOD 2017, LNCS 10710, pp. 88–99, 2018.
https://doi.org/10.1007/978-3-319-72926-8_8

The NE approach is also predictive. While there are competing kinetic-based models of metabolic pathways, these methods generally require a large number of parameters that cannot be directly measured and must be determined through model regression. Thus kinetic approaches are correlative and not predictive, particularly for experimental conditions that deviate from the training data used to determine model parameters.

1.1 A Nash Equilibrium Approach to Metabolic Pathways

In this sub-section, only a brief summary of the Nash Equilibrium approach to modelling metabolic pathways is presented. The reader is referred to [1,13] for details. The key ideas behind the NE approach to metabolic pathway analysis are to (1) represent the network using first principles rigorous chemical reaction equilibrium and element mass balances and (2) view enzymes as players in a multi-player game, in which each enzyme minimizes the change in Gibbs free energy for the biochemical reaction it catalyzes subject to appropriate elemental mass balances (i.e., conservation of mass of carbon, hydrogen, oxygen, nitrogen, phosphorous and sulfur). This leads to the representation of any metabolic network as a set of N nonlinear programming sub-problems (NLPs), where the network objective function is defined by:

$$\frac{G(v)}{RT} = \sum_{j=1}^{N} \min \frac{G_j(v_j)}{RT} \tag{1}$$

where G is the Gibbs free energy, v denotes the vector of metabolic fluxes, R is the universal gas constant, T is absolute temperature and j denotes the jth NLP sub-problem. The details of the Gibbs free energy and heat of formation data required in the NE formulation and a description of the cellular fluid model can be found in [1,13].

1.2 Element Mass Balances and Charge Balancing

The NE approach provides a natural way to ensure that atomic mass balances are satisfied within and across the metabolic network. In particular, appropriate element mass balances are included within each NLP sub-problem. Movement from one NLP sub-problem to the next automatically guarantees that element mass balances are satisfied since the outputs of one reaction are typically some or all of the inputs to the next reaction(s) in the network. For example, consider the first reaction in the Krebs (TCA) cycle given by:

$$C_4H_2O_5^{-2} + C_{23}H_{34}N_7O_{17}P_3S^{-4} + H_2O$$
$$\rightleftharpoons C_6H_5O_7^{-3} + C_{21}H_{32}N_7O_{16}P_3S^{-4} + H^+ \tag{2}$$

in which oxaloacetate and acetyl-CoA combine with water to form citrate, co-enzyme A, and hydrogen ions in the presence of the enzyme citrate synthase. The corresponding element balances for the reaction in Eq. 2 are the hydrogen, nitrogen, oxygen, and carbon balances and are represented by the matrix-vector equation:

$$A_{ik,j}\, v_{k,j} = M_{i,j} \tag{3}$$

where the index i corresponds to the individual elements, index j denotes sub-problem j (here $j = 1$ since the reaction under consideration is the first reaction in the network), index k corresponds to individual metabolites/cofactors, vector $v_{k,j} = (v_{1,1}, v_{2,1}, v_{3,1}, v_{4,1}, v_{5,1})^T$, vector $M_{i,j} = (H_1, N_1, O_1, C_1)^T$ and H_1, N_1, O_1, C_1 represent the molar amounts of hydrogen, nitrogen, oxygen, and carbon in that order for subproblem $j = 1$. The $k = 5$ independent fluxes (chemical species) in Eq. 3 are water, acetyl-CoA, co-enzyme A, oxaloacetate, and citrate respectively. H^+ is a dependent flux. Additionally, the full matrix $A_{ik,j}$ is

$$A_{ik,j=1} = \begin{pmatrix} 2 & 34 & 32 & 2 & 5 & 1 \\ 0 & 7 & 7 & 0 & 0 & 0 \\ 1 & 17 & 16 & 5 & 7 & 0 \\ 0 & 23 & 21 & 4 & 6 & 0 \end{pmatrix} \tag{4}$$

Note that while phosphorous and sulfur are also present, these elements, along with nitrogen, are fixed in the ratio $N_7 P_3 S$ so only one of the element mass balances in the subset N, P, S is linearly independent and can be used as a constraint.

In the second reaction in the TCA cycle, the citrate from reaction 1 binds with the enzyme aconitase to form isocitrate. Considering only the overall metabolite reaction, we have

$$C_6 H_5 O_7^{-3} \rightleftharpoons iC_6 H_5 O_7^{-3} \tag{5}$$

Note that there is only a single independent element balance for this second reaction since it is simply an isomerization reaction.

The key points here are that:

1. If the element balances for the first reaction are satisfied, then the amount of citrate that is available for the second reaction preserves element mass balances.
2. Element balancing automatically accounts for correct charge balancing.

2   Explicitly Incorporating Enzyme-Substrate Reactions

To our knowledge there is no approach to metabolic network modeling and analysis that explicitly includes the binding and unbinding of substrates with enzymes and therefore no methodology capable of predicting enzyme concentrations and activities or their impact. The inclusion of enzyme-substrate binding/unbinding reactions opens up a wide range of possibilities that can provide important quantitative information such as:

1. The amount of a given enzyme needed to catalyze a given reaction.
2. The impact of changes in enzymatic activity on the steady-state behavior of metabolic networks.

Including enzyme-substrate reactions within the Nash Equilibrium formulation of a metabolic network is straightforward. However, it is well known that enzyme-substrate binding (also called protein docking) exhibits many minima and saddle points on the energy surface. To obtain this data, molecular length scale protein docking software (e.g., AutoDock [14], which was used exclusively in this work) can be used to create a look-up table of ranked enzyme-substrate binding energies from histograms and this information can be easily up-scaled to the cellular length scale for use in the NE calculations. In general, information for conformations (or docking solutions) with the lowest Gibbs free energy are up-scaled to and used at the cellular length scale - unless there are reasons for choosing a different conformation.

2.1   Enzyme-Substrate Reactions

General enzyme-substrate reactions can be described using the simple two-reaction sequence:

$$E + S \rightleftharpoons E - S \tag{6}$$

$$E - S \rightleftharpoons E + P \tag{7}$$

which represent, respectively, the binding of the enzyme, E, and substrate, S, to form a stable complex, denoted by $E - S$, followed by rearrangement, cleaving, or some other interaction and then subsequent unbinding to regenerate enzyme and produce product, P. Within the Nash Equilibrium framework, binding and unbinding are considered to reach chemical equilibrium.

2.2   An Example of Binding and Unbinding Reactions

Consider the simple example of the binding of citrate (shown in blue and red) with the iron sulfate complex (orange and yellow) of the enzyme aconitase as shown in Fig. 1. A sample of the output produced by AutoDock is given in Appendix A.1. Note that binding takes place at sites in the large binding pocket containing the iron sulfate complex, which is surrounded by α helix and β sheet portions of the enzyme. Moreover, this large pocket is the 'correct' binding site and the one that results in the production of isocitrate.

2.3   Multiple Minima from Protein Docking

It is well known that protein-ligand docking is a multi-minima problem, in which there are a large number of minima and saddle point solutions. For example, for the aconitase-citrate illustration, using just twenty-five (25) random starting points AutoDock located twenty-five different solutions or conformations in three separate binding pockets. Figure 2 shows three key solutions, which have corresponding Gibbs free energies of binding of -11.38, -6.72, and -6.22 kcal/mol respectively. Solution 1 (top left) is the global minimum and the one that leads to the conversion of citrate to isocitrate. Solutions 2 (top right) and 3 (bottom)

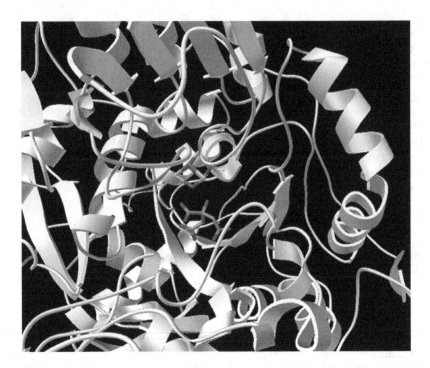

Fig. 1. Docking of citrate with aconitase (Protein Data Bank (PDB) ID: 1C96). (Color figure online)

represent conformations in which citrate binds to sites that are above the large binding pocket in the center of the enzyme and, as a result, do not convert citrate to isocitrate. These solutions can be ranked based on their respective Gibbs free energies and clustering information from the corresponding histogram (see Appendix A.2) can be used to define cluster efficiencies given by the rule

$$k_{ijm} = \frac{n_{ijm}}{\sum_{l=1}^{M} n_{ijl}}, \quad i = 1,...,n_S; \; j = 1,...,n_E; \; m = 1,...,M \tag{8}$$

where i is a substrate index, n_S is the number of substrates, j denotes the enzyme index, n_E is the number of enzymes, m is the cluster index, and M is the total number of clusters. In the illustrative example, there is one substrate (citrate), one enzyme (aconitase) and three clusters, which gives

$$k_{111} = \frac{23}{25} = 0.92; \; k_{112} = \frac{1}{25} = 0.04; \; k_{113} = \frac{1}{25} = 0.04 \tag{9}$$

respectively for solutions 1, 2 and 3 shown in Fig. 2. Normally, the cluster with the highest efficiency is chosen unless there is reason to choose a different cluster efficiency. There are a few key points to note:

1. The total time for docking simulations is not prohibitive. For this small example, the total time to compute all twenty-five solutions was ~10 min on a

laptop. This clearly indicates that generating a database of enzyme-substrate binding energies, clusters, and a cluster ranking is tractable - even for a large number of enzymes and substrates. Once this data is determined for a given enzyme-substrate pair, it never has to be computed again.

2. Our proposed approach for including molecular length scale enzyme-substrate information relies on the Protein Data Bank (PDB) and a tool for computing relative binding Gibbs free energies and cluster information (e.g., AutoDock).

3. In a more general sense, ranked enzyme-substrate efficiencies open up many possibilities, not the least of which is the capability to include behavior such as changes in enzyme activity due to genetic modifications, misfolding, ageing, and so on, provided of course the structural changes in the enzyme resulting from these modifications can be determined (e.g. via protein folding calculations, molecular dynamics, etc.).

Fig. 2. Multiple binding solutions for aconitase-citrate docking.

2.4   A Multi-scale Methodology for Including Enzyme-Substrate Reactions

It is instructive to illustrate for the reader the way in which enzyme-substrate reactions are included in the NE framework for the purpose of determining enzyme activities and concentrations. Here again we use the example of citrate conversion to isocitrate to illustrate. The conversion of citrate to isocitrate can be treated as a two-reaction sequence, in which the first reaction is given by

$$C_6H_5O_7^{-3} + E \;\rightleftharpoons\; E - C_6H_5O_7^{-3} \tag{10}$$

where E = aconitase. The corresponding NLP sub-problem for this reaction is

$$\min \frac{G_1(v_1, v_2, v_3)}{RT} \tag{11}$$

subject to element balances

$$5v_1 + 0v_2 + 5v_3 = H \tag{12}$$

$$0v_1 + 1v_2 + 1v_3 = E \tag{13}$$

where H and E represent the amount of hydrogen and aconitase in the initial pool and the subscripts 1, 2 and 3 correspond to citrate, aconitase, and the aconitase-citrate complex, respectively. It is important for the reader to understand that enzymes are treated by assuming they undergo no change in mass. Only the substrates or metabolites undergo chemical change. As a result, element balancing of the enzyme is unnecessary, which avoids scaling and other complicating issues due to the typically large number of residues and corresponding molecular weight of enzymes.

The second reaction, Eq. 7, is the unbinding of isocitrate from aconitase and results in the NLP sub-problem given by

$$\min \frac{G_1(v_1, v_2, v_3)}{RT} \tag{14}$$

subject to the same set of mass balance constraints (i.e., Eqs. 12 and 13). The only difference here is that subscript 1 in Eqs. 12 and 13 now represents isocitrate. Note that charge balancing associated with the overall conversion of citrate to isocitrate remains unchanged in the presence of enzyme-substrate reactions.

2.5   Enzyme Activity

One way to get a measure of enzyme activity is to plot the rate of reaction as a function of substrate concentration. This leads to the simple expression for the rate of conversion of citrate to isocitrate (here for $E.\ coli$) given by

$$V_0 = \frac{v_P}{V} \tag{15}$$

where v_P represents the steady-state flux of product, V is the reaction volume of the appropriate compartment of the cell (e.g., cytosol in $E.\ coli$), and V_0 is the rate of catalysis. Biochemists usually express the rate of catalysis in terms of Michaelis-Menton kinetics using an equation of the form

$$V_0 = V_{max} \frac{[S]}{[S] + K_M} \tag{16}$$

where V_{max} is the maximum rate of catalysis, $[S]$ is the substrate concentration, and K_M is the Michaelis constant, which is defined as the substrate concentration that gives a reaction velocity equal to $V_{max}/2$.

Another important metric of enzyme activity is <u>turnover number</u>. The turnover number, k_{cat}, is the reaction rate constant associated with the conversion of enzyme-substrate complex to enzyme plus product (i.e., the rate constant associated with Eq. 7). The turnover number can be computed using the expression

$$k_{cat} = \frac{V_{max}}{[E]_T} = \frac{V_{max}}{[E] + [E-S]} \tag{17}$$

where $[E]_T$ is the total enzyme concentration.

3   Numerical Results

Numerical results for the inclusion of enzyme-substrate reactions within a Nash Equilibrium formulation are presented. To make the presentation clear, we focus on the citrate-aconitase-isocitrate example for *E. coli* from the previous sections (Eqs. 10–14) to provide proof-of-concept. Of specific interest is the quantitative determination of enzyme concentrations and activity metrics. All computations were performed on a Dell Inspiron laptop with the Lahey-Fujitsu LF95 compiler.

Table 1. Aconitase conversion of citrate to isocitrate at $25\,^{\circ}\mathrm{C}$.

Enzyme/Substrate	Flux (mmol/s)	Concentration (mM)
Citrate	0.066900	2.79040
Aconitase	0.106647	4.44828
Aconitase-citrate	0.022484	0.93782
Isocitrate	0.033813	1.41034

Table 1 shows the steady-state fluxes and concentrations for substrates, enzyme and enzyme-substrate complex for the conversion of citrate to isocitrate in the presence of aconitase for an initial pool of 0.1 mmol of citrate and 0.13 mmol of aconitase and temperature of 25 °C. Table 1 also shows that under the given conditions ∼33% of the citrate is converted to isocitrate and that ∼82% of the aconitase is regenerated.

Figure 3 shows the rate of isocitrate as a function of substrate concentration for 0.13 mM/s of aconitase and initial citrate concentrations ranging from 0.2 mM to 4 M at 25 °C and a cytosolic volume for *E. coli* of $V = 1\,\mu\mathrm{m}^3$ (Fig. 1 in [15]).

Note that the reaction velocity increases as substrate concentration increases until the aconitase is saturated. At that point there are no more active sites available and the reaction velocity (enzyme activity) reaches a maximum rate of 0.091 M/s.

Figure 4, on the other hand, is an enlargement of Fig. 3 at low substrate concentration, which is necessary to graphically determine the Michaelis constant, K_M. From the value $V_{max}/2 = 0.0455$ M/s, the Michaelis constant is $K_M = 4.91$ mM. Using Eq. 17 and the saturated enzyme concentration of 5.386 mM, the predicted turnover number is 16.90 s^{-1}. Both metrics, $K_M = 4.91$ mM

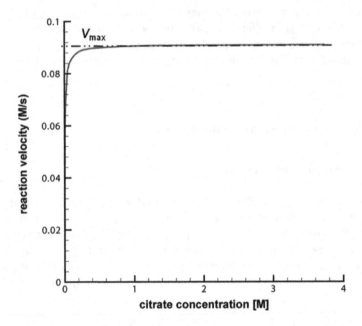

Fig. 3. Reaction velocity as a function of citrate concentration.

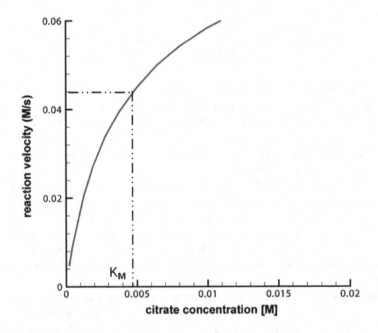

Fig. 4. Enlargement of V_0 as a function of low citrate concentration.

and $k_{cat} = 16.90$ s^{-1} match published experimental data (i.e. $K_M = [1.16 - 11]$ mM in [16] and $k_{cat} = 13.5$ s^{-1} in [17]) quite well.

Finally, from the histogram information, we use an enzyme efficiency of 92% and adjust the initial pool of active enzyme from 0.13 mmol to 0.1196 mmol. For the illustration in Table 1, this simply gives a slightly lower conversion of citrate to isocitrate (32 vs. 33%), slightly lower enzyme concentration (4.05 vs. 4.45 mM), and shifts the curve in Fig. 3 downward yielding a lower value of V_{max} (0.083 vs. 0.091 M/s) and a lower K_M (4.69 vs. 4.91 mM).

4 Conclusions

The inclusion of enzymatic reactions in a Nash Equilibrium framework for metabolic pathway analysis was presented. Results for a simple illustration of the conversion of citrate to isocitrate in the presence of aconitase clearly show that the proposed approach can be used to predict key metrics used to describe enzyme activity as well as enzyme and enzyme-substrate complex concentrations. The results presented in this work easily generalize to any enzymatic reaction and can be used for strain development via genetic modification, understanding epigenetics, therapeutics, and other biological tasks.

Appendix A.1

See Fig. 5.

```
      LOWEST ENERGY DOCKED CONFORMATION from EACH CLUSTER
      _____


      Keeping original residue number (specified in the input PDBQ
      file) for outputting.

      MODEL        21
      USER     Run = 21
      USER     Cluster Rank = 1
      USER     Number of conformations in this cluster = 23
      USER
      USER     RMSD from reference structure       = 0.503 A
      USER
      USER       Estimated Free Energy of Binding  =  -11.38 kcal/mol
      [=(1)+(2)+(3)-(4)]
      USER       Estimated Inhibition Constant, Ki =    4.52 nM
      (nanomolar)  [Temperature = 298.15 K]
      USER
      USER     (1) Final Intermolecular Energy     =  -12.88 kcal/mol
      USER         vdW + Hbond + desolv Energy     =   -8.69 kcal/mol
      USER         Electrostatic Energy            =   -4.18 kcal/mol
      USER     (2) Final Total Internal Energy     =   +1.43 kcal/mol
      USER     (3) Torsional Free Energy           =   +1.49 kcal/mol
      USER     (4) Unbound System's Energy  [=(2)] =   +1.43 kcal/mol
      USER
      USER
      USER
```

Fig. 5. Sample output from protein-ligand docking software.

Appendix A.2

See Fig. 6.

```
 Clus | Lowest   | Run | Mean     | Num | Histogram
 -ter | Binding  |     | Binding  | in  |
 Rank | Energy   |     | Energy   | Clus|   5    10    15    20    25
      |          |     |          |     |   :___|____:___|___:__
   1  |  -11.38  | 21  |  -10.27  | 23  |#######################
   2  |   -6.72  | 13  |   -6.72  |  1  |#
   3  |   -6.22  |  6  |   -6.22  |  1  |#
```

Fig. 6. Sample AutoDock histogram for protein-ligand docking.

References

1. Lucia, A., DiMaggio, P.A.: A Nash equilibrium approach to metabolic network analysis. In: Pardalos, P.M., Conca, P., Giuffrida, G., Nicosia, G. (eds.) MOD 2016. LNCS, vol. 10122, pp. 45–58. Springer, Cham (2016). https://doi.org/10.1007/978-3-319-51469-7_4

2. Varma, A., Palsson, B.O.: Metabolic flux balancing: basic concepts, scientific and practical use. Nat. Biotechnol. **12**, 994–998 (1994)

3. Kauffman, K.J., Prakash, P., Edwards, J.S.: Advances in flux balance analysis. Curr. Opin. Biotechnol. **14**, 491–496 (2003)

4. Holzhutter, H.G.: The principles of flux minimization and its application to estimate stationary fluxes in metabolic networks. Eur. J. Biochem. **271**, 2905–2922 (2004)

5. Julius, A.A., Imielinski, M., Pappas, G.J.: Metabolic networks analysis using convex optimization. In: Proceedings of the 47th IEEE Conference on Decision and Control, p. 762 (2008)

6. Smallbone, K., Simeonidis, E.: Flux balance analysis: a geometric perspective. J. Theor. Biol. **258**, 311–315 (2009)

7. Murabito, E., Simeonidis, E., Smallbone, K., Swinton, J.: Capturing the essence of a metabolic network: a flux balance analysis approach. J. Theor. Biol. **260**(3), 445–452 (2009)

8. Lee, S., Phalakornkule, C., Domach, M.M., Grossmann, I.E.: Recursive MILP model for finding all the alternate optima in LP models for metabolic networks. Comput. Chem. Eng. **24**, 711–716 (2000)

9. Henry, C.S., Broadbelt, L.J., Hatzimanikatis, V.: Thermodynamic metabolic flux analysis. Biophys. J. **92**, 1792–1805 (2007)

10. Mahadevan, R., Edwards, J.S., Doyle, F.J.: Dynamic flux balance analysis in diauxic growth in Escherichia coli. Biophys. J. **83**, 1331–1340 (2002)

11. Patane, A., Santoro, A., Costanza, J., Nicosia, G.: Pareto optimal design for synthetic biology. IEEE Trans. Biomed. Circuits Syst. **9**(4), 555–571 (2015)

12. Angione, C., Costanza, J., Carapezza, G., Lio, P., Nicosia, G.: Multi-target analysis and design of mitochondrial metabolism. PLoS One **9**, 1–22 (2015)

13. Lucia, A., DiMaggio, P.A., Alonso-Martinez, D.: Metabolic pathway analysis using Nash equilibrium. J. Optim. (2017, in press)

14. Morris, G.M., Huey, R., Lindstrom, W., Sanner, M.F., Belew, R.K., Goodsell, D.S., Olson, A.J.: Autodock4 and AutoDockTools4: automated docking with selective receptor flexibility. J. Comput. Chem. **16**, 2785–2791 (2009)
15. Milo, R.: What is the total number of protein molecules per cell volume? A call to re-think some published values. BioEssays **35**, 1050–1055 (2013)
16. Schomburg, I., Hofmann, O., Bänsch, C., Chang, A., Schomburg, D.: Enzyme data and metabolic information: BRENDA, a resource for research in biology, biochemistry, and medicine. Gene Funct Dis. **3**, 109–118 (2000)
17. Villafranca, J.J., Mildvan, A.S.: The mechanism of aconitase action: I. Preparation, physical properties of the enzyme, and activation by iron (II). J. Biol. Chem. **246**, 772–779 (1971)

A Comparative Study on Term Weighting Schemes for Text Classification

Ahmad Mazyad[✉], Fabien Teytaud, and Cyril Fonlupt

LISIC, Université du Littoral Côte d'Opale,
50 Rue Ferdinand Buisson, 62100 Calais, France
Ahmad.Mazyad@univ-littoral.fr

Abstract. Text Classification (or Text Categorization) is a popular machine learning task. It consists in assigning categories to documents. In this paper, we are interested in comparing state of the art classifiers and state of the art feature weights. Feature weight methods are classic tools that are used in text categorization. We extend previous studies by evaluating numerous term weighting schemes for state of the art classification methods. We aim at providing a complete survey on text classification for fair benchmark comparisons.

1 Introduction

Nowadays, at the time of the rapid growth of the internet, the volume of text documents becomes more and more important. Consequently, effective document retrieval may be a really hard task, especially without any organization. Text classification has become a state of the art solution to this problem. Over time, several classification methods appear [5], such as k-nearest neighbor [16], Naïve Bayes [8], decision trees [1], neural networks [9], boosting methods [11] and Support Vector Machines [2]. In this paper we are interested in finding a good term weighting method for state of the art classification algorithms. The paper is organized as follows: in Sect. 2 we present text classification and in particular state of the art term weighting method definitions. In Sect. 3 we present the state of the art classifiers used in our study. In Sect. 4, we compare the different term weighting methods applied to 3 famous text categorization benchmarks. Finally, we discuss and present future works in Sect. 5.

2 Text Classification

Text Classification (TC) aims at automatically assigning a set of predefined categories to a text document. Depending on the text corpus being classified, each document can be in one or multiple categories. This task is achieved by using a classifier learned on a training set of labeled documents.

A fundamental step in learning a classifier is to represent text documents in a suitable format recognizable by this classifier. In Vector Space Model (VSM), each text document is represented as a vector of index terms in which each term

© Springer International Publishing AG 2018
G. Nicosia et al. (Eds.): MOD 2017, LNCS 10710, pp. 100–108, 2018.
https://doi.org/10.1007/978-3-319-72926-8_9

is associated with a weight (score) that measures how informative/discriminative the correspondent term is. The method which assigns a weight to a term is called Term Weighting Scheme (TWS).

To the best of our knowledge, no complete survey exists on how effective TWS performs with different state of the art classifiers. In this paper, we focus on this comparison in order to have a fair and complete study.

One of the most famous (TWS) is $tf.idf$ proposed by Jones in [12] and stands for term frequency-inverse document frequency. $tf.idf$ is an unsupervised term weighting method. It is the product of the Term Frequency component (TF) by the Collection Frequency component (CF): Term Frequency (tf) and Inverse Document Frequency (idf) respectively. We use the logarithmically scaled tf defined as:

$$tf_{t,d} = 1 + \log(f_{t,d}).$$

where $f_{t \in d}$ stands for the occurrence of term t in the document d. And idf defined as:

$$idf(t) = \log \frac{|D|}{|\{d' \in D | t \in d'\}|}.$$

where $|D|$ is the total number of documents and $|\{d' \in D | t \in d'\}|$ is the number of documents that contains the term t.

TC is a supervised learning task, such that document membership (class information) is known in advance. We call Supervised Term Weighting (STW), the term weighting that incorporates the class information. In that context, researchers proposed various supervised term weighting methods that replace the unsupervised collection frequency component idf by a supervised component. For instance, Chi-square (χ^2) is a test of independence between two variables and it was first used as a TWS in text categorization in [4,5]. Gain ratio (gr) in [4], odds ratio (or) in [5], relevance frequency (rf) was proposed by Lan et al. in [7], inverse category frequency (icf) proposed by Wang et al. in [14], and term relevance ratio (trr) by Youngjoong in [17].

Thus, the general formula for the different TWS in this paper, could be defined as:

$$w_{t,d} = tf_{t,d} \times CF(t).$$

Table 1 shows all the CF included in this study that are used in almost all TC works. All these TWS are used in classic machine learning tools for TC. We present some of these tools in the next section.

3   Classifiers

To study the effect of each Supervised Term Weighting (STW) on classification tasks, we use five known learning algorithms: Support Vector Machine (SVM), Passive-Aggressive (PA), Stochastic Gradient Descent (SGD), Nearest Centroid (NC), C4.5 (C4.5). However it's important to note that we are studying the effectiveness of STW rather than the performance of the learning algorithms.

Table 1. Collection Frequency Components. Given a term t and a category c, N stands for the total number of documents, $|C|$ is the total number of categories and $|C_t|$ is the number of categories where the term t occurs, w is the number of documents that contain t and belong to category c, x is the number of documents that contain t and do not belong to c, y is the number of documents that do not contain t and belong to cj, z is the number of documents that do not contain t and do not belong to c.

CF	Formula
idf	$\log(N/(w+y))$
χ^2	$N \times ((w \times z - x \times y)^2)/((w+y)(x+z)(w+x)(y+z))$
ig	$((w/N) \times \log(w \times N)/((w+x)(w+y))) + ((y/N) \times \log(y \times N)/((y+z)(w+y)))$ $+((x/N) \times \log(x \times N)/((w+x)(b+z))) + ((z/N) \times \log(z \times N)/((y+z)(b+z)))$
gr	$ig/((-(w+y)/N)(\log(w+y)/N) - ((x+z)/N)(\log(x+z)/N))$
or	$\log(2 + (w*z)/(x*y))$
rf	$\log(2 + (w/\max(1,x)))$
icf	$\log_2(C/C_i)$

SVM is a supervised machine learning algorithm used for both classification and regression. SVM has been proposed by Cortes and Vapnik in [2]. Joachims [6] was the first to use SVM for text categorization in which he shows the superiority of SVM over other traditional learning methods.

PA introduced in [3] is an online learning algorithm for large scale dataset. The algorithm watches a stream of instances. Once a new instance is received, the algorithm outputs a prediction. Later, the instance true label is uncovered and the algorithm updates its prediction function.

SGD [18] is another learning algorithm for large scale classification task. It is used to learn linear models such as linear SVM, by minimizing its objective function.

NC [13] is a simple neighborhood-based classification algorithm. The algorithm computes a centroid for each class. It then outputs the label of the nearest centroid to the test instance as the predicted label.

C4.5 [10] is a supervised tree-based learning algorithm. In 2008, C4.5 has received a considerable amount of attention after being ranked first in the *Top 10 Algorithms in Data Mining* [15].

4   Results and Discussion

4.1   Experiments

Three widely-used datasets are used to evaluate the classifiers: Reuters, Oshumed and 20 Newsgroups.

Reuters-21578[1] is one of the most used test collection for TC research. We use the "ModApte" split which contains 90 categories.

---

[1] http://disi.unitn.it/moschitti/corpora.htm.

The second dataset is extracted from the Oshumed[2] collection compiled by William Hersh.

The last test collection used in our experiment is the 20 Newsgroups. The dataset "20news-bydate"[3] is sorted by date and splitted into training set (about 60%) and test set (about 40%). Duplicates are removed. Newsgroup-identifying headers (Xref, Newsgroups, Path, Followup-To, Date) are also removed.

In all three test collections, we applied lower case transformation, word stemming and stop word removal. No additional preprocessing steps or feature selection is performed.

Reuters-21578 and Oshumed are multi-labelled datasets. 20Newsgroups is a multi-class dataset. In all cases, we transform the task into multiple binary single label tasks using the one-vs.-all transformation strategy aka one-vs.-rest.

Table 2 shows some statistics on the three collections.

Table 2. Statistics on the three test collections (train data/test data).

	Reuters	Oshumed	Newsgroups
# documents	7769/3019	6286/7643	11314/7532
# terms	26000	30198	101322
# categories	90	23	20
Size of the smallest category	1/1	65/70	377/251
Size of the largest category	2877/1087	1799/2153	600/399

4.2   Evaluation

To assess the performance of STW, we use the standard F1 measure. The F1 score considers both precision (true positive over true positive plus false positive) p and recall (true positive over true positive plus false negative) r and can be formally defined as: $F1(p, r) = \frac{2rp}{r+p}$. We also report the precision and recall. The precision and recall results of the multiple binary tasks are averaged using the micro-(μ) and macro-(m) averaged measures.

Tables 3, 4, 5, 6, 7, 8, 9, 10 and 11 show the μ/m-averaged precision, recall and f-score, for reuters, oshumed and 20newsgroups datasets, respectively. In these tables, the highest μ/m score over a column is underlined, and the best pair of μ/m scores considering all classifiers and all TWS are bolded. The pair that have the highest average is choosen as the best.

4.3   Results

In Tables 3, 4 and 5, we present the μ/m-averaged precision, recall, and f-score, respectively, for Reuters-21578 dataset. In Table 5, NC shows the lowest performance, considering both μ and m scores. PA have the highest μ-score (87.22%)

[2] http://disi.unitn.it/moschitti/corpora.htm.
[3] http://qwone.com/~jason/20Newsgroups/.

Table 3. μ/m-averaged precision results (%) on Reuters-21578 dataset using different weighting methods.

	PA	SVM	SGD	NC	C4.5
tf	91.50/62.69	94.37/56.75	94.40/56.64	39.22/30.28	82.17/57.17
tfchi2	91.35/63.23	94.37/56.75	94.46/55.54	39.22/30.28	82.18/56.23
tfgr	91.26/61.37	94.37/56.75	94.48/57.21	39.22/30.28	82.44/55.26
tficf	<u>93.21</u>/64.03	94.95/<u>57.31</u>	<u>94.69</u>/<u>61.25</u>	48.87/<u>50.00</u>	81.64/55.34
tfidf	**93.12**/**64.14**	**95.17**/56.95	94.45/58.85	<u>63.40</u>/47.57	81.82/56.65
tfig	91.56/62.63	94.37/56.75	94.48/56.68	39.22/30.28	<u>82.45</u>/<u>58.53</u>
tfor	91.73/63.42	94.37/56.75	94.47/56.62	39.22/30.28	82.07/56.24
tfrf	91.51/60.75	94.37/56.75	94.45/55.52	39.22/30.28	81.93/55.63

Table 4. μ/m-averaged recall results (%) on Reuters-21578 dataset using different weighting methods.

	PA	SVM	SGD	NC	C4.5
tf	82.27/<u>42.74</u>	<u>78.85</u>/33.51	79.73/35.08	**89.93**/**61.76**	81.62/<u>53.79</u>
tfchi2	81.76/42.64	<u>78.85</u>/33.51	79.62/34.82	**89.93**/**61.76**	81.41/52.91
tfgr	82.27/41.55	<u>78.85</u>/33.51	79.54/34.81	**89.93**/**61.76**	81.62/51.81
tficf	79.59/39.44	75.27/30.64	77.19/33.20	86.75/52.96	80.26/51.91
tfidf	82.02/41.81	78.37/<u>33.60</u>	<u>80.02</u>/<u>36.29</u>	87.55/55.60	80.80/53.65
tfig	<u>82.61</u>/41.93	<u>78.85</u>/33.51	79.51/34.81	**89.93**/**61.76**	81.70/53.51
tfor	82.10/42.59	<u>78.85</u>/33.51	79.46/34.66	**89.93**/**61.76**	81.68/53.32
tfrf	82.00/41.15	<u>78.85</u>/33.51	79.57/34.75	**89.93**/**61.76**	<u>81.97</u>/52.97

Table 5. μ/m-averaged f-score results (%) on Reuters-21578 dataset using different weighting methods.

	PA	SVM	SGD	NC	C4.5
tf	86.64/48.48	85.91/39.74	86.45/41.15	54.61/34.75	81.90/53.63
tfchi2	86.29/<u>48.51</u>	85.91/39.74	86.41/40.80	54.61/34.75	81.79/53.24
tfgr	86.53/47.14	85.91/39.74	86.37/41.14	54.61/34.75	82.03/51.79
tficf	85.87/46.42	83.97/37.77	85.05/40.28	62.52/46.43	80.94/52.05
tfidf	**87.22**/**48.20**	<u>85.95</u>/<u>40.32</u>	<u>86.64</u>/<u>42.73</u>	<u>73.55</u>/<u>47.05</u>	81.31/53.36
tfig	86.86/47.76	85.91/39.74	86.35/40.97	54.61/34.75	**82.08**/**54.24**
tfor	86.65/48.48	85.91/39.74	86.32/40.85	54.61/34.75	81.87/52.82
tfrf	86.49/46.74	85.91/39.74	86.37/40.76	54.61/34.75	81.95/52.82

Table 6. μ/m-averaged precision results (%) on Oshumed dataset using different weighting methods.

	PA	SVM	SGD	NC	C4.5
tf	71.13/73.55	78.77/81.13	79.85/<u>82.42</u>	39.22/35.64	57.09/53.40
tfchi2	64.72/61.40	72.81/71.56	71.57/69.70	47.34/45.23	<u>58.22</u>/<u>56.02</u>
tfgr	<u>76.17</u>/78.21	**81.04/80.14**	<u>80.67</u>/79.39	58.65/58.85	56.72/52.89
tficf	74.27/75.04	80.80/81.07	77.92/77.81	<u>69.32</u>/<u>68.58</u>	55.63/53.09
tfidf	75.76/<u>78.26</u>	80.83/80.36	80.48/79.11	54.40/53.06	57.54/53.61
tfig	76.14/77.84	**81.04/80.14**	80.81/79.49	58.65/58.85	56.84/53.65
tfor	74.25/76.45	79.74/81.91	79.44/81.19	53.58/53.61	57.34/54.62
tfrf	74.08/76.38	80.29/<u>83.20</u>	80.39/82.11	52.24/52.12	57.64/54.63

Table 7. μ/m-averaged recall results (%) on Oshumed dataset using different weighting methods.

	PA	SVM	SGD	NC	C4.5
tf	52.91/44.32	46.21/35.96	47.27/37.76	**68.04/66.55**	56.08/51.73
tfchi2	56.50/51.22	<u>54.83</u>/48.33	50.77/45.94	64.82/65.07	56.70/52.42
tfgr	54.89/47.80	48.61/40.01	52.08/44.79	66.60/64.62	56.89/52.70
tficf	45.57/40.16	35.50/29.46	42.02/36.32	51.58/47.07	<u>57.43</u>/52.71
tfidf	53.55/45.80	46.82/37.43	50.19/42.00	67.71/65.54	56.35/52.50
tfig	54.76/47.84	48.61/40.01	51.96/44.73	66.60/64.62	57.36/<u>53.67</u>
tfor	<u>58.15</u>/<u>53.84</u>	54.68/<u>48.36</u>	<u>56.53</u>/<u>51.26</u>	66.14/66.29	55.89/51.46
tfrf	56.38/50.68	52.32/44.55	53.69/46.72	66.32/65.98	55.53/51.79

Table 8. μ/m-averaged f-score results (%) on Oshumed dataset using different weighting methods.

	PA	SVM	SGD	NC	C4.5
tf	60.68/53.95	58.25/47.02	59.39/48.78	49.76/44.48	56.58/52.42
tfchi2	60.33/55.51	62.55/55.27	59.40/52.01	54.72/51.83	<u>57.45</u>/<u>53.88</u>
tfgr	63.80/58.11	60.77/51.71	63.29/56.05	<u>62.37</u>/<u>60.16</u>	56.80/52.65
tficf	56.48/51.32	49.33/41.93	54.60/48.32	59.15/55.25	56.51/52.67
tfidf	62.75/56.42	59.30/49.08	61.83/53.41	60.33/57.43	56.94/52.88
tfig	63.71/58.12	60.77/51.71	63.25/56.02	<u>62.37</u>/<u>60.16</u>	57.10/53.47
tfor	**65.22/62.37**	64.87/<u>58.78</u>	66.05/<u>60.57</u>	59.20/57.43	56.60/52.76
tfrf	64.03/60.08	63.36/55.52	64.38/57.19	58.44/56.05	56.56/53.00

Table 9. μ/m-averaged precision results (%) on 20newsgroups dataset using different weighting methods.

	PA	SVM	SGD	NC	C4.5
tf	63.91/63.77	66.94/66.58	61.05/62.78	55.91/62.22	44.07/44.12
tfchi2	58.55/60.54	60.26/60.35	59.33/59.51	47.73/60.20	38.20/38.16
tfgr	68.43/68.37	69.69/69.41	**70.19/70.06**	62.85/71.44	43.07/43.37
tficf	68.14/68.23	69.15/69.23	69.24/68.85	59.43/<u>71.87</u>	<u>49.19</u>/<u>51.77</u>
tfidf	68.31/68.10	69.69/69.29	61.26/66.59	<u>64.27</u>/69.19	43.65/43.67
tfig	<u>68.97</u>/<u>68.86</u>	<u>70.14</u>/<u>69.79</u>	<u>70.26</u>/69.85	63.64/71.56	44.16/44.40
tfor	68.57/68.19	69.80/69.26	69.54/69.24	56.44/69.03	45.13/44.77
tfrf	56.00/55.42	57.73/56.95	56.57/55.36	36.56/46.22	42.22/42.58

Table 10. μ/m-averaged recall results (%) on 20newsgroups dataset using different weighting methods.

	PA	SVM	SGD	NC	C4.5
tf	63.91/62.87	66.94/65.81	61.05/59.69	55.91/55.17	44.07/43.01
tfchi2	58.55/57.05	60.26/58.74	59.33/57.82	47.73/47.05	38.20/37.17
tfgr	68.43/67.33	69.69/68.52	70.19/68.92	62.85/62.10	43.07/42.07
tficf	68.14/66.96	69.15/67.90	69.24/67.95	59.43/58.62	<u>49.19</u>/<u>48.15</u>
tfidf	68.31/67.20	69.69/68.48	61.26/59.95	<u>64.27</u>/<u>63.32</u>	43.65/42.73
tfig	<u>68.97</u>/<u>67.86</u>	<u>70.14</u>/<u>68.93</u>	**70.26/68.94**	63.64/62.78	44.16/43.15
tfor	68.57/67.42	69.80/68.52	69.54/68.26	56.44/55.81	45.13/44.05
tfrf	56.00/54.86	57.73/56.38	56.57/55.10	36.56/36.06	42.22/41.33

Table 11. μ/m-averaged f-score results (%) on 20newsgroups dataset using different weighting methods.

	PA	SVM	SGD	NC	C4.5
tf	63.91/63.06	66.94/65.85	61.05/60.38	55.91/56.97	44.07/43.18
tfchi2	58.55/56.89	60.26/58.18	59.33/57.21	47.73/50.62	38.20/36.91
tfgr	68.43/67.55	69.69/68.60	70.19/<u>68.98</u>	62.85/64.45	43.07/42.36
tficf	68.14/67.18	69.15/68.10	69.24/67.97	59.43/61.74	<u>49.19</u>/<u>49.08</u>
tfidf	68.31/67.37	69.69/68.49	61.26/62.38	<u>64.27</u>/<u>64.90</u>	43.65/42.86
tfig	<u>68.97</u>/<u>68.05</u>	<u>70.14</u>/<u>68.96</u>	**70.26/68.93**	63.64/65.09	44.16/43.39
tfor	68.57/67.52	69.80/68.51	69.54/68.30	56.44/59.19	45.13/43.97
tfrf	56.00/54.69	57.73/56.18	56.57/54.47	36.56/38.41	42.22/41.47

and the second highest m-score (48.51%) preceded only by C4.5 with a m-score of 54.24%. Regarding TWS, even though, $tf.idf$ shows higher scores, the results are very close.

Tables 6, 7 and 8 shows the μ/m-averaged precision, recall, and f-score, respectively, for Oshumed dataset. Considering both precision and recall scores in Table 8, PA shows the best performance, followed by SGD, SVM. Strangely C4.5 shows the lowest performance.

Regarding TWS, $tf.or$ outperforms clearly all other methods except when used in conjunction NC. $tf.rf$, $tf.gr$ and $tf.ig$ have close results, and come second, followed by tf and $tf.idf$. $tf.icf$ performs poorly.

For these two datasets, we can note that, in comparison with the other algorithms, NC have a very high recall scores in Tables 4 and 7. However, NC reports the lowest precision scores in Tables 3 and 6.

Scores for Newsgroups dataset are presented in Tables 9, 10 and 11. $tf.ig$ and $tf.gr$ record the best scores (70%/69%) in conjunction with both SVM and SGD. Overall, in this dataset, SVM performs the best, followed by SGD and PA. C4.5 records very low scores. As for TWS, $tf.ig$ and $tf.gr$ give the best results, followed closely by $tf.or$, $tf.idf$ and $tf.icf$. $tf.rf$ shows the lowest scores.

In contrast to the high recall scores and low precision scores registered by NC algorithm on Reuters-21578 and Oshumed datasets, NC registered approximately equal results on both precision and recall.

Concerning C4.5, we can note that precision and recall results are approximately equal on the three datasets.

Overall, in our study, we find that $tf.or$ is the best TWS. $tf.idf$, $tf.gr$ and $tf.ig$ are also good choices for weighting features. $tf.\chi^2$, $tf.icf$ and $tf.rf$ are the worst methods.

5   Conclusion

The aim of this paper is to give an insight into the different TWS available for TC. These schemes are used in conjunction with five classifiers tested on Reuters-21578, Oshumed and 20newsgroups datasets. Our work aims at extending previous surveys and establishing a clean and fair basis for TC benchmarks.

To sum up, we find that the superiority of supervised term weighting methods over unsupervised methods is still not clear. Even though, in our experiment, $tf.or$ gives better results than $tf.idf$, we find no consistent superiority. In addition, $tf.idf$ is shown to be superior that the three supervised methods $tf.rf$, $tf.\chi^2$ and $tf.icf$. We find also that, alongside with $tf.or$ which gave the best results, $tf.idf$, $tf.gr$ and $tf.ig$ are good choices for weighting features.

References

1. Apte, C., Damerau, F., Weiss, S., et al.: Text mining with decision rules and decision trees. Citeseer (1998)
2. Cortes, C., Vapnik, V.: Support-vector networks. Mach. Learn. **20**, 273–297 (1995). Springer

3. Crammer, K., Dekel, O., Keshet, J., Shalev-Shwartz, S., Singer, Y.: Online passive-aggressive algorithms. J. Mach. Learn. Res. **7**, 551–585 (2006)
4. Debole, F., Sebastiani, F.: Supervised term weighting for automated text categorization. In: Sirmakessis, S. (ed.) Text Mining and its Applications. Studies in Fuzziness and Soft Computing, vol. 138, pp. 81–97. Springer, Heidelberg (2004). https://doi.org/10.1007/978-3-540-45219-5_7
5. Deng, Z.-H., Tang, S.-W., Yang, D.-Q., Li, M.Z.L.-Y., Xie, K.-Q.: A comparative study on feature weight in text categorization. In: Yu, J.X., Lin, X., Lu, H., Zhang, Y. (eds.) APWeb 2004. LNCS, vol. 3007, pp. 588–597. Springer, Heidelberg (2004). https://doi.org/10.1007/978-3-540-24655-8_64
6. Joachims, T.: Text categorization with support vector machines: learning with many relevant features. In: Nédellec, C., Rouveirol, C. (eds.) ECML 1998. LNCS (LNAI), vol. 1398, pp. 137–142. Springer, Heidelberg (1998). https://doi.org/10.1007/BFb0026683
7. Lan, M., Tan, C.L., Su, J., Lu, Y.: Supervised and traditional term weighting methods for automatic text categorization. IEEE Trans. Pattern Anal. Mach. Intell. **31**(4), 721–735 (2009)
8. McCallum, A., Nigam, K., et al.: A comparison of event models for Naive Bayes text classification. In: Workshop on Learning for Text Categorization, AAAI 1998, vol. 752, pp. 41–48. Citeseer (1998)
9. Ng, H.T., Goh, W.B., Low, K.L.: Feature selection, perceptron learning, and a usability case study for text categorization. In: ACM SIGIR Forum, vol. 31, pp. 67–73. ACM (1997)
10. Quinlan, J.R.: C4.5: Programs for Machine Learning. Morgan Kaufmann Publishers Inc., San Francisco (1993)
11. Schapire, R.E., Singer, Y.: Boostexter: a boosting-based system for text categorization. Mach. Learn. **39**(2–3), 135–168 (2000)
12. Sparck Jones, K.: A statistical interpretation of term specificity and its application in retrieval. J. Documentation **28**(1), 11–21 (1972)
13. Tibshirani, R., Hastie, T., Narasimhan, B., Chu, G.: Diagnosis of multiple cancer types by shrunken centroids of gene expression. Proc. Natl. Acad. Sci. **99**(10), 6567–6572 (2002)
14. Wang, D., Zhang, H.: Inverse category frequency based supervised term weighting scheme for text categorization. Preprint arXiv:1012.2609v4 (2013)
15. Wu, X., Kumar, V., Quinlan, J.R., Ghosh, J., Yang, Q., Motoda, H., McLachlan, G.J., Ng, A., Liu, B., Philip, S.Y., et al.: Top 10 algorithms in data mining. Knowl. Inf. Syst. **14**(1), 1–37 (2008)
16. Yang, Y.: Expert network: effective and efficient learning from human decisions in text categorization and retrieval. In: Croft, B.W., van Rijsbergen, C.J. (eds.) SIGIR 1994, pp. 13–22. Springer, Cham (1994). https://doi.org/10.1007/978-1-4471-2099-5_2
17. Youngjoong, K.: A study of term weighting schemes using class information for text classification. In: Proceedings of the 35th International ACM SIGIR Conference on Research and Development in Information Retrieval, pp. 1029–1030. ACM (2012)
18. Zhang, T.: Solving large scale linear prediction problems using stochastic gradient descent algorithms. In: Proceedings of the Twenty-first International Conference on Machine Learning, ICML 2004, pp. 919–926. Omnipress (2004)

Dual Convergence Estimates for a Family of Greedy Algorithms in Banach Spaces

S. P. Sidorov$^{(\boxtimes)}$, S. V. Mironov, and M. G. Pleshakov

Saratov State University, Saratov, Russian Federation
sidorovsp@info.sgu.ru
http://www.sgu.ru

Abstract. The paper examines four weak relaxed greedy algorithms for finding approximate sparse solutions of convex optimization problems in a Banach space. First, we present a review of primal results on the convergence rate of the algorithms based on the geometric properties of the objective function. Then, using the ideas of [16], we define the duality gap and prove that the duality gap is a certificate for the current approximation to the optimal solution. Finally, we find estimates of the dependence of the duality gap values on the number of iterations for weak greedy algorithms.

Keywords: Greedy algorithms · Nonlinear optimization · Sparsity

1 Introduction

Let X be a Banach space with norm $\| \cdot \|$. Let E be a convex function defined on X. The problem of convex optimization is to find an approximate solution to the problem

$$E(x) \to \min_{x \in X}. \tag{1}$$

A set of elements $\mathcal{D}$ from the space X is called a *dictionary* (see, e.g. [24]) if each element $g \in \mathcal{D}$ has norm bounded by one, $\|g\| \leq 1$, and the closure of span $\mathcal{D}$ is X, i.e. $\overline{\mathrm{span}\mathcal{D}} = X$. A dictionary $\mathcal{D}$ is called symmetric if $-g \in \mathcal{D}$ for every $g \in \mathcal{D}$. In this paper we assume that the dictionary $\mathcal{D}$ is symmetric.

Many problems in machine learning can be reduced to the problem (1) with E as a loss function [4]. In many real applications it is required that the optimal solution x^* of (1) should have a simple structure, e.g. be a *finite* linear combination of elements from a dictionary $\mathcal{D}$ in X. In other words, x^* should be a sparse element with respect to the dictionary $\mathcal{D}$ in X. Of course, one can substitute the requirement of sparsity by a constraint on cardinality (i.e. the limit on the number of elements used in linear combinations of elements from the dictionary $\mathcal{D}$ to

This work was supported by the Russian Fund for Basic Research, projects 16-01-00507, 18-01-00408.

G. Nicosia et al. (Eds.): MOD 2017, LNCS 10710, pp. 109–120, 2018.
https://doi.org/10.1007/978-3-319-72926-8_10

construct a solution of the problem (1)). However, in many cases the optimiza-tion problems with cardinality-type constraint are NP-complete. By this reason, practitioners and researchers in real applications choose to use greedy methods. By that design, greedy algorithms are capable of producing sparse solutions.

We are interested in finding the solutions of the problem (1), that are sparse with respect to $\mathcal{D}$, i.e. we are looking for solving the following problem:

$$E(x) \to \inf_{x \in \Sigma_m(\mathcal{D})}, \tag{2}$$

where $\Sigma_m(\mathcal{D})$ is the set of all m-term polynomials with respect to $\mathcal{D}$:

$$\Sigma_m(\mathcal{D}) = \left\{ x \in X \; : \; x = \sum_{g \in \Lambda} c_g g, \; \#(\Lambda) = m, \; \Lambda \subset \mathcal{D} \right\}. \tag{3}$$

One of the apparent choices among constructive methods for finding the best m-term approximations are greedy algorithms. The design of greedy algo-rithms allows us to obtain sparse solutions with respect to $\mathcal{D}$. Perhaps, the Frank-Wolfe method [11], which is also known as the "conditional gradient" method [19], is one of the most prominent greedy algorithms for finding optimal solutions of constrained convex optimization problems. Important contributions to the development of Frank-Wolfe type algorithms can be found in [5,12,16]. The review of gradient methods in Banach spaces can be found in [7,23]. The paper [16] provides general primal-dual convergence results for Frank-Wolfe-type algorithms by extending the duality concept presented in the work [5]. One can find recent convergence results for greedy algorithms in the works [1,6,8–10,13,15,17,18,22,25,27].

This paper examines four greedy algorithms for finding solutions of a con-vex optimization problem, which are sparse with respect to some dictionary, in Banach spaces:

- basic greedy algorithm (BGA),
- weak greedy algorithm (WGA),
- weak greedy algorithm with free relaxation (WGAFR),
- weak greedy algorithm with free relaxation and error δ (WGAFR(δ)).

Primal convergence results for BGA, WGA, WGAFR, WGAFR(δ) algo-rithms were obtained in [10,24]. Extending the ideas of [5,16] we force into application the notion of the duality gap to obtain dual convergence estimates for sparse-constrained convex optimization problems of type (2) by means of BGA, WGA, WGAFR, WGAFR(δ) algorithms.

It should be noted, that the paper by Temlyakov [24] shows that the greedy algorithms (WGAFR, WGAFR(δ)) for finding the solutions of (2) with respect to the dictionary $\mathcal{D}$ solve the problem (1) as well. Following [24], we examine the problem in an infinite dimensional space setting, since in many real appli-cations the dimension of the space X even if finite, it is too large. Therefore, our interest lies in obtaining estimates on the rate of convergence not depending on the dimension of X. Obviously, the results for infinite Banach spaces provide such estimates on the convergence rate. Moreover, in recent years some learning schemes in Banach spaces have been developed and justified using semi-inner-products or reproducing kernel Banach space approach (see e.g. [14,26]).

2   Greedy Algorithms

'For a functional $F \in X^*$ and an element $f \in X$ in this paper we will use an appropriate bracket notation $F(f) = \langle F, f \rangle$.

Let $\Omega := \{x \in X : E(x) \leq E(0)\}$ and suppose that Ω is bounded. We will suppose that function E is Fréchet differentiable on Ω. We note that it follows from convexity of E that for any $x, y \in \Omega$

$$E(y) \geq E(x) + \langle E'(x), y - x \rangle,$$

where $E'(x)$ denotes Fréchet differential of E at x.

Let $A_1(\mathcal{D})$ denote the closure (in X) of the convex hull of $\mathcal{D}$.

We analyze the family of greedy algorithms in a Banach space which use the Fréchet differential to choose a steepest descent direction at each iteration.

For optimization problem (2), the simplest iterative optimizer is described in Algorithm 1, which for each $m \geq 1$ defines the next point G_m by induction using the current point G_{m-1} and element ϕ_m that is obtained in the gradient greedy step. The gradient greedy step maximizes a certain functional determined by the gradient information from the previous steps of the algorithm. The algorithm is the Frank-Wolfe type method, since at each current point G_{m-1} it uses the linearization of the objective function E, and moves towards a minimizer of this function taken over the dictionary $\mathcal{D}$.

Algorithm 1. BASIC GREEDY ALGORITHM (BGA)

---

begin
    · Let $G_0 := 0$;
    for each $m \geq 1$ **do**
        · *(Gradient greedy step)* Find the element $\phi_m = \phi_m \in \mathcal{D}$ such that
        $\langle -E'(G_{m-1}), \phi_m \rangle = \sup_{s \in \mathcal{D}} \langle -E'(G_{m-1}), s \rangle$;
        · *(Line-search step)* Find a real number λ_m such that
        $E((1 - \lambda_m)G_{m-1} + \lambda_m \phi_m) = \inf_{\lambda} E((1 - \lambda)G_{m-1} + \lambda \phi_m)$;
        · *(Update step)* Define $G_m = G_m^{r,\tau} = (1 - \lambda_m)G_{m-1} + \lambda_m \phi_m$;
end

We would like to note that the gradient greedy step of greedy algorithms is looking for supremum over the dictionary $\mathcal{D}$ (not its convex hull $A_1(\mathcal{D})$), since points from $A_1(\mathcal{D})$ are mostly linear combinations of infinite number of the dictionary elements. Thus, the optimal solution obtained this way is not obliged to be sparse with respect to $\mathcal{D}$.

Solving the subproblem $\sup_{s \in \mathcal{D}} \langle -E'(G_{m-1}), s \rangle$ at Gradient greedy step exactly can be too expensive (or even impossible) for many real problems. Let $\tau := \{t_m\}_{m=1}^{\infty}$ be a given sequence of nonnegative numbers $t_m \leq 1$, $m = 1, 2, \ldots$. Algorithm 2 uses the sequence τ in the gradient greedy step to find approximate minimizer ϕ_m instead, which has approximation (multiplicative) quality at least t_m in step m. That is why the algorithm is called "weak" and the sequence τ is called "weakness" sequence.

Algorithm 2. WEAK GREEDY ALGORITHM (WGA)

---

begin
 · Let $G_0 = 0$;
 for each $m \geq 1$ **do**
  · *(Gradient greedy step)* Find the element $\phi_m \in \mathcal{D}$ such that
  $\langle -E'(G_{m-1}), \phi_m \rangle \geq t_m \sup_{s \in \mathcal{D}} \langle -E'(G_{m-1}), s \rangle$;
  · *(Line-search step)* Find a real number λ_m such that
  $E\left((1 - \lambda_m)G_{m-1} + \lambda_m \phi_m\right) = \inf_{\lambda} E\left((1 - \lambda)G_{m-1} + \lambda \phi_m\right)$;
  · *(Update step)* Define $G_m = (1 - \lambda_m)G_{m-1} + \lambda_m \phi_m$;

end

After we found ϕ_m at the gradient greedy step we can update the current state G_{m-1} using different ways. Some variants of choosing G_m are used in optimization algorithms like gradient method, reduced gradient method, conjugate gradients, gradient pursuits (see, for instance, [2,3,11,21]). Algorithms 1 and 2 at the line-search step proceeds by choosing the best point on the line connecting ϕ_m and the current point G_{m-1}.

In Algorithm 3 we use so called free relaxation and choose the element G_m from span(G_{m-1}, ϕ_m) which gives the infimum of E over all linear combinations of G_{m-1} and ϕ_m.

We assume that there exists an element (not necessarily unique) x^* in the Banach space X where the minimum E^* is attained, $E(x^*) = E^*$. It is obvious that the set of all such optimal elements x^* where the minima is attained is convex. Moreover, the minimum E^* is obliged to attain on the set Ω.

The paper [10] pointed out that very often we cannot calculate values of E exactly. Moreover, in many applications we are not able to find the exact value of the $\inf_{\lambda, \omega} E\left((1 - \omega)G_{m-1} + \lambda \phi\right)$ in the search with free relaxation step of Algorithm 3. Therefore, the paper [10] studies the following Algorithm 4 that is a modification of WGAFR with changing the step of Algorithm 3.

Comparing to the BGA, the idea behind using weakness sequence τ in the WGA, the free relaxation search in the WGAFR or the error δ in WGAFR(δ) is that it will hopefully make more progress per iteration, and may result in a better sparsity of the final solution. Obviously, it has its price, since the optimization problem in each iteration can now have the same complexity as the original optimization problem.

Algorithm 3. WEAK GREEDY ALGORITHM WITH FREE RELAXATION (WGAFR)

---

begin
   · Let $G_0 = 0$;
   for each $m \geq 1$ **do**
      · *(Gradient greedy step)* Find the element $\phi_m \in \mathcal{D}$ such that
      $\langle -E'(G_{m-1}), \phi_m \rangle \geq t_m \sup_{s \in \mathcal{D}} \langle -E'(G_{m-1}), s \rangle$;
      · *(Search with free relaxation)* Find real numbers ω_m and λ_m, such that
      $E\left((1 - \omega_m)G_{m-1} + \lambda_m \phi_m\right) = \inf_{\omega, \lambda} E\left((1 - \omega)G_{m-1} + \lambda \phi_m\right)$;
      · *(Update step)* Define $G_m = (1 - \omega_m)G_{m-1} + \lambda_m \phi_m$;
end

Thus, on the search step of Algorithm 4 we try to solve the *bivariate* convex optimization problem with respect to λ and ω with an error δ. The book [20] gives fast algorithms to solve such problems approximately (see also [10]). In the search step of Algorithm 3 the existence of optimal ω_m and λ_m is assumed. However, Algorithm 4 uses approximate implementation which avoids this assumption.

Algorithm 4. WEAK GREEDY ALGORITHM WITH FREE RELAXATION AND ERROR δ (WGAFR(δ))

---

begin
   · Let $G_0 = 0$ and $\delta > 0$ be a fixed real;
   for each $m \geq 1$ **do**
      · *(Gradient greedy step)* Find the element $\phi_m \in \mathcal{D}$ such that
      $\langle -E'(G_{m-1}), \phi_m \rangle \geq t_m \sup_{s \in \mathcal{D}} \langle -E'(G_{m-1}), s \rangle$;
      · *(Search with free relaxation and error δ)* Find real numbers ω_m and λ_m, such that
      $E\left((1 - \omega_m)G_{m-1} + \lambda_m \phi_m\right) \leq \inf_{\omega, \lambda} E\left((1 - \omega)G_{m-1} + \lambda \phi_m\right) + \delta$;
      · *(Update step)* Define $G_m = (1 - \omega_m)G_{m-1} + \lambda_m \phi_m$;
end

---

3   Primal Convergence Results

The modulus of smoothness of function E on the bounded set Ω is defined as follows:

$$\rho(E, u) = \frac{1}{2} \sup_{x \in \Omega, \|y\|=1} |E(x + uy) + E(x - uy) - 2E(x)|. \tag{4}$$

E is called uniformly smooth function on Ω if $\lim_{u \to 0} \rho(E, u)/u = 0$.

Let $E^* := \inf_{x \in X} E(x) = \inf_{x \in \Omega} E(x)$, $E(x^*) = E^*$. Let

$$\epsilon_m := \inf\{\epsilon : A_\epsilon^q m^{1-q} \le \epsilon\}. \tag{5}$$

Exploiting the geometric properties of the function E, papers [10,24] show the following estimate of the convergence rate of BGA, WGA and WGAFR.

Theorem 1 (*Primal Convergence for BGA, WGA and WGAFR*). *Let E be a uniformly smooth convex function with modulus of smoothness $\rho(E, u) \le \gamma u^q$, $1 < q \le 2$. Then, for a weakness sequence $\tau = \{t_m\}_{m=1}^{\infty}$, $0 < t_m \le 1$, $m = 1, 2, \ldots$, we have for BGA (with $t_m = 1$, $m = 1, 2, \ldots$), WGA and WGAFR*

$$E(G_m) - E^* \le C(E, q, \gamma)\epsilon_m, \tag{6}$$

where $C(E, q, \gamma)$ is a positive constants not depending on m.

Let $A_0 := \inf\{M : x^* \in \mathcal{L}_M\}$. Notice that since $x^* \in \mathcal{L}_{A_0}$ then (6) can be rewritten in the simpler form:

$$E(G_m) - E^* \le C(E, q, \gamma)A_0^q m^{1-q}.$$

Lemma 1 (*See, e.g. Lemma 1.1 of [24]*). *Let E be a Fréchet differentiable and convex on Ω. Then for all $x \in \Omega$*

$$0 \le E(x + uy) - E(x) - u\langle E'(x), y\rangle \le 2\rho(E, u\|y\|).$$

The following lemma is proved in [24] as Lemma 2.2.

Lemma 2. *Let F be a bounded linear functional and let $\mathcal{D}$ be a dictionary. Then*

$$\sup_{s \in \mathcal{D}} \langle F, s\rangle = \sup_{s \in A_1(\mathcal{D})} \langle F, s\rangle.$$

Lemma 3. *Let nonnegative $a_1, \ldots, a_N$ be such that*

$$a_m \le a_{m-1} + \inf_\lambda(-\lambda v a_{m-1} + B\lambda^q) + \delta, \ B > 0, \ \delta \in [0, 1],$$

for $m \le K := [\delta^{-1/q}]$, $q \in (1, 2]$. Then $a_m \le C(q, v, B)m^{1-q}$, $m \le K$.

Denote

$$\mathcal{L}_M := \{s \in X : s/M \in A_1(\mathcal{D})\},$$

$$A_\epsilon := A(E, \epsilon) = \inf\{M : \exists y \in \mathcal{L}_M \text{ s.t. } E(y) - E^* \le \epsilon\}.$$

The following lemma was proved in [24] (Lemma 4.1).

Lemma 4. *Let E be uniformly smooth with modulus of smoothness $\rho(E, u)$ on Ω. Let f_ϵ be from $\mathcal{L}_{A_\epsilon}$. Then for WGAFR(δ) we have*

$$E(G_m) \le E(G_{m-1}) + \inf_{\lambda \ge 0}(-\lambda t_m A_\epsilon^{-1}(E(G_{m-1}) - E(f)) + 2\rho(E, C_0\lambda)) + \delta,$$

for $m = 1, 2, \ldots, \delta^{-1/q}$.

We present the following estimate of the convergence rate of the WGAFR(δ) based on the geometric properties of the function E.

Theorem 2 *(Primal Convergence for WGAFR(δ)). Let E be a uniformly smooth convex function with modulus of smoothness $\rho(E, u) \leq \gamma u^q$, $1 < q \leq 2$. Then, for a weakness sequence $\tau = \{t_k\}_{k=1}^{\infty}$, $0 < t_k \leq 1$, $k = 1, 2, \ldots$, we have*

$$E(G_m) - E^* \leq C(E, q, \gamma)\epsilon_m, \quad m \leq \delta^{-\frac{1}{q}}, \tag{7}$$

where $C(E, q, \gamma)$ is positive constants not depending on m.

Proof. The proof is from the work [10]. We present it for completeness. Since

$$E(G_m) \leq E(G_{m-1}) + \delta \tag{8}$$

and $m \leq \delta^{-1/q}$, we have $E(G_m) \leq E(0) + 1$, and therefore, $G_m \in \Omega_1 := \{x \in X : E(x) \leq E(0) + 1\}$ for $m \leq \delta^{-1/q}$. Let $a_n := E(G_n) - E(f)$. It follows from Lemma 4 that $a_m \leq a_{m-1} + \inf_{\lambda>0}(-\lambda t_m A_\epsilon^{-1} a_{m-1} + 2\gamma(C_0\lambda)^q) + \delta$. If a_m are nonnegative then we apply Lemma 3 with $v = t_m A_\epsilon^{-1}$ and $B = 2\gamma C_0^q$.

Denote $K := [\delta^{-1/q}]$. Let n' be the smallest from $[1, K]$ such that $a_{n'} < 0$. Then it follows from (8) and $m \leq \delta^{-1/q}$ that $a_m \leq Cm^{1-q}$ for all $n' \leq m \leq K$.

4 Duality Gap and Convergence Result

In this section we present dual convergence results for BGA, WGA, WGAFR and WGAFR(δ).

Definition 1. Let us define *the duality gap* at element $G \in \Omega$ and error ϵ by

$$g(G) = g(G, \epsilon) =: A_\epsilon \sup_{s \in \mathcal{D}} \langle E'(G), A_\epsilon^{-1}G - s \rangle. \tag{9}$$

The useful property of the duality gap is described in the following proposition.

Proposition 1. Let E be a convex function defined on Banach space X. Then for any $x \in \Omega$ we have

$$E(x) - E(x^*) \leq g(x, \epsilon) + \epsilon.$$

Proof. Let x_ϵ be such that $E(x_\epsilon) - E(x^*) < \epsilon$ and $x_\epsilon/A_\epsilon \in A_1(\mathcal{D})$, i.e. $x_\epsilon \in \mathcal{L}_{A_\epsilon}$. Let us first prove that

$$E(x) - E(x_\epsilon) \leq g(x).$$

Since E is convex on X, for any $x \in \Omega$ we have

$$\begin{aligned}
E(x_\epsilon) &\geq E(x) + \langle E'(x), x_\epsilon - x \rangle \\
&\geq E(x) - \sup_{s \in \mathcal{L}_{A_\epsilon}} \langle E'(x), x - s \rangle \quad (\text{since } x_\epsilon \in \mathcal{L}_{A_\epsilon}) \\
&= E(x) - A_\epsilon \sup_{s \in \mathcal{L}_{A_\epsilon}} \langle E'(x), xA_\epsilon^{-1} - sA_\epsilon^{-1} \rangle \\
&= E(x) - A_\epsilon \sup_{s' \in A_1(\mathcal{D})} \langle E'(x), xA_\epsilon^{-1} - s' \rangle \quad (\text{since} \mathcal{L}_{A_\epsilon} = A_\epsilon A_1(\mathcal{D})) \\
&= E(x) - A_\epsilon \sup_{s' \in \mathcal{D}} \langle E'(x), xA_\epsilon^{-1} - s' \rangle, \tag{10}
\end{aligned}$$

where we have used Lemma 2 (or Lemma 2.2 in [24]). Then it follows from (10) that

$$E(x) - E(x_\epsilon) \leq A_\epsilon \sup_{s' \in \mathcal{D}} \langle E'(x), x A_\epsilon^{-1} - s' \rangle =: g(x), \tag{11}$$

and we get Lemma (since $E(x_\epsilon) - E(x^*) < \epsilon$).   □

Thus, the usefulness of the duality gap is based on the fact that the duality gap $g(x)$ is a certificate for the current approximation $E(x)$ to the optimal solution $E(x^*)$.

We can state the following dual result for BGA, WGA and WGAFR.

Theorem 3. *Let E be a uniformly smooth convex function defined on Banach space X. Let $\rho(E, u)$ be the modulus of smoothness of E and suppose that $\rho(E, u) \leq \gamma u^q$, $1 < q \leq 2$. Let $\tau = \{t_m\}_{m=1}^\infty$, $0 < \theta < t_k < 1$, $k = 1, 2, \ldots,$ be a weakness sequence. Assume that BGA or WGA or WGAFR is run for $N > 2$ iterations. Then there are an iterate $1 \leq \tilde{m} \leq N$ and $\beta > 0$ such that*

$$g(G_{\tilde{m}}) \leq \beta C(E, q, \gamma) \epsilon_N. \tag{12}$$

As it can be seen it is sufficient to prove the dual result for WGAFR(δ) and then Theorem 3 follows immediately from the corresponding result for WGAFR(δ).

We need some preliminary results to prove the main theorem.

Lemma 5. *Let E be a uniformly smooth convex function defined on Banach space X. Let $\rho(E, u)$ denote the modulus of smoothness of E. Then the following inequality holds for the WGAFR(δ):*

$$E(G_m) \leq E(G_{m-1}) + \inf_{\lambda \geq 0} \left(-\lambda t_m A_\epsilon^{-1} g(G_{m-1}) + 2\rho(E, C_0 \lambda) \right) + \delta, \quad m = 1, 2, \ldots,$$

where C_0 does not depend on m.

Proof. The proof is a modified version of the proof of Lemma 4.1 in [24]. From the definition of G_m in Step 3 of WGAFR(δ) we have

$$G_m = (1 - \omega_m) G_{m-1} + \lambda_m \phi_m.$$

Step 2 of WGAFR(δ) implies

$$E(G_m) \leq \inf_{\lambda \geq 0, \omega} E(G_{m-1} - \omega G_{m-1} + \lambda \phi_m) + \delta. \tag{13}$$

It follows from Lemma 1 that

$$E(G_{m-1} - \omega G_{m-1} + \lambda \phi_m) \leq E(G_{m-1})$$
$$+ \lambda \langle -E'(G_{m-1}), \phi_m \rangle - \omega \langle -E'(G_{m-1}), G_{m-1} \rangle + 2\rho(E, \|\lambda \phi_m - \omega G_{m-1}\|). \tag{14}$$

We have from Step 1 of WGAFR(δ)

$$\langle -E'(G_{m-1}), \phi_m \rangle \geq t_m \sup_{s \in \mathcal{D}} \langle -E'(G_{m-1}), s \rangle. \tag{15}$$

Let us take $\omega = \lambda t_m A_\epsilon^{-1}$ then

$$E(G_{m-1} - \omega G_{m-1} + \lambda \phi_m)) \leq E(G_{m-1})$$
$$- \lambda t_m \langle -E'(G_{m-1}), \phi_m - G_{m-1} A_\epsilon^{-1} \rangle + 2\rho(E, \lambda \| \phi_m - G_{m-1} A_\epsilon^{-1} \|). \tag{16}$$

Using (15) and the definition of the duality gap (9), we have

$$\langle -E'(G_{m-1}), \phi_m - G_{m-1} A_\epsilon^{-1} \rangle$$
$$\geq t_m \sup_{s \in \mathcal{D}} \langle -E'(G_{m-1}), s - G_{m-1} A_\epsilon^{-1} \rangle = t_m A_\epsilon^{-1} g(G_{m-1}). \tag{17}$$

It follows from (13), (16) and (17) that

$$E(G_m) \leq E(G_{m-1}) + \inf_{\lambda \geq 0} \left(-\lambda t_m A_\epsilon^{-1} g(G_{m-1}) + 2\rho(E, \lambda \| \phi_m - G_{m-1} A_\epsilon^{-1} \|) \right) + \delta.$$

It follows from $E(G_{m-1}) \leq E(0)$ that $G_{m-1} \in \Omega$. Our assumption on boundness of Ω implies that there exists a constant C_1 such that $\|G_{m-1}\| \leq C_1$. Since $\phi_m \in \mathcal{D}$, we have $\|\phi_m\| \leq 1$. Thus,

$$\|A_\epsilon^{-1} G_{m-1} - \phi_m \| \leq A_\epsilon^{-1} C_1 + 1 =: C_0.$$

This completes the proof of Lemma. $\square$

Lemma 6. *Let $0 < \mu < 1$ be a real and M be an integer. Then $\epsilon_{m_0} \leq \mu^{1-q} \epsilon_M$, where ϵ_m defined in Theorem 2 and $m_0 = \lceil \mu M \rceil$.*

Proof. It follows from (5) that

$$\inf \left\{ \epsilon : m^{1-q} < \frac{\left(\frac{\epsilon}{\mu^{1-q}} \right)}{A \left(\frac{\epsilon}{\mu^{1-q}} \right)} \right\} = \mu^{1-q} \epsilon_m.$$

Let $\epsilon^* := \frac{\epsilon}{\mu^{1-q}}$, i.e. $\epsilon^* \mu^{1-q} = \epsilon$. We note that $\mu^{1-q} > 1$ and $\frac{\epsilon}{\mu^{1-q}} < \epsilon$. We get

$$A \left(\frac{\epsilon}{\mu^{1-q}} \right) \geq A_\epsilon. \tag{18}$$

It follows from (18) that

$$\epsilon_{m_0} = \inf \left\{ \epsilon : M^{1-q} < \frac{\epsilon^*}{A_\epsilon^q} \right\} \leq \inf \left\{ \epsilon^* : M^{1-q} < \frac{\epsilon^*}{A_{\epsilon^*}^q} \right\} = \mu^{1-q} \epsilon_M.$$

$\square$

Theorem 4. *Let E be a uniformly smooth convex function defined on Banach space X. Let $\rho(E, u)$ be the modulus of smoothness of E and suppose that $\rho(E, u) \leq \gamma u^q$, $1 < q \leq 2$. Let $\tau = \{t_m\}_{m=1}^{\infty}$, $0 < \theta < t_k < 1$, $k = 1, 2, \ldots$, be a weakness sequence. Assume that WGAFR(δ) is run for $0 < N \leq \delta^{-\frac{1}{q}}$ iterations. Then there are an iterate $1 \leq \tilde{m} \leq N$ and $\beta > 0$ such that*

$$g(G_{\tilde{m}}) \leq \beta C(E, q, \gamma) \epsilon_N. \tag{19}$$

Proof. It follows from Theorem 2 that

$$E(G_m) - E^* \leq C(E, q, \gamma) \epsilon_m, \quad m \leq \delta^{-\frac{1}{q}}, \ \epsilon_m := \inf\{\epsilon : A_\epsilon^q m^{1-q} \leq \epsilon\}.$$

Let us suppose that

$$g(G_m) > \beta C(E, q, \gamma) \epsilon_N \tag{20}$$

for all $[\mu N] \leq m \leq N$, $0 < \mu < 1$ (μ is fixed and will be chosen later).

It follows from Lemma 5 with $\lambda = \epsilon_m$,

$$E(G_{m+1}) - E^* \leq E(G_m) - E^* - \epsilon_m t_m A_\epsilon^{-1} g(G_m) + 2\gamma(C_0 \epsilon_m)^q + \delta. \tag{21}$$

Using our assumption (20), the inequality (21) can be rewritten in the form

$$E(G_{m+1}) - E^* \leq E(G_m) - E^* - \epsilon_m t_m A_\epsilon^{-1} \beta C(E, q, \gamma) \epsilon_N + 2\gamma(C_0 \epsilon_m)^q + \delta. \tag{22}$$

Now we are going to use the following inequalities: $\theta \leq t_k \leq 1$, $k = 1, 2, \ldots$; $\epsilon_{m_0} \leq \mu^{1-q} \epsilon_N$ (Lemma 6); Since $[\mu N] \leq m \leq N$, we have $\epsilon_{[\mu N]} \geq \epsilon_m \geq \epsilon_N$. Then (22) gives

$$E(G_{m+1}) - E^* \leq E(G_m) - E^* - A_\epsilon^{-1} \beta \theta C(E, q, \gamma) \epsilon_N^2 + 2\gamma(C_0)^q \mu^{q(1-q)} \epsilon_N^q + \delta. \tag{23}$$

If we write the chain of inequalities for all $m = m_0, \ldots, N$, $m_0 := [\mu N]$, we get

$$E(G_{m+1}) - E^* \leq E(G_{m_0}) - E^*$$
$$- (N - m_0) \epsilon_N \left[A_\epsilon^{-1} \beta \theta C(E, q, \gamma) \epsilon_N - 2\gamma(C_0)^q \mu^{q(1-q)} \epsilon_N^{q-1} + \frac{\delta}{\epsilon_N} \right]$$
$$\leq C(E, q, \gamma) \epsilon_{m_0}$$
$$- N(1 - \mu) \epsilon_N \left[A_\epsilon^{-1} \beta \theta C(E, q, \gamma) \epsilon_N - 2\gamma(C_0)^q \mu^{q(1-q)} \epsilon_N^{q-1} + \frac{\delta}{\epsilon_N} \right]$$
$$\leq C(E, q, \gamma) \mu^{1-q} \epsilon_N$$
$$- N(1 - \mu) \epsilon_N \left[A_\epsilon^{-1} \beta \theta C(E, q, \gamma) \epsilon_N - 2\gamma(C_0)^q \mu^{q(1-q)} \epsilon_N^{q-1} + \frac{\delta}{\epsilon_N} \right]$$
$$= \epsilon_N \Bigg(C(E, q, \gamma) \mu^{1-q}$$
$$- N(1 - \mu) \left[A_\epsilon^{-1} \beta \epsilon_N \theta C(E, q, \gamma) - 2\gamma(C_0)^q \mu^{q(1-q)} \epsilon_N^{q-1} + \frac{\delta}{\epsilon_N} \right] \Bigg). \tag{24}$$

If we take sufficiently big β,

$$\beta > \frac{\frac{C(E,q,\gamma)\mu^{1-q}}{N(1-\mu)} + 2\gamma(C_0)^q \mu^{q(1-q)} \epsilon_N^{q-1} - \frac{\delta}{\epsilon_N}}{A_\epsilon^{-1} \epsilon_N \theta C(E,q,\gamma)}$$

then we get $E(G_m) - E^* < 0$ which is impossible. The parameter μ can be chosen as follows:

$$\mu := \arg \min_{0 \leq \mu \leq 1} \left(\frac{C(E,q,\gamma)\mu^{1-q}}{N(1-\mu)} + 2\gamma(C_0)^q \mu^{q(1-q)} \epsilon_N^{q-1} - \frac{\delta}{\epsilon_N} \right).$$

$\square$

5   Conclusion

Theorems 1 and 2 give small primal errors for the weak greedy algorithms with free relaxation. However, since in real application problems the optimum value $E(x^*)$ and the constant γ in the modulus of smoothness of E are usually unknown, estimates for the current approximation quality are strongly desired. The duality gap $g(G)$ defined in (9) (estimates of which are obtained in Theorems 3 and 4) is computed at the gradient greedy step of the week greedy algorithms and it is an appropriate quality measure for the primal error $E(G) - E(x^*)$, since it is natural upper bounds for the primal error.

This paper examines the weak greedy algorithms with free relaxation. Another problem we would like to address is the problem of extending the results to the Chebyshev-type greedy algorithms which use so called Chebyshev-type search and choose the element G_m from span$\{\phi_i\}_{i=1}^m$ which gives the infimum of E over all linear combinations of ϕ_i, $i = 1, 2, \ldots, m$.

The problem of quantifying the number of iterations, that are necessary to reduce to a certain value the error, may be also addressed to the future work.

References

1. Barron, A.R., Cohen, A., Dahmen, W., DeVore, R.A.: Approximation and learning by greedy algorithms. Ann. Stat. **36**(1), 64–94 (2008)
2. Blumensath, T., Davies, M.E.: Gradient pursuits. IEEE Trans. Signal Process. **56**, 2370–2382 (2008)
3. Blumensath, T., Davies, M.: Stagewise weak gradient pursuits. IEEE Trans. Signal Process. **57**, 4333–4346 (2009)
4. Bubeck, S.: Convex optimization: algorithms and complexity. Found. Trends Mach. Learn. **8**(3–4), 231–358 (2015)
5. Clarkson, K.L.: Coresets, sparse greedy approximation, and the Frank-Wolfe algorithm. ACM Trans. Algorithms **6**(4), 1–30 (2010)
6. Davis, G., Mallat, S., Avellaneda, M.: Adaptive greedy approximation. Constr. Approx. **13**, 57–98 (1997)
7. Demyanov, V., Rubinov, A.: Approximate Methods in Optimization Problems. American Elsevier Publishing Co., New York (1970)

8. Dereventsov, A.V.: On the approximate weak Chebyshev greedy algorithm in uniformly smooth banach spaces. J. Math. Anal. Appl. **436**(1), 288–304 (2016)

9. DeVore, R.A., Temlyakov, V.N.: Some remarks on greedy algorithms. Adv. Comput. Math. **5**, 173–187 (1996)

10. DeVore, R.A., Temlyakov, V.N.: Convex optimization on Banach spaces. Found. Comput. Math. **16**(2), 369–394 (2016)

11. Frank, M., Wolfe, P.: An algorithm for quadratic programming. Naval Res. Logis. Quart. **3**, 95–110 (1956)

12. Freund, R.M., Grigas, P.: New analysis and results for the Frank-Wolfe method. Math. Program. **155**(1), 199–230 (2016)

13. Friedman, J.: Greedy function approximation: a gradient boosting machine. Ann. Stat. **29**(5), 1189–1232 (2001)

14. Georgiev, P.G., Sánchez-González, L., Pardalos, P.M.: Construction of pairs of reproducing kernel Banach spaces. In: Demyanov, V., Pardalos, P., Batsyn, M. (eds.) Constructive Nonsmooth Analysis and Related Topics. SOIA, vol. 87, pp. 39–57. Springer, New York (2014). https://doi.org/10.1007/978-1-4614-8615-2_4

15. Huber, P.J.: Projection pursuit. Ann. Statist. **13**, 435–525 (1985)

16. Jaggi, M.: Revisiting Frank-Wolfe: projection-free sparse convex optimization. In: Proceedings of the 30th International Conference on Machine Learning (ICML-13), pp. 427–435 (2013)

17. Jones, L.: On a conjecture of Huber concerning the convergence of projection pursuit regression. Ann. Statist. **15**, 880–882 (1987)

18. Konyagin, S.V., Temlyakov, V.N.: A remark on greedy approximation in Banach spaces. East J. Approx. **5**(3), 365–379 (1999)

19. Levitin, E.S., Polyak, B.T.: Constrained minimization methods. USSR Comp. Math. M. Phys. **6**(5), 1–50 (1966)

20. Nemirovski, A.: Optimization II: Numerical methods for nonlinear continuous optimization. Lecture Notes, Israel Institute of Technology (1999)

21. Nesterov, Y.: Introductory Lectures on Convex Optimization: A Basic Course. Kluwer Academic Publishers, Boston (2004)

22. Nguyen, H., Petrova, G.: Greedy strategies for convex optimization. Calcolo **41**(2), 1–18 (2016)

23. Polyak, B.T.: Introduction to Optimization. Optimization Software Inc., New York (1987)

24. Temlyakov, V.N.: Greedy approximation in convex optimization. Constr. Approx. **41**(2), 269–296 (2015)

25. Temlyakov, V.N.: Dictionary descent in optimization. Anal. Mathematica **42**(1), 69–89 (2016)

26. Zhang, H., Zhang, J.: Learning with reproducing Kernel Banach spaces. In: Dang, P., Ku, M., Qian, T., Rodino, L.G. (eds.) New Trends in Analysis and Interdisciplinary Applications. TM, pp. 417–423. Springer, Cham (2017). https://doi.org/10.1007/978-3-319-48812-7_53

27. Zhang, Z., Shwartz, S., Wagner, L., Miller, W.: A greedy algorithm for aligning DNA sequences. J. Comput. Biol. **7**(1–2), 203–214 (2000)

Nonlinear Methods for Design-Space Dimensionality Reduction in Shape Optimization

Danny D'Agostino[1,2], Andrea Serani[1], Emilio F. Campana[1],
and Matteo Diez[1(✉)]

[1] CNR-INSEAN, National Research Council–Marine
Technology Research Institute, Rome, Italy
`matteo.diez@cnr.it`
[2] Department of Computer, Control, and Management Engineering "A. Ruberti",
Sapienza University of Rome, Rome, Italy

Abstract. In shape optimization, design improvements significantly depend on the dimension and variability of the design space. High dimensional and variability spaces are more difficult to explore, but also usually allow for more significant improvements. The assessment and breakdown of design-space dimensionality and variability are therefore key elements to shape optimization. A linear method based on the principal component analysis (PCA) has been developed in earlier research to build a reduced-dimensionality design-space, resolving the 95% of the original geometric variance. The present work introduces an extension to more efficient nonlinear approaches. Specifically the use of Kernel PCA, Local PCA, and Deep Autoencoder (DAE) is discussed. The methods are demonstrated for the design-space dimensionality reduction of the hull form of a USS Arleigh Burke-class destroyer. Nonlinear methods are shown to be more effective than linear PCA. DAE shows the best performance overall.

Keywords: Shape optimization · Hull-form design
Nonlinear dimensionality reduction · Kernel methods
Deep autoencoder

1 Introduction

The simulation-based design (SBD) paradigm has demonstrated its capability of supporting the design decision process, providing large sets of design options and reducing time and costs of the design process. The recent development of high performance computing (HPC) systems has driven the SBD towards its integration with optimization algorithms, moving the SBD paradigm further, to automatic SBD optimization (SBDO). In shape optimization, SBDO consists of three main elements: (i) a simulation tool, (ii) an optimization algorithm, and (iii) a shape modification tool, which need to be integrated efficiently and robustly. In this context, design improvements significantly depend on the dimension and extension of the design space: high dimensional and variability spaces

© Springer International Publishing AG 2018
G. Nicosia et al. (Eds.): MOD 2017, LNCS 10710, pp. 121–132, 2018.
https://doi.org/10.1007/978-3-319-72926-8_11

are more difficult and computationally expensive to explore but, at the same time, potentially allow for bigger improvements. The assessment and breakdown of the design-space dimensionality and variability are therefore a key element for the success of the SBDO [1].

Online linear dimensionality reduction techniques have been developed, requiring the evaluation of the objective function or its gradient. As an example, principal component analysis (PCA) or proper orthogonal decomposition (POD) methods have been applied for reduced-dimensionality local representations of feasible design regions [2]. A PCA/POD based approach is used in the active subspace method (ASM) [3] to discover and exploit low-dimensional and monotonic trends in the objective function, based on the evaluation of its gradient. Online methods improve the shape optimization efficiency by basis rotation and/or dimensionality reduction. Nevertheless, they do not provide an assessment of the design space and the associated shape parametrization before optimization is performed or objective function and/or gradient are evaluated.

Offline linear methodologies have been developed with focus on design-space variability and dimensionality reduction for efficient optimization procedures. A method based on the Karhunen-Loève expansion (KLE) has been formulated for the assessment of the shape modification variability and the definition of a reduced-dimensionality global model of the shape modification vector in [1]. No objective function evaluation nor gradient is required by the method. The KLE is applied to the continuous shape modification vector, requiring the solution of a Fredholm integral equation of the second kind. Once the equation is discretized, the problem reduces to the PCA of discrete data. Offline linear methods improve the shape optimization efficiency by reparametrization and dimensionality reduction, providing the assessment of the design space and the shape parametrization before optimization and/or performance analysis are performed. The assessment is based on the geometric variability associated to the design space of the shape optimization. Although linear methods have been successfully applied for a wide range of problems, they may be not efficient when complex non linear relationship are involved in the performance analysis and optimization.

In the last years researchers have developed nonlinear methods for data dimensionality reduction. Nonlinear dimensionality reduction (NLDR) methods generalize linear methods to address data with nonlinear structures. Kernel PCA (KPCA) solves a PCA eigenproblem in a new space (called feature space) by using kernel methods [4]. Local PCA (LPCA) divides the initial design space in k clusters and a PCA is applied for each of them, supposing that the data in each cluster has an approximate linear structure. LPCA techniques [5] may be differentiated based on the clustering method, which may follow k-means [6] or spectral approaches [7]. Artificial neural networks (ANN) have been also used to reduce data dimensionality [8], by performing both encoder and decoder tasks (the method is also known as autoencoder).

The objective of the present work is to combine NLDR techniques with shape parametrization in SBDO for ship hydrodynamics. Specifically KPCA, LPCA with k-means (LPCA-KM), LPCA with spectral clustering (LPCA-SC), and

Deep Autoencoder (DAE) are used to build a reduced-dimensionality design-space, resolving at least the 95% of the original design variability based on the concept of geometric variance [1]. The methods are demonstrated for the design-space dimensionality reduction of the hull form of USS Arleigh Burke-class destroyer, namely the DTMB 5415 model, an early and open to public version of the DDG-51. The effectiveness of the NLDR techniques is shown and discussed, comparing the results to the linear KLE/PCA method from earlier work [1].

2   Dimensionality Reduction Methods

General definitions and assumptions for the current problem are presented in the following, along with linear and nonlinear dimensionality reduction methods.

2.1   General Definitions and Assumptions

Consider a geometric domain $\mathcal{G}$ (which identifies the initial shape) and a set of coordinates $\mathbf{x} \in \mathcal{G}$.

Assume that $\mathbf{u} \in \mathcal{U}$ is the design variable vector, which defines a continuous shape modification vector $\boldsymbol{\delta}(\mathbf{x}, \mathbf{u})$. Consider the design variables $\mathbf{u}$ as a random field defined over a domain $\mathcal{U}$, with associated probability density function $p(\mathbf{u})$. The associated mean shape modification is evaluated as

$$\langle \boldsymbol{\delta} \rangle = \int_{\mathcal{U}} \boldsymbol{\delta}(\mathbf{x}, \mathbf{u}) p(\mathbf{u}) d\mathbf{u} \tag{1}$$

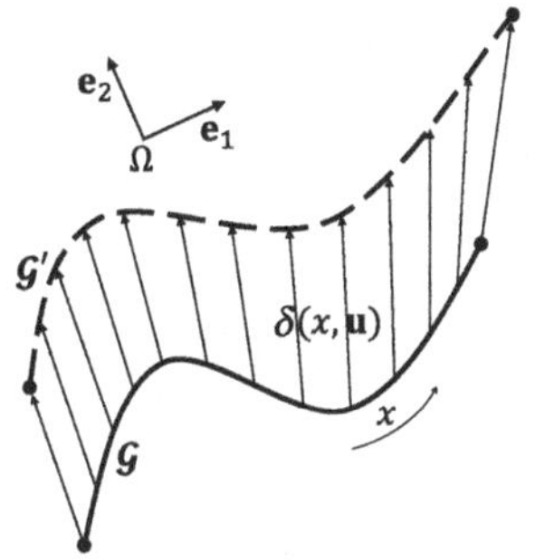

If one defines the internal product in $\mathcal{G}$ as

$$(\mathbf{f}, \mathbf{g}) = \int_{\mathcal{G}} \mathbf{f}(\mathbf{x}) \cdot \mathbf{g}(\mathbf{x}) \, d\mathbf{x} \tag{2}$$

with associated norm $\|\mathbf{f}\| = (\mathbf{f}, \mathbf{f})^{1/2}$, the variance associated to the shape modification vector (geometric variance) may be defined as

$$\sigma^2 = \left\langle \|\hat{\boldsymbol{\delta}}\|^2 \right\rangle = \int_{\mathcal{U}} \int_{\mathcal{G}} \hat{\boldsymbol{\delta}}(\mathbf{x}, \mathbf{u}) \cdot \hat{\boldsymbol{\delta}}(\mathbf{x}, \mathbf{u}) p(\mathbf{u}) d\mathbf{x} d\mathbf{u} \tag{3}$$

Fig. 1. Scheme and notation for the current formulation, showing an example for $n = 1$ and $m = 2$

where $\hat{\boldsymbol{\delta}} = \boldsymbol{\delta} - \langle \boldsymbol{\delta} \rangle$, and $\langle \cdot \rangle$ denotes the ensemble average over $\mathbf{u}$. Generally, $\mathbf{x} \in \mathbb{R}^n$ with $n = 1, 2, 3$, $\mathbf{u} \in \mathbb{R}^M$ with M number of design variables, and $\boldsymbol{\delta} \in \mathbb{R}^m$ with $m = 1, 2, 3$ (with m not necessarily equal to n). Figure 1 shows an example with $n = 1$ and $m = 2$. Ensemble averages $\langle \cdot \rangle$ over $\mathbf{u} \in \mathcal{U}$ may be evaluated by Monte Carlo (MC) sampling using a statistically convergent number of random realizations S, $\{\mathbf{u}_k\}_{k=1}^{S} \sim p(\mathbf{u})$. These are collected in a $[S \times L]$ matrix

$$\mathbf{D} = \left[\mathbf{d}(\mathbf{u}_1) \middle| \quad \dots \quad \middle| \mathbf{d}(\mathbf{u}_S) \right]^T \tag{4}$$

representing the (MC sampled) original design space, where $\mathbf{d}(\mathbf{u}_k) = \{d_q(\mathbf{u}_k)\}_{q=1}^m$ is the deviation from the mean of the shape modification vector and its q-th component is evaluated at discrete coordinates $\mathbf{x}_t$, $t = 1 \dots, T$, as

$$d_q(\mathbf{u}_k) = \left\{ \begin{array}{c} \delta_q(\mathbf{x}_1, \mathbf{u}_k) \\ \vdots \\ \delta_q(\mathbf{x}_T, \mathbf{u}_k) \end{array} \right\} - \frac{1}{S} \sum_{k=1}^{S} \left\{ \begin{array}{c} \delta_q(\mathbf{x}_1, \mathbf{u}_k) \\ \vdots \\ \delta_q(\mathbf{x}_T, \mathbf{u}_k) \end{array} \right\} \tag{5}$$

with $\delta_q = \boldsymbol{\delta} \cdot \mathbf{e}_q$, where $\{\mathbf{e}_q\}_{q=1}^m \in \mathbb{R}^m$ is a basis of orthogonal unit vector. Note that $L = mT$.

A reduced-dimensionality representation of $\mathbf{D}$ is sought after for later use in the SBDO.

2.2   Principal Component Analysis

PCA allows to reduce the input dimensionality of the data, performing a projection of the points in a new linear subspace, defined by the eigenvectors of the $[L \times L]$ covariance matrix $\mathbf{C} = \mathbf{D}^T \mathbf{D}/S$. These eigenvectors have the properties to maximize the variance of points projected on them and to minimize the mean squared distance between the original points and the relative projections [9]. The principal components are defined by the solution of the eigenproblem

$$\mathbf{C}\mathbf{z} = \lambda \mathbf{z} \tag{6}$$

The solutions $\{\mathbf{z}_i\}_{i=1}^L$ of the Eq. 6 are used to build a reduced-dimensionality space for the shape modification vector $\mathbf{d}$ as

$$\mathbf{d} \approx \sum_{i=1}^{N} \alpha_i \mathbf{z}_i = \hat{\mathbf{d}} \tag{7}$$

where α_i is the i-th component of the new design variable vector $\boldsymbol{\alpha} \in \mathbb{R}^N$. Equation 7 may be truncated to the N-th order, preserving a desired level of confidence β ($0 < \beta \le 1$), provided that

$$\sum_{i=1}^{N} \lambda_i \ge \beta \sum_{i=1}^{L} \lambda_i = \beta \sigma^2 \tag{8}$$

assuming $\lambda_i \ge \lambda_{i+1}$. Only M eigenvalues are expected to be non zeros.

2.3   Kernel Principal Component Analysis

The kernel PCA (KPCA) method [4] is a nonlinear extension of PCA. It finds directions of maximum variance in a higher (possibly infinite) dimensional feature space $\mathcal{F}$, mapping the points from the input space $\mathcal{I}$ by a possible nonlinear function $\Phi : I \to \mathcal{F}$ as

$$\mathbf{d}_k \to \Phi(\mathbf{d}_k), \qquad \forall k = 1, \ldots, S \tag{9}$$

where, for the sake of simplicity, the $\mathbf{d}(\mathbf{u}_k)$ of Eq. 4 is here simplified in $\mathbf{d}_k$. Then PCA is computed in the feature space $\mathcal{F}$. Assuming that $\sum_k \Phi(\mathbf{d}_k) = 0$, the kernel principal component $\{\mathbf{z}_p\}_{p=1}^P$ can be find solving the eigenproblem

$$\Sigma_\Phi \mathbf{z}_p = \lambda_p \mathbf{z}_p \tag{10}$$

where Σ_Φ is the $[P \times P]$ covariance matrix in the feature space $\mathcal{F}$, defined as

$$\Sigma_\Phi = \frac{1}{S} \sum_{k=1}^S \Phi(\mathbf{d}_k)\Phi(\mathbf{d}_k)^T \tag{11}$$

KPCA allows the solution of Eq. 10 without computing explicitly the Eq. 9, since it appears only within an inner product [10], which can be computed efficiently by a kernel function $K(\mathbf{d}_i, \mathbf{d}_k) = \Phi(\mathbf{d}_i)^T \Phi(\mathbf{d}_k)$. Defining $\mathbf{z}_p$ as a linear expansion of $\Phi(\mathbf{d}_k)$

$$\mathbf{z}_p = \sum_{k=1}^S c_{pk}\Phi(\mathbf{d}_k) \tag{12}$$

the Eq. 10 can be recasted as

$$\mathbf{K}\mathbf{c}_p = \lambda_p S \mathbf{c}_p \tag{13}$$

where $\mathbf{K}$ is the symmetric and positive-semidefinite $[S \times S]$ kernel matrix, with $\mathbf{K}_{ik} = K(\mathbf{d}_i, \mathbf{d}_k)$. The length of the S-component vector $\mathbf{c}_p$ is chosen such that $\mathbf{z}_p^T \mathbf{z}_p = \lambda_p S \mathbf{c}_p^T \mathbf{c}_p = 1$. Once the eigenproblem in Eq. 13 is solved, the new design variables can be found projecting $\Phi(\mathbf{d})$ on $\mathbf{z}_p$ as

$$\alpha = \Phi(\mathbf{d})\mathbf{z}_p = \sum_{k=1}^S c_{pk}\Phi(\mathbf{d})^T \Phi(\mathbf{d}_k) = \sum_{k=1}^S c_{pk} K(\mathbf{d}, \mathbf{d}_k) \tag{14}$$

The reconstruction of the original data from the feature space $\mathcal{F}$ in KPCA is more problematic than PCA, since it needs to find, for every point $\Phi(\mathbf{d}_k)$, the relative pre-image $\mathbf{d}_k$ in the input space $\mathcal{I}$. In this paper, approximate pre-images technique proposed in [11] is used.

2.4   Local Principal Component Analysis

Local PCA (LPCA) performs a PCA for every different disjoint region of the input space $\mathcal{I}$, assuming that, if the local regions are small enough, the data manifold will not curve much over the extent of the region and the linear model will be a good fit [5].

The first step in LPCA is to cluster the data in k sets, applying a clustering algorithm, such that $\mathbf{D} = \{\mathbf{D}_1, \ldots, \mathbf{D}_i\}_{i=1}^k$. Herein, LPCA is performed with two clustering techniques: the k-means (LPCA-KM) algorithm [6] and a spectral clustering (LPCA-SC) [12]. The k-means clustering algorithm is described in Algorithm 1.

Algorithm 1. k-means clustering algorithm

Require: Random k centroids as representative points of each cluster $\mathbf{D}_i$ $\forall i = 1, \ldots, k$.
1: **repeat**
2:     Assign each point $\mathbf{d}_j$ to the nearest centroid μ_i using the Euclidean distance as similarity measure.
3:     Update the centroids according to: $\mu_i = \frac{1}{|\mathbf{D}_i|} \sum_{\mathbf{d}_j \in \mathbf{D}_i} \mathbf{d}_j$
4: **until** μ_i $\forall i = 1, \ldots, k$ remains unchanged

---

One issue in k-means is that using the euclidean distance as similarity measure assumes a convex shape to the underlying clusters [13].

Spectral clustering can be effective even if the clusters shape are more complex. There are several versions of the spectral clustering algorithms, the main difference is in which graph Laplacian is used [7]. Herein, the symmetric normalized Laplacian $\mathbf{A}_{\text{sym}} = \mathbf{I} - \mathbf{B}^{-\frac{1}{2}} \mathbf{W} \mathbf{B}^{-\frac{1}{2}}$ [12] is used and the corresponding algorithm is summarized in Algorithm 2 [7].

After the data are partitioned in k clusters, a PCA is performed on them solving k PCA eigenproblem

$$\mathbf{C}_i \mathbf{z}_i = \lambda_i \mathbf{z}_i \qquad \forall i = 1, \ldots, k \tag{15}$$

LPCA results are highly dependent by the clustering procedure and specially by the number of clusters used. Moreover, the number of clusters k should be set carefully to avoid extensive computation.

Algorithm 2. Normalized Spectral Clustering

Require: Let k the number of clusters to identify, build a similarity graph as:

- K-nearest neighbor graphs: fix K, $\mathbf{d}_i$ is connected to a point $\mathbf{d}_j$ if it is among the K-nearest neighbor of $\mathbf{d}_i$ or viceversa.

1: Compute the adjacency matrix $\mathbf{W}$ of the graph and the diagonal degree matrix $\mathbf{B}$, where each element is equal to $b_{ii} = \sum_{j=1}^S w_{ij}$.
2: Compute the symmetric normalized Laplacian $\mathbf{A}_{\text{sym}}$.
3: Find the first k eigenvector $\mathbf{v}_1, \ldots, \mathbf{v}_k$ corresponding to the k smallest eigenvalues of $\mathbf{A}_{\text{sym}}$.
4: Construct a $[S \times k]$ matrix $\mathbf{V}$ with the eigenvectors as columns.
5: Normalize the rows of matrix $\mathbf{V}$ by $\hat{v}_{ij} = v_{ij} / (\sum_k v_{ik}^2)^{\frac{1}{2}}$
6: Run k-means on matrix $\mathbf{V}$.

2.5 Deep Autoencoders

An autoencoder (AE) is an ANN that performs two main tasks [8]: (1) an encoder function $\mathcal{E}$ maps the data $\mathbf{d}$ to compress data $\boldsymbol{\alpha}$; (2) a decoder function $\mathcal{D}$ maps from the compressed data $\boldsymbol{\alpha}$ back to $\hat{\mathbf{d}}$. This operation is performed setting the same number of neurons L in the input and output layer and constraining the hidden layer to have $N < M$ neurons.

Consider a single hidden layer AE (see Fig. 2), if the new design variable $\boldsymbol{\alpha}$ can be written as

$$\boldsymbol{\alpha} = \mathcal{E}(\mathbf{H}^{(1)}\mathbf{d} + \mathbf{b}^{(1)}) \tag{16}$$

where $\mathbf{H}$ is a relative weight matrix, $\mathbf{b}$ the bias vector, and the apex "(1)" represent the hidden layer, then the reconstruction vector $\hat{\mathbf{d}}$ from $\boldsymbol{\alpha}$ can be expressed as

$$\hat{\mathbf{d}} = \mathcal{D}(\mathbf{H}^{(2)}\boldsymbol{\alpha} + \mathbf{b}^{(2)}) \tag{17}$$

where the apex "(2)" represent the output layer. The network parameters $\mathbf{H}$ and $\mathbf{b}$, are evaluated minimizing the reconstruction error

$$E(\mathbf{H}^{(1)}, \mathbf{b}^{(1)}, \mathbf{H}^{(2)}, \mathbf{b}^{(2)}) = \frac{1}{2}\sum_{k=1}^{S}||\mathbf{d}_k - \hat{\mathbf{d}}_k||^2 \tag{18}$$

$$= \frac{1}{2}\sum_{k=1}^{S}||\mathbf{d}_k - \mathcal{D}(\mathbf{H}^{(2)}\mathcal{E}(\mathbf{H}^{(1)}\mathbf{d}_k + \mathbf{b}^{(1)}) + \mathbf{b}^{(2)})||^2$$

If $\mathcal{E}$ and $\mathcal{D}$ are linear then the Eq. 18 has a unique global minimum, in which the weights in the hidden layer span the same subspace as the first N-principal components of the data [14,15]. AE with nonlinear activation functions and more hidden layers (called deep autoencoder, DAE) provides a nonlinear generalization of the PCA [16], but in this case the error function (Eq. 18) becomes non convex and the optimization algorithm may get

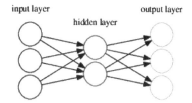

Fig. 2. Example of AE with one hidden layer with $L = 3$ and $N = 2$

stuck in poor local minima. Moreover, the intrinsic dimensionality of the data (the number of neurons N in the hidden layer) cannot be known a priori and have to be fixed respect to the reconstruction error.

3 Shape Modification of a Destroyer Hull

The DTMB 5415 model is an open-to-public early concept of the DDG-51, a USS Arleigh Burke-class destroyer, widely used for both towing tank experiments [17] and hull-form SBDO [18]. Figure 3 shows its geometry and body surface grid used to discretize the shape modification domain.

The offline design-space assessment and dimensionality reduction of the DTMB 5415 hull form (assuming full-scale with a length between perpendiculars $L_{\text{pp}} = 142$ m) is presented as a pre-optimization study of the following problem

Minimize $f(\mathbf{u})$
subject to $g_a(\mathbf{u}) = 0,$ with $a = 1, \ldots, A$ (19)
and to $h_e(\mathbf{u}) \leq 0,$ with $e = 1, \ldots, E$

where f is the objective function related to the ship performance (*i.e.* resistance, seakeeping, etc.) and $\mathbf{u}$ are the (original) design variables. Geometrical equality constraints, g_a, include fixed length between perpendicular (L_{pp}) and displacement (∇), whereas geometrical inequality constraints, h_e, include 5% maximum variation of beam and draught and reserved volume for the sonar in the bow dome, corresponding to 4.9 m diameter and 1.7 m length (cylinder).

Shape modifications $\boldsymbol{\delta}(\mathbf{x}, \mathbf{u})$ are applied directly on the Cartesian coordinates $\mathbf{g}$ of the computational body surface grid, as per

$$\mathbf{g}(\mathbf{u}) = \mathbf{g}_0 + \boldsymbol{\delta}(\mathbf{x}, \mathbf{u}) \qquad (20)$$

where $\mathbf{g}_0$ represents the original grid.

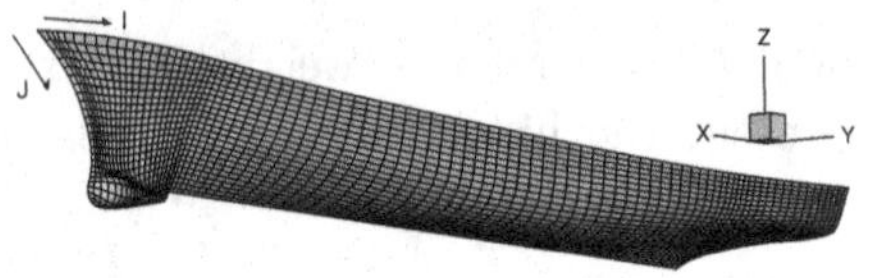

Fig. 3. DTMB 5415 geometry and body surface discretization

The shape modification is defined using a linear combination of $M = 27$ vector-valued functions of the Cartesian coordinates $\mathbf{x}$ over a hyper-rectangle embedding the demi hull [18]

$$\boldsymbol{\psi}_i(\mathbf{x}) \; : \; \mathcal{V} = [0, L_{x_1}] \times [0, L_{x_2}] \times [0, L_{x_3}] \in \mathbb{R}^3 \longrightarrow \mathbb{R}^3 \qquad (21)$$

with $i = 1, \ldots, M$, as

$$\boldsymbol{\delta}(\mathbf{x}, \mathbf{u}) = \sum_{i=1}^{M} u_i \, \boldsymbol{\psi}_i(\mathbf{x}) \qquad (22)$$

where the coefficients $u_i \in \mathbb{R}$ $(i = 1, \ldots, M)$ are the (original) design variables,

$$\boldsymbol{\psi}_i(\mathbf{x}) := \prod_{j=1}^{3} \sin\left(\frac{a_{ij}\pi x_j}{L_{x_j}} + r_{ij}\right) \mathbf{e}_{q(i)} \qquad (23)$$

and the following orthogonality property is imposed:

$$\int_{\mathcal{V}} \boldsymbol{\psi}_i(\mathbf{x}) \cdot \boldsymbol{\psi}_k(\mathbf{x}) \mathrm{d}\mathbf{x} = \delta_{ik} \qquad (24)$$

In Eq. 23, $\{a_{ij}\}_{j=1}^{3} \in \mathbb{R}$ define the order of the function along j-th axis; $\{r_{ij}\}_{j=1}^{3} \in \mathbb{R}$ are the corresponding spatial phases; $\{L_{x_j}\}_{j=1}^{3}$ are the hyper-rectangle edge lengths; $\mathbf{e}_{q(i)}$ is a unit vector. Modifications are applied along x_1, x_2, or x_3, with $q(i) = 1, 2$, or 3 respectively. The parameter values used here are taken from [18].

Fixed L_{pp} and ∇ are satisfied by automatic geometric scaling, while geometries exceeding the constraints are not considered.

4   Numerical Results

The results obtained by linear PCA and the nonlinear methods (KPCA, LPCA-KM, LPCA-SC, and DAE) are presented in the following subsections. Two evaluation metrics are used to assess the methods' performance and compare them.

4.1 Evaluation Metrics

The methods are assessed by the portion of original geometric variance resolved ($\hat{\beta}$) and the root mean square error (RMSE) of matrix reconstruction $\hat{\mathbf{D}}$, defined as

$$\hat{\beta} = \frac{\frac{1}{S}\sum_{j=1}^{L}\sum_{k=1}^{S}(\hat{d}_{jk}-\hat{\mu}_j)^2}{\frac{1}{S}\sum_{j=1}^{L}\sum_{k=1}^{S}(d_{jk}-\mu_j)^2} \quad\text{and}\quad \text{RMSE} = \sqrt{\frac{1}{S}\sum_{k=1}^{S}||\mathbf{d}_k-\hat{\mathbf{d}}_k||^2} \quad (25)$$

where $\hat{\mu}_j$ is the mean value of $\hat{\mathbf{D}}$ j-th column.

4.2 Evaluation of Design-Space Dimensionality Reduction Capabilities

In assessing the methods' performance, a cubic polynomial kernel is used for the KPCA, a number of cluster $k = 32$ and 24 is used for LPCA-KM and LPCA-SC respectively, a seven hidden layer DAE (composed by 300-150-50-N-50-150-300 neurons) with hyperbolic tangent (as activation function) is used and trained with Adam optimization algorithm [19].

The design space ($M = 27$) is sampled using a uniform random distribution of $S = 1,000$ hull-form designs. For each dimensionality-reduction method, Fig. 4a shows the geometric variance ($\hat{\beta}\%$) resolved by a N-dimensional design space, whereas Fig. 4b shows the corresponding reconstruction error (RMSE). The nonlinear methods result to be more effective than the linear PCA in terms of both $\hat{\beta}\%$ and RMSE.

Table 1. Numerical results

Method	N [–]	$\hat{\beta}\%$	RMSE/L_{pp}
PCA	24	95.0	1.12E−1
KPCA	18	100	0.00E+0
LPCA-KM	12	95.0	1.12E−1
LPCA-SC	15	95.4	1.08E−1
DAE	5	97.8	9.60E−2

Specifically, in order to reduce the design-space dimensionality while resolving at least the 95% of the original geometric variance, $N = 24$ is required by PCA, whereas $N = 18, 12, 15,$ and 5 are needed by KPCA, LPCA-KM, LPCA-SC, and

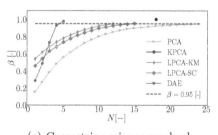

(a) Geometric variance resolved

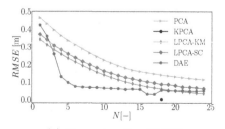

(b) Reconstruction RMSE

Fig. 4. Convergence of dimensionality-reduction methods in terms of $\hat{\beta}\%$ (a) and RMSE (b) versus the reduced-dimensionality N

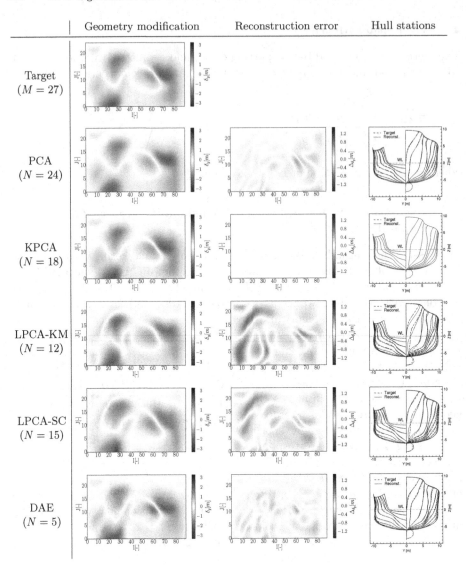

Fig. 5. Reconstruction of the geometry modification vector δ_y, reconstruction error, and corresponding hull stations of target geometry (original input)

DAE, respectively. The results are summarized in Table 1. It is worth noting that KPCA requires $N = 18$, but resolves the 100% of the original variance and shows a reconstruction error equal to zero. In the current study, it was not possible to reduce N further, due to numerical issue associated to the computation of pre-images.

Finally, Fig. 5 shows the shape modification (δ_y) and the reconstruction error ($\Delta\delta_y$) versus grid-node index (I, J), and the corresponding hull stations for a design originally included in the data matrix **D**. For this design, LPCA shows the largest reconstruction error. PCA and DAE produce a close reconstruction to the target, whereas KPCA reproduce the target exactly. With only $N = 5$, DAE is the most efficient overall.

5   Conclusions and Future Work

Four nonlinear methods for design-space dimensionality-reduction in shape optimization have been presented and compared. Specifically, kernel PCA (KPCA), local PCA with k-means and spectral clustering (respectively LPCA-KM and LPCA-SC), and deep autoencoder (DAE) have been used for an offline pre-optimization dimensionality-reduction of the hull-form parametrization of the DTMB 5415 model hull. A linear PCA method from earlier studies has been also included in the analysis, for comparison.

The original shape parametrization was defined by $M = 27$ design variables. The reduced-dimensionality space is required to resolve at least the 95% of the original design variability, based on the concept of geometric variance. The linear PCA achieved a reduction of 11.2% of the original design dimensionality (requiring a number of design variables $N = 24$). All nonlinear methods outperform the linear PCA. Specifically, a 33.4% dimensionality reduction is achieved by KPCA ($N = 18$), 55.5% by LPCA-KM ($N = 12$), 44.4% by LPCA-SC ($N = 15$), and finally a remarkable 81.5% by DAE ($N = 5$). Nonlinear methods have shown their superior effectiveness in terms of both variance resolved and reconstruction error, compared to linear PCA. DAE have shown the best performance overall.

The analysis of some specific behavior of the methods presented, such as the assessment of the clusters used by the LPCA, will be addressed in future work. Moreover, in order to investigate further on the methods' effectiveness, future work will include the optimization of the DTMB 5415 using the reduced-dimensionality space produced by linear and nonlinear methods, with comparison of objective function improvement and convergence to the optimum. Also, combined geometry and physics based design variability studies [20,21] will be addressed using current nonlinear methods.

Acknowledgments. The work is supported by the US Office of Naval Research Global, NICOP grant N62909-15-1-2016, under the administration of Dr. Woei-Min Lin, Dr. Salahuddin Ahmed, and Dr. Ki-Han Kim, and by the Italian Flagship Project RIT-MARE. The research is performed within NATO STO Task Group AVT-252 Stochastic Design Optimization for Naval and Aero Military Vehicles. The authors wish to thank Prof. Frederick Stern and Dr. Manivannan Kandasamy of The University of Iowa for inspiring the current research on nonlinear dimensionality reduction methods.

References

1. Diez, M., Campana, E.F., Stern, F.: Design-space dimensionality reduction in shape optimization by Karhunen-Loève expansion. Comput. Method. Appl. Mech. Eng. **283**, 1525–1544 (2015)
2. Raghavan, B., Breitkopf, P., Tourbier, Y., Villon, P.: Towards a space reduction approach for efficient structural shape optimization. Struct. Multi. Optim. **48**, 9871000 (2013)

3. Lukaczyk, T., Palacios, F., Alonso, J.J., Constantine, P.: Active subspaces for shape optimization. In: Proceedings of the 10th AIAA Multidisciplinary Design Optimization Specialist Conference, 13–17 January 2014, National Harbor, Maryland, USA (2014)
4. Schölkopf, B., Smola, A., Müller, K.R.: Nonlinear component analysis as a kernel eigenvalue problem. Neural Comput. 10(5), 1299–1319 (1998)
5. Kambhatla, N., Leen, T.K.: Dimension reduction by local principal component analysis. Neural Comput. 9(7), 1493–1516 (1997)
6. Lloyd, S.: Least squares quantization in PCM. IEEE Trans. Inf. Theor. 28(2), 129–137 (1982)
7. Von Luxburg, U.: A tutorial on spectral clustering. Stat. Comput. 17(4), 395–416 (2007)
8. Hinton, G.E., Salakhutdinov, R.R.: Reducing the dimensionality of data with neural networks. Science 313(5786), 504–507 (2006)
9. Bishop, C.M.: Pattern Recognition and Machine Learning (Information Science and Statistics). Springer, New York (2006)
10. Smola, A.J., Schölkopf, B.: Learning with Kernels. Citeseer (1998)
11. Bakır, G.H., Weston, J., Schölkopf, B.: Learning to find pre-images. Adv. Neural Inf. Process. Syst. 16, 449–456 (2004)
12. Ng, A.Y., Jordan, M.I., Weiss, Y., et al.: On spectral clustering: analysis and an algorithm. In: NIPS, vol. 14, pp. 849–856 (2001)
13. Aggarwal, C.C., Reddy, C.K.: Data Clustering: Algorithms and Applications. Chapman and Hall/CRC (2013)
14. Bourlard, H., Kamp, Y.: Auto-association by multilayer perceptrons and singular value decomposition. Biol. Cybern. 59(4), 291–294 (1988)
15. Baldi, P., Hornik, K.: Neural networks and principal component analysis: learning from examples without local minima. Neural Netw. 2(1), 53–58 (1989)
16. LeCun, Y., Bengio, Y., Hinton, G.: Deep learning. Nature 521(7553), 436–444 (2015)
17. Stern, F., Longo, J., Penna, R., Olivieri, A., Ratcliffe, T., Coleman, H.: International collaboration on benchmark CFD validation data for surface combatant DTMB model 5415. In: Proceedings of the Twenty-Third Symposium on Naval Hydrodynamics, 17–22 September 2000, Val de Reuil, France (2000)
18. Serani, A., Fasano, G., Liuzzi, G., Lucidi, S., Iemma, U., Campana, E.F., Stern, F., Diez, M.: Ship hydrodynamic optimization by local hybridization of deterministic derivative-free global algorithms. Appl. Ocean Res. 59, 115–128 (2016)
19. Kingma, D., Ba, J.: Adam: A method for stochastic optimization. arXiv preprint arXiv:1412.6980 (2014)
20. Diez, M., Serani, A., Stern, F., Campana, E.F.: Combined geometry and physics based method for design-space dimensionality reduction in hydrodynamic shape optimization. In: Proceedings of the 31st Symposium on Naval Hydrodynamics, Monterey, CA, USA (2016)
21. Serani, A., Campana, E.F., Diez, M., Stern, F.: Towards augmented design-space exploration via combined geometry and physics based Karhunen-Loève expansion. In: 18th AIAA/ISSMO Multidisciplinary Analysis and Optimization Conference (MA&O), AVIATION 2017, 5–9 June 2017, Denver, USA (2017)

A Differential Evolution Algorithm to Develop Strategies for the Iterated Prisoner's Dilemma

Manousos Rigakis$^{(\boxtimes)}$, Dimitra Trachanatzi, Magdalene Marinaki,
and Yannis Marinakis

School of Production Engineering and Management, Technical University of Crete,
University Campus, Chania, Greece
{mrigakis,dtrachanatzi}@isc.tuc.gr, magda@dssl.tuc.gr,
marinakis@ergasya.tuc.gr

Abstract. This paper presents the application of the Differential Evolution (DE) algorithm in the most known dilemma in the field of Game Theory, the Prisoner's Dilemma (PD) that simulates the selfish behavior between rational individuals. This study investigates the suitability of the DE to evolve strategies for the Iterated Prisoner's Dilemma (IPD), so that each individual in the population represents a complete playing strategy. Two different approaches are presented: a classic DE algorithm and a DE approach with memory. Their results are compared with several benchmark strategies. In addition, the Particle Swarm Optimization (PSO) and the Artificial Bee Colony (ABC) that have been implemented in the same framework are compared with the DE approaches. Overall, the strategies developed by DE outperform all the others. Also, it has been observed over iterations that when the DE algorithm is used the player manages to learn his opponent, therefore, DE converges with a quick and efficient manner.

Keywords: Differential evolution · Game theory
Iterated Prisoner's Dilemma

1 Introduction

The Prisoner's Dilemma (PD) is a game between two rational and mutually interdependent players with conflicting interests. PD has been discussed extensively by game theorists and, also, finds application in diverse areas ranging from business, finance to sociology. This study focuses on the PD's variation, the Iterated Prisoner's Dilemma (IPD) and presents an algorithmic scheme to develop well-performing strategies regarding the latter game. Since Axelrod's original work [1], in which he evolved a population of strategies using a genetic algorithm, a number of other works have been published [3,7]. A detailed review relevant to the IPD approaches for generating strategies and their representation can be found in [5]. The basic component of this approach is the Differential Evolution (DE) algorithm, which has a purpose of generating a binary M-bit

© Springer International Publishing AG 2018
G. Nicosia et al. (Eds.): MOD 2017, LNCS 10710, pp. 133–145, 2018.
https://doi.org/10.1007/978-3-319-72926-8_12

length, decision vector that prevails to his opponent after a number of repeated PD games. To improve the developed strategies, a variety of opponents has been used, such as man-made strategies, denoted as Benchmark and, also, strategies evolved from other algorithms, the Particle Swarm Optimization (PSO) and the Artificial Bee Colony (ABC). In order to enhance the solution of the DE algorithm, another approach was implemented, which incorporates the attribute of memory over the executions, denoted as version 2. From the experimental results, presented in the following, it is evident that both versions of the DE approach provide efficient strategies. An additional deriving conclusion, is that, when the DE approach with memory faces an approach of the ABC with memory, the strategies co-evolve and eventually, after a number of iterations the DE's evolved strategies manage to learn their opponent. The rest of this paper is organized as follows. Firstly, a short overview of the DE algorithm is given in Sect. 2 and in sequence the theoretical description of the PD game is presented in Sect. 3. The latter section includes a short outline of the Iterated PD and of several Benchmark strategies. An extended analysis of the proposed solution algorithm based on the DE is presented in Sect. 4, regarding both versions. In Sect. 5, the experimental results of this research are illustrated and analysed. Finally, Sect. 6 summarizes this paper and provides suggestions for future research.

2   Differential Evolution: A Short Overview

Differential Evolution (DE) is a stochastic, real-parameter optimization algorithm proposed by Storn and Price [9], designed for continuous-optimization problems. DE is a population-based search method, which includes processes such as mutation, crossover and selection. One of the DE advantages is the small number of control parameters, the population size (NP), the mutation rate (F) and the crossover rate (Cr). The main idea is the perturbation of a vectors population through a number of generations, that incorporates vector differences and recombination. Initially, a randomly disturbed population of NP individuals is generated. Each one is a D-dimensional real vector x_{ij}, where $i \in \{1, \cdots, NP\}$ and $j \in \{1, \cdots, D\}$. The first evolutionary process that takes place in every generation is the mutation. During mutation, three vectors are randomly chosen, a *base* vector $(i_1 \neq i)$ and two others $(i \neq i_1 \neq i_2 \neq i_3)$. The difference of x_{i_2} and x_{i_3} is amplified by the mutation rate F, which is a real, constant value between 0 and 2. The scaled difference is added to the base vector, see Eq. 1, in order to form the mutant vector $v_{ij}(t)$, for each target vector of the population, in t generation, i.e. for each individual.

$$v_{ij}(t+1) = x_{i_1 j}(t) + F * (x_{i_2 j}(t) - x_{i_3 j}(t)) \tag{1}$$

Afterwards, the crossover process occurs, which is a recombination of each target and its corresponding mutant and generates the trial vector. There are two common kinds of crossover methods, exponential and binomial. The binomial crossover will be used in this research and it is implemented as follows, for every target vector in the population. Through Eq. 2, it is determined what parameters will be inherited to the trial vector from the mutant and what from the target vector. The crossover is controlled by the Cr parameter and its value is decided by the user, within the range $[0,1]$. For each parameter j, a random number ϕ and a random index j_{rand} are generated such as $\phi, j_{rand} \in [0,1]$. If the random number ϕ is less or equal to Cr, or if the parameter's index equals to j_{rand}, the trial vector inherits the corresponding element from the mutant vector and otherwise from the target vector. The j_{rand} ensures that at least one parameter will be forwarded to the trial vector from the mutant vector.

$$u_{ij}(t+1) = \begin{cases} v_{ij}(t+1), & \text{if } \phi \leq C_r \text{ or } j = j_{rand} \\ x_{ij}(t), & \text{otherwise} \end{cases} \tag{2}$$

Both processes, mutation and crossover, increase the diversity of the population and thereby they carry out the exploration phase of the search. In order to involve an exploitation phase of the search and to retain the size of the population, a selection procedure is performed. Thus, subsequent to crossover process, one of the correlated vectors, target and trial, have to remain in the population and the other one has to be discarded. DE uses a greedy technique as selection and the vector with the highest fitness value will survive over the other and will be included in the next generation's population, see Eq. 3.

$$x_{ij}(t+1) = \begin{cases} u_{ij}(t+1), & \text{if } f(u_{ij}(t+1)) \leq f(x_{ij}(t)) \\ x_{ij}(t), & \text{otherwise} \end{cases} \tag{3}$$

A number of variations to the basic DE algorithm have been developed over the years. Different DE strategies have a general notation $DE/x/y/z$, where x is cited to the way that a target vector is selected, y is the number of difference vectors used and z refers to the crossover scheme. The presented research adopts the $DE/rand/1/bin$ variation, thus the vector to be mutated is chosen randomly, the difference of one pair of vectors is involved to Eq. 1 and the binomial crossover is implemented.

3   Prisoner's Dilemma

Prisoner's Dilemma is a non-zero-sum, non-cooperative game and was first formalized by Tucker in 1950 [10]. In a non-zero-sum game, when one player wins, the loss of his opponent is not strictly implied. In terms of non-cooperative, the communication between players prior to the game is forbidden and as a result they are not able to make any kind of agreement. An analytical description of the game is given: two crime suspects have been arrested and detained in separate cells without communication. The prosecutor offers them two possible choices: either to cooperate (C) or to defect (D). Cooperation denotes that the prisoner conspires with his/her associate to remain silent, while defection means that the prisoner acts selfish, accusing the other one, in order to make a deal with the authorities. Combining these two choices of each individual, three different scenarios may emerge. If simultaneously one prisoner chooses to defect and the other one to cooperate, then the defector will be released and the cooperator will be jailed for m years. If simultaneously both decide to cooperate, then, both will be jailed for n years, $n < m$. If simultaneously both decide to defect, then, both will be jailed for r years, $n < r < m$. The game can be presented in a matrix form, as shown in Table 1. The values, R, S, T, P, inside the cells are related to each player's payoff, depending on his/her decision. The first value is related to the payoff of player I and the second to the payoff of player II. Specifically, the letter R denotes the reward payoff in case of mutual cooperation, S refers to the sucker's payoff, because in a conflicting-decision situation the player cooperates. In contrary, T is the temptation payoff that player receives when he defects against the other's cooperation. Finally, P expresses the punishment payoff of mutual defection. Regarding the payoffs, the following constrains have to be satisfied: $T > R > P > S$ and $2R > S + T$.

3.1   Iterated PD and Benchmark Strategies

PD is a single game, since each player has to take a decision only once, in order to maximize his/her payoff. The Iterated Prisoner's Dilemma (IPD), is merely a PD game played by the same participants, repeatedly. The key element of the IPD is that the number of iterations have to be unknown for both parties. Thus, players of IPD are in position to form a strategy in order to maximize the total payoff over the repeated games. Since Axelord's original work [1,2], researchers have developed various efficient IPD strategies. In the presented research the following strategies, denoted as benchmark, will be used to evaluate the ones developed by the DE approach.

1. Random: A random sequence of decisions, either cooperation of defection.
2. Always Cooperate (AC): Unconditionally cooperation in every game.
3. Pavlov: A decision is repeated if it was beneficial (i.e. the corresponding payoff was T or R). Otherwise, the next decision is opposite to the previous one (i.e. the corresponding payoff was P or S).
4. Tit-for-tat (TFT): The first decision is cooperation and, afterwards, the player imitates the last decision of his opponent.
5. Evil tit-for-tat (ETFT): The first decision is defection and, afterwards, the player imitates the last decision of his opponent.

4   DE Develops IPD Strategies

In this study, an implementation of the Differential Evolution algorithm is applied to develop strategies for the IPD. Thus, each developed strategy is represented by a solution vector that includes binary values (0 or 1). In this case, 1 denotes cooperation and 0 stands for defection. Following the DE's algorithmic scheme, an initial population is generated of N random decision vectors x_{ij} (where N, the number of players is equivalent to the population size NP). Each individual i has j parameters that correspond to the player's decision in every game, $j \in [1, \cdots, M]$, therefore M represents the total number of PD games.

After the generation of the initial population, each individual has to be evaluated. On that account, each developed strategy has to be assessed and since each element represents one decision (one PD game), the fitness function is merely the dilemma itself. In more detail, the fitness function's value per strategy is the total payoff achieved from all the PD games, when a player with that strategy plays with all the players in the population. A simple way to express it, is that all players have to compete against all others, with respect to their developed strategies and summarize their achieved payoffs per dilemma/per opponent.

As mentioned above, the equation of DE algorithm that mutates the solution vectors, see Eq. 1, requires continuous values, therefore, the binary vectors should be converted to continuous-valued vectors. Consequently, the sigmoid function [6] is applied for this trans-

Table 1. Payoff matrix of fitness function

	Player II	
	Cooporate	Defect
Player I Cooporate	**3,3** (R,R)	**0,5** (S,T)
Defect	**5,0** (T,S)	**1,1** (P,P)

formation, on every parameter of the population, through Eq. 4. In sequence, the converted solution vector of every one in the population, is mutated and the crossover process takes place, see Eq. 2, as described in Sect. 2 and the trial vectors are generated. To evaluate the new solution vectors and to perform the selection process between each trial and target vector, the fitness value is need. Thus, the trial vectors have to be transformed in binary ones, through Eq. 5, which is controlled by a random generated value $\phi_1 \in [0,1]$.

$$Sig(x_{ij}) = \frac{1}{1 + exp(x_{ij})}, i \in [1, \cdots, N], j \in [1, \cdots, M] \qquad (4)$$

$$u_{ij}(t+1) = \begin{cases} 1, & \text{if } \phi_1 < u_{ij}(t+1) \\ 0, & \text{if } \phi_1 \geq u_{ij}(t+1) \end{cases} \qquad (5)$$

In this way, a set of new solution vectors (strategies) is formed and in effort to calculate their fitness value, a tournament all-against-all is implemented, as previously described. Finally, only one of the N strategies emerges, the one with the highest value of payoff, the most productive strategy of the current generation. All the above processes are repeated for a number of iterations L, concluding to an efficient strategy. The algorithm has been executed several number of times, W. An overview of the DE approach steps are presented below in Algorithm 1. In order to test the quality of the strategy, that is evolved by the DE approach, another Iterated Prisoner's Dilemma will be implemented. At this stage, one of the players is the DE's evolved strategy and the other party is a player that follows one of the benchmark strategies, which are described in Sect. 3.1. To be more thorough, the authors have compared the DE's strategy with others, that have been developed by two nature inspired algorithms, the Artificial Bee Colony (ABC) [4] and the Particle Swarm Optimization (PSO) algorithm [6], based on the previous work of Rigakis et al. [8].

Algorithm 1. Solution Algorithm for the IPD (version 1)

 1: Define number of executions (W), iterations (L), games (M)
    Differential Evolution Algorithm
 2: *Initialization*
 3: Define values of the DE parameters (F) and (Cr)
 4: Define population size (N)
 5: Randomly create (N) decision vectors x_{ij}
 6: Tournament all-against-all
 7: Calculate each strategy's payoff, according to Table 1
 8: *Main Phase*
 9: **while** number of iterations is not equal to L **do**
10:    Transform the decision vectors $x_{ij}(t)$ to continuous-valued, Eq. (4)
11:    Create mutant $v_{ij}(t+1)$ for each target vector, Eq. (1)
12:    Create trial vectors $u_{ij}(t+1)$ for each mutant, Eq. (2)
13:    Transform the trial vectors to binary, Eq. (5)
14:    Tournament all-against-all
15:    Calculate each strategy's payoff, according to Table 1
16:    Select trial or target vector, Eq.(3)
17: **end while**
18: Save the most efficient strategy of the population (maximum payoff)
    END Differential Evolution Algorithm
19: Employ the best strategy for M games against each of the 5 Benchmark strategies/
    PSO algorithm/ ABC algorithm
20: Return to line 2, until W executions are completed

4.1   The DE Approach with Memory

After close examination of the presented DE approach for IPD games, the authors have decided to embed memory in it. So far, the evolved strategies are being improved through the DE's iterations and by merely facing each other. The idea is to make those strategies, also, relevant to their opponents. Thus, a new approach is formed, the DE algorithm remains mostly intact, but instead of one best strategy, five strategies with high payoff, will play against their rivals. Each of the "elite-five" will counter Benchmark strategies, PSO and ABC evolved strategies. The "global best" one, which will achieve the highest payoff, will be memorized. In sequence, the global best strategy will replace one of the random solution vectors of the initial population in the next execution, see Algorithm 2. The result of this replacement is to enhance the developed strategies with beneficial traits, regarding those opposing to them.

5   IPD Experiments

The experimental results of our research are stated in this section and are described succinctly. The following are divided to subsections, where each one contains the results of DE algorithm against other methods in Iterated Prisoner's Dilemma, for different parametrization. The original DE control parameters F

Algorithm 2. Solution Algorithm for the IPD with Memory (version 2)

1: Define number of executions (W), iterations (L), games (M)
2: Define population size (N)
3: *Initialization*
4: **if** It is the first execution **then**
5:   Randomly create (N) decision vectors x_{ij}
6: **else**
7:   Create $N-1$ random decision vectors
8:   Add the global best strategy of the previous execution
9: **end if**
10: Tournament all-against-all
11: Calculate each strategy's payoff, according to Table 1
12: *Main Phase*
13: **while** number of iterations not equal to L **do**
14:   **DE** algorithm's steps to evolve strategies, see Algorithm 1, lines 10–16
15: **end while**
16: Save the five most efficient strategies of the population
17: Employ them for M games against each of the 5 Benchmark strategies/ PSO algorithm/ ABC algorithm
18: Save the global best strategy (maximum payoff)
19: Return to line 5, until W executions are completed

---

and Cr are both equal to 0.5, regarding to all the conducted experiments. In the figures presented below, the vertical axis shows the payoff achieved for every strategy while in the horizontal axis, the number of execution appears.

Results with Small Number of Iterations. Initially, the DE algorithm was executed with small values of the control parameters. More precisely, for $N = 5$ players, $L = 5$ iterations, $M = 5$ games and $W = 20$ executions of the DE algorithm. It should be mentioned that the W executions are considered as different procedures and are not compared with each other. As it was mentioned previously, with each execution of the algorithm, a player's strategy is created ready to face all the Benchmark strategies. Furthermore, a comparison is made between the described approach and the ABC algorithm presented by [8] to evolve strategies for the IPD. Thus, to have a fair comparison, the control parameters (L, M, N, W) are selected to be equal to the ones chosen in [8]. Figure 1 shows the payoff that each player developed by the DE algorithm (red cross) gains, against to the AC strategy (green circle). The AC strategy is the most innocent one, therefore, it is predictable and easy to be dominated by both versions (with memory and without memory) of DE. It is obvious from Fig. 2 that the DE algorithm in version 1 (without memory) provides competitive strategies against the most unpredictable opponent (the random strategy). In version 2 (with memory) the quality of the results is deteriorated as we excepted, since the memory is inefficient against a constantly changing behaviour. As seen in Fig. 3, DE algorithm adapts to his opponent, which follows the Pavlov strategy and both versions provide sufficient results. Specifically, in version 1, DE wins 18 out of the 20 executions and achieves two draws. Version 2 improves the results and prevails by 100%. Figure 4 corresponds to the game between DE algorithm and the most known from Axelrod's experiments benchmark strategy, TFT. It is evident, that DE's evolved strategies achieve higher payoff at every execution, regarding both algorithm's versions. A strategy which manages to fully compete DE algorithm and most times overcomes it, is ETFT. In Fig. 5, the results for a small number of iterations can be observed. Specifically, in version 1 the best result that DE algorithm obtains is one draw (1/20). It is obvious that version 2 with memory has improved the results (6/20 draws), but the number of iterations ($L = 5$) seems to be not enough in order to learn the selfish behaviour of his opponent. In order to evaluate the performance of the DE algorithm against each Benchmark strategy, the following equation: $Perc_{opponent} = \frac{payoff_{DE} - payoff_{opponent}}{payoff_{opponent}}$ is used to calculate the percentage difference of the respectively achieved total payoffs from W executions. Moreover, the same efficiency measure is presented about the approach of ABC algorithm and the results are showed in Table 2.

Results with Large Number of Iterations. In an effort to improve the results and to determine more productive parameters' values, the following changes are

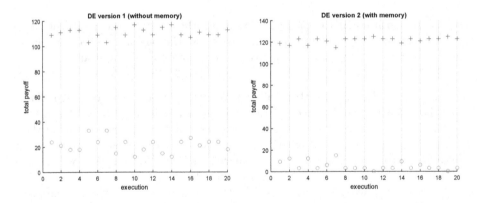

Fig. 1. DE vs AC (DE: red cross, AC: green circle) (Color figure online)

Table 2. Percentage differences of payoffs (%) between the two versions of DE and ABC algorithm, for $W = 20$, $N = 5$, $M = 5$, $L = 5$

Strategies	DE (Version 1)	DE (Version 2)	ABC (Version 1)	ABC (Version 2)
AC	3,7747	13,4043	1,6774	2,2627
Random	0,5455	−0,5173	−0,0647	−0,7724
Pavlov	0,6408	4,7582	−0,1213	1,9606
TFT	0,4013	0,9320	0,2255	0,3461
ETFT	−0,1786	−0,1519	−0,2215	−0,2944

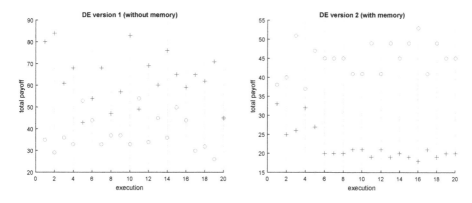

Fig. 2. DE vs RANDOM (DE: red cross, RANDOM: green circle) (Color figure online)

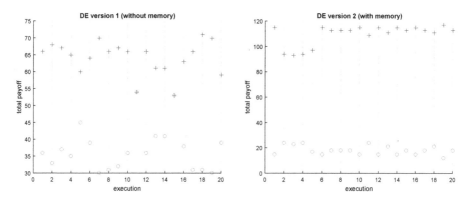

Fig. 3. DE vs PAVLOV (DE: red cross, PAVLOV: green circle) (Color figure online)

implemented. The number of iterations was increased to $L = 40$, the number of players to $N = 20$ and the number of decisions exchanged amongst the players to $M = 10$. The executions W are independent runs of the algorithm and the corresponding value is chosen to remain stable at $W = 20$, since a different value would not affect the algorithm's performance. After the increase in control parameter's values, both versions of DE behaved almost similar as previously described (for $N, L, M = 5$). Most interesting were the results obtained with the Evil-Tit-For-Tat strategy, thus they are analysed bellow. From Fig. 6, it is observed that a larger number of iterations ($L = 40$) is more effective against ETFT strategy, as both versions of DE succeed more draws. Very interesting is the analysis of the DE's second version against ETFT strategy, which is the most competitive strategy. It has been observed that by holding all variables fixed and gradually increasing only the number of iterations, the results show a significant improvement. Figure 7 illustrates the experimental results of the DE approach with memory for $L = 60$ and $L = 600$, where 11/20 and 20/20 draws are demonstrated respectively. For $L = 600$ the outcome was remarkable, since the DE algorithm managed to learn

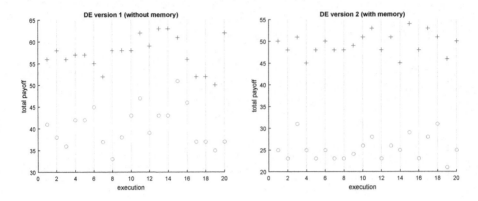

Fig. 4. DE vs TFT (DE: red cross, TFT: green circle) (Color figure online)

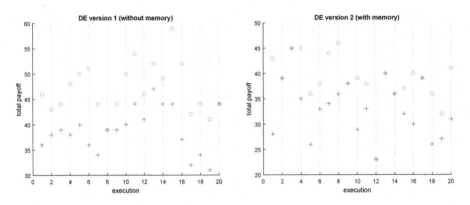

Fig. 5. DE vs ETFT (DE: red cross, ETFT: green circle) (Color figure online)

the behaviour of his/her opponent. Table 3 summarizes the experiments and the percentage difference is presented regarding the different L values (60, 120, 240, 600) that have been tested. Additionally, in order to determine whether a DE approach is able to develop efficient IPD strategies, another set of experiments has been implemented. In particular, DE's evolved strategy plays against one developed by PSO algorithm. It is concluded that both DE versions (without and with memory) outperform the PSO algorithm and the corresponding percentage differences are 1,4720 and 2.3341, respectively. Comparisons are also made between the strategies evolved by conducting games between the approaches, DE versus ABC and DE with memory versus ABC with memory (Fig. 8). DE provides more sufficient strategies than ABC algorithm and the percentage differences are 1,5380 and 2.4631, respectively.

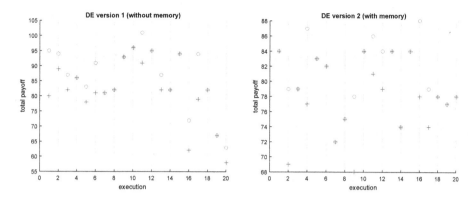

Fig. 6. DE vs ETFT (DE: red cross, ETFT: green circle) (Color figure online)

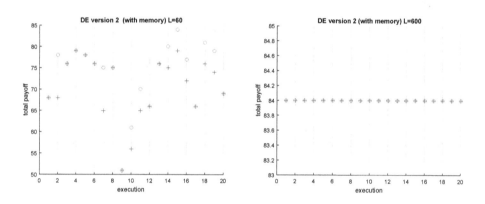

Fig. 7. DE vs ETFT (DE: red cross, ETFT: green circle) (Color figure online)

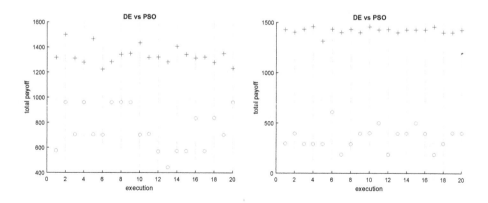

Fig. 8. DE vs PSO (DE: red cross, PSO: green circle) (Color figure online)

Table 3. Percentage differences of payoffs (%) between DE Version 2 and ETFT.

DE with memory versus Evil-Tit-For-Tat strategy			
W=20, N=20, M=10, with : L=**60** L=**120** L=**240** L=**600**			
-0,0386 -0,0349 -0,0167 0,0			

6 Conclusions

In the presented research, an approach is introduced based on the DE algorithm, to evolve strategies for the 2-person IPD. It is concluded by the experimental results, that both approaches are in position to develop productive strategies, in terms of their achieved payoff, against a variety of opponents. The experiments included several Benchmark strategies and others generated by approaches of the PSO algorithm and of the ABC algorithm. The presented DE approach exhibits superior behaviour in comparison to the other tested algorithms, not only regarding their performance against the Benchmark strategies, but also, in terms of facing each others' developed strategies. Moreover, it would be interesting to expand the solution approach to the N-person IPD (more than 2 players). Finally, future work will include other known problems from the field of Game Theory, such as the Battle of Sexes.

Acknowledgments. This work was partially financed by the School of Production Engineering and Management of the Technical University of Crete, as postgraduate research.

References

1. Axelrod, R.: The Evolution of Cooperation. Basic Books, New York (1984)
2. Axelrod, R.: The evolution of strategies in the iterated prisoner's dilemma. In: Davis, L. (ed.) Genetic Algorithms and Simulated Annealing, pp. 32–41. Morgan Kaufman, Los Altos, CA (1987)
3. Haider, S.A., Bukhari, A.S.: Using genetic algorithms to develop strategies for Prisoner's dilemma. Asian J. Inf. Technol. 5(8), 866–871 (2006)
4. Karaboga, D., Basturk, B.: On the performance of artificial bee colony (ABC) algorithm. Appl. Soft Comput. 8(1), 687–697 (2008)
5. Kendall, G., Yao, X., Chong, S.Y.: The Iterated Prisoners' Dilemma: 20 Years On, vol. 4. World Scientific, Singapore (2007). 261 p
6. Kennedy, J., Eberhart, R.C.: Particle swarm optimization. In: Proceedings of the IEEE International Conference on Neural Networks, pp. 1942–1948. IEEE Service Center, Piscataway (1995)
7. Mittal, S., Deb, K.: Optimal strategies of the Iterated Prisoner's Dilemma problem for multiple conflicting objectives. IEEE Trans. Evol. Comput. 13(3), 554–565 (2009)

8. Rigakis, M., Trachanatzi, D., Marinaki, M., Marinakis, Y.: Artificial bee colony optimization approach to develop strategies for the iterated prisoners dilemma. In: 7th International Conference of Bionspired Methods and Their Applications, Bled, Slovenia, pp. 18–20 (2016)
9. Storn, R., Price, K.: Differential evolution a simple and efficient heuristic for global optimization over continuous spaces. J. Glob. Optim. **11**(4), 341–359 (1997)
10. Tucker, A.W.: A two-person dilemma. In: Readings in Games and Information (1950)

Automatic Creation of a Large and Polished Training Set for Sentiment Analysis on Twitter

Stefano Cagnoni, Paolo Fornacciari, Juxhino Kavaja, Monica Mordonini,
Agostino Poggi, Alex Solimeo, and Michele Tomaiuolo[✉]

Dipartimento di Ingegneria e Architettura, Università di Parma,
Parco Area delle Scienze 181/A, 43124 Parma, Italy
{cagnoni,fornacciari,kavaja,monica,poggi,solimeo,tomamic}@ce.unipr.it,
michele.tomaiuolo@unipr.it

Abstract. Within the field of sentiment analysis and emotion detection applied to tweets, one of the main problems related to the construction of an automatic classifier is the lack of suitable training sets. Considering the tediousness of manually annotating a training set, and the noise present in data collected directly from the social web, in this paper we propose an iterative learning approach, which combines distant supervision with dataset pruning techinque. In particular, following the "eat your own dogfood" idea, we have applied a classifier, trained on raw data obtained from different Twitter channels, to the same original dataset, for removing the most doubious instances automatically. This kind of approach has been used to obtain a more polished training set for emotion classification, based on Parrot's model of six basic emotions. On the basis of the achieved results, we argue that the automatic filtering of training sets can make the application of the distant supervision approach more effective in many use cases.

Keywords: Social media · Emotion detection · Distant supervision
Machine learning

1 Introduction

Several institutions have always been interested in obtaining informations about the emotional state of people involved, directly or indirectly, in their activities. With the diffusion of social media, this task has been enormously favoured by people sharing large quantities of data related to their feelings. Sentiment analysis uses this data for the automatic detection of theirs authors' emotional state. Within this field, however, one of the main problems related to the construction of an automatic classifier is the lack of suitable training sets. On the one hand, the manual annotation of training sets is very difficult to apply, as it requires much time and attention from many people to overcome the subjectiveness of evaluation. On the other hand, a pure distant supervision approach tends to produce noisy datasets. Thus, in this paper we propose an approach combining distant supervision with a dataset pruning techinque.

© Springer International Publishing AG 2018
G. Nicosia et al. (Eds.): MOD 2017, LNCS 10710, pp. 146–157, 2018.
https://doi.org/10.1007/978-3-319-72926-8_13

This approach has been applied for the creation of a single "flat" seven-output classifier based on Parrot's emotion categorization (joy, love, surprise, fear, anger, sadness) with the addition of the "objective" class, used to label sentences without emotion. For creating the necessary dataset, we relied on the fact that Twitter users, when expressing emotions, add specific hashtags corresponding to their sentiments. In order to understand which hashtags used to express a given sentiment are the most popular, we downloaded a set of tweets starting directly from tags related to primary and secondary emotions. Then we manually searched among them for more hashtags, used in a consistent way. We decided to download more hashtags for each emotion, in order to possibly represent all its different facets.

A classifier has been trained over this raw dataset, with classes inferred directly from the hashtags. Following the "eat your own dogfood" principle, the same classifier has been then applied to the same raw dataset, to filter out the more doubious instances from the training set, automatically. Finally a classifier, trained on the dataset collected and filtered with this approach, has been tested on a test set derived from the EmoTweet-28 [1] dataset, leading to higher accuracy than the classifier trained without dataset pruning. The results obtained show the importance of combining dataset pruning techinques with the distant supervision approach, in order to remove, as much as possible, spurious instances that are unavoidably affect data collected from the social web.

The paper is organized according to the following structure. Section 2 describes some related research work, in the fields of social media analysis, sentiment analysis, emotion detection, and distant supervision. Then, Sect. 3 describes the methodology for the acquisition of data and the creation of a classification model. Section 4 presents and discusses the obtained results. Finally, some concluding remarks are presented.

2   Related Work

The individual behavior and decision-making process depend on the user's emotions with respect to a fact or a product. The growth of online social networks extends and improves the benefits, both for an individual and organizations, coming from the interactions among the users (the so called social capital [2]) and much work has been done to try to model complex systems like social networks efficiently [3,4]. The ability to retrieve and analyze large amounts of data, in particular the chance to predict the collective decision by automatic data classification [5,6], has attracted the interest of marketing and politics.

The automatic classification of human activities is a well-known problem in different research areas [7,8]. In the case of social-network analysis, Sentiment Analysis(SA) techniques [9,10], as well as the study of the dissemination of information [11,12], have been applied to the users belonging to a given network [13,14].

In recent years, some tools providing more specific classifications than the simple positive or negative polarity of the classical SA, have been developed [15–18]. In [19] emotion analysis on brand tweets are conducted using both approaches of SentiWordNet [20] and NRC Hashtag Emotion Lexicon [21], without relying on any a-priori knowledge. On the other hand, it is possible to capture some a-priori knowledge by using a hierarchical classification system, in which first the subjectivity, then the polarity, and finally the particular emotion of a text are detected. In [22], a comparison between hierarchical and flat classification of emotions in text are reported. In [23], a comparison between the two approaches is conducted on two datasets of tweets, coming directly from some Twitter channels without any manually interaction in their composition: the tweets are filtered using an automated procedure. The tweets are categorized according to Parrot's classification [24], in which the number of positive and negative emotions are balanced. In [25], Plutchick's wheel of emotion is used [26] to treat the inherently multi-class problem of emotion detection as a binary problem, for four opposing emotion pairs.

In this work, we have applied the distant supervision method, which has been shown to be an effective way to overcome the need for a large set of manually labeled data to produce accurate classifiers [27,28]. Distant supervision is a semi-supervised method to retrieve noisy data which are used to train traditional supervised systems. In [29] these methods are used to remove noisy data from automatically generated datasets of text (mentions) with good results. A survey of dataset pruning methods for distant supervision in sentiment analysis is exposed in [30].

3   Methodology

The performances of an automatic system for emotion analysis are mainly affected by the quality of the dataset used to train it. A few publicly available, reliable and manually annotated datasets, to be used for sentiment analysis of tweets, are described in the scientific literature, but they only address valence (polarity) classification. As a result, for the purpose of creating an emotional classifier based on Parrot's model of six basic emotions, we had to create our own training set. Given the high costs required for manually annotating a training set, we decided to use a distant supervision approach. This approach was easily implemented because different users of Twitter tend to label their emotional states with specific hashtags. In the following section we describe which hashtags we have chosen and how we have used the Twitter REST API for downloading the instances of our training set. The distant supervision approach, as mentioned before, has the advantage of allowing the collection of a dense training set, in short time. However, its main disadvantage is the lack of control over the way people decide to label their tweets, resulting in noisy data. Because of this, we have combined distant supervision with an automatic dataset pruning technique, that will be described in the following sections. In order to evaluate the effectiveness of the dataset pruning phase we trained, in the same way but with different training sets, a number of seven-outputs "flat" classifiers:

- **Raw classifier**: trained on the training set collected **using distant supervision without applying dataset pruning**;
- **A set of improved classifiers**: trained on the training sets obtained from the **dataset pruning phase** executed with different thresholds;

and compared them on the same manually-annotated test set. We underline the importance of having a manually-annotated test set, in order to actually measure the validity of our approach.

3.1   Training Set Creation

To implement the distant supervision approach, we used the Twitter REST API for downloading tweets containing some given hashtags, corresponding to Parrot's primary sentiments, and other terms selected by an empirical study of tweets we had collected tweets. In creating this dataset we relied on the fact that Twitter's users, when expressing emotions, add specific hashtags corresponding to their emotions. In order to identify the most popular hashtags used to express a given sentiment, we downloaded a set of tweets and manually searched among them for hashtags used in a consistent way. We decided to download more hashtags for each emotion to represent all possible different facets. The selected hashtags for each emotion are presented in Table 1.

Table 1. Hashtags selected for each sentiment.

Sentiment	Hashtags
Joy	#joy, #happiness, #happy, #joyful, #blessed, #smile, #goodvibes, #proud
Love	#love, #loveofmylife, #fiance
Surprise	#surprisesurprise, #wtf, #omg
Anger	#fuckyou, #pissedoff, #angry, #furious, #fuckoff, #annoyed, #stfu
Sadness	#sad, #sadness, #sosad, #disappointed
Fear	#terror, #scared

Since the objective class is considered in the task of polarity classification, and considering that there are publicly available datasets for this field, we have decided to collect the instances relative to the "objective" class from these sets. The datasets we have chosen is "SemEval-2013 Task #2": Sentiment Analysis on Twitter [31] and "Emotweet-28" [1].

With the name of **raw training set** we will refer to the set of tweets downloaded using the hashtags presented in Table 1, those collected for the objective class and the corresponding labels obtained as previously described. It has been essential to proceed with a pre-processing stage:

- tweets are cleared from elements with no emotional meaning, such as hashtags, user references, punctuation or retweet information;
- tweets are cleared from links;
- repetition of tweets are removed;
- emoticons and contractions are replaced with their textual extension;
- keys used to download the tweets are removed;

After these operations, the raw training set is composed by 42533 instances equally distributed whithin each class.

In the following subsections, we will present the approach used to train our classifiers and the dataset pruning technique we used. Starting from the raw training set, we will describe the algorithms and tools used to derive our raw classifier. Then, we will describe how this classifier has been used to derive the **filtered training sets**, starting from the raw training set.

3.2   Classification

In this work, in order to evaluate the validity of the dataset pruning technique, we have obtained different classifiers: one from the raw training set and many others from the filtered training set, in a scheme that we familiarly call "dogfood learning". In fact, following the "eat your own dogfood" principle, the classifier obtained from the raw training set has been then applied to the same initial raw dataset, to filter out the more doubious instances automatically. As will be described later, we have been able to filter out doubious instances at different levels, and hence obtain a different classifier for each of the "cleaning levels" considered. All classifiers have been trained with the same approach, which is described in detail in this section.

Our classifiers have been trained using the *Naive Bayes Multinomial* algorithm (in particular, the implementation provided by Weka). In order to define the features of our training set, we have used the *String to Word Vector* algorithm, that turns a string into a set of attributes representing word occurrences. However, it is important to use not only uni-grams (single word) but to extend the representation to n-grams (set of maximum "n" words). To select the features that are more relevant for our training sets, we have used the *Information Gain* algorithm (in particular, the implementation provided by Weka).

For each training set, a preliminary phase has been dedicated to optimizing the parameters representing the number of features and n-grams to be used. We started from a grid of pairs *(n-grams, number of features)* and used *cross-validation* to estimate the quality of classifiers configured with the parameters defined by these pairs. Then, we used the pair that returned best results. Figure 1 shows the case of the **raw classifer**; it can be noted that accuracy peak corresponds to n-grams = 2 and number of features = 6760.

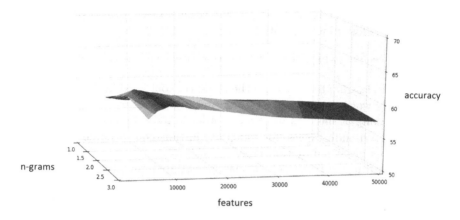

Fig. 1. Parameters optimization.

3.3 Dataset Pruning

The basic assumption underlying our dataset pruning scheme is that the most uncertain instances, contained in the raw training set, represent only a fraction of the ones that are correctly classified. This hypothesis has been considered true since the results of the work described in [32] show that the instances obtained by distant supervision have similar quality to annotations of trained human judges.

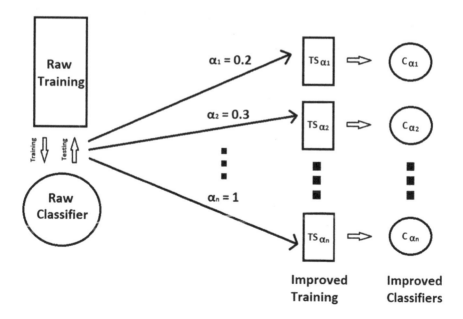

Fig. 2. Representation of the dataset pruning scheme.

Figure 2 summarizes the dataset pruning scheme used in this work, which is composed of the following sequential steps:

1. Train a classifier (which we call the "raw classifier") using the whole raw training set.
2. Use the raw classifier to classify all the instances contained in the raw training set. In other words, use the raw training set in place of a test set.
3. Save the results of this classification in order to have, for each instance:
   - The corresponding class, for which the instance has been downloaded;
   - The class predicted by the classifer;
   - The **confidence factor** of the classifier in predicting the class.
     The Naive Bayes algorithm classifies a given instance based on the class with the maximum posterior probability distribution given the observation. So this probability is used as confidence factor in predicting a given class.
4. Remove, from the previously saved dataset, the incorrectly classified instances.
5. Obtain different training sets $TS_{\alpha 1}$, $TS_{\alpha 2}$, ..., $TS_{\alpha n}$ by applying a variable threshold α from 0.2 to 1. The threshold is used to remove all the instances that have been classified correctly but with a confidence factor lower than the threshold value.

We have decided to produce also another training set obtained, just by removing the incorrectly classified instances, without applying any threshold to the correctly classified instances: this training set is called T_0.

All the resulting training sets have been used to train different classifiers $C_{\alpha 1}$, $C_{\alpha 2}$, ..., $C_{\alpha n}$, whose parameters have been selected by the optimization process previously described. Table 2 shows the parameters n-gram and number of features used for each classifiers.

Table 2. Parameters optimization results.

Classifier	N-Gram (max)	Features
Raw	2	6760
C_0	2	4800
$C_{0.2}$	2	4800
$C_{0.3}$	2	4800
$C_{0.4}$	2	4800
$C_{0.5}$	2	4760
$C_{0.6}$	2	4760
$C_{0.7}$	2	4840
$C_{0.8}$	2	4760
$C_{0.9}$	2	4400
C_1	2	4000

4   Results

In this section we present the results obtained, on a common test set, by the different classifiers. Since all the classifiers have been trained in the same way but with different training sets, we have been able to assess the effectiveness of the dataset pruning technique introduced and to evaluate which threshold allows one to obtain the best performance.

4.1   Test Set

The choice of the test set is a critical element, for evaluating the performance of a classifier. In this work we have derived a test set from the EmoTweet-28 dataset [1]. This dataset consists of tweets manually classified according to 28 different emotions. Since, for our work, we only need a subset of these emotions, we have defined some classes of EmoTweet-28 emotions, that can be associated to each of our primary sentiments. In Table 3 we summarize this process:

Table 3. EmoTweet-28 classes used as representative of Parrot's primary sentiments. The tweets corresponding to classes of EmoTweet-28 not reported in the table have not been included in the test set.

Macro-categories	EmoTweet-28 emotions
Joy	"Amusement", "Excitement", "Happiness", "Inspiration", "Pride"
Love	"Fascination", "Love"
Surprise	"Surprise"
Anger	"Anger", "Hate", "Jealousy"
Fear	"Fear"
Sadness	"Sadness", "Regret", "Sympathy"
Objective	"none"

Since many of these tweets are labeled with more than one emotion, we decided to maintain only the tweets with associated emotions of the same macro-categories according to Table 3. Further, we applied the pre-processing stage as described in Sect. 3.1. Finally, as mentioned in the previous chapter, considered the large amounts of objective tweets, we decided to remove some of these from the test set and insert them in the raw training set.

As result, the test set has 10499 instances subdivided for each class as follows:

- Joy: 2781;
- Love: 447;
- Surprise: 15;
- Anger: 1221;
- Sadness: 98;
- Fear: 204;
- Objective: 5733

4.2 Accuracy

In this section we compare the results obtained on the test set, previously described, by the original raw classifier and the improved classifiers.

Figure 3 shows the accuracy for each classifier. The raw classifier has an accuracy of 39,00% and all the other classifiers, obtained using the different filtered training sets, improve the accuracy to some degree. More in detail, note that even the classifier C_0, from which only wrongly classified instances have been removed and no threshold application, allows to boost the accuracy of the results.

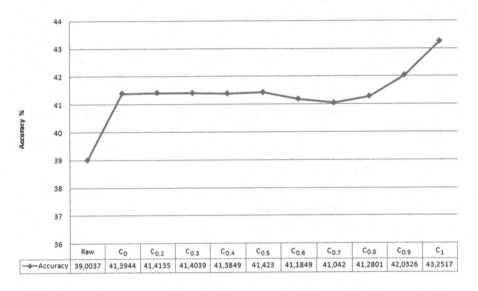

Fig. 3. Visualization of the accuracy obtained by the different classifiers on the given test set.

In order to have a better feeling of the performances of the classifiers, we present in Fig. 4 the F-measures obtained by each classifier. The figure shows that the impact of the data pruning technique is not the same on all classes, possibly because of the different average certainty degree of the different classes, which may cause the filter to alter the balance of the original dataset. However if one considers the average F-measure over the 7 classes, a steady increment in the global performance can be observed. It should be noticed that this measure is independent of the a priori distribution of the test data among the 7 classes.

At this point two important observation have to be made:

- Even if the C_1 classifier produces a small increment of the F-measure in relation to the "surprise" class, the low F-measures of the class **Surprise**, obtained by all the classifiers, are probably related to the lack of a suitable

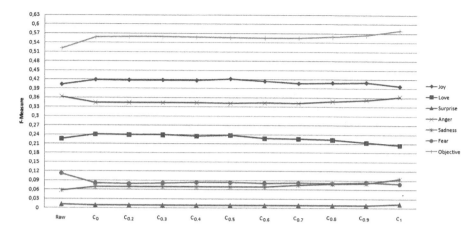

Fig. 4. F-measure trends for each classifier.

number of instances in the test set. EmoTweet-28 contains many tweets associated with the Surprise label; however, many of these were ignored since Surprise was not the only label assigned to them. This caused very few instances of that classes to be included in the test set.

- The improved classifiers couldn't obtain an increment of the "fear" F-measure. The reason is probably releted to the fact that Twitter users are hesitant on sharing their real fears. It follows that the distant supervision approach is not effective with this type of class. So, the reduction of the trend can be explained by the fact that, for this particular class, the hypothesis of applicability of our dataset pruning technique are not verified, since the percentage of spurious instances is superior to the percentage of correct ones.

5   Conclusion

The automatical analysis of the social network users' emotional state is of increasing importance. For the creation of a classifier for emotion detection, it is of utmost importance to collect a proper training set with low costs and efforts. In this work we propose an approach for automatically deriving a training set from Twitter, using a distant supervision approach combined with a dataset pruning technique. Even if it has been proven that training sets obtained with distant supervision correspond well to annotation of human judges [32], in this paper we show that is possible to increment the quality of the training set using a simple and automated dataset pruning technique.

References

1. Yan, J.L.S., Turtle, H.R., Liddy, E.D.: EmoTweet-28: a fine-grained emotion corpus for sentiment analysis. In: Proceedings of the 10th International Conference on Language Resources and Evaluation. LREC 2016, pp. 1149–1156 (2016)
2. Franchi, E., Poggi, A., Tomaiuolo, M.: Social media for online collaboration in firms and organizations. Int. J. Inf. Syst. Model. Des. (IJISMD) **7**, 18–31 (2016)
3. Sani, L., et al.: Efficient search of relevant structures in complex systems. In: Adorni, G., Cagnoni, S., Gori, M., Maratea, M. (eds.) AI*IA 2016. LNCS (LNAI), vol. 10037, pp. 35–48. Springer, Cham (2016). https://doi.org/10.1007/978-3-319-49130-1_4
4. Amoretti, M., Ferrari, A., Fornacciari, P., Mordonini, M., Rosi, F., Tomaiuolo, M.: Local-first algorithms for community detection. In: 2nd International Workshop on Knowledge Discovery on the WEB, KDWeb 2016 (2016)
5. Ducange, P., Pecori, R., Mezzina, P.: A glimpse on big data analytics in the framework of marketing strategies. Soft. Comput. **21**, 1–18 (2017)
6. Bollen, J., Mao, H., Pepe, A.: Modeling public mood and emotion: Twitter sentiment and socio-economic phenomena. In: ICWSM, vol. 11, pp. 450–453 (2011)
7. Ugolotti, R., Sassi, F., Mordonini, M., Cagnoni, S.: Multi-sensor system for detection and classification of human activities. J. Ambient Intell. Humaniz. Comput. **4**, 27–41 (2013)
8. Matrella, G., Parada, G., Mordonini, M., Cagnoni, S.: A video-based fall detector sensor well suited for a data-fusion approach. In: Assistive Technology from Adapted Equipment to Inclusive Environments. Assistive Technology Research Series, vol. 25, pp. 327–331 (2009)
9. Liu, B.: Sentiment analysis and opinion mining. Synth. Lect. Hum. Lang. Technol. **5**, 1–167 (2012)
10. Mohammad, S.M.: Sentiment analysis: detecting valence, emotions, and other affectual states from text. In: Emotion Measurement (2015)
11. Fornacciari, P., Mordonini, M., Tomauiolo, M.: Social network and sentiment analysis on Twitter: towards a combined approach. In: 1st International Workshop on Knowledge Discovery on the WEB, KDWeb 2015 (2015)
12. Mislove, A., Lehmann, S., Ahn, Y.Y., Onnela, J.P., Rosenquist, J.N.: Pulse of the nation: US mood throughout the day inferred from twitter. Northeastern University (2010)
13. Allisio, L., Mussa, V., Bosco, C., Patti, V., Ruffo, G.: Felicittà: Visualizing and estimating happiness in italian cities from geotagged tweets. In: ESSEM@ AI* IA, pp. 95–106 (2013)
14. Healey, C., Ramaswamy, S.: Visualizing Twitter sentiment (2010). Accessed 17 Jun 2016
15. Strapparava, C., Valitutti, A., et al.: Wordnet affect: an affective extension of wordnet. In: LREC, vol. 4, pp. 1083–1086 (2004)
16. Strapparava, C., Mihalcea, R.: Learning to identify emotions in text. In: Proceedings of the 2008 ACM symposium on Applied computing, pp. 1556–1560. ACM (2008)
17. Poria, S., Cambria, E., Winterstein, G., Huang, G.B.: Sentic patterns: dependency-based rules for concept-level sentiment analysis. Knowl. Based Syst. **69**, 45–63 (2014)

18. Kao, E.C., Liu, C.C., Yang, T.H., Hsieh, C.T., Soo, V.W.: Towards text-based emotion detection a survey and possible improvements. In: 2009 International Conference on Information Management and Engineering, ICIME 2009, pp. 70–74. IEEE (2009)

19. Al-Hajjar, D., Syed, A.Z.: Applying sentiment and emotion analysis on brand tweets for digital marketing. In: 2015 IEEE Jordan Conference on Applied Electrical Engineering and Computing Technologies (AEECT), pp. 1–6. IEEE (2015)

20. Baccianella, S., Esuli, A., Sebastiani, F.: Sentiwordnet 3.0: an enhanced lexical resource for sentiment analysis and opinion mining. In: LREC. Vol. 10, pp. 2200–2204 (2010)

21. Mohammad, S.M., Kiritchenko, S.: Using hashtags to capture fine emotion categories from tweets. Comput. Intell. **31**, 301–326 (2015)

22. Ghazi, D., Inkpen, D., Szpakowicz, S.: Hierarchical versus flat classification of emotions in text. In: Proceedings of the NAACL HLT 2010 Workshop on Computational Approaches to Analysis and Generation of Emotion in Text, pp. 140–146. Association for Computational Linguistics (2010)

23. Angiani, G., Cagnoni, S., Chuzhikova, N., Fornacciari, P., Mordonini, M., Tomaiuolo, M.: Flat and hierarchical classifiers for detecting emotion in tweets. In: Adorni, G., Cagnoni, S., Gori, M., Maratea, M. (eds.) AI*IA 2016. LNCS (LNAI), vol. 10037, pp. 51–64. Springer, Cham (2016). https://doi.org/10.1007/978-3-319-49130-1_5

24. Parrott, W.G.: Emotions in Social Psychology: Essential Readings. Psychology Press, Philadelphia (2001)

25. Suttles, J., Ide, N.: Distant supervision for emotion classification with discrete binary values. In: Gelbukh, A. (ed.) CICLing 2013, Part II. LNCS, vol. 7817, pp. 121–136. Springer, Heidelberg (2013). https://doi.org/10.1007/978-3-642-37256-8_11

26. Plutchik, R., Kellerman, H.: Emotion: Theory, Research and Experience. Academic press, New York (1986)

27. Go, A., Bhayani, R., Huang, L.: Twitter sentiment classification using distant supervision. CS224N Project Report, Stanford 1 (2009)

28. Purver, M., Battersby, S.: Experimenting with distant supervision for emotion classification. In: Proceedings of the 13th Conference of the European Chapter of the Association for Computational Linguistics, pp. 482–491. Association for Computational Linguistics (2012)

29. Intxaurrondo, A., Surdeanu, M., De Lacalle, O.L., Agirre, E.: Removing noisy mentions for distant supervision. Procesamiento del lenguaje natural **51**, 41–48 (2013)

30. Roth, B., Barth, T., Wiegand, M., Klakow, D.: A survey of noise reduction methods for distant supervision. In: Proceedings of the 2013 Workshop on Automated Knowledge Base Construction, pp. 73–78. ACM (2013)

31. Nakov, P., Kozareva, Z., Ritter, A., Rosenthal, S., Stoyanov, V., Wilson, T.: Semeval-2013 task 2: Sentiment analysis in twitter (2013)

32. Mohammad, S.M.: # Emotional tweets. In: Proceedings of the First Joint Conference on Lexical and Computational Semantics - Volume 1: Proceedings of the Main Conference and the Shared Task, and Volume 2: Proceedings of the Sixth International Workshop on Semantic Evaluation. SemEval 2012, pp. 246–255. Association for Computational Linguistics, Stroudsburg (2012)

Forecasting Natural Gas Flows
in Large Networks

Mauro Dell'Amico[1], Natalia Selini Hadjidimitriou[1(✉)], Thorsten Koch[2],
and Milena Petkovic[2]

[1] University of Modena and Reggio Emilia,
Via Amendola 2, Pad. Morselli, 42122 Reggio Emilia, Italy
selini@unimore.it
[2] Zuse Institute Berlin, Takustr. 7, 14195 Berlin, Germany

Abstract. Natural gas is the cleanest fossil fuel since it emits the lowest amount of other remains after being burned. Over the years, natural gas usage has increased significantly. Accurate forecasting is crucial for maintaining gas supplies, transportation and network stability. This paper presents two methodologies to identify the optimal configuration o parameters of a Neural Network (NN) to forecast the next 24 h of gas flow for each node of a large gas network.

In particular the first one applies a Design Of Experiments (DOE) to obtain a quick initial solution. An orthogonal design, consisting of 18 experiments selected among a total of 4.374 combinations of seven parameters (training algorithm, transfer function, regularization, learning rate, lags, and epochs), is used. The best result is selected as initial solution of an extended experiment for which the Simulated Annealing is run to find the optimal design among 89.100 possible combinations of parameters.

The second technique is based on the application of Genetic Algorithm for the selection of the optimal parameters of a recurrent neural network for time series forecast. GA was applied with binary representation of potential solutions, where subsets of bits in the bit string represent different values for several parameters of the recurrent neural network.

We tested these methods on three municipal nodes, using one year and half of hourly gas flow to train the network and 60 days for testing. Our results clearly show that the presented methodologies bring promising results in terms of optimal configuration of parameters and forecast error.

Keywords: Machine learning · Neural networks · Genetic algorithm
Simulated annealing · Design Of Experiments (DOE)
Time series forecast

1 Introduction

Over the years, lower prices and better infrastructure led to a significant increase of natural gas usage in transportation and consumption in residential and

© Springer International Publishing AG 2018
G. Nicosia et al. (Eds.): MOD 2017, LNCS 10710, pp. 158–171, 2018.
https://doi.org/10.1007/978-3-319-72926-8_14

industrial sectors. Accurate forecasting of natural gas flows is crucial for maintaining gas supplies, transportation and network stability.

In the past, several methods have been used to predict daily gas demands [1–3]. The most commonly used among these models are ARIMA models which assume a linear relationship between data. However, in real-life applications a lot of time series do not have linear relationships. Neural Networks (NN) can model both linear and nonlinear relationships. This, as well as the capability of NN to extract the relationship between the inputs and outputs of a process, are the reasons why NNs have become one of the methods frequently used in time series analysis in recent years.

Usually, the process of constructing the NN model for time series forecasting is based on trial and error heuristics. When designing a NN model several parameters combinations (i.e. neurons, training algorithms, delays...) are set. For each of them, training, validation and test errors are obtained, and one of the configuration with better generalization capability is selected to forecast the future values. An improved procedure is to use an automatic NN design with hyperparameters optimization techniques.

The goal of this paper is to develop accurate methods capable to automatically determine hyperparameters of NN for forecasting gas flows in large networks.

The paper is organized as follows. Section 2 presents a literature review on hyperparameters optimization techniques. Section 3 explains two methodologies: Recurrent Neural Network (RNN) with Design Of Experiments (DOE) and Simulated Annealing and RNN with Genetic Algorithm for forecasting the next 24 h of gas flows. Section 4 reports the conclusions.

1.1   Literature Review

Neural Networks have been extensively used to forecast time series. In [4], the authors underline that the application of neural networks to forecast time series started already in the fifties. After a decline in the use of Neural Networks, the introduction of the backpropagation algorithm for network training and weight optimization together with the increased processing power of computers, have created the basis for a wider deployment of Neural Networks for several types of applications. The connection between autoregressive models and neural networks is shown by [6] who demonstrate that NARMA models (Nonlinear Autoregressive Moving Average) are a particular type of recurrent neural network. Since the prediction is influenced by the outliers, they propose a filtering algorithm which is integrated into the training process. In [5], a recurrent neural network is trained with a hybrid algorithm and it is shown that the combination of two global optimization algorithms allows to predict missing values of the time series. They also add that further research should be focused on the parameters settings.

Several approaches have been proposed to find the optimal hyperparameters of Neural Networks. The simplest one is the random search which consists in the initialization of the parameters based on a distribution. After each iteration, the solution is updated until a stop criterion is met. Similarly, the grid search consists of dividing the space of solutions into a grid and to systematically explore each area. If the subset of optimal parameters is known and an acceptance function is defined, [7] showed that the random search performs better than the grid search for classification problems.

There are numerous researchers that proposed Evolutionary NN for different problems (see for instance [21–23]). Several authors use genetic algorithms to optimize the neural networks' hyperparameters (i.e. [11,13–15]). In [10], the initial population consists of a range of parameters initialized with a set of potential solutions. The couples of chromosomes (solutions) are evaluated based on a fitness function. The best chromosomes are then selected by considering the value of the cumulative distribution.

The genetic algorithm has been implemented in combination with the Design of Experiments (DOE). For instance, a hybrid genetic algorithm has been proposed by [9] to optimize a neural network architecture and its hyperparameters. The traditional genetic algorithm is combined with the Taguchi orthogonal design by inserting the result of the DOE between crossover and mutation to select the best genes.

Other approaches to optimize the configuration of NNs are based on the combination of different solutions. For instance, ensembles consists of combining several networks outputs. For prediction problems, NN are often trained individually and the prediction results are averaged or weighted averaged [40]. In [12], the authors proposed a technique which is based on genetic algorithm to select and ensemble several neural networks. This technique consists of training a number of neural networks; assigning random weights to the network and running a genetic algorithm to improve the fitness of the ensemble model. The result is a smaller network with an increased generalization capacity.

Sequential Model-Based Optimization (SMBO) is a methodology that consists of the space of parameters' evaluation based on a function [8]. The configuration space is described as a probability distribution or as a set of discrete values. [16], for instance, combined algorithm selection and hyperparameter optimization by considering the set of Waikato Environment for Knowledge Analysis (WEKA) algorithms [17]. The objective was to identify the optimal combination of algorithms and parameters that minimize the cross-validation loss. Irace [18] is an iterated racing procedure for automated parameters configuration. The method consists of three steps: the sampling of possible configurations according to a distribution, the selection of the best configuration and the update of the distribution based on the best selected solution. SMAC [8] is a sequential-based optimization method for parameter optimization that can be seen as an extension of the iterated racing algorithm. The method improves the intensification procedure and provides the possibility to handle categorical variables and multiple instances. ParamILS [19] is a stochastic local search algorithm for parameter configuration. The general framework consists of starting with an initial configuration, modifying it and accepting the new configuration as soon as the performance improves. Finally, CALABRIA [20] combines the Taguchi design with the local search to narrow the space of parameters.

The next section describes the dataset deployed to forecast gas flow, the network and how the forecast results are evaluated.

1.2   The Data Set

In this paper we observed a data set of hourly gas flow time series from three Municipal nodes with 17.520 h, that is twenty three months, from 1/10/2013 to 30/9/2015. The objective is to forecast the demand for the next 24 h to plan the optimal distribution of flows over a large gas network. We used one year and 9 months (1/10/2013–25/7/2015) to train the network and we performed the hourly daily forecast for each of the 60 remaining days (26/7/2015–30/9/2015).

1.3   Input Features

The network receives a fixed set of features: gas flow at the same time of the previous day, flow at the same time of the second previous day, .., and so on, up to the flow at the same time of same day of the previous week. The average flow registered at the same time in the previous seventh days is also computed and used as input to the model. Finally, since the gas flow of the three nodes is influenced by the temperature and its use depends on whether it is holiday or weekend, these information are included in the set of input features. The input data are normalized and the predictions are transformed back to evaluate the forecast results.

1.4   The Network

The network that we want to optimize is a simple Recurrent Neural Network. This type of network was developed with the aim to model the structure of time and it was introduced by [35]. More specifically, it is a layer recurrent neural network with a similar structure as the feedforward neural network with the difference that it includes tap delays that connect past time steps to the current output. The number of past time steps to be deployed to forecast the next 24 h is one of the parameters that is optimized.

1.5   Evaluation

The evaluation is based on the Root Mean Square Error (RMSE) that measures the difference between the predicted and observed values. The network produces the 24-h forecast based on the previous data. The average daily RMSE is computed for the 24-h forecast for each of the 60 days and the average forecast is recorded. The cross validation is performed on the 30% of the dataset.

2   Recurrent Neural Network (RNN) with Design of Experiments (DOE) and Simulated Annealing

The Design Of Experiments (DOE) allows to separate and study the main effects of the different components on the final outcome of a process or system [28]. The development of this methodology has been guided by the industry as there was the need for new strategies aimed at reducing the number of experiments to determine the optimal combinations of parameters. The DOE is an iterative process which consists in the execution of a subset of experiments to move towards the optimal condition based on the analysis of the average and variance of the experiments' responses.

The most common method of DOE is called fractional factorial design defined as the methodology that allows reducing the size of the experiment by limiting the loss of critical information [29]. The basic principle of these experiments relies on the possibility to estimate the effect of each factor independently from the others and this is possible when the design is orthogonal. [25] describe an orthogonal array of m factors of level s and N experiments, $OA_N(s^m)$, as a matrix $N * m$ in which each factor can be evaluated independently. In the case of mixed levels design, an orthogonal array is denoted by $OA_N(s^m * t^n)$ where s and t $(s \neq t)$ denote the cardinality of two different levels' set and n and m $(n \neq m)$ indicate the number of factors for each level.

Taguchi has proposed a set of orthogonal designs in order to further reduce the number of experiments needed to identify the optimal combination of parameters (see [24,27,39]). The main concept introduced by Taguchi is the loss function which expresses the cost of quality loss. The hypothesis is that the losses are caused by the deviation from the target, therefore:

- losses are 0 at the target values;
- when the deviance from the target increases, the loss also increases.

Practically, any deviation from the target causes an increase of loss. There is, therefore, a margin of tolerability which depends on the type of problem.

DOE have already being used to configure NNs. In particular, [26] applied the Taguchi method to optimize a Feedforward Neural Network by varying several factors such as the number of hidden layers, the number of neurons, the size of the training set and the learning rate.

2.1   The Experiment

This work presents three experiments executed on three different nodes of a gas network according to the Taguchi's L_{18} $(3^6 * 6^1)$ mixed levels orthogonal array consisting of one factor with six levels (6^1), six factors with three levels (3^6) and 18 experiments. Each factor is a parameter of the Neural Network. The levels are fixed values assigned to the parameters.

The experiment runs 18 separate trial conditions selected among a total of 4.374 combinations of seven parameters (training algorithm, transfer function, regularization, learning rate, lags, and epochs). The average daily RMSE of the 24-h forecast for 60 days is recorded. The selection of the design is based on the hypothesis that the train function is the main source of variability of a Neural Network, as such more training functions should be tested. The other parameters (transfer function, number of hidden neurons, lags, etc.) were selected based on the experience and the existing literature.

Description of the Selected Factors. Training a neural network means finding the weights and biases such that, given a set of input values, the output of the network is as close as possible to the target values. The most common algorithm for training a neural network is the backpropagation. According to this approach, the error is minimized by computing the partial derivatives of the error function with respect to the weights. This allows computing the steepest descent, the sign of which provides an indication of how much and towards which direction the weights have to be updated such that the difference between the output and the target is minimized. As noted by [30], the backpropagation algorithm does not guarantee the convergence to the global optimum because complex problems may have several local optima, it is difficult to understand how the algorithm behaves with different number of neurons and layers and the convergence is often slow. For this reason, there is not a standard technique that guides in the selection of the most appropriate training function.

In the context of the selected design (L_{18} ($3^6 * 6^1$)), six different training functions have been included in the experiment. Five of them are based on the backpropagation algorithm:

- The Levenberg-Marquardt backpropagation (LM) training process combines the conjugate gradient until it is closed to the optimum; after that it performs a quadratic approximation according to the Gauss-Newton algorithm [32].
- The Scaled conjugate gradient backpropagation (SCG) combines the Levenberg-Marquardt with the conjugate gradient approach. The Scaled conjugate gradient determines the search direction for the weights; and based on the direction, it defines the step size [31].
- The Resilient Backpropagation (RP) considers only the sign of the partial derivative, without taking into account the size of the weights, to decide the direction of the weights update.
- Gradient Descent backpropagation (GD) adapts the weights and bias according to the gradient descent.
- Gradient Descent with momentum backpropagation (GDM) is a batch steepest descent which utilize the entire training set to compute the gradient of the cost function with reference to the parameters.
- Broyden-Fletcher-Goldfarb-Shanno Quasi-Newton (BFGS) is an iterative method for solving unconstrained nonlinear optimization problem [33].

The factors together with their corresponding levels are shown in Table 1.

The *transfer function* allows the non-linear transformation of the input into the output. The *neurons* are the number of nodes included in each layer of the

Table 1. Factors/Parameters

Level	Training Fnc	Transfer Fnc	Neurons	Lags	Regularization	Learning rate	Epochs
1	LM	logsig	10	2	0.5	0.001	100
2	SCG	purelin	14	3	0.7	0.01	300
3	RP	tansig	18	4	0.9	0.1	600
4	GD	-	-	-	-	-	-
5	GDM	-	-	-	-	-	-
6	BFGS	-	-	-	-	-	-

network, that is the number of non linear transformations of the input into the output based on the transfer function. The *lags* are the number of previous data to be used to forecast the next 24-h. The *regularization* is a parameter that allows avoiding overfitting by removing a certain percentage of connections between neurons. The *learning rate* allows to set the time needed to reach the convergence. If the learning speed is high, the algorithm converges very fast but it could fail finding the optimal solution. Finally, the *epochs* are the presentation of the training set to the network. The higher is the number of epochs, the higher is the accuracy but the higher is the time needed to reach the convergence.

The results of the experimental design for three nodes of the gas network is shown in Table 2 where the 18 combinations of factors and the corresponding RMSE are presented according to the selected Taguchi design.

2.2 Optimal Design with Simulated Annealing

This section presents a procedure to optimize the design with Simulated Annealing. The algorithm, proposed by [34], was inspired by the behaviour of the atoms in thermal equilibrium in case of temperature variation. The algorithm starts with an initial configuration and an initial temperature. At each iteration, the temperature changes and a new candidate configuration is generated by a perturbation of the initial configuration. The difference between the two objective functions is then evaluated. If the value of the objective function improves, the new configuration is accepted, otherwise it is updated.

In our implementation, the objective is to find the optimal combination of NN parameters that minimize the RMSE based on the quick initial solution provided by the DOE. More specifically, the problem consists of minimizing the RMSE and finding the integer variables that specify the value of the NN parameters in terms of their position in the vector of parameters.

To this aim, the range of parameters to be evaluated is extended to 89.100 possible combinations of seven parameters. The levels of each parameter are reported here below:

- Train Function: LM, SCG, RP, BFGS, GDM, GD

Table 2. Results of the DOE

Run	Lags	Neurons	Learn. Rate	Transfer Fnc	Regular	Training Fnc	Epochs	NodeA-RMSE	NodeB-RMSE	NodeC-RMSE
1	3	10	0.01	logsig	0.9	RP	600	**6.9**	98.6	18.8
2	3	18	0.1	logsig	0.5	GDM	300	136.5	493.1	104.5
3	2	10	0.001	logsig	0.5	LM	100	7.4	96.5	**18.4**
4	3	14	0.001	purelin	0.5	BFGS	600	7.3	101.1	20.3
5	4	18	0.01	logsig	0.7	BFGS	100	10.5	103.8	26.5
6	2	10	0.1	tansig	0.9	BFGS	300	8.8	102.6	21.2
7	2	18	0.001	purelin	0.7	RP	300	7.5	101.6	20.0
8	2	14	0.01	purelin	0.9	GDM	100	346.7	3232.9	607.6
9	4	10	0.01	purelin	0.5	SCG	300	7.4	101.0	19.8
10	4	18	0.1	purelin	0.9	LM	600	7.3	103.1	19.9
11	3	14	0.01	tansig	0.7	LM	300	7.2	100.3	18.7
12	2	18	0.01	tansig	0.5	GD	600	73.9	717.3	355.2
13	3	18	0.001	tansig	0.9	SCG	100	7.2	100.9	18.7
14	4	10	0.001	tansig	0.7	GDM	600	69.8	691.3	154.5
15	3	10	0.1	purelin	0.7	GD	100	298.3	1691.6	582.9
16	2	14	0.1	logsig	0.7	SCG	600	7.9	100.0	19.2
17	4	14	0.001	logsig	0.9	GD	300	172.1	2110.4	412.5
18	4	14	0.1	tansig	0.5	RP	100	7.0	**98.1**	19.4

- Transfer Function: logsig, tansig, purelin
- Neurons: 10, 11, 12, 13, 14, 15, 16, 17, 18, 19, 20
- Lags: 2, 3, 4
- Regularization: 0.5, 0.6, 0.7, 0.8, 0.9
- Learning Rate: 0.001, 0.01, 0.1
- Epochs: 100, 200, 300, 400, 500, 600, 700, 800, 900, 1000

The best combination of parameters obtained with the Taguchi design in terms of minimum RMSE is selected as initial solution of the simulated annealing. The lower and upper bounds of each integer variable of the optimization problem is set. Since the value of the variables corresponds to the position in the vector of possible values of the parameters, the upper bound corresponds to the number of levels of each variable.

The algorithm stops if the average change in the objective function, after 300 iterations, is below 1. The best solution of the Taguchi experiment is selected as initial solution.

The results of the simulated annealing for the three nodes are reported in Table 3. For each node, the optimal configuration of parameters is shown together with the value of the objective function, the RMSE. The results indicate that, for all nodes, the optimal number of neurons is 19. Furthermore, the number of

Table 3. Results of SA

Parameters	NodeA	NodeB	NodeC
Training Fnc	LM	RP	SCG
Transfer Fnc	logsig	tansig	tansig
Number of neurons	19	19	19
Lags	2	3	3
Regularization	0.9	0.9	0.9
Learning Rate	0.1	0.01	0.001
Epochs	700	300	600
RMSE	5.99	94.04	18.04

previous data to be deployed to forecast the next 24 h, is limited to 2 or 3 observations (lags). The optimal regularization rate is rather high for all three nodes and it is included between 0.8 and 0.9. The optimal number of epochs ranges between 600 and 700, except for NodeB which uses an RP training function with a learning rate of 0.01.

3 Recurrent Neural Network (RNN) with Genetic Algorithm (GA)

This section describes the application of genetic algorithm for selecting the optimal parameters' configuration of a recurrent neural network for time series forecast. Genetic algorithm (GA) is a robust evolutionary optimization method based on elementary mechanisms of evolution. It was originally proposed by John Holland [36] and further improved by numerous researchers (i.e. [37,38]). This evolutionary algorithm is based on a population of individuals where every individual represents a potential solution. Each candidate solution has a set of properties which can be mutated and altered, and a measure of adjustment, i.e. the value of optimization function. Using operators like selection, crossover and mutation GA attempts, through iterations, to achieve the optimal value of the fitness function.

The proposed GA algorithm for optimal configuration of RNN parameters consists of the following steps:

1. **Setting.** In the first step, the RMSE defined according to Sect. 1.5 is set as objective function for the optimization procedure. The maximum number of iterations and fitness function tolerance (minimal average relative change in the fitness function value over iterations) are set in advance as termination criteria to the values of 20 and 10^{-3} respectively. The population of potential solutions is initialized with 20 individuals. The crossover and mutation rates are set to 0.8 and 0.001 respectively. Finally, the data set is divided into training and test sets according to Sect. 1.2.

2. **Encoding.** The procedure for encoding consists of representing each NN parameter that is going to be optimized with a certain number of bits with fixed length. The main idea is to use binary code for every parameter and its level according to Table 1. Every potential solution consists of 15 bits long string. First three bits are used to identify the training function which has six levels, based on the following relation $(\sum_{i=0}^{n} g(i) * 2^{n-i}) + 1$ where $g(i) = 0$ or 1 and n is number of bits for each parameter. For example, when the value of the first three bits of the string is *000*, the string corresponds to the first level of the training function (LM training function). All the other parameters are expressed by two bits because the possible levels are only three. In case of binary codes that correspond to higher levels of the parameter reported in Table 1, infeasible solutions are discarded as soon they are generated. All the other parameters are encoded with 2 bits since they have 3 levels.

3. **Decoding and training the network.** The first generation of potential solutions, expressed as string of bits, are decoded into the parameter of the network reported in Table 1.

4. **Genetic operations.** For every potential solution in the current iteration, the value of the objective function is calculated. The reproduction, crossover and mutation are then performed to create the new generation. In the reproduction, two strategies are applied: elite and selection. In elite strategy, potential solutions with the best value of the objective function are stored as elite solutions among all the others in that iteration. For the individual elite, all bits are kept unchanged in the next iteration. In the selection strategy, some candidate solutions with better objective function value are chosen for crossover. In the crossover, two of all candidate solutions are randomly selected and some bits of each parent individual are interchanged. Mutation means that randomly selected bits of parent individual change their values in order to avoid that search ends in local optima. If one of the termination criteria is satisfied, the procedure terminates; otherwise the algorithm is repeated from step 4.

Table 4. GA results

Parameters	NodeA	NodeB	NodeC
Training Fnc	LM	LM	LM
Transfer Fnc	tansig	tansig	logsig
Number of neurons	18	18	18
Lags	2	2	3
Regularization	0.7	0.9	0.5
Learning rate	0.01	0.01	0.01
Epochs	600	600	600
RMSE	6.6	101.09	18.05

Table 5. Robustness of the solutions

Node	Method	Training Fnc	Transfer Fnc	Neurons	Lags	Regularization	Learning rate	Epochs	RMSE
NodeA	DOE - SA	trainlm	logsig	19	2	0.9	0.1	700	6.7
		trainrp	tansig	19	3	0.9	0.01	300	7.0
		trainscg	tansig	19	3	0.9	0.001	600	7.1
	GA	trainlm	tansig	18	2	0.7	0.01	600	6.8
		trainlm	tansig	18	2	0.9	0.01	600	6.6
		trainlm	logsig	18	3	0.5	0.01	600	6.5
NodeB	DOE - SA	trainlm	logsig	19	2	0.9	0.1	700	99.1
		trainrp	tansig	19	3	0.9	0.01	300	97.1
		trainscg	tansig	19	3	0.9	0.001	600	97.1
	GA	trainlm	tansig	18	2	0.7	0.01	600	97.4
		trainlm	tansig	18	2	0.9	0.01	600	100.7
		trainlm	logsig	18	3	0.5	0.01	600	97.6
NodeC	DOE - SA	trainlm	logsig	19	2	0.9	0.1	700	18.5
		trainrp	tansig	19	3	0.9	0.01	300	18.8
		trainscg	tansig	19	3	0.9	0.001	600	19.2
	GA	trainlm	tansig	18	2	0.7	0.01	600	19.2
		trainlm	tansig	18	2	0.9	0.01	600	18.8
		trainlm	logsig	18	3	0.5	0.01	600	18.7

The optimal configuration of parameters and value of the RMSE obtained with the genetic algorithm are reported in Table 4. The results of the GA are very similar to the ones obtained with the SA both in terms of optimal configuration and RMSE. The obtained results, one side confirm that the considered range of parameters lead to a certain amount of errors. In this particular case, the prediction errors also depend on the dimension of the gas flow (gas flow of NodeB is higher compared to gas flow of NodeA). On the other side, the results evidences that additional characteristics should be considered when looking for the optimal configuration of a neural network such as the selection of the input features and the evaluation of different neural network architectures.

4   Conclusion

In the present work, two different methodologies are proposed to optimize the configuration of the neural network parameters for time series forecast. The DOE allows to perform a subset of experiments based on the selection of an orthogonal design. The results of the experiment are deployed as initial solution to optimize the design with the Simulated Annealing.

The second methodology is based on the implementation of a Genetic Algorithm to select optimal combination of parameters of a recurrent neural network. The parameters are represented using strings of bits and are the same parameters deployed for the design of experiment.

Based on the results obtained thanks to the application of the two techniques and the evaluation of the three similar nodes of the gas network, it is possible to make some considerations on the selection of parameters. For instance, it is generally recommended to set the number of neurons equal to the number of input features, while the results clearly show that the optimal number of neurons is higher for all three nodes. All optimal configurations are tested on all nodes and the results are shown in Table 5. The table shows that the solutions found by the two approaches are robust as there is a little variation in terms of RMSE.

The main limitation of the two approaches relies on the time needed to run the algorithms which includes the time needed to train 60 times the network. The final RMSE is, in fact, an average of 60 days 24-h forecast errors. This increases substantially the running time that also depends on the parameter setting. For instance, when low values of the learning rates (i.e. 0.001) are set, the running time further increases.

Future research may introduce in the optimization process a methodology to select features with the final aim to identify the features with better prediction capability.

With reference to the performance of the two algorithms, the results show that the Simulated Annealing provides different training functions for the three nodes. The GA provides, instead, more stable configuration in terms of parameters selection. This may be an indication of the fact that the GA evaluates less solutions compared to the SA or that the solution provided by the GA is more robust.

Future research may consider the comparison of the two proposed approaches with more mature configuration optimizers. Furthermore, more complex neural network architectures might be considered as well as a wider range of parameters to be optimized.

Acknowledgments. The authors would like to acknowledge networking support by the COST Action TD1207.

References

1. Marx, B.: Fitting a continuous profile to hourly natural gas flow data, Ph.D. thesis, Marquette University Department of Electrical and Computer Engineering, Milwaukee, WI (2007)
2. Lyness, F.K.: Gas demand forecasting. Statistician **33**(1), 9–21 (1984)
3. Lim, H.L., Brown, R.H.: Hourly gas load forecasting model input factor identification using a genetic algorithm. In: Proceedings of 44th IEEE Midwest Symposium on Circuits and Systems, Dayton, OH, pp. 670–673 (2001)
4. Zhang, G., Patuwo, B.E., Hu, M.Y.: Forecasting with artificial neural networks: the state of the art. Int. J. Forecast. **14**, 35–62 (1998)
5. Cai, X., Zhang, N., Venayagamoorthy, G.K., Wunsch, D.C.: Time series prediction with recurrent neural networks trained by a hybrid PSO-EA algorithm. Neurocomputing **70**(13), 2342–2353 (2007)
6. Connor, J.T., Martin, R.D., Atlas, L.E.: Recurrent neural networks and robust time series prediction. IEEE Trans. Neural Netw. **5**(2), 240–254 (1994)

7. Bergstra, J., Bengio, Y.: Random search for hyper-parameter optimization. J. Mach. Learn. Res. **13**, 281–305 (2012)
8. Hutter, F., Hoos, H.H., Leyton-Brown, K.: Sequential model-based optimization for general algorithm configuration. In: Coello, C.A.C. (ed.) LION 2011. LNCS, vol. 6683, pp. 507–523. Springer, Heidelberg (2011). https://doi.org/10.1007/978-3-642-25566-3_40
9. Tsai, J.-T., Chou, J.-H., Liu, T.-K.: Tuning the structure and parameters of a neural network by using hybrid Taguchi-genetic algorithm. IEEE Trans. Neural Netw. **17**(1), 69–80 (2006)
10. Leung, Y.W., Wang, Y.: An orthogonal genetic algorithm with quantization for global numerical optimization. IEEE Trans. Evol. Comput. **5**(1), 41–53 (2001)
11. Yao, X.: Evolving artificial networks. Proc. IEEE **87**, 1423–1447 (1999)
12. Zhou, Z.-H., Wu, J., Tang, W.: Ensembling neural networks: many could be better than all. Artif. Intell. **137**(1), 239–263 (2002)
13. Koza, R., Rice, J.P.: Genetic generation of both the weights and architecture for a neural network. In: Proceedings of International Joint Conference on Neural Networks, Seattle, WA, pp. 397–404 (1991)
14. Richards, N., Moriarty, D.E., Miikkulainen, R.: Evolving neural networks to play go. Appl. Intell. **8**, 85–96 (1997)
15. Curran, D., O'Riordan, C.: Applying Evolutionary Computation to Designing Neural Networks: A Study of the State of the Art, Department of Information Technology, National University of Ireland, Galway, Ireland, Technical report NUIG-IT-111 002 (2002)
16. Kotthoff, L., Thornton, C., Hoos, H., Hutter, F., Leyton-Brown, K.: Auto-WEKA 2.0: Automatic model selection and hyperparameter optimization in WEKA. J. Mach. Learn. Res. **18**(25), 1–5 (2017)
17. Eibe, F., Hall, M.A., Witten, I.H.: The WEKA Workbench. Online Appendix for Data Mining: Practical Machine Learning Tools and Techniques, 4th edn. Morgan Kaufmann, San Francisco (2016)
18. López-Ibáñez, M., Dubois-Lacoste, J., Pérez Cáceres, L., Stützle, T., Birattari, M.: The irace package: iterated racing for automatic algorithm configuration. Oper. Res. Perspect. **3**, 43–58 (2016)
19. Hutter, F., Holger, H.H., Leyton-Brown, K., Stützle, T.: ParamILS: an automatic algorithm configuration framework. J. Artif. Res. **36**, 267–306 (2009)
20. Adenso-Díaz, B., Laguna, M.: Fine-tuning of algorithms using fractional experimental designs and local search. Oper. Res. **54**(1), 99–114 (2006)
21. Cortez, P., Rocha, M., Neves, J.: Time series forecasting by evolutionary neural networks. In: Rubuñal, J., Dorado, J. (eds.) Artificial Neural Networks in Real-Life Applications, Chapter III, pp. 47–70. Idea Group Publishing, Hershey (2006)
22. Niska, H., Hiltunena, T., Karppinenb, A., Ruuskanena, J., Kolehmainena, M.: Evolving the neural network model for forecasting air pollution time series. Eng. Appl. Artif. Intell. **17**(2), 159–167 (2004)
23. Chena, Y.-H., Chang, F.-J.: Evolutionary artificial neural networks for hydrological systems forecasting. J. Hydrol. **367**(1–2), 125–137 (2009)
24. Taguchi, G., Chowdhury, S., Taguchi, S.: Robust Engineering. McGraw Hill, New York (1999)
25. Bose, R.C., Bush, K.A.: Orthogonal arrays of strength two and three. Ann. Math. Stat. **23**, 508–524 (1952)
26. Khaw, J.F.C., Lim, B.S., Lim, L.E.N.: Optimal design of neural networks using the Taguchi method. Neurocomputing **7**(3), 225–245 (1995)

27. Kackar, R.N., Lagergren, E.S., Fillben, J.J.: Taguchi's orthogonal arrays are classical designs of experiments. J. Res. Natl. Inst. Stand. Technol. **96**, 577–591 (1991)
28. Montgomery, D.C.: Experimental design for product and process design and development. J. R. Stat. Soc. Ser. D (The Statistician) **48**(2), 159–177 (1999)
29. Gunst, R.F., Mason, R.L.: Fractional factorial design. WIREs Comput. Stat. **1**, 234–244 (2009)
30. Sanger, T.D.: Optimal unsupervised learning in a single-layer linear feedforward neural network. Neural Netw. **2**, 459–473 (1989)
31. Møller, M.F.: A scaled conjugate gradient algorithm for fast supervised learning. Neural Netw. **6**(4), 525–533 (1993)
32. Yu, H., Wilamowski, B.M.: Levenberg-Marquardt Training, in Industrial Electronics Handbook. 5 Intelligent Systems. CRC Press, Boca Raton (2011). 12-1–12-15
33. Fletcher, R.: Practical Methods of Optimization, 2nd edn. Wiley, New York (1987)
34. Kirkpatrick, S., Gelatt, C.D., Vecchi, M.P.: Optimization by simulated annealing. Science **220**(4598), 671–680 (1983)
35. Elman, J.L.: Finding structure in time. Cogn. Sci. **14**, 179–211 (1990)
36. Holland, J.H.: Adaptation in Natural and Artificial Systems. MIT Press, Cambridge (1992)
37. Haupt, R., Haupt, S.E.: Practical Genetic Algorithms. Wiley, New York (1998)
38. Michalewicz, Z.: Genetic Algorithms + Data Structures = Evolution Programming, 3rd edn. Springer, Berlin (1999)
39. Roy, R.K.: A Primer on Taguchi Method. Van Nostrand Reinhold, New York (1990)
40. Krogh, A., Vedelsby, J.: Neural network ensembles, cross validation and active learning. In: Tesauro, G., Touretzky, D.S., Leen, T.K. (eds.) Proceedings of the 7th International Conference on Neural Information Processing Systems (NIPS 1994), pp. 231–238. MIT Press, Cambridge (1994)

A Differential Evolution Algorithm to Semivectorial Bilevel Problems

Maria João Alves[1,3(✉)] and Carlos Henggeler Antunes[2,3]

[1] CeBER and Faculty of Economics, University of Coimbra, Coimbra, Portugal
mjalves@fe.uc.pt
[2] DEEC, University of Coimbra, Polo 2, Coimbra, Portugal
ch@deec.uc.pt
[3] INESC Coimbra, Coimbra, Portugal

Abstract. Semivectorial bilevel problems (SVBLP) deal with the optimization of a single function at the upper level and multiple objective functions at the lower level of hierarchical decisions. Therefore, a set of nondominated solutions to the lower level decision maker (the follower) exists and should be exploited for each setting of decision variables controlled by the upper level decision maker (the leader). This paper presents a new algorithmic approach based on differential evolution to compute a set of four *extreme* solutions to the SVBLP. These solutions capture not just the optimistic vs. pessimistic leader's attitude but also possible follower's reactions more or less favorable to the leader within the lower level nondominated solution set. The differential evolution approach is compared with a particle swarm optimization algorithm. In this experimental comparison we draw attention to pitfalls associated with the interpretation of results and assessment of the performance of algorithms in SVBLP.

Keywords: Semivectorial bilevel problems · Differential evolution
Particle swarm optimization · Optimistic/pessimistic frontiers
Optimistic/deceiving solutions · Pessimistic/rewarding solutions

1 Introduction

A semivectorial bilevel problem (SVBLP) is an optimization problem with a single objective function at the upper (leader's) level and multiple objective functions at the lower (follower's) level of hierarchical non-cooperative decisions. Hence, a multiobjective (MO) optimization problem contributes to define the feasible region to the leader's problem, in the sense that a lower level nondominated region exists for each setting of upper level variables. Thus, when solving his/her optimization problem, the leader must anticipate the follower's choice of a nondominated solution embodying a trade-off between the lower level multiple objectives. The follower's reaction may strongly affect the leader's optimal solution, depending on the follower's preference

© Springer International Publishing AG 2018
G. Nicosia et al. (Eds.): MOD 2017, LNCS 10710, pp. 172–185, 2018.
https://doi.org/10.1007/978-3-319-72926-8_15

structure vis-à-vis the nondominated region established by the instantiation of the leader's decision variables. Therefore, it is useful for the leader to have an overview of possible optimal solutions resulting from different attitudes (optimistic or pessimistic) in face of his/her expectation of the more or less favorable follower's choice. In addition to the intrinsic theoretical and computational difficulty in computing solutions to the SVBLP, the leader does not have a-priori information about the nondominated solution the follower will choose according to his/her (unknown) preferences.

In this setting, this paper presents an algorithmic approach intertwining single and MO versions of Differential Evolution (DE) for the upper level and lower level problems, which is aimed at computing a set of extreme solutions to the SVBLP. These extreme solutions are: the *optimistic* solution offering the leader the best objective function value when the follower's decision for each setting of upper level variables is the best for the leader; the *deceiving* solution when the leader adopts an optimistic approach but the follower's reaction is the worst for the leader; the *pessimistic* solution offering the best objective function value for the leader when the follower's decision for each setting of upper level variables is the worst for the leader; and the *rewarding* solution when the leader adopts a pessimistic approach but the follower's reaction is the most favorable to the leader.

The algorithmic approach introduces new concepts of optimistic and pessimistic frontiers and adapts DE mechanisms to combine the search at both levels with the population split between orientations to each frontier. This approach is compared with a Particle Swarm Optimization (PSO) algorithm we have previously developed [1], which has been extended herein to compute these four extreme solutions. The algorithms are tested on a set of benchmark problems for multiobjective bilevel (MOBL) optimization (considering only one of the objective functions in the upper level). We were able to determine analytically the exact solutions to these problems, which enable to assess the quality of the solutions obtained by the algorithms. A thorough analysis of the computational results allowed us to unveil pitfalls associated with the interpretation of results and assessment of the algorithm performance in SVBL and MOBL optimization. This paper also aims at drawing attention to these pitfalls.

In Sect. 2, the SVBLP is presented and the definitions of the extreme (*optimistic*, *deceiving*, *pessimistic* and *rewarding*) solutions are introduced. Algorithmic approaches to deal with the SVBLP are also briefly reviewed in this section. The concepts of optimistic and pessimistic frontiers are presented and illustrated in Sect. 3. In Sect. 4, the Semivectorial Bilevel Differential Evolution (SVBLDE) algorithm is proposed. Computational results are presented and discussed in Sect. 5. Concluding remarks are presented in Sect. 6.

2   The SVBLP: Optimistic vs. Pessimistic Approaches

The SVBLP is a bilevel optimization problem with a single objective function at the upper level $F(x, y)$ and multiple objective functions $f_k(x, y)$, $k = 1, \ldots, m$ at the lower level.

$$\underset{x \in X}{'\min'} \quad F(x, y)$$

$$\text{s.t.} \quad G(x, y) \leq 0 \tag{1}$$

$$y \in \underset{y \in Y}{\arg \min} \{ (f_1(x, y), \ldots, f_m(x, y)) : g(x, y) \leq 0 \}$$

with $X \subseteq \Re^{n_1}$ and $Y \subseteq \Re^{n_2}$, which impose bounds (box constraints) on the upper level variables x (which are controlled by the leader) and on the lower level variables y (which are controlled by the follower), respectively. $G(x, y) \leq 0$ and $g(x, y) \leq 0$ are general constraints, respectively in the upper and the lower level problems.

Since the decision process is sequential and the leader decides first, x assumes a constant vector in the optimization of $f_k(x, y)$, $k = 1, \ldots, m$. For each $x \in X$ there is a set of efficient (Pareto optimal or nondominated) solutions to the lower level problem represented by $\Psi_{Ef}(x)$. Let $Y(x) = \{ y \in Y : g(x, y) \leq 0 \}$.

Thus, $\Psi_{Ef}(x) = \{ y \in Y : (\text{there is no } y' \in Y(x) | f(x, y') \prec f(x, y)) \}$ where $\prec$ denotes the dominance relation, i.e., $f(x, y') \prec f(x, y)$ iff $f_j(x, y') \leq f_j(x, y)$ for all $j = 1, \ldots, m$, and $f_j(x, y') < f_j(x, y)$ for at least one j.

Since there is not, in general, a single efficient solution to the lower level problem for each x, problem (1) is ambiguous. This is the reason for the quotation marks in the upper level objective function. Two main approaches have been suggested in the literature to address the problem – the optimistic and the pessimistic approaches – leading to two reformulations of (1). As in the single objective bilevel problem with non-unique optimal solutions to the lower level problem, the optimistic formulation of the SVBLP is much simpler to tackle and has therefore been the most investigated.

The optimistic approach assumes that the leader is able to influence the choice of the follower. Thus, the upper level optimization can be taken with respect to x and y to determine the optimal optimistic solution. This means that, for a given upper level decision x, the lower level decision y is the one that presents the minimum $F(x, y)$ among the efficient solutions to the lower level problem for that x, which also satisfy upper level constraints (if there are upper level constraints coupled with lower level variables, i.e. $G(x, y) \leq 0$). The optimal optimistic solution will be called just *optimistic* solution and is defined as follows:

- the *optimistic* solution, (x^o, y^o), is given by

$$\underset{x \in X, y \in Y}{\min} \{ F(x, y) : y \in \Psi_{Ef}(x), G(x, y) \leq 0 \}$$

In the pessimistic approach the leader prepares for the worst case. The leader chooses the x that leads to a feasible solution with minimum F in view of the follower's decisions y worst for the leader. The optimal pessimistic solution will be called just *pessimistic* solution and is defined as follows:

- the *pessimistic* solution, (x^p, y^p), is given by

$$\min_{x \in X} \left\{ \max_{y \in Y} \{F(x,y) : y \in \Psi_{Ef}(x)\} : G(x,y) \leq 0 \right\}$$

A failed optimistic approach leads to the deceiving solution. This means that the leader chooses x according to the optimistic approach but the follower does not react accordingly and takes the decision with worst value for the leader's objective function. Thus, given the optimistic upper level decision x^o,

- the *deceiving* solution is $(x^d, y^d) = (x^o, y^d)$ where y^d is given by

$$\max_{y \in Y} \{F(x^o, y) : y \in \Psi_{Ef}(x^o)\}$$

According to the above definition, the *deceiving* solution may be infeasible to the leader, i.e. infeasible for the SVBLP. Knowing whether the deceiving follower's reaction is feasible or infeasible to the upper level problem is also a useful information to the leader.

A successful pessimistic approach leads to the *rewarding* solution. Thus, given the pessimistic upper level decision x^p, the *rewarding* solution can be defined as the feasible $(x^r, y^r) = (x^p, y^r)$ such that y^r is given by

$$\min_{y \in Y} \{F(x^p, y) : y \in \Psi_{Ef}(x^p), G(x^p, y) \leq 0\}$$

Bonnel [2] and Bonnel and Morgan [3] firstly addressed the SVBLP by providing necessary optimality conditions [2] and a penalty function method [3] for determining the *optimistic* solution. Other methods based on penalty functions to compute the *optimistic* solution were developed by Ankhili and Mansouri [4], Zheng and Wan [5] and Ren and Wang [6] for the SVBLP with a MO linear problem in the lower level. Calvete and Galé [7] focused on the same problem and proposed an exact method and a genetic algorithm, considering the optimistic approach. Liu et al. [8] developed necessary optimality conditions for the *pessimistic* solution and Lv and Chen [9] proposed a discretization iterative algorithm to compute the *pessimistic* solution to a SVBLP without upper level variables in the lower level constraints. Alves et al. [1] firstly introduced the concept of *deceiving* solution and proposed an algorithm based on PSO to approximate the *optimistic, pessimistic* and *deceiving* solutions to the SVBLP. The *rewarding* solution was introduced in [10], where illustrative examples of these four types of extreme solutions were presented. In the present paper we propose a new algorithm based on DE to compute these four extreme solutions and extend the algorithm in [1] to compute also the *rewarding* solution.

3   Optimistic and Pessimistic Frontiers

Let us now define two new concepts to be used in the algorithm proposed in the next section, which are the *Optimistic* and the *Pessimistic frontiers*.

The *Optimistic frontier* (O) consists of the feasible solutions (x, y'), such that y' is the follower's efficient solution $y' \in \Psi_{Ef}(x)$, $G(x, y') \leq 0$, that provides the minimum (best) F for that $x \in X$:

$$O = \left\{ (x, y') : x \in X, y' \in \arg\min_{y \in Y} \{ F(x, y) : y \in \Psi_{Ef}(x), G(x, y) \leq 0 \} \right\}$$

The *optimistic* solution (x^o, y^o) to the SVBLP is the solution $(x, y') \in O$ with minimum F.

The *Pessimistic frontier* (P) consists of the solutions (x, y'') such that y'' is the follower's efficient solution $y'' \in \Psi_{Ef}(x)$ that provides the maximum (worst) F for that $x \in X$:

$$P = \left\{ (x, y'') : x \in X, y'' \in \arg\max_{y \in Y} \{ F(x, y) : y \in \Psi_{Ef}(x) \} \right\}$$

The *pessimistic* solution (x^p, y^p) to the SVBLP is the feasible solution $(x, y'') \in P$, $G(x, y'') \leq 0$, with minimum F.

The *deceiving* solution (x^d, y^d) is the solution in P with $x^d = x^o$.

The *rewarding* solution (x^r, y^r) is the solution in O with $x^r = x^p$.

In the example in Fig. 1, there is a significant difference between the *optimistic* and the *deceiving* solutions for the leader's objective function. Therefore, if the leader opts for an optimistic approach he/she takes a high risk, since the *deceiving* solution is very bad. Conversely, there is a small difference between the *pessimistic F* and the *rewarding* one, being the *F* value in the *rewarding* solution close to the *optimistic F*.

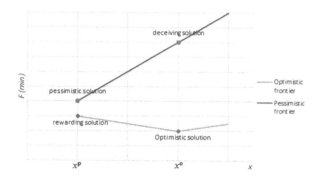

Fig. 1. *F* values in the Optimistic and Pessimistic efficient frontiers of a SVBL linear problem with one upper level variable (*x*) and two objective functions at the lower level.

Since the *deceiving* solution is obtained from the *Pessimistic frontier* using an optimistic approach and the *rewarding* solution is obtained from the *Optimistic frontier* using a pessimistic approach, both frontiers should be simultaneously explored by an algorithm aimed at computing these four extreme solutions.

4   A Differential Evolution Algorithm for the SVBLP

In the SVBLDE algorithm proposed below, the population Pop of individuals is divided into two sub-populations Pop' and Pop'' which share the upper level x vectors. Let Nu be the number of upper level individuals. $Pop = Pop' \cup Pop''$ where $Pop' = \left\{ (x_1, y_1'), (x_2, y_2'), \ldots, (x_{Nu}, y_{Nu}') \right\}$ and $Pop'' = \left\{ (x_1, y_1''), (x_2, y_2''), \ldots, (x_{Nu}, y_{Nu}'') \right\}$. The individuals of Pop' aim at approximating the *Optimistic frontier* while the individuals of Pop'' aim at approximating the *Pessimistic frontier*. DE operations are employed to evolve the population of the upper level problem and, for each upper level vector x, a lower level DE algorithm (DE_LOWERLEVEL_O_P) is used to determine (x, y') and (x, y''). Below, $\mathcal{F}$ denotes the mutation scaling factor and CR the crossover rate in the DE operations. Let Tu be the number of upper level generations. The DE upper level search is described in Algorithm 1. We have used $\mathcal{F} = 0.7$ and CR = 0.9.

Algorithm 1: Upper Level Search

1. $t \leftarrow 1$
2. Create a random initial population of Nu upper level vectors: $x_{i,t} \in X, i = 1, \ldots, Nu$
3. **For** $i = 1$ **to** Nu **do**
4.   DE_LOWERLEVEL_O_P($x_{i,t}, y'_{i,t}, y''_{i,t}$)
5.   Insert $(x_{i,t}, y'_{i,t})$ into Pop'_t and $(x_{i,t}, y''_{i,t})$ into Pop''_t
6. **End For** i
      Initialize the incumbent solutions:
7. $(x^o, y^o) \leftarrow$ arg min $\{F(x,y): (x,y) \in Pop'_t, G(x,y) \le 0\}$ //optimistic solution
      Let $i1$ be the index of (x^o, y^o) in Pop'_t such that $(x^o, y^o) = (x_{i1,t}, y'_{i1,t})$
8. $(x^p, y^p) \leftarrow$ arg min $\{F(x,y): (x,y) \in Pop''_t, G(x,y) \le 0\}$ //pessimistic solution
      Let $i2$ be the index of (x^p, y^p) in Pop''_t such that $(x^p, y^p) = (x_{i2,t}, y''_{i2,t})$
9. $(x^d, y^d) \leftarrow (x_{i1,t}, y''_{i1,t}) \in Pop''_t$ //deceiving solution
10. $(x^r, y^r) \leftarrow (x_{i2,t}, y'_{i2,t}) \in Pop'_t$ //rewarding solution
11. **For** $t = 1$ **to** Tu **do**
12.   **For** $i = 1$ **to** Nu **do**
13. ◆   Select r_1, r_2 and r_3 // selection dependent on the DE variant
14. $j_{rand} = \text{randint}(1, n_1)$ //n_1 is the number of UL variables
15.   **For** $j = 1$ **to** n_1 **do**
16. ◆
$$u_{i,j,t+1} = \begin{cases} x_{r3,j,t} + \mathcal{F}(x_{r1,j,t} - x_{r2,j,t}) & \text{if rand}(0,1) < CR \text{ or } j = j_{rand} \\ x_{i,j,t} & \text{otherwise} \end{cases}$$
17.   **End For** j
18.   DE_LOWERLEVEL_O_P($u_{i,t+1}, w'_{i,t+1}, w''_{i,t+1}$)
       Update the incumbent solutions:
19.   **If** $F(u_{i,t+1}, w'_{i,t+1}) < F(x^o, y^o)$ **and** $G(u_{i,t+1}, w'_{i,t+1}) \le 0$ **then**
20. $(x^o, y^o) \leftarrow (u_{i,t+1}, w'_{i,t+1})$ //update the optimistic
21. $(x^d, y^d) \leftarrow (u_{i,t+1}, w''_{i,t+1})$ // and deceiving solutions
22.   **If** $F(u_{i,t+1}, w''_{i,t+1}) < F(x^p, y^p)$ **and** $G(u_{i,t+1}, w''_{i,t+1}) \le 0$ **then**
23. $(x^p, y^p) \leftarrow (u_{i,t+1}, w''_{i,t+1})$ //update the pessimistic
24. $(x^r, y^r) \leftarrow (u_{i,t+1}, w'_{i,t+1})$ // and rewarding solutions
25. ◆   **If** ACCEPT $u_{i,t+1}$ **then**   // criterion dependent on the DE variant
26. $x_{i,t+1} = u_{i,t+1}$
27.   Insert $(u_{i,t+1}, w'_{i,t+1})$ into Pop'_{t+1} and $(u_{i,t+1}, w''_{i,t+1})$ into Pop''_{t+1}
28.   **Else**
29. $x_{i,t+1} = x_{i,t}$
30.   Insert $(x_{i,t}, y'_{i,t})$ into Pop'_{t+1} and $(x_{i,t}, y''_{i,t})$ into Pop''_{t+1}
31.   **End For** i
32. **End For** t
     Output: $(x^o, y^o), (x^d, y^d), (x^p, y^p), (x^r, y^r)$

In Step 16, if $u_{i,j,t+1}$ does not satisfy the bounds defined by X, then it is projected into the closest bound.

We consider two DE variants: *DE/rand/1/bin* (the original version, which obtained good results in the comparative study of DE variants for global optimization in [11]) and *DE/best/1/bin* (the variant with highest performance in the same study). The steps marked with ◆ change from one variant to the other. In Step 13, *DE/rand/1/bin* randomly selects indexes $r_1 \neq r_2 \neq r_3$ from $\{1, \ldots, Nu\}$, while *DE/best/1/bin* randomly selects indexes $r_1 \neq r_2$ for x_{r1} and x_{r2} but an x_{best} is used in Step 16 to replace x_{r3}. The *DE/best/1/bin* variant divides the population into two equal parts: the first half is mainly oriented towards the *optimistic* solution, so $x_{best} = x^o$, and the second half of the population is mainly oriented towards the *pessimistic* solution, so $x_{best} = x^p$. In addition, r_1 and r_2 are randomly selected from $\{1, \ldots, Nu/2\}$ for $i \leq Nu/2$ and from $\{Nu/2 + 1, \ldots, Nu\}$ otherwise.

The criterion to decide whether $u_{i,t+1}$ is accepted or not in Step 25 (ACCEPT) also depends on the DE variant. Steps 25–30 define the population for the next generation. In *DE/rand/1/bin*, if (**a**) the new individual obtained for approximating the *Optimistic* frontier $(u_{i,t+1}, w'_{i,t+1})$ improves the current one in Pop'_t, i.e. $F(u_{i,t+1}, w'_{i,t+1}) < F(x_{i,t}, y'_{i,t})$, or (**b**) the new individual obtained for approximating the *Pessimistic* frontier $(u_{i,t+1}, w''_{i,t+1})$ improves the current one in Pop''_t, i.e. $F(u_{i,t+1}, w''_{i,t+1}) < F(x_{i,t}, y''_{i,t})$, then the new upper level individual $u_{i,t+1}$ is accepted and Steps 26–27 are performed. Otherwise, the previous individual is kept and Steps 29–30 are performed. In *DE/best/1/bin*, the acceptance criterion in the first half of the population (oriented to the *optimistic* solution) only considers condition (**a**) to decide whether $u_{i,t+1}$ is accepted or not, whereas in the second half of the population only condition (**b**) is considered.

The DE_LOWERLEVEL_O_P algorithm aims at computing two extreme efficient solutions to the lower level problem for a given x, one belonging to the *Optimistic* frontier and the other belonging to the *Pessimistic* frontier: $(x, y') \in O$ and $(x, y'') \in P$.

Let Tl be the number of lower level generations and Nl (an even number) the size of the lower level population. The algorithm attempts to converge to a population $Popy$ of efficient solutions to the lower level problem polarized to the extreme values of the upper level objective function F (the maximum and the minimum). The first $Nl/2$ individuals of $Popy$ are oriented to converge to y' while the remaining $Nl/2$ individuals are oriented to converge to y''.

Algorithm 2: DE_LOWERLEVEL_O_P $(x \downarrow, \; y' \uparrow, \; y'' \uparrow)$

---

1. $t \leftarrow 1$
2. Create a random initial population of Nl lower level vectors $Popy_t = \{y_{i,t} \in Y, i = 1,...,Nl\}$ and sort $Popy_t$ by increasing order of F.
3. Define Eff with the solutions in $Popy_t$ that satisfy $g(x, y_{it}) \leq 0$ and are not dominated by any other solution regarding the lower level objective functions $f_1,... f_m$.
4. Initialize y' and y'':

$$y' \leftarrow \arg \min_{y} \{F(x, y) : y \in Eff, G(x, y) \leq 0\} \quad //y': \text{ solution on the Optimistic frontier}$$

$$y'' \leftarrow \arg \max_{y} \{F(x, y) : y \in Eff\} \quad // y'': \text{ solution on the Pessimistic frontier}$$

5. **For** $t = 1$ **to** Tl **do**
6.   **For** $i = 1$ **to** Nl **do**
7.   **If** $i \leq Nl/2$ **then**
8. ◆   Randomly select $r_1 \neq r_2 \in \{1,...,Nl/2\}$ and select r_3 //depends on the DE variant
9.   **Else**
10. ◆   Randomly select $r_1 \neq r_2 \in \{Nl/2+1,...,Nl\}$ and select r_3
11. $j_{rand} = \text{randint}(1, n_2)$ //n_2 is the number of LL variables
12.   **For** $j = 1$ **to** n_2 **do**

13. ◆ $v_{i,j,t+1} = \begin{cases} y_{r3,j,t} + \mathcal{F}(y_{r1,j,t} - y_{r2,j,t}) & \text{if rand}(0,1) < CR \text{ or } j = j_{rand} \\ y_{i,j,t} & \text{otherwise} \end{cases}$

14.   **End For** j
15.   **End For** i
16.   Insert in Eff the mutually nondominated $(x, v_{i,t+1})$, $i = 1,...,Nl$ that satisfy $g(x, v_{i,t+1}) \leq 0$ and are not dominated by any member of Eff. Delete solutions that become dominated in Eff
17.   **For** $i = 1$ **to** Nl **do**
18.   **If** ACCEPT_LL $v_{i,t+1}$ **then**   //depends on x, $v_{i,t+1}$, $y_{i,t}$, i, Eff
19. $y_{i,t+1} \leftarrow v_{i,t+1}$
20.   **Else**
21. $y_{i,t+1} \leftarrow y_{i,t}$
22.   Insert $y_{i,t+1}$ into $Popy_{t+1}$
23.   Update the incumbent solutions y' and y'' as in Step 4
24. **End For** t
    Output: y' and y''

In step 2, for each new $y_{i,t} \in Y$ randomly generated, the lower level constraints $g(x, y_{i,t}) \leq 0$ are checked; if the constraints are violated then another $y_{i,t} \in Y$ is drawn. If the first and the second trials are infeasible, the solution with smaller overall violation of constraints g is selected. In step 13, if $v_{i,j,t+1}$ does not satisfy the bounds defined by Y, then it is projected into the closest bound.

As in Algorithm 1, the steps marked with ♦ change from one DE variant to the other. In Steps 8 and 10, *DE/rand/1/bin* randomly selects $r_3 \in \{1, \ldots, Nl/2\}$ for $i \leq Nl/2$ and $r_3 \in \{Nl/2 + 1, \ldots, Nl\}$ for $i > Nl/2$. The *DE/best/1/bin* variant defines $y_{best} = y'$ for $i \leq Nl/2$ and $y_{best} = y''$ for $i > Nl/2$; y_{best} is used in Step 13 to replace y_{r3}.

The acceptance criterion in step 18 (ACCEPT_LL) determines whether the new individual $v_{i,t+1}$ is accepted or not to replace $y_{i,t}$ in the next population. The acceptance criterion firstly observes whether the solutions $(x, v_{i,t+1})$ and $(x, y_{i,t})$ satisfy the lower level constraints $g(x, y) \leq 0$ (*g*-feasibility), privileging the feasible solution if one of them is infeasible. If both are *g*-feasible, then it is checked whether they are non-dominated w.r.t. to the current set of solutions *Eff*. If one of the solutions $v_{i,t+1}$ or $y_{i,t}$ is nondominated (i.e., it belongs to *Eff*) and the other is dominated, the nondominated solution is selected. If both solutions have the same status, the selection is based upon the upper level objective function value: for $i \leq Nl/2$ (sub-population oriented to the *Optimistic* frontier) the individual with lowest F is selected; for $i > Nl/2$ (sub-population oriented to the *Pessimistic* frontier) the individual with highest F is selected. It is worthwhile to note that, in an initial version of the algorithm, we did not use the set *Eff* in the acceptance criterion of $v_{i,t+1}$. The algorithm only compared the two candidate solutions, $v_{i,t+1}$ and $y_{i,t}$, checking whether one dominated the other or both were nondominated w.r.t. to each other. However, the algorithm revealed a very poor convergence of the population to nondominated solutions, which was overcome with the current strategy.

5   Computational Experiment

The SVBLDE algorithm has been compared with the PSO algorithm in [1], which was extended to compute also the *rewarding* solution as this algorithm had been originally designed to determine the other three extreme solutions. Below we shortly designate the *optimistic, pessimistic, deceiving* and *rewarding* solutions by *sol.o*, *sol.p*, *sol.d* and *sol.r*, respectively (with F^o, F^p, F^d and F^r being the respective upper level objective values).

To test and compare the algorithms we have considered two sets of problems. The first set includes 4 problems – *Prob.1* to *Prob.4* – whose formulations and *sol.o*, *sol.p* and *sol.d* are presented in [1]; these problems were adapted from the MOBL problems in [12] by considering only one upper level objective function. All the problems have one upper level variable and two lower level objective functions. Below we briefly describe these problems by indicating the number of lower level variables (n_2) and showing the values of F^o, F^p, F^d and F^r.

Prob.1 – $n_2 = 2$; *sol.o* $\neq$ *sol.d* $\neq$ *sol.p* with $F^o = 0.5$, $F^d = 1.25$, $F^p = 1$; *sol.r* = *sol.p*, so $F^r = 1$.

Prob.2 – generalization of *Prob.1* with $n_2 = k$. We consider $k = 14$. The extreme solutions have the same characteristics as in *Prob.1* and the same upper/lower level objective values.

Prob.3 and *Prob.4* have $n_2 = 2$ and differ from each other in the upper level objective function. They include an upper level constraint G depending on lower level variables, which increases their difficulty. *Prob.3*: *sol.o* = *sol.r* with $F = -2$ and *sol.d* = *sol.p* with $F = -1$. *Prob.4*: this problem admits alternative *pessimistic* solutions (all with $F^p = 0$) but with different outcomes for the corresponding *rewarding* solution (with $F = -\alpha$, $0 \le \alpha \le 1$). The best *rewarding* solution corresponds to *sol.p* = *sol.d*, $F^p = F^d = 0$, being the *rewarding* solution *sol.r* = *sol.o* with $F^r = F^o = -1$.

The second group of test problems are the MOBL problems *DS1* to *DS5* in [13], originally proposed in [14]. We consider only F_1 for the upper level objective function in our problems. This is a set of scalable problems with a variety of complex features to the algorithms. Problems *DS1-DS3* have k upper level and k lower level variables – we consider $k = 5$. Problems *DS4* and *DS5* have one upper level variable and $k + l$ lower level variables – we consider $k = 3$ and $l = 2$. All the other parameters were set as in [13]. The corresponding values of F^o, F^p, F^d and F^r are presented in Table 1.

Table 1. Median and interquartile range of F in 30 independent runs for each algorithm.

		SVBLDE		PSO algorithm		Exact F	M-W test
		Median F	IQR F	Median F	IQR F		
Prob.1	*Sol.o*	**0.497384**	0.000708	0.496248	0.001489	**0.5**	+
	Sol.p	0.993762	7.04E−05	0.993742	9.36E−05	**1**	−
	Sol.d	1.246284	0.01238	1.246038	0.016857	**1.25**	−
	Sol.r	0.993713	0.004252	0.988769	0.021896	**1**	−
Prob.2	*Sol.o*	**0.487397**	0.002296	0.407539	0.026634	**0.5**	+
	Sol.p	**0.999220**	0.011627	0.991885	0.000691	**1**	+
	Sol.d	**1.250138**	0.035337	1.202307	0.069874	**1.25**	+
	Sol.r	0.908306	0.057767	**0.98603**	0.007036	**1**	+
Prob.3	*Sol.o*	−2	0.006734	−1.99995	4.01E−05	**−2**	−
	Sol.p	−0.99985	0.001206	−0.99984	0.000215	**−1**	−
	Sol.d	−1.00214	0.003443	−1.00296	0.001645	**−1**	−
	Sol.r	−1.95307	0.101921	**−1.99036**	0.002566	**−2**	+
Prob.4	*Sol.o*	−0.99694	0.007655	−0.99995	5.12E−05	**−1**	−
	Sol.p	**−0.00356**	0.001833	−0.00606	0.000695	**0**	+
	Sol.d	−0.00334	0.001618	−0.00391	0.001138	**0**	−
	Sol.r	**−0.96020**	0.092859	−0.89689	0.111896	**−1**	+
DS1	*Sol.o*	**2,51E−05**	3,1E−05	5,61E−05	3,11E−05	**0**	+
	Sol.p	0,07746	0,056167	**0,099769**	0,000128	**0.1**	+
	Sol.d	0,092602	0,061951	**0,099981**	0,000168	**0.1**	+
	Sol.r	**3,23E−05**	0,000107	0,000179	0,000193	**0**	+
DS2	*Sol.o*	**−0,25977**	0,013058	−0,34826	0,035846	**−0.238773**	+
	Sol.p	−0,23876	8,07E−06	**−0,23877**	2,68E−06	**−0.238773**	+
	Sol.d	−0,23873	0,000247	**−0,23877**	9,51E−07	**−0.238773**	+
	Sol.r	−0,23876	1,36E−05	**−0,23878**	0,091723	**−0.238773**	+

(*continued*)

Table 1. (*continued*)

		SVBLDE		PSO algorithm		Exact F	M-W test
		Median F	IQR F	Median F	IQR F		
DS3	Sol.o	**1.85E−07**	2.64E−07	5.34E−05	9.29E−05	0	+
	Sol.p	1.84E−07	1.03E−07	**0.200086**	1.65E−05	0.2	+
	Sol.d	1.86E−07	1.83E−07	**0.200299**	0.000192	0.2	+
	Sol.r	**1.99E−07**	2.65E−07	0.001461	0.004658	0	+
DS4	Sol.o	**0**	0	0	0.845635	0	+
	Sol.p	102	0	102	0	102	−
	Sol.d	**204**	0	204	100.5451	204	+
	Sol.r	**1.000245**	0.000347	2.388914	0.519674	1	+
DS5	Sol.o	**0.760132**	0.000174	2.01667	0.021934	0.76	+
	Sol.p	102	0	102	0	102	−
	Sol.d	**188.9164**	27.86941	`102	0	167.3	+
	Sol.r	**1.000139**	0.000134	2.268107	0.36943	1	+

We have considered the following parameters for both algorithms, which were tuned through experimentation: Nu-N_l-Tu-T_l equal to 20-60-50-100 for the first set of problems except *Prob.2*; 20-100-50-100 for *Prob.2, DS4* and *DS5*, which also have one upper level variable but more than 2 lower level variables; 100-100-100-100 for *DS1* to *DS3*, which have a higher number of upper level variables. Specific parameters of the PSO algorithm were set as in [1]. We performed 30 independent runs of each algorithm in each problem.

Concerning the DE variants of the SVBLDE algorithm, we observed that the results of *DE/rand/1/bin* were not statistically different from the results of *DE/best/1/bin* in about half of the cases; however, *DE/rand/1/bin* provided very poor results in a few other cases. Therefore, and due to space reasons, we omit herein the results of that variant. Table 1 presents the median and the interquartile range IQR of the F values obtained for the four extreme solutions over the 30 runs using the variant *DE/best/1/bin* of SVBLDE and the PSO algorithm. We also include the exact values of F (obtained analytically), which are very useful to assess the quality of the results obtained. The non-parametric Mann-Whitney test has been applied to assess whether the differences of the F values obtained with the two algorithms are statistically significant, considering a confidence level of 95%. The best result for each solution is highlighted in bold if the difference is statistically significant ('+' in the last column).

It is noteworthy that there are several difficulties in evaluating results to SVBLP. These difficulties can easily lead to pitfalls in the interpretation of results, which may be very difficult to avoid in general problems for which the exact solutions are not known. We draw attention to some of these pitfalls:

- Only efficient (Pareto optimal) solutions to the lower level problem are feasible to the SVBLP. Therefore, an algorithm may yield apparently better solutions (for any of the four extreme solutions), i.e. with lower F values, but the solutions are invalid because they are not efficient to the lower level problem.

- Even if only efficient (Pareto optimal) solutions to the lower level problem are obtained, other difficulties arise in assessing the *pessimistic* and *deceiving* solutions. Solutions with lower F values (i.e., which seem to be better) may be false because they are not in the *Pessimistic* frontier, i.e., they are not the worst for the leader for that setting of x. We can observe this situation in Table 1 for several *sol.d* and *sol.p* (e.g., *Prob.2, Prob.4, DS1, DS3* or *DS5*).

From Table 1, we observe that SVBLDE outperformed the PSO algorithm in 17 out of the 36 cases (4 extreme solutions to 9 problems) while the PSO algorithm outperformed SVBLDE in 9 cases (the differences in the other 10 cases were not statistically significant). Therefore, SVBLDE seems to perform slightly better than the PSO algorithm. We can also observe that SVBLDE is very effective in approximating the *optimistic* solution, being always better or equal to the PSO algorithm, but SVBLDE reveals more difficulty in attaining the real *pessimistic* and *deceiving* solutions in several cases.

6   Conclusions

We presented a new DE algorithm to compute the optimistic/deceiving and pessimistic/rewarding solutions to the SVBLP. These four extreme solutions capture the optimistic vs. pessimistic leader's attitude and possible follower's reactions more or less favorable to the leader. The DE approach seems to perform slightly better than the PSO-based approach, but the results do not evidence a clear performance advantage of the SVBLPDE algorithm with respect to PSO. The experiments unveiled some pitfalls associated with the interpretation of results and assessment of the algorithm performance in SVBLP. These pitfalls could be avoided because we were able to determine analytically the exact solutions to the problems tested. Research is underway on techniques to mitigate these pitfalls in general problems, which are nevertheless intrinsic to this kind of problems and cannot be entirely avoided.

Acknowledgment. This work was supported by projects UID/MULTI/00308/2013 and SAICTPAC/0004/2015-POCI-01-0145-FEDER-016434.

References

1. Alves, M.J., Antunes, C.H., Carrasqueira, P.: A PSO approach to semivectorial bilevel programming: pessimistic, optimistic and deceiving solutions. In: Proceedings of the Genetic and Evolutionary Computation Conference (GECCO 2015), pp. 599–606 (2015)
2. Bonnel, H.: Optimality conditions for the semivectorial bilevel optimization problem. Pac. J. Optim. **2**, 447–468 (2006)
3. Bonnel, H., Morgan, J.: Semivectorial bilevel optimization problem: penalty approach. J. Optim. Theor. Appl. **131**, 365–382 (2006)

4. Ankhili, Z., Mansouri, A.: An exact penalty on bilevel programs with linear vector optimization lower level. Eur. J. Oper. Res. **197**, 36–41 (2009)
5. Zheng, Y., Wan, Z.: A solution method for semivectorial bilevel programming problem via penalty method. J. Appl. Math. Comput. **37**, 207–219 (2011)
6. Ren, A., Wang, Y.: A novel penalty function method for semivectorial bilevel programming problem. Appl. Math. Model. **40**, 135–149 (2016)
7. Calvete, H., Galé, C.: On linear bilevel problems with multiple objectives at the lower level. Omega **39**, 33–40 (2011)
8. Liu, B., Wan, Z., Chen, J., Wang, G.: Optimality conditions for pessimistic semivectorial bilevel programming problems. J. Inequal. Appl. **2014**, 41 (2014)
9. Lv, Y., Chen, J.: A discretization iteration approach for solving a class of semivectorial bilevel programming problem. J. Nonlinear Sci. Appl. **9**, 2888–2899 (2016)
10. Alves, M.J., Antunes, C.H.: An illustration of different concepts of solutions in semivectorial bilevel programming. In: 2016 IEEE Symposium on Computational Intelligence (SSCI) (2016)
11. Mezura-Montes, E., Velázquez-Reyes, J., Coello Coello, C.A.: A comparative study of differential evolution variants for global optimization. In: Proceedings of the 8th Annual Conference on Genetic and Evolutionary Computation, pp. 485–492 (2006)
12. Deb, K., Sinha, A.: Solving bilevel multi-objective optimization problems using evolutionary algorithms. In: Ehrgott, M., Fonseca, C.M., Gandibleux, X., Hao, J.-K., Sevaux, M. (eds.) EMO 2009. LNCS, vol. 5467, pp. 110–124. Springer, Heidelberg (2009). https://doi.org/10.1007/978-3-642-01020-0_13
13. Sinha, A., Malo, P., Deb, K.: Approximated set-valued mapping approach for handling multiobjective bilevel problems. Comput. Oper. Res. **77**, 194–209 (2017)
14. Deb, K., Sinha, A.: Constructing test problems for bilevel evolutionary multi-objective optimization. In: 2009 IEEE Congress on Evolutionary Computation, pp. 1153–1160 (2009)

Evolving Training Sets for Improved Transfer Learning in Brain Computer Interfaces

Jason Adair[✉], Alexander Brownlee, Fabio Daolio, and Gabriela Ochoa

Computing Science and Mathematics, University of Stirling, Stirling, Scotland, UK
{jad,sbr,fda,goc}@cs.stir.ac.uk

Abstract. A new proof-of-concept method for optimising the performance of Brain Computer Interfaces (BCI) while minimising the quantity of required training data is introduced. This is achieved by using an evolutionary approach to rearrange the distribution of training instances, prior to the construction of an Ensemble Learning Generic Information (ELGI) model. The training data from a population was optimised to emphasise generality of the models derived from it, prior to a recombination with participant-specific data via the ELGI approach, and training of classifiers. Evidence is given to support the adoption of this approach in the more difficult BCI conditions: smaller training sets, and those suffering from temporal drift. This paper serves as a case study to lay the groundwork for further exploration of this approach.

Keywords: Optimisation · Machine learning · Ensemble
Brain-computer interface · P300 · Evolutionary computation
Transfer learning

1 Introduction

Brain Computer Interfaces (BCI) are applications in which neurological recordings are utilised for the control of digital systems. Uses for BCI range from manipulation of prosthetic limbs, psychological interventions, and assisted communication devices [1]. Approaches to obtain these recordings can be separated into two main groupings; invasive and non-invasive. While invasive recordings can allow exceptional spatial and temporal resolutions, they involve sub-cranial surgery with potentially severe health risks and prohibitive financial costs [2]. For these reasons, non-invasive approaches have garnered significant interest. These include *electroencephalography (EEG)*; a technique that involves placing electrodes on the surface of the scalp to measure electrical fields produced by the underlying neurons. While this technique comes with little or no health risks, it lacks high resolutions [3], and is subject to noise from muscle movements (*electromyography*), cardiac rhythms (*electrocardiography*), eye movements (*electrooculography*), and environmental electrical sources [4].

Due to the aforementioned issues, a large quantity of training data is often required for each individual to calibrate the classifiers. As training sessions are

© Springer International Publishing AG 2018
G. Nicosia et al. (Eds.): MOD 2017, LNCS 10710, pp. 186–197, 2018.
https://doi.org/10.1007/978-3-319-72926-8_16

often supervised by a health or technical expert, this increased training time not only comes at a financial cost, but proves frustrating and stressful for the participant which, in turn, introduces further noise into the training set.

For these reasons, it is deemed imperative to minimise the amount of training data by exploiting all available data, from all potential sources.

We propose a novel method for the optimisation of the distribution of instances within a database of sets recorded from previous participants, in a manner that ensures that they can be used to create an ensemble that is maximally general to the population. This database is then used to seed a previously established method (Ensemble Learned for Generic Information) that recombines instances obtained from different participants with small quantities of participant-specific data, to create a robust participant-specific ensemble. This should allow for the creation of a BCI that requires only a small amount of training data, and should retain accuracy over time in a way that a traditional BCI does not. This is achieved by moving instances between previously obtained datasets via a random mutation hill-climber.

The structure of the paper is as follows: A brief literature introduction to Transfer Learning in the BCI field is given (Sect. 2), and algorithms described, with a hypothesis based on the new technique (Sect. 2.2). This leads us to the paradigm, dataset, and methodology used for experimentation (Sects. 3 and 4). Finally, the results are presented (Sect. 5) and discussed (Sect. 6).

2 Related Work on Transfer Learning in BCI

As described in Sect. 1, BCIs are difficult to calibrate due to recordings having a low signal to noise ratio. This is further compounded by the non-stationary nature of brain signals: neural patterns not only differ between participants, but are also subject to *temporal drift*, where data obtained from a single participant changes drastically over time [5]. *Zero Training systems*, trained exclusively on participants from previous sessions, are an ideal goal, but this non-stationarity means highly accurate zero training systems may not be possible. Consequently, we must instead focus on minimising the participant-specific training information required by maximising the effectiveness of the data available.

Sufficient data from an individual for the creation of an accurate system comes with significant costs, so utilising databases from other participants offers an attractive avenue to alleviate this burden. Transfer Learning has been employed in a number of domains containing multiple sources to allow data inference to unseen sources. For a more in-depth discussion of the wider field, [6] provides a recent, thorough survey. More specifically, BCI literature typically reports *domain adaptation* approaches [5], the most popular of which being *Common Spatial Patterns* [7]. This involves creating a transformation of the data that will allow a single classification rule to be applied across all instances. A much less commonly explored approach is '*Rule Adaptation*' [5], in which a number of rules are created from the existing datasets, and then applied to the new instances. Both cases however, rely upon the natural distribution of the data as grouped by their original participant. Some attempts have been made to group datasets by known variants such as gender [8], and others using the information extracted from the trained models [9]; but little has been done in regards to instance selection for each model.

2.1 Ensembles

One method for incorporating data from other domains is the use of *ensembles*. Ensembles typically consist of an array of different classifiers trained with the same dataset. Each classifier makes predictions on a test set, and these are collated in a voting process. This allows multiple different relationships to be detected for the classification process, many of which may not be obvious, even to a domain expert. Another approach is to use multiple instances of the same classifier, trained with different initial datasets.

Ensembles have been used in a number of different BCI applications to increase accuracy and reduce the amount of training data required for participants. Arguably, the most well known *P300-Speller ensemble* is [10] in which an ensemble of SVMs were used to reduce variability in signal inputs by averaging classifier outputs, but relied on a substantial quantity of subject-specific data. This, like most BCI ensembles [11], used naive partitioning in which the instances were divided by their associated labels, whether it be by source domain or by stimuli. This proves useful for weighting classifiers within the ensembles; allowing information regarding the appropriateness of each model and the test-domain to be extracted [9]. It was demonstrated in [11] that overlapping these naive divisions can actually increase accuracy, suggesting that having the same training data duplicated amongst the classifiers can benefit the overall performance.

2.2 ELGI

In 2015, Xu et al. [12] introduced the *Ensemble Learning Generic Information (ELGI)* approach. Rather than using the small amount of training data to train a classifier, or for weighting the models within a larger ensemble trained on the data of other participants, ELGI combines the participant-dependent data with participant-independent data to form a hybrid ensemble. This is achieved by splitting the datasets of each existing patient within the database into target and non-target sets. The removed missing instance class (target or non-target) is then replaced by a copy of the corresponding class from the participant-specific training data. This results in an ensemble consisting of $2n - 1$ classifiers, where n is the number of participants within the database.

This paper proposes a new technique in which the database containing the previously recorded participants' datasets are optimised to create an ensemble that is maximally generalised for the population, prior to the combination process of ELGI. The procedure is outlined fully in Sect. 4.

3   Methodology

This section defines the BCI Paradigm used in the work and describes the datasets. It then goes on to describe the offline filtering applied to the data and finally defines the algorithms to be compared in the experiments. In particular, we focus on how the datasets are initially derived for the eELGI approach, but background information on the application is summarised here for convenience and the interested reader can find more detail in [12].

3.1   P300 Speller Paradigm

A promising application of BCI systems are their use in communication assistive technologies. Some conditions, such as *Amyotrophic Lateral Sclerosis (ALS)*, cause degradation of physical movement [2], rendering patients unable to communicate with the outside world. Detection of neural activity can allow patients to control computers, and produce synthetic speech [1]. A common form of the system involves using a computer screen displaying a 6×6 grid of alphanumeric characters. The user concentrates on the character they wish to select, and all columns and rows are flashed in a randomised sequence. When the target character's row and columns flash, a fluctuation in neural patterns can be observed. This is known as a *P300 wave*. The goal of a BCI system in this context is to identify the P300 wave among the many other detected signals.

3.2   Dataset Recordings

The dataset used in this paper was obtained from [13], in which the P300 Speller Paradigm was adapted to present 6 images to the participant; each eliciting a response by increasing the brightness of the image. It included EEG recordings for 4 disabled participants and 4 able-bodied PhD students. Participants 1, 2 and 4 were able to speak with some dysarthria, but participant 3 was unable to communicate verbally due to their late stages of amyotrophic lateral sclerosis. All 4 disabled participants were wheelchair users, with limited to no control over their upper limbs. Each participant attended 4 recording sessions, each consisting of, on average, 810 trials; resulting in approximately 3240 trials per participant.

3.3   Prefiltering

The data underwent prefiltering as described in [13]. In summary, the procedure for this was: referencing against the mastoid electrodes, Butterworth bandpass filtering between 1 and 12 Hz, downsampling to 32 Hz, and Windsorizing to mitigate EoG and EMG artifacts.

3.4   Classifier

A Bayesian Linear Discriminate Analysis classifier (as in [13]) was used. Each stimuli presentation was treated as a binary problem, and the Bayesian probability of the prediction was recorded. Due to the paradigm structure, every subdivision of 6 stimuli presentations has 1 target and 5 non-target. These groupings are deemed as a 'round'. A prediction is made based on the highest probability within each round. In each run, 20 rounds of all 6 stimuli are presented. This allows the Bayesian probabilities of each round to be summed with previous predictions, increasing predictive accuracy over the course of the run.

3.5   Conditions

The complex nature of BCI allows a number of different factors to be considered:

Quantity of Participant-Specific Data. As a primary aim in BCI is minimising the required participant-specific training data, the impact of training set size was explored. The datasets follow a common hierarchical structure; each participant recording 4 sessions of 6 runs. All models were trained with data from the first session and 3 training set sizes were used: 3, 4, and 5 runs.

Time Between Testing Sessions. A major challenge in BCI, other than between-participant transference, is between-session transference in single participants. As neural drift occurs over time, highly fitted models tend to lose accuracy. All models were tested on data acquired from 3 sessions, recorded over 2 days; session 2 on the same day as the training data, and 3 and 4 on a day no more than 2 weeks later.

3.6   Compared Algorithms

Three approaches were compared in our experiments, two taken from the literature and the proposed new method:

Standard Learning Individual Information (SLII). A Bayesian LDA model trained using participant-specific data exclusively. The highest probability in each round was selected as the target, and the rest, assumed to be non-targets [12].

Ensemble Learning Generic Information (ELGI). The ELGI [14] creates an array of classifiers by utilising the participant-specific and participant-independent datasets in the following manner:

$$[C_{2N}] = \sum_{i=1}^{N} [C(P_i^T + P_k^{NT}), C(P_i^{NT} + P_k^T)]$$

The training data P from each participant P_i is split into two subgroups; target T and non-target NT. A copy of the target instances from the test-participant

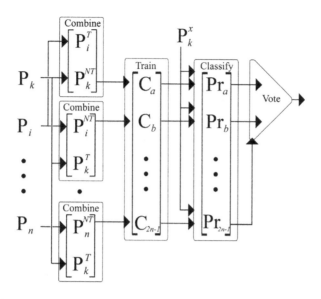

Fig. 1. ELGI approach displaying that 2 classifiers are trained for every participant in the database P_i by a splitting and recombination of their target P_i^T and non-target P_i^{NT} instances with the corresponding instances from the test-participant's training data P_k. These classifiers are then used to make predictions on the test-participants unseen data P_k^x. Finally, these predictions are collated via voting.

k (P_k^T) are then added to the non-target subgroup P_i^{NT}, and conversely, a copy of the test-participant's non-target instances P_k^{NT} are added to the target subgroup P_i^T. Each of these new subgroups are used to train an ensemble of classifiers C. Predictions Pr are made by each classifier in the ensemble based on the unseen data from the test-participant P_k^x, and these predictions are collated. This is done using the Sum Rule voting method where the Bayesian posterior probabilities are summed for each class. This is further depicted in Fig. 1.

Evolved Ensemble Learning Generic Information (eELGI). The novel proposed approach of this paper, as described in Sect. 4. In this, we assume that the natural grouping of instances by participant is not optimal. Instead, an evolutionary algorithm transplants instances between datasets taken from each participant, aiming to maximise the generalisability of each set in reference to other previously recorded participants, prior to their combination with participant-specific data via the ELGI.

4   Evolved ELGI Ensemble

We propose a new approach whereby the database containing the previous participants' datasets is optimised, with the goal of creating an ELGI ensemble that better generalises to the population. This is achieved by a leave-one-out technique in which a participant's bin, that is the subset containing all data from that participant, is selected at random, and a portion of the instances obtained from that participant are moved into the bin of another randomly selected participant. Two models are then trained; one using the data from the bin that was selected for transfer, and one from the bin that was selected as the destination.

These models make predictions on the data in the remaining unselected bins. The resulting overall predictive accuracy is used as the fitness function for a random mutation hill climber. This seeks the allocation of training data to bins that maximises the predictive accuracy within the database.

We now describe the implementation in more detail. The procedure is given formally in Algorithm 1. The search is seeded with a solution consisting of 7 bins; each consisting of an individual's data, but excluding any information from the new participant, as in the Zero Training Model. A 500 iteration Hillclimber was then applied with the following mutation operator and fitness function.

Mutation (Move Operator). The move operator selects a target bin a and a destination bin b at random from the training set bins; a subset m with 10% of the target bin's instances are moved into the destination bin. Subsets P_{ea} and P_{eb} are created by removing subset m from P_a and appending it to P_b, respectively.

Fitness Function. To assess the fitness of the candidate solution, 2 classifiers C_{ea} and C_{eb} were trained from the subsets P_{ea} and P_{eb}. These were then used to make predictions on the remaining instances within all subsets P, excluding the participant datasets selected for mutation (P_a and P_b). The average round accuracy over all the non-selected bins was calculated for both models affected by the mutation (f_{ea} and f_{eb}); a solution was deemed successful if the fitnesses obtained were an increase over the fitness (f_a and f_b) of both models created from the incumbent solution (C_a and C_b). The mutation was rejected if it caused a decrease in accuracy within either model.

This evolved dataset was then used to seed the original ELGI from [12].

5   Results

Figure 2 presents the performance of the SLII, ELGI and eELGI algorithms averaged across all 8 participants. Rows 1, 2 and 3 show performance of models with 3, 4 and 5 runs (see Sect. 3.5) of training data available, respectively. Columns display performance over 3 different testing sessions. While the confidence intervals of the different approaches vary due to differing sample sizes, the SLII and ELGI are almost indiscernible. The mean line of the eELGI is typically higher than that of the other algorithms, with its smaller confidence interval often visibly higher. The instances in which notable improvements are made are in the extremity conditions: low availability of participant specific data (row 1) and the testing session farthest from the training session (column 3).

The Round Accuracy is presented in Fig. 3 for the SLII, ELGI and eELGI algorithms. It is displayed by participant with each point representing the accuracy achieved with 3, 4 and 5 runs of training data provided for training.

Algorithm 1. Evolution of instances in eELGI

Input: Initial solution is $P = \mathcal{P}(P_i)$
Output: Final solution is Modified $P = \mathcal{P}(P_i)'$

1: **for** $x = 1 \rightarrow 500$ **do**
2:      Choose a and b from $1 : N$ where N is the $|P|$
3:      Create $m \subset P_a$
4:      $P_{ea} \leftarrow P_a$ with m removed
5:      $P_{eb} \leftarrow P_b$ appended with m
6:      Train classifiers C_a and C_b with P_a and P_b
7:      Train classifiers C_{ea} and C_{eb} with P_{ea} and P_{eb}
8:      $f_a = 0$, $f_b = 0$, $f_{ea} = 0$, $f_{eb} = 0$
9:      **for** $i = 1 \rightarrow N$ **do**
10:        **if** $i \neq a$ && $i \neq b$ **then**
11:          $f_a = f_a + C_a(P_i)$, $f_b = f_b + C_b(P_i)$
12:          $f_{ea} = f_{ea} + C_{ea}(P_i)$, $f_{eb} = f_{eb} + C_{eb}(P_i)$
13:        **end if**
14:     **end for**
15:     **if** $f_a < f_{ea}$ && $f_b < f_{eb}$ **then**
16:        $P_a = P_{ea}, P_b = P_{eb}$
17:     **end if**
18: **end for**

Increases in the quantity of participant-specific training data increases the predictive accuracy in each participant except 6. Participant 5 is the outlier in terms of variance; increases in participant-specific training data makes a much more substantial change to this classifier's accuracy than others. When considering overall round accuracies across differing training set sizes, eELGI performed better than the SLII and ELGI in 62.5% of cases, and obtained the second best results in the remainder. In no cases was eELGI the worst performer.

Figure 4 demonstrates each algorithm's resilience to neural drift over time. The round accuracy of the SLII, ELGI and eELGI over each of the testing sessions is given. A decrease in predictive accuracy was observed between session 2 and session 3 in 62.5% of the cases, and a decrease between session 3 and 4 in 58.3%. Overall, a decrease in predictive accuracy between session 2 and 4 was observed in 79.2% of the cases, as expected due to temporal neural drift. For 5 of the 8 participants, the eELGI retained the highest round accuracy after 2 weeks, while still maintaining relativity high accuracy in the remaining 3.

To analyse the differences between each algorithm's effectiveness in mitigating the effects of neural drift over time, hierarchical linear models were used as recommended in [15]. The results of these are given in Figs. 5a and b. In Fig. 5a, lines show the expected average behaviour when considering the variation across participants, with points representing the residual deviation of each participant from the estimated common behaviour. Although no statistical significance can be claimed here, the trends suggest that in all 3 testing sessions, the eELGI performed better than both the SLII and ELGI. It should also be noted that there appears to be less variance within and between testing sets for the eELGI.

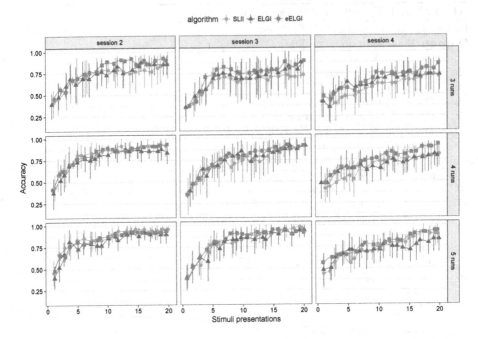

Fig. 2. Algorithm performance by number of stimuli presentations, with differing quantities of participant-specific training data available. Error bars show the confidence intervals around the means. Horizontal jitter has been added to improve discernibility.

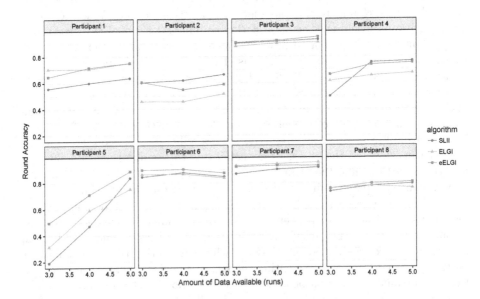

Fig. 3. Round Accuracy over all testing sets displayed for each quantity of participant-specific training data, separated for each participant.

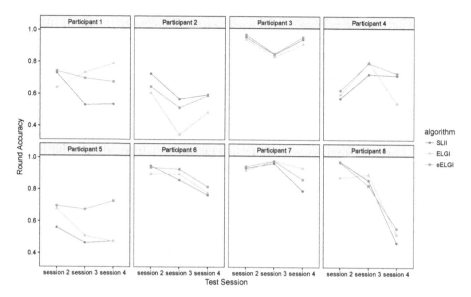

Fig. 4. Round Accuracy over all quantities of training data for each testing set, separated for each participant.

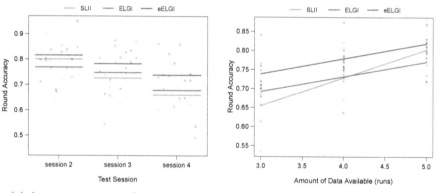

(a) Accuracy over time from training.   (b) Accuracy over available training data.

Fig. 5. Fit of hierarchical linear models, with random effects for each participant, estimating (a) the overall Round Accuracy per testing set and (b) the change in Round Accuracy over training set size.

This suggests that the eELGI not only performs better than the other algorithms, but is also less susceptible to neural drift over time.

As seen in Fig. 5b, the round accuracy of all 3 algorithms increases with the amount of participant-specific data available. The SLII is most dependent on the quantity of participant-specific data, with ELGI performing much better when fewer training instances are available. However, this advantage is lost as volume of training data increases. The eELGI line has a similar slope to the ELGI (0.0402 and 0.0394, respectively) but with a higher intercept (0.618 to 0.574), resulting in better overall performance than both the SLII and ELGI in all 3 conditions. In fact, a post-hoc Tukey's comparison of the model estimates, averaging over algorithm-data interactions [16], showed that the eELGI produced a statistically significant increase in round accuracy over the SLII ($p = 0.0387$) while the ELGI did not ($p = 0.1483$). Therefore, with respect to the ELGI, the effect of evolving the base dataset appears to increase the intercept, without having any adverse affects to the rate of improvement seen when increasing participant-specific data.

6   Discussion and Conclusion

This paper served as a case study for the proposed eELGI approach. However, statistical significance can be difficult to determine with small datasets. This being said, even with small samples, we have demonstrated that there is a visible advantage to optimisation of the participant database for use in transfer learning techniques. We can see that an evolved database has 3 primary advantages:

1. *A higher classification accuracy*, regardless of quantity of training data. As seen in Fig. 3, 62.5% of cases see eELGI performing better than ELGI and SLII, with the remaining still close to the optimal. In Fig. 5b we can notice, in the majority of cases, a marked improvement over the non-evolved ELGI.
2. *A reduction in variance* in performance across not only sessions, but participants as well. When comparing sessions in Fig. 5a, and training set size in Fig. 5b, the groupings of round accuracies are noticeably more dense. Figure 2, is perhaps the most dramatic demonstration of this. By including all participants over all test sets, the error bars for both the SLII and ELGI are substantial, while the eELGI provides a modest difference.
3. *A means for protection against temporal drift*. Figure 5b demonstrates that the traditional BCI approach (SLII) is highly susceptible to the neural drift seen over time. While ELGI alleviates that to a degree, eELGI provides a much more linear, and slower degradation in predictive accuracy over the testing sessions.

As this paper focused on a small dataset, with an equal number of able and disabled patients, further work should investigate the effects of optimising different base datasets. For example, it should contain substantially more participants, and, in more commonly observed situations, contain disproportionately more able bodied participants. In terms of algorithms; while a simple hillclimber has provided some promising results, it would be prudent to apply more complex heuristics to the problem. A potentially promising direction would be utilisation of a genetic algorithm with an encoding that would allow oversampling of the more prototypical instances.

Acknowledgements. This research was funded by the ESPRC through the DAASE project [grant number EP/J017515/1]. The authors are grateful for the assistance of Kate Howie in preparating the statistical analyses.

Data Access Statement. The dataset and source code used in this paper are available on request from the lead author.

References

1. Waytowich, N.R., Lawhern, V.J., Bohannon, A.W., Ball, K.R., Lance, B.J.: Spectral transfer learning using information geometry for a user-independent brain-computer interface. Front. Neurosci. **10** (2016)
2. Nicolas-Alonso, L.F., Gomez-Gil, J.: Brain computer interfaces, a review. Sensors **12**, 1211–1279 (2012)
3. Schwartz, A.B., Cui, X.T., Weber, D.J., Moran, D.W.: Brain-controlled interfaces: movement restoration with neural prosthetics. Neuron **52**(1), 205–220 (2006)
4. Khatwani, P., Tiwari, A.: A survey on different noise removal techniques of EEG signal. Int. J. Adv. Res. Comput. Commun. Eng. **2**(2), 1091–1095 (2013)
5. Jayaram, V., Alamgir, M., Altun, Y., Scholkopf, B., Grosse-Wentrup, M.: Transfer learning in brain-computer interfaces. IEEE Comp. Intell. Mag. **11**(1), 20–31 (2016)
6. Weiss, K., Khoshgoftaar, T.M., Wang, D.: A survey of transfer learning. J. Big Data **3**, 9 (2016). Springer
7. Blankertz, B., Tomioka, R., Lemm, S., Kawanabe, M., Müller, K.R.: Optimizing spatial filters for robust EEG single-trial analysis. IEEE Signal Process. Mag. **25**(1), 41–56 (2008)
8. Cantillo-Negrete, J., Gutierrez-Martinez, J., Carino-Escobar, R., Carrillo-Mora, P., Elias-Vinas, D.: An approach to improve the performance of subject-independent BCIs-based motor imagery allocating subjects by gender. Biomed. Eng. **13**(1), 158 (2014)
9. Lotte, B.F.: To minimize or suppress calibration time in oscillatory activity-based brain computer interfaces. Proc. IEEE **103**(6), 871–890 (2015)
10. Rakotomamonjy, A., Guigue, V.: BCI competition III: dataset II- ensemble of SVMs for BCI P300 speller. IEEE Trans. Biomed. Eng. **55**(3), 1147–1154 (2008)
11. Onishi, A., Natsume, K.: Overlapped partitioning for ensemble classifiers of P300-based brain-computer interfaces. PLoS One **9**(4), e93045 (2014)
12. Xu, M., Liu, J., Chen, L., Qi, H., He, F., Zhou, P., Cheng, X., Wan, B., Ming, D.: Inter-subject information contributes to the ERP classification in the P300 speller. In: International IEEE/EMBS Conference on Neural Engineering 2015 July, pp. 206–209 (2015)
13. Hoffmann, U., Vesin, J., Ebrahimi, T., Diserens, K.: An efficient P300-based brain-computer interface for disabled subjects. J. Neurosci. Methods **167**, 115–125 (2008)
14. Xu, M., Liu, J., Chen, L., Qi, H., He, F., Zhou, P., Wan, B., Ming, D.: Incorporation of inter-subject information to improve the accuracy of subject-specific P300 classifiers. Int. J. Neural Syst. **26**(3), 1–12 (2016)
15. Locascio, J.J., Atri, A.: An overview of longitudinal data analysis methods for neurological research. Dement Geriatr. Cogn. Dis. Extra **1**(1), 330–357 (2011)
16. Hothorn, T., Bretz, F., Westfall, P.: Simultaneous inference in general parametric models. Biometrical J. **50**(3), 346–363 (2008)

Hybrid Global/Local Derivative-Free Multi-objective Optimization via Deterministic Particle Swarm with Local Linesearch ·

Riccardo Pellegrini[1], Andrea Serani[1], Giampaolo Liuzzi[2], Francesco Rinaldi[3], Stefano Lucidi[4], Emilio F. Campana[1], Umberto Iemma[5], and Matteo Diez[1(✉)]

[1] CNR-INSEAN, National Research Council,
Marine Technology Research Institute, Rome, Italy
`matteo.diez@cnr.it`
[2] CNR-IASI, National Research Council,
Institute for Systems Analysis and Computer Science, Rome, Italy
[3] Department of Mathematics, University of Padua, Padua, Italy
[4] Department of Computer, Control and Management Engineering "A. Ruberti",
Sapienza University, Rome, Italy
[5] Department of Engineering, Roma Tre University, Rome, Italy

Abstract. A multi-objective deterministic hybrid algorithm (MODHA) is introduced for efficient simulation-based design optimization. The global exploration capability of multi-objective deterministic particle swarm optimization (MODPSO) is combined with the local search accuracy of a derivative-free multi-objective (DFMO) linesearch method. Six MODHA formulations are discussed, based on two MODPSO formulations and three DFMO activation criteria. Forty five analytical test problems are solved, with two/three objectives and one to twelve variables. The performance is evaluated by two multi-objective metrics. The most promising formulations are finally applied to the hull-form optimization of a high-speed catamaran in realistic ocean conditions and compared to MODPSO and DFMO, showing promising results.

Keywords: Hybrid global/local optimization
Multi-objective optimization · Particle swarm optimization
Linesearch method · Derivative-free optimization
Deterministic optimization

1 Introduction

Simulation-based design optimization (SBDO) supports the design of complex engineering systems. The process consists in the evaluation of several numerical simulations to the aim of exploring and assessing design opportunities with improved performance for a set of often conflicting objectives. Multi-objective optimization algorithms drive the search for the best compromise among all design objectives, which are generally provided in the form of Pareto solutions.

© Springer International Publishing AG 2018
G. Nicosia et al. (Eds.): MOD 2017, LNCS 10710, pp. 198–209, 2018.
https://doi.org/10.1007/978-3-319-72926-8_17

In this context, objectives may be noisy and/or their derivatives are often not provided by the simulation tool. Therefore, derivative-free optimization algorithms are preferred as a viable option for the SBDO.

Global or local optimization algorithms are used, whether a fine search region is or is not known a priori. Global methods explore the whole design domain, providing approximate solutions to the decision problem. Local algorithms investigate accurately a limited domain region, also providing proof of convergence (generally not available for global methods). Hybrid global/local algorithms combine the global search capability of global methods with the accuracy and convergence properties of local algorithms. Examples of hybrid methods in the context of multi-objective optimization can be found in [1,2].

Among other derivative-free global methods, particle swarm optimization [3] has been successfully applied in SBDO [4] and extended to hybrid global/local formulations for both single- [5,6] and multi-objective [7–13] problems. Most algorithms are stochastic, requiring extensive numerical campaigns to achieve statistically significant results. Often this is not attainable in SBDO, especially if CPU-time expensive simulations provide directly objectives and constraints. Therefore, deterministic methods have been developed and assessed [4,14].

The objective of the present work is to introduce and assess a novel multi-objective deterministic hybrid algorithm (MODHA), which combines the global exploration capabilities of multi-objective deterministic particle swarm optimization (MODPSO [14]) with the local search accuracy of a deterministic derivative-free multi-objective (DFMO [15]) linesearch method.

Six formulations are proposed, based on two MODPSO formulations [14] and three DFMO activation criteria. Two of these are based on the particle velocity and one on the hypervolume metric [16]. A comparative study is performed using 45 analytical test problems, with a number of objective functions ranging from two to three and a number of variables from one to twelve. The DFMO activation criterion is investigated along with the number of function evaluations assigned to the local search. A full-factorial combination of formulations and setting parameters is investigated through more than 14,000 optimization runs. Two multi-objective performance metrics are assessed, namely the number of solutions found and the hypervolume bounded by the solution set.

The most promising formulations are applied to the reliability-based robust design optimization (RBRDO) of a high-speed catamaran in realistic ocean environment, sailing in head waves in the North Pacific Ocean with stochastic sea state and speed [17]. A comparison with MODPSO and DFMO is provided.

2   Optimization Problem Formulation

The multi-objective minimization problem can be formulated as

$$
\begin{aligned}
\text{minimize} \quad & \mathbf{f}(\mathbf{x}) = \{f_m(\mathbf{x})\}, \quad \text{with} \quad m = 1, \ldots, N_{\text{of}} \\
\text{subject to} \quad & z_i(\mathbf{x}) \leq 0, \quad \text{with} \quad i = 1, \ldots, I \\
\text{and to} \quad & h_j(\mathbf{x}) = 0, \quad \text{with} \quad j = 1, \ldots, J \\
\text{and to} \quad & \mathbf{l} \leq \mathbf{x} \leq \mathbf{u}
\end{aligned}
\tag{1}
$$

where $\mathbf{x} \in \mathbb{R}^{N_{dv}}$ is the vector collecting the N_{dv} variables, N_{of} is the number of objective functions f_m, z_i are the inequality constraints, h_j are the equality constraints, and $\mathbf{l}$ and $\mathbf{u}$ are the lower and upper bound for $\mathbf{x}$, respectively.

Defining the feasible solution set as $\mathcal{X} = \{\mathbf{x} \in \mathbb{R}^{N_{dv}} \mid [\cap_i^I z_i(\mathbf{x}) \leq 0] \wedge [\cap_j^J h_j(\mathbf{x}) = 0 \wedge [\mathbf{l} < \mathbf{x} < \mathbf{u}]\}$, the solution of Eq. 1 is the locus of non dominated feasible solutions represented in the variable space by the Pareto solution set $\mathcal{PS} = \{\mathbf{x} \in \mathcal{X} \mid \mathbf{f}(\mathbf{x}) \prec \mathbf{f}(\mathbf{y}), \forall \mathbf{y} \in \mathcal{X}\}$. In the objective function space, the locus is represented by the Pareto front $\mathcal{PF} = \{\mathbf{f}(\mathbf{x}) : \mathbf{x} \in \mathcal{PS}\}$. In the following, the approximate solution set $\mathcal{S}$ (set of non dominated solutions represented either in the variable or function space) achieved by the optimizer at a specific iteration n is indicated by $\mathcal{S}^n = \{(\mathbf{x}, \mathbf{s}) : \mathbf{s} = \mathbf{f}(\mathbf{x}) \prec \mathbf{f}(\mathbf{y}), \forall \mathbf{y}\}$. Similarly, the approximate Pareto front (assessed by numerical experiments and used as a reference solution set for the performance analysis) is defined as $\mathcal{R} = \{(\mathbf{x}, \mathbf{r}) \in \cup_{i=1}^{N_s} \mathcal{S}_i : \mathbf{r} = \mathbf{f}(\mathbf{x}) \prec \mathbf{f}(\mathbf{y}), \forall \mathbf{y}\}$, where N_s is the number of solution sets available, provided by different algorithm formulations/setups.

3 Performance Metrics

The algorithm performance is evaluated in terms of *capacity* (related to the number of Pareto solutions $\mathcal{S}$), *convergence* (related to the distance between $\mathcal{S}$ and $\mathcal{R}$), and *diversity* (related to how $\mathcal{S}$ is wide). Here, the following two metrics are used. The Ratio of Reference Point Found (C1$_R$, [18])

$$\text{C1}_R = \frac{|\mathcal{S} \cap \mathcal{R}|}{|\mathcal{R}|} \tag{2}$$

is used as capacity metric, whereas a normalized version of the hypervolume (HV) [16] is used as a convergence-diversity metric, defined as

$$\text{NHV} = \frac{\text{HV}(\mathcal{S}, \mathcal{R})}{\text{HV}(\mathcal{R}, \mathcal{R})}, \quad \text{with} \quad \text{HV}(\mathcal{S}, \mathcal{R}) = \text{volume} \left(\bigcup_{i=1}^{|\mathcal{S}|} v_i \right) \tag{3}$$

where $\text{HV}(\mathcal{S}, \mathcal{R})$ gives the (hyper) volume dominated by the solution set $\mathcal{S}$, evaluated using as a reference the anti-ideal point of $\mathcal{R}$ [19].

Additionally, the relative variability $\sigma_{r,k}^2$ [20] is used to assess the impact of the k-th setting parameter on the algorithm performance.

4 Hybrid Global/Local Deterministic Algorithm

The selected global and local algorithms are described in the following along with their hybridization.

4.1 MODPSO

PSO algorithm [3] is based on the social-behaviour metaphor of a flock of birds or a swarm of bees searching for food and belongs to the class of metaheuristic algorithms for single-objective derivative-free global optimization. Pinto et al. [21] proposed a multi-objective deterministic version of PSO as

$$
\begin{cases}
\mathbf{v}_i^{n+1} = \chi \left[\mathbf{v}_i^n + c_1 \left(\mathbf{p}_i - \mathbf{x}_i^n \right) + c_2 \left(\mathbf{g}_i - \mathbf{x}_i^n \right) \right] \\
\mathbf{x}_i^{n+1} = \mathbf{x}_i^n + \mathbf{v}_i^{n+1}
\end{cases}
\tag{4}
$$

where $\mathbf{v}_i^n$ and $\mathbf{x}_i^n$ are the velocity and the position of the i-th particle at the n-th iteration, χ is a constriction factor, c_1 and c_2 are the cognitive and social learning rate, and $\mathbf{p}_i$ and $\mathbf{g}_i$ are the cognitive and social attractor.

In this work two MODPSO formulations are selected from [14], namely:

- MODPSO1, where $\mathbf{p}_i$ is the closest point to the i-th particle of the personal solution set $\mathcal{S}_{\mathrm{p},i}^n$ (i.e., the set of all non dominated solutions ever visited by the i-th particle) and $\mathbf{g}_i$ is the closest point to the i-th particle of the solution set $\mathcal{S}^n$;
- MODPSO3, where $\mathbf{p}_i$ is the personal minimizer of the aggregated objective function $F(\mathbf{x}_i) = \sum_{m=1}^{N_{\mathrm{of}}} f_m(\mathbf{x}_i)$ and $\mathbf{g}_i$ is the closest point to the i-th particle of the solution set $\mathcal{S}^n$.

4.2 DFMO

It is a derivative-free algorithm for constrained (possibly) non-smooth multi-objective problems [15], representing a so-called "a posteriori" method in the sense that it is able to approximate the entire $\mathcal{PF}$ by producing in output a set of non dominated points. More in particular, at every iteration, the algorithm produces (or updates) a set of non dominated points (rather than a single point, as it is common in the single-objective case). As the iteration count grows, these sets of points tend to the $\mathcal{PF}$ of the problem.

Other relevant features of DFMO are: (i) a linesearch approach that takes into account the presence of multiple objectives; (ii) an exact penalty approach for dealing with the nonlinear constraints. At each iteration, for each point in $\mathcal{S}$, DFMO starts a linesearch along a suitably generated direction d_j. If such a direction is able to guarantee "sufficient" decrease, then a "sufficiently" large movement λ along the direction is performed. This allows to (possibly) improve $\mathcal{S}$. Detail of algorithm formulation and implementation can be found in [15].

4.3 MODHA

A critical issue when combining MODPSO and DFMO is to define when and where from the local search starts. Here, three approaches are defined: two are based on the velocity of the particle and one on the HV metric.

The velocity-based formulation starts a local search if the normalized speed of the i-th particle drops under a threshold value β, namely when

Algorithm 1. MODHA pseudo-code

```
 1: Initialize a swarm of N_p particles
 2: while (n < Max number of iterations) do                    ▷ MODPSO begins
 3:     for i = 1, N_p do
 4:         Evaluate f(x_i^n)
 5:         Compute S_{p,i}^n
 6:     end for
 7:     Compute S^n
 8:     for i = 1, N_p do
 9:         Identify cognitive attractor p_i
10:         Identify social attractor g_i
11:         Update particle velocities v_i^{n+1}
12:         Update particle positions x_i^{n+1}
13:     end for
14:     Evaluate condition for performing DFMO based on hybridization scheme
15:     if condition for performing DFMO is true then
16:         Define N_dv coordinate directions d_j
17:         Identify N_l starting points for DFMO based on hybridization scheme
18:         for i = 1, N_l do                                   ▷ DFMO begins
19:             Set N_DFMO to zero
20:             for j = 1, N_dv do
21:                 while N_DFMO < max. allowed (depending on α) and λ > λ_min do
22:                     Perform one step equal to λ along d_j from the starting point i
23:                     Evaluate f
24:                     Set N_DFMO to N_DFMO + 1
25:                     if At least one objective function decreases "sufficiently" then
26:                         Update DFMO solution set
27:                         go to 20
28:                     else
29:                         Reduce λ
30:                     end if
31:                 end while
32:             end for
33:         end for                                            ▷ DFMO ends
34:         Update S^n (and S_{p,i}^n if required) with DFMO solution set
35:     end if
36: end while                                                 ▷ MODPSO ends
37: Output S^n
```

$||\mathbf{v}_i|| / ||(\mathbf{u} - 1)|| < \beta$. The local search starts either from the current particle position (PP) or from the particle social attractor (SA). The hypervolume-based formulation starts the local search from each point of the current solution set (SS) if $HV(\mathcal{S}^n, \mathcal{S}^n) < \gamma HV(\mathcal{S}^{n-1}, \mathcal{S}^n)$. $HV(\mathcal{S}^n, \mathcal{S}^n)$ is the hypervolume associated to $\mathcal{S}^n$, γ is the threshold coefficient, and $HV(\mathcal{S}^{n-1}, \mathcal{S}^n)$ is the hypervolume associated to $\mathcal{S}^{n-1}$.

The number of problem evaluations (N_{DFMO}) performed at each call of the local algorithm is defined as $N_{DFMO} = \alpha N_{dv} N_{of}$ and $N_{DFMO} = \alpha N_p$ for velocity- and hypervolume-based formulations, respectively. Algorithm 1 shows the pseudo code for the current hybrid formulations.

4.4 Algorithm Parameters and Setup

The MODPSO1 and MODPSO3 setups are defined as in [14]. The number of particles N_p is set to $8N_{of}N_{dv}$, initialized using a Hammersley sequence sampling [22] over variable domain and boundary. The coefficients are set as proposed by

Clerc [23], with $\chi = 0.721$ and $c_1 = c_2 = 1.655$. A semi-elastic wall-type approach is used for box constraints [4].

Threshold values for local search activation are set to $\beta = \{0.1, 1.0, 10\}$ and $\gamma = \{1.0, 1.1, 1.2\}$. The budget of local search evaluations (for each call) is set by $\alpha = \{1, 5, 10\}$. The linesearch step is reduced by a factor of two until it reaches a minimum step size $\lambda_{\min} = 1E-9$, starting from a maximum value equal to the 10% of the design variables space dimension.

The number of problem evaluations (N_{peval}), where one problem evaluation involves one evaluation of each objective function, is assessed by $N_{\text{peval}} = \nu N_{\text{of}} N_{\text{dv}}$ where $\nu = 125 \cdot 2^c$, $c \in \mathbb{N}[0, 4]$ therefore N_{peval} ranges between $125 N_{\text{of}} N_{\text{dv}}$ and $2000 N_{\text{of}} N_{\text{dv}}$.

5   Numerical Results

A preliminary study on analytical benchmark problems is used to identify the most promising MODHA formulation and setup. The MODHA formulations under analysis are summarized in the following:

- PP1 and PP3 perform $\alpha N_{\text{dv}} N_{\text{of}}$ local search for each call, starting from the current particle position $\mathbf{x}_i^n$, and are activated by the velocity threshold β;
- SA1 and SA3 perform $\alpha N_{\text{dv}} N_{\text{of}}$ local search for each call, starting from the particle social attractor $\mathbf{g}_i$, and are activated by the velocity threshold β;
- SS1 and SS3 perform αN_{p} local search for each call, starting from the current solution set $\mathcal{S}^n$, and are activated by the HV threshold value γ.

"1" and "3" indicate the MODPSO formulation. The most promising MODHA formulations are finally applied to the RBRDO of the high-speed catamaran and compared with MODPSO1, MODPSO3, and DFMO.

5.1   Analytical Benchmark Problems

A number of 45 benchmark problems [14] is used, including convex and non-convex, continuous and discontinuous Pareto fronts, with $N_{\text{of}} = 2, 3$ and $1 \leq N_{\text{dv}} \leq 12$.

In order to provide a proper comparison between different problems with different codomain size, each solution set $\mathcal{S}$ is normalized with the function range, therefore $\mathbf{s}_i \in [0, 1]$ and the reference point HV is $\{1\}_{i=1}^{N_{\text{of}}}$. The computation of HV is performed with the code provided in [24].

Figure 1 shows the relative variability of $C1_R$ and NHV, conditional to the setup parameters. Considering both metrics, the velocity-based formulations (PP1, PP3, SA1, and SA3) are mainly affected by the velocity threshold β, whereas the hypervolume-based formulation is mainly influenced by the coefficient α, but SS3.

Figure 2 compares $C1_R$ and NHV provided by global, local, and hybrid global/local algorithms. Although DFMO achieves the highest $C1_R$, hybrid methods provide significantly larger NHV values. In general, within the same

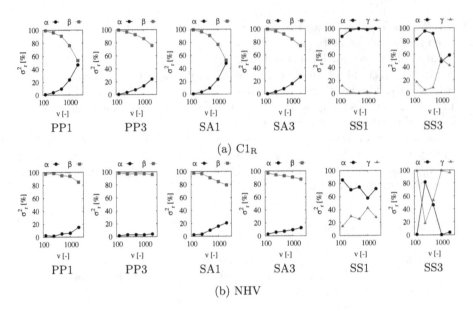

(a) C1$_R$

(b) NHV

Fig. 1. Analytical test problems, relative variability $\sigma^2_{r,k}$ conditional to the formulation parameters

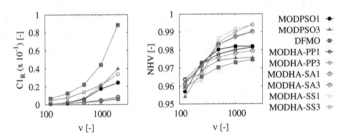

Fig. 2. Analytical test problems, comparison of MODPSO1, MODPSO3, DFMO, and most promising setup of MODHA formulations

hybridization approach, MODPSO1 and MODPSO3 achieve similar performances. It is worth noting that the velocity-based hybrid formulation, that start the local search from the current particle position (PP1 and PP3), are not able to outperform the corresponding global algorithms.

Table 1 summarizes the most promising setup for each hybrid formulation, based on budget-averaged NHV. MODHA-SS3 with an activation threshold $\gamma = 1.0$ and a coefficient $\alpha = 10$ for the DFMO problem evaluations is the best performing overall (on average).

Finally, Fig. 3 shows illustrative examples of the solution achieved by global, local, and hybrid (SS1 and SS3) algorithms for the *Sch1* problem [25] with $N_{of} = 2$ and $N_{dv} = 1$. The hybrid algorithms show a more accurate approximation of the Pareto front than local and global algorithms.

Table 1. Most promising MODHA setup based on budget-averaged NHV

MODHA formulation	β	γ	α	C1$_R$	NHV
PP1	0.1	–	1	2.981E−3	0.9750
PP3	0.1	–	5	3.764E−3	0.9786
SA1	0.1	–	1	2.920E−3	0.9739
SA3	0.1	–	5	3.700E−3	0.9785
SS1	–	1.0	10	1.633E−2	0.9789
SS3	–	1.0	10	1.632E−2	**0.9797**

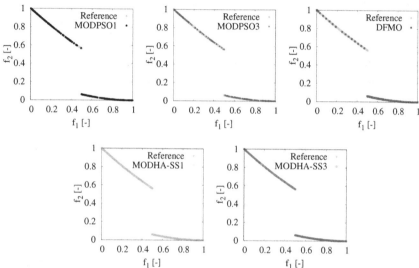

Fig. 3. Global, local, and hybrid algorithm solution for the *Sch1* problem with $2000 N_{\text{of}} N_{\text{dv}}$ problem evaluations

5.2 High-Speed Catamaran Optimization

A reliability-based robust design optimization of a 100 m high-speed catamaran is solved for realistic conditions, associated to the North Pacific Ocean including stochastic sea state and speed [17]. The multi-objective problem aims at the reduction of the expected value of the mean total resistance in irregular waves (φ_1) and the increase of the ship operability referring to a set of motion-related constraints (φ_2). The design optimization problem is formulated as

$$
\begin{aligned}
\text{minimize} \quad & \{\varphi_1(\mathbf{x}), -\varphi_2(\mathbf{x})\}^T \\
\text{subject to} \quad & \mathbf{l} \leq \mathbf{x} \leq \mathbf{u} \\
\text{and to} \quad & \varphi_1 \leq 0; \quad \varphi_2 \geq 0
\end{aligned}
\tag{5}
$$

The problem is solved by means of stochastic radial-basis function interpolation [26] of high-fidelity URANS simulations. The inequalities in Eq. 5 are handled by a linear penalty function, so that $\varphi_k = \varphi_k + 100 \sum_{j=1}^{N_{dv}} \max(x_j - u_j, 0) + 100 \sum_{j=1}^{N_{dv}} |\min(l_j - x_j, 0)|$ if domain bounds violation occurs and $\varphi_k = 10000\varphi_k$ if $\varphi_1 > 0$ or $\varphi_2 < 0$. Four design variables ($N_{dv} = 4$) control global shape modifications of the catamaran hull, based on the Karhunen-Loève expansion of the shape modification vector. Details may be found in [17]. A total number of 16,000 problem evaluations are performed and used to compute the reference non dominated solution set $\mathcal{R}$.

Figure 4 shows the solution obtained by MODPSO1 and 3, DFMO, and the hybrid algorithms SS1 and SS3 with the most promising parameter set summarized in Table 1. The hybrid algorithms are able to cover the reference solution, outperforming the global and local algorithms. It is worth noting that the hybrid algorithms are able to accurately identify the upper right section of $\mathcal{R}$. The solution provided by SS3 is more accurate than that provided by SS1.

Table 2 summarizes C1$_R$ and NHV percentage values achieved by each algorithm, conditional to the budget parameter ν. Both metrics confirm the results depicted in Fig. 4. The hybrid algorithms perform better than the global and local algorithms and provide more dense solutions. In particular, SS3 is found to

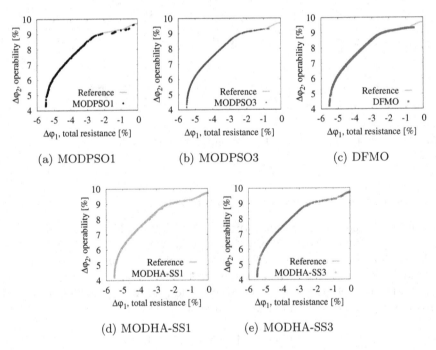

Fig. 4. Global, local, and hybrid algorithm solution for the catamaran problem with $2000 N_{of} N_{dv}$ problem evaluations

Table 2. Catamaran problem, summary of the optimization results

ν	MODPSO1		MODPSO3		DFMO		MODHA-SS1		MODHA-SS3	
	$C1_R$	NHV	$C1_R$	NHV	$C1_R$	NHV	$C1_R$	NHV	$C1_R$	NHV
125	0.000E+0	0.9977	0.000E+0	0.9223	0.000E+0	0.9983	0.000E+0	0.9969	0.000E+0	0.9983
250	1.388E−3	0.9983	0.000E+0	0.9687	6.246E−3	0.9984	3.470E−4	0.9978	4.511E−3	0.9986
500	8.675E−3	0.9984	7.634E−3	0.9826	2.325E−2	0.9984	6.246E−3	0.9980	1.410E−1	0.9986
1000	3.088E−2	0.9994	5.274E−2	0.9873	7.911E−2	0.9984	4.580E−2	0.9988	3.540E−1	0.9999
2000	3.227E−2	0.9995	1.620E−1	0.9889	2.866E−1	0.9984	1.117E−1	0.9999	**4.060E−1**	**0.9999**

be the best formulation, achieving higher values of $C1_R$ and NHV and providing more dense solutions than SS1.

6   Conclusions and Future Work

A multi-objective deterministic hybrid algorithm (MODHA) has been presented, combing two multi-objective deterministic particle swarm formulations with a local derivative-free multi-objective linesearch algorithm. Three hybridization schemes have been studied: two are based on the particle velocity and one on the hypervolume metric. The velocity-based formulation starts the local search when the particle velocity drops under a threshold value (β) and use as a starting point either the current particle position (PP) or the particles social attractor (SA). The hypervolume-based formulation starts the local search when the HV associated to the current solution set does not improve sufficiently (by a factor equal to γ) compared to the previous iteration. In this case, a local search is performed starting from each point of the current solution set (SS). These hybridization schemes are combined to both MODPSO1 and MODPSO3, resulting in six MODHA formulations.

A comparative study has been performed using 45 analytical test problems, with a number of objective functions ranging from two to three and a number of variables from one to twelve, varying the activation criterion and the number of problem evaluations for the local search. A full-factorial combination of formulations and parameters has been investigated through more than 14,000 optimization runs. Two multi-objective performance metrics ($C1_R$ and NHV) have been evaluated and discussed.

Velocity-based formulations depend significantly on the local search activation threshold, whereas the hypervolume-based formulation is affected mainly by the coefficient related to the number of evaluations reserved for the local search. Hybrid formulations based on the hypervolume show the best performance. Specifically, MODHA-SS3 with $\alpha = 10$ and $\gamma = 1.0$ is found the most promising on average. Hypervolume formulations have been applied to the hull-form optimization of a high-speed catamaran (aimed at reducing the resistance and increasing the operability in realistic ocean conditions), showing better results than global and local algorithms. Also for the catamaran, MODHA-SS3 provides the best performance.

Current results are promising and motivate further investigations of metrics-based formulations, with focus on the method for the selection of local search starting points. Future work includes the development and assessment of a hybrid version of the crowding-distance based MOPSO [27] and the use of the crowding distance to select the local search starting points. The effects of the local search stop criterion on the overall performance will be included in the analysis. Finally, novel strategies for the approximation of the Pareto front (*e.g.* [28]) will be considered to enhance the exploration capabilities of the MODHA formulations.

Acknowledgements. The work is supported by the US Office of Naval Research Global, NICOP grant N62909-15-1-2016, under the administration of Dr. Woei-Min Lin, Dr. Salahuddin Ahmed, and Dr. Ki-Han Kim, and by the Italian Flagship Project RITMARE, founded by the Italian Ministry of Education.

References

1. Qian, C., Yu, Y., Zhou, Z.H.: Pareto ensemble pruning. In: Proceedings of the Twenty-Ninth AAAI Conference on Artificial Intelligence, AAAI 2015, pp. 2935–2941. AAAI Press (2015)
2. Qian, C., Tang, K., Zhou, Z.-H.: Selection hyper-heuristics can provably be helpful in evolutionary multi-objective optimization. In: Handl, J., Hart, E., Lewis, P.R., López-Ibáñez, M., Ochoa, G., Paechter, B. (eds.) PPSN 2016. LNCS, vol. 9921, pp. 835–846. Springer, Cham (2016). https://doi.org/10.1007/978-3-319-45823-6_78
3. Kennedy, J., Eberhart, R.: Particle swarm optimization. In: Proceedings of the IEEE International Conference on Neural Networks, vol. 4, pp. 1942–1948 (1995)
4. Serani, A., Leotardi, C., Iemma, U., Campana, E.F., Fasano, G., Diez, M.: Parameter selection in synchronous and asynchronous deterministic particle swarm optimization for ship hydrodynamics problems. Appl. Soft Comput. **49**, 313–334 (2016)
5. Serani, A., Diez, M., Campana, E.F., Fasano, G., Peri, D., Iemma, U.: Globally convergent hybridization of particle swarm optimization using line search-based derivative-free techniques. In: Yang, X.S. (ed.) Recent Advances in Swarm Intelligence and Evolutionary Computation. SCI, vol. 585, pp. 25–47. Springer, Cham (2015). https://doi.org/10.1007/978-3-319-13826-8_2
6. Serani, A., Fasano, G., Liuzzi, G., Lucidi, S., Iemma, U., Campana, E.F., Stern, F., Diez, M.: Ship hydrodynamic optimization by local hybridization of deterministic derivative-free global algorithms. Appl. Ocean Res. **59**, 115–128 (2016)
7. Liu, D., Tan, K.C., Goh, C.K., Ho, W.K.: A multiobjective memetic algorithm based on particle swarm optimization. IEEE Trans. Syst. Man Cybern. Part B (Cybern.) **37**(1), 42–50 (2007)
8. Kaveh, A., Laknejadi, K.: A novel hybrid charge system search and particle swarm optimization method for multi-objective optimization. Expert Syst. Appl. **38**(12), 15475–15488 (2011)
9. Cheng, S., Zhan, H., Shu, Z.: An innovative hybrid multi-objective particle swarm optimization with or without constraints handling. Appl. Soft Comput. **47**, 370–388 (2016)
10. Santana-Quintero, L.V., Ramírez, N., Coello, C.C.: A multi-objective particle swarm optimizer hybridized with scatter search. In: Gelbukh, A., Reyes-Garcia, C.A. (eds.) MICAI 2006. LNCS (LNAI), vol. 4293, pp. 294–304. Springer, Heidelberg (2006). https://doi.org/10.1007/11925231_28

11. Izui, K., Nishiwaki, S., Yoshimura, M., Nakamura, M., Renaud, J.E.: Enhanced multiobjective particle swarm optimization in combination with adaptive weighted gradient-based searching. Eng. Optim. **40**(9), 789–804 (2008)
12. Mousa, A., El-Shorbagy, M., Abd-El-Wahed, W.: Local search based hybrid particle swarm optimization algorithm for multiobjective optimization. Swarm Evol. Comput. **3**, 1–14 (2012)
13. Xu, G., Yang, Y.Q., Liu, B.B., Xu, Y.H., Wu, A.J.: An efficient hybrid multiobjective particle swarm optimization with a multi-objective dichotomy line search. J. Comput. Appl. Math. **280**, 310–326 (2015)
14. Pellegrini, R., Serani, A., Leotardi, C., Iemma, U., Campana, E.F., Diez, M.: Formulation and parameter selection of multi-objective deterministic particle swarm for simulation-based optimization. Appl. Soft Comput. **58**, 714–731 (2017)
15. Liuzzi, G., Lucidi, S., Rinaldi, F.: A derivative-free approach to constrained multiobjective nonsmooth optimization. SIAM J. Optim. **26**(4), 2744–2774 (2016)
16. Zitzler, E., Thiele, L.: Multiobjective optimization using evolutionary algorithms - a comparative case study. In: Eiben, A.E., Bäck, T., Schoenauer, M., Schwefel, H.-P. (eds.) PPSN 1998. LNCS, vol. 1498, pp. 292–301. Springer, Heidelberg (1998). https://doi.org/10.1007/BFb0056872
17. Diez, M., Campana, E.F., Stern, F.: Development and evaluation of hull-form stochastic optimization methods for resistance and operability. In: Proceedings of the 13th International Conference on Fast Sea Transportation (FAST 2015) (2015)
18. Czyzak, P., Jaszkiewicz, A.: Pareto simulated annealing-a metaheuristic technique for multiple-objective combinatorial optimization. J. Multi-criteria Decis. Anal. **7**(1), 34–47 (1998)
19. Jiang, S., Ong, Y.S., Zhang, J., Feng, L.: Consistencies and contradictions of performance metrics in multiobjective optimization. IEEE Trans. Cybern. **44**(12), 2391–2404 (2014)
20. Campana, E.F., Diez, M., Iemma, U., Liuzzi, G., Lucidi, S., Rinaldi, F., Serani, A.: Derivative-free global ship design optimization using global/local hybridization of the DIRECT algorithm. Optim. Eng. **17**(1), 127–156 (2015)
21. Pinto, A., Peri, D., Campana, E.F.: Multiobjective optimization of a containership using deterministic particle swarm optimization. J. Ship Res. **51**(3), 217–228 (2007)
22. Wong, T., Luk, W., Heng, P.: Sampling with Hammersley and Halton points. J. Graphics Tools **2**(2), 9–24 (1997)
23. Clerc, M.: Stagnation analysis in particle swarm optimization or what happens when nothing happens (2006). http://clerc.maurice.free.fr/pso
24. Fonseca, C.M., Paquete, L., Lòpez-Ibàñez, M.: An improved dimension - sweep algorithm for the hypervolume indicator. In: Proceedings of the Congress on Evolutionary Computation (CEC 2006), pp. 1157–1163. IEEE (2006)
25. Huband, S., Hingston, P., Barone, L., While, L.: A review of multiobjective test problems and a scalable test problem toolkit. IEEE Transa. Evol. Comput. **10**(5), 477–506 (2006)
26. Volpi, S., Diez, M., Gaul, N., Song, H., Iemma, U., Choi, K.K., Campana, E.F., Stern, F.: Development and validation of a dynamic metamodel based on stochastic radial basis functions and uncertainty quantification. Struct. Multidisciplinary Optim. **51**(2), 347–368 (2015)
27. Raquel, C.R., Naval Jr., P.C.: An effective use of crowding distance in multiobjective particle swarm optimization. In: Proceedings of the 7th Annual Conference on Genetic and Evolutionary Computation, pp. 257–264. ACM (2005)
28. Žilinskas, A.: Visualization of a statistical approximation of the pareto front. Appl. Math. Comput. **271**, 694–700 (2015)

Artificial Bee Colony Optimization to Reallocate Personnel to Tasks Improving Workplace Safety

Beatrice Lazzerini and Francesco Pistolesi[✉]

Department of Information Engineering, University of Pisa,
Largo Lucio Lazzarino 1, 56122 Pisa, Italy
{b.lazzerini,f.pistolesi}@iet.unipi.it

Abstract. Worldwide, just under 5,800 people go to work every day and do not return because they die on the job. The groundbreaking Industry 4.0 paradigm includes innovative approaches to improve the safety in the workplace, but Small and Medium Enterprises (SMEs) – which represent 99% of the companies in the EU – are often unprepared to the high costs for safety. A cost-effective way to improve the level of safety in SMEs may be to just reassign employees to tasks, and assign hazardous tasks to the more cautious employees. This paper presents a multi-objective approach to reallocate the personnel of a company to the tasks in order to maximize the workplace safety, while minimizing the cost, and the time to learn the new tasks assigned. Pareto-optimal reallocations are first generated using the Non-dominated Sorting artificial Bee Colony (NSBC) algorithm, and the best one is then selected using the Technique for Order of Preference by Similarity to Ideal Solution (TOPSIS). The approach was tested in two SMEs with 11 and 25 employees, respectively.

Keywords: Bee colony algorithm · Occupational safety and health
Multi-objective optimization · Personnel reallocation
Risk perception · TOPSIS

1 Introduction

Improving the safety of work environments is key. Every 15 s, a worker dies as a consequence of occupational injuries and accidents [1]. New technologies helps save a countless number of lives today, but workplace fatalities are not diminishing enough.

In economic terms, occupational illnesses and accidents at work result in costs up to 6% of GDP, in country estimates [1]. Statistics also say that up to 80% of the accidents are caused by workers' actions or omissions [9]. It is thus crucial to study both the employees' behavior and personality when assigning tasks. Workers are indeed characterized by the so-called *human factors*, i.e., individual aspects and organizational, environmental and job factors that modify a worker's behavior in a way that can influence occupational safety and health (OSH) [9].

© Springer International Publishing AG 2018
G. Nicosia et al. (Eds.): MOD 2017, LNCS 10710, pp. 210–221, 2018.
https://doi.org/10.1007/978-3-319-72926-8_18

Human factors affect *risk perception*, i.e., how one understands characteristics and level of danger in the presence of hazards [3,4,16,18,19]. Human factors include, e.g., age, past experience and health status, social and cultural aspects, psychological traits, trust in risk management institutions, optimism bias [7,11] and locus of control [10,11]. These factors have been investigated in the Sociology and Psychology literature, but the way they affect human behavior when in the presence of risk remains vague.

Risk awareness courses help employees achieve appropriate risk awareness and enforce safety guidelines be observed. Employees regularly undergo risk awareness training, with high capital investments for companies. However, fatalities and accidents at work are too frequent. Novel techniques were proposed in [12,13] to profile workers depending on their sensitivity to risk. This can help provide employees tailored risk awareness courses. Also, the Smart Manufacturing approach, a part of the groundbreaking Industry 4.0 paradigm, has the workplace safety improvement as a primary objective. Limited economic resources make safety hard to manage for Small and Medium Enterprises (SMEs), which represent ∼99% of the companies in the EU and employ 65 million people [2]. If human factors were included into OSH procedures, accidents at work could be reduced [9]. To help SMEs achieve a low-cost workplace safety increase, it may be sufficient to reallocate the personnel to tasks analyzing the employees': (i) human factors; (ii) ability to learn new tasks; (iii) behavior when exposed to the hazards of the tasks [14,15].

This paper presents a multi-objective approach to reallocate the personnel of a company to the tasks, in order to improve the workplace safety, while keeping low both the costs and the time to learn the new tasks assigned. The learning time that an employee takes to learn a new task is predicted using his/her past jobs where the employee performed that task. Risk-free practical tests help estimate the learning time in the case the worker has never performed the task before.

A neural network-based system [6] calculates every employee's level of caution towards each task, starting from the human factors and behavior when in the presence of the risks of the task. An employee's behavior while performing a task is expressed on the basis of the precautions taken during the task execution.

The multi-objective problem is solved by generating an approximation of the whole Pareto front using the Non-dominated Sorting Bee Colony optimization algorithm (NSBC). The most appropriate Pareto-optimal personnel (re)allocation is selected using the TOPSIS algorithm. Experiments were carried out involving two footwear companies with 11 and 25 employees, respectively.

The paper is organized as follows: Sect. 2 contains a background on multi-objective optimization and the description of the NSBC algorithm; Sect. 3 describes TOPSIS; Sect. 4 contains the details on how an employee's level of caution towards a task is computed; Sect. 5 gives the problem formulation; in Sect. 6 the experiments are discussed; Sect. 7 draws the conclusions.

2   Multi-objective Optimization

Multi-objective optimization (MOO) problems deal with the optimization of multiple objectives, typically conflicting [8]. An MOO problem can be written as Minimize$_{x \in \mathcal{X}}$ $\mathbf{f}(\mathbf{x}) = [f_1(\mathbf{x}), \ldots, f_k(\mathbf{x})]$, where $\mathcal{X} = \{x \in \mathbb{R}^p : g_i(\mathbf{x}) \leq 0 \; \forall i = 1, \ldots, m, h_j(\mathbf{x}) = 0 \; \forall j = 1, \ldots, n\}$. The vector function $\mathbf{f} : \mathbb{R}^p \to \mathbb{R}^k$ contains the objective functions. In general, any solution does not minimize all the objective functions at the same time. Thus, *Pareto dominance* and *Pareto-optimality* are introduced. A solution $\mathbf{x}^1$ dominates $\mathbf{x}^2$ if $f_i(\mathbf{x}^1) \leq f_i(\mathbf{x}^2) \; \forall i \in \{1, \ldots, k\}$, and $f_j(\mathbf{x}^1) < f_j(\mathbf{x}^2)$ for at least one $j \in \{1, \ldots, k\}$. Pareto-optimal solutions map to the *Pareto front*, in the objective space.

2.1   Non-dominated Sorting Bee Colony Optimization

Overview. The Non-dominated Sorting Bee Colony (NSBC) algorithm is a popular algorithm inspired by the foraging behavior of bees [17]. NSBC encodes a solution using the position of a food source and its nectar amount (i.e., the fitness of the solution). NSBC divides a bee colony into *onlookers, employed bees* and *scouts*. Onlookers stand on a dance area waiting to decide for a quality food source; employed bees are associated with food sources; scouts perform a random search. The total number of employed and onlooker bees is equal to the number of candidate solutions.

The NSBC Optimization Algorithm. NSBC initializes a population of N food sources (candidate solutions) of dimension D. Food source $\mathbf{X}_i(t) = (x_{i,1}, \ldots, x_{i,D})$ of population $P_{t=0}$ is randomly initialized in the range $[\mathbf{X}_{min}, \mathbf{X}_{max}]$, with $\mathbf{X}_{min} = (x_1^{min}, \ldots, x_D^{min})$ and $\mathbf{X}_{max} = (x_1^{max}, \ldots, x_D^{max})$. The j-th component of $\mathbf{X}_i(t = 0)$ is $x_{i,j}(t = 0) = x_j^{min} + \mathcal{U}(0, 1) \cdot (x_j^{max} - x_j^{min})$, where $\mathcal{U}(0, 1)$ is a random number uniformly distributed in $[0, 1]$. The k-th component $f_k(\mathbf{X}_i(t = 0))$, i.e., the fitness of every food source $\mathbf{X}_i(t = 0)$, is computed for each $i = 1, \ldots, N$.

Each employed bee looks for a new food source $\mathbf{X}'_i(t) = (x_{i,1}, \ldots, x'_{i,j}, \ldots, x_{i,D})$ changing the j-th component, selected randomly. The new value is:

$$x'_{i,j}(t) = x_{i,j}(t) + \mathcal{U}(-1, 1) \cdot (x_{i,j}(t) - x_{k,j}(t)) \tag{1}$$

where $k \neq i$ and $\mathcal{U}(-1, 1)$ is a random number uniformly distributed in $[-1, 1]$. If $\mathbf{X}_i(t)$ is dominated by $\mathbf{X}'_i(t)$, the bee substitutes the previous food source with $\mathbf{X}'_i(t)$. Otherwise, the bee maintains both solutions in memory. This step iterates for every food source. The population so obtained (size $N \leq N' \leq 2N$) is sorted according to non-domination. Non-dominated food sources take rank 1 (first front). These food sources are then neglected to find the second front, etc.

A parent population P'_t of size N is built for the onlooker bee phase, according to the ascending order of the non-domination ranking. The food sources in the last front that can be inserted in P_t are sorted in descending order of crowding distance, i.e., the sum of the distances from a food source to its closest food source along each objective. Let $\mathcal{D}_i$ be the set of the food sources dominated by $\mathbf{X}_i(t)$. Each food source $\mathbf{X}_i(t = 0)$, where $i = 1, \ldots, N$, is associated with a probability to be selected by the onlooker bee. This probability is equal to $\pi_i = \frac{|\mathcal{D}_i|}{N}$, where $| \cdot |$ denotes the cardinality of a set. Onlooker bees evaluate the fitness of each food source from all employed bees and select a food source $\mathbf{X}_i(t)$ on the basis of probability π_i. Onlooker bees change the position of the food source in their memory in accordance to Eq. (1), and test the fitness of the new food source.

Population P'_t (size $N \leq N' \leq 2N$) stems from evaluating the Pareto dominance between the neighborhood and the previous food sources. As employed bees do, using the crowding distance non-domination sorting, the non-dominated food sources in P'_t are found to build population P_{t+1} of size N.

Finally, when a food source is not improved throughout a certain number of epochs, it is replaced by a randomly position found by the scouts. NSBC iterates until a stop condition is met.

3 Technique for Order of Preference by Similarity to Ideal Solution (TOPSIS)

TOPSIS is an MCDM approach [5]. Considering a decision problem characterized by n alternatives and m criteria, TOPSIS requires an $n \times m$ decision matrix $\mathbf{H} = [h_{ij}]$, where $i \in \{1, \ldots, n\}$ and $j \in \{1, \ldots, m\}$. The goodness of alternative i w.r.t. criterion j is measured by element h_{ij}. In addition, criteria must be prioritized by assigning them weights. Let these weights be contained in a vector $\boldsymbol{\omega} = (\omega_1, \ldots, \omega_m)$, with $\sum_{j=1}^{m} \omega_j = 1$. TOPSIS first calculates the weighted normalized decision matrix $\mathbf{V} = [v_{ij}] = \omega_j h_{ij} / \sqrt{\sum_{i=1}^{n} h_{ij}^2}$, then it finds the ideal best (IB) and worst (IW) solutions. The indices in Ω_B and Ω_C indicate benefit and cost criteria, respectively. Let $IB = (a_1^+, \ldots, a_m^+)$ and $IW = (a_1^-, \ldots, a_m^-)$, where $a_j^+ = \max_i v_{ij}$ for $j \in \Omega_B$ or $a_j^+ = \min_i v_{ij}$ for $j \in \Omega_C$, and $a_j^- = \min_i v_{ij}$ for $j \in \Omega_B$ or $a_j^- = \max_i v_{ij}$ for $j \in \Omega_C$. TOPSIS measures the Euclidean distance of every single alternative from IB, namely, $D_i^+ = \sqrt{\sum_{j=1}^{m} (v_{ij} - a_j^+)^2}$, and IW, i.e., $D_i^- = \sqrt{\sum_{j=1}^{m} (v_{ij} - a_j^-)^2}$. TOPSIS eventually measures the relative closeness coefficient of every alternative to IB as $RCL_i^+ = D_i^- / (D_i^+ + D_i^-)$: the higher RCL_i^+ the better the alternative. The alternative $k = \arg\max_i RCL_i^+$ is chosen and results to be the best.

4 Worker's Risk Perception and Caution

Consider a set of tasks $\mathcal{T} = \{t_1, \ldots, t_{|\mathcal{T}|}\}$ and a set of employees $\mathcal{E} = \{e_1, \ldots, e_{|\mathcal{E}|}\}$. An employee is assumed to be exposed to a set $\mathcal{R}_i$ of risks when performing task t_i. Let the set of risks of the workplace be $\mathcal{R} = \bigcup_{i=1}^{|\mathcal{T}|} \mathcal{R}_i$. Each employee innately takes specific precautions when exposed to the risks of a task.

Formally, let set $\mathcal{A}_k = \{a_{k,1}, \ldots, a_{k,|\mathcal{A}_k|}\}$ contain *preventive actions*, i.e., precautions that an employee can take to mitigate a risk $r_k \in \mathcal{R}$, where $k \in \{1, \ldots, |\mathcal{R}|\}$. Preventive actions can mitigate a risk, i.e., they decrease the risk occurrence and/or its impact. Depending on this extent, each preventive action is associated with a *level of prevention* in $\mathcal{L} = \{1, \ldots, L\}$. The more the action makes a risk less likely and/or mitigates its impact, the higher the level of prevention. Experts in risk assessment assign the levels of prevention to preventive actions.

Consider a set $\mathcal{H} = \{h_1, \ldots, h_{|\mathcal{H}|}\}$ of *human factors* (or *factors*). Each h_v takes values in a domain $\mathcal{D}_v$. Set $\mathcal{H}$ is made of factors that relate to the worker's past history and work experience, and factors related to the task. The first group is composed of P *personal* factors. The second contains T *task-related* factors. The *risk perception personal level* $pers\_perc_j$ of employee e_j stems from the set $\mathcal{P}_j = \bigcup_{v=1}^{P} d_{v,j}$, which contains the values $d_{v,j} \in \mathcal{D}_v$ of each personal factor h_v.

A function $\varphi_{PERSONAL}$ such that $\mathcal{P}_j \mapsto \varphi_{PERSONAL}(\mathcal{P}_j) = pers\_perc_j$ exists, and the perception level $task\_perc_{i,j}$ of e_j for the risks of task t_i is established by $\mathcal{T}_j = \bigcup_{v=P+1}^{P+T} d_{v,j}$. Here, $d_{v,j}$ are the values of task-related factors h_v for employee e_j. The risk perception personal level $pers\_perc_j$ of w_j also influences $task\_perc_{i,j}$. Thus, there exists a function φ_{TASK} such that $(\mathcal{T}_j, pers\_perc_j) \mapsto \varphi_{TASK}(\mathcal{T}_j, pers\_perc_j) = task\_perc_{i,j}$.

For each risk r_k and employee e_j, the caution of e_j for r_k is measured on the basis of the number of preventive actions per level of prevention that e_j performs when exposed to r_k: this is the *behavior* of e_j towards r_k. Let us denote the number of $\bar{\ell}$-level preventive actions that e_j performs when exposed to r_k as $\#\mathcal{A}_{k,\ell=\bar{\ell},j}$. A function ρ_k such that $(\#\mathcal{A}_{k,\ell=1,j}, \ldots, \#\mathcal{A}_{k,\ell=L,j}) \mapsto \rho_k(\#\mathcal{A}_{k,\ell=1,j}, \ldots, \#\mathcal{A}_{k,\ell=L,j})$ $= risk\_caution_{k,j}$ can thus be configured for each $k = 1, \ldots, |\mathcal{R}|$.

For each task t_i and employee e_j, the caution of employee e_j when performing task t_i therefore depends on $risk\_caution_{k,j}, \forall k \in \mathcal{R}_i$. For this reason, a group of functions τ_i, one for each $i = 1, \ldots, |\mathcal{T}|$, such that $\bigcup_{r_k \in \mathcal{R}_i} risk\_caution_{k,j} \mapsto \tau_i \left(\bigcup_{r_k \in \mathcal{R}_i} risk\_caution_{k,j}\right) = task\_caution_{i,j}$, computes the level of caution of employee e_j for each task t_i, given the employee's levels of caution for the risks involved. A tuple $\theta_j = \{\bigcup_{v=1}^{P+T} d_{v,j}, \bigcup_{k=1}^{|\mathcal{R}|} \bigcup_{\lambda=1}^{L} \#\mathcal{A}_{k,l=\lambda,j}\}$ thus represents the employee e_j in the model. It is important to point out that $|\mathcal{H}| = P + T$, and that $v \in \{1, \ldots, P\}$ refers to personal factors, whereas task-related factors are referred to as $v \in \{P+1, \ldots, P+T\}$.

Given tuple θ_j, the levels of risk perception and caution of each employee towards every task are determined in this paper by using the neural network-based system whose architecture and training process are described in detail in [6].

5   Problem Formulation

5.1   Objectives

Consider a decision variable $x_{ij} \in \{0, 1\}$. Let $x_{ij} = 1$ if employee e_j is assigned to task t_i, and let $x_{ij} = 0$ if not, where $i \in \{1, \ldots, |\mathcal{T}|\}$ and $j \in \{1, \ldots, |\mathcal{E}|\}$. The vector $\mathbf{x} \in \{0,1\}^{|\mathcal{T}| \times |\mathcal{E}|}$ is a *personnel assignment*, and has decision variables x_{ij} as elements, in lexicographic order. The three objectives considered in the optimization approach described in this paper are formalized in the next sections.

Cost. Assigning task t_i to employee e_j results in a cost that depends on the employee's work seniority for task t_i and his/her need to be trained to perform t_i. The longer the overall time during which e_j has performed t_i in life, the lower the cost for the training. The cost c_{ij} of assigning employee e_j to task t_i includes the cost for the training and what the employer pays for salary and benefits. The overall cost objective function $COST(\mathbf{x}) : \{0,1\}^{|\mathcal{T}| \times |\mathcal{E}|} \to \mathbb{R}^+$ to minimize is modeled as:

$$COST(\mathbf{x}) = \sum_{i=1}^{|\mathcal{T}|} \sum_{j=1}^{|\mathcal{E}|} c_{ij} x_{ij}. \tag{2}$$

Learning Time. In general, more experienced workers are preferred to be assigned to tasks. Consider the average number of days AVG_TIME_i typically required to employees to be properly trained for task t_i. If an employee has never performed the task before, this number of days is assumed to be required to train the worker. If the employee has a past experience for that task, let $\mathcal{P}_{i,j}$ be the set of the past jobs where employee e_j performed task t_i. The experience of employee e_j for task t_i is estimated as

$$experience_{i,j} = \frac{|\mathcal{P}_{i,j}|}{\sum_{u \in \mathcal{P}_{i,j}} duration_u^{-1}}, \tag{3}$$

where $duration_u$ is the duration (in days) of past job u. The harmonic mean is used in Eq. (3) as it mitigates (intensifies) the impact of large (small) outliers. The time $T_{i,j}$ that an experienced employee e_j takes to be trained for t_i is estimated as:

$$T_{i,j} = \begin{cases} AVG\_TIME_i & \text{if } experience_{i,j} \geq k \cdot AVG\_TIME \\ a_{i,j} \cdot AVG\_TIME_i & \text{otherwise} \end{cases} \tag{4}$$

where the parameter $k > 0$ is set by experts in the field, and $a_{i,j} > 0$ results from risk-free practical tests where experts in the field evaluate how skilled employee e_j is in executing task t_i. The overall learning time is estimated through the mean to variance ratio

$$LEARNING(\mathbf{x}) = \frac{T_{MEAN}}{\sum_{i=1}^{|\mathcal{T}|} \sum_{j=1}^{|\mathcal{E}|} (T_{i,j} x_{ij} - T_{MEAN})^2},$$

where T_{MEAN} is the average learning time for the tasks assigned, defined as $T_{MEAN} = \frac{1}{|\mathcal{T}|} \sum_{i=1}^{|\mathcal{T}|} \sum_{j=1}^{|\mathcal{E}|} T_{i,j} x_{ij}$.

Caution. Let us define the average level of caution for the tasks assigned as: $C_{MEAN} = \frac{1}{|\mathcal{T}|} \sum_{i=1}^{|\mathcal{T}|} \sum_{j=1}^{|\mathcal{E}|} task\_caution_{i,j} x_{ij}$. The overall level of caution for the tasks assigned $CAUTION(\mathbf{x}) : \{0,1\}^{|\mathcal{T}|\times|\mathcal{E}|} \to \mathbb{R}^{+}$ to maximize, is defined here as the mean to variance ratio of the level of caution of each employee towards the task assigned:

$$CAUTION(\mathbf{x}) = \frac{C_{MEAN}}{\sum_{i=1}^{|\mathcal{T}|} \sum_{j=1}^{|\mathcal{E}|} (caution_{i,j} x_{ij} - C_{MEAN})^2}. \tag{5}$$

5.2 Problem Formulation

Consider a set of tasks $\mathcal{T}$ and a set of employees $\mathcal{E}$ where $|\mathcal{T}| = |\mathcal{E}|$. Each task must be (re)assigned to one worker and vice versa. The optimization problem is:

$$\underset{\mathbf{x}}{\text{Minimize}} \ \ \mathbf{f}(\mathbf{x}) = [COST(\mathbf{x}), LEARNING(\mathbf{x}), -CAUTION(\mathbf{x})] \tag{6a}$$

subject to:

$$\sum_{i=1}^{|\mathcal{T}|} x_{ij} = 1, \quad \forall j = 1, \dots, |\mathcal{E}| \tag{6b}$$

$$\sum_{j=1}^{|\mathcal{E}|} x_{ij} = 1, \quad \forall i = 1, \dots, |\mathcal{T}| \tag{6c}$$

$$x_{ij} \in \{0,1\}, \quad \forall i = 1, \dots, |\mathcal{T}|, \forall j = 1, \dots, |\mathcal{E}|. \tag{6d}$$

Equation (6a) is the objective function $\mathbf{f}(\mathbf{x}) : \{0,1\}^{|\mathcal{T}|\times|\mathcal{E}|} \to \mathbb{R}_{+}^{2} \times \mathbb{R}_{-}$ whose components are the overall cost, the average learning time, the overall level of caution (inverted in sign) towards the tasks assigned of assignment (i.e., personnel reallocation) $\mathbf{x} \in \{0,1\}^{|\mathcal{T}|\times|\mathcal{E}|}$. Constraints (6b) force each worker be assigned to one task. Constraints (6c) let instead each task of the factory be (re)assigned to one worker. Equation (6d) express the integer constraint.

6 Experiments and Discussion

The proposed approach was applied to two scenarios ("Scenario A" and "Scenario B") based on two real-world case studies related to two footwear companies. The optimization approach was implemented in MATLAB.

6.1 Dataset

A website was implemented to collect information about the employees: the values of their human factors and behavior. The employees were required to fill out a questionnaire through the website. Data were collected in compliance with the privacy laws. For each employee e_j, the questionnaire collects:

- the values of the human factors in order to compute $task\_perc_{i,j}$;
- data relating the past jobs to estimate the learning times;
- data relating the behavior towards each risk r_k of every task t_i on the basis of the preventive actions that the employee chooses from a predefined set of actions. The actions chosen let $task\_caution_{i,j}$ for each task t_i be computed.

The dataset consists of 36 interviews: 11 interviews relate to the first company, the other 25 interviews relate to the second company. Due to privacy laws and ethical issues, the dataset cannot be made public domain.

6.2 Setup of the Parameters

The system was implemented in MATLAB. By means of a trial and error approach, the NSBC algorithm was run for 1000 generations and the population
size was set to 150. To find this configuration, a total of 30 trials were run using
different values for population size and number of generations. These values were
determined on the basis of heuristic considerations on the problem.

6.3 Optimization Results

Personnel Assignment Strategy in the Involved Companies. In SMEs,
managers determine how suitable an employee is to perform a task on the basis
of his/her experience: the more the experience, the more suitable is the employee.
As many tasks of the footwear industry are handmade and require the use of
dangerous machines, workers are continuously exposed to serious risks including
crushing injuries, burns and amputation. However, risk management is typically
carried out by SMEs by assigning the more dangerous tasks to the more experienced employees, mainly because SMEs are unprepared to make important
capital investments. This is tremendously dangerous because more experienced
workers typically have higher locus of control and this can decrease the risk
awareness [10].

The Shoe Making Process. Making a shoe is a complex process, with many
handmade operations. The process starts with cutting pieces of leather using
cutting machines and knives to prepare some of the parts of a shoe, i.e., uppers,
linings, reinforcements and insoles. Die cutters are used to prepare other parts,
such as welts, vamps, soles, heels. The thickness of the leather is made uniform
using milling cutters while preparing the upper. Ornaments are then applied to
the shoe. By sewing all the parts above, the upper is assembled: this phase is
called stitching. A pounding phase lets possible folds of the leather be smoothed.
The upper is then mounted on the last (i.e., a sculpture of the shoe) using a
lasting machine, and is finally joined with the insole. The sole is applied by
using sanding machines, through glueing, sewing or welts. Die casting or nails
are used to fix the heel. Heels are typically coated with leather or wrapped with
a material similar to the one the upper is made of. A press fixes the upper to the
block made of sole and heel, in the case of rubber soles. The bottom of the shoe
is finished by: (i) sanding heel and sole by using rotating machines; (ii) waxing
and coloring the sole contour with rotating tools; (iii) polishing sole and heel.
    The process ends with embellishment steps that include cleaning the upper
with solvents/brushes, waxing-up the sole, polishing and starching. The shoes
are eventually put in pairs into shoeboxes to be stored into the warehouse.

Proposed Strategy for Personnel Reallocation. The proposed approach
for personnel reallocation started asking the management to prioritize the objectives. Preventive actions were then classified into three prevention levels: low,

medium, high. Each employee's data about human factors and behavior were collected by means of the website described in Sect. 6.1. Each employee's *task_perc* and *risk_caution* towards each risk were computed using these data. The neural system referred to in Sect. 4 computed each employee's levels of *task_caution* and *task_perc* towards every task. The Pareto front was approximated by means of NSBC. The best solution selected by TOPSIS is in Table 1, for both scenarios.

Discussion of Scenario A. In this scenario, the company has 11 employees. The company aims to improve the workplace safety with a low increase in cost. The weights of the objectives are $(0.35, 0.2, 0.45)$, in the order they appear in Eq. (6a). Cost and caution are thus the most important objectives. The left-hand side of Fig. 1 shows the Pareto front obtained by means of the NSBC algorithm.

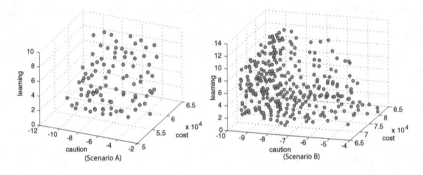

Fig. 1. Pareto front obtained for Scenario A (left) and Scenario B (right).

The current assignment has a cost of 30,180 € and shows an overall level of caution of -5.15, as summarized in Table 1. Learning is equal to zero because all employees can perform their task correctly.

The proposed solution reassigns 5 tasks (highlighted with colored cells in Table 1) and guarantees $\sim$120% improvement in the overall level of caution at the expense of $\sim$18% increase in cost. Cost is paramount for the involved company, and a percentage like this seems unreasonable. Anyway, that is not what it seems like because the increase in cost is just temporary, as it is due to the cost of training the reallocated workers.

Data on the employees' behavior w.r.t. every task cannot be reported. Consider that each task of a footwear company exposes an employee to five risks on average. Each risk can in turn be prevented by four actions per level of prevention, on average. Also, three levels of prevention (low, medium and high) are considered. This causes an explosion of the number of preventive actions per employee to report. The discussion is thus based on intuitive considerations.

The proposed approach, if implemented, would guarantee that $\sim$75% of the employees would deal with the task assigned with high-level preventive actions only. The remaining percentage would have behaviors between a poor level of caution (3 low-level actions) and a good level of caution (2 medium-level actions and 2 high-level actions).

It is important to note that the most unsafe behavior (i.e., just 1 low-level action) stems from workers who stamp the insoles or make supervision. As can be understood, these tasks are almost risk-free. Workers may experience just muscle tightness and eyestrain.

Instead, in the current assignment, 2 safety-critical tasks are performed by employees with behavior uniquely made of 1 low-level preventive action. These tasks involve die cutters and pounding machines, which can be a serious threat for the health if used unsafely. For instance, employees may experience grazes, crushing injuries to the hands, and amputation. The employees currently assigned to these tasks are highly likely to get hurt due to their poor level of caution.

Assigning safety-critical tasks to certain people just because they have a high number of years of experience is dangerous. These workers, of course, perform the task better than others, but the management is highly wrong if neglects their behavior when in the presence of risk.

In the current assignment, safety-critical tasks are assigned to employees with, on average, 15 years of experience for the task. However, their level of caution is poor. As one can imagine, people become familiar with something that is performed every day in life. And this causes a decrease in risk awareness. Accordingly, the employees that the management assigned to safety-critical tasks have become familiar with the task, and they may have inadequate awareness of the risks they take when performing the task.

Regarding the learning time, the proposed reallocation of personnel guarantees a fast training, ~3 days, on average, as can be seen from Table 1.

Table 1. Current and proposed personnel assignment for Scenarios A and B.

		TASKS																								
		1	2	3	4	5	6	7	8	9	10	11	12	13	14	15	16	17	18	19	20	21	22	23	24	25
A	Current	5	2	10	3	7	4	11	6	1	9	8	-	-	-	-	-	-	-	-	-	-	-	-	-	-
	Proposed	5	9	11	3	2	4	10	6	1	7	8	-	-	-	-	-	-	-	-	-	-	-	-	-	-
B	Current	23	3	8	21	17	6	25	9	11	19	16	5	24	7	18	14	1	4	10	15	13	2	20	12	22
	Proposed	23	3	10	21	17	22	25	9	7	19	16	5	2	11	18	14	20	4	8	15	13	24	1	12	6

Discussion of Scenario B. In this scenario, the weights of the objectives are $(0.4, 0.1, 0.5)$ The management thus wants to reallocate the personnel to improve the safety, keeping low the costs.

The current assignment of personnel to tasks is characterized by an overall caution equal to -4.78 (see Table 2). By reallocating the personnel as suggested by the proposed approach (see Table 1), the overall level of safety is more than doubled, at the expense of a temporary increase in cost of ~10%. The Pareto

Table 2. Values of the objectives for Scenarios A and B.

		COST [€]	LEARNING	−CAUTION
A	Current	30,180.00	0	−5.15
	Proposed	35,720.00	3.22	−11.41
B	Current	68,995.00	0	−4.78
	Proposed	76,150.00	5.38	−9.94

front is in Fig. 1. The behaviors that highly impact on safety are discussed in the following.

As an example, it is fundamental to observe that the safest behaviors, i.e., the ones made of high-level preventive actions alone, pass from 5 (in the current personnel assignment) to 9 thanks to the proposed reallocation of personnel to tasks. Also, with the proposed solution no employee behaves showing low-level preventive actions only. In the current assignment, there are instead 6 employees characterized by behaviors like this, the two most hazardous of them involve a task where employees may experience crushing injuries while using die cutters and severe excoriations, respectively. This is another SME where the management chooses more experienced employees for the most dangerous tasks.

Finally, the time required by the employees to learn the new tasks assigned is estimated to be a bit longer than 5 days (see Table 2). This is a short amount of time if one thinks that 10 tasks (colored cells in Table 1) are reassigned. The time required to learn the new tasks assigned is thus compliant with the importance assigned by the management to the corresponding objective.

7 Conclusion

This paper has presented a MOO approach for personnel reallocation whose aim is to improve the workplace safety in SMEs, while keeping low both the costs and the time required to train the employees. Each employee's level of caution for every task is determined by a neural network-based system, based on some human factors and the precautions he/she takes when performing the task. NSBC and TOPSIS are used to find a Pareto-optimal personnel reallocation. The approach was tested in two footwear companies. A high improvement of the caution was obtained in both scenarios, with a low and temporary increase in cost. Risks thus become less harmful and less likely as tasks are assigned to more appropriate employees w.r.t. their level of caution while performing the tasks.

Acknowledgments. This research was supported by the PRA 2016 project "Analysis of Sensory Data: from Traditional Sensors to Social Sensors", funded by the University of Pisa.

References

1. International Labour Organization Statistics Database (ILOSTAT). http://www.ilo.org
2. Bandini, S., Manzoni, S., Sartori, F.: Case-based reasoning to support work and learning in small and medium enterprises. In: 21st IEEE International Conference on Tools with Artificial Intelligence, pp. 253–260 (2009)
3. Bouyer, M., Bagdassarian, S., Chaabanne, S., Mullet, E.: Personality correlates of risk perception. Risk Anal. **21**(3), 457–465 (2001)
4. Chauvin, B., Hermand, D., Mullet, E.: Risk perception and personality facets. Risk Anal. **27**(1), 171–185 (2007)
5. Hwang, C.-L., Yoon, K.: Multiple Attribute Decision Making. Springer, Heidelberg (1981). https://doi.org/10.1007/978-3-642-48318-9
6. Cococcioni, M., Lazzerini, B., Pistolesi, F.: A semi-supervised learning-aided evolutionary approach to occupational safety improvement. In: IEEE Congress on Evolutionary Computation (2016)
7. Costa-Font, J., Mossialos, E., Rudisill, C.: Optimism and the perception of new risks. J. Risk Res. **12**(1), 27–41 (2009)
8. Deb, K., Kalyanmoy, D.: Multi-Objective Optimization Using Evolutionary Algorithms. Wiley, Chichester (2001)
9. Health, Safety Executive: Reducing error and influencing behaviour. HSE Books (1999)
10. Horswill, M.S., McKenna, F.P.: The effect of perceived control on risk taking. J. Appl. Soc. Psychol. **29**(2), 377–391 (1999)
11. Klein, C.T.F., Helweg-Larsen, M.: Perceived control and the optimistic bias: a meta-analytic review. Psychol. Health **17**(4), 437–446 (2002)
12. Lazzerini, B., Pistolesi, F.: Profiling risk sensibility through association rules. Expert Syst. Appl. **40**(5), 1484–1490 (2013)
13. Lazzerini, B., Pistolesi, F.: Classifying workers into risk sensibility profiles: a neural network approach. In: 8th European Modelling Symposium, pp. 33–38 (2014)
14. Lazzerini, B., Pistolesi, F.: An integrated optimization system for safe job assignment based on human factors and behavior. IEEE Syst. J. (2017). https://doi.org/10.1109/JSYST.2016.2646843
15. Lazzerini, B., Pistolesi, F.: Multiobjective personnel assignment exploiting workers' sensitivity to risk. IEEE Trans. Syst. Man Cybern. Syst. (2017). https://doi.org/10.1109/TSMC.2017.2665349
16. Peters, E., Slovic, P.: The role of affect and worldviews as orienting dispositions in the perception and acceptance of nuclear power. J. Appl. Soc. Psychol. **26**, 1427–1428 (1996)
17. Rakshit, P., Sadhu, A.K., Bhattacharjee, P., Konar, A., Janarthanan, R.: Multi-robot box-pushing using non-dominated sorting bee colony optimization algorithm. In: Panigrahi, B.K., Suganthan, P.N., Das, S., Satapathy, S.C. (eds.) SEMCCO 2011. LNCS, vol. 7076, pp. 601–609. Springer, Heidelberg (2011). https://doi.org/10.1007/978-3-642-27172-4_71
18. Sjöberg, L.: Factors in risk perception. Risk Anal. **20**(1), 1–12 (2000)
19. Slovic, P.: Perception of risk. Science **236**(4799), 280–285 (1987)

Multi-objective Genetic Algorithm for Interior Lighting Design

Alice Plebe[(✉)] and Mario Pavone

Department of Mathematics and Computer Science,
University of Catania, V.le A. Doria 6, 95125 Catania, Italy
alice.plebe@gmail.com, mpavone@dmi.unict.it

Abstract. This paper proposes a novel system to help in the design of interior lighting. It is based on multi-objective optimization of the key criteria involved in lighting design: the respect of a given target level of illuminance, uniformity of lighting, and electrical energy saving. The proposed solution integrates the 3D graphic software Blender, used to reproduce the architectural space and to simulate the effect of illumination, and the genetic algorithm NSGA-II. This solution offers advantages in design flexibility over previous related works.

Keywords: Lighting design · Genetic algorithm · Blender

1 Introduction

The design of interior lighting is the crucial and complex process of integrating luminaries into the fabric of architecture [11,17]. The goal is to select the lighting equipment and their placement in the interior environment, that result in a comfortable and pleasant visual experience. The design process should take into account several aspects, such as the type of occupants and the type of activities in the given space, or the interior surface finishes and furnishings.

In addition, in the last decades increasing attention has been paid to the issue of energy savings. In U.S. the energy consumed for lighting accounts for about 30% of the total energy consumed by commercial buildings, and in the European Union the yearly consumption is over 170 TWh. Therefore, the concept of *sustainable* lighting design has become central in architectural strategies [23].

A well established aids offered by computational tools to the designer is by photorealistic architectural rendering, simulating in computer graphics the effect of a lighting solution on a model of the interior environment [15]. Mathematically, this is the solution of the *direct lighting problem*. The drawback of direct lighting tools is that, if the achieved illumination is not satisfactory, it is not easy to infer which modifications to the current solution may lead to improvements. Very likely, the final solution chosen by the designer over a collection of trials, will be far from optimal.

As discussed in Sect. 2, a more effective assistance would be given by computational tools implementing the *inverse lighting problem*: the determination

© Springer International Publishing AG 2018
G. Nicosia et al. (Eds.): MOD 2017, LNCS 10710, pp. 222–233, 2018.
https://doi.org/10.1007/978-3-319-72926-8_19

of the lighting equipment and their placement from specifications on the illuminance. This is the line of research undertaken by this paper. The inverse lighting problem lacks reliable analytic solutions, therefore it is often formulated as an optimization problem. The proposed methodology aims to optimize the main common requirements of interior lighting design, chiefly the desired level of average luminous intensity, the uniformity of light in the interior space, taking into account energy consumption. Due to the clashing of these multiple factors, the resulting problem is multi-objective in nature, as discusses in Sect. 3, where previous proposals for interior lighting design are compared and contrasted with the present one.

Our methodology, detailed in Sect. 4, is based on the combination of a 3D graphic software providing a rendering engine for direct illumination, and genetic algorithms for solving the multi-objective inverse illumination optimization. Results on a variety of interior environments are shown in Sect. 5.

2   The Inverse Lighting Problem

In a nutshell, lighting design for interior spaces is the process of integration of artificial light sources in architectural complexes – be it industrial, public or private [11]. Since the discovery of the electric light system by Thomas Edison in 1879, lighting design has experienced several significant revolutions, such as fluorescent lamps in 1938 and, more recently, solid-state lighting.

Traditionally, illumination design has been seen as a blend of art and practice, where all the challenges are left to the creativity and the experience of the design architect. Given the aesthetic nature of the task, lighting design may seems difficult to formally model. Nevertheless, the design process has been lately considered as a mathematical and physical problem to be solved with optimization techniques.

The *inverse lighting problem* [2,14,20,24] is the problem of determining potential light sources satisfying a set of given illumination requirements, for a pre-defined interior space. Conversely, the *direct lighting problem* refers to the computation of radiance distribution in an environment that is completely known a priori, including its lighting parameters. In the inverse problem, lighting configurations are inferred from the desired illumination requirements, taking into account positions, kinds of luminaries, intensities and number of light sources. Energy efficiency is often considered as well.

Given the many feasible solutions possible, the application of optimization methods still allows the designer to have a degree of freedom and creativity in the final choice of the lighting configuration, from one of the optimal solutions obtained.

2.1   Blender as Direct Engine

A system facing the inverse lighting problem must provide two fundamental components:

- a three-dimensional environment able to accurately reproduce the architectural space and its spectral reflectometric properties;
- a physical simulation platform for illumination calculation in sample points of the architectural space.

Several approaches were considered in order to satisfy these requirements. Lightsolve [1] is an interactive dedicated environment for daylight design, with a performance-driven decision support system. The system lacks a detailed architectural reproduction, and the inclusion of interior furniture is difficult to manage. In [10] the 3d models of building facades are obtained with the simple modeling tool Google SketchUp, which offers a quick and easy way to outline an architectural space, but resulting in a low level of realism. Conversely, the popular software Radiance, widely used in the field of optimal lighting design [9,15,19], consists of a sophisticated physically-correct rendering engine for illumination calculation, and it allows architectural spaces reproduction at arbitrary levels of detail. Nevertheless, it is a non-interactive system composed by a collection of command-line programs, and all architectural specifications have to be coded into configuration files. Attempts have been made to unify those extremes, for example Painting With Light is an integration environment for Rhinoceros, a commercial CAD software, and Radiance [4].

This paper investigates the adoption of the 3D graphic software Blender as a unified solution to the two requirements stated above. Firstly, Blender is the most comprehensive open-source computer graphic tool available, it is particularly suitable for modeling architectural interiors, with the possibility of importing components from CAD files. Secondly, Blender provides a physically-based rendering engine, named Cycles, able to exhaustively evaluate lighting configurations needed for solving the inverse lighting problem. Moreover, Blender embeds a Python interpreter which can run scripts supplied by the user, in order to extend its functionalities, and is known for its remarkable software integrity [12]. Thanks to its intrinsic versatility, Blender has already been applied to a number of different problems, including industrial applications [21].

3   Multi-objective Optimization

Interior illumination design involves multiple and often conflicting factors, therefore the resulting problem is multi-objective in nature. In contrast to a single-optimization problem, where there is usually a single optimal solution, a multi-objective optimization finds a set of solutions that satisfies all conflicting criteria.

Multi-objective optimization methods have been widely used in architectural and lighting design, and plenty of them are nature-inspired. A particle swarm optimization algorithm was developed to design curtain wall facades for office buildings and to achieves low energy consumption [22]. A study has adopted multicriteria ant colony optimization to design paneled building envelopes, optimizing lighting performance and cost criteria [25]. Harmony search algorithms have been applied in the field of civil engineering several times, such as structural design optimization [16], and residential buildings design with low-emission and energy-efficient requirements [7].

But most of all, genetic algorithms have been proven to be successfully useful in a variety of architectural tasks, including lighting. The *GENE_ARCH* tool [3,4] is a popular generative design system for energy-efficient and sustainable architectural solutions, based on a Pareto genetic algorithm. In [26] genetic algorithms and parametric modeling are combined to explore the morphology of a dome, taking into account structural performance and daylight transmittance. A micro-genetic algorithm is used in [10] to explore facade designs based on illuminance and glare criteria. A study has applied genetic programming to design decorative wall of lights, and to create stained-glass window for large public spaces [19].

For the problem in hand, we used two objectives treated as multi-objective, and we deem this is the appropriate value for real cases. In fact we might have more requirements for illumination design, for example in our experiments we used as illumination quality both the deviation from a target value, and the overall uniformity. However, it seems that requirements can always be unified in two combined fitness only, which are significantly conflicting: one the sum up to lighting quality, and a contrasting one that expresses the cost for achieving quality.

3.1   Previous Related Works

There is a number of multi-objective genetic formulations of the inverse lighting problem that shares similarities with the one here proposed. A variant of genetic algorithm called *generalized extremal optimization* is used in [5] to minimize the deviation of lighting to a desired target, and the energy consumption. Our algorithm takes into account also the uniformity of lighting, but the main difference is that the methodology proposed by Cassol et al. is customized to a rectangular enclosure formed by surfaces that are perfectly diffuse, while our system is fully flexible in the geometry of the interior space, and the properties of the surfaces.

In [27] one of the criteria to satisfy, named *suitable office lighting*, is derived by interpolating subjective data obtained from psycho-visual tests, while the other criteria is energy savings. The optimization is solved using genetic algorithm, but it affects only the relative dimming of two fixed light sources.

In [18] a genetic algorithm was employed for simultaneously minimizing the power consumption and the uniformity of the illuminance. Our algorithm takes into account, in addition to the uniformity, the adherence of the average illumination to a given target. But the most important difference is that in Madias et al. the location of the light sources is assumed constant, and only their dimming is variable, while in our strategy there is full flexibility in light selection, placement, and dimming.

3.2 NSGA-II

The Non-dominated Sorting Genetic Algorithm II (NSGA-II), introduced by Deb et al. [6], is an elitist multi-objective genetic algorithm that performs well with real world problems, producing Pareto-optimal solutions to the optimization problem. The evaluation of the population of solutions takes into account the dominance and the crowding distance of the individuals. The first criterion is used to sort the population into different fronts of non-dominated individuals, while the second criterion gives preference to solutions that are less crowded. Non-dominated individuals belonging to a high-rank front with a larger crowding distance are selected to reproduce more than others. The offspring are generated through the genetic operators of crossover and mutation. The next-generation population is then selected among the best individuals from both the offspring and the parent population, ensuring elitism. The result of the algorithm is the set of non-dominated solutions of the whole final population, namely the Pareto front.

One of the key working principles of the genetic algorithm is the chromosomal representation of a solution. The algorithm works with a coding of decision variables, instead of the variable themselves, and choosing the right representation scheme is crucial to its performance [13]. The most traditional approach is to code the decision variables in a binary string of fixed length, which is a natural translation of real-life genetic chromosomes. Such strings are directly manipulated by the genetic operators, crossover and mutation, to obtain a new (and hopefully better) set of individuals. Another well established method is the floating point representation of chromosomes, where each solution is coded as a vector of floating point numbers, and crossover and mutation operators are adapted to handle real parameter values.

For the algorithm presented in this paper, we developed a novel chromosomal representation of solutions, specifically tailored for lighting design optimization. Each individual represents a possible illumination configuration, and it is coded as a vector of variable length containing a set of lamp specifications. A lamp specification is the set of features describing the luminaries in the 3D environment, including position and orientation, intensity, color temperature of light, and model of light fixture (wall or ceiling mount). Special operators of crossover and mutation are implemented to handle this peculiar chromosomal representation. The design of such operators is, however, facilitated by the transparency of the representation itself. Therefore, our approach is introduced especially to deal with representation of complex structured individuals, and it ensures more flexibility with respect to previous proposals.

3.3   Fitness Evaluation and Constraint Handling

The goal of the proposed model is to find the lighting configuration that best satisfies the most common and compelling requirements faced by the lighting designer. In accordance to what stated in the Introduction, there are goals directly related with the quality of the lighting, and an additional goal of energy saving. We adopt as goal for the light quality the combination of two objectives: achieving an illuminance level closest as possible to the given target, and obtaining light distribution uniform enough in the given space. The evaluation of light quality is performed on samplers S, horizontal surfaces distributed in the interior space, captured by virtual cameras placed in Blender over each sampler. The system allows two placement methods: one automatic that locate as much as evenly as possible the sample in the space, or a manual placing, more convenient in the case of complex spaces, or when key portions of the space, that require the best quality, are known in advance. Compliance with the target level of light, and degree of uniformity, are combined in a single fitness f_1 of the individual I, with the following computations:

$$t\left(I\right) = \frac{1}{M} \sum_{i=0}^{M} |S_i - T| \tag{1}$$

$$u\left(I\right) = \sqrt{\frac{1}{M} \sum_{i=0}^{M} \left(S_i - \overline{S}\right)^2} \tag{2}$$

$$f_1\left(I\right) = wt(I) + (1 - w)u(I) \tag{3}$$

where S_i is the illumination measured on the i-th sampler produced by the lighting configuration of individual I, and M is the number of samplers S. Note that treating $t(I)$ and $u(i)$ as separate fitness in multi-objective optimization would be incorrect, because they are not conflicting. It can be easily verified in the limit case of an individual $\hat{I}$ that illuminates all samplers exactly at target level T, from Eqs. (1) and (2) we obtain $t(\hat{I}) = u(\hat{I}) = 0$. The weight w control the balance between the desired compliance with the target level of light and uniformity, the default value used in all reported results is 0.5.

Energy consumption represents the second fitness and it is quantified as the overall power consumption of the lamps (measured in Watt) divided by the volume of the room:

$$f_2\left(I\right) = \frac{\sum_{i=0}^{N} C_i}{V} \tag{4}$$

where C_i is the amount of Watts consumed by the i-th lamp of the individual I, V the volume of the interior environment in m^3, and N the number of lamps composing the solution.

In the presented problem of lighting optimization there are some conditions on the design process to be satisfied, therefore a constraint handling method has to be considered as well. The constrains in question concern positioning the lamps inside the interior environment:

- a lamp must be placed inside the room and in contact with the room surface;
- two lamps can not be placed in the same location;
- a lamp should be mounted on the walls or on the ceiling in accordance with its model of light fixture;

- depending on the room design, there might be some areas where the lamp placement is not allowed, for example in presence of windows, pillars, or supporting beams.

The constraint specifications are provided to the system within the 3D model of the environment itself. The walls and ceiling are structured as a discrete grid of vertices, each representing a feasible position for a lamp. With this approach, the set of constraints can be effortlessly reformulated for different experiments, ensuring absolute flexibility in the design process.

Since the satisfaction of the above constraints is mandatory for the problem, they can be referred as *hard constraints*. To handle them, we adopted a method based on preserving feasibility of solutions. In this approach, two feasible solutions, after crossover and mutation operation, will create two feasible offspring. Nevertheless, it can happen that crossover produces an individual composed of exactly the same lamps of another solution. In that case, the duplicated solution is discarded.

4 The Proposed Strategy

The algorithm presented in this paper has been implemented in the form of a Blender script, composed by 9 main Python modules. The simulation environment set-up is performed by the first group of modules, which rely on Blender's modeling features. The architectural interior scene of interest is represented inside the computer graphics software by means of geometric meshes and material shaders. The room structure (walls, floors, ceiling) and its furnishings are defined by the meshes, while colors, textures and reflectivity properties of the objects are specified through the shaders.

When evaluating the fitness of a solution, the 3D scene is enriched with further supporting elements: the proposed lamps illuminating the environment, and basic 3D structures employed to perform individual lighting measurements at locations of interest. Using a sophisticated ray-tracing render engine, Blender executes an accurate simulation of illumination, taking into account a variety of environmental factors. The obtained rendered images are processed by the second group of python modules to extract light intensity values and their distribution across the interior space.

These outputs are used, in the third group of modules, by the genetic algorithm to compute the actual fitness values of a solution. After evaluating the entire current population and selecting the mating pool, the genetic operators of crossover and mutation are applied to generate the offspring. The operators are specifically implemented for the presented case problem, as mentioned in Sect. 3.2, with the support of an evolutionary computation python framework named DEAP [8], which allows to freely customize any component of the genetic algorithm workflow.

At the end of the execution of the algorithm, the obtained result is the Pareto front of the final population, namely the set of non-dominated solutions, each one of them representing an optimal lighting configuration for the given interior

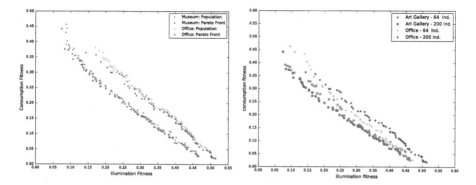

Fig. 1. On the left the final populations and the Pareto fronts in the two case studies. On the right the comparison of Pareto fronts for executions with 64 and 200 individuals.

environment. Optionally, a photorealistic rendering of the illuminated scene can be generated.

5   Results

We evaluated empirically our lighting optimization algorithm on two case studies. As discussed in the Introduction, a satisfactory lighting quality is highly dependent on the visual tasks that are to be performed in the interior space, and on specific requirements of visual interest within the space. These specifications are passed to the model with the placement of the samplers and fixing the target illumination level. All genetic parameters of the model have been tuned in a preliminary phase on simpler and smaller rooms, and these settings did not required further tweaking in the two case studies. The chosen case environments are both complex architectural interiors, with irregular and non-convex planimetries, demonstrating that there are no limitations in the flexibility of application of the presented system.

5.1   Art Gallery

The first case study is an art gallery environment hosting temporary exhibitions, its dimensions are $24 \times 12 \times 4.5$ m. The architecture of this room is characterized by a wide open space with high ceilings, a supporting beam, and two load-bearing columns placed in the middle of the room. A temporary wallboard is also placed as support for hanging canvas painting, and other ground stands are used for various sculpture exhibitions.

A total of 16 samplers have been used to evaluate illumination levels, placed in key areas where light should create visual interest. An illumination target of 0.95 has been selected, since the overall lighting level needed for art exhibitions is slightly lower than typical. The genetic algorithm has been run with a population of 200 individuals, the final population after 30 generations is shown on the left in

Fig. 2. Two renderings, a plan view and an interior view, of two different optimal light configurations in the art gallery environment.

Fig. 1, where it is possible to appreciate how the solutions smoothly span a large Pareto front of the two fitness. The right plot in Fig. 1 shows that the algorithm with an initial population of 64 individuals and 20 generations only already provides an acceptable approximation of the best Pareto front, obtained with 200 individuals. The Fig. 2 shows photorealistic renderings of two of the solutions belonging to the Pareto front, the first gives more importance to the quality of illumination, while the second privileges optimal energy consumption. A more qualitative evaluation of the quality of light is given by the isophotes plotted at 1 m level in the room, in Fig. 4 on the left. This case study demonstrates how the presented algorithm can be a suitable tool to effectively design light configuration for a frequently changing environment, a temporary art gallery, with minimum effort from the user.

5.2 Office

The second case study is a typical open-space office, with dimensions of $29 \times 13 \times 3$ m, composed by a reception area connected to an hallway leading to the main office area. The space is suitable for 20 work stations, and it also includes a separated private area serving as meeting room or as lounge room. The architecture is even more complicated by the presence of a curved wall in the reception, a supporting column and a full window wall in the office area. A total of 12 samples have been used, evenly spaced in the working area, and an illumination target of 1.0 has been specified, since office work requires standard lighting level. Apart from number of samples and illumination target, all parameters of the algorithm are the same as in the Art Gallery case. As in the previous

Fig. 3. Two renderings, a plan view and an interior view, of two different optimal light configurations in the open-space office.

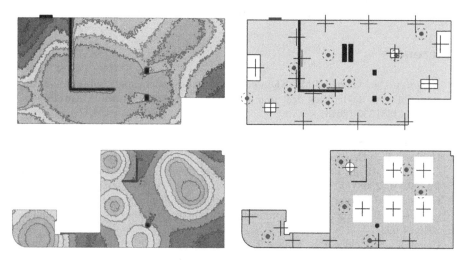

Fig. 4. On the left plots of isophotes. On the right configurations of the solutions, dots and a dashed circles represent lamps, crosses represent samplers.

case study, there is a wide and smooth coverage of the Pareto front. However, as can be seen in Fig. 1, the Pareto front of this case study did not reach the same optimal level in the illumination fitness as the previous one. This result can be explained by the greater complexity of the planimetry of the office, a narrow and long hallway near to a large spacious room appears to be more challenging to illuminate uniformly. Nonetheless, the visual results are rather satisfying, as shown in the photorealistic renderings of two optimal solutions in Fig. 3, the first one preferring light quality and uniformity, the second one considering higher level of energy saving.

The final results of both cases are single executions, it is not practical to perform several runs with different seeds. The timing, on a iMac Intel Core i7 4 GHz, is of 377 min for the Art Gallery with 200 individuals and 30 generations, and 305 min for the Office, in both cases 97% of the time is spent in the rendering of light.

6 Conclusions

This paper proposed a system for inverse design of interior lighting based on the integration between the 3D graphic software Blender and a multi-objective genetic algorithm. The system takes as input an arbitrary interior environment, including realistic furniture and materials, with the description of the lighting requirements in terms of desired average illumination. It produces a Pareto front of solutions minimizing the compliance with the target illumination level, the uniformity of light distribution in the interior space, and the consumption of electric power. The cases presented demonstrate the effectiveness of the system in helping the process of lighting design in complex architectural interiors.

References

1. Andersen, M., Gagne, J.M., Kleindienst, S.: Interactive expert support for early stage full-year daylighting design: a user's perspective on Lightsolve. Autom. Constr. **35**, 338–352 (2013)
2. Baltes, H. (ed.): Inverse Source Problems in Optics. Princeton University Press, Princeton (1978)
3. Caldas, L.: Generation of energy-efficient architecture solutions applying GENE_ARCH: an evolution-based generative design system. Adv. Eng. Inform. **22**, 59–70 (2008)
4. Caldas, L.: Painting with light: an interactive evolutionary system for daylighting design. Building and Environment (2016). https://doi.org/10.1016/j.buildenv. 2016.07.023
5. Cassol, F., Schneider, P.S., França, F.H., Neto, A.J.S.: Multi-objective optimization as a new approach to illumination design of interior spaces. Build. Environ. **46**, 331–338 (2011)
6. Deb, K., Agrawal, S., Pratap, A., Meyarivan, T.: A fast elitist non-dominated sorting genetic algorithm for multi-objective optimization: NSGA-II. In: International Conference on Parallel Problem Solving From Nature, pp. 849–858 (2000)
7. Fesanghary, M., Asadi, S., Geem, Z.W.: Design of low-emission and energy-efficient residential buildings using a multi-objective optimization algorithm. Build. Environ. **49**, 245–250 (2012)
8. Fortin, F.A., De Rainville, F.M., Gardner, M.A., Parizeau, M., Gagné, C.: DEAP: evolutionary algorithms made easy. J. Mach. Learn. Res. **13**, 2171–2175 (2012)
9. Futrell, B., Ozelkan, E.C., Brentrup, D.: Optimizing complex building design for annual daylighting performance and evaluation of optimization algorithms. Energy Build. **92**, 234–245 (2014)

10. Gagne, J., Andersen, M.: A generative facade design method based on daylighting performance goals. J. Build. Performance Simul. **5**, 141–154 (2012)
11. Gordon, G.: Interior Lighting for Designers. Wiley, New York (2014)
12. Grasso, G., Plebe, A.: Conceptual integrity without concepts. In: International Conference on Software Engineering and Knowledge Engineering, pp. 422–427. KSI Research Inc. and Knowledge Systems Institute, Pittsburgh (PA) (2016)
13. Janikow, C.Z., Michalewicz, Z.: An experimental comparison of binary and floating point representations in genetic algorithms. In: Proceedings of the 4th International Conference on Genetic Algorithms, pp. 31–36 (1991)
14. Kawai, J., Painter, J.S., Cohen, M.F.: Radioptimization: goal based rendering. In: Proceedings of the 20th Annual Conference on Computer Graphics and Interactive Techniques, pp. 147–154 (1993)
15. Larson, G.W., Shakespeare, R.: Rendering with Radiance: The Art and Science of Lighting Visualization. Morgan Kaufmann, San Francisco (1997)
16. Lee, K.S., Geem, Z.W.: A new structural optimization method based on the harmony search algorithm. Comput. Struct. **82**, 781–798 (2004)
17. Livingston, J.: Designing with Light: The Art, Science, and Practice of Architectural Lighting Design. John Wiley, New York (2015)
18. Madias, E.N.D., Kontaxis, P.A., Topalis, F.V.: Application of multi-objective genetic algorithms to interior lighting optimization. Energy Build. **125**, 66–74 (2016)
19. Moylan, K., Ross, B.J.: Interior illumination design using genetic programming. In: Johnson, C., Carballal, A., Correia, J. (eds.) EvoMUSART 2015. LNCS, vol. 9027, pp. 148–160. Springer, Cham (2015). https://doi.org/10.1007/978-3-319-16498-4_14
20. Patow, G., Pueyo, X.: A survey of inverse rendering problems. Comput. Graphics Forum **22**, 663–687 (2003)
21. Plebe, A., Grasso, G.: Particle physics and polyedra proximity calculation for hazard simulations in large-scale industrial plants. In: American Institute of Physics Conference Proceedings, pp. 090003-1–090003-4 (2016)
22. Rapone, G., Saro, O.: Optimisation of curtain wall facades for office buildings by means of PSO algorithm. Energy Build. **45**, 189–196 (2012)
23. Sansoni, P., Farini, A., Mercatelli, L. (eds.): Sustainable Indoor Lighting. Springer, Berlin (2015). https://doi.org/10.1007/978-1-4471-6633-7
24. Schoeneman, C., Dorsey, J., Smits, B., Arvo, J., Greenberg, D.: Painting with light. In: Proceedings of the 20th Annual Conference on Computer Graphics and Interactive Techniques, pp. 143–146 (1993)
25. Shea, K., Sedgwick, A., Antonuntto, G.: Multicriteria optimization of paneled building envelopes using ant colony optimization. In: Smith, I.F.C. (ed.) EG-ICE 2006. LNCS (LNAI), vol. 4200, pp. 627–636. Springer, Heidelberg (2006). https://doi.org/10.1007/11888598_56
26. Turrin, M., von Buelow, P., Stouffs, R.: Design explorations of performance driven geometry in architectural design using parametric modeling and genetic algorithms. Adv. Eng. Inform. **25**, 656–675 (2011)
27. Villa, C., Labayrade, R.: Multi-objective optimisation of lighting installations taking into account user preferences - a pilot study. Lighting Res. Technol. **45**, 176–196 (2013)

An Elementary Approach to the Problem of Column Selection in a Rectangular Matrix

Stéphane Chrétien[1(✉)] and Sébastien Darses[2]

[1] National Physical Laboratory, Hampton Road, Teddington TW11 0LW, UK
stephane.chretien@npl.co.uk
[2] Aix Marseille Université, CNRS, Centrale Marseille,
I2M UMR 7373, 13453 Marseille, France
sebastien.darses@univ-amu.fr

Abstract. The problem of extracting a well conditioned submatrix from any rectangular matrix (with e.g. normalized columns) has been a subject of extensive research with applications to machine learning (rank revealing factorization, sparse solutions to least squares regression problems, clustering, $\cdots$), optimisation (low stretch spanning trees, $\cdots$), and is also connected with problems in functional and harmonic analysis (Bourgain-Tzafriri restricted invertibility problem).

In this paper, we provide a deterministic algorithm which extracts a submatrix X_S from any matrix X with guaranteed individual lower and upper bounds on each singular value of X_S. We are also able to deduce a slightly weaker (up to a log) version of the Bourgain-Tzafriri theorem as an immediate side result.

We end the paper with a description of how our method applies to the analysis of a large data set and how its numerical efficiency compares with the method of Spieman and Srivastava.

Keywords: Bourgain Tzafriri theorem · Restricted invertibility
Column selection problems

1 Introduction

Let $X \in \mathbb{R}^{n \times p}$ be a matrix such that all columns of X have unit euclidean ℓ_2-norm. We denote by $\|x\|_2$ the ℓ_2-norm of a vector x and by $\|X\|$ (resp. $\|X\|_F$) the associated operator norm (resp. the Frobenius norm). Let X_T denote the submatrix of X obtained by extracting the columns of X indexed by $T \subset \{1, \ldots, p\}$. For any real symmetric matrix A, let $\lambda_k(A)$ denote the k-th eigenvalue of A, and we order the eigenvalues as $\lambda_1(A) \geq \lambda_2(A) \geq \cdots$. We also write $\lambda_{\min}(A)$ (resp. $\lambda_{\max}(A)$) for the smallest (resp. largest) eigenvalue of A. We finally write $|S|$ for the size of a set S.

The problem of well conditioned column selection that we consider here consists in finding the largest subset of columns of X such that the corresponding submatrix has all singular values in a prescribed interval $[1 - \varepsilon, 1 + \varepsilon]$. The one-sided problem of finding the largest possible T such that $\lambda_{\min}(X_T^t X_T) \geq 1 - \varepsilon$ is

© Springer International Publishing AG 2018
G. Nicosia et al. (Eds.): MOD 2017, LNCS 10710, pp. 234–243, 2018.
https://doi.org/10.1007/978-3-319-72926-8_20

called the Restricted Invertibility Problem and has a long history starting with the seminal work of Bourgain and Tzafriri [3]. Applications of such results are well known in the domain of harmonic analysis [3]. The study of the condition number is also a subject of extensive study in statistics and signal processing [15].

In data science and machine learning applications, the matrix X represents n objects that are described using p features. The column subset selection problem corresponds to choosing the most relevant features among the available ones and is a nice alternative to PCA analysis, a very highly popular method which changes the representation basis without constraining the new features to have an interpretable meaning with respect to the data. An alternative to PCA is the well known Sparse-PCA [6] method. However, Sparse-PCA relies on solving Semi-Definite Programming problems and therefore may not be scalable to high dimension problems such as encountered in "Big Data"-like settings. On the other hand, several methods have been proposed in the literature for column selection; see for instance [2,4,5], etc., and the references therein. In these works, the problem considered is the one of approximating X in a given norm and the methods. Moreover, the methods often involve randomisation. Greedy approaches seem to be suitable for large dimensional problem such as described in [7].

The goal of the present work is to propose a simple greedy approach to the column subset selection problem with the objective to extract independent features. The constraint given before hand is the one of selecting the columns in such a way that the smallest singular value of the extracted matrix is above a pre-specified level. This problem is intimately related to the Bourgain-Tzafriri restricted invertibility problem [3]. Our approach is completely elementary and readily implementable. As a side feature, our algorithm allows at once to also recover the abstract Bourgain-Tzafriri bound up to a log term in the number of selected columns. Our main results are based on two simple ingredients:

1. Choosing recursively $y \in \mathcal{V}$, the set of remaining columns of X, satisfying

$$Q(y) \le \frac{1}{|\mathcal{V}|} \sum_{x \in \mathcal{V}} Q(x),$$

where Q is a relevant quantity depending on the previous chosen vectors;
2. Analysing the steps using a well-known equation (sometimes called *secular equation*) whose roots are the eigenvalues of a square matrix after appending a row and a line.

1.1   Historical Background

Concerning the Restricted Invertibility problem, Bourgain and Tzafriri [3] obtained the following result for square matrices:

Theorem 1.1 ([3]). *Given a $p \times p$ matrix X whose columns have unit ℓ_2-norm, there exists $T \subset \{1, \ldots, p\}$ with $|T| \ge d\dfrac{p}{\|X\|^2}$ such that $C \le \lambda_{\min}(X_T^t X_T)$, where d and C are absolute constants.*

See also [14] for a simpler proof. Vershynin [17] generalized Bourgain and Tzafriri's result to the case of rectangular matrices and the estimate of $|T|$ was improved as follows.

Theorem 1.2 ([17]). *Given a $n \times p$ matrix X and letting $\widetilde{X}$ be the matrix obtained from X by ℓ_2-normalizing its columns. Then, for any $\varepsilon \in (0,1)$, there exists $T \subset \{1, \ldots, p\}$ with*

$$|T| \geq (1 - \varepsilon) \frac{\|X\|_F^2}{\|X\|^2}$$

such that $C_1(\varepsilon) \leq \lambda_{\min}(\widetilde{X}_T^t \widetilde{X}_T) \leq \lambda_{\max}(\widetilde{X}_T^t \widetilde{X}_T) \leq C_2(\varepsilon)$.

Recently, Spielman and Srivastava proposed in [13] a deterministic construction of T which allows them to obtain the following result.

Theorem 1.3 ([13]). *Let X be a $p \times p$ matrix and $\varepsilon \in (0,1)$. Then there exists $T \subset \{1, \ldots, p\}$ with $|T| \geq (1 - \varepsilon)^2 \frac{\|X\|_F^2}{\|X\|^2}$ such that $\varepsilon^2 \frac{\|X\|^2}{p} \leq \lambda_{\min}(X_T^t X_T)$.*

The technique of proof relies on new constructions and inequalities which are thoroughly explained in the Bourbaki seminar of Naor [10]. Using these techniques, Youssef [18] improved Vershynin's result as:

Theorem 1.4 ([18]). *Given a $n \times p$ matrix X and letting $\widetilde{X}$ be the matrix obtained from X by ℓ_2-normalizing its columns. Then, for any $\varepsilon \in (0,1)$, there exists $T \subset \{1, \ldots, p\}$ with $|T| \geq \frac{\varepsilon^2}{9} \frac{\|X\|_F^2}{\|X\|^2}$ such that $1 - \varepsilon \leq \lambda_{\min}(\widetilde{X}_T^t \widetilde{X}_T) \leq \lambda_{\max}(\widetilde{X}_T^t \widetilde{X}_T) \leq 1 + \varepsilon$.*

1.2 Our Contribution

We provide a deterministic algorithm that extracts a submatrix Y_r from the matrix X with guaranteed individual lower and upper bounds on each singular value of Y_r.

Consider the set of vectors $V_0 = \{x_1, \ldots, x_p\}$, where the x_i are the columns of X. At step $r = 1$, choose $y_1 \in V_0$. By induction, let us be given $y_1, \ldots, y_r$ at step r. Let Y_r denote the matrix whose columns are $y_1, \ldots, y_r$ and let v_k be an unit eigenvector of $Y_r^t Y_r$ associated to $\lambda_{k,r} := \lambda_k(Y_r^t Y_r)$.

We say that $u(\cdot, \cdot)$ satisfies the hypothesis (H) if u satisfies for $r \geq 1$:

$$0 \leq u(k, r) \leq u(k+1, r+1), \quad k \in \{0, \cdots, r\}; \tag{1.1}$$

$$0 \leq u(k+1, r) \leq u(1, r) < u(0, r) \quad k \in \{1, \cdots, r-1\}. \tag{1.2}$$

We now introduce the "potential" associated to $u(\cdot, \cdot)$ satisfying (H):

$$Q_r(x) = \sum_{k=1}^r \frac{(v_k^t Y_r^t x)^2}{u(0, r) - u(k, r)}, \quad x \in V_0.$$

We then choose $y_{r+1} \in V_r := \{x_1, \ldots, x_p\} \setminus \{y_1, \ldots, y_r\}$ so that

$$Q_r(y_{r+1}) \leq \frac{1}{p-r} \sum_{x \in V_r} Q_r(x) = \frac{1}{p-r} \sum_{k=1}^{r} \frac{\sum_{x \in V_r} (v_k^t Y_r^t x)^2}{u(0,r) - u(k,r)}. \tag{1.3}$$

The following result, for which we propose a short and elementary proof, gives a control on *all* singular values in the column selection problem.

Theorem 1.5. *Let u satisfies Hypothesis (H). Set $R \leq p/2$. Then, we can extract from X some submatrices Y_r such that for all r and k with $1 \leq k \leq r \leq R$, we have*

$$1 - \delta_R \ u(r - k + 1, r)\sqrt{\lambda_{1,r}} \ \leq \ \lambda_{k,r} \ \leq \ 1 + \delta_R \ u(k,r)\sqrt{\lambda_{1,r}}, \tag{1.4}$$

where

$$\delta_R = \sqrt{\frac{2\|X\|^2}{p} \sup_{1 \leq r \leq R} \sum_{k=1}^{r} \frac{u(0,r)^{-1}}{u(0,r) - u(k,r)}}. \tag{1.5}$$

In particular,

$$\lambda_{1,r} \leq 1 + 2\delta_R \ u(1,r).$$

2   Proof of Theorem 1.5

2.1   Suitable Choice of the Extracted Vectors

Consider the set of vectors $V_0 = \{x_1, \ldots, x_p\}$. At step 1, choose $y_1 \in V_0$. By induction, let us be given $y_1, \ldots, y_r$ at step r. Let Y_r denote the matrix whose columns are $y_1, \ldots, y_r$ and let v_k be an unit eigenvector of $Y_r^t Y_r$ associated to $\lambda_{k,r} := \lambda_k(Y_r^t Y_r)$. Let us choose $y_{r+1} \in V_r := \{x_1, \ldots, x_p\} \setminus \{y_1, \ldots, y_r\}$ so that

$$\sum_{k=1}^{r} \frac{(v_k^t Y_r^t y_{r+1})^2}{u(0,r) - u(k,r)} \leq \frac{1}{p-r} \sum_{x \in V_r} \sum_{k=1}^{r} \frac{(v_k^t Y_r^t x)^2}{u(0,r) - u(k,r)}$$

$$= \frac{1}{p-r} \sum_{k=1}^{r} \frac{\sum_{x \in V_r} (v_k^t Y_r^t x)^2}{u(0,r) - u(k,r)}. \tag{2.6}$$

Lemma 2.1. *For all $r \geq 1$, y_{r+1} satisfies*

$$\sum_{k=1}^{r} \frac{(v_k^t Y_r^t y_{r+1})^2}{u(0,r) - u(k,r)} \leq \frac{\lambda_{1,r}\|X\|^2}{p-r} \sup_{1 \leq j \leq r} \sum_{k=1}^{j} \frac{1}{u(0,j) - u(k,j)}.$$

Proof. Let X_r be the matrix whose columns are the $x \in V_r$, i.e. $X_r X_r^t = \sum_{x \in V_r} xx^t$. Then

$$\sum_{x \in V_r} (v_k^t Y_r^t x)^2 = \text{Tr}\left(Y_r v_k v_k^t Y_r^t X_r X_r^t\right) \leq \text{Tr}(Y_r v_k v_k^t Y_r^t)\|X_r X_r^t\| \leq \lambda_{k,r}\|X\|^2,$$

which yields the conclusion by plugging in into (2.6) since $\lambda_{k,r} \leq \lambda_{1,r}$.

In practice, the next y_{r+1} can be chosen by selecting random candidates among the remaining columns.

2.2   Controlling the Individual Eigenvalues

It is clear that (1.4) holds for $r = 1$ since then, 1 is the only singular value because the columns are supposed to be normalized.

Assume the induction hypothesis (H_r): for all k with $1 \leq k \leq r < R$, (1.4) holds.

Let us then show that (H_{r+1}) holds. By Cauchy interlacing theorem, we have

$$\lambda_{k+1,r+1} \leq \lambda_{k,r}, \quad 1 \leq k \leq r$$
$$\lambda_{k+1,r+1} \geq \lambda_{k+1,r}, \quad 0 \leq k \leq r - 1.$$

We then deduce, due to the induction hypothesis (H_r) and Assumption (H),

$$\lambda_{k+1,r+1} \leq 1 + \delta_R u(k,r)\sqrt{\lambda_{1,r}} \leq 1 + \delta_R u(k+1,r+1)\sqrt{\lambda_{1,r+1}}, \quad 1 \leq k \leq r, \tag{2.7}$$

$$\lambda_{k+1,r+1} \geq 1 - \delta_R u(r-k,r)\sqrt{\lambda_{1,r}}$$
$$\geq 1 - \delta_R u(r+1-(k+1)+1, r+1)\sqrt{\lambda_{1,r+1}}, \quad 0 \leq k \leq r-1. \tag{2.8}$$

It remains to obtain the upper estimate for $\lambda_{1,r+1}$ and the lower one for $\lambda_{r+1,r+1}$. We write

$$Y_{r+1}^t Y_{r+1} = \begin{bmatrix} y_{r+1}^t \\ Y_r^t \end{bmatrix} \begin{bmatrix} y_{r+1} \ Y_r \end{bmatrix} = \begin{bmatrix} 1 & y_{r+1}^t Y_r \\ Y_r^t y_{r+1} & Y_r^t Y_r \end{bmatrix}, \tag{2.9}$$

and it is well known that the eigenvalues of $Y_{r+1}^t Y_{r+1}$ are the zeros of the secular equation:

$$q(\lambda) := 1 - \lambda + \sum_{k=1}^{r} \frac{(v_k^t Y_r^t y_{r+1})^2}{\lambda - \lambda_{k,r}} = 0. \tag{2.10}$$

We first estimate $\lambda_{1,r+1}$ which is the greatest zero of q, and assume for contradiction that

$$\lambda_{1,r+1} > 1 + \delta_R u(0,r)\sqrt{\lambda_{1,r}}. \tag{2.11}$$

From (H_r), we then obtain that for $\lambda \geq 1 + \delta_R u(0,r)\sqrt{\lambda_{1,r}}$,

$$q(\lambda) \leq 1 - \lambda + \frac{1}{\delta_R \sqrt{\lambda_{1,r}}} \sum_{k=1}^{r} \frac{(v_k^t Y_r^t y_{r+1})^2}{u(0,r) - u(k,r)} := g(\lambda).$$

Let λ^0 be the zero of g. We have $g(\lambda_{1,r+1}) \geq q(\lambda_{1,r+1}) = 0 = g(\lambda^0)$. But g is decreasing, so

$$\lambda_{1,r+1} \leq \lambda^0 = 1 + \frac{1}{\delta_R \sqrt{\lambda_{1,r}}} \sum_{k=1}^{r} \frac{(v_k^t Y_r^t y_{r+1})^2}{u(0,r) - u(k,r)}.$$

Thus, using Lemma 2.1, the equality (1.5) and noting that $r \leq p/2$, we can write:

$$\lambda_{1,r+1} \leq 1 + \frac{2}{\delta_R} \frac{\sqrt{\lambda_{1,r}} \|X\|^2}{p} \sum_{k=1}^{r} \frac{1}{u(0,r) - u(k,r)} \leq 1 + \delta_R u(0,r)\sqrt{\lambda_{1,r}},$$

(2.12)

which yields a contradiction with the inequality (2.11). Thus, we have

$$\lambda_{1,r+1} \leq 1 + \delta_R u(0,r)\sqrt{\lambda_{1,r}} \leq 1 + \delta_R u(1,r+1)\sqrt{\lambda_{1,r+1}}. \qquad (2.13)$$

This shows that the upper bound in (H_{r+1}) holds.

Finally, to estimate $\lambda_{r+1,r+1}$ which is the smallest zero of q, we write

$$q(\lambda) \geq 1 - \lambda - \frac{1}{\delta_R \sqrt{\lambda_{1,r}}} \sum_{k=1}^{r} \frac{(v_k^t Y_r^t y_{r+1})^2}{u(0,r) - u(k,r)} := \tilde{g}(\lambda).$$

By means of the same reasoning as above, we show that the lower bound in (H_{r+1}) holds.

2.3 Controlling the Greatest Eigenvalue

Set $\mu_{1,r} = \lambda_{1,r} - 1 \geq 0$.

Since $u(1,r) \leq u(1,R) \leq u(0,R)$, we can write

$$\mu_{1,r} \leq \delta_R \sqrt{\mu_{1,r} + 1}.$$

Hence, using that $x \leq A\sqrt{1+x}$ implies $x \leq 2A$, we reach the upper estimate for $\lambda_{1,r}$.

This concludes the proof of Theorem 1.5.

2.4 Two Simple Examples

Let us choose $u(k,r) = \frac{2r-k}{\sqrt{r}}$. Using $(r+1)(2r-k)^2 \leq r(2r+1-k)^2$ and $(r+1)(r+k)^2 \leq r(r+1+k)^2$, we thus deduce that u satisfies Hypothesis (H). Applying Theorem 1.5, we obtain that we can extract a submatrix with R columns and $\lambda_{1,R} \leq 1 + \varepsilon$, provided that

$$R \log R \leq \frac{\varepsilon^2}{8} \frac{p}{\|X\|^2},$$

which is a slightly weaker bound than the one known from [3].

One can also check that $u(k,r) = \sqrt{r-k}$ satisfies Hypothesis (H) and yields a similar bound.

Algorithm 1. Greedy column selection

1: **procedure** GREEDY COLUMN SELECTION
2:     Set $r = 1$ and choose a random singleton $T = \{j^{(1)}\} \subset \{1, \ldots, p\}$. Fix $\varepsilon > 0$.
    Fix R such that $R \log(R) \leq \varepsilon^2 p/(8\|X\|^2)$. Set $r = 1$.
3:     **while** $r \leq R$ **do**
4:         Choose $y_{r+1} \in \mathcal{V}_r := \{x_1, \ldots, x_p\} \setminus \{y_1, \ldots, y_r\}$ so that

$$\sum_{k=1}^{r} \frac{(v_k^t Y_r^t y_{r+1})^2}{u(0, r) - u(k, r)} \leq \frac{1}{p - r} \sum_{k=1}^{r} \frac{\sum_{x \in \mathcal{V}_r} (v_k^t Y_r^t x)^2}{u(0, r) - u(k, r)}. \tag{3.14}$$

        and let $j^{(r+1)}$ denote the index of the selected column.
5:         Set $T = T \cup \{j^{(r+1)}\}$.
6:         $r \leftarrow r + 1$

---

3 Computational Considerations

3.1 A Simple Algorithm

The method studied in this paper can be summarised as follows.

Finding a column $y_{r+1} = x_{j^{(r+1)}}$ in the matrix X that satisfies (3.14) may take some time if we scan through the set of columns of X at each iteration of the method. Yet another option is to draw new candidates uniformly at random. Another idea is to draw new candidates among the remaining columns in X with respect to a probability distribution that could be taken proportional to the maximum absolute scalar product with the vectors already selected at the previous iterations. This approach is reminiscent of the K-means++ method for clustering [1]. In practice, we observed that uniform sampling was often the most efficient solution, but the non-uniform sampling approach based on absolute scalar products may be relevant for very difficult problems with extremely many almost co-linear columns.

3.2 Scalability vs Accuracy

The method described in the present paper is constructive and allows to extract a set of almost orthogonal columns from any given matrix $X \in \mathbb{R}^{n \times p}$. The problem of extracting highly non collinear columns from a data matrix is a crucial in many applications such as

- Machine Learning: rank revealing factorization, sparse solutions to least squares regression problems, clustering, $\cdots$
- Optimisation: low stretch spanning trees, sketching for SemiDefinite programming, $\cdots$

The problem of computing such extraction in polynomial time has recently been addressed in a series of impressive papers [10, 13, 16, 18] $\ldots$

Different criteria have also been proposed in the literature such as the one of maximizing the volume of the polytope generated by the extracted columns such as in the very interesting [11,12]. However, volume maximisation was proved to be NP-hard and one usually resort to randomized algorithms. Notice that in [11], an interesting variant of the Bourgain-Tzafriri theorem is obtained and a quick proof for this theorem is provided.

From the computational viewpoint, the approach of [13] needs to perform recursive invertions of intermediate matrices. This results in a very difficult computational burden, especially when applied to Big Data examples. Our approach avoids this difficulty while maintaining a remarkable accuracy when compared against [13] for the Bourgain-Tzafriri problem, since we only loose a log factor of R for the gain of a much better scalability.

3.3   Extracting Representative Images from a Dataset

Extracting representative objects in a dataset is of great importance in data analytics. It can be used to detect outliers or clusters. In this example, we applied our technique to the Yale Faces database[1] shown in Fig. 1. In order to cluster the set of images, we performed a preliminary scattering transform [8,9] of the images in the dataset. We then reshaped the resulting scattering transform matrices into column vectors that we further concatenated into a single matrix X. We selected 11 faces using our column selection algorithm and we obtained the result shown in Fig. 2. The total time for this computation was .22 s.

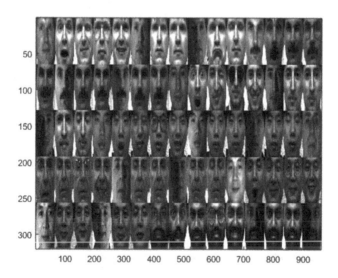

Fig. 1. Faces from the Yale database

We then used the method for extracting representative images from a large data set of 10000 images. We applied Mallat's scattering transform to each of

---

[1] http://www.cad.zju.edu.cn/home/dengcai/Data/Yale/Yale_64x64.mat.

Fig. 2. Faces selected by our algorithm

these images and then ran our method on the resulting matrix in the same manner as in the previous subsection. Our method took less that 15 min in order to extract 20 columns. In comparison, we stopped the algorithm implementing the method of Spielman and Srivastava before convergence, after several hours of computations.

4   Conclusion

In the present paper, we presented and analysed an efficient column extraction method for feature extraction with a view towards Big Data analytics. We discussed computational experiments which confirm that the method has practical advantages over existing methods. Our method is fully deterministic in nature and as such, is different from most existing alternative approaches. Randomisation can be used in order to accelerate the method.

We leave the following question open for further research: does there exist a function u satisfying Hypothesis (H) and allowing to reach the optimal bound known in the Bourgain Tzafriri theorem [3] via our new algorithm?

References

1. Arthur, D., Vassilvitskii, S.: k-means++: the advantages of careful seeding. In: Proceedings of the Eighteenth Annual ACM-SIAM Symposium on Discrete Algorithms, pp. 1027–1035. Society for Industrial and Applied Mathematics (2007)
2. Avron, H., Boutsidis, C.: Faster subset selection for matrices and applications. SIAM J. Matrix Anal. Appl. **34**(4), 1464–1499 (2013)
3. Bourgain, J., Tzafriri, L.: Invertibility of "large" submatrices with applications to the geometry of Banach spaces and harmonic analysis. Israel J. Math. **57**(2), 137–224 (1987)
4. Boutsidis, C., Drineas, P., Magdon-Ismail, M.: Near-optimal column-based matrix reconstruction. SIAM J. Comput. **43**(2), 687–717 (2014)
5. Boutsidis, C., Drineas, P., Mahoney, M.: On selecting exactly k columns from a matrix (2008, in press)

6. d'Aspremont, A., Ghaoui, L.E., Jordan, M.I., Lanckriet, G.R.: A direct formulation for sparse PCA using semidefinite programming. In: Advances in Neural Information Processing Systems, pp. 41–48 (2005)
7. Farahat, A.K., Elgohary, A., Ghodsi, A., Kamel, M.S.: Greedy column subset selection for large-scale data sets. Knowl. Inform. Syst. **45**(1), 1–34 (2015)
8. Mallat, S.: Group invariant scattering. Commun. Pure Appl. Math. **65**(10), 1331–1398 (2012)
9. Bruna, J., Mallat, S.: Invariant scattering convolution networks. IEEE Trans. Pattern Anal. Mach. Intell. **35**(8), 1872–1886 (2013)
10. Naor, A.: Sparse quadratic forms and their geometric applications [following Batson, Spielman and Srivastava]. Séminaire Bourbaki: Vol. 2010/2011. Exposés 1027–1042. Astérisque No. 348 (2012), Exp. No. 1033, viii, 189–217
11. Nikolov, A.: Randomized rounding for the largest simplex problem. In: Proceedings of the Forty-Seventh Annual ACM on Symposium on Theory of Computing, pp. 861–870 (2015)
12. Nikolov, A., Singh, M.: Maximizing determinants under partition constraints. In: STOC 2016, pp. 192–201 (2016)
13. Spielman, D.A., Srivastava, N.: An elementary proof of the restricted invertibility theorem. Israel J. Math. **190**, 83–91 (2012)
14. Tropp, J.A.: The random paving property for uniformly bounded matrices. Studia Math. **185**(1), 67–82 (2008)
15. Tropp, J.A.: Norms of random submatrices and sparse approximation. C. R. Acad. Sci. Paris, Ser. I **346**, 1271–1274 (2008)
16. Tropp, J.A.: Column subset selection, matrix factorization, and eigenvalue optimization. In: Proceedings of the Twentieth Annual ACM-SIAM Symposium on Discrete Algorithms, pp. 978–986. Society for Industrial and Applied Mathematics (2009)
17. Vershynin, R.: John's decompositions: selecting a large part. Israel J. Math. **122**, 253–277 (2001)
18. Youssef, P.: A note on column subset selection. Int. Math. Res. Not. IMRN **23**, 6431–6447 (2014)

A Simple and Effective Lagrangian-Based Combinatorial Algorithm for S^{3}VMs

Francesco Bagattini$^{(\boxtimes)}$, Paola Cappanera, and Fabio Schoen

DINFO, Università degli Studi di Firenze, Firenze, Italy
fbagattini@unifi.it

Abstract. Many optimization techniques have been developed in the last decade to include the unlabeled patterns in the Support Vector Machines formulation. Two broad strategies are followed: *continuous* and *combinatorial*. The approach presented in this paper belongs to the latter family and is especially suitable when a fair estimation of the proportion of positive and negative samples is available. Our method is very simple and requires a very light parameter selection. Experiments on both artificial and real-world datasets have been carried out, proving the effectiveness and the efficiency of the proposed algorithm.

Keywords: Semi-supervised learning · Support vector machines Lagrangian combinatorial heuristics

1 Introduction and Related Work

1.1 The Semi-supervised Scenario

The process of manually labeling instances, essential to a supervised classifier, can be expensive and time-consuming. In such a scenario the *semi-supervised* approach, which makes use of the unlabeled patterns when building the decision function, is a more appealing choice. Indeed, large amounts of unlabeled samples often can be easily obtained. *Semi-supervised support vector machines*, (S^{3}VMs) extend the well-known SVMs classifiers [17] to the semi-supervised scenario. In addition to using the labeled part of the training set to maximize the margin between classes, these classifiers take advantage of the unlabeled patterns by forcing the decision function to traverse through low density areas. This approach implements the so called *cluster assumption*, which states that samples that are close each other are likely to have the same label.

Let us consider the linear binary classification problem. We are given ℓ labeled samples, $\{\mathbf{x}_i, y_i\}_{i=1}^{\ell}$ and a set of u unlabeled ones, $\{\mathbf{x}_i\}_{i=\ell+1}^{n}$, where each $\mathbf{x}_i$ is a d-dimensional vector, $y_i \in \{-1, 1\}$ and $n = \ell + u$. When dealing with S^{3}VMs the objective function one needs to optimize depends on both the separation hyperplane parameters $(\mathbf{w}, b)$ and the unknown labels $y_{i=\ell+1}^{n}$,

$$P(\mathbf{w}, b, y_{i=l+1}^{n}) := \frac{1}{2}||\mathbf{w}||^2 + C \sum_{i=1}^{\ell} V(y_i, \alpha_i) + C^* \sum_{i=\ell+1}^{n} V(y_i, \alpha_i), \qquad (1)$$

© Springer International Publishing AG 2018
G. Nicosia et al. (Eds.): MOD 2017, LNCS 10710, pp. 244–254, 2018.
https://doi.org/10.1007/978-3-319-72926-8_21

where $\alpha_i = \mathbf{w}^T \mathbf{x}_i + b$ are the linear predictions. The *Hinge* loss

$$V(y_i, \alpha_i) = \max\{0, 1 - \alpha_i y_i\}^p \tag{2}$$

with p usually chosen as 1 or 2 (from now on, we will assume $p = 1$) is a common choice for V, while the hyper-parameters C and C^* are used to give more importance respectively to the labeled or the unlabeled error term. A *balance constraint* is frequently added to the formulation mainly to avoid trivial solutions (with all the examples being classified as belonging to a single class), in particular when the number of labeled examples is significantly smaller than that of the unlabeled ones. This is achieved by letting the user specify a desired ratio r of unlabeled samples to be classified as positive, which leads to the following constraint:

$$\frac{1}{u} \sum_{i=\ell+1}^{n} \max\{y_i, 0\} = r. \tag{3}$$

The above equation is equivalent to

$$\frac{1}{u} \sum_{i=\ell+1}^{n} y_i = 2r - 1, \tag{4}$$

as explained in [5].

1.2 Continuous vs Combinatorial Approach

The optimization techniques that have been recently developed to optimize (1) follow one of two broad strategies, the *continuous* and the *combinatorial* ones. The former approach gets rid of the variables $y_{i=\ell+1}^n$ by replacing them with the expression of their prediction $sgn(\mathbf{w}^T \mathbf{x}_i + b)$, obtaining a non convex function that depends only on $(\mathbf{w}, b)$: the non convexity (that violates one of the nicest properties of supervised SVMs) is due to the unlabeled part of the objective; this is what the continuous approach pays by removing the dependency of P on $y_{i=\ell+1}^n$. Another drawback of substituting the unknown labels with their predictions is that of getting a non linear balance constraint: a common approach to tackle this issue is that of working with a linear relaxation of the constraint (see [3]). Due to the loss of convexity, as pointed out in [5], off-the-shelf dual based SVM software cannot be used directly to optimize (1): for this reason, non linear decision functions are often obtained by implementing the *kernel trick* on the primal formulation, as in [3,8]. An exception to this common practice is [6]. Further details on this family of approaches can be found in [5].

Let us switch to the combinatorial approach, which our method belongs to. Once the unknown labels $y_{i=\ell+1}^n$ are fixed, what we get is a standard SVM formulation. Defining

$$I(y_{i=\ell+1}^n) := \min_{\mathbf{w}, b} P(\mathbf{w}, b, y_{i=l+1}^n), \tag{5}$$

combinatorial approaches aim to minimize $I(y_{i=\ell+1}^n)$ over the set of binary variables $y_{i=\ell+1}^n$. The same applies to the non linear case, where $\mathbf{w} = \sum_{i=1}^n c_i y_i \mathbf{x}_i$ and the real variable $\mathbf{c}$ to be optimized is that of the Lagrangian dual of P. The first S³VM implementation (S³VM$^{\texttt{light}}$, [10]) belongs to this family of strategies. It alternates minimizations in $(\mathbf{w}, b)$ obtained by training a standard SVM on the current labeled set at disposal and a heuristic labeling process of the unlabeled set, in which the ru unlabeled samples $\mathbf{x}_i$ with highest $\mathbf{w}^T \mathbf{x}_i$ are classified as positive (so to satisfy the balance constraint), while the others are assigned to the negative class. An additional label switching phase is carried out to improve the value of the objective function: since only two labels per iteration are switched, the overall training process is very slow. The combinatorial approach itself is known to be intractable when u is large, due to the huge number of possible labeling combinations of the unlabeled patterns. Among others, we mention [4,7,15] as methods belonging to the combinatorial family.

An interesting discussion on the scenarios where the use of unlabeled samples can be *unsafe* can be found in [12]. Some real-world applications of semi-supervised learning and particularly of S³VMs are [11] (spoken dialog systems evaluation), [18] (satellite image classification) and [16] (health-care).

2   Lagrangian S³VM

2.1   Dealing with Hyper-parameters

The labeled samples which a classifier is trained on are vital when tuning the hyper-parameters of its learning algorithm. When only a few patterns in a dataset have a label, cross-validation techniques are likely to pick bad hyper-parameter settings, due to the very small size of the validation sets that can be built. Thus, the more hyper-parameters a method needs to be fine-tuned, the less robust the method is. In addition, even with a very fast decision function learning process, the initial hyper-parameter tuning phase could be very time-consuming. In Sect. 2.4 we will put emphasis on the way our algorithm tackles these issues.

2.2   Balance Constraint as a Guide

In many classification scenarios, a reasonable confidence on the percentage of examples to be classified as positive (r in the literature) is available. Let us think about a medical procedure in which we have to distinguish between patients who are likely to contract a particular disease and those who are not: the overall population incidence of a disease is often well-known or can be fairly well estimated from historical data. Moreover, semi-supervised approaches are often used when a large amount of unlabeled data is available: this renders our estimation of r more likely to be a fair approximation of the *true* ratio of positive samples. This insight can be usefully plugged in a S³VM by means of the balance constraint. Our method carefully takes advantage of this information to draw the separating hyperplane.

2.3 Inductive vs Transductive S³VMs

S³VMs have been introduced by [17] as *Transductive SVMs*. This definition was due to the fact that the unlabeled samples that were used alongside the labeled ones to draw the separating hyperplane were the same the authors would like to have labeled. Conversely, the approach of building a classification rule on the entire input space (which is the classical approach of SVMs) is known in the literature as *inductive*. In [5] the authors elaborate on this distinction and carry out an empirical analysis. As we will point out in Sect. 3, our experimental setting refers to the inductive approach.

2.4 Method Details

The method we present here is a decomposition algorithm that uses a standard SVM as subroutine. For the sake of simplicity, we are going to use linear classification to elaborate on the algorithm's idea. The extension to the kernel formulation is straightforward: the presented method needs to be aware only of the distances between the unlabeled samples and the separating hyperplane.

At first, we initialize the decision function, that is, optimizing (1) in $(\mathbf{w}, b)$ using the labeled part of the training set. At each iteration we give a label to the unlabeled patterns $\{\mathbf{x}_i\}_{i=\ell+1}^{n}$ by taking into account both the current decision function and the balance constraint, that is handled by a Lagrangian technique[1]; this labeling process is extremely fast. The inner SVM is trained again on the extended labeled set, comprised of the patterns in $\{\mathbf{x}_i\}_{i=1}^{\ell}$ and those just labeled by the Lagrangian heuristic; the weight C^* is increased at each iteration, until it is assigned the same value of C. This global optimization approach, known as *annealing*, is often used in the literature (see [5,8]) and aims at giving increasing importance to the (initially) unlabeled patterns, as the learning process evolves.

Let us elaborate on how the labeling Lagrangian heuristic works. Once $(\mathbf{w}, b)$ have been computed, the variable part of (1) remains

$$U(y_{i=l+1}^{n}) := \sum_{i=\ell+1}^{n} \max\{0, 1 - y_i(\mathbf{w}^T\mathbf{x}_i + b)\}, \tag{6}$$

with the balance constraint (4). Then, the idea is to relax the constraint by means of the Lagrangian multiplier λ and solve the corresponding dual problem, which takes the form:

$$\max_{\lambda} \min_{y_{i=l+1}^{n}} L(\lambda, y_{i=l+1}^{n}) :=$$

$$\max_{\lambda} \min_{y_{i=l+1}^{n}} \sum_{i=\ell+1}^{n} \max\{0, 1 - \alpha_i y_i\} + \lambda(\sum_{i=\ell+1}^{n} y_i - \beta),$$

---

[1] The idea of employing Lagrangian techniques to relax the balance constraint is proposed also in [15], but for a totally different formulation.

where α_i is the prediction $\mathbf{w}^T \mathbf{x}_i + b$ and the constant term β is defined as $(2r - 1)u$. We can write, equivalently:

$$L(\lambda, y_{i=l+1}^n) = \underbrace{\sum_{i=\ell+1}^{n} \max\{\lambda y_i, 1 - \alpha_i y_i + \lambda y_i\} - \lambda\beta.}_{:=F(\lambda, \mathbf{y})} \qquad (7)$$

If we fix the Lagrange multiplier to a starting value λ_0, the optimization problem becomes separable in the y variables and each component of $\mathbf{y}^* :=$ $\arg\min_{\mathbf{y}} F(\lambda_0, \mathbf{y})$ can be independently computed:

$$y_i^* = \arg\min_{y_i \in \{-1, +1\}} \max\{0, 1 - \alpha_i y_i\} + \lambda_0 y_i, \quad i = l + 1 \ldots n. \qquad (8)$$

We can then plug $\mathbf{y}^*$ in (7) and update the multiplier value. To this aim, it can be easily shown that

$$L(\lambda) = \min_{y_{i=l+1}^n} L(\lambda, y_{i=l+1}^n) \qquad (9)$$

is a concave function of λ: we can take advantage of this property to obtain an updated value λ^{next} for the subsequent iteration of the Lagrangian heuristic. This iterative approach, called *cutting plane* (see [1]), is a typical non-differentiable optimization method used to solve the Lagrangian dual. Let $(\lambda^a, \mathbf{y}^a)$ and $(\lambda^b, \mathbf{y}^b)$ be a pair of dual solutions, such that

$$\sum_{i=\ell+1}^{n} y_i^a - \beta < 0 \quad \text{and} \quad \sum_{i=\ell+1}^{n} y_i^b - \beta > 0, \qquad (10)$$

and λ^{next} the multiplier value for which

$$L(\lambda^{next}, \mathbf{y}^a) = L(\lambda^{next}, \mathbf{y}^b). \qquad (11)$$

Then, let y^{next} be the labeling obtained by plugging λ^{next} in (8): we can now set $y^a = y^{next}$ or $y^b = y^{next}$ according to the sign of the constraint violation $\sum_{i=\ell+1}^n y_i^{next} - \beta$, and iterate the process until convergence. During the very first iterations of the heuristic, when $(\lambda^a, \mathbf{y}^a)$ and $(\lambda^b, \mathbf{y}^b)$ are not yet available, the multiplier λ is updated (coherently with the constraint violation) as prescribed by the *sub-gradient method* (see [1]). When the Lagrangian dual is polyhedral (the point-wise minimum of a finite number of affine functions), as it happens in our case, the cutting plane method terminates finitely (see Proposition 6.3.2 of [1]).

As described above, the Lagrangian heuristic returns with a labeling $y_{i=l+1}^n$ when the balance constraint is exactly satisfied. During the subsequent annealing iteration of Lagrangian-S^{3}VM a new separation function is computed, taking into account the enhanced labeled set, as described above. Algorithm 1 reports the pseudo-code of our semi-supervised method; a few snapshots of the evolution of the decision function built by the method are depicted in Fig. 1.

Algorithm 1. Lagrangian-S³VM

Input: $\{\mathbf{x}_i, y_i\}_{i=1}^n$, r, a SVM and an annealing rule

$C \leftarrow$ cross-validate the SVM on the labeled set $\{\mathbf{x}_i, y_i\}_{i=1}^\ell$

$C^* \leftarrow$ a fraction of C

$\mathcal{H} \leftarrow$ separating function obtained by SVM on $\{\mathbf{x}_i, y_i\}_{i=1}^\ell$

while $C^* \leq C$ **do**

> $\boldsymbol{\alpha} \leftarrow$ distances of $\{\mathbf{x}_i\}_{i=\ell+1}^n$ from $\mathcal{H}$
>
> $y_{i=\ell+1}^n \leftarrow$ labeling computed wrt $\boldsymbol{\alpha}$ and r as in (6)–(11)
>
> $\mathcal{H} \leftarrow$ separating function obtained by SVM on the enhanced labeled set,
> with C^* weighting the patterns which initially had no label
>
> $C^* \leftarrow$ update C^* according to the annealing rule

end

return the trained classifier

(a)   (b)   (c)   (d)

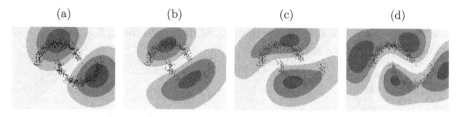

Fig. 1. The snapshots of some selected annealing iterations of Lagrangian-S³VM. The plot (a) shows the contours of the surface that separates the two labeled examples (the blue big circle and the red big triangle). After this initialization the unlabeled patterns are taken into account increasingly, and the decision function evolves until it reaches a balanced solution in (d). (Color figure online)

Let us explain more in detail our parameter selection strategy. Almost everywhere in the literature all of the SVM parameters (typically, C and the kernel parameter γ) are validated in addition to C^*. All the latter need to be fine tuned by the experimenter (their choice widely influences the accuracy of the S³VM approach), resulting in a very time-consuming validation phase. Our choice is totally different: our implementation does nothing but validate the internal supervised solver's parameter C (γ is kept fixed) on the (usually very limited) labeled set. This is done to initialize the SVM and is a very quick process; in addition, the experimenter interaction is not required, being an automated step. Finally, C^* is handled by a standard annealing sequence.

The choice of r, which is the only actual parameter of our method, comes, as is common in the literature, from the knowledge of the problem domain. In addition, differently from C, C^* and γ, this parameter's value has a clear and intelligible meaning and can be easily used by the experimenter to feed our algorithm. In the next section we will show how C and C^* have been chosen to carry out our experiments.

3 Experiments

In this Section we introduce our experimental setup and show the performance of the proposed algorithm, in terms of classification accuracy and execution time.

3.1 Algorithms

We have compared our method with a standard SVM classifier and two semi-supervised ones. In particular, we used the Python sklearn implementation of SVM (based on the LIBSVM library, [2]) as supervised method, while we chose QN-S^{3}VM ([8]) and Well-SVM ([13]) as semi-supervised ones; the latter are the most accurate S^{3}VM implementations available on-line. It is worth to notice that both these implementations belong to the continuous family: in fact, in the last decade, combinatorial methods have been increasingly put aside, due to their poor efficiency and bad scalability. One of the aims of this paper is to show the potential efficiency of combinatorial methods, through a smart use of optimization techniques and heuristics.

3.2 Datasets

To assess its effectiveness and efficiency, we have tested our method with three artificial datasets, 2moons (Fig. 1 shows an instance), 2gauss and 4gauss (see [8] for construction details) and two real-world datasets, usps ([9]) and coil20 ([14]). For each dataset we rescaled the features such that each value lies in [0, 1]; for coil20, we also rescaled each picture to 20×20 pixels. It can be noticed that many classification tasks can be derived from each real-world dataset: we denote with _(i,j) the binary classification task of distinguish between object i and j (pictures of everyday objects in coil20 and handwritten digits in usps); we have taken into account those objects that are more difficult to recognize (e.g., pictures of very similar toy cars from coil20).

3.3 Model Selection

The setting in which we have compared our algorithm with the other solvers is the following: for the inner SVM model we have chosen a Gaussian kernel $K(\mathbf{x}, \mathbf{x}') = \exp(-\gamma||\mathbf{x} - \mathbf{x}'||^2))$, and the supervised solver is internally preset (using the labeled patterns) by cross-validating over a very small set of values for C, while the kernel parameter γ is kept fixed at $1/d$, where d is the number of features. More in detail, with reference to Algorithm 1, C is selected from $\{2^i, i \in [0, 5]\}$ (line 2) and C^* take values $\frac{1}{10}C, \frac{1}{4}C, \frac{1}{2}C, C$, resulting in four annealing iterations (lines 3, 5 and 9). Table 1 recaps how the hyper-parameters' values have been chosen for (all) our experiments. Dealing with the semi-supervised solvers, we refer to the respective papers for a deeper look to the parameter selection strategy, which we followed in our tests. For all our experiments we used a 3-fold cross-validation approach to validate algorithms' parameters. The only exception is the moons dataset: when $l \le 10$, we used $C = 1$ for Lagrangian-S^3VM and default values for the other solvers. It is worth to notice, as already mentioned above, that our validation process is very light; in fact, we only have to choose a value for C to initialize the inner SVM classifier. Differently from the literature, where the hyper-parameters are selected by cross-validating the (parameters of the) whole semi-supervised method, we just validate the inner supervised routine to pick a starting value for C, selecting it from a small set. Nevertheless, in order to keep the comparison as fair as possible, all the execution time analyses we have carried out do not take into account the model selection phase, but only the training steps.

For what concerns r, it has been set (for all the compared algorithms) to the ratio of positive samples in the whole dataset. We are thus assuming to have a good confidence on this measure; note that in the training and test sets this percentage could be different. Doing this we add some uncertainty to the value of r we pick and render its selection more fair. Finally, it is common in the literature to make use of a surrogate function to approximate the (non differentiable) *Hat* loss $H(t) = exp(-st^2)$: the latter arises in continuous approaches when replacing the unknown labels with the expression of their prediction (see Sect. 1.2). QN-S^3VM uses this approximation, and we set $s = 3$ as the authors did in [8].

Table 1. A recap of the hyper-parameter selection.

C	Picked in $\{2^i, i \in [0,5]\}$ by validating SVM on $\{\mathbf{x}_i\}_{i=1}^{\ell}$
C^*	$\{\frac{1}{10}C, \frac{1}{4}C, \frac{1}{2}C, C\}$, resulting in four annealing iterations
γ	Fixed at $1/d$
r	Ratio of positive samples on $\{\mathbf{x}_i\}_{i=1}^{n}$

3.4 Experimental Results

In our first experiment, we have compared algorithms' classification accuracy on the artificial and real-world datasets. Following the experimental setup of [8], two different ratios of labeled examples are used for each classification task. Table 2 reports the mean classification error (that is, the percentage of test patterns being misclassified, scaled between 0.0 and 100.0) and its standard deviation of ten different splits of each dataset configuration; ℓ, u and t denote respectively the number of labeled, unlabeled and test samples, while the best score for each configuration is marked in bold. Our experimental setting is inductive (see Sect. 2.3): we use two separated unlabeled sets, respectively for training and testing. In other words, we employ ℓ labeled and u unlabeled samples for training a classification rule, and use the latter to label the t samples in the test set; this holds for all the experiments in this section. Looking at Table 2, it is easy to notice that the semi-supervised approach outperforms standard SVM classification everywhere. For what concerns the semi-supervised methods, Lagrangian-S^3VM is the most accurate 72% of the time, which confirms its effectiveness.

Table 2. Experimental results. ℓ, u and t denote respectively the number of labeled, unlabeled and test samples; for each configuration we report the mean and the standard deviation of the classification error on ten different dataset splits. `Lagrangian-S`^{3}`VM`, `QN-S`^{3}`VM` and `Well-SVM` are referred to respectively as `lagr`, `qn` and `well`. Best results are in bold.

Dataset	ℓ	u	t	svm	lagr	qn	well
moons	2	498	500	14.92 ± 7.21	8.66 ± 9.01	11.22 ± 5.22	$\mathbf{5.44 \pm 1.66}$
moons	3	497	500	29.92 ± 8.31	7.98 ± 9.68	11.24 ± 3.61	$\mathbf{5.50 \pm 1.26}$
moons	5	495	500	20.10 ± 15.52	7.08 ± 8.80	10.60 ± 2.55	$\mathbf{5.08 \pm 1.58}$
moons	10	490	500	11.02 ± 4.74	$\mathbf{1.36 \pm 1.67}$	3.28 ± 3.19	4.28 ± 0.93
moons	20	480	500	6.02 ± 3.13	$\mathbf{1.10 \pm 1.48}$	2.08 ± 3.46	1.22 ± 1.87
2gauss	25	225	250	23.48 ± 14.10	$\mathbf{3.04 \pm 1.12}$	3.68 ± 2.76	4.56 ± 1.30
2gauss	50	200	250	8.68 ± 2.32	$\mathbf{2.40 \pm 0.88}$	2.76 ± 0.83	5.12 ± 1.41
4gauss	25	225	250	17.52 ± 4.68	15.60 ± 18.46	8.60 ± 9.83	$\mathbf{8.44 \pm 4.22}$
4gauss	50	200	250	8.64 ± 3.17	3.84 ± 2.78	$\mathbf{3.12 \pm 1.63}$	5.20 ± 1.72
usps(2,5)	16	806	823	9.74 ± 6.04	$\mathbf{3.49 \pm 0.39}$	4.25 ± 1.32	4.69 ± 1.11
usps(2,5)	32	790	823	5.58 ± 1.08	$\mathbf{3.62 \pm 0.33}$	3.88 ± 1.14	4.34 ± 1.02
usps(2,7)	17	843	861	3.25 ± 0.95	$\mathbf{1.41 \pm 0.29}$	1.79 ± 0.60	2.75 ± 0.89
usps(2,7)	34	826	861	2.46 ± 1.02	$\mathbf{1.38 \pm 0.35}$	1.85 ± 0.48	1.94 ± 0.42
usps(3,8)	15	751	766	9.65 ± 2.70	$\mathbf{6.12 \pm 1.60}$	7.22 ± 2.51	8.17 ± 2.33
usps(3,8)	30	736	766	6.76 ± 1.29	$\mathbf{4.84 \pm 1.39}$	5.16 ± 1.88	5.69 ± 1.44
usps(8,0)	22	1,108	1,131	4.76 ± 2.12	$\mathbf{1.67 \pm 0.65}$	2.40 ± 1.19	2.88 ± 1.08
usps(8,0)	45	1,085	1,131	3.53 ± 1.20	$\mathbf{1.51 \pm 0.52}$	1.95 ± 0.83	2.11 ± 0.76
coil(3, 6)	15	100	29	16.90 ± 14.91	22.76 ± 17.46	$\mathbf{15.17 \pm 14.40}$	17.59 ± 12.19
coil(3, 6)	25	90	29	6.21 ± 5.30	3.45 ± 4.88	$\mathbf{2.07 \pm 3.52}$	7.59 ± 7.52
coil(5, 9)	15	100	29	29.31 ± 13.20	21.38 ± 10.09	$\mathbf{15.86 \pm 11.25}$	15.86 ± 11.56
coil(5, 9)	25	90	29	14.14 ± 6.97	10.00 ± 9.44	$\mathbf{6.90 \pm 9.87}$	9.66 ± 7.52
coil(6, 19)	15	100	29	7.59 ± 10.77	$\mathbf{6.55 \pm 12.09}$	8.62 ± 17.73	8.97 ± 17.39
coil(6, 19)	25	90	29	5.52 ± 9.53	$\mathbf{00.00 \pm 00.00}$	$\mathbf{00.00 \pm 00.00}$	0.34 ± 1.03
coil(18, 19)	15	100	29	10.34 ± 15.73	$\mathbf{0.69 \pm 2.07}$	2.07 ± 6.21	3.79 ± 4.74
coil(18, 19)	25	90	29	1.72 ± 2.78	0.69 ± 2.07	$\mathbf{00.00 \pm 00.00}$	4.48 ± 4.64

Our second experiment aims at comparing the execution time of the selected semi-supervised solvers. To do so we have varied the training set size $n = \ell + u$ of two different classification tasks of `usps` from 100 up to 1000 samples. The execution time (averaged over ten different dataset splits) is reported in Fig. 2: in both tasks, `Lagrangian-S`^{3}`VM` and `QN-S`^{3}`VM` are the most efficient methods, growing linearly with the size of the training set; conversely, `Well-SVM` turned out to be the worst scalable among the compared algorithms.

3.5   Technical Details

We have implemented our algorithm in `Python 2.7`, with the `sklearn` implementation of SVM as the internal supervised classifier. All execution time analyses have been performed on a desktop computer with an `Intel`® i7 CPU at 2.93 GHz, running `Ubuntu 14.04 LTS`.

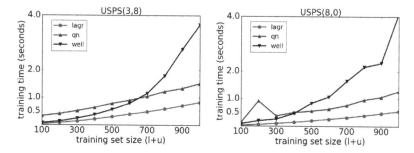

Fig. 2. Execution time (in seconds) with growing training set sizes (labeled and unlabeled samples, with ℓ being fixed at 25) on `usps(3,8)` and `usps(8,0)`. Results are averaged over ten different splits.

4   Conclusion and Remarks

The supervised approach to classification is not reliable when labeled data are scarce. Involving the unlabeled data when training a classifier can help in improving the classification accuracy in such a scenario. However, directly optimizing the unknown labels (combinatorial approach) can be intractable; on the other hand, expressing these variables in terms of their predictions (continuous approach) renders the objective non convex. Recently, several methods have been proposed to tackle these two main drawbacks of semi-supervised classification. A common weak point of these methods is the large number of hyper-parameters they need to be cross-validated on a usually very small validation set. Our approach faces this issue by implementing an automated and very lightweight validation phase. An additional drawback of the continuous methods lies in the need of linearly relaxing the balance constraint. Directly involving the balance constraint in the optimization problem has proved to be a good choice to outperform state-of-the-art solvers' accuracy on most datasets. The presented algorithm is also very efficient, thanks to the quick labeling process guided by a Lagrangian combinatorial heuristic, which renders our approach suitable for larger scale scenarios; future work should broaden the execution time analysis on datasets having a large number of both features and instances. Of course, our method is sensitive to the ratio r of unlabeled examples to be classified as positive, and should be used in a scenario in which there is enough confidence on the value of this parameter.

References

1. Bertsekas, D.P.: Nonlinear Programming. Athena Scientific (2003)
2. Chang, C.C., Lin, C.J.: LIBSVM: a library for support vector machines. ACM Trans. Intell. Syst. Technol. **2**, 27:1–27:27 (2011)
3. Chapelle, O., Chi, M., Zien, A.: A continuation method for semi-supervised SVMs. In: ICML 2006, pp. 185–192. Max-Planck-Gesellschaft, ACM Press, New York, June 2006
4. Chapelle, O., Sindhwani, V., Keerthi, S.S.: Branch and bound for semi-supervised support vector machines. Technical report, Max Plank Institute (2006)
5. Chapelle, O., Sindhwani, V., Keerthi, S.S.: Optimization techniques for semi-supervised support vector machines. J. Mach. Learn. Res. **9**, 203–233 (2008)
6. Collobert, R., Sinz, F., Weston, J., Bottou, L.: Large scale transductive SVMs. J. Mach. Learn. Res. **7**, 1687–1712 (2006)
7. De Bie, T., Cristianini, N.: Semi-supervised Learning Using Semi-definite Programming. MIT Press, Cambridge (2006)
8. Gieseke, F., Airola, A., Pahikkala, T., Kramer, O.: Fast and simple gradient-based optimization for semi-supervised support vector machines. Neurocomputing **123**, 23–32 (2014). Contains Special issue articles: Advances in Pattern Recognition Applications and Methods
9. Hastie, T., Tibshirani, R., Friedman, J.: The Elements of Statistical Learning. Springer Series in Statistics. Springer, New York (2001). https://doi.org/10.1007/978-0-387-84858-7
10. Joachims, T.: Transductive inference for text classification using support vector machines. In: Proceedings of the Sixteenth International Conference on Machine Learning, ICML 1999, pp. 200–209. Morgan Kaufmann Publishers Inc., San Francisco (1999)
11. Li, B., Yang, Z., Zhu, Y., Meng, H., Levow, G., King, I.: Predicting user evaluations of spoken dialog systems using semi-supervised learning. In: 2010 IEEE Spoken Language Technology Workshop (SLT) (2010)
12. Li, Y.F., Zhou, Z.H.: Towards making unlabeled data never hurt. IEEE Trans. Pattern Anal. Mach. Intell. **37**(1), 175–188 (2015)
13. Li, Y., Tsang, I.W., Kwok, J.T., Zhou, Z.: Convex and scalable weakly labeled SVMs. CoRR abs/1303.1271 (2013)
14. Nene, S.A., Nayar, S.K., Murase, H.: Columbia Object Image Library (COIL-100). Technical report, February 1996
15. Sindhwani, V., Keerthi, S.S., Chapelle, O.: Deterministic annealing for semi-supervised kernel machines. In: Proceedings of the 23rd International Conference on Machine Learning, ICML 2006, pp. 841–848. ACM, New York (2006)
16. Timsina, P., Liu, J., El-Gayar, O., Shang, Y.: Using semi-supervised learning for the creation of medical systematic review: an exploratory analysis. In: 2016 49th Hawaii International Conference on System Sciences (HICSS) (2016)
17. Vapnik, V.N., Sterin, A.: On structural risk minimization or overall risk in a problem of pattern recognition. Autom. Remote Control **10**(3), 1495–1503 (1977)
18. Yang, W., Yin, X., Xia, G.S.: Learning high-level features for satellite image classification with limited labeled samples. IEEE Trans. Geosci. Remote Sens. **53**(8), 4472–4482 (2015)

A Heuristic Based on Fuzzy Inference Systems for Multiobjective IMRT Treatment Planning

Joana Dias[1,2(✉)], Humberto Rocha[1,2], Tiago Ventura[3], Brígida Ferreira[4], and Maria do Carmo Lopes[3,5]

[1] Faculdade de Economia, CeBER, Universidade de Coimbra, 3004-512 Coimbra, Portugal
`joana@fe.uc.pt, hrocha@mat.uc.pt`
[2] INESC-Coimbra, Rua Sílvio Lima, Pólo II, 3030-290 Coimbra, Portugal
[3] Serviço de Física Médica, IPOC-FG, EPE, 3000-075 Coimbra, Portugal
`{tiagoventura,mclopes}@ipocoimbra.min-saude.pt`
[4] School for Allied Health Technologies, Porto, Portugal
`bcf@estsp.ipp.pt`
[5] Departamento de Física, I3N, Universidade de Aveiro, 3810-193 Aveiro, Portugal

Abstract. Radiotherapy is one of the treatments used against cancer. Each treatment has to be planned considering the medical prescription for each specific patient and the information contained in the patient's medical images. The medical prescription usually is composed by a set of dosimetry constraints, imposing maximum or minimum radiation doses that should be satisfied. Treatment planning is a trial-and-error time consuming process, where the planner has to tune several parameters (like weights and bounds) until an admissible plan is found. Radiotherapy treatment planning can be interpreted as a multiobjective optimization problem, because besides the set of dosimetry constraints there are also several conflicting objectives: maximizing the dose deposited in the volumes to treat and, at the same time, minimizing the dose delivered to healthy cells. In this paper we present a new multiobjective optimization procedure that will, in an automated way, calculate a set of potential non-dominated treatment plans. It is also possible to consider an interactive procedure whenever the planner wants to explore new regions in the non-dominated frontier. The optimization procedure is based on fuzzy inference systems. The new methodology is described and it is applied to a head-and-neck cancer case.

Keywords: Multiobjective · Radiotherapy planning · Fuzzy inference systems

1 Introduction

Radiotherapy is one of the possible treatments used against cancer, possibly combined with surgery and chemotherapy. In a radiotherapy treatment (RT), the patient is immobilized in a couch, and receives radiation from a linear accelerator, mounted on a gantry that can rotate along a central axis parallel to the couch. The rotation of the couch and gantry allows radiation to be delivered from almost any direction (angle) around the tumor. However, the equidistant coplanar angle configuration (radiation beams equally spaced lay on the plane of rotation of the linear accelerator) is usually used. There are different RT

© Springer International Publishing AG 2018
G. Nicosia et al. (Eds.): MOD 2017, LNCS 10710, pp. 255–267, 2018.
https://doi.org/10.1007/978-3-319-72926-8_22

modalities, sharing essentially the same workflow. First all volumes to treat (Planning Target Volumes – PTV) and radiosensitive structures to spare (Organs at Risk – OAR) are delineated using the patient's 3D medical images. Then, the medical prescription is defined, imposing lower and/or upper radiation doses to be deposited, or maximum/minimum volumes that should receive a given maximum/minimum radiation dose. This medical prescription has then to be translated into a plan configuration. In the last stage, the quality of the proposed treatment plan is analyzed through dose-volume statistic tools and dose distribution inspection. The present work is focused on the process that leads to a treatment plan to be delivered to the patient. In current clinical practice, this process is done by resorting to a computer assisted trial-and-error time consuming procedure using dedicated dose calculation software (Treatment Planning System-TPS). TPS asks the planner to introduce weights, bounds or other parameters that will be used in the TPS optimization procedure, that is seen by the planner as a *blackbox*. With these parameters fixed, the TPS will run an optimization procedure, generating a dose distribution that will be compared with the desired dose distribution defined by the medical prescription. The planner will iteratively change the TPS dependent parameters, trying to comply with the medical prescription. The procedure is repeated until the planner is satisfied, runs out of time, or does not find other ways of improving the treatment plan. Depending on the complexity of the case, this interactive process can take from several hours to several days for a single patient, and the optimality of the solution is not guaranteed. Moreover, the planner will have to deal with many difficult decisions and tradeoffs. It is not possible to guarantee that a solution satisfying all the dosimetry constraints even exists. If this is the case, the planner will have to try to satisfy the constraints "as much as possible", being difficult to define in a rigorous way this concept. The planner will also have to consider the existence of tradeoffs between the doses delivered to different structures, since RT planning is inherently a multiobjective problem: the maximization of the dose delivered to PTV versus minimization of the dose in OAR.

In this work, Intensity Modulated Radiation Therapy (IMRT) is considered, although the developed methodology can be easily extended to other RT modalities. In IMRT the head of the linear accelerator is composed of pairs of individual leaves that can move independently (multileaf collimator). These leaves will block radiation, and different configurations allow the conformal shaping of the treatment beams to the tumor shape and the possibility of having different radiation intensity profiles. Each radiation beam is interpreted as a set of individual beamlets. In clinical practice, the planner will usually determine *a priori* the number of beams to use and their directions. For each set of TPS parameters, an optimization procedure is run (IMRT Fluence Map Optimization – FMO) that will generate the optimal radiation intensity associated with each beamlet from each of the angles to be used in the treatment (fluence maps). The dose deposited in each voxel (measured in Gy) can then be calculated. In this paper, we present an approach where the trial-and-error procedure is replaced by an automated procedure that optimizes fluence maps by using Fuzzy Inference Systems (FIS). The procedure will consider different sets of angles, and will calculate a set of potential non-dominated solutions that can then be presented to the planner. Solutions are called "potential" non-dominated because it is not possible to know for sure if they are indeed Pareto optimal solutions. The presented methodology can thus be seen as a heuristic procedure. The paper is organized as follows: in Sect. 2, a brief review of the literature is presented, focusing on fuzzy logic and multiobjective approaches applied to RT planning. Section 3 describes the mathematical optimization problem and the FMO problem. Section 4 describes the developed approach. Section 5 describes an application to a head-and-neck cancer case. Section 6 presents some conclusions and directions for future research.

2 Brief Review of the Literature

The rules that guide the planner in the interactive process of changing TPS dependent parameters can be hard to represent in a mathematical formal way. They usually are simple rules that can be written in natural language, and one of the ways of representing this kind of information is resorting to fuzzy numbers and fuzzy logic. The methodology presented in this paper is an adaptation of an algorithm previously developed by the authors [1]. All the model parameters are iteratively and automatically changed by resorting to a FIS system, without any type of human intervention. The algorithm considers how far the present treatment plan is from what is desired by the medical prescription, and uses common-sense rules of the form "*if the spinal cord is not being spared enough then increase the importance of this structure in the optimization process*", translated into fuzzy rules, to automatically tune the TPS parameters. One of the drawbacks of the methodology is that it asks the planner to define priorities associated with each structure (that can all have the same value), calculating a single solution based on those priorities. The method is capable of delivering high quality plans within reasonable computational times. Fuzzy logic has been applied to RT planning before. Li and Yin [2] apply fuzzy logic for determining the best prescription for the normal tissue. Yan et al. [3, 4] consider the changing of weights assigned to each structure through the use of a FIS composed of eight rules. The authors extend this work [5] by developing a neuro-FIS using a trained neural network to determine the parameters of the fuzzy inference system.

The multiobjective inherently nature of RT planning problems have been recognized by several different authors. It has been demonstrated that multiobjective optimization can help planners, especially the less trained ones, to improve the quality of the treatment plans, with a reduction of planning time [6, 7]. Romeijn et al. [8] present several results showing that under some conditions several non-convex objectives usually used in RT planning can be transformed in convex ones, preserving the set of non-dominated treatment plans. In [9, 10] a database of treatment plans is created for *a posteriori* navigation, under the condition that the multiobjective optimization problem is convex. In [11] the authors analyze two different navigation algorithms, and conclude that only a limited number of plans is needed during navigation. In [12] the authors tackle the problem of non-convexity whenever different sets of beam angles are considered, by developing a methodology that allows the navigation between different convex Pareto surfaces. Teichert et al. [13] present a methodology to compare two convex Pareto sets considering two different sets of beam angles. Metaheuristics have also been applied. Holdsworth et al. [14, 15] present a hierarchical evolutionary algorithm for IMRT plan generation. The higher level population represents parameters that are used in the fitness function calculation for the lower level deterministic optimization algorithm. Aubry et al. [16] present a simulated annealing approach, where different objective functions are iteratively chosen to guide the algorithm. Lexicographic approaches to radiotherapy planning have been developed [17], considering a pre-determined ordered list of objectives and constraints. At the present moment, there is no automated procedure for treatment planning that explicitly considers the multiobjective nature of this problem.

3   Multiobjective Optimization Problem

The multiobjective optimization problem is determined by the medical prescription. The defined constraints are inherently linked to the desired objectives. The type of restrictions and objectives to consider will be patient dependent, but they usually consist of dose-volume restrictions that relate the dose delivered with the volume that receives that dose and that one wishes to maximize or minimize, according to the specific structure. One of the main tools to assess the quality of a RT plan is the Dose Volume Histogram (DVH), so including in the optimization process restrictions and objectives related with points in the DVH has several advantages. However, these type of constraints are usually considered as being very difficult to include in FMO problems [18–20], because they present the drawback of creating a non-convex feasibility space, with many local minima. It can also be useful to consider the mean-tail-dose rather than conventional dose-volume constraints [21] (mean dose of a hottest or coldest fractional volume). Consider the medical prescription defined in Table 1.

Table 1. Prescribed doses for each structure considered

Structure	Type of constraint		Limit
Spinal cord	Maximum dose	Lower than	45 Gy
Brainstem	Maximum dose	Lower than	54 Gy
Left parotid	Mean dose	Lower than	26 Gy
Right parotid	Mean dose	Lower than	26 Gy
PTV_{70}	$D_{95\%}$	Greater than	66.5 Gy
PTV_{70}	Maximum dose	Lower than	74.9 Gy
PTV_{59}	$D_{95\%}$	Greater than	56.4 Gy
PTV_{59}	$V_{107\%}$	Lower than	Percentage of PTV_{70} volume inside PTV_{59} plus a 10% margin
Body	Maximum dose	Lower than	80 Gy

This medical prescription considers five structures that should be spared (spinal cord, brainstem, left and right parotids, body), and two PTVs that have different dose requirements: PTV_{59} that should receive 59.4 Gy and PTV_{70} that should receive 70 Gy (the prescribed doses). In this particular case, PTV_{70} is inside PTV_{59}. In an ideally situation 100% of PTV_{70} voxels would receive 70 Gy, and 100% of PTV_{59} voxels (except those belonging to PTV_{70}) would receive 59.4 Gy. It is not possible to guarantee this complete coverage, so different types of constraints are imposed. This medical prescription can be interpreted as defining the following set of constrains:

- No voxel belonging to the spinal cord should receive more than 45 Gy;
- No voxel belonging to the brainstem should receive more than 54 Gy;
- The mean dose in the parotids should not exceed 26 Gy;
- 95% of the voxels in PTV_{70} should receive at least 66.5 Gy ($D_{95\%}$);
- No voxel belonging to PTV_{70} should receive more than 74.9 Gy;
- 95% of the voxels in PTV_{59} should receive at least 56.4 Gy ($D_{95\%}$);
- The percentage of voxels in PTV_{59} that are allowed to receive more than 107% of the prescribed dose ($V_{107\%}$) are limited to the percentage of PTV_{70} volume inside PTV_{59} plus a 10% margin.

These constraints are related with optimization objectives:

- The maximum dose received by the spinal cord should be as low as possible;
- The maximum dose received by the brainstem should be as low as possible;
- The mean dose received by the parotids should be as low as possible;
- $D_{95\%}$ for PTV_{70} should be as close to 70 Gy as possible;
- $D_{95\%}$ for PTV_{59} should be as close to 59.4 Gy as possible;
- The percentage of voxels in PTV_{59} that receive more than 107% of the prescribed dose ($V_{107\%}$) should be minimized.

It is not expected that a single treatment plan will be able to simultaneously optimize all these objectives. If we consider that the set of beam angles is fixed *a priori* then the multiobjective problem that has to be solved is the FMO problem considering simultaneously several objectives. Let V represent the number of voxels, N the number of beamlets and D the dose matrix, such that D_{ij} represents the contribution of beamlet j to the total dose deposited in voxel i. The total dose received by voxel i can be calculated as $\sum_{j=1}^{N} D_{ij} w_j$ with w_j representing the intensity of beamlet j. For this particular case, the FMO model can then be defined as follows:

$$f_1(w) = Min \underset{i \in \text{Spinal cord}}{Max} \sum_{j=1}^{N} D_{ij} w_j \tag{1}$$

$$f_2(w) = Min \underset{i \in \text{Brainstem}}{Max} \sum_{j=1}^{N} D_{ij} w_j \tag{2}$$

$$f_3(w) = Min \frac{\sum_{i \in \text{Right parotid}} \sum_{j=1}^{N} D_{ij} w_j}{\#\{i : i \in \text{Right parotid}\}} \tag{3}$$

$$f_4(w) = Min \frac{\sum_{i\in \text{Left parotid}} \sum_{j=1}^{N} D_{ij}w_j}{\#\{i:i \in \text{Left parotid}\}} \tag{4}$$

$$f_5(w) = Min\,Max\{0, 70 - D_{95\%}(PTV_{70})\} \tag{5}$$

$$f_6(w) = Min\,Max\{0, 59.4 - D_{95\%}(PTV_{59})\} \tag{6}$$

$$f_7(w) = Min\,V_{107\%}(PTV_{70}) \tag{7}$$

Subject to:

$$\sum_{j=1}^{N} D_{ij}w_j \le 45, \forall i \in \text{spinal cord} \tag{8}$$

$$D_{95\%}(PTV_{70}) \ge 66.5 \tag{9}$$

$$\sum_{j=1}^{N} D_{ij}w_j \le 54, \forall i \in \text{brainstem} \tag{10}$$

$$\sum_{j=1}^{N} D_{ij}w_j \le 74.9, \forall i \in PTV_{70} \tag{11}$$

$$\frac{\sum_{i\in \text{Right parotid}} \sum_{j=1}^{N} D_{ij}w_j}{\#\{i:i \in \text{Right parotid}\}} \le 26 \tag{12}$$

$$D_{95\%}(PTV_{59}) \ge 56.4 \tag{13}$$

$$\frac{\sum_{i\in \text{Left parotid}} \sum_{j=1}^{N} D_{ij}w_j}{\#\{i:i \in \text{Left parotid}\}} \le 26 \tag{14}$$

$$\sum_{j=1}^{N} D_{ij}w_j \le 74.9, \forall i \in PTV_{70} \tag{15}$$

$$V_{107\%}(PTV_{59}) \le \text{Percentage of } PTV_{70} \text{ volume inside } PTV_{59} \text{ plus a 10\% margin} \tag{16}$$

$$w_j \ge 0, j = 1,\dots,N \tag{17}$$

4 Heuristic Procedure Based on FIS

In order to optimize this nonlinear multiobjective optimization problem, a much simpler problem will be iteratively solved. Let U_i and L_i be upper/lower bounds associated with voxel i. $\underline{\lambda}_i$ and $\bar{\lambda}_i$ are penalty weights. The FMO model is defined as:

$$f(w) = Min_w \sum_{i=1}^{V} \left[\underline{\lambda}_i \max \left\{ 0, L_i - \sum_{j=1}^{N} D_{ij} w_j \right\}^2 + \bar{\lambda}_i \max \left\{ 0, \sum_{j=1}^{N} D_{ij} w_j - U_i \right\}^2 \right] \quad (18)$$

$$\text{s.t. } w_j \geq 0, \, j = 1, \dots, N \quad (19)$$

The objective function considered does not have any clinical meaning whatsoever. This optimization problem will only be used as a tool for finding RT plans satisfying all the defined constraints. This problem will be iteratively solved, having its parameters (both weights and bounds) automatically changed resorting to FIS. Structures violating the respective constraints will have their importance increased in (18), either by changing the corresponding bounds, or weights, or both. Let d_S represent the distance between the dosimetry values of the current solution and the bounds defined by the violated constraints for S. The fuzzy rules considered are of the form: if d_S is *large* then increase (decrease) L_S (U_S) by a *large* amount; if d_S is *medium* then increase (decrease) L_S (U_S) by a *medium* amount; if d_S is *small* then increase (decrease) L_S (U_S) by a *small* amount, where concepts like *small*, *medium* or *large* are defined by fuzzy membership functions and the change in the right hand side of the constraint is determined by FIS. A detailed description of this procedure can be found in [1]. The algorithm tries to find a solution that satisfies all constraints. If this is not possible, then it will relax some of the constraints, also using FIS and considering the distance between each one of the dose metrics and the desired values (the greater the distance, the greater the relaxation, by changing the right hand side values of the constraints (8)–(16)). When a feasible solution is reached, the algorithm tries to improve this solution by being more demanding regarding the dosimetry constraints. The right hand side values are, once again, changed by using FIS. In [1] the multiobjective nature of the problem was not explicitly considered, and the planner is asked to assign priorities to all structures. These priorities would define (again using a FIS) how the right hand side values of the restrictions would be changed: the algorithm would give more importance to structures with higher priorities, meaning that it would be more demanding with these structures, accepting worse values in the other ones. The algorithm will stop when it is not possible to improve the current treatment plan further. In this work no priority list is considered and equal importance is given to all the structures. A set of solutions is calculated. The algorithmic approach is described next:

1. Choose a set of radiation beam angles. *improve*←0;
2. Initialize all the model's parameters; *it*←0.
3. Solve the FMO with the current parameters; *it*←*it* +1.
4. Do the dosimetry calculations. *Admissible*←*true*.

5. For each structure S
   (a) If S is violating a constraint then change the upper/lower bounds associated with S according to FIS. *Admissible←false*.
   (b) If the upper/lower bound associated with S has reached a predetermined threshold, then change the corresponding weight according to FIS.
6. If *Not Admissible* go to 7, else go to 9.
7. If $it \leq Nmax$ then go to 3, else go to 8.
8. If *improve* then go to 9. Else relax some of the violated constraints using FIS. $it←0$. Go to 3.
9. *improve←1*. For each structure S and for each objective function f involving S
   (a) Change the right hand side of the constraint related with S and f, by using FIS.
   (b) Execute 2 to 8.
   (c) Save the current solution to a set *SOL*.
10. If every set of angles was already considered, then go to 11. Else, select a different set of angles and go to 2.
11. Analyse set *SOL* and identify all the potential non-dominated solutions.

The algorithm begins by considering a given set of beam angles, and tries to find an admissible solution (steps 2 to 8). If it is not possible to find an admissible solution, the algorithm relaxes some of the constraints (step 8). When a solution is finally calculated (step 9), then the algorithm will consider a structure S and one objective function related with that structure at a time. The right hand side of the corresponding constraint will be more demanding (step 9a). This is interpreted as a new problem, that is again solved by steps 2 to 8. The procedure is repeated for all pairs of structures and objectives. All the solutions that are calculated along the process are saved. When all sets of beam angles have been tried, this pool of solutions is analyzed so that only the non-dominated solutions are kept. These solutions are non-dominated considering this set, but it is not possible to assure that they are indeed non-dominated for the original problem. This algorithm can thus be interpreted as a heuristic procedure that approximates the non-dominated Pareto frontier. Step 9a should be further explained. It is motivated by a well known result by Ross and Soland [22], where they show that it is possible to find non-dominated solutions for a linear multiobjective mixed integer programming problem by simply using a weighted objective function and additional constraints, one for each objective. Changing the right hand side of these constraints and optimizing the problem will lead to non-dominated solutions. Although we are not in the presence of a mixed integer linear multiobjective programming problem, the idea is the same: changing the right hand side of constraints that are related with the objective functions will trigger the discovery of new solutions. This change is done looking at how far the current solution is from the upper/lower bounds defined by the constraint associated with the objective function. Simple fuzzy rules are considered, assuming that if the current solution is fulfilling the current constraint by a large amount (the slack is high) the algorithm can be more demanding. On the contrary, if they are barely fulfilling the constraint, then the change has to be only slight.

After generating a set of non-dominated solutions, it is still possible that the planner wants to calculate other solutions different from the ones already available. It is possible to consider an interactive procedure where the planner chooses two known solutions.

Bounds based on these two solutions and his preferences can be defined. To calculate this new solution, two different situations have to be considered: if both solutions were generated using the same beam angles set, then it is possible to simply consider a linear combination of the corresponding fluence maps [12, 23] to find a new admissible solution, taking no more than a few seconds of computational time. If they were generated considering two different sets of beam angles, then it is no longer possible to consider a linear combination of fluence maps directly. The algorithm will consider one beam angle set at a time and will look for a solution generated by that set that is as close as possible to the solution generated with the other beam angle set. A linear combination is then considered. This means that two new solutions are generated, one for each beam angle set. If at least one of the solutions satisfies the new bounds, then the solution is presented to the planner. If not, the algorithm has to be executed again, considering only the two beam angle sets and the new bounds. The computational time is expected to be in the order of 4 to 30 min (according to computational experiments made). Consider the example depicted in Table 2, where two solutions have been found, but the planner wants to calculate another one. The type of existing constraints will determine the new bounds to consider (where $\varepsilon \rightarrow 0$).

Table 2. Calculating a new solution

Structure	Type of constraint	Solution 1	Solution 2	New bounds
Spinal cord	Maximum dose	44.5	38.1	$44.5 - \varepsilon$
Brainstem	Maximum dose	53.7	51.3	$53.7 - \varepsilon$
Left parotid	Mean dose	21.9	20.9	$21.9 - \varepsilon$
Right parotid	Mean dose	21.9	22.5	$21.9 - \varepsilon$
PTV$_{70}$	$D_{95\%}$	67.1	66.5	$66.5 + \varepsilon$
PTV$_{70}$	Maximum dose	74.9	74.9	74.9
PTV$_{59}$	$D_{95\%}$	57.5	56.9	$56.9 + \varepsilon$

5   Illustration of the Application of the Procedure

The algorithm was applied to one head-and-neck cancer case where proper PTV coverage and OAR sparing was difficult to obtain in clinical practice (Fig. 1). The OARs and PTVs considered are defined in Table 1, as well as the medical prescription.

In clinical practice, most of the times, these cases are treated with 5 up to 11 beam angles. In this paper 9 beam angle plans are considered, and every equidistant beam angle solution with 5° discretization was tried. Tests were performed on an Intel Core i7 CPU 2.8 GHz computer with 4 GB RAM and Windows 7. CERR 3.2.2 [24] and MATLAB 7.4.0 (R2007a) were used. The dose was computed using CERR's pencil beam algorithm (QIB), with corrections for heterogeneities. The sample rate for Body was 32 and for the remaining structures was 4. The FMO problem was solved using a trust-region-reflective algorithm (fmincon). FIS made use of the Fuzzy Sets Toolbox. The algorithm was initialized as described in [1]. It found a total of 78 different potential non-dominated solutions in approximately 12 h of computational time. Figure 2 shows

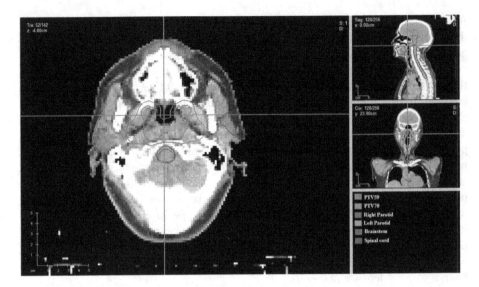

Fig. 1. Contoured structures in one CT slice for the considered case.

the distribution of the dosimetry values for each structure using boxplots. Figure 3 shows a heatmap created by considering the dosimetry values of the solutions set. It can be seen that there are not many differences in the PTV coverage, and choices have to be made regarding the irradiation of parotids and spinal cord. Figure 4 shows a line chart considering the subset of solutions that are in the quartile with higher doses delivered to PTVs. Dominated solutions would be represented by a line that would be always under at least one other line. It is possible to observe that they are all non-dominated between themselves.

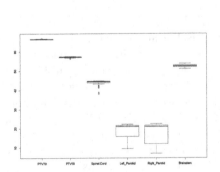

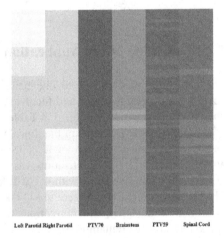

Fig. 2. Box-plot of the dosimetry values for each structure.

Fig. 3. Heatmap illustrating all the non-dominated solutions found.

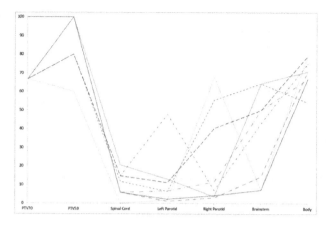

Fig. 4. Line chart considering dosimetry values scaled into [0–100] where 0 is the worst value for the structure and 100 is the best.

6 Conclusions

In this paper a new methodology based on FIS that is able to calculate sets of non-dominated solutions for RT planning is described. This set is built without requiring human intervention. The *a priori* calculation of this set could then support an interactive navigation procedure, where the planner can explore the existing tradeoffs and choose the best treatment plan according to his preferences. It is also possible to consider an interactive procedure, where new plans are calculated if the planner wants to explore new regions of the Pareto frontier. The optimal design of such a decision support system, and the exploration of new visualization tools that allow the user to simultaneously deal with more than three objectives, is out of the scope of this paper and is an interesting path of research. The analysis of the number of solutions calculated with each set of beam angles and the corresponding tradeoffs can also provide valuable insights for the integration of a proper beam angle optimization in a multiobjective framework.

Acknowledgments. This work has been supported by the Fundação para a Ciência e a Tecnologia (FCT) under project grant UID/MULTI/00308/2013.

References

1. Dias, J., Rocha, H., Ventura, T., Ferreira, B., Lopes, M.C.: Automated fluence map optimization based on fuzzy inference systems. Med. Phys. **43**, 1083–1095 (2016)
2. Li, R.-P., Yin, F.-F.: Optimization of inverse treatment planning using a fuzzy weight function. Med. Phys. **27**, 691–700 (2000)
3. Yan, H., Yin, F.-F., Guan, H., Kim, J.H.: Fuzzy logic guided inverse treatment planning. Med. Phys. **30**, 2675–2685 (2003)

4. Yan, H., Yin, F.-F., Willett, C.: Evaluation of an artificial intelligence guided inverse planning system: clinical case study. Radiother. Oncol. **83**, 76–85 (2007)
5. Stieler, F., Yan, H., Lohr, F., Wenz, F., Yin, F.-F.: Development of a neuro-fuzzy technique for automated parameter optimization of inverse treatment planning. Radiat. Oncol. **4**, 39 (2009)
6. Kierkels, R.G.J., Visser, R., Bijl, H.P., Langendijk, J.A., van't Veld, A.A., Steenbakkers, R.J.H.M., Korevaar, E.W.: Multicriteria optimization enables less experienced planners to efficiently produce high quality treatment plans in head and neck cancer radiotherapy. Radiat. Oncol. **10**, 87 (2015)
7. Thieke, C., Kufer, K.H., Monz, M., Scherrer, A., Alonso, F., Oelfke, U., Huber, P.E., Debus, J., Bortfeld, T.: A new concept for interactive radiotherapy planning with multicriteria optimization: First clinical evaluation. Radiother. Oncol. **85**, 292–298 (2007)
8. Romeijn, H.E., Dempsey, J.F., Li, J.G.: A unifying framework for multi-criteria fluence map optimization models. Phys. Med. Biol. **49**, 1991–2013 (2004)
9. Craft, D., Halabi, T., Shih, H.A., Bortfeld, T.: An approach for practical multiobjective IMRT treatment planning. Int. J. Radiat. Oncol. Biol. Phys. **69**, 1600–1607 (2007)
10. Craft, D.L., Halabi, T.F., Shih, H.A., Bortfeld, T.R.: Approximating convex Pareto surfaces in multiobjective radiotherapy planning. Med. Phys. **33**, 3399–3407 (2006)
11. Craft, D., Richter, C.: Deliverable navigation for multicriteria step and shoot IMRT treatment planning. Phys. Med. Biol. **58**, 87 (2013)
12. Craft, D., Monz, M.: Simultaneous navigation of multiple Pareto surfaces, with an application to multicriteria IMRT planning with multiple beam angle configurations. Med. Phys. **37**, 736–741 (2010)
13. Teichert, K., Süss, P., Serna, J.I., Monz, M., Küfer, K.H., Thieke, C.: Comparative analysis of Pareto surfaces in multi-criteria IMRT planning. Phys. Med. Biol. **56**, 3669 (2011)
14. Holdsworth, C., Kim, M., Liao, J., Phillips, M.H.: A hierarchical evolutionary algorithm for multiobjective optimization in IMRT. Med. Phys. **37**, 4986–4997 (2010)
15. Holdsworth, C., Kim, M., Liao, J., Phillips, M.: The use of a multiobjective evolutionary algorithm to increase flexibility in the search for better IMRT plans. Med. Phys. **39**, 2261–2274 (2012)
16. Aubry, J.-F., Beaulieu, F., Sevigny, C., Beaulieu, L., Tremblay, D.: Multiobjective optimization with a modified simulated annealing algorithm for external beam radiotherapy treatment planning. Med. Phys. **33**, 4718–4729 (2006)
17. Breedveld, S., Storchi, P.R.M., Voet, P.W.J., Heijmen, B.J.M.: iCycle: integrated, multicriterial beam angle, and profile optimization for generation of coplanar and noncoplanar IMRT plans. Med. Phys. **39**, 951–963 (2012)
18. Deasy, J.O.: Multiple local minima in radiotherapy optimization problems with dose–volume constraints. Med. Phys. **24**, 1157 (1997)
19. Zarepisheh, M., Shakourifar, M., Trigila, G., Ghomi, P.S., Couzens, S., Abebe, A., Noreña, L., Shang, W., Jiang, S.B., Zinchenko, Y.: A moment-based approach for DVH-guided radiotherapy treatment plan optimization. Phys. Med. Biol. **58**, 1869–1887 (2013)
20. Scherrer, A., Yaneva, F., Grebe, T., Küfer, K.-H.: A new mathematical approach for handling DVH criteria in IMRT planning. J. Glob. Optim. **61**, 407–428 (2014)
21. Romeijn, H.E., Ahuja, R.K., Dempsey, J.F., Kumar, A.: A new linear programming approach to radiation therapy treatment planning problems. Oper. Res. **54**, 201–216 (2006)
22. Ross, T., Soland, R.: A multicriteria approach to the location of public facilities. Eur. J. Oper. Res. **4**, 307–321 (1980)

23. Hoffmann, A.L., Siem, A.Y.D., den Hertog, D., Kaanders, J.H.A.M., Huizenga, H.: Derivative-free generation and interpolation of convex Pareto optimal IMRT plans. Phys. Med. Biol. **51**, 6349 (2006)
24. Deasy, J.O., Blanco, A.I., Clark, V.H.: CERR: a computational environment for radiotherapy research. Med. Phys. **30**, 979–985 (2003)

Data-Driven Machine Learning Approach for Predicting Missing Values in Large Data Sets: A Comparison Study

Ogerta Elezaj[1], Sule Yildirim[1], and Edlira Kalemi[2(✉)]

[1] Faculty of Computer Science and Media Technology,
Norwegian University of Science and Technology, 2815 Gjøvik, Norway
ogertae@stud.ntnu.no, sule.yildirim@ntnu.no
[2] Faculty of Engineering and Physical Sciences, University of Surrey, Guildford GU2 7XH, UK
e.kalemi@surrey.ac.uk

Abstract. Pre-processing of large scale datasets in order to ensure data quality is a very important task in data mining. One of the serious threats to data quality is the lack of data collected during field experiments, which negatively affects the data quality. The missing data usually have significant effects in many real-life pattern classification scenarios, especially when it leads to biased parameter estimates but also disqualify for analysis purposes. The process of filling in the missing data based on other valid values of rest of the variables of a data set is known as the imputation process. In this paper, we present a new data-driven machine learning approach for imputing the missing data. Even though Machine Learning methods are used in order to impute missing data in the literature, it is difficult to decide on a single method to apply on a given data set for imputation. This is because imputation process is not considered as science but as art that focuses on choosing the best method with the least biased value. For this reason, we compare different machine learning methods, such as decision tree (C4.5), Bayesian network, clustering algorithm and artificial neural networks in this work. The comparison of the algorithms indicates that, for predicting categorical and numerical missing information in large survey data sets, clustering method is the most efficient out of the others methods found in literature. A hybrid method is introduced which combines unsupervised learning methods with supervised ones based on the missing ratio, for achieving a data imputation with higher accuracy. Additionally, some statistical imputation methods such as Mean\Mode, Hot-Deck have been applied emphasizing their limitations in large scale datasets in comparison to the machine learning methods. A comparison of all above mentioned methods, traditional statistical methods and machine learning methods has been made and conclusions are drawn for achieving data imputation with higher accuracy in data sets of large scale survey. Also, another objective of these experiments is to discover the effect of balancing the training data set in the performance of classifiers. All methods are tested to a real world data set, population and housing census.

Keywords: Imputation · Data pre-processing · Clustering · Decision tree
Neural networks · Hot-Deck

© Springer International Publishing AG 2018
G. Nicosia et al. (Eds.): MOD 2017, LNCS 10710, pp. 268–285, 2018.
https://doi.org/10.1007/978-3-319-72926-8_23

1   Introduction

Information plays an important role in today's modern society and it is considered as a valuable asset for companies and organizations as it can be used for future prediction or decision-making processes. The rapid growth of information coming from different sources and in different formats (such as text, numeric, image, video audio and so on) requires interdisciplinary knowledge to be processed and used for decision making. Nowadays, machine learning techniques are playing an important role in knowledge discovery in databases which is a process that aims to identify valid data models. Identifying valid data models is complex and cannot be managed only based on experience or intuition. The successful implementation of this process requires the application of technologies and techniques that are based on analysis and research. Predictive methods such as classification, regression and time series analyses are used to learn models that are used for classification, i.e. predicting a certain class. Descriptive methods such as clustering, association or link analysis deal with extracting interesting, understandable and interpretable knowledge in order to discover regularities in the data and to uncover patterns. In the discipline of knowledge discovery, the term "process" is always used to indicate that knowledge discovery consists of several steps. There are 5 main steps in this process: (1) understanding the domain and selection of relevant data sets; (2) data pre-processing; (3) data transformation (4) data mining; (5) interpretation and evaluation [1]. Data pre-processing is a complex process that faces a number of challenging aspects. Most of these challenges are common such as scalability, dimensionality and heterogeneous data, data quality, data ownership and distribution and data privacy preservation. In this paper, we will be focus on applying machine learning techniques to preprocessing process for dealing with the data quality challenge. The real-world data sets are considered as dirty data sets due to inconsistence among values, missing values or being noisy data sets. The data can be incomplete when the data are not collected, when there are differences between the time when the data are collected and when the data are analyzed or because of human error or hardware/software malfunction during data collection and data capturing. A data set is considered as noisy when it contains errors for different reasons such as faulty data collection instruments, data entry errors or data transmission errors. In these conditions, when working with real world data set, lacking quality data means lacking quality in mining results and consequently misleading decision making based on these results. Most of the work done by organizations when they implement their data warehouse includes data extraction, data cleaning and data transformation. The missing data can be discovered by searching for null values in a data set but also wrong data or outliers are considered and treated as missing data during the data pre-processing phase.

Missing or partially answered questions, in the field of statistics are referred to as item non-response in contrast to unit non-response. The unit non-response can happen when the person refuses to be part of a survey or when he cannot be contacted during the field work phase (Fig. 1). Weighting procedures are used for handling the unit non-response. Item non- responses occur frequently in large scale surveys and in this kind of surveys it can never be prevented totally. They occur in different domains due to different reasons such as partial refusal of respondent, malfunctions or failures of systems or sensors used to collect and process data, hardware unavailability, etc.

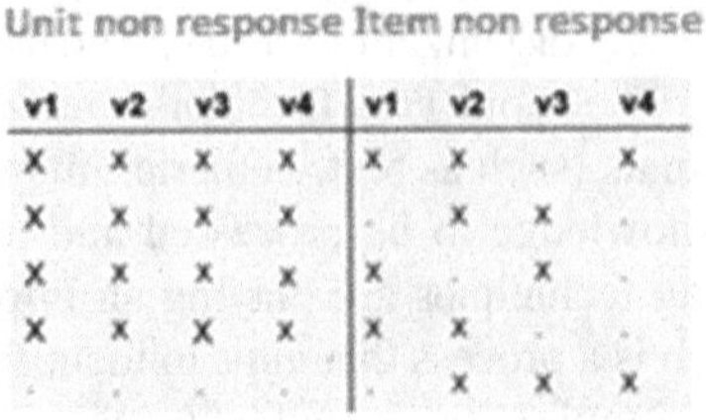

Unit non response Item non response

v1	v2	v3	v4	v1	v2	v3	v4
X	X	X	X	X	X		X
X	X	X	X		X	X	
X	X	X	X	X		X	
X	X	X	X	X	X		
				X	X	X

Fig. 1. Unit non response and item non response

In this situation, researchers have to handle the missing data through understanding them and imputing them in a way to achieve unbiased parameter estimates and accurate standard errors. The amount of missing data is considered to be an important quality indicator for survey data processing [2]. Also, analysis of the effect of imputation is undertaken as part of assessing quality of the data collection in EUROSTAT Quality Assurance Framework [3]. Many techniques have been developed to handle the imputation process. Deletion techniques are the most traditional techniques used to handle missing data by statisticians and they are based on the approach to discard the missing cases from analyses. These techniques are easy and they are recommended to be used in the cases when the data set contains a small number of missing records. In contrast, most of real world data sets contain a considerable amount of missing records and applying these methods can lead to biased parameter estimates. So, in these data sets different techniques should be applied to predict the missing information based on the available information. Several statistical techniques are reported for imputation such as mean/mode, regression methods, and multiple imputation methods. Also, machine learning methods are used to impute the missing data such as decision trees, Bayesian networks, K-Nearest Neighbor, multilayer perceptron and clustering techniques. The decision about what methods to use is not easy and depends on many factors such as the data sets itself, the missing data mechanism, missing patterns and data types that are going to be imputed. This research presents a comparison of some statistical methods and machine learning techniques used for imputation in real world data sets with artificially created missing records. In this paper, we use different missing ratio in the data from 25% to 0.5% for the attributes of the selected data set. This paper is divided into following sections such as introduction, related work, system design, performance measurements and results, and conclusions.

2   Related Work

In a data set, there can be non-observed values or unknown values called as missing values. Most of the missing data occurs during the data collection phase where there can be a lack of information due to uncontrolled system failure or the data are not reported due to privacy or refusal issues. Also, in some cases the values can be lost during the data processing phase. As the presence of missing data in knowledge discovery in databases process is harmful, leading to wrong knowledge extraction and wrong decision making, procedures to handle them should be in place.

The problem of missing data in an experimental design was firstly introduced in 1933 by Allan and Wishart [4]. Their solution was based on estimating the missing data based on iteration methods. This approach is not useful when the number of missing observations is high due to computational time needed during iteration process. In 1937, it was Bartlett who used the analyses of covariance among variables to impute the missing data based on the degree of relationship [5]. Since then, many researchers have proposed different methods to solve the problem of missing data. The selection of the methods is based on missing patterns, missing mechanisms (MCAR, MAR, MNAR) and data set characteristics (numerical or categorical).

The missing pattern of a data set gives information about the structure of the missing data. There are two missing patterns, monotone and arbitrary pattern. In the arbitrary pattern, the missing data are interspersed among full data values while in monotone patterns the missing data are at the end, from left to right and there are not gaps between the missing data and full data, illustrated in Fig. 2 [6]. A monotone missing pattern is easier and more flexible in the selection of the techniques for imputation of data. The first step that a researcher does when she wants to start an imputation process is to analyze the missing patterns via available procedures.

Missing monotone				Missing arbitrarily			
v1	v2	v3	v4	v1	v2	v3	v4
X	x	x	x	x	x	.	x
X	x	x	x	.	x	x	.
X	x	x	.	x	.	x	.
X	x	.	.	x	x	.	.
X	.	.	.	.	x	x	x

Fig. 2. Missing monotone and missing arbitrary (v: variable, missing, x: existing variable value)

Missing data mechanism, classified as Missing Completely at Random (MCAR), Missing At Random (MAR) or Not Missing At Random (NMAR), is seen as a key factor for developing a successful imputation procedure. MCAR is a mechanism where the probability of a record to have a missing value is independent of the existing observed data or the missing one. This case is the highest level of randomness and suggested approach is not to take into account the records with missing values during data processing. In case that the number of records with missing value is too high, using this approach can reduce the data set and end up with loss of information. MAR mechanism occurs when the missing data depends on the observed data but it does not depend on the missing value itself. Most of the real-world data sets follow this mechanism and different imputation methods exist to predict the missing information based on the existing data. NMAR occurs when the probability of having missing data depends on the value of the missing attribute [7, 8].

Another factor to take into consideration during the imputation process is the type of the variables that we are imputing. As there are many imputation methods, some of them can handle only categorical variables (nominal or ordinal), others only numerical variables (discrete or continues) and some other methods can handle both type of variables, numerical and categorical ones. Also, during the imputation process an important aspect to take into account is the capability of methods to deal with complex data sets, in terms of number of variables and number of records.

In state of the art, various imputation techniques are available. Traditional imputation techniques are based on statistical methods and are model based. Handling missing data in surveys using statistical imputation techniques means defining models based on the non missing data sets and doing inferences based on probability distribution of the model [9]. The simplest method to impute missing values is the mean/mode approach [10]. Also, other methods such as Hot-Deck and machine learning are well used for the imputation. However, as the focus of this paper is to find out the best method to be used for imputing survey database such as census, the evaluation measurement should be done based on the accuracy of classification for each categorical target variable and the ability of each imputation strategy in restoring the original marginal distribution of the numerical target variable. After a good overview of some recent research in imputation field, we found out different machine learning algorithms used for imputation and the comparison among them is based on univariate analyses of data, putting apart the preservation of joint distribution.

Bayesian networks are considered as efficient methods based on some previous studies. In 2004, Di Zio suggested the use of Bayesian networks for imputing missing values. He applied this methodology in a subset of 1991 UK population census [11] and concluded that this method improves the consistency and preservation of joint distribution.

Recently, in [12] authors experimented with several clustering algorithms to impute a financial data set and concluded that the k means algorithm suit the imputation process in larger data sets better in terms of achieving a higher accuracy compared to other algorithms.

To the best of our knowledge, there is a lack of comparing supervised and unsupervised machine learning methods with statistical methods used in imputation process in terms of univariate data characteristics and preservation of marginal distribution. Despite the considerable efforts that have been done by researchers in the imputation filed, there is not a broad consensus among them regarding the most appropriate training model or the parameters to use for these models, in order to improve the accuracy of the estimated values. From our review, we find out that there is not a machine learning algorithm that outperforms the others. This happens because the performance of imputation does not only depend on the amount of missing data, but also depends on the missing data structure, the missing data mechanism, the type of variables that are going to be imputed and the nature of the data set that is going to be imputed.

We conducted many experiments with population data estimating the missing values with most used models in this field, to find the most promising model for our situation.

3   System Design

As indicated previously, in this study, we impute the missing values of a census data set using different machine learning methods in order to compare their performance. In this section, we describe the processes of data collection, preparation and the machine learning algorithms we selected to test.

3.1   Data Source and Data Preparation

In this work, as an experiment dataset, we use the Albanian Population and Housing Census, data collected for the entire population of Albania in 2011. This is the main statistical activity carried out by a country with the aim to count all the residents collecting accurate demographical, economic and social data, providing a clear picture of the social-economic situation in the country [13]. This data set is very important for the policy making in the country because it gives changes in population from a social-economic perspective. Meanwhile, as it is a complex and large-scale activity, the reliability of the data depends on the whole data processing steps. As the scope of this paper is the process of imputation, we will be focused only on handling missing data and not on the other type of errors.

The data set in total contains 2,800,138 records which is the total population of the country in 2011, the latest enumeration process. In our study, we have selected only female aged over 15 years old and we have chosen two variables to impute, a categorical and a numerical one (Table 1). This paper deals with how to impute a numerical variable and a categorical variable through its dependence on other numerical and/or categorical variables.

Table 1.   The number of records by gender and age

Total population	Female	Female >15 age old
2,800,138	1,397,079	1,094,429

We decided to impute two different type of variables due to the fact that most of the ML algorithms are data type dependent (Fig. 3). The categorical variable is "having or not any live-born children" and the numerical one is "number of children live-born".

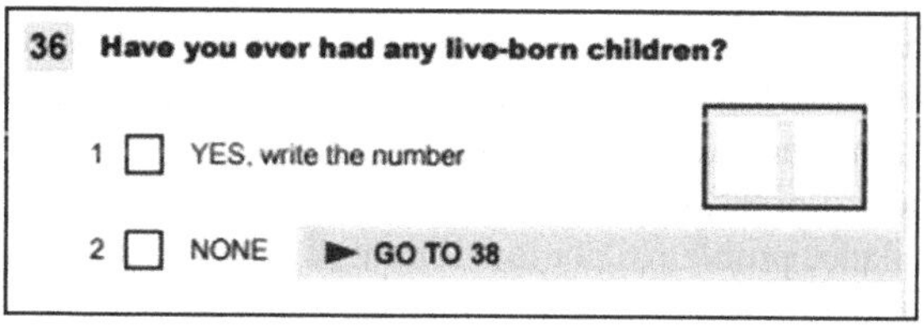

36   Have you ever had any live-born children?
1 ☐   YES, write the number
2 ☐   NONE   ► GO TO 38

Fig. 3.   Question 36 of the population and housing census questionnaire

In surveys, the missing data follows a missing at random mechanism (MAR) and not missing completely at random (MCAR) due to the fact that some people are more or less likely to have a valid response. For example, in censuses some questions had a higher probability to be left blank for example number of children live-born for younger women. The selection of variables that are used during the imputation phase are crucial for the entire process, as the main desirable features of the process is to reduce biases in the final results arising from non responses in the data sets. A set of variables that have a good explanatory power with respect to fertility are used in all the selected machine learning algorithms. They are: woman age, marital status, urban/rural area, employment status, level of education, woman disability status. The data set is anonymized to prevent direct/indirect identifications of persons. All the simulations are conducted using the same data set and the same variables. The data set which contains 1,094,430 records is a final data set without missing values. For our experiment, we have created artificially missing values which later are imputed with different methods. Since the original value of the artificially missing data created are known values, we can evaluate the performance of the machine learning algorithms and do a comparison among them. As the imputation procedure depends not only on the missing mechanism, but also on the amount of missing, data sets with different missing ratios (50%, 20% and 1%) are created as shown in Table 2. These data sets are generated as follows: the original data set is ordered randomly and random missing values are generated for the two variables that are going to be imputed. The missing mechanism of the data sets is MCAR. The missing model used is Uniformly Distributed (UD) where each variable has equal number of missing values.

Table 2. Settings of missing data simulations

Missing mechanism	No of attributes having missing value	Missing ratio	Missing model	No of records per data set
MCAR	2	50%–25% per variable	Uniformly Distributed	1,094,429
		20%–10% per variable		
		1%–0.5% per variable		

Usually in the problem of class prediction, the real-world data set are unbalanced which has an impact in the performance of the classifier during the training phase. The problem of unbalanced data is a crucial problem in many domains such as in medical data sets when rare disease is predicted, in fraud detection, network intrusions etc.

To handle the unbalance class distribution data set problem, external techniques are required during the pre-processing phase. In literature, there are two strategies to deal with the class imbalance problem, data-level strategy and algorithm level strategy [15].

The methods at data level approach adjust the data sets with adding or removing records in order to reduce the discrepancy among classes, while the algorithm level strategy is focused on tuning classifier algorithms to improve the learning process in respect to the minority class. The most known data-level strategy are over-sampling and down-sampling which both has the focus to reduce the imbalance ratio of the data set used in the training phase. When we apply down-sampling we remove part of the records belonging to the majority class until each category is represented by the same number of records, and when we apply up-sampling we duplicate records from the minority class [16]. In the case when we apply down-sampling approach, we reduce the chance of over-fitting but there is a high risk to lose potential information from the training data set as considerable amount of records are excluded during the learning phase.

When we apply the second approach, the duplication of the records of minority class do not provide additional information about the class, so it is not dealing with the lack of the data from minority class.

For doing the experiment, we have split the data set into two, one with missing record and one without missing records. For taking into account the problem of unbalance classes for the imputation of categorical variables, we have used three different data sets, down sampling, no sampling and over sampling for each of the missing patterns.

The principle of analysis is shown in Fig. 4.

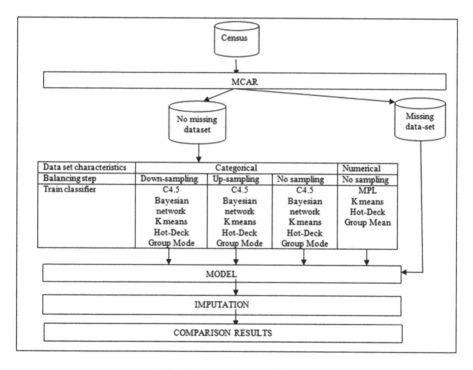

Fig. 4. Comparison architecture

3.2 Methods for Imputation of Missing Values

In literature, there exist various imputation methods starting from traditional one to sophisticated model-based imputation techniques. The methods used in this paper are three unsupervised imputation algorithms: Mean/Mode, Hot-Deck and clustering, and three supervised machine learning algorithms: decision trees, Bayesian Network and neural networks. All these methods are briefly introduced in this section.

Mean/Mode imputation
When this method is applied, the missing values are imputed with the mean/mode of the corresponding variable [19]. So, the imputed value y_i^* for a missing value of y_i is calculated by the observed mean for the numerical variables:

$$y_i^* = \frac{\sum_{k \in obs} y_k}{p} \tag{1}$$

where y_k is the observed value of kth record and p is the total number of non missing values of target variable y. The main limitation of this method is that it does not take into account auxiliary information from other variables during the imputation. For that reason, in this paper we applied Group Mean/Mode imputation which takes into account the variation among groups. When this method is applied, the data are grouped into similar groups based on auxiliary variables and a missing value is replaced by the mean/mode value of the corresponding group.

Hot-Deck imputation
It is based on the nearest neighbor imputation (NNI) and we will introduce this method with a simple case. Let's consider a data set $(x_1, y_1), (x_2, y_2), \ldots, (x_n, y_n)$ and let's suppose that from n y-values there are p observed values and the other values are missing $(m = n - p)$. Let's suppose that the missing ones are $y_{p+1}, y_{p+2}, \ldots, y_n$. The Hot-Deck method imputes the missing value y_i, where $p + 1 <= i <= n$ by y_j where $1 <= j <= p$. The nearest neighbor methods consider y_j as the nearest neighbor of y_i, if j satisfies:

$$\left| x_j - x_i \right| = min \left| x_k - x_i \right| \tag{2}$$

where $1 <= k <= p$. In the case where there is more than one nearest neighbor of i, one of them is randomly selected.

This procedure is repeated until all the missing values are imputed.

The most widely used method for imputing statistical surveys is donor Hot-Deck imputation which for each missing value m a donor record d with similar characteristics is searched in the dataset [20]. This method can be applied for numerical and categorical variables and allows multiple imputations of all the missing variables of a record with information provided by the corresponding donor. The donor pool is defined based on minimizing a distance function (nearest neighbor Hot-Deck).

Decision tree
A decision tree composed by nodes and leaves is a supervised classifier that can handle both numeric and categorical variables used to split the instance space. In this study, we used C4.5 algorithm which is an evolution of ID3 and this algorithm is considered to be a good one for treating missing values using gain ratio which is calculated using entropy based measurement [21]. Using this method, the original data set is divided into two

sets, one with missing values and the other one without missing values. A decision tree is created with the data set without missing values and a missing variable is considered a target variable (class variable). The main limitation is that the target variable has to be a categorical variable and the tree is sensitive to the number of records used to split the nodes.

Bayesian network

A Bayesian network is a directed acyclic graph containing nodes and arcs. A node corresponds to a variable and an arc to the casual relationship among variables. Bayesian networks are used in imputation of missing values because of their ability to deal with the problem of preservation of the variable distribution in large scale data sets. This network can be seen as a hierarchical ordering of variables associated with their conditional probability. This hierarchical ordering allows us to impute a missing value conditionally based only on the variables that are directly related to it.

Clustering

Clustering consists in grouping similar objects into clusters, where objects belonging to a class are similar among them and dissimilar to the objects belonging to the other cluster. As the amount of data to be processed is increased rapidly, clustering is becoming a powerful technique for drawing useful patterns. There has been many clustering algorithms proposed in literature but k means is the most popular. K means used in our study, is a centroid based algorithm [22]. At the initial phase of the algorithm, the k number of clusters should be specified and for each cluster k centroids are randomly selected as initial centers.

Neural networks

Neural networks are created by small units called neurons which try to imitate the neural brain system. All the neurons of a network are connected among them via connections that are weighted. This is considered to be an adaptive system by learning to estimate the parameters based on the training records. During the training phase the network tries to adjust the connection weights to improve the class prediction based on the input signal. The neural networks are widely used when the data sets are high dimensional, when they contain noisy data or when there are complex and hidden patterns among variables. So, because of their ability to present non linear models, they are recently used in imputation of missing data. The neurons of a neural network are organized in layers. The first layer is known as the input layer and the last one as the output layer. There are different neural networks such as perceptron, multi layer perceptron, Kohonen features maps, SOM etc. In this study, we selected the multilayer perceptron as it is the most popular type of neural network. It contains multi layers and is a feed-forward network because cycles or loops within the same layer are not allowed and during the training phase; the weights are updated by mapping inputs to outputs [22].

As a summary, Table 3 shows a summary of methods used in imputation process illustrating their main benefits and limitations.

Table 3. Methods used for imputation process

Methods	Main benefit	Main limitation	Variable type	Computational cost
Mean/Mode	Valid when the amount of missing is low	Variance is reduced artificially	Categorical/ Numerical	Low
Hot-Deck	Variance is not reduced artificially	All possible combinations need to be computed	Categorical/ Numerical	Medium
Decision tree (C4.5)	Use all information for construction of the tree	Does not offer a complete data table	Categorical	High
Bayesian network	Use all information for construction of the tree	Does not offer a complete data table	Categorical	High
Artificial neural network	Maintain complex non linear relationship	Poor generalization in complex data sets	Categorical/ Numerical	Very high
Clustering (k means)	Performs well in large data sets	Parameters in the initialization phase	Categorical/ Numerical	High

4   Performance Measurements and Results

The proposed approach is implemented in Weka 3.8 and executed in a PC with Intel® Core i5 processor with 2.7 GHz speed and 8 GB of RAM. SAS software, Version 9.2 has also been used for processing. Hot-Deck imputation is done using CONCORD JAVA (CONtrollo e CORrezione dei Dati version with Java interface) software developed for data editing and imputation, and IDEA (Indices for Data Editing Assessment) software is used for calculating similarity indexes. These are open source software developed by Italian National Institute of Statistics (ISTAT) [18].

4.1   Algorithms Tuning

All the above mentioned methods were evaluated with 10-fold cross-validation within the Weka data mining library. A ten-way cross-validation approach was selected where the dataset was partitioned randomly into ten subsets of equal size. Nine of these datasets were used as the training set, and the induced tree was used for predicting on the tenth subset (test set).

C4.5 was used to learn and predict values for the target variables. Since this method is a supervised classifier, during the training phase the missing records were not included. We learn a pruned tree, using a confidence threshold of 0.25 with a minimum number of 2 instances per leaf. For K-means algorithm, we select Manhattan distance metric to compute the distance between any two data objects, and the numbers of clusters k is specified using Elbow rule ($k = 10$).

Regarding the neural networks, we used the default heuristic values in all of our experiments because tuning them seemed to have little impact on the final results.

4.2   Evaluation Measures

The evaluation of all the machine learning algorithms has been done by comparing the original value and the imputed value by each algorithm. Imputation performance for the categorical variable is evaluated using different evaluation metrics such as classifier accuracy and root mean square error (RMSE). Accuracy and RMSE are defined as follow:

Accuracy or recognition rate: percentage of test set records that are correctly classified

$$\text{Accuracy} = \frac{(\text{TruePositive} + \text{TrueNegative})}{\text{All}} \tag{3}$$

The RMSE explores the average difference of actual values with the imputed values

$$\text{RMSE} = \sqrt{\frac{\sum_{i=1}^{N} (P_i - O_i)^2}{N}} \tag{4}$$

Where P_i is the imputed value of ith missing value ($1 <= i <= N$), O_i is the actual value of ith artificially created missing value, and N number of artificially created missing values.

As imputing a numerical variable is not a classification problem like categorical variables, the evaluation measurements should be done at two levels: micro level preserving single values for each variable and at macro level preserving marginal distribution of variables. In our study, we used the following measurements:

- micro level: univariate analyses of the variable before and after imputation (mean, mode, firs quartile, median, third quartile, minimum, maximum);
- macro level: Kolmogorov-Smirnov Distance(KS) index for measuring similarities of distribution of variable before and after imputation.

KS index is calculated based on the following cumulative distribution functions:

$$F_{\overline{O}_n}(t) = \frac{1}{N} \sum_{i=1}^{N} I\left(\overline{O}_i \leq t\right) \tag{5}$$

$$F_{P_n}(t) = \frac{1}{N} \sum_{i=1}^{N} I(P_i \leq t) \tag{6}$$

Before and after imputation are computed the value:

$$KS = \max_t \left(\left| F_{\overline{O}_n}(t) - F_{P_n}(t) \right| \right) \tag{7}$$

The KS distance is equal to 0 when the distributions of the variable are equal before and after imputation and takes the maximum value to 1 when there is a maximum difference between the two distributions.

In cases when the experiment is large, which means that there is a huge number of records to be processed from real world data sets, Kovahi in [14] proposed to use 10-fold cross-validation. In the following supervised algorithm, we used this method, which randomly partitions the data set in 10 equal size subsamples, where each of the subsamples is used as a validation data.

As our dataset is not balanced (Table 4) within classes, another objective of these experiments is to discover the effect of balancing the training data set in the performance of classifiers.

Table 4. Frequency of classes

		Frequency	Percent
Class	1	798972	73.0
	2	295457	27.0
	Total	1094429	100.0

4.3 Results and Considerations

In this section, we are discussing the experimental results of the census data set, and some considerations are drawn.

For both variables imputed and for each missing data simulation conducted in the experiments Tables 3 and 4 give the performance measurements for all the algorithms applied in the imputation phase. Table 3 illustrates the performance measurements of the categorical variable whereas Table 5, the univariate characteristics and KS index are given.

Table 5. Performance of algorithms on census data set

		C4.5		Bayesian		K means		Hot-Deck		Group Mode	
Missing ratio	Balance approach	Accuracy %	RMSE	Accuracy %	RMSE	Accuracy %	RMSE	Accuracy %	RMSE	Accuracy %	RMSE
25%	no sampling	93.8552	0.2264	93.4441	0.2306	95.92721	0.2017	86.88	0.3622	83.84	0.2481
	down sampling	92.2622	0.2442	92.2322	0.2469	96.8758	0.1919	86.99	0.3606	83.77	0.2451
	up sampling	90.0247	0.2813	89.9771	0.2872	96.3174	0.1965	85.45	0.3814	82.99	0.2647
10%	no sampling	93.9672	0.2259	93.4229	0.2312	96.4695	0.1878	87.25	0.3570	87.93	0.2463
0.50%	no sampling	93.8526	0.2269	93.29	0.2348	91.5057	0.2914	88.91	0.3330	93.54	0.2541

It is clear that the best overall results based on accuracy and RMSE measurements are achieved when we use clustering algorithm K means where the accuracy is higher and RMSE is lower. The worst case is obtained when we apply traditional statistical methods for imputation such as Hot-Deck. Regarding mode imputation method, we did not apply it because the data set is unbalanced. If this method is applied in an unbalanced data set, all the records are imputed with the value of the class that has the highest frequency. To avoid this problem, we applied the group mode method when we impute the missing class based on the available information from other variables. Due to the fact that we have selected variables that are highly correlated to fertility, the group mode method outperforms the Hot-Deck method in all the cases of missing ratios. Usually the Hot-Deck technique is considered as a better statistical method compared to the group mode because it allows multiple imputation based on the information coming from donors and also it is better in preserving the condition distribution of the imputed variable. When the record missing rate is very low, the mean method fits well in comparison to the other methods. In the other case, where the missing rate is 25% (273,449 records in our data set), the machine learning algorithms performs better in terms of accuracy of prediction. To test the effect of balancing the data over unbalanced data, we applied all the methods in three datasets, down sampling, up sampling and no sampling. We selected the case with highest missing ratio because in this case it makes more sense to test the balancing approach. When we compare the results of the methods used to impute it is evident that the model on unbalance data is more accurate compared to the balanced data sets for mostly of the methods. Also, the confusion matrix of all the machine learning algorithms shows that false positive and false negative values are lower when we use the unbalanced data set to predict the missing values. Only when we apply clustering methods, a very slight improvement in accuracy is evident in the down sampling approach. Our experiment concludes the fact that a 50:50 (down-sampling or over-sampling) balance ratio between classes in the training dataset does not improve the classifier performance. Also, other studies discovered the same finding, when the data set is large, balancing the data is not beneficial as a consequence of the fact that the size of samples for both classes is big enough containing significant information for each of them [17]. Also, as a conclusion from the results, we can find out that down-sampling approach may be more suitable in comparison to up-sampling due to the fact that the false positive cases are minimized. In particular, the k means imputation provides superior results across all the methods when the missing ratio is 25%. This method improved the classification accuracy by 2% compared to the Bayesian network and by 12% compared to group mode imputation (Table 6).

Table 6. Preservation of distribution and aggregates, Kolmogorov-Smirnov Index

MPL	25%					10%					0.5%				
	Original	MPL	K means	Hot-Deck	Group Mean	Original	MPL	K means	Hot-Deck	Group Mean	Original	MPL	K means	Hot-Deck	Group Mean
Mean	2.34	1.89	2.38	2.5	2.347	2.35	2.95	2.31	2.7	2.3473	2.34	2.7616	2.35	3.1	2.3724
STD	2.174	1.616	1.762	3.45	1.676	2.172	1.611	1.835	3.4	1.67649	2.178	1.82487	1.769	3.3	1.6780
Max	16	6	6	20	6	20	6	6	20	6	16	6	6	20	6
Min	0	0	0	0	0	0	0	0	0	0	0	0	0	0	0
Q1	0	1	2	1	1	0	1	1	1	1	0	2	1	1	1
Q2	2	2	2	2	2	2	3	2	2	2	2	3	2	2	2
Q3	3	3	3	5	4	3	4	3	5	4	3	4	4	5	4
Mode	0	0	2	2	2	0	1	3	2	2	0	3	2	2	2
KS	-	0.151	0.013	0.167	0.066	-	0.157	0.075	0.155	0.068	-	0.188	0.104	0.099	0.073

According to the KS index, which measures the ability to restore the original marginal distribution, it seems that all the methods give good results (the index values do not exceed 0.188). The lowest index value is achieved using the K means algorithms (0.013) when the missing ratio is 25%. We observe that K means performs better on a larger data set. When the missing data set is small, the traditional method of mean imputation performs better in terms of a smaller KS index and a mean closer to the original mean. Related to the mean preservation, multi layer perceptron algorithm produces negative bias for all the different missing percentage. The Group Mean method and K-means produce a slight bias on means.

An important aspect to analyze during imputation is the variability of the variable distribution measured by the standard deviation (STD). From the results, it is very evident the fact that all the methods produce consistent reduction of variability, except the Hot-Deck method. This can be explained by the fact that the other methods were not able to reproduce extreme original values. The maximum value imputed from the methods is 6 while the original data set has higher maximum values. All the imputation methods preserve variability better when the data sets are smaller due to the fact that, it is easier to control variations in these data sets. Regarding the use of neural networks for imputing the numerical variables, based on our experiment, we can conclude that they come with a greater computation and training complexity compared to all the other methods.

5   Proposed Imputation Approach

Based on the results, the proposed imputation approach is as shown in figure below (Fig. 5):

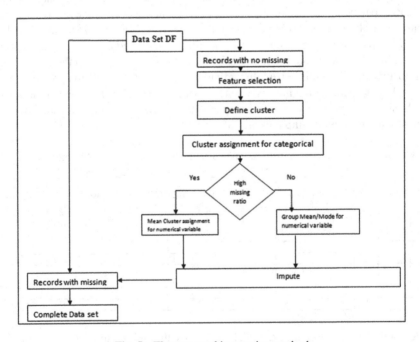

Fig. 5. The proposed imputation method

When dealing with data sets where there are missing values first a detailed analyze should be done regarding the missing mechanism and missing patterns. The data set has to be split in two sets, one without missing records and the other one with missing records. The data set without missing records will be used in the learning phase of the machine learning algorithms. The data sets with missing records will be used in the testing phase. Important step when dealing with imputation is the feature selection, which consists in analyzing the correlation among variables to find out the ones that have higher correction with the variables that are going to be imputed. The selected variables ate this step will be used during the clustering process. Similar records are grouped into same cluster and the cluster information is used to categorical variables. Predicting the missing value of a categorical variable is seen as classification process. When it comes to numerical variable imputation, it has to be taken into account the missing ratio in the data set. When this ratio is high based on our experiment results, imputing the missing values with cluster mean when the missing records belongs to, the imputed value is closer to the original value. In case that the missing ratio is low, imputing with group mean method we achieve better results. More specifically, the clustering approach we apply shows better results in terms of RMSE and KS index. In terms of RMSE, our clustering method exhibits better the missing value prediction for categorical variables, where the RMSE is 0.201. On the other hand, imputing the numerical variables when the amount of missing data is considerable, using clustering approach we can achieve a KS index 0.013, a value closer to 0 is achieved when the distributions of the variable are almost equal before and after imputation.

6   Conclusions

In this paper, the performance of some statistical and machine learning imputation methods has been evaluated through an experimental application on a real-life census data set. Starting from a complete data set we artificially created missing data sets which gives us the possibility to evaluate the performance of the imputation process by comparing the original value with the estimated one. The analyses of results suggested that all methods produce good results in terms of preservation of the marginal distribution for the numerical target variable, but the best result is achieved when we apply the clustering technique in the data set where the missing ratio is higher. As expected, we found that simple imputation methods such as group mean imputation performed just as well as or more than the machine learning methods in data sets where the missing amount of records is low. We found similar performance with the use of decision trees and Bayesian networks for imputing categorical variables. K means is the method which performs better even in predicting the missing values of categorical and numerical variables.

In general, results confirm that when the data sets have a higher missing ratio, the machine learning algorithms perform better compared to the statistical methods such as group mean or Hot-Deck. The results confirm as well that these methods produce satisfactory results related to univariate characteristics of data and preservation of marginal distribution. So we can conclude that machine learning algorithms provide better performance in reproducing the original value, producing significantly better estimates for statistical parameter such as mean, mode, standard deviation. Based on the proposed approach, in surveys for a data set containing a large amount of missing data, clustering methods should be the first methods to be considered. We discover that the basic K-means algorithm outperforms the mean substitution method or other statistical methods, which are considered as simple and common approaches for missing data imputation. As the imputation process is much related to the data set characteristics, missing patterns and mechanism, different methods should be considered for finding the method that better fits the needs. As this process is very crucial, after it there is a high need to always check the result of the imputation.

Further studies are needed in order to measure the performance of these methods when more covariates containing missing data are used, when the number of variables containing missing data is higher, and where the missing mechanism is not missing completely at random. The experiments should also be performed in other domain data sets such as health or financial ones.

References

1. Mannila, H.: Data mining: machine learning, statistics, and databases. In: 8th International Conference on Scientific and Statistical Database Management (SSDBM 1996), p. 2 (1996)
2. Groves, R.M.: Survey Errors and Survey Costs. Wiley, New York (1989)
3. http://ec.europa.eu/eurostat/documents/64157/4372717/Eurostat-Quality-Assurance-Framework-June-2013-ver-1-1-EN.pdf/352234ca-77a0-47ca-93c7-d313d760bbd6
4. Allan, F.G., Wishart, J.: A method of estimating the yield of missing plot in field experiments. J. AgricSci. **20**, 399–406 (1930)
5. Barlett, M.S.: Some examples of statistical methods of research in agriculture and applied biology. J. R. Stat. Soc. B **4**, 137–185 (1973)
6. Berglund, P.A.: An Introduction to Multiple Imputation of Complex Sample Data using SAS® v9.2. SAS Global Forum 2010-Statistics and Data Analysis (2010)
7. Rubin, D.B.: Inference and missing data. Biometrika **63**, 581–592 (1976)
8. Rubin, D.B.: Multiple Imputation for Nonresponse in Surveys. Wiley, New York (1987)
9. Little, R.J., Rubin, D.B.: Statistical Analysis with Missing Data, 2nd edn. Wiley, New York (2002)
10. Schneider, T.: Analysis of incomplete climate data: estimation of mean values and covariance matrices and imputation of missing values. J. Clim. **14**, 853–871 (2001). American Meteorological Society
11. Di Zio, M., Scanu, M., Coppola, L., Luzi, O., Ponti, P.: Bayesian networks for imputation. J. R. Stat. Soc. Ser. A **167**(2), 309–322 (2004a)
12. Purwar, A., Singh, S.K.: Empirical evaluation of algorithms to impute missing values for financial dataset. IEEE (2014)
13. http://www.instat.gov.al/en/census/census-2011.aspx
14. Kohavi, R.: A study of cross-validation and bootstrap for accuracy estimation and model selection. In: Proceedings of the Fourteenth International Joint Conference on Artificial Intelligence, vol. 2, pp. 1137–1143 (1995)
15. Li, Y., Sun, G., Zhu, Y.: Data imbalance problem in text classification. In: 2010 Third International Symposium on Information Processing (ISIP). IEEE (2010)

16. Chawla, N.V., et al.: SMOTE: synthetic minority over-sampling technique. arXiv preprint arXiv:1106.1813 (2002)
17. Visa, S.: Fuzzy classifiers for imbalanced data sets. Department of Electrical and Computer Engineering and Computer Science, University of Cincinnati, Cincinnati (2006)
18. https://www.istat.it/en/tools/methods-and-it-tools/processing-tools/concordjava
19. Han, J., Kamber, M.: Data Mining: Concepts and Techniques. Morgan Kaufmann Publishers, San Francisco (2001)
20. Liu, Y., Salvendy, G.: Visualization support to better comprehend and improve decision tree classification modeling process: a survey and appraisal. Theor. Issues Ergon. Sci. **8**(1), 63–92 (2007)
21. Fujikawa, Y., Ho, T.B.: Cluster-based algorithms for dealing with missing values. In: Chen, M.-S., Yu, P.S., Liu, B. (eds.) PAKDD 2002. LNCS (LNAI), vol. 2336, pp. 549–554. Springer, Heidelberg (2002). https://doi.org/10.1007/3-540-47887-6_54
22. Westin, L.: Missing data and the preprocessing perceptron. Department of Computing Science, Umeå University (2002)

Mineral: Multi-modal Network Representation Learning

Zekarias T. Kefato$^{(\boxtimes)}$, Nasrullah Sheikh, and Alberto Montresor

University of Trento, Trento, Italy
{zekarias.kefato,nasrullah.sheikh,alberto.montresor}@unitn.it

Abstract. Network representation learning (NRL) is a task of learning an embedding of nodes in a low-dimensional space. Recent advances in this area have achieved interesting results; however, as there is no solution that fits all kind of networks, NRL algorithms need to be specialized to preserve specific aspects of the networks, such as topology, information content, and community structure. One aspect that has been neglected so far is how a network reacts to the diffusion of information. This aspect is particularly relevant in the context of social networks. Studies have found out that diffusion reveals complex patterns in the network structure that are otherwise difficult to be discovered by other means. In this work, we describe a novel algorithm that combines topology, information content and diffusion process, and jointly learns a high quality embedding of nodes. We performed several experiments using multiple datasets and demonstrate that our algorithm performs significantly better in many network analysis tasks over existing studies.

Keywords: NRL · Diffusion patterns · Cascades

1 Introduction

Network representation learning (NRL) is the task to embed nodes of a network into a low-dimensional space, while preserving important aspects of the original network. This strategy is an invaluable tool to tackle a variety of subsequent network analysis problems, such as node classification, link prediction, and visualization. It is not only a hard and daunting task to manually engineer high-quality features for the aforementioned problems, but also the resulting features lack the capability of being applicable across different problems. For example, features that are engineered for node classification might not be suitable for link prediction or vice versa; therefore, one has to develop a new set of features for almost every new task.

Automatic network embedding approaches [1–9], however, are highly effective in capturing interesting patterns that are applicable to a range of tasks. They are well-suited for learning features that are otherwise difficult to find even

---

Z. T. Kefato and N. Sheikh contributed equally to this work.

© Springer International Publishing AG 2018
G. Nicosia et al. (Eds.): MOD 2017, LNCS 10710, pp. 286–298, 2018.
https://doi.org/10.1007/978-3-319-72926-8_24

for experts. Such techniques have been employed in multiple disciplines, such as speech recognition and signal processing [10] and object recognition [11,12], improving previous state-of-the-art solutions by several orders of magnitude [13].

Recent studies in representation learning through neural networks have achieved remarkable results [10–12]. An interesting aspect that makes these model attractive is that different components of the model, called neurons, are activated while detecting different kinds of patterns. In other words, the learned embedding has a set of discriminative features that are shared among different tasks [13]. This is one of the main reasons that made the representations learned using this technique applicable across multiple tasks [13].

There have been a plethora of studies [1–8] that apply neural networks to NRL. The goal of such studies is usually to learn a representation that preserves one or more of the following properties of nodes: (i) neighborhood structure, (ii) content/attribute information, (iii) community affiliation.

First of all, a high-quality embedding should enable to effectively reconstruct the original network. Therefore, preserving the structural information is of paramount importance. A second aspect to be considered is that approaches that incorporate content/attribute information and enforce a constraint on an embedding algorithm to preserve it, achieve higher-quality embeddings compared to content-oblivious approaches [4,6], sometimes by over an order of magnitude.

While significant improvements over traditional techniques have been obtained, there are still several aspects of information networks that reveal interesting properties of the network. For example, it has been observed that the dynamics of diffusion of influence and information (cascades) unveil complex patterns of the network that are effective in identifying groups of users [14,15].

To complement existing studies of NRL, in this study we propose a novel algorithm that learns an embedding of the network that preserves the topology, the content information, as well as the dynamics of diffusion cascades.

Our approach integrates content and diffusion information into the network structure, without requiring any additional data structure. Based on this, we propose a novel algorithm called MINERAL (Multi-modal Network Representation Learning).

Given that in some datasets, only a fraction of nodes are included in cascades, while in other datasets cascade information is completely missing, we simulate a diffusion process that enables to capture complex local and global network structures. Then, we acquire context information of nodes related to their local neighborhood (directly connected neighbors) and global neighborhood (community membership).

Our contribution can be summarized as follows:

- we combine different aspects of a network that enable learning an effective network embedding;
- we propose a novel scalable algorithm for NRL;
- we perform several experiments using multiple datasets and across multiple network analysis tasks.

The rest of the paper is organized as follows. Section 2 introduces the basic concepts and notations and presents the problem statement. Section 3 discusses the proposed algorithm. Section 4 reports the experiments and results. Finally, Sect. 5 discusses related works; the paper is concluded in Sect. 6.

2   Preliminary

We start by providing definitions of the data models and describe our problem.

Definition 1. *We consider a network $G = (V, E)$, where V is a set containing n nodes and E is a set containing m edges.*

As in social networks, we assume that the nodes are involved in two types of activities: (i) generating their own content (e.g. posts) and (ii) consuming/spreading others' content. Given a node u, $A(u)$ contains all the pieces of content generated or consumed by u. Content is assumed to be textual; in case of multimedial information, metadata and tags could be used instead. One way to incorporate content information is to add a separate node for each piece of content. However, given that often the goal of incorporating content is to better identify similarities between nodes in the representation learning process, we simply introduce a similarity function π on the edges that is defined as follows.

Definition 2. *We consider a similarity function $\pi : E \rightarrow [0, 1]$, such that for any $(u, v) \in E$, $\pi(u, v)$ is equal to the Jaccard similarity between u and v:*

$$\pi(u, v) = \frac{|A(u) \cap A(v)|}{|A(u) \cup A(v)|}$$

If the content is textual, one can easily compute π. For example, consider a user u that actively tweets about *politics* and *religion* and a user v tweeting about *sport* and *politics*. One can construct $A(u)$ and $A(v)$ from the set of keywords extracted from their posts and estimate π. This modeling is simple and efficient, as it requires no additional structure with respect to the existing network; it only associates weights to edges. Unless there is a particular benefit one can gain from adding independent nodes for content, which could be expensive, we argue that such modeling is sufficient.

The final piece of our data model is a set of finite cascades $\mathcal{C}$:

Definition 3. *We consider a set of cascades $\mathcal{C} = \{C_1, \ldots, C_c\}$ of size c, where a cascade $C = [u_1, u_2, \ldots, u_{|C|}]$ is a sequence of finite events, each of them representing the infection of a user by a given contagion.*

We use $C(i) = u_i$ to denote the i-th node of the cascade C. We say that a node u is infected before node v in a cascade C, and we write $u \prec_C v$, if and only if $u = C(i)$, $v = C(j)$ and $i < j$. Given a node u and a context size s, we define the *left-hand side infection context* $C(u; s)^{\prec}$ and the *right-hand side infection context* $C(u; s)^{\succ}$:

$$C(u; s)^{\prec} = \{v : v = C(i) \wedge u = C(j) \wedge j - s \leq i \leq j - 1\}$$
$$C(u; s)^{\succ} = \{v : v = C(i) \wedge u = C(j) \wedge j + 1 \leq i \leq j + s\}$$

Definitions 1–3 represents the input of our problem:

Problem 1. Given a network G, a set of cascades $\mathcal{C}$, a similarity function π, and a dimensional number d, we seek to learn a representation of the network specified by $\varPhi : V \to \mathbb{R}^d$, provided that $\varPhi$ preserves as much as possible (i) the network structure, (ii) the similarity between nodes and (iii) the node infection context.

3   Mineral

In this section, we present a detailed description of MINERAL, which exploits two sources of information: in SPC-Mode (Structure+Content-Mode), it uses structural information (the network G) as well as content information associated to nodes (the function π). In CSD-Mode (Cascade-Mode), it utilizes the observed diffusion information (the set of cascades C).

Thanks to function π, the network G can be considered as a weighted graph. Hence without requiring additional structures, we can design an effective algorithm to learn the representation of the network that preserves both structural and content similarity between nodes. One strategy that has proved to be effective for NRL is to use a similar approach to word representation learning in natural language documents. In word representation, the basic idea is to learn a representation of words by predicting their context. Nonetheless, unlike words in a document where their context is obvious as a result of their linear structure, we do not have a straightforward way to deduce the context of nodes in a network. Several strategies have been developed in the literature to address this problem.

In this work, we extend existing approaches based on random walks [1,2] by considering instead a diffusion process. It has been observed that the dynamics of diffusion processes reveal complex local and global structural patterns of the network. Therefore we simulate the diffusion of influence or information using the independent cascade (IC) model [16] to obtain context information for nodes. The cascades generated by simulating IC are merged with actual (observed) cascades, when available.

Algorithm 1 shows the high-level steps required to generate cascades. For each node $u \in V$, r cascades are generated starting from u, based on the IC model and using the content similarity π as an unnormalized probability of infection.

When SIMULATEDIFFUSION(G, π, u, h) is invoked, a cascade of size h is generated starting from u. Let $\mathcal{I}_t$ denote the set of nodes infected at time t; the diffusion process works as follows:

1. At time $t = 0$, a cascade sequence is initiated by infecting the current root, i.e. $C = [u]$, i.e., $\mathcal{I}_0 = \{u\}$.

2. At time $t > 0$, each node $v \in \mathcal{I}_{t-1}$ makes a single attempt to infect each of its outgoing neighbor $w \in out(v)$ that is not already infected (i.e., $w \notin C$). The infection succeeds with a probability proportional to $\pi(v, w)$; in such case, w is appended to C and it is included in $\mathcal{I}_t$.
3. Repeat the process starting from step 2 while $|C| \leq h$.

We restrict the size of cascades (the number of infected nodes) to be at most h nodes, because large, viral cascades (unlike non-viral ones) usually do not capture any relevant local or global structural relation of nodes [14, 15].

Generated cascades, together with existing ones if available, are thus used to learn embeddings. Since cascades are sequences of nodes, we borrow the SKIP-GRAM [17] model for word representation learning to perform network representation learning. For the purpose of being self-contained, we briefly describe the SKIPGRAM [17] model in our context.

CASCADEGENERATOR(G, π, r, h)

---

1 $C = \emptyset$
2 for $u \in V$ do
3   repeat r times
4          $C = $ SIMULATEDIFFUSION (G, π, u, h)
5          $\mathcal{C}$.insert(C)

6 return $\mathcal{C}$

SKIPGRAM. Given a center node $u \in C$, this model maximizes the log probability of observing context nodes $v \in C(u; s)^{\preceq}$ and $w \in C(u; s)^{\succeq}$ within a window size s. Based on the assumption that the likelihood of observing each context node given a center node is independent, more formally the SKIPGRAM model optimizes the objective in Eq. 1 with respect to the model parameter Φ.

$$\max_{\Phi} \sum_{u \in V} \log Pr(C(u; s)^{\preceq} \mid \Phi(u)) + \log Pr(C(u; s)^{\succeq} \mid \Phi(u)) \tag{1}$$

$$\log Pr(C(u; s)^D \mid \Phi(u)) = \sum_{v \in C(u; s)^D} \log Pr(v | \Phi(u)) \tag{2}$$

where D is either $\preceq$ or $\succeq$, and $\Phi(u) \in [0, 1]^d$ is a d-dimensional representation of u. The right-hand side term in Eq. 2 is specified using the softmax function:

$$Pr(v | \Phi(u)) = \frac{\exp(\Phi(v)^T \cdot \Phi(u))}{\sum_{w \in N} \exp(\Phi(w)^T \cdot \Phi(u))} \tag{3}$$

Nonetheless, directly estimating the conditional probability in Eq. 3 is expensive, because of the normalization constant that needs to be computed for every node. For this reason, different approximation strategies have been suggested in the literature; in this work, we adopt the "Negative Sampling" strategy [17] that characterizes a good model by its power to discriminate appropriate context nodes from noise. Then, the computation of $\log Pr(v|\Phi(u))$ using the negative sampling strategy is shown in Eq. 4.

$$\log Pr(v|\Phi(u)) = \log \sigma(\Phi(v)^T \Phi(u)) + neg(u; l) \qquad (4)$$

σ is the logistic function, and we need our model to effectively differentiate v from the l negative examples drawn from some noise distribution $\mathcal{N}(u)$ of u, where $neg(u; l)$ is the noise model and is defined as:

$$neg(u; l) = \sum_{i=1}^{l} \mathbf{E}_{w_i \sim \mathcal{N}(u)}[-\log \sigma(\Phi(w_i)^T \Phi(u))] \qquad (5)$$

Numerically, a good model should produce a small expected probability for the noise model and larger probability for the data model (the first term on the right-hand-side of Eq. 4).

Finally, we employ the stochastic gradient descent algorithm to minimize the negative log-likelihood of the objective in Eq. 1 based on the negative sampling strategy in Eqs. 4 and 5 and obtain the complete model parameters $\Phi \in V \rightarrow [0, 1]^d$.

Table 1. Summary of the datasets

Dataset	$|\mathbf{V}|$	$|\mathbf{E}|$	$|\mathcal{C}|$	Number of labels	Type of labels
Twitter	595,460	14,273,311	397,681	5	top-5 communities
Memetracker	3,836,314	15,540,787	71,568	5	top-5 communities
Flickr	80,513	5,899,882	–	195	Groups
Blogcatalog	10,312	333,983	–	39	Interests

4   Experiments and Results

In order to demonstrate the effectiveness of our algorithm, we have carried out several experiments across multiple network analysis problems using multiple datasets, listed below. A brief summary of the characteristics of the datasets is given in Table 1.

- Twitter [14]: a dataset containing the follower network of Twitter users and cascade information of hashtags. Each time a user adopts a hashtag (by creating a new or using an existing one), it is added to the set of her keywords. A cascade is constructed by sorting the users according to their first use of a particular hashtag.

– Memetracker [18]: a dataset containing the interaction history between different news media and blog web pages during a year. Each page is associated with a set of memes, which are considered as its keywords. Memes are grouped into clusters, and we consider each cluster id as a contagion that has infected every page that has mentioned a meme that belongs to the cluster. Similar to Twitter, cascades are built by sorting the users of a contagion according to the time of first use.
– Blogcatalog [19]: a dataset containing a network of bloggers. There are 39 topic categories which are considered as content information for each author.
– Flickr [19]: a photo sharing site paired to a social network. Users place their pictures under a set of predefined categories which can be considered content information.

For Twitter and Memetracker, users are labeled based on their communities. First we identify the (non-overlapping) community to which a user belongs using [20], and then we associate it as her label. We utilize both SPC and CSD modes for these datasets, since information regarding structure, content, and cascades is available. In addition, in all the experiments we have used $h = 500$ for Twitter and Memetracker, $h = 200$ for Flickr and $h = 50$ for Blogcatalog.

4.1   Baselines

Existing methods [4,6] that consider content information are usually based on matrix factorization, which makes them unscalable for large networks. For this reason, we only consider the following two content-oblivious approaches as baseline methods:

1. DEEPWALK [1]: is a method that utilizes truncated random walks for network embedding, where each step of a walk is chosen uniformly at random. Equivalent to the current work, they use the SKIPGRAM model and it is trained using the walks.
2. LINE [3]: is a proximity based approach, trained by concatenating two independently trained models based on the notions of first-order and second-order similarity of nodes. In other words, in the first phase they train a model that preserves the undirected link structure between nodes; in the second phase, they train a model that preserves the directed or undirected 2-hop link structure of the network.

Table 2. Result for the link prediction task on the Twitter dataset

Algorithm	P@100	P@500	P@1000	P@5000	P@10000	p@50000	p@100000	p@500000
MINERAL	**99.9**	99.8	99.8	99.8	**99.8**	**99.8**	**99.8**	**99.0**
DEEPWALK	96.6	97.0	97.1	97.1	97.1	97.1	97.1	96.9
Line	99.3	99.8	**99.9**	99.8	99.7	98.5	94.5	71.0

Table 3. Result for the link prediction task on the Memetracker dataset

Algorithm	P@100	P@500	P@1000	P@5000	P@10000	p@50000	p@100000	p@500000
MINERAL	**100.0**	**99.9**	**99.9**	**99.6**	**99.5**	**99.5**	**99.4**	98.6
DEEPWALK	99.1	99.0	99.0	99.1	99.0	99.0	99.0	**99.0**
Line	91.2	92.2	89.9	85.2	83.3	72.8	68.9	65.4

Table 4. Result for the link prediction task on the Flickr dataset

Algorithm	P@50	P@100	P@500	P@1000	p@5000	p@10000	p@50000	p@100000
MINERAL	**99.2**	**99.6**	**99.6**	**99.6**	**99.4**	**99.2**	97.4	94.9
DEEPWALK	96.6	96.6	97.4	97.5	97.5	97.5	97.4	**97.1**
LINE	54.4	61.0	61.6	58.8	51.6	48.9	44.2	42.5

4.2 Link Prediction

Link prediction is one of the most important network analysis problems. There are three main techniques solving it, based on node similarity, topology, and social theory [21]. Very often, such techniques rely on experts to craft informative features that enable us to effectively predict links, and this makes them expensive. Instead of manually-crafted features, we use here the learned embeddings to perform link prediction. Towards this end, we randomly sampled 15% of the existing edges from the network; we also randomly sampled the same amount of node pairs that are not in the edge set. We then used the learned embedding to effectively predict the links. That is, given a pair of nodes $\{u, v\} \subseteq V$, we compute the probability $p(u, v)$ of an edge existing between the two nodes as:

$$p(u, v) = \frac{1}{1 + e^{-(\Phi(u)^T \cdot \Phi(v))}}$$

Then we sort the predicted edges according to $p(u, v)$ in descending order and evaluate the performance of an embedding in correctly predicting the edges using the precision-at-K (P@K) score. P@K measures the fraction of correctly predicted edges on the top-K results, i.e. what percent of the top-K edges are true edges from the randomly sampled edges. For each K value we perform the experiments 10 times and report the average. Tables 2, 3 and 4 show the results for the Twitter, Memetracker and Flickr datasets; MINERAL performs as good as or better than the baselines.

4.3 Node Label Classification

The second problem we addressed is label classification. We consider two instance of it, namely multi-class and multi-label classifications. For the Twitter and Memetracker datasets, we tackled the multi-class classification problem, because–as shown in Table 1–labels are communities and each node belongs to

Table 5. Node classification accuracy on different levels of labeled training set ratio for the Twitter dataset

Algorithm	Training ratio								
	0.9	0.8	0.7	0.6	0.5	0.4	0.3	0.2	0.1
MINERAL	**98.19**	**98.05**	**97.97**	**97.98**	**97.95**	**97.91**	**97.74**	**97.51**	**96.93**
DEEPWALK	97.78	97.76	97.86	97.67	97.61	97.45	97.42	97.02	96.01
LINE	84.19	85.74	85.02	85.11	85.18	84.69	84.06	82.20	76.19

Table 6. Node classification accuracy on different levels of labeled training set ratio for the Memetracker dataset

Algorithm	Training ratio								
	0.09	0.08	0.07	0.06	0.05	0.04	0.03	0.02	0.01
MINERAL	**85.25**	**85.24**	**85.23**	**85.21**	**85.19**	**85.14**	**85.08**	**84.98**	**84.58**
DEEPWALK	80.74	80.60	80.70	80.83	80.84	80.73	80.77	80.73	80.41
LINE	53.56	53.54	53.50	53.47	53.44	53.41	53.32	53.12	52.58

just a single community. In the other datasets, given that multiple labels are present, we performed multi-label classification. To evaluate the effectiveness of a model in the classification task, we adopt the same evaluation metrics as in previous studies, and hence we use Accuracy, F1-Micro and F1-Macro metrics.

The Multi-class classification results for the Twitter and Memetracker datasets are reported in Tables 5 and 6, respectively. Similar to previous studies, we performed these experiments on different fractions of labeled training sets (Training Ratio $\in \{10\%, 20\%, 30\%, 40\%, 50\%, 60\%, 70\%, 80\%, 90\%\}$). Under this setting, accuracy is the evaluation metric; and as shown in the tables, MINERAL performs slightly better than DEEPWALK and significantly better than LINE.

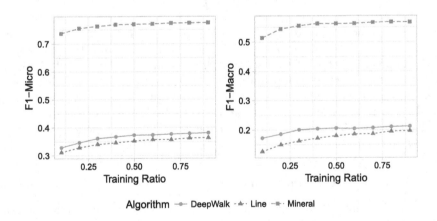

Fig. 1. Multi-label classification (using one-vs-rest logistic regression classifier) on the Blogcatalog dataset

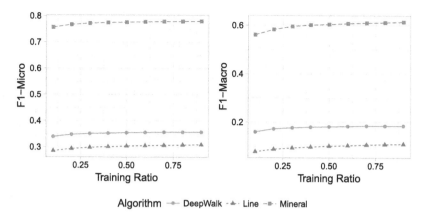

Fig. 2. Multi-label classification (using one-vs-rest logistic regression classifier) on the Flickr dataset

For the other datasets, however, MINERAL significantly outperforms both base-lines in multi-label classification. Figures 1 and 2 report the results on different training ratios (x-axis) using F1-Micro and F1-Macro measures (y-axis).

4.4 Network Visualization

The last but not the least application of NRL is network visualization. We use the Twitter dataset for this task, and the visualization is performed using t-Distributed Stochastic Neighbor Embedding (t-SNE) [22]. Given a set of q communities, an informative visualization should maintain a knit cluster for members of the same community and maintain clear boundaries between different communities. As shown in Fig. 3, MINERAL's visualization gives the best result. Members of each community are densely clustered and are far from members of other communities.

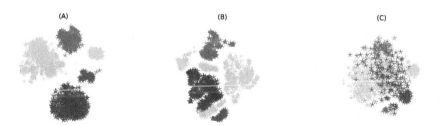

Fig. 3. Visualization of top-5 communities with atmost 2000 users in the Twitter Dataset using (A) MINERAL (B) DEEPWALK and (C) Line

4.5 Parameter Sensitivity

Now we turn into analyzing the sensitivity of the virality controlling parameter, which is h. In Sect. 3 we have argued that "viral" or large cascades do not capture any meaningful dependency between infected nodes of the cascades. To empirically prove that such is the case, we have performed experiments over different values of $h \in \{50, 100, 250, 500, 1000\}$ on the Blogcatalog dataset. As shown in Fig. 4, the precision@k significantly drops as we increase the size of h. For example, for a fixed $k = 10$, the precision@k is $P@k = 0.86$ for $h = 50$, $P@k = 0.6$ for $h = 100$, $P@k = 0.29$ for $h = 500$, and $P@k = 0.15$ for $h = 1000$.

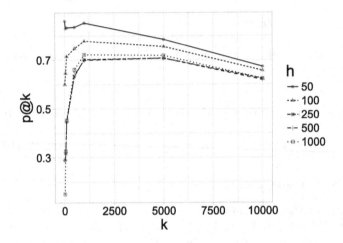

Fig. 4. Sensitivity of the parameter h using the link prediction task on Blogcatalog

5 Related Work

Recent advances in neural network models have attracted researches from several communities such as computer vision, NLP, and social network analysis. In the last two communities in particular, a seminal work of Mikolov et al. [17] in representation learning (embedding) of words in documents using a shallow neural network model has inspired studies [1,2] in network representation learning. Among the approaches introduced for word embedding, the Skip-Gram model [17] is the one that has been most largely used for network representation learning. The Skip-Gram model is used to learn a representation of words by way of predicting context words. The context of a node in a network, however, does not have a straightforward definition. Studies have introduced different strategies of capturing nodes context, for example using random walks [1,2], pair-wise proximities [3,5], and community structures [7,8]. Once a context is formalized, different neural network (based on either shallow or deep models) are employed for the representation learning task. Then the learned representations are utilized for downstream network analysis tasks.

Studies such as [6] propose a NRL algorithm based on matrix factorization. Such techniques, however, are computationally expensive and not scalable for large networks.

6   Conclusion

This study presents MINERAL, a novel algorithm for network representation learning (NRL) that leverages three network aspects: topology, node content, and diffusion. The algorithm efficiently encodes content information associated with nodes into a similarity function between pairs of connected nodes. Then it combines the network and similarity information with natural (observed) or simulated cascades, and acquires context information of nodes. Finally, we combine everything as a set of cascades and employ the SKIPGRAM model to learn an embedding that preserves structural, content, and diffusion context of nodes.

We performed several experiments using multiple datasets across several network analysis problems, and compared the performance of our approach with existing NRL baseline methods. Our results show that MINERAL significantly outperforms the baselines specially in multi-label classification and network visualization. It also performs slightly better than the baselines in link prediction. Even though our data modeling is effective in capturing many kinds of content information, in this study we have focused on textual information.

Acknowledgements. This research was partially supported by EIT Digital Project Sensemaking Service: Entity Linking for Big Linked Data - Act. #17151 - 2017.

References

1. Perozzi, B., Al-Rfou, R., Skiena, S.: DeepWalk: online learning of social representations. In: Proceedings of the 20th ACM International Conference on Knowledge Discovery and Data Mining (KDD 2014), pp. 701–710. ACM (2014)
2. Grover, A., Leskovec, J.: Node2vec: scalable feature learning for networks. In: Proceedings of the 22nd ACM International Conference on Knowledge Discovery and Data Mining (KDD 2016), pp. 855–864. ACM (2016)
3. Tang, J., Qu, M., Wang, M., Zhang, M., Yan, J., Mei, Q.: LINE: large-scale information network embedding. CoRR, vol. abs/1503.03578 (2015)
4. Huang, X., Li, J., Hu, X.: Label informed attributed network embedding. In: Proceedings of the Tenth ACM International Conference on Web Search and Data Mining (WSDM 2017), pp. 731–739. ACM (2017)
5. Wang, D., Cui, P., Zhu, W.: Structural deep network embedding. In: Proceedings of the 22nd ACM International Conference on Knowledge Discovery and Data Mining (KDD 2016), pp. 1225–1234. ACM (2016)
6. Yang, C., Liu, Z., Zhao, D., Sun, M., Chang, E.Y.: Network representation learning with rich text information. In: Proceedings of the 24th International Conference on Artificial Intelligence (IJCAI 2015), pp. 2111–2117. AAAI Press (2015)
7. Tu, C., Wang, H., Zeng, X., Liu, Z., Sun, M.: Community-enhanced network representation learning for network analysis. CoRR, vol. abs/1611.06645 (2016)

8. Wang, X., Cui, P., Wang, J., Pei, J., Zhu, W., Yang, S.: Community preserving network embedding. In: AAAI (2017)

9. Xie, R., Liu, Z., Jia, J., Luan, H., Sun, M.: Representation learning of knowledge graphs with entity descriptions. In: Proceedings of the Thirtieth AAAI Conference on Artificial Intelligence (AAAI 2016), pp. 2659–2665. AAAI Press (2016)

10. Dahl, G.E., Ranzato, M., Mohamed, A.-R., Hinton, G.: Phone recognition with the mean-covariance restricted Boltzmann machine. In: Proceedings of the 23rd International Conference on Neural Information Processing Systems (NIPS 2010), pp. 469–477. Curran Associates Inc. (2010)

11. Ciregan, D., Meier, U., Schmidhuber, J.: Multi-column deep neural networks for image classification. In: 2012 IEEE Conference on Computer Vision and Pattern Recognition, pp. 3642–3649, June 2012

12. Szegedy, C., Liu, W., Jia, Y., Sermanet, P., Reed, S., Anguelov, D., Erhan, D., Vanhoucke, V., Rabinovich, A.: Going deeper with convolutions. In: 2015 IEEE Conference on Computer Vision and Pattern Recognition (CVPR), pp. 1–9, June 2015

13. Bengio, Y., Courville, A., Vincent, P.: Representation learning: a review and new perspectives. IEEE Trans. Pattern Anal. Mach. Intell. **35**, 1798–1828 (2013)

14. Weng, L., Menczer, F., Ahn, Y.-Y.: Virality prediction and community structure in social networks. Sci. Rep. **3**(2522) (2013)

15. Weng, L., Menczer, F., Ahn, Y.-Y.: Predicting successful memes using network and community structure. In: Adar, E., Resnick, P., Choudhury, M.D., Hogan, B., Oh, A.H. (eds.) ICWSM. The AAAI Press (2014)

16. Kempe, D., Kleinberg, J., Tardos, E.: Maximizing the spread of influence through a social network. In: Proceedings of the Ninth ACM International Conference on Knowledge Discovery and Data Mining (KDD 2003), pp. 137–146. ACM (2003)

17. Mikolov, T., Sutskever, I., Chen, K., Corrado, G., Dean, J.: Distributed representations of words and phrases and their compositionality. In: Proceedings of the 26th International Conference on Neural Information Processing Systems (NIPS 2013), pp. 3111–3119. Curran Associates Inc. (2013)

18. Leskovec, J., Backstrom, L., Kleinberg, J.: Meme-tracking and the dynamics of the news cycle. In: Proceedings of the 15th ACM International Conference on Knowledge Discovery and Data Mining (KDD 2009), pp. 497–506. ACM (2009)

19. Tang, L., Liu, H.: Relational learning via latent social dimensions. In: Proceedings of the 15th ACM International Conference on Knowledge Discovery and Data Mining (KDD 2009), pp. 817–826. ACM (2009)

20. Blondel, V.D., Guillaume, J.-L., Lambiotte, R., Lefebvre, E.: Fast unfolding of communities in large networks. J. Stat. Mech. Theor. Exp. **2008**(10), P10008 (2008)

21. Wang, P., Xu, B., Wu, Y., Zhou, X.: Link prediction in social networks: the state-of-the-art. Sci. China Inf. Sci. **58**(1), 1–38 (2015)

22. van der Maaten, L., Hinton, G.: Visualizing high-dimensional data using t-SNE. J. Mach. Learn. Res. **9**, 2579–2605 (2008)

Visual Perception of Mixed Homogeneous Textures in Flying Pigeons

Margarita Zaleshina[1], Alexander Zaleshin[1(✉)], and Adriana Galvani[2]

[1] Moscow Institute of Physics and Technology, Moscow, Russia
terbiosorg@gmail.com
[2] University of Bologna, Bologna, Italy
adriana.galvani@unibo.it

Abstract. In this study, we simulated the visual perception of the terrain in flying pigeons over combined homogeneous terrain with multiple textures – forest and grassland, water surface and seacoast. The surfaces along the pigeon's flight trajectory were considered as mixed textures observed from a bird's eye view. In the proposed method, the main structural elements for the analyzed textures were selected and then statistically homogeneous characteristics of the texture were determined. The textural characteristics and their changes during flight were recorded in the form of distinct "event channels". For different types of terrain, the frequency characteristics of visual perception were calculated and compared. In addition, we considered the possibility of comparing the frequency characteristics of the textures with data regarding the pigeon's rhythmic brain activity. Spatial data—open-access remote sensing datasets—were processed using the geographical information system QGIS. Our results show that recognizing mixed landscape textures can help solve navigation tasks when flying over terrain with sparse landmarks.

Keywords: Visual perception · Spatial navigation · Brain activity

1 Introduction

Here, we simulated a pigeon's flight over natural terrain. The aim of the study was to determine whether visual properties of mixed homogenous textures can be used to solve navigation tasks in motion.

In this simulation, the distinguishing feature of homogeneity is defined as homogenous distribution of texture elements by the spatial frequency. The elements of the texture are defined as visually recognizable items that repeat along the surface. During motion, these elements provide visual information for perception.

We propose criteria by which individual elements of textures can be considered "events along the route". In addition, we demonstrate that repeating of similar events can determine the frequency of an extensive texture.

Moments of observations of textural elements can also be recorded in the event channels. By taking into account flight speed, the frequency characteristics of different

© Springer International Publishing AG 2018
G. Nicosia et al. (Eds.): MOD 2017, LNCS 10710, pp. 299–308, 2018.
https://doi.org/10.1007/978-3-319-72926-8_25

types of terrains can be calculated and compared. The pigeon perception in flight corresponds with visual flicker threshold about 75 Hz.

Records in visual event channels can be compared with other simultaneously recorded data, including the bird's precise spatial position (using GPS data) and the bird's brain activity (using EEG data).

Figure 1 shows the typical examples of homogenous and multiple textures (A) and change in attention of flying pigeon (B).

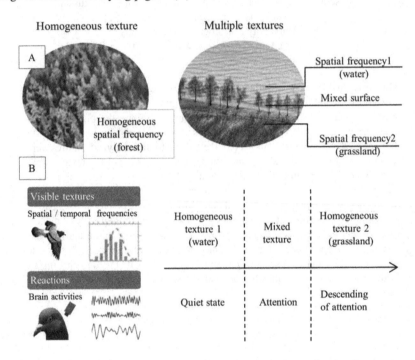

Fig. 1. Homogeneous and multiple textures (A) and typical change in pigeon attention when flying from sea to coast (B)

This paper is structured as follows. In Sect. 2, we provide a brief review of the published literature regarding the following topics: *(i)* the perception of textures in animals and humans; *(ii)* the study of pigeon's flight using GPS-based navigation; and *(iii)* the principles of recognizing textures and textural elements while in motion. In Sect. 3, we describe spatial data processing methods for natural terrain images obtained at the bird's flight altitude. In Sect. 4, we present the "event channels" and compare the frequency parameters of the observed textures with the data obtained regarding the rhythmic brain activity in flying pigeon. In Sect. 5 we conclude that spatial frequency recognition in textures can help when flying over mixed homogeneous terrain.

2   Background and Related Works

2.1   Visual Perception of Textures by Animals and Humans

Spatial analysis is associated to the mind which organizes the observations, and the connections; the analysis is able to modulate new perceptions into different organizations. The brain is in fact a geographer: there is the external space, the brain observes it, the brain interprets it; describes it; interiorises it, and finally uses it, and modifies it.

Visual perception of the textures in the surrounding world is a continuously ongoing process. The ability to perceive, distinguish, and respond rapidly to minor changes in textures can be very high—and highly varied—in both animals and humans [1, 2].

When observed visually, a texture can be perceived either as its separate elements or as a generalized surface. This has been illustrated most clearly in terms of the orientation of insects. For example, bees can orient themselves using separate repeated local objects, but they can also produce a global, holistic perception of the visual picture [3].

Leonhardt et al. [4] showed the difference between orientation tasks in homogenous terrain (for example, a forest) and orientation tasks in disturbed terrains with many notable landmarks and scenery fragments ("visually/structurally homogeneous terrains, and disturbed terrains with many prominent landmarks and fragmented"). In their study, Leonhardt et al. found that homogeneity and heterogeneity of the landscape can differentially affect the bees' homing.

In addition to responding to the visual scenes themselves, the brain also reacts to the rate of change in the scenes. Neural mechanisms of speed perception have been described [5]. For example, the hippocampus can respond to deviations from the correct route, thereby supporting the pigeon's directional adjustment in the event of a navigation error when returning home [6].

Primates can recognize textures with high accuracy, as demonstrated by the response of the primary (V1) and secondary (V2) visual cortex [7, 8]. Presentation of the direction maps in V1, V2 brain areas were showed in [9].

The perception is different in birds and in primates. For example, primates have mechanisms of visual perception that include both saccades and smooth pursuit eye movements, which allows a primate to trace the texture elements during motion [10]. In contrast, birds have characteristic forward-and-backward head movements while walking, running, or landing after flight [11].

Another difference between birds and primates is their flicker threshold value (i.e., the flicker frequency at which separate moving stimuli appear to be continuous), which is several times higher in pigeons than in humans [1, 2]. Visual acuity is also much higher in pigeons, enabling them to distinguish objects at a distance of up to three kilometers.

Yet another obvious difference is that most birds can fly, which allows them to rapidly change their spatial position and to refine their visual information.

2.2 Studying a Pigeon in Flight Using GPS and Brain Activity Loggers

Jimenez Ortega et al. [11] reported that pigeons occasionally fly home in a straight line, but rather fly along familiar highways, preferring to be guided by familiar landmarks. In addition, Biro et al. [12] argued that pigeons rely on both visual reference points and their internal sense of direction. Schiffner et al. [13] analyzed pigeons' GPS tracks and discussed whether a pigeon requires visual perception in order to accurately determine its location.

Vyssotski et al. [14] analyzed a pigeon's flight using sensors (Neurologger and a GPS logger). In their paper, authors described the differences between flight over the open sea and flight over land, taking into account variations in visibility. They also registered and analyzed EEG activity during flight over various types of terrains and under various levels of visibility (i.e., poor, moderate, and good visibility).

Mann et al. [15] studied the influence of mixed terrain on the pigeon's navigational behavior, including the effect of extensive linear landmarks such as roads and rivers. They concluded that pigeons navigate best in areas in which the terrain's complexity is neither too high nor too low.

The visual perception of terrain textures typically occurs in pigeon's flight at an altitude of 100–500 m at an average speed of 60 km/h. During flight, the pigeon can change its speed, direction, and altitude based on the varying information.

2.3 The Recognition of Texture and Texture Elements in Motion

Texture consists of a combination of separate elements united by a common principle (or set of principles). The basic approaches of texture properties were proposed by Haralick et al. in 1973 [16]. The authors confirmed that a texture can be used for classification and segmentation, and they showed that texture and luminance contrast are always present in images, and they are independent. In 1990, Webster et al. noted "spatial frequency of patterns defined by luminance variations" [17].

To use dynamic texture properties for navigation, it is necessary to identify the most relevant attributes. Zaccolo [18] showed that texture attributes suitable for tracking must be repeatable frame-to-frame in a visual sequence, and they must be clearly distinguishable from the noise. Based on repeated attributes, a texture's temporal and spatial frequency can be described.

The internal tracking ability is activated in the presence of structural and organizational components of the space, which are perceived as sets of visual stimuli. Exogenous attention depends on the second order of the stimulus spatial frequency [19].

To classify remote sensing data the methods of the texture recognition can be used, even for low-density textures [20]. However, a lot of methods are used primarily for the recognition of stationary images. In practice, in order to recognize separate elements, it is possible to use elements' dictionaries for satellite and aerial imageries [21]. In his review, Blaschke [22] compared methods for selecting objects of interest. In another review, Du et al. [23] showed that optimal results can be obtained using a combination of image analysis methods, including both the spectral and textural characteristics of images. Skowronek et al. [24] showed that the most frequently used methods yield strikingly similar results. Indeed, the three most frequently used methods (maximum entropy, biased support vector machines, and boosted regression trees) yield highly similar results. Joseph et al. [25] used artificial sets of textures to highlight the optimum set of scales for observing texture, based on the fact that "As the texture is scaled down, increasing the number of checks within the fixed display size, performance increases while the efficiency decreases."

Simple elements used in detection the texture frequency should be:

- perceptible in motion (the texture frequency should be lower than the flicker threshold for visual perception);
- large enough to be in the detection range (the elements observation scale should correspond to visual acuity);
- small enough to be perceived independently (the spatial frequency should be present within the visual field);
- visually distinguishable (should be differ from noise and other texture content);
- statistically significant (should be appear repeatedly enough inside the texture); and
- relatively periodic.

At the same time, mixed homogenous textures can be represented by several sets of elements, repeated with different frequencies (for example, mixed growth of bushes and trees). In addition, the orientation of some texture elements can be taken into account; for example, Kingdom and Keeble [26] showed double modulation for differently oriented grids with the same spatial frequency.

The speed and accuracy of detecting the frequency characteristics of the texture can be increased by performing a layered separation of different types of elements.

3   Materials and Methods

3.1   Spatial Datasets

In this study open remote sensing data were applied. The source data layers were added using the OpenLayers Plugin (http://openlayers.org) in QGIS, which allows to obtain Google Maps, Bing Maps and another open layers. The original coordinate system is WGS 84/Pseudo-Mercator (EPSG:3857). For processing, we used grayscale images.

Calculations were performed for the following cases:

- Flying over homogeneous forest;
- Flying over combined terrain: from the sea to the coast.

Case 1. Homogeneous Terrain: Forest
Case 1 shows the spatial texture frequency calculation for pigeon's flight over homogeneous terrain.

View the primary data source in Google Maps is available at: https://www.google.com/maps/@60.499723,5.1349364,500m/data=!3m1!1e3.

Case 2. Mixed Terrain: Flying from the sea to the coast

Case 2 shows the example of flight over mixed homogeneous coastal terrain: water surface, coast and forest; and generalization the "event channels" for different spatial texture frequencies.

View the primary data source in Google Maps is available at: https://www.google.com/maps/@44.0549291,9.94026,300m/data=!3m1!1e3 or http://ant.dev.openstreetmap.org/bingimageanalyzer/?lat=44.054929161689764&lon=-350.0599533319473&zoom=18&l=bing.

In addition, variations in 12–60 Hz EEG power obtained in similar experiments on pigeon [14] and records of "event channels" are compared.

3.2 Methods

This work was performed in the following steps:

Step 1 - Calculating and analyzing spatial texture frequencies for pigeon's flight over homogeneous surface (Cases 1, 2):

- Load data from OpenLayers. Generate raster layer from the appropriate texture location.
- Create shapefile of pigeon's track.
- Create vector buffer around pigeon's tracks (Fig. 2). This buffer simulates the area of special attention in pigeon.

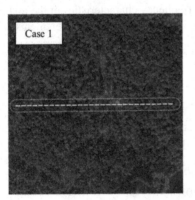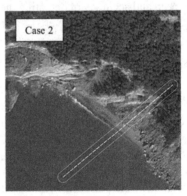

Fig. 2. Case 1. Homogeneous terrain. Case 2. Mixed terrain: flying from the sea to the coast.

- Calculate sampling intervals of texture. Generate vector contour lines at calculated sampling intervals.
- Generate "event channels" in accordance with spatial texture frequencies. Compute basic texture statistics and statistics for vector polygons.
- Generate histogram of spatial texture frequency distribution.
- Create a histogram of line directions — rose diagram (weighted using the line segment lengths).

Step 2 - Calculating and analyzing spatial and temporal texture frequencies for pigeon's flight over mixed surface (Cases 2):

– Generate multiple "event channels" in accordance with the sets of spatial texture frequencies.
– Comparing spatial and temporal texture frequencies and EEG data obtained in similar experiments on pigeon [14, 27].
– Calculating spatial texture frequencies for different texture scaling.

3.3   QGIS Plugins

The data were processed using the open source software program QGIS (http:// qgis.org), including additional analysis plugins: QGIS geoalgorithms, SAGA, GDAL/OGR and GRASS.

4   Results

4.1   Analysis of Typical Texture Frequencies

We calculated and analyzed the texture frequencies for different textures using the geographical information system QGIS.

Calculated results (contour extraction, event channel and line direction histogram) for Case 1 are shown in Fig. 3.

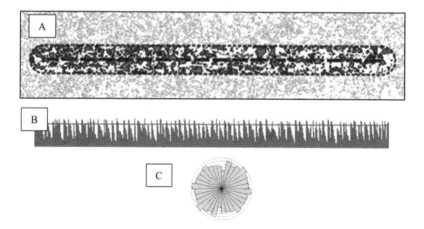

Fig. 3. A. Contour extraction and selection data for statistical analysis. B. Event channel. C. Line direction histogram shows that the difference in texture directions is insignificant.

Contour extraction was made using the GdalTools plugin (see http://planet.qgis.org/ planet/tag/contours). Creation of Polygon Centroids was made using QGIS Geometry Tools/Polygon Centroids. The event channel was made based on the principle that each centroid means one event point in the buffer zone.

4.2 Comparison Texture Frequencies and Brain Activity in Flying Pigeon

We calculated the sets of frequencies of textures based on the data along the pigeon's flight trajectory from sea to coastal area. We also calculated the temporal frequency parameters of the observed textures, taking into account the typical pigeon's speed. The events along the flight track were registered in distinct channels.

We compared obtained spatial and temporal texture frequencies in Case 2 with typical EEG power [14]. At this work authors found that the pigeon's brain activity differs depending on the type of landscape over which the pigeon is flying. EEG power in the 12–60 Hz frequency range was higher over mixed ground landscape than over sea.

Figure 4 summarizes result of this comparison.

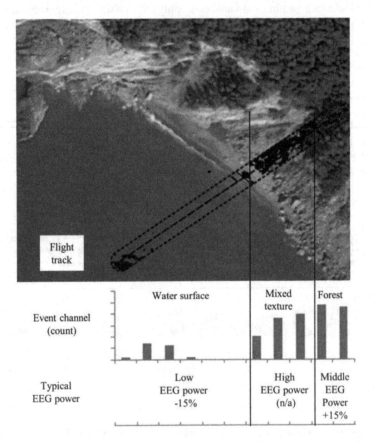

Fig. 4. Data comparison in distinct "event channels". Data about typical EEG power is reproduced from [14].

It should be noted that texture frequency cannot be used in classification tasks directly, since different textures can be similar in frequencies.

5   Conclusions

Our simulation methods can be applied to study the navigational mechanisms in pigeon flying over a mixed homogeneous terrain: forest, grassland, water surface or coast.

As future work, precise coordinate definition of texture characteristics could be compared with GPS location of pigeon and with EEG records of brain activity. To improve accuracy, one should use high-resolution imagery and take into account 3d and parallax effects.

References

1. Bovet, D., Vauclair, J.: Picture recognition in animals and humans. Behav. Brain Res. **109**, 143–165 (2000)
2. D'Eath, R.B.: Can video images imitate real stimuli in animal behaviour experiments? Biol. Rev. **73**, 267–292 (1998). https://doi.org/10.1111/j.1469-185X.1998.tb00031.x
3. Avargues-Weber, A., Dyer, A.G., Ferrah, N., Giurfa, M.: The forest or the trees, preference for global over local image processing is reversed by prior experience in honeybees. Proc. Biol. Sci. **282**, 20142384 (2015)
4. Leonhardt, S.D., Kaluza, B.F., Wallace, H., Heard, T.A.: Resources or landmarks, which factors drive homing success in Tetragonula carbonaria foraging in natural and disturbed landscapes? J. Comp. Physiol. A Neuroethol. Sens. Neural Behav. Physiol. **202**, 701–708 (2016)
5. Krekelberg, B., van Wezel, R.J.A.: Neural mechanisms of speed perception, transparent motion. J. Neurophysiol. **110**, 2007–2018 (2013)
6. Gagliardo, A., Ioale, P., Savini, M., Dell'Omo, G., Bingman, V.P.: Hippocampal-dependent familiar area map supports corrective re-orientation following navigational error during pigeon homing: a GPS-tracking study. Eur. J. Neurosci. **29**, 2389–2400 (2009)
7. Freeman, J., Ziemba, C.M., Heeger, D.J., Simoncelli, E.P., Movshon, J.A.: A functional and perceptual signature of the second visual area in primates. Nat. Neurosci. **16**, 974–981 (2013)
8. Ziemba, C.M., Freeman, J., Movshon, J.A., Simoncelli, E.P.: Selectivity and tolerance for visual texture in macaque V2. Proc. Natl. Acad. Sci. 201510847 (2016)
9. Lu, H.D., Chen, G., Tanigawa, H., Roe, A.W.: A motion direction map in macaque V2. Neuron **68**, 1002–1013 (2010). https://doi.org/10.1016/j.neuron.2010.11.020
10. Horiuchi, T.K., Koch, C.: Analog VLSI-based modeling of the primate oculomotor system. Neural Comput. **11**, 243–265 (1999)
11. Jimenez Ortega, L., Stoppa, K., Gunturkun, O., Troje, N.F.: Vision during head bobbing, are pigeons capable of shape discrimination during the thrust phase? Exp. Brain Res. **199**, 313 (2009). https://doi.org/10.1007/s00221-009-1891-5
12. Biro, D., Freeman, R., Meade, J., Roberts, S., Guilford, T.: Pigeons combine compass and landmark guidance in familiar route navigation. Proc. Natl. Acad. Sci. U. S. A. **104**, 7471–7476 (2007)
13. Schiffner, I., Siegmund, B., Wiltschko, R.: Following the Sun, a mathematical analysis of the tracks of clock-shifted homing pigeons. J. Exp. Biol. **217**, 2643–2649 (2014)
14. Vyssotski, A.L., Dell'Omo, G., Dell'Ariccia, G., Abramchuk, A.N., Serkov, A.N., Latanov, A.V., et al.: EEG responses to visual landmarks in flying pigeons. Curr. Biol. **19**, 1159–1166 (2009)

15. Mann, R.P., Armstrong, C., Meade, J., Freeman, R., Biro, D., Guilford, T.: Landscape complexity influences route-memory formation in navigating pigeons. Biol. Lett. **10**, 20130885 (2014)
16. Haralick, R.M., Shanmugam, K., Dinstein, I.: Textural features for image classification (1973)
17. Webster, M.A., De Valois, K.K., Switkes, E.: Orientation and spatial-frequency discrimination for luminance and chromatic gratings. J. Opt. Soc. Am. A **7**, 1034–1049 (1990)
18. Zaccolo, M.: Good features to track. Methods Mol. Biol. **178**, 255–258 (2002)
19. Barbot, A., Landy, M.S., Carrasco, M.: Differential effects of exogenous and endogenous attention on second-order texture contrast sensitivity. J. Vis. **12**, 1–15 (2012)
20. Song, B., Li, P., Li, J., Plaza, A.: One-class classification of remote sensing images using kernel sparse representation. IEEE J. Sel. Top. Appl. Earth Obs. Remote Sens. **9**, 1613–1623 (2016)
21. Moody, D.I., Brumby, S.P., Rowland, J.C., Altmann, G.L.: Land cover classification in multispectral satellite imagery using sparse approximations on learned dictionaries. In: Huang, B., Chang, C.-I., Lopez, J.F. (eds.) Proceedings of the SPIE, vol. 9124, p. 91240Y (2014)
22. Blaschke, T.: Object based image analysis for remote sensing. ISPRS J. Photogramm. Remote Sens. **65**, 2–16 (2010). https://doi.org/10.1016/j.isprsjprs.2009.06.004
23. Du, P., Xia, J., Zhang, W., Tan, K., Liu, Y., Liu, S.: Multiple classifier system for remote sensing image classification. Sensors **12**, 4764–4792 (2012)
24. Skowronek, S., Asner, G.P., Feilhauer, H.: Performance of one-class classifiers for invasive species mapping using airborne imaging spectroscopy. Ecol. Inform. **37**, 66–76 (2017)
25. Joseph, J.S., Victor, J.D., Optican, L.M.: Scaling effects in the perception of higher-order spatial correlations. Vis. Res. **37**, 3097–3107 (1997)
26. Kingdom, F.A.A., Keeble, D.R.T.: Luminance spatial frequency differences facilitate the segmentation of superimposed textures. Vis. Res. **40**, 1077–1087 (2000)

Estimating Dynamics of Honeybee Population Densities with Machine Learning Algorithms

Ziad Salem[1]([✉]), Gerald Radspieler[1], Karlo Griparić[2], and Thomas Schmickl[1]

[1] Artificial Life Lab at the Institute for Zoology, Karl-Franzens-University Graz,
Universitätsplatz 2, 8010 Graz, Austria
{ziad.salem,gerald.radspieler,thomas.schmickl}@uni-graz.at
[2] LARICS Lab at the Faculty of Electrical Engineering and Computing,
University of Zagreb, Unska 3, 10000 Zagreb, Croatia
karlo.griparic@fer.hr

Abstract. The estimation of the density of a population of behaviourally diverse agents based on limited sensor data is a challenging task. We employed different machine learning algorithms and assessed their suitability for solving the task of finding the approximate number of honeybees in a circular arena based on data from an autonomous stationary robot's short range proximity sensors that can only detect a small proportion of a group of bees at any given time. We investigate the application of different machine learning algorithms to classify datasets of pre-processed, highly variable sensor data. We present a new method for the estimation of the density of bees in an arena based on a set of rules generated by the algorithms and demonstrate that the algorithm can classify the density with good accuracy. This enabled us to create a robot society that is able to develop communication channels (heat, vibration and airflow stimuli) to an animal society (honeybees) on its own.

Keywords: Machine learning · Data mining
Classification algorithms · Density estimation · Robots · Honeybees

1 Introduction

The availability of a precise estimation of the population density is an important prerequisite for the establishment of a mixed society of interacting bees and robots, which is the aim of the ASSISI|$_{bf}$ (*Animal and robot Societies Self-organise and Integrate by Social Interaction (bees and fish)*) project. The main concept of the project is to generate a mixed society of honeybees and autonomous robots, which aims to establishing a robotic society that is able to develop communication channels to animal societies (bees and fish swarms) on its own [20]. The robots will adapt by evolutionary algorithms until they have learned to interact with animals in a desired way. Honeybees are an established and widely used model organism in the field of collective behaviour and swarm intelligence due to their social nature [23]. The project's long-term objective is to

© Springer International Publishing AG 2018
G. Nicosia et al. (Eds.): MOD 2017, LNCS 10710, pp. 309–321, 2018.
https://doi.org/10.1007/978-3-319-72926-8_26

integrate bio-inspired robots and biological agents to form an interactive mixed society. The experiments focus on limited interaction between stationary robots and bees in an experimental arena. These robots are implemented as *CASU*s (Combined Actuator Sensor Units) that communicate with the bees via their actuators (heat, vibration and airflow) [21]. The bees communicate with the CASUs via their mere presence (detected by the CASUs' six proximity sensors). The task of estimating the population density of a group of agents, based on information gathered by sensors with severely limited surveillance range, poses a great challenge for the development of automated realtime solutions. This is especially true for agents such as honeybees with highly variable and unpredictable behaviour. Supervised machine learning methods are the most promising candidates for a reliable and time efficient solution for density estimation. The new agent density estimation approach poses a new challenge when selecting a good learning algorithm and its parametrization. Therefore, this aspect had high priority in our work. To our knowledge, agents density estimation based on machine learning has not been applied before to solve such a problem.

1.1   State of the Art

Researchers use machine learning classification algorithms to generate decision trees or rulesets as descriptors of datasets, which represent the problem to be solved [18]. The algorithm is trained by the training dataset where the generated rules are tested with a test dataset. The algorithms either split the dataset into separate sets of test and training data, or use cross-validation for generating the classifiers. The produced classifiers can be evaluated by different measures such as accuracy, robustness, speed and scalability [11]. For this work, we used accuracy, calculated as the number of correct predictions per total number of predictions made while using bees as agents. In prior works, Salem and Schmickl [19] used *bristle-bots*, simple micro robots propelled by vibrating the slanted bristles they rest on [7], as substitutes for groups of bees. In this case, an algorithm was used to learn how to derive the number of bristle bots in a circular arena from the sensor activities of a CASU at the center of the arena. The resulting rules were induced by an algorithm trained with datasets collected during the experiments. The work showed that the set of rules was able to predict the number of bristle-bots with satisfying accuracy. While this study was valuable for the development of the project, there are important differences between bristle-bots and bees that required an extended study with bees. For instance, the bees' locomotion patterns are modulated by the environment and dependent on communication between individuals. The work reported in this paper aims to enable the CASUs to determine the number of bees in an arena with a good accuracy by employing different algorithms. By determining the best suited algorithm and its parametrization we extend the scope of machine learning applications to the field of bio-hybrid societies, where they will be implemented on different layers of control and evaluation software. In additional related work an artificial neural network based on LSTM architecture was designed and trained for bee density estimation [16].

1.2   Research Questions

The work presented in this paper was devoted to answering the following research questions:

1. Does the number and type of attributes correlate with the classification accuracy?
2. Does the number of classes correlate with the accuracy and number of generated rules?
3. Does the number of examples affect the accuracy and number of generated rules?
4. Does the number of generated rules correlate with the accuracy?
5. Is there a specific experimental setup that allows any or all of the algorithms to perform exceptionally well?
6. Are there differences between the three tested algorithms regarding the achieved accuracy and number of generated rules?
7. Is there an algorithm that performs consistently better or worse than the others in all experimental setups?

2   Material and Methods

The data is generated by the sensors mounted on the hexagonal top part of a CASU. The CASU is located at the center of a circular arena ($d = 12.5$ cm), which is equipped with a wax floor and a plastic wall. The infra-red proximity sensors are triggered by objects at a distance of up to 1.5 cm. The values of each sensor are logged at a rate of $10s^{-1}$, thus producing data with six features. For each animal experiment, single or groups of young (up to one day old) European honeybees (*Apis mellifera* sp. [10]) were released into the arena and left to walk freely for a specified time. The experiments were conducted in an infrared lit environment which is essentially dark for the bees. The number of bees in the arena was varied in different steps between experiments. The resulting log files were processed to extract various attributes relevant for the learning process, which we combined to constitute the actual datasets. We conducted several series of experiments on these datasets with different selections of attributes and evaluated the accuracy achieved with each combination along with the corresponding Kappa coefficient [4]. The aim of the learning process was to assign to each group size one of several group size classes, which were also subject to change between iterations of the series of experiments. In order to prevent the adverse effects of an excessive number of classes on the accuracy, we grouped the population sizes into larger and fewer bins for some experiments and assessed the impact of this measure on the accuracy by repeating the experiment four times with different numbers of classes. In this work we used different algorithms from the *Weka package*[1] (*Waikato Environment for Knowledge Analysis*) [22]. We focussed on three different methods which process combined training and

---

[1] Available for download at http://www.cs.waikato.ac.nz/ml/weka/.

test datasets that store pre-existing knowledge about the classification (the output, in our case the group size) inline with the attributes and their values used for the training (the input) [12].

2.1 Hardware Description

The hardware system that delivers the data to learn from is a custom stationary robot (CASU) developed to facilitate interactions with nearby bees (*Bee-CASU system*) [8]. The CASU is mounted beneath the arena, with its hexagonal top part ($d = 1.5$ cm) above the aluminium ring protruding through the arena floor into the arena. This part hosts six lateral proximity sensors (VCNL4010) with an I2C communication interface to detect nearby honeybees. The sensors are fully integrated and implement an independent distance measurement procedure that resorts to a built-in infra-red emitter and a photo diode to detect reflected infrared light. The sensors are able to detect bees at a distance of up to 1.5 cm and thus do not allow to directly determine the total number of honeybees in the arena. During the experiments, the values reported by the proximity sensors and other relevant status information of the CASU were logged by the control software at a rate of $10s^{-1}$. One or more bees are detected by a sensor whenever its value reaches or exceeds its threshold value. The sensor specific threshold is assumed at 3% above the minimum value encountered by the sensor during the entire experiment. The short detection range of the sensors is the primary reason why we had to develop a method to (spatially and temporally) integrate the sparse information retrieved from the individual sensors in order to get an estimation of the total number of honeybees in the arena.

2.2 Learning Algorithms

All learning experiments were performed using three algorithms, one of them based on decision trees (J48) and two based on classification rules (JRip and PART).

J48 Decision Tree. J48 is a Java implementation of the C4.5 decision tree algorithm [17]. J48 is an extension of ID3 algorithm and is often referred to as a statistical classifier.

JRip Rules Classifier. JRip is a fast and efficient RIPPER algorithm [2]. Classes are examined in increasing size and an initial set of rules for each class is generated using incremental reduced error pruning [5]. JRip proceeds by treating all the examples of a particular judgement in the training data as a class, and finding a set of rules that cover all the members of that class.

PART Algorithm. PART is a partial decision tree algorithm, which is the developed version of C4.5 and RIPPER algorithms [15]. However, decision trees are sometimes more problematic due to the larger size of the tree which could be oversized and might perform badly for classification problems [3].

2.3   Dataset Description

In order to create the datasets for or our learning experiments, we pre-processed the data logs for the sensors including six integer values that reflect the status of each proximity sensor, and extracted a number of attributes that integrate the information of one or several sensors. Depending on the experiment series, we averaged the raw data over a period of 3 s (30 values) or 1 min (600 values) and employed different sub-sets of the defined attributes[2].

2.4   Learning Experiments

The classical supervised learning problem is to construct a classifier that can correctly predict the classes of new objects given training examples of old objects [14]. If the classifier classifies most cases in the test examples correctly, we can assume that it works accurately also on the future data. If the classifier makes too many errors (misclassifications) in the test examples, we can assume that it was a wrong model. A better model can be searched after modifying the data, changing the settings of the learning algorithm, or by using another classification method [9]. In order to identify the best combination of algorithm and dataset setup, we conducted several series experiments to test the variability of rules generated using different setups. This approach resulted in five main experiment series (s1–s5) with different population sizes and different population size granularity. s1 was complemented with four additional sub-series (s1.1–s1.4). The main reason to iterate over different experiment series was to determine the response to altering the number of bees, definition of group sizes, attributes or experiment duration on the learning algorithm for the aim of concluding to the best setup. For this purpose, we processed the averaged or summed attributes with each of the algorithms to test. The dataset based on sums was tested in its original version (examples sorted by population size) and in a randomly shuffled version (examples in random order). In **s1**, we used a very simple dataset setup with 14 attributes and 14 classes, where every population size of bees is considered as a class. We derived the attributes from both averaged (real) and summed (integer) value in order to test for a possible impact of the data type on the performance of the algorithm, which proved to be marginal. In s1.1–s1.4 we used different grouping schemes with different population numbers in every sub-series while sticking with the same attributes. In **s2**, we introduced new attributes based on the original sensor readings in addition to the attributes used earlier. We also added a new class for *no bees*. In this case the algorithm can learn from a mixture of real and logical values with more information, which can be beneficial for generating better results. In this and all subsequent series, we averaged the

---

[2] A sample dataset is available at https://doi.org/10.5281/zenodo.824923.

sensor values over 600 sensor log records, which correspond to 1 min of record-ings. This method reduces the number of records in the dataset and also reduces the impact of the large variance in the instantaneous data. In **s3**, we introduced normalized versions of several attributes (sum and standard deviation of sensor values) in order to better compensate for differences between sensors, to reduce the data range and to complement the logical and integer values with real values, thus creating a more diverse dataset, which is expected to have a positive impact on the algorithm's search space. In **s4**, the duration of the bee experiments was shortened to 10 min. In **s5** we modified the setup of the experiments to decrease the population size granularity. We cycled through all population sizes from 1 to 32 bees and repeated each experiment twice. For details on experiment setups in the different series of learning experiments, consult Table 1.

Table 1. Setups of experiment series. The number of attributes, examples and classes and the duration (τ) and number of repetitions (n_{runs}) of the bee experiments are shown for the different series. Each example consists of the indicated number of attributes and resolves to the respective number of classes.

Series	$n_{examples}$	$n_{attributes}$	$n_{classes}$	$\tau[min]$	n_{runs}
1	1008	14	14	30	2
1.1	1008	14	2	30	2
1.2	1008	14	3	30	2
1.3	1008	14	4	30	2
1.4	1008	14	4	30	2
2	1080	25	5	30	2
3	1080	25	5	30	2
4	360	25	5	10	2
5	320	25	4	10	1

3   Results

The J48 and PART algorithms achieved a similar accuracy (differences between 3–6%) while JRip typically performed slightly lower (see Fig. 1). For all algo-rithms, the lowest accuracy (47.4% and 46.7% for J48 and PART) was achieved in the s1 experiments (no categorization) and the highest (97.8% and 98.3% for J48 and PART) in s1.1 experiments (categorization into the lowest number of classes). The Kappa statistics demonstrate a sufficient (s1) to excellent (s2, s3) agreement between predicted and actual classes (see Table 3). The accuracy achieved while learning from the different types of dataset (based on averages, sums and shuffled sums), which were provided in the s1 experiments, was almost identical for the three dataset types (see Fig. 1). While the number of rules generated by JRip slightly changed depending on the dataset (25, 22 and 24 rules for averages, sums and shuffled sums), the PART algorithm generated the same number of rules (90) with all datasets (see Table 2). During the average based learning of s1 experiments, missing values (zeros) occurred in 68% of all examples for 4 of the 14 attributes. During the s2 experiments, missing values occurred in 10% of all examples for 3 out of 25 attributes. While the states were evenly distributed in the s1 datasets (72 examples per state), the introduction of classes resulted in an imbalanced distribution of the dataset examples over the different states. For example, of the five classes in the datasets used for the s2 and s3 experiments, two were covered by 72, two by 360 and one by 216 examples. The results of categorization into different numbers of nominal classes (see Fig. 2a) corroborate the negative correlation between number of classes and achieved accuracy for all three algorithms (compare s1 experiments of Fig. 1). However, for all tested numbers of classes the achieved accuracy was higher than for learning from uncategorized (continuous) states. Over all numbers of classes and with uncategorized states, J48 and PART achieved a higher accuracy than JRip. There is a direct correlation between the number of classes and the number of rules generated by the JRip and PART algorithms (see Fig. 2b and Table 2). This correlation is approximately linear in the investigated range of 2–15 classes. PART produced more rules than JRip over the entire range. In contrast to the number of classes, the number of attributes does not have an influence on the number of rules generated by JRip or PART (see Table 2). It follows from the correlation between number of classes and number of generated rules and the negative correlation between number of classes and accuracy, that there is also a negative correlation between the number of generated rules and the achieved accuracy (see Fig. 3).

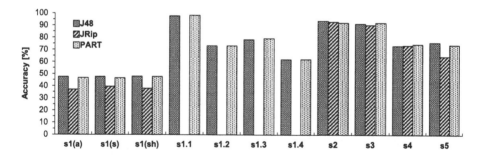

Fig. 1. Overview over the accuracy achieved in the different learning experiment series. The setups of different series differ in data reduction method and number of states (classes). The dataset for s1(a) is based on averages, for s1(s) on sums and for s1(sh) on shuffled sums. For s5, (evenly) redistributed classes were used. Consult Table 1 for numbers of classes and attributes used in each series. The JRip algorithm was not used in experiment series s1.1–s1.4.

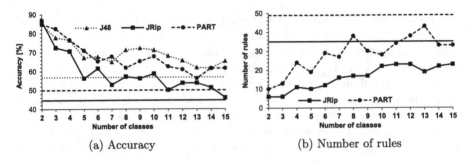

(a) Accuracy   (b) Number of rules

Fig. 2. The dependence of accuracy and number of generated rules on the number of classes. The learning experiments were conducted for different numbers of distinguished classes (lines with markers; group sizes of 1–32 honeybees categorized into 2–15 nominal classes) and for uncategorized group sizes (markerless lines). The learning experiments were conducted with the setup of the s5 experiments. (a) The accuracy achieved by all three algorithms for different numbers of classes. (b) The number of rules generated by the JRip and PART algorithms for different numbers of classes

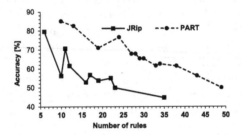

Fig. 3. The accuracy achieved by the JRip and PART algorithms for different numbers of generated rules

4   Discussion

In this paper, the primary benchmark for the quality of the algorithms is the accuracy of the predictions made by a model trained on the test dataset. The lowest accuracy was achieved in the s1 experiments, which were based on datasets with uncategorized states. This finding is explicable by the scattering of the dataset over 14 possible states (3 to 42 bees in steps of 3), which has a negative effect on the ability of the searching mechanism to find the best rules to represent the whole dataset. A considerable improvement was achieved by categorizing the states into classes (i.e. by grouping populations; compare experiment series s1.1–s1.4), which reduces the amount of scattering. In accordance with the negative relationship between number of classes and accuracy, the highest accuracy was achieved in the s1.1 experiments, in which only two classes were distinguished. The method of calculating the attribute values (summation or averaging) and the randomization of the order of examples did not have any impact on the accuracy. Some of the 14 attributes used in the s1 and s1.1–s1.4 experiments assumed zero-values in many dataset examples. Frequent occurrences of such missing values can affect the quality of the classifiers trained on these datasets [1] as they tend to inflate the ruleset. Due to the interdependencies between number of classes, number of rules and accuracy, it is not possible to imply from these experiments a direct impact of zero values on the accuracy. The new attributes and data reduction methods introduced with s2 had a dampening effect on the number of zero values encountered in the datasets, as both the number of occurrences and the number of affected attributes decreased. On the other hand, the introduction of classes with s1.1 brought along the problem of uneven class distribution, which can seriously affect the performance of the learning algorithms both during the training and the evaluation phase [6]. For example, in the s1.1 experiments, one of the two classes comprised populations of 3 bees while the other comprised populations of 6–42 bees, thus causing an extremely imbalanced class coverage. Despite this factor, the learning from categorized states, which was employed in s1.1 and higher, proved to be beneficial to the accuracy. Therefore, the effect of uneven class distribution appears to be outbalanced by the negative toll taken by the large number of states to differentiate in datasets without categorization. In experiment series s2–s5, additional features (attributes) were introduced to the dataset, bringing their number from 14 to 25 and thus providing the algorithms with more information to learn from (see Table 1). Additionally, some of the newly introduced attributes directly reflect low level sensor activities rather than the values derived from logical combinations of sensor values. Both of these changes had a beneficial effect on the performance of the algorithms. The setup of the s2 experiments proved to be especially beneficial to the accuracy of the learning algorithms. In this series with categorization into 5 classes, the J48 and PART algorithms performed better than in the s1.2, s1.3 and s1.4 experiments that only differentiated 3, 4 and 4 classes, respectively, and the JRip algorithm performed on level with J48 and PART. However, J48 and PART didn't achieve the same accuracy as in s1.1 experiments, in which only two classes were differentiated. Given the persistence of other parameters, this improvement can only be ascribed to the increase in number and qualitative changes of the attributes and the resulting improvement of the algorithms' database. The changes introduced in s3–s5 experiments did not improve the performance of the algorithms compared to s2. While they performed comparably well in s3 experiments, the setup changes in s4 and s5 had a detrimental effect on the classification accuracy of all three algorithms. Since the number of classes is the same (5) in s4 as in s2 and s3 and even lower (4) in s5 and the attributes are identical as well, this effect appears to be due to the lower number of examples in the respective datasets (360) compared to those used for s2 and s3 (1080). In this case, the shorter duration of the biological experiments was reflected by smaller datasets, which in turn provided less information for the algorithms to learn from. The comparison of the accuracy achieved by the three algorithms shows that JRip typically has a lower performance than the other two algorithms, which perform approximately equal. This is especially obvious when considering the accuracy in correlation with the number of distinct classes. However, in experiment series

Table 2. The number of rules generated by the JRip and PART algorithm in all learning experiment series. For s1, the values are given for attributes constructed of averages (a), sums (s) and shuffled sums (sh).

Algorithm	s1(a)	s1(s)	s1(sh)	s1.1	s1.2	s1.3	s1.4	s2	s3	s4	s5
JRip	25	22	24	-	-	-	-	16	18	8	9
PART	90	90	90	9	8	30	46	20	25	26	15

Table 3. Kappa statistics. The kappa coefficients calculated during the learning process of the three algorithms are shown for the major experiment series.

Algorithm	s1(a)	s1(s)	s1(sh)	s2	s3	s4	s5
PART	0.4263	0.4241	0.437	0.8957	0.8945	0.6693	0.738
Jrip	0.3216	0.3494	0.3323	0.9057	0.8701	0.6595	0.6847
J48	0.4338	0.4348	0.438	0.9148	0.8859	0.6517	0.6728

s2–s4, all three algorithms have a comparable performance. The number and quality of generated rules are an important factor that determines the classification accuracy both during the learning process and upon application of the ruleset to a classification problem, where a lower number of rules is beneficial for the applicability of the classifier. In this regard, JRip clearly outperforms PART as it manages to achieve a comparable accuracy with fewer rules in all tested setups. Generally it can be said that the setups of the experiments played a much greater role for the accuracy than the selected algorithm.

5   Conclusions

Our approach of detecting the approximate number of bees in an arena by employing machine learning algorithms to evaluate highly variable data from a limited detection system has shown promising results. The algorithms applied to the problem were able to classify the bees population in the arena with good accuracy and a reasonable number of generated rules (where applicable). It is obvious that the choice of the algorithm and a proper configuration of dataset and algorithm is crucial to the quality of the classification results. However, finding the optimal parameters for all applied algorithms and datasets requires a large amount of resources [13] as each algorithm has to be tuned individually for each dataset. Although there are automatic methods for setting parameters, we adhered to the default parametrization for all machine learning algorithms employed in our experiments, so that some potential remains to further improve the accuracy of our classification results. The comparison of the different algorithms employed in our experiments showed that PART is the best suited algorithm for our specific problem as it achieves the highest classification accuracy with the smallest rulesets. However, all three algorithms achieved a comparable accuracy when processing datasets configured according to the setup of the s2 experiments. This setup proved to be the best suited among the different setups we tested due to the high accuracy it allowed the algorithms to achieve while differentiating a reasonable number of classes. Due to this compromise between accuracy and a useful number of classes, this setup will continue to be used in further experiments. Therefore, the research questions on which the work presented in this paper was focused, can be answered as follows:

1. The number of attributes and the procedure of deriving their values from the raw sensor data have an influence on the classification accuracy. However, the method of calculating their values (averages, sums or shuffled sums) does not influence the accuracy.
2. The classification accuracy correlates negatively and the number of generated rules positively with the number of classes.
3. A low number of examples can have a negative effect on the classification accuracy if the learning algorithm is trained on too little data to achieve its potential.
4. The number of generated rules is interdependent with the achieved accuracy, but it is not clear from our results whether this is a real or spurious correlation.
5. All three algorithms showed their best performance when trained on datasets configured according to the s2 and s3 setup. However, there is potential for further improvements.
6. J48 and PART perform similarly in terms of classification accuracy while JRip falls behind. PART also generates fewer rules than JRip.
7. Due to its prevalence over the competing algorithms in terms of accuracy and ruleset size, PART shows the best overall performance during all experiments. Judged by the same criteria, JRip underperformed consistently.

Therefore, our approach to determine the approximate number of bees in an arena using machine learning algorithms was successful. The proportion of correct classifications was excellent, so that the method can be implemented in future experimentation, where it will for example provide stationary robots with the information required to take control over a swarm of bees or other animals. At the same time, the flexibility of the method and its implementation will allow for easy adaptation to different scenarios or environments. This wide applicability gives the method relevance for various fields of research.

6   Future Work

The density detection mechanisms developed in this work will be implemented in future experiments conducted in the framework of the $ASSISI_{bf}$ project. The pre-generated rulesets will be used by the CASU control program to estimate the number of bees in the arena. The control program will be able to integrate information from several CASUs and thus classify the group sizes with higher accuracy. We will complement this top-down approach by the implementation of a learning algorithm directly in each CASUs' control logic. This will enable the CASUs to independently estimate the number of bees in the arena and to adapt to new scenarios by dynamically modifying the classification rulesets. The methods developed in our work could be used to monitor bees in their hive in order to continuously assess the colony size or to detect abnormal behaviour. This could provide beekeepers with a valuable new tool to survey the health of their colonies.

Acknowledgments. This study was supported by the EU FP7 FET-Proactive project $ASSISI_{bf}$, grant no. 601074.

References

1. Acuña, E., Rodriguez, C.: The treatment of missing values and its effect on classifier accuracy. In: Banks, D., McMorris, F.R., Arabie, P., Gaul, W. (eds.) Classification, Clustering, and Data Mining Applications. Studies in Classification, Data Analysis, and Knowledge Organisation, pp. 639–647. Springer, Heidelberg (2004). https://doi.org/10.1007/978-3-642-17103-1_60
2. Cohen, W.W.: Fast effective rule induction. In: Proceedings of the Twelfth International Conference on Machine Learning, pp. 115–123. Morgan Kaufmann, San Francisco (1995)
3. Duin, R.P.: A note on comparing classifiers. Pattern Recogn. Lett. **17**(5), 529–536 (1996)
4. Foody, G.M.: Thematic map comparison: evaluating the statistical significance of differences in classification accuracy. Photogram. Eng. Remote Sens. **70**(5), 627–633 (2004)
5. Fürnkranz, J., Widmer, G.: Incremental reduced error pruning. In: Proceedings of the 11th International Conference on Machine Learning (ML-1994), pp. 70–77 (1994)
6. Ganganwar, V.: An overview of classification algorithms for imbalanced datasets. Int. J. Emerg. Technol. Adv. Eng. **2**(4), 42–47 (2012)
7. Giomi, L., Hawley-Weld, N., Mahadevan, L.: Swarming, swirling and stasis in sequestered bristle-bots. In: Proceedings of the Royal Society a Mathematical Physical and Engineering, vol. 469, p. 20120637. The Royal Society (2013)
8. Griparić, K., Haus, T., Miklić, D., Bogdan, S.: Combined actuator sensor unit for interaction with honeybees. In: 2015 IEEE Sensors Applications Symposium (SAS), pp. 1–5. IEEE (2015)
9. Hämäläinen, W., Vinni, M.: Classifiers for educational data mining. In: Handbook of Educational Data Mining, pp. 57–74 (2010)
10. Heinrich, B.: The Hot-Blooded Insects: Strategies and Mechanisms of Thermoregulation. Springer, Heidelberg (2013). https://doi.org/10.1007/978-3-662-10340-1
11. Jordan, M., Mitchell, T.: Machine learning: Trends, perspectives, and prospects. Science **349**(6245), 255–260 (2015)
12. Kotsiantis, S.B.: Supervised machine learning: a review of classification techniques. Informatica - Int. J. Comput. Inform. **31**(3), 249–268 (2007)
13. Kotthoff, L., Gent, I.P., Miguel, I.: A preliminary evaluation of machine learning in algorithm selection for search problems. In: Fourth Annual Symposium on Combinatorial Search (2011)
14. Mitchell, T.M.: Machine Learning, 1st edn. McGraw-Hill, Inc., New York (1997)

15. Mohamed, W.N.H.W., Salleh, M.N.M., Omar, A.H.: A comparative study of reduced error pruning method in decision tree algorithms. In: IEEE International Conference on Control System, Computing and Engineering (ICCSCE), pp. 392–397. IEEE (2012)

16. Polić, M., Salem, Z., Griparić, K., Bogdan, S., Schmickl, T.: Estimation of moving agents density in 2D space based on LSTM neural network. In: IEEE Conference on Evolving and Adaptive Intelligent Systems (EAIS), pp. 1–8. IEEE (2017)

17. Quinlan, J.R.: C4.5: Programs for Machine Learning. Elsevier, San Francisco (2014)

18. Salem, Z.: Enhanced computer algorithms for machine learning. Ph.D. thesis, Intelligent system Research laboratory. Cardiff University, Wales, UK (2002)

19. Salem, Z., Schmickl, T.: The efficiency of the rules-4 classification learning algorithm in predicting the density of agents. Cogent Eng. **1**(1), 986262 (2014)

20. Schmickl, T., et al.: ASSISI: mixing animals with robots in a hybrid society. In: Lepora, N.F., Mura, A., Krapp, H.G., Verschure, P.F.M.J., Prescott, T.J. (eds.) Living Machines 2013. LNCS (LNAI), vol. 8064, pp. 441–443. Springer, Heidelberg (2013). https://doi.org/10.1007/978-3-642-39802-5_60

21. Schmickl, T., Szopek, M., Bodi, M., Hahshold, S., Radspieler, G., Thenius, R., Bogdan, S., Miklić, D., Griparić, K., Haus, T., Kernbach, S., Kernbach, O.: Assisi: charged hot bees shakin'in the spotlight. In: 2013 IEEE 7th International Conference on Self-Adaptive and Self-Organizing Systems, pp. 259–260. IEEE (2013)

22. Witten, I.H., Frank, E., Hall, M.A., Pal, C.J.: Data Mining: Practical Machine Learning Tools and Techniques, 4th edn. Morgan Kaufmann, San Francisco (2016)

23. Zahadat, P., Hahshold, S., Thenius, R., Crailsheim, K., Schmickl, T.: From honeybees to robots and back: division of labor based on partitioning social inhibition. Bioinspiration & Biomimetics **10**(6), 066005 (2015)

SQG-Differential Evolution for Difficult Optimization Problems under a Tight Function Evaluation Budget

Ramses Sala[✉], Niccolò Baldanzini, and Marco Pierini

Department of Industrial Engineering, University of Florence, Florence, Italy
rsala.unifi@gmail.com

Abstract. In the context of industrial engineering, it is important to integrate efficient computational optimization methods in the product development process. Some of the most challenging simulation-based engineering design optimization problems are characterized by: a large number of design variables, the absence of analytical gradients, highly non-linear objectives and a limited function evaluation budget. Although a huge variety of different optimization algorithms is available, the development and selection of efficient algorithms for problems with these industrial relevant characteristics, remains a challenge. In this communication, a hybrid variant of Differential Evolution (DE) is introduced which combines aspects of Stochastic Quasi-Gradient (SQG) methods within the framework of DE, in order to improve optimization efficiency on problems with the previously mentioned characteristics. The performance of the resulting derivative-free algorithm is compared with other state-of-the-art DE variants on 25 commonly used benchmark functions, under tight function evaluation budget constraints of 1000 evaluations. The experimental results indicate that the new algorithm performs excellent on the "difficult" (high dimensional, multi-modal, inseparable) test functions. The operations used in the proposed mutation scheme, are computationally inexpensive, and can be easily implemented in existing differential evolution variants or other population-based optimization algorithms by a few lines of program code as an non-invasive optional setting. Besides the applicability of the presented algorithm by itself, the described concepts can serve as a useful and interesting addition to the algorithmic operators in the frameworks of heuristics and evolutionary optimization and computing.

Keywords: Meta-heuristics · Derivative-free optimization
Evolutionary computing · Differential evolution · Black box optimization
Stochastic Quasi-Gradient Descend · SQG-DE

1 Introduction

The combination of computational optimization with modeling and simulation is becoming increasingly important in the modern development processes of complex engineering products and systems. During the last decades, a huge variety of heuristic and meta-heuristic search techniques have been developed [1, 2] and applied to real-world industrial problems [3, 4]. In the quest for product and process efficiency, an important question is: How to select efficient optimization methods for a particular

© Springer International Publishing AG 2018
G. Nicosia et al. (Eds.): MOD 2017, LNCS 10710, pp. 322–336, 2018.
https://doi.org/10.1007/978-3-319-72926-8_27

problem? The extension of the conservation law of generalization of performance [5], and the "no free lunch" (NFL) theorems for machine learning to the field of search and optimization [6], identified that often "generalization is a zero sum enterprise" [5], and that therefore: the search for a universal best performing optimization algorithm is futile. The still standing challenge is: to develop and identify efficient optimization algorithms for particular optimization problems, or classes of optimization problems, taking into account the available resources in the context of their application.

In this communication, we target optimization problems under strict function evaluation budget constraints, which often occur in the context of industrial optimization problems which involve the simulation responses of complex dynamic systems. Industrial applications of such problems are for example: simulation based crashworthiness optimization of vehicle structures [9–11], and Computational Fluid Dynamics (CFD) based optimization [7, 8]. The optimization problems of such complex system responses are often characterized by: a large number of design variables, the absence of analytical gradient information, highly non-linear system responses, and computationally expensive function evaluations resulting in a limited function evaluation budget. The need to adapt the engineering development and optimization process to products and systems with increasing complexity make the research for efficient optimization algorithms, for non-convex optimization problems, under tight function evaluation constraints of great relevance in engineering [3, 4, 10–13].

Despite recent and ongoing research on the theoretical performance analysis of heuristic search algorithms on fixed budget problems [14], the performance analysis of complex problems and optimization algorithm is in practice still restricted to numerical comparative tests. The optimization algorithm performance comparisons in the literature are however often w.r.t. algorithm convergence behavior using a large number (hundreds of thousands to millions) of function evaluations. For engineering optimization problems which involve computationally expensive simulations, the function evaluation budget is often orders of magnitudes smaller, such that true optimization near to the global optimum is often infeasible [10, 24]. When the function evaluation budget strongly constraints the optimization, different aspects of the optimization algorithms are of practical relevance.

In this communication, we present a new variant of the well-known and widely used Differential Evolution (DE) algorithm [15, 16], by introducing a novel mutation operator inspired by concepts of Stochastic Quasi-Gradient (SQG) methods [25, 26]. The new hybrid algorithm targets to improve the search efficiency in the setting of optimization under tight function evaluation budget constraints. To investigate the effect of the new DE mutation operator, the performance of the hybrid algorithms is compared with "classical" DE and several state-of-the-art DE variants [17–21], on a commonly used set of test functions of various structure and complexity, under budget constraints of 1000 function evaluations. Although the mutation operator could also be used in other similar algorithm classes such as Particle Swarm Optimization (PSO) [29], or Evolutionary Strategies (ES), the focus of this first study will be limited to the implementation and performance comparison of the SQG-mutation operator in the framework of Differential Evolution and several of its state-of-the-art variants. In the context of budget limited optimization problems in structural and multidisciplinary optimization, DE was identified and recommended as an efficient algorithm for car-body optimization problems involving computationally expensive crashworthiness responses [9–12]. DE algorithms are also used for optimization of aircraft engines [7], wind turbines [8], and many other applications [22, 23]. Optimization problems in the "expensive" function evaluation setting can also benefit from meta-modelling or surrogate model based optimization techniques such as e.g. [27, 28]. In-depth investigations on the interactions between optimization algorithm operators and different meta-models and control parameters require however an extended scope. Nevertheless even in the present limited scope, the obtained results indicate that the new algorithm, could already be an efficient alternative to several state-of-the-art DE variants, for difficult problems under a tight function evaluation budget.

2 Description of the SQG-DE Algorithm

2.1 Conceptual Description

The objective in the design of heuristic and meta-heuristic optimization algorithms is to obtain a beneficial compromise between efficiency and accuracy (or optimality), taking into account the available resources. The here presented hybrid method is developed to improve the efficiency of DE for a variety of problem types under strict evaluation budgeted constraints. This is achieved by means of a new mutation operator. While in conventional DE, the mutation operator uses a sum of random vector differences from the DE population, in the new mutation operator the perturbation directions of the new population members are constructed by a weighted sum approach, using the weights dependent of respective fitness differences. This concept for the perturbation directions was inspired by the stochastic quasi-gradient estimations [25, 26] used in the SQG method. Whereas in SQG-descend the stochastic gradient estimations are based on vector differences of small stochastic perturbations, the here described method applies the concept to vector differences of the DE population.

2.2 Quantitative Description

The new hybrid SQG-DE algorithm uses the framework of the conventional original Differential Evolution (DE) algorithm. For the description of the relatively well-known DE algorithm we refer to [15, 16], while for an overview of variants we refer to the reviews in [22, 23]. For the here proposed hybrid method a new mutation operator was developed, which will be described in this section. The new mutation operator was inspired by the Stochastic Quasi-Gradient (SQG) method (initially introduced as: "search by means of statistical gradients"). A detailed description of stochastic quasi-gradient methods is given in [26]. For the sake of clarity and briefness, the description here is limited to concepts relevant for the new mutation operator.

In SQG, the search direction is the stochastic gradient approximation $\xi(x)$ of a function $f(x)$ at point x. This direction is proportional to the following expression:

$$\xi(x) \sim \sum_{k=1}^{r} \left(\frac{\left(f\left(x + \Delta z_k\right) - f(x)\right)}{\Delta} \right) * z_k \tag{1}$$

Where $z_k \in [-1, 1]^D$ are uniform random perturbation vectors of current trial vector x in dimension D. For a sufficiently large r, and sufficiently small values of Δ this approximation converges in probability to the direction of the gradient ∇f. SQG is however often applied using r significantly smaller than the problem dimension D, leading to coarse and "inexpensive" gradient approximations. Compared to other gradient based methods that require finite difference gradient approximations, SQG often however performs surprisingly well on local search problems, considering the efficiency in terms of the total amount of function evaluations. The idea of approximating the gradient direction d at point x, by means finite perturbations of the trial vector x, can be generalized to a sum of differences between pairs of distinct vectors x_a and x_b, in a sufficiently small neighborhood ε of x with: $\|x - x_a\| < \varepsilon$, and $\|x - x_b\| < \varepsilon$ by:

$$d \sim \frac{1}{w} \sum_{k=1}^{w} \frac{\left(f\left(x_{a,k}\right) - f\left(x_{b,k}\right)\right)}{\|x_{a,k} - x_{b,k}\|} * \left(x_{a,k} - x_{b,k}\right) \tag{2}$$

The key concept of the proposed hybrid DE method is the extension of this concept for gradient estimation to the application of finding new mutation vectors based on an existing differential evolution population. This extension thus omits the neighborhood constraint on a sufficiently small ε, and uses the differences between members of the population at a given iteration of a population based algorithm.

The mutation operator for the originally proposed "DE/rand/1/bin" version of DE [15] is determined by:

$$v_i = x_a + F(x_b - x_c) \tag{3}$$

where v_i are the mutant vectors for the next generation, x_q are parent vectors of the current population generation, with mutually exclusive $(a \neq b \neq \cdots \neq q)$ random permutation indices $a, b, \ldots, q \in \{1, 2, .., P\}$ to population members, in a population of size P. The scaling factor $F \in [0, 2]$ controls the amplification or step size of the differential variation. Later DE versions were introduced [16] such as "DE/best/2/bin", in which the mutation operator was based on a sum of more vector differences:

$$v_i = x_{best} + F\left(\left(x_a - x_b\right) + \left(x_c - x_d\right)\right) \tag{4}$$

In which x_{best} is the best member of the population as opposed to a random member as in (3).

Combining the previous considerations we introduce the SQG-DE hybrid scheme "SQG-DE/best/w/bin" with the SQG-mutation operator defined as:

$$v_i = x_{best} - F * \varphi * \sum_{k=1}^{w} \frac{(y_{b,k} - y_{c,k})}{\|x_{b,k} - x_{c,k}\|} (x_{b,k} - x_{c,k}) \tag{5}$$

where $y_{q,k}$ refers to the fitness or function evaluation value corresponding to the parent vector $x_{q,k}$ as $y_{q,k} = f(x_{q,k})$, and w is the number of mutually exclusive vector pairs used. To preserve the population self-adaptivity of the DE algorithm, a scaling factor φ on the perturbation magnitude is included. This factor is chosen as:

$$\varphi = \frac{1/w \left\| \sum_{k=1}^{w} (x_{b,k} - x_{c,k}) \right\|}{\left\| \sum_{k=1}^{w} \frac{(y_{b,k} - y_{c,k})}{\|x_{b,k} - x_{c,k}\|} (x_{b,k} - x_{c,k}) \right\|} \tag{6}$$

In which the denominator normalizes the magnitude of the perturbation direction, while the numerator scales the perturbation magnitude to a similar magnitude as the mutation operator in the original algorithm (3). The mutation formulation in Eq. (5) is similar to the original concept of vector differences, with the difference that now a sum of weighted vector differences are used, with a particular choice for the weights. The weights for the vector differences are calculated according to the fitness differences between the corresponding population members, such that in a high-dimensional setting, directions with larger directional "differences" are prioritized over directions for which the fitness differences are smaller.

Fitness differences are also used implicitly in the context of PSO, where single point pair differences between the global and local best-known locations are used to "guide" the search directions resulting from the mutation operator. In contrast, the basic SQG-mutation operator uses fitness difference based weighed sums of point pair differences between any population members, to guide the randomization of the search points. While the SQG concept is originally aimed at local gradient approximation, this is not the primary aim of the SQG-mutation operator. The extension of the SQG concept from perturbation points in a local neighborhood to random mutually exclusive population members, is likely to result in inaccurate local gradient approximations since the distance between the population members can be relatively large, and the fitness functions are generally non-linear. Since in "classical" DE the mutation operator is however only intended to introduce randomization of the population in its original implementation, there is no mechanism to favor any particular search direction. New non-descending SQG-mutation search points based on inaccurate gradient estimates are therefore also not problematic in the context of DE. However, for problems in which there are global trend directions, the SQG-mutations are statistically biased towards global descend directions. The key idea of the new method is however that SQG-mutations, tend to favor search directions along which the fitness differences are larger, over directions with small differences. This "direction-screening" is particularly relevant when, not all problem dimensions are equally important. For the mutation operator, parameter values of w between 2 and 5 give very satisfactory results, based on our current experience. It should be noted that for very small population sizes, high values of w should be avoided to maintain sufficient variance in future populations.

Although Eq. (5) is more complex than Eq. (3) only a relatively modest number of scalar and vector operations is required for the new mutation operator, while no additional function evaluations are required. The mutation operator can be easily implemented as an optional setting in existing code of DE or other population-based optimization algorithm implementations. On request, a MATLAB/Octave or other implementation of the algorithm is available from the authors.

3 A Comparison of Algorithm Performance on a Constrained Function Evaluation Budget

To investigate the performance of the new hybrid algorithm, a comparative assessment of SQG-DE, with two of the original DE algorithms, the SQG algorithm, and 5 state-of-the-art DE variants is performed, on a set of 25 test functions, using a budget of maximum 1000 function evaluations.

3.1 Algorithms and Test Functions

For the comparative assessment the following optimization algorithms are used:

1. DE - original differential evolution "rand/1/exp" [15]
2. DE2 - "best/2/bin" differential evolution [16]
3. jDE - self-adapting differential evolution [17]
4. JADE - adaptive differential evolution [20]
5. SaDE - strategy adaptation differential evolution [18]
6. epsDE - ensemble parameters differential evolution [19]
7. CoDe - composite trial vector strategy differential evolution [21]
8. SQG - Stochastic Quasi-Gradient search [25, 26]
9. SQG-DE - Stochastic Quasi-Gradient based Differential Evolution

To assess and compare the performance of the different algorithms the 25 test functions of the CEC 2005 benchmark [30] were used. Although many more benchmark sets have been developed since, these test functions are widely used in the optimization community (more than 1500 citations at present). The test function set is composed of optimization problems in 4 categories. One of these categories is of particular interest in the context of this work: The 4th category of "difficult" inseparable complex multimodal rotated functions, of which many are even hard to solve with a large function evaluation budget. Although these functions are usually used in the conventional context of global optimization, without tight function evaluation limits (typical budgets of hundreds of thousands to millions of objective function evaluations), they are also of interest as surrogate-test problems, in the context of algorithm performance assessment for complex industrial problems under tight function evaluation constraints. In this assessment the number of function evaluations per optimization run is constrained to a maximum of 1000. The test functions were evaluated for dimensions 30 and 50. Table 1 gives an overview of the test function descriptions and the problem categories. For a more detailed description of the test functions we refer to the description in [30].

Table 1. List of test functions

Function nr.	Test function description from [30]
1 Unimodal Functions (5):	
$f_1(x)$	Shifted sphere function
$f_2(x)$	Shifted Schwefel's Problem 1.2
$f_3(x)$	Shifted Rotated High Conditioned Elliptic Function
$f_4(x)$	Shifted Schwefel's Problem 1.2[a]
$f_5(x)$	Schwefel's Problem 2.6[b]
2 Multimodal Basic Functions (7):	
$f_6(x)$	Shifted Rosenbrock's Function
$f_7(x)$	Shifted Rotated Griewank's Function
$f_8(x)$	Shifted Rotated Ackley's Function[b]
$f_9(x)$	Shifted Rastrigin's Function
$f_{10}(x)$	Shifted Rotated Rastrigin's Function
$f_{11}(x)$	Shifted Rotated Weierstrass Function
$f_{12}(x)$	Schwefel's Problem 2.13
3 Multimodal Expanded Functions (2):	
$f_{13}(x)$	Expanded Extended Griewank's plus Rosenbrock's Function
$f_{14}(x)$	Shifted Rotated Expanded Scaffer's F6
4 Multimodal Hybrid Composition Functions (11):	
$f_{15}(x)$	Hybrid Composition Function
$f_{16,18,21,24}(x)$	Rotated Hybrid Composition Functions
$f_{17}(x)$	Rotated Hybrid Composition Function[a]
$f_{20}(x)$	Rotated Hybrid Composition Function[c]
$f_{19}(x)$	Rotated Hybrid Composition Function[b]
$f_{22}(x)$	Rotated Hybrid Composition Function[d]
$f_{23}(x)$	Non-Continuous Rotated Hybrid Composition Function
$f_{25}(x)$	Rotated Hybrid Composition Function

[a]With noise in fitness function
[b]With the global optimum on the bounds
[c]With a narrow basin for the global optimum
[d]With a high condition number matrix

For the comparison, the optimization runs of each algorithm were independently repeated with different random seeds for 100 times, for each test function to obtain statistically significant performance results. For all algorithms except SQG, the initial population size was set to 100, for all problems and dimension as was also done in previous works [31, 32]. For SQG a "warm" start was provided by choosing the best start point from a pseudo-random sample set of equal size as the population size of the other algorithms. The control parameters for DE, DE2 were F = 0.8, CR = 0.8, and in addition w = 5 for and SQG-DE, and SQG. For the five other algorithms with adaptive parameters, the control parameters were as described in the corresponding references [17–21]. For the adaptive parameter algorithms, the implementations as available in [34] were used.

3.2   Performance Measures

A commonly used optimization algorithm performance metric is Expected Running Time (ERT) [33], which can be defined as:

$$ERT(f_{target}) = mean\left(T_{f_{target}}\right) + \left(\left(1 - p_s\right)/p_s\right)T_{max} \tag{7}$$

Where f_{target} is a reference threshold value, $T_{f_{target}}$ is the number of function evaluations to reach an objective value better than f_{target}, T_{max} is the maximum number of function evaluations per optimization run, and p_s is the success rate defined as: $p_s = N_{succes}/N_{total}$, where N_{succes} is the number of successful runs (where the best obtained objective value is better than f_{target}). If the experiments result in no successful runs for a particular algorithm such that ($p_s = 0$), then expression (7) is undefined, in that case the information available on the ERT is that: $ERT(f_{target}) > T_{max} * N_{total}$.

ERT can be interpreted as the expected number of function evaluations of an algorithm to reach an objective function threshold for the first time. For the ERT performance measure, a threshold or success criterion is required. For conventional optimization performance studies this criterion is often related to reaching the value of the known global optimum, within a specified tolerance. For the optimization of difficult problems under tight budget constraints the probability of coming close to the global optimum is usually statistically negligible, therefore an alternative success criterion is required. To compare qualitative performance using ERT it is necessary that all compared algorithms meet the success criterion at least a few times. For the optimization performance assessment under tight budget evaluation restrictions, we define the success criterion as reaching a target value which corresponds to the expected value of the best objective function value obtained from uniform random sampling with the given function evaluation budget (1000 samples). For a test function f_k we will refer to the expected objective value as $E_{f_k}^{RSE}$. The estimation of $E_{f_k}^{RSE}$ is based on the same number of repetitions as is used to measure the performance of the other algorithms (100 in this case). We will refer to the ERT w.r.t. this objective function value limit as Random Sampling Equivalent-Expected Run Time (ERT_{RSE}).

Besides the fixed target performance measure ERT, a further (more intuitive) way to compare the performance of the optimization algorithms on the test functions is by means of diagrams on which the Best Function Value (BFV) of the objective functions, is plotted against the number of algorithm iterations or Function evaluations. For the diagrams in the results section, the BFV has been normalized (BNFV) with $E_{f_k}^{RSE}$, for the corresponding test function. The advantage of these diagrams is that they give an intuitive picture of performance for both, fixed-cost, and fixed-target scenarios.

Except for the ERT reference value, and an increased number of repetitions, the experimental set up of the study followed the benchmark description in [30]. To assess the significance of the overall performance test results, the non-parametric Wilcoxon signed rank test [36] was applied pairwise between the results of the algorithms, with the best algorithm as the reference (see also [35]). The null hypothesis of this test is: a zero difference of the median between two results sets. The conventional significance threshold of 0.05 is used to indicate that the null hypothesis cannot be rejected with sufficient certainty.

3.3 Results Comparison

All test problems in this comparison are minimization problems. Good optimization algorithm performance is thus related to reaching a low BNFV in few function evaluations. For the 50-dimensional problem set, BNFV diagrams comparing the algorithm performance for the 9 algorithms are displayed in Fig. 1. The SQG-DE algorithm ranked as the best algorithm in terms of BNFV performance after 1000 function evaluations in 16 out of the 25 test problems and was the winner in all of the test problems of the 4th

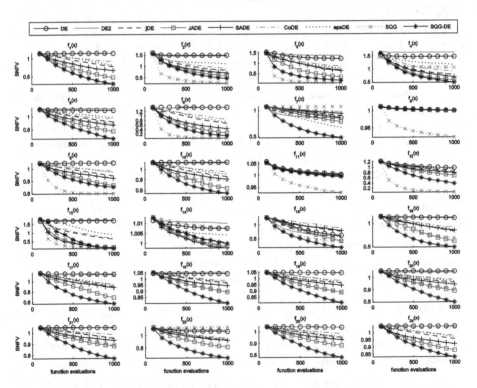

Fig. 1. Evolution of the best normalized function value (BNFV) for increasing function evaluations, functions 1–24 (D = 50).

category (Multi-modal Hybrid composition functions). The results for the 30-dimensional problem set were similar, but not reported in a figure due to space constraints.

An overview of algorithm performance in terms of the ERT_{RSE} for all 30 and 50 dimensional problems is given in Table 2. An ERT_{RSE}-value of for example 300 means that the corresponding algorithms requires 300 function evaluations to obtain a function evaluation better than the threshold, (which was defined as the expected best objective value for 1000 uniform random samples in the problem domain). Table 2 shows that SQG-DE achieves the best ERT performance in 30 out of the 50 test problems. The new hybrid algorithm performed also with respect to the ERT measure as the best in all of the test problems of the 4th category.

Table 2. Algorithm performance in terms of ERT_{RSE}, for all test functions

Test functions	Dimension 30									Dimension 50								
	DE	DE2	jDE	JADE	SADE	Code	epsDE	SQG	SQG-DE	DE	DE2	jDE	JADE	SADE	Code	epsDE	SQG	SQG-DE
1 Unimodal																		
$f_1(x)$	11657	24049	491	264	329	904	606	202	**168**	19030	9054	439	205	254	674	579	192	**178**
$f_2(x)$	7401	11567	739	309	379	894	3653	**137**	318	9055	8145	572	316	388	908	2135	**127**	296
$f_3(x)$	6536	32763	606	356	519	1148	1045	**135**	210	3819	10317	514	289	455	764	566	**121**	145
$f_4(x)$	6182	13326	570	**285**	342	662	1712	694	310	19059	10175	538	341	434	920	2931	1332	**310**
$f_5(x)$	4342	4766	536	311	527	871	564	459	**160**	9173	19040	568	294	366	964	1418	451	**214**
2 Multimodal Basic																		
$f_6(x)$	7404	11728	491	234	315	807	603	**124**	168	11557	13327	491	235	244	831	609	**132**	182
$f_7(x)$	256	124	482	310	596	879	169	32365	**107**	394	187	443	289	536	884	231	24041	**109**
$f_8(x)$	2877	3011	1783	2084	2673	2720	2082	**132**	2771	1732	2347	2098	1792	2724	2522	2740	**117**	1960
$f_9(x)$	15735	24058	467	269	380	818	726	**126**	193	15719	19060	482	245	295	845	738	**123**	194
$f_{10}(x)$	11658	24054	474	256	320	850	630	201	**172**	24071	24059	456	235	306	739	661	**169**	188
$f_{11}(x)$	2414	1555	1967	1502	1321	1541	1979	**368**	1949	2024	2568	2187	1324	1814	1825	1675	**347**	1331
$f_{12}(x)$	1072	934	509	342	416	931	594	**128**	225	1183	1263	560	340	418	1132	774	**119**	224
3 Multimodal Expanded																		
$f_{13}(x)$	15709	10150	524	224	231	723	898	**161**	289	7381	9057	514	227	193	604	1139	**145**	336
$f_{14}(x)$	7497	19174	2735	1735	**1509**	2366	6379	6899	1658	7476	15824	1792	1310	**1198**	1296	3482	4706	1808
4 Multimodal Hybrid																		
$f_{15}(x)$	437	373	668	366	565	857	458	1016	**185**	341	291	557	354	556	852	331	205	**173**
$f_{16}(x)$	2999	11671	490	335	615	689	616	219	**179**	7506	11568	495	320	487	738	593	260	**156**
$f_{17}(x)$	6925	11670	514	362	572	845	693	9096	**178**	11585	32369	640	327	515	794	809	5743	**192**
$f_{18}(x)$	1017	1036	501	314	460	741	341	263	**143**	10264	19135	631	319	380	1071	892	352	**226**
$f_{19}(x)$	1271	1045	533	317	551	836	389	259	**148**	24351	7387	511	290	389	866	728	307	**210**
$f_{20}(x)$	1190	1098	491	315	545	1039	384	284	**154**	19307	13407	533	321	372	915	794	380	**227**
$f_{21}(x)$	9466	24251	498	253	327	901	597	301	**175**	13571	9089	486	243	320	871	605	325	**176**
$f_{22}(x)$	2433	2889	489	281	389	730	602	13320	**202**	3311	5068	564	291	431	755	724	19048	**153**
$f_{23}(x)$	13650	24194	474	254	325	802	586	664	**181**	7468	7544	483	248	333	880	590	537	**170**
$f_{24}(x)$	7785	4449	544	272	376	911	573	10183	**187**	15709	13335	498	231	245	784	688	8170	**201**
$f_{25}(x)$	220	115	445	304	554	855	167	5321	**109**	304	163	488	315	552	1038	210	13368	**109**

Table 3 provides a summary of algorithm performance in terms of ERT_{RSE}, divided by test function category, and averaged overall performance. The results in Table 3 indicate that: SQG-DE had the best average performance over all test problems, with an ERT of approximately 10% less w.r.t. the second overall best algorithm JADE. More remarkable is that in the 4th category with the hardest test functions, SQG-DE obtained ERT scores which are about 40% better than the second-best algorithm (JADE). The Wilcoxon signed rank test indicated that the all the results were statistically significant, except for the small test problem group 3. Closer inspection revealed that this was exclusively caused by test function 14, for which all of the investigated algorithms performed worse than random sampling, which indicates that f_{target} was rarely reached.

Table 3. Overview of algorithm performance in terms of ERT_{RSE}, by test function category

Test function groups	Dimension	DE	DE2	jDE	JADE	SADE	Code	epsDE	SQG	SQG-DE	p 1st. rank
1 Unimodal	D=30	7223	17294	588	305	419	896	1516	325	**233**	p<0.01
f_{1-5}	D=50	12027	11346	526	289	379	846	1526	444	**228**	p<0.01
2 Multimodal Basic	D=30	5917	9352	882	714	860	1221	969	4778	798	p<0.01
f_{6-12}	D=50	8097	8973	960	637	905	1254	1061	3578	**598**	p<0.01
3 Multimodal Expanded	D=30	11603	14662	1630	980	870	1544	3638	3530	974	p=0.197
f_{13-14}	D=50	7429	12440	1153	768	696	950	2311	2426	1072	p=0.014
4 Multimodal Hybrid	D=30	4309	7526	513	307	480	837	491	3721	**167**	p<0.01
f_{15-25}	D=50	10338	10851	535	296	416	869	633	4427	**181**	p<0.01
Overal performance	D=30	5925	10562	721	474	605	1013	1082	3322	**422**	p<0.01
f_{1-25}	D=50	9816	10551	702	428	568	979	1066	3233	**379**	p<0.01

The results from this benchmark indicate that for hard high dimensional multimodal, problems under a tight function evaluation budget the new hybrid algorithm performs significantly better than the original DE, SQG, and the state-of-the-art DE variants tested.

4   Discussion and Outlook

The performance comparison results show efficiency gains of SQG-DE ranging up to 40%, w.r.t. the next best algorithm in the category of Multimodal Hybrid Composition functions. Overall the performance benefits of SQG-DE w.r.t. the "parent" algorithms (SQG and DE2 "best/2/bin") indicates a useful synergy effect, which already could be exploited to solve complex budget constrained optimization problems, in its present state.

The remarkable results also call for further activities and investigations, such as: further performance comparisons against optimization algorithms other than DE; implementing the SQG-mutation operator in other DE variants; control parameter tuning; implementation of suitable self-adaptive parameters strategies; and hybridization of SQG with other population-based meta-heuristic algorithms such as ES and PSO.

In the present study the conventional control parameters settings, according to the recommendations in the respective literature were used for the optimization algorithms.

The best choice for the control parameters, is however both problem and budget dependent. Budget dependent control parameter tuning or optimization, for a particular test function or set of test functions is possible. For computationally expensive industrial optimization problems such parameter tuning is however several orders of magnitude more expensive than an optimization run, such that direct control parameter optimization on real-world problems is often infeasible. Although it is possible to optimize or tune the algorithms settings on conventional synthetic test or benchmark problems, it is important in the context of industrially relevant problems to know or estimate the algorithm performance correlations between the synthetic test problems, and a given real-world problem. The industrial relevance of detailed comparative studies including new algorithms, operators or tuned control parameters, are relative to the quantifiability of performance correlations with real-world problems, or by the gained theoretical insights. We are however obliged to note that performance on most of the conventional synthetic benchmark functions (including those used for this study) is difficult to relate (or quantitatively correlate) to performance on particular real-world optimization problems, which is thus a strong limiting factor for direct practical relevance. Also the theoretical insights and generalizability of the results are limited by the lack of systematic relations among the conventional test functions. These strong limitations apply to the presented study, as well as to most of the work in the literature which is based on conventional synthetic test problems and benchmark sets.

In order to obtain systematic results that could lead to insights of theoretical value, and improved optimization performance in real-world problems, most of the earlier mentioned plans for further investigations on the SQG-mutation operator will be performed using test functions with parameterized function characteristics such as presented in [37], benchmarks based on engineering design optimization applications, and new synthetic test approaches such as representative surrogate problems [12]. Important open questions are: How are the performance of SQG-DE and other meta-heuristic optimization algorithms related to particular problem characteristics? How do the control parameters interact with problem characteristics in terms of algorithm performance? Further investigations and insights are required to address these questions.

5   Conclusions

A new SQG inspired mutation operator is introduced in the framework of DE, resulting in a new hybrid algorithm "SQG-DE". The algorithm is compared with conventional DE and several state-of-the-art DE variants, w.r.t. optimization performance under strict function evaluation budget constraints. The results of the comparison indicate that the new algorithm excels the other compared algorithms on average by 10% in overall performance, on the investigated benchmark problems. The new algorithm performs particularly well on high dimensional multi-modal composite test problems (of the 4th test problem category), where w.r.t. fixed target performance measures, averaged function evaluation savings of about 40% are achieved. The results are promising, and the displayed optimization efficiency could be of relevance for Industrial real-world problem settings, which involve a strict function evaluation budget. The described mutation operator is computationally inexpensive, easy to implement and could therefore also be used in other population-based meta-heuristic optimization approaches. On request, an implementation of the SQG-DE algorithm is available from the authors.

Acknowledgments. This work was partially funded by the GRESIMO project grant agreement no. 290050 by the European community 7th Framework program. We would like to thank the anonymous reviewers for their remarks to improve the manuscript. Furthermore, we like to express our gratitude to Qingfu Zhang and all other cited authors who made code of their algorithms and test benches publicly available on their websites.

References

1. Floudas, C.A., Gounaris, C.E.: A review of recent advances in global optimization. J. Global Optim. **45**(1), 3–38 (2009)
2. Rios, L.M., Sahinidis, N.V.: Derivative-free optimization: a review of algorithms and comparison of software implementations. J. Global Optim. **56**(3), 1247–1293 (2013)
3. Sobieszczanski-Sobieski, J., Haftka, R.T.: Multidisciplinary aerospace design optimization: survey of recent developments. Struct. Optim. **14**(1), 1–23 (1997)
4. Venkataraman, S., Haftka, R.T.: Structural optimization complexity: what has Moore's law done for us? Struct. Multidisc. Optim. **28**(6), 375–387 (2004)
5. Schaffer, C.: A conservation law for generalization performance. In: Proceedings of the 11th International Conference on Machine Learning, pp. 259–265 (1994)
6. Wolpert, D.H., Macready, W.G.: No free lunch theorems for optimization. IEEE Trans. Evol. Comput. **1**(1), 67–82 (1997)
7. Aissa, M.H., Verstraete, T., Vuik, C.: Aerodynamic optimization of supersonic compressor cascade using differential evolution on GPU. In: Simos, T., Tsitouras, C. (eds.) AIP Conference Proceedings, p. 480077 (2016)
8. Carrigan, T.J., Dennis, B.H., Han, Z.X., Wang, B.P.: Aerodynamic shape optimization of a vertical-axis wind turbine using differential evolution. ISRN Renew. Energy **2012**, 1–16 (2012)
9. Kiani, M., Yildiz, A.R.: A comparative study of non-traditional methods for vehicle crashworthiness and NVH optimization. Arch. Comput. Meth. Eng. **23**(4), 723–734 (2016)
10. Duddeck, F.: Multidisciplinary optimization of car bodies. Struct. Multidisc. Optim. **35**(4), 375–389 (2008)
11. Sala, R., Pierini, M., Baldanzini, N.: Optimization efficiency in multidisciplinary vehicle design including NVH criteria. In: Proceedings of the 26th International Conference on Noise and Vibration Engineering, ISMA, pp. 1571–1585 (2014)
12. Sala, R., Baldanzini, N., Pierini, M.: Representative surrogate problems as test functions for expensive simulators in multidisciplinary design optimization of vehicle structures. Struct. Multidisc. Optim. **54**(3), 449–468 (2016)
13. Haftka, R.T., Watson, L.T.: Multidisciplinary design optimization with quasiseparable subsystems. Optim. Eng. **6**(1), 9–20 (2005)
14. Jansen, T., Zarges, C.: Fixed budget computations: a different perspective on run time analysis. In: Proceedings of the 14th Annual Conference on Genetic and Evolutionary Computation, pp. 1325–1332. ACM (2012)
15. Storn, R., Price, K.: Differential evolution - a simple and efficient adaptive scheme for global optimization over continuous spaces, Technical report TR-95-012, ICSI (1995)

16. Storn, R., Price, K.: Differential evolution–a simple and efficient heuristic for global optimization over continuous spaces. J. Global Optim. **11**(4), 341–359 (1997)
17. Brest, J., Greiner, S., Boskovic, B., Mernik, M., Zumer, V.: Self-adapting control parameters in differential evolution: a comparative study on numerical benchmark problems. IEEE Trans. Evol. Comput. **10**(6), 646–657 (2006)
18. Qin, A.K., Huang, V.L., Suganthan, P.N.: Differential evolution algorithm with strategy adaptation for global numerical optimization. IEEE Trans. Evol. Comput. **13**(2), 398–417 (2009)
19. Mallipeddi, R., Suganthan, P.N., Pan, Q.K., Tasgetiren, M.F.: Differential evolution algorithm with ensemble of parameters and mutation strategies. Appl. Soft Comput. **11**(2), 1679–1696 (2011)
20. Zhang, J., Sanderson, A.C.: JADE: adaptive differential evolution with optional external archive. IEEE Trans. Evol. Comput. **13**(5), 945–958 (2009)
21. Wang, Y., Cai, Z., Zhang, Q.: Differential evolution with composite trial vector generation strategies and control parameters. IEEE Trans. Evol. Comput. **15**(1), 55–66 (2011)
22. Das, S., Suganthan, P.N.: Differential evolution: a survey of the state-of-the-art. IEEE Trans. Evol. Comput. **15**(1), 4–31 (2011)
23. Das, S., Mullick, S.S., Suganthan, P.N.: Recent advances in differential evolution–an updated survey. Swarm Evol. Comput. **27**, 1–30 (2016)
24. Knowles, J., Corne, D., Reynolds, A.: Noisy multiobjective optimization on a budget of 250 evaluations. In: Ehrgott, M., Fonseca, C.M., Gandibleux, X., Hao, J.-K., Sevaux, M. (eds.) EMO 2009. LNCS, vol. 5467, pp. 36–50. Springer, Heidelberg (2009). https://doi.org/10.1007/978-3-642-01020-0_8
25. Ermoliev, Y.M.: Methods of solution of nonlinear extremal problems. Cybernetics **2**(4), 1–14 (1966)
26. Ermoliev, Y.M.: Stochastic quasigradient methods and their application to system optimization. Stochast. Int. J. Probab. Stochast. Process. **9**(1–2), 1–36 (1983)
27. Wang, G.G., Shan, S.: Review of metamodeling techniques in support of engineering design optimization. J. Mech. Des. **129**(4), 370–380 (2007)
28. Krityakierne, T., Ginsbourger, D.: Global optimization with sparse and local Gaussian process models. In: Pardalos, P., Pavone, M., Farinella, G.M., Cutello, V. (eds.) MOD 2015. LNCS, vol. 9432, pp. 185–196. Springer, Cham (2015). https://doi.org/10.1007/978-3-319-27926-8_16
29. Kennedy, J., Eberhart, R.: Particle swarm optimization. In: Proceedings of the IEEE International Conference on Neural Networks, vol. 4, pp. 1942–1948. IEEE (1995)
30. Suganthan, P.N., Hansen, N., Liang, J.J., Deb, K., Chen, Y.P., Auger, A., Tiwari, S.: Problem definitions and evaluation criteria for the CEC 2005 special session on real-parameter optimization. KanGAL report, 2005005 (2005)
31. Yao, X., Liu, Y., Lin, G.: Evolutionary programming made faster. IEEE Trans. Evol. Comput. **3**(2), 82–102 (1999)
32. Brest, J., Zamuda, A., Bošković, B., Greiner, S., Žumer, V.: An analysis of the control parameters' adaptation in DE. In: Chakraborty, U.K. (ed.) Advances in Differential Evolution, vol. 143, pp. 89–110. Springer, Heidelberg (2008)
33. Auger, A., Hansen, N.: Performance evaluation of an advanced local search evolutionary algorithm. In: Proceedings of the IEEE Congress on Evolutionary Computation, vol. 2, pp. 1777–1784. IEEE (2005)
34. Qingfu Zhang's Homepage. http://dces.essex.ac.uk/staff/qzhang/code/codealgorithm/. Accessed 5 Apr 2017

35. García, S., Molina, D., Lozano, M., Herrera, F.: A study on the use of non-parametric tests for analyzing the evolutionary algorithms' behaviour: a case study on the CEC'2005 special session on real parameter optimization. J. Heuristics **15**(6), 617–644 (2009)
36. Wilcoxon, F.: Individual comparisons by ranking methods. Biometrics Bull. **1**(6), 80–83 (1945)
37. Sala, R., Baldanzini, N., Pierini, M.: Global optimization test problems based on random field composition. Optim. Lett. **11**(4), 699–713 (2017)

Age and Gender Classification of Tweets Using Convolutional Neural Networks

Roy Khristopher Bayot[(✉)] and Teresa Gonçalves

LISP - Laboratório de Informática, Sistemas e Paralelismo,
Universidade de Évora, Évora, Portugal
d11668@alunos.uevora.pt, tcg@uevora.pt

Abstract. Determining age and gender from a series of texts is useful for areas such as business intelligence and digital forensics. We explore the use of convolutional neural networks together with word2vec word embeddings for this task in comparison to handcrafted features. The network constructed consists of five layers and is trained using adadelta. It starts with an embedding layer where a word is represented by a vector, followed by a convolutional layer composed of three filters, each with 100 feature maps. It is followed by a max-over-time pooling layer which is done on each map and the resulting features are concatenated before a dropout layer and a softmax layer. The network was trained to classify age and gender for English and Spanish tweets. The predictions per tweet were aggregated using the majority prediction as the final prediction for the user who gave the tweets. The results outperform previous experiments. The highest English age and gender classification accuracy obtained are 49.6% and 72.1% respectively. The highest Spanish age and gender classification accuracy obtained on the other hand are 56.0% and 69.3% respectively.

Keywords: Author profiling · Twitter · Word vectors · Word2vec
Convolutional neural networks

1 Introduction

Social media has grown rapidly in the recent years, especially with the advent of sites like Facebook, Instagram, Twitter, and Snapchat. With it comes new communication models. Logging worker tasks and productivity for instance could be done through Yammer while Slack or Telegram is used for team communications. These communication models still face problems such as fake profiles or to a lesser extent, incomplete information about the person writing the content. Authorship analysis then becomes a way to possibly deal with such a problem. One facet of authorship analysis is author profiling wherein a person's traits are determined using the text they created.

Our work tries to solve author profiling for age and gender in English and Spanish with twitter text using convolutional neural networks. It follows the work of Kim [8] with some minor modifications. Aside from using convolutional

© Springer International Publishing AG 2018
G. Nicosia et al. (Eds.): MOD 2017, LNCS 10710, pp. 337–348, 2018.
https://doi.org/10.1007/978-3-319-72926-8_28

neural networks with word vectors, the work also aims to observe the effect of the size of the vector dimensions to the accuracy. It also aims to observe if fine-tuning pre-trained vectors would improve the result. Finally, the work compares the current accuracy results to the results from previous experiments.

Although the current work is focused on age and gender, it could possible be extended to other traits. It could also be applied to business intelligence for targeted advertising or product reviews and analysis wherein brands can suggest products to people who gave incomplete profiles. The application could also be geared towards profiling users who circulate certain kinds of content such as fake news.

The paper is organized as follows. Section 2 covers related literature where it initially discusses previous author profiling endeavors, then followed by methods in PAN, followed by an explanation on word2vec, the uses of convolutional neural networks, and finally a short summary on the previous experiment to which this work was compared. Section 3 describes the methodology, beginning with the creation of word2vec vectors, to the dataset, to details of the convolutional neural network architecture, the different variations, and then how it was evaluated. Section 4 gives the results and discussion while Sect. 5 gives the conclusion and recommendations.

2   Related Literature

In previous author profiling research, most of the work is centered on handcrafted features as well as content-based and style-based ones. For instance, in the work of Argamon et al. in [2] where texts were categorized based on gender, age, native language, and personality, different content-based features and style-based features were used. Content-based features used were the 1000 words that appear frequently in the corpus which has the highest information gain to differentiate between classes. The style-based features included the nodes of a taxonomic tree made from systemic functional linguistics [7] with each value giving the frequency of the node's occurence normalized by the number of words in the text. Another example is the work of Schler et al. in [25] where writing styles in blogs are related to age and gender. Stylistic and content features were extracted from 71,000 different blogs and a Multi-Class Real Winnow was used to learn the models to classify the blogs. Stylistic features included parts-of-speech tags, function words, hyperlinks, and non-dictionary words; content features included word unigrams with high information gain.

2.1   PAN Editions

One particular initiative dealing with author profiling is PAN. In the first edition of PAN [20] in 2013, the task was age and gender profiling for English and Spanish blogs. In PAN 2014 [20], the task was profiling authors with text from four different sources - social media, twitter, blogs, and hotel reviews. In PAN 2015 [21], the task was limited to tweets but expanded to different tasks with age and gender classification and a personality dimension. The languages include English, Spanish, Italian, and Dutch. There were 5 different personality dimensions - extroversion, stability, agreeableness, conscientiousness, and openness. The more recent edition, PAN 2016 [22] deals with cross-genre evaluation where classifiers are trained on English, Spanish, and Dutch tweets while tested on other genres such as blogs, reviews, and other forms of social media.

Most of the methods include extracting content-based features such as bag of words, named entities, dictionary words, slang words, contractions, sentiment words, and emotion words; another would be stylistic features such as frequencies, punctuations, POS, HTML use, readability measures, and other various statistics. There are also features that are n-grams based, IR-based, and collocations-based; named entities, sentiment words, emotion words, contractions and words with character flooding were also considered. Some variations would be that of Maharjan et al. [13], where n-grams were used with stopwords, punctuations, and emoticons, and idf count was also used before placed into a classifier. In [29], different features were used that were related to length (number of characters, words, sentences), information retrieval (cosine similarity, okapi BM25), and readability (Flesch-Kincaid readability, correctness, style). Another approach was to use term vector model representation as in [28]. On the other hand, Marquardt et al. in [14], used a combination of content-based features (MRC, LIWC, sentiments) and stylistic features (readability, html tags, spelling and grammatical error, emoticons, total number of posts, number of capitalized letters number of capitalized words). Classifiers also varied for this edition; there was the use of logistic regression, multinomial Naïve Bayes, liblinear, random forests, Support Vector Machines, and decision tables.

In most of the editions, the work of Lopez-Monroy et al. in [11] provided a framework that works best for most tasks in most editions. They placed second for both English and Spanish in 2013 where they used second order representation based on relationships between documents and profiles. The work of Meina et al. [15] used collocations and placed first for English while the work of Santosh et al. in [24] worked well with Spanish using POS features in the same year. In the following year, the work of Lopez-Monroy et al. in [12] which uses the same method as the previous year [11] gave the best result with an average accuracy of 28.95% on all corpus-types and languages.

In 2015, the work of Alvarez-Carmona et al. [1] gave the best results on English, Spanish, and Dutch; their work used second order profiles as in the previous years as well as LSA. On the other hand, the work of Gonzales-Gallardo et al. [6] gave the best result for Italian; this used stylistic features represented by character n-grams and POS n-grams.

2.2 Word2vec

The overall theme is that hand-crafted features are extracted from the text and used into a classifier to predict. However there is a more recent trend where the system learns suitable filters at run time and uses the learned filters to generate a feature representation suitable for classification. This approach begins with learning word embeddings that captures semantic information between words that could be later leveraged. One of the more prominent embeddings is word2vec by Mikolov in [16,17]. Essentially, words from a dictionary of a given corpus are initially represented with a vector of random numbers. A word's vector representation is learned by predicting it through its adjacent words; the basis for the words order is in a large corpus. Obtaining the vector can be done in two different ways - skip grams and continuous bag of words (CBOW). In CBOW, the word vector is predicted given the context of adjacent words; in skip grams, the context words are predicted given a word. The word vectors are then updated after all the predictions and will result in vectors that are not just random but have some semantic relation to each other.

2.3 Previous System

We used word embeddings in age and gender classification on the same dataset in our previous paper [3]. We experimented with using word2vec and support vector machines, where one training example is the average of the word vectors taken from all the tweets made by one user. We compared accuracy results between *tfidf* against 100 dimension word2vec vectors trained on continuous bag of words. Word2vec performs better than the usual *tfidf* for the given task. Additional experiments also showed that vectors which used the skip-grams method performed better than that which used continuous bag of words.

2.4 Convolutional Neural Networks

Averaging the vectors still seemed somewhat crude since all the vectors are given the same weight and filters are not learned to see which feature is necessary. We then look into neural network architectures that uses word vectors to find suitable filters for classification tasks. One such architecture is that of LeCun in [9]. The original paper works on images however it was adapted to work on text. For instance, convolutional neural networks were used for semantic parsing in the work of Yih et al. in [31] while Shen et al. used it for search query retrieval in [26]. The work done in this paper however closely follows that of Kim [8]. He used word vectors together with convolutional neural network on multiple benchmarks: movie reviews with one sentence per review in [19], Stanford Sentiment Treebank (neutral reviews removed and only binary labels), subjectivity dataset in [18], TREC question dataset in [10], customer reviews in [10], and opinion polarity detection of the MPQA dataset in [30]. This approach was able to improve the state of the art on five out of seven benchmarks - everything except TREC question dataset and the subjectivity datasets. The details of the network he used is described in Sect. 3.4.

3 Methodology

The Fig. 1 given below shows an overview description of the system from how the dataset is manipulated before fed into the convolutional neural network and how it is evaluated. The details are described in the following subsections.

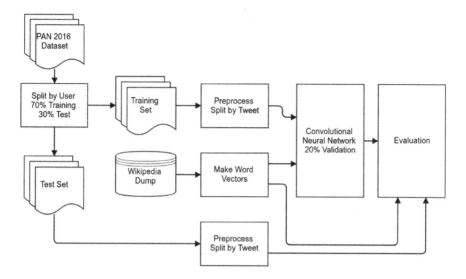

Fig. 1. Illustration of the flow of the data from the split, to preprocessing, to feeding to the network, and to the evaluation.

3.1 Pre-trained Vectors

Before training begins, word embeddings need to be created. The wikipedia dump from February 05, 2016 was used. The English wikipedia dump at that time was 11.8 Gb compressed with bzip. The Spanish dump on the other hand had 2.2 Gb compressed. This dump was then extracted and transformed such that everything was turned into lowercase and entries are in one file. This was then used as input to the word2vec implementation of gensim [23] to generate our own vectors. Regarding word2vec parameters, no lemmatization was done, and 5 was the window size used. We also used skip grams instead of continuous bag of words and finally, the output dimensions were 100 and 300.

3.2 Dataset

After creating the vectors, the dataset is processed. The dataset comes from PAN 2016 Author Profiling task [22]. It is composed of tweets from English, Spanish, and Dutch with profiling elements of age and gender. The categories for age classification are 18–24, 25–34, 35–49, 50–64, and 65 and above. Dutch does not have age information so we will not be using it. The Table 1 show information about the dataset. The dataset was then split with 70% to be used for training while the remaining 30% was held out for testing.

3.3 Preprocessing

All XMLs files from each user are read for both the training and test set. The tweets taken from each user are extracted to form one training example. The

Table 1. Age and gender distribution for the number of users in the PAN 2016 dataset.

		English	Spanish
Gender	Male	218	125
	Female	218	125
Age	18–24	28	16
	25–34	139	64
	35–49	182	126
	50–64	80	38
	65–xx	6	6

Table 2. Basic statistics for PAN 2016 dataset

	English		Spanish	
	Train	Test	Train	Test
Total number of users	306	130	175	75
Total number of tweets	194953	82839	151362	57258
Total number of nans removed	1963	1371	1986	55
Max number of tokens in tweet	69	70	98	47
Min number of tokens in tweet	1	1	1	1
Average tweet length	13.15	13.15	13.84	13.92
Standard deviation	6.21	6.32	6.25	6.19
Mode	11	10	18	16

examples are transformed by putting them all in lower case. No stop words are removed. Hash tags, numbers, mentions, shares, and retweets were not processed or transformed to anything else. They were retained as is and will correspond to another item in the dictionary of words. The test set will be set aside for the final evaluation while the training set will be used to train the network.

3.4   Model

The model architecture is shown in Fig. 2. This is similar to the architecture of Kim [8] which is a variant of the architecture given by Collobert et al. [5] and is implemented in Keras [4] with a Theano [27] backend ran on an NVIDIA Tesla K20c GPU.

All words in the training set are turned into number indices that corresponds to a word vector. Each training example will be represented by a sequence of numbers. The sequence length will vary. The total number of indices in the sequence is held at 59 and padding is done to ensure it. We then feed the sequence into the system. Each number will be looked up in the embedding layer and converted to a word vector according to the pre-trained word vectors previously discussed. The whole sequence will then form a matrix. Feature maps are then

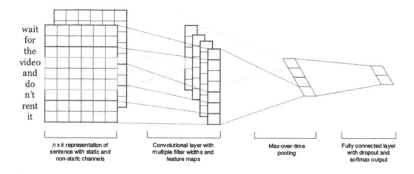

Fig. 2. Kim's architecture.

created by convolving filters to the matrix and using a non-linearity after the convolution. There are three filter windows for this experiment - 3, 4, and 5. Each filter window has 100 feature maps. Note that the coefficients of the filters are initially random and then updated while training. The non-linearity used in our experiments is the *tanh* function. After making a feature map, max over time pooling is performed. This means that from each feature map, only the maximum is recorded. Therefore, there will be a total of 300 features after max over time pooling is done. Then a dropout layer is added. Our dropout probability is 0.5. We finally add a softmax layer as the final layer with the weight vectors constrained to an l_2-norm.

Training is done through stochastic gradient descent over shuffled minibatches with the Adadelta update rule where each mini-batch is made of 3000 examples. The dev set is comprised of 20% of the training set. We also kept the number of epochs to 30 and to provide for early stopping.

3.5   Model Variations

We experimented on two aspects. The first is dimension varying between 100 and 300 both from skip-grams. The second is the difference between fine-tuning or not towards the accuracy. *CNN-static*, indicates that word vectors were taken from pre-trained word2vec but kept static. *CNN-non-static* is the same as the first but it was tuned while training.

3.6   Evaluation

We set aside 30% of the dataset for final evaluation. After the training is done, we apply the model on the tweets we set aside. After getting a prediction, we group the tweets that belong to the same user and get the majority prediction from all the tweets gathered for that user. We used the majority prediction as a final prediction for the user and base our accuracy off of that.

4   Results and Discussion

To recap, we did experiments on two different languages (English and Spanish), on two different classification tasks (age and gender), using two different vector dimensions (100 and 300), and two different treatments for the vectors (static and non-static). This gives a total of 16 experiments. Table 3 show the CNN accuracy results for English and Spanish. We can observe different patterns with these accuracy results.

We first look at the difference between accuracy over tweets and accuracy over users. The accuracy over tweets mean that each tweet is regarded as an example for which we evaluate the accuracy. Accuracy over users means that the tweets are aggregated first by the user who sent the tweet and uses the majority prediction as the final prediction for the user. The accuracy is evaluated on the user. We can see that using the majority predicted class for a tweet to predict the user generally improves the result except for Spanish gender evaluation.

Looking at the accuracy per user, we observe the effects of dimensionality as well as the effect of treating vectors static or non-static. Increasing the dimensions gives different effects. In three cases, it diminishes the accuracy. In two cases, the result is the same. And in the final three cases, the accuracy improves. However, the magnitudes in the times the accuracy improved is bigger than the magnitude of the times when the accuracy decreased. The effect also varies when treating the vectors as static or non-static. We have 4 cases where the accuracy increases, 3 cases where it decreases, and 1 case where it does not change. However, the magnitudes of increased accuracy are much more significant than those of the decreased ones. This is possibly because the vectors get finely tuned with more training data. There is an increase of 1.5% and 7.7% for English gender classification using 100 and 300 dimensions respectively. There's also a 1.3% increase for Spanish age classification using 100 dimensions while using 300 dimensions did not yield any difference. There's also an increase of 8.0% for Spanish gender classification using 300 dimensions. The other instances however lower the result when tuning the vectors. The biggest decrease is 4.0% which comes from gender classification using 100 dimensions.

We look at Table 4 which compares the accuracy from the best settings for convolutional neural networks that we were able to obtain against the accuracy from the best settings from previous experiments. In the previous work [3], we have results comparing the accuracy of an SVM classifier trained on *tfidf* against another SVM classifier trained on average of word vectors. In addition to what was done in the previous paper, we also experimented with skip-grams as well as varying the dimensions from 100 to 300 for skip-grams. The tasks and the datasets are the same and the test set was also the same for the previous experiment and the experiment detailed on this paper. We found that using skip-grams with 300 dimensions got a better result for the previous system. We then adapted the previous system to do a majority vote on tweets as a prediction for the user so that it would be comparable to the current system. We can see a better accuracy from the convolutional neural networks after comparing the two

Table 3. Accuracy comparison between evaluation by tweets and evaluation by user

	English				Spanish			
	Age		Gender		Age		Gender	
	Static	Non-static	Static	Non-static	Static	Non-static	Static	Non-static
100								
Tweet	0.407	0.397	0.618	0.623	0.541	0.481	0.561	0.551
User	0.481	0.473	0.651	0.667	0.547	0.560	0.557	0.550
300								
Tweet	0.410	0.409	0.626	0.613	0.538	0.467	0.693	0.653
User	0.496	0.473	0.643	0.721	0.547	0.547	0.507	0.587

Table 4. Comparison between the best results of the CNN and previous work

	English		Spanish	
	Age	Gender	Age	Gender
CNN	**0.473**	**0.721**	0.547	**0.587**
Past-work	0.415	0.600	**0.560**	0.533

Table 5. Comparison of accuracy results based on different number of epochs for training

	English		Spanish	
	Age	Gender	Age	Gender
Dim = 100				
Epoch = 30	0.481	0.651	0.547	0.557
Epoch = 200	0.496	0.636	0.547	0.640
Dim = 300				
Epoch = 30	0.496	0.643	0.547	0.507
Epoch = 200	0.481	0.643	0.560	0.560

systems. The only time the previous system fares better is with English gender classification and only by 0.013.

Another aspect to look at is that the number of epochs for training. Looking at the learning rates from previous results, learning does not seem to plateau since the early stopping callback did not take into effect. We then ran the same experiment of age and gender classification on English and Spanish with 100 and 300 vector dimensions but only for static. The difference this time is that training was ran on 200 epochs to see if the accuracy would improve. The comparison is given in Table 5. We can see that aside from Spanish gender classification, the accuracy improvements are marginal or none at all.

5   Conclusion and Recommendations

To summarize, we were able to use word vectors in conjunction to a convolutional neural network using Kim's architecture [8] as basis. We observed that using a bigger word vector dimension for English tasks improves the result but the same is not true for Spanish tasks. We observed that the effect also varies when tuning or not-tuning vectors. It generally improves the performance when dimensions are increased. There is a general tendency to have a better accuracy result for users instead tweets. And finally, we are able to report a better accuracy score for all four tasks as compared to previous results.

However this work has a lot of hyperparameters that were either fixed based on Kim's architecture or decided based on previous experiments. Some of it might be sub-optimal. For instance, we used $f(x) = tanh(x)$ as our activation function instead of $f(x) = relu(x)$ according to Kim's architecture. Another thing could be the number of feature maps. Sequence size is also an important parameter that was overlooked. This work was left to 59 based on Kim's work but the highest token count was 98. Another main concern is the use of vectors trained on wikipedia instead of twitter. Preprocessing also does not account for the fact that hyperlinks and twitter mentions be queried as a separate vector. This could give important information. Other things that could be experimented more are the number of layers, padding, the dropout, and even the regularization that was at the final layers. Other architectures such as LSTM and Bidirectional LSTM could also be used for further study and comparison. These are possible things to do for future work.

Acknowledgments. The authors would like to thank FCT, Fundação de Ciências e Tecnologia under LISP research center (UID/CEC/4668/2016) for partially supporting this research.

References

1. Álvarez-Carmona, M.A., López-Monroy, A.P., Montes-y Gómez, M., Villaseñor-Pineda, L., Jair-Escalante, H.: INAOE's participation at PAN 2015: author profiling task. In: Working Notes Papers of the CLEF 2015 Evaluation Labs (2015)
2. Argamon, S., Koppel, M., Pennebaker, J.W., Schler, J.: Automatically profiling the author of an anonymous text. Commun. ACM **52**(2), 119–123 (2009)
3. Bayot, R., Gonçalves, T.: Author profiling using SVMs and word embedding averages-notebook for PAN at CLEF 2016. In: Balog et al. [22]
4. Chollet, F.: keras, (2015). https://github.com/fchollet/keras
5. Collobert, R., Weston, J., Bottou, L., Karlen, M., Kavukcuoglu, K., Kuksa, P.: Natural language processing (almost) from scratch. J. Mach. Learn. Res. **12**(Aug), 2493–2537 (2011)
6. González-Gallardo, C.E., Montes, A., Sierra, G., Antonio Núñez-Juárez, J., Salinas-López, A.J., Ek, J.: Tweets classification using corpus dependent tags, character and POS N-grams. In: Proceedings of CLEF (2015)
7. Halliday, M., Matthiessen, C.M., Matthiessen, C.: An Introduction to Functional Grammar. Routledge, Abingdon (2014)

8. Kim, Y.: Convolutional neural networks for sentence classification. arXiv preprint arXiv:1408.5882 (2014)
9. LeCun, Y., Bottou, L., Bengio, Y., Haffner, P.: Gradient-based learning applied to document recognition. Proc. IEEE **86**(11), 2278–2324 (1998)
10. Li, X., Roth, D.: Learning question classifiers. In: Proceedings of the 19th International Conference on Computational Linguistics-Volume 1, pp. 1–7. Association for Computational Linguistics (2002)
11. Lopez-Monroy, A.P., Montes-y Gomez, M., Escalante, H.J., Villasenor-Pineda, L., Villatoro-Tello, E.: Inaoe's participation at PAN 2013: author profiling task. In: CLEF 2013 Evaluation Labs and Workshop (2013)
12. López-Monroy, A.P., Montes-y Gómez, M., Escalante, H.J., Villaseñor Pineda, L.: Using intra-profile information for author profiling. In: CLEF (Working Notes), pp. 1116–1120 (2014)
13. Maharjan, S., Shrestha, P., Solorio, T.: A simple approach to author profiling in mapreduce. In: CLEF (Working Notes), pp. 1121–1128 (2014)
14. Marquardt, J., Farnadi, G., Vasudevan, G., Moens, M.-F., Davalos, S., Teredesai, A., De Cock, M.: Age and gender identification in social media. In: Proceedings of CLEF 2014 Evaluation Labs (2014)
15. Meina, M., Brodzinska, K., Celmer, B., Czoków, M., Patera, M., Pezacki, J., Wilk, M.: Ensemble-based classification for author profiling using various features. Notebook Papers of CLEF (2013)
16. Mikolov, T., Chen, K., Corrado, G., Dean, J.: Efficient estimation of word representations in vector space. arXiv preprint arXiv:1301.3781 (2013)
17. Mikolov, T., Sutskever, I., Chen, K., Corrado, G.S., Dean, J.: Distributed representations of words and phrases and their compositionality. In: Advances in Neural Information Processing Systems, pp. 3111–3119 (2013)
18. Pang, B., Lee, L.: A sentimental education: sentiment analysis using subjectivity summarization based on minimum cuts. In: Proceedings of the 42nd annual meeting on Association for Computational Linguistics, p. 271. Association for Computational Linguistics (2004)
19. Pang, B., Lee, L.: Seeing stars: exploiting class relationships for sentiment categorization with respect to rating scales. In: Proceedings of the 43rd Annual Meeting on Association for Computational Linguistics, pp. 115–124. Association for Computational Linguistics (2005)
20. Rangel, F., Rosso, P., Koppel, M.M., Stamatatos, E., Inches, G.: Overview of the author profiling task at pan 2013. In: CLEF Conference on Multilingual and Multimodal Information Access Evaluation, pp. 352–365. CELCT (2013)
21. Rangel, F., Rosso, P., Potthast, M., Stein, B., Daelemans, W.: Overview of the 3nd author profiling task at pan: In: Cappellato, L., Ferro, N., Gareth, J., San Juan, E. (eds) CLEF 2015 Labs and Workshops, Notebook Papers, vol. 1391 (2015)
22. Rangel, F., Rosso, P., Verhoeven, B., Daelemans, W., Pottast, M., Stein, B.: Overview of the 4th author profiling task at PAN 2016. In: Balog, K., Cappellato, L., Ferro, N., Macdonald, C. (eds) Working Notes Papers of the CLEF 2015 Evaluation Labs. CEUR Workshop Proceedings, vol. 1609, pp. 750–784. CLEF and CEUR-WS.org, September 2016
23. Řehůřek, R., Sojka, P.: Software framework for topic modelling with large corpora. In: Proceedings of the LREC 2010 Workshop on New Challenges for NLP Frameworks, Valletta, Malta, pp. 45–50. ELRA, May 2010. http://is.muni.cz/publication/884893/en
24. Santosh, K., Bansal, R., Shekhar, M., Varma, V.: Author profiling: predicting age and gender from blogs. In: Notebook Papers of CLEF (2013)

25. Schler, J., Koppel, M., Argamon, S., Pennebaker, J.W.: Effects of age and gender on blogging. In: AAAI Spring Symposium: Computational Approaches to Analyzing Weblogs, vol. 6, pp. 199–205 (2006)
26. Shen, Y., He, X., Gao, J., Deng, L., Mesnil, G.: Learning semantic representations using convolutional neural networks for web search. In: Proceedings of the 23rd International Conference on World Wide Web, pp. 373–374. ACM (2014)
27. Theano Development Team. Theano: A Python framework for fast computation of mathematical expressions. arXiv e-prints, abs/1605.02688, May 2016
28. Villena Román, J., González Cristóbal, J.C.: Daedalus at pan 2014: guessing tweet author's gender and age (2014)
29. Weren, E.R.D., Moreira, V.P., de Oliveira, J.P.M.: Exploring information retrieval features for author profiling. In: CLEF (Working Notes), pp. 1164–1171 (2014)
30. Wiebe, J., Wilson, T., Cardie, C.: Annotating expressions of opinions and emotions in language. Lang. Resour. Eval. **39**(2–3), 165–210 (2005)
31. Yih, W.-T., He, X., Meek, C.: Semantic parsing for single-relation question answering. In: ACL, vol. 2, pp. 643–648. Citeseer (2014)

Approximate Dynamic Programming with Combined Policy Functions for Solving Multi-stage Nurse Rostering Problem

Peng Shi[✉] and Dario Landa-Silva

ASAP Research Group, School of Computer Science,
The University of Nottingham, Nottingham, UK
{peng.shi,dario.landasilva}@nottingham.ac.uk

Abstract. An approximate dynamic programming that incorporates a combined policy, value function approximation and lookahead policy, is proposed. The algorithm is validated by applying it to solve a set of instances of the nurse rostering problem tackled as a multi-stage problem. In each stage of the problem, a weekly roster is constructed taking into consideration historical information about the nurse rosters in the previous week and assuming the future demand for the following weeks as unknown. The proposed method consists of three phases. First, a pre-process phase generates a set of valid shift patterns. Next, a local phase solves the weekly optimization problem using value function approximation policy. Finally, the global phase uses lookahead policy to evaluate the weekly rosters within a lookahead period. Experiments are conducted using instances from the Second International Nurse Rostering Competition and results indicate that the method is able to solve large instances of the problem which was not possible with a previous version of approximate dynamic programming.

Keywords: Dynamic programming · Approximation function
Policy function · Nurse scheduling problem

1 Introduction

This paper investigates the ability of approximate dynamic programming using a combined policy function to tackle a multi-stage nurse rostering problem. Approximate dynamic programming (ADP) is designed to tackle the Markov Decision Process that dynamic programming is unable to solve in practice [1]. ADP aims to learn the selection of the optimal policy for mapping the state space into the action space. The purpose of policies in ADP is to determine decisions. The technique presented here is a hybrid approach that combines the lookahead policy and the value function approximation policy. The *lookahead policy* makes decisions now by explicitly optimizing over some time horizon by combining some approximation of future information while the *value function*

© Springer International Publishing AG 2018
G. Nicosia et al. (Eds.): MOD 2017, LNCS 10710, pp. 349–361, 2018.
https://doi.org/10.1007/978-3-319-72926-8_29

approximation policy refers to an approximation of the value of being in a future state as a result of a decision made now [2].

The Nurse Rostering Problem (NRP) is an NP-Hard problem that consists in constructing rosters for a number of nurses over a time horizon of typically no more than a few weeks. Constructing a roster involves assigning shifts types of each nurse for each day in order to fulfill daily duty requirements plus satisfying a number of soft and hard constraints [3]. In this paper, the NRP is tackled as a multi-stage optimisation problem is used to test the proposed technique because it is a widely investigated problem and presents an interesting challenge to ADP. Tackling the NRP as a multi-stage problem was proposed by [4].

Solving the NRP with dynamic programming is impractical due to the curse of dimensionality [2,5]. Our previous work investigated ADP to solve NRP, where a value function approximation based method was proposed to tackle various instances of the NRP [6]. However, the computation time required for constructing solution samples and the memory space required for recording rewards increased exponentially for larger problem instances. Hence, that shortfall has motivated the present work. A number of ADP practical issues related to the complexity of the environment, in particular when dealing with large state or action space, are reported in the literature [5]. The technique proposed in this paper enhances the ability of ADP to solve NRP as a multi-stage problem by combining two policy functions, value function approximation to solve the weekly problem, and lookahead policy to evaluate weekly rosters with artificially constructed future demand within a given lookahead period.

The contribution of this paper is an enhanced approximate dynamic programming approach that takes advantage of tackling the NRP in multiple stages and is able to tackle instances of this problem with longer planning horizons. The rest of paper is structured as follows. Section 2 describes NRP used in this investigation and its modelling as a Markov Decision Process. Section 3 explains the details of the proposed algorithm. Section 4 presents the experimental results. Section 5 concludes the paper and outlines future work.

2   The Multi-stage Nurse Rostering Problem

In the multi-stage nurse rostering problem the planning horizon is seen as multiple non-overlapping stages, nurse rosters should be selected one stage at a time. A stage is a part of the planning period for which the demands are completely known at its start [7]. In this paper, the Second International Nurse Rostering Competition (INRC-II) instances are used for experimentation. In these instances, each stage is a week under the competition setting. This section outlines the problem and its modelling as a Markov Decision Process proposed in a previous paper [6].

2.1   Problem Description

An instance in the INRC-II consists of three data parts, *global information, week requirement* and *history data*. The global constraints, listed below, are those that are the same for each stage of the problem and those that are applicable to the last stage only.

H1 A nurse can be assigned at most one working shift per day.

H3 Two consecutive shifts of a nurse must follow a legal shift type successor, for example a late shift could not be followed by a early shift.

H4 A shift of a given skill must be fulfilled by a nurse having that skill.

S5 Each nurse is required to either work or rest on both days of weekends.

S6 For the whole planning period, each nurse has a minimum and maximum total number of working assignments.

S7 For the whole planning period, each nurse works a maximum number of weekends.

Week requirement is a list of specific hard or soft constraints in each week:

H2 For each day, shift or skill combination, the assigned number of nurses must cover the minimum requirement.

S1 The number of nurses for each shift with each skill must be equal to the optimal requirement.

S2 Maximum and minimum number of consecutive assignment per shift or day.

S3 Maximum and minimum number of consecutive days off.

S4 Respect to the specific shift requirement for each nurse.

History data is a summary of the actual roster for the previous stage which is required when tackling the problem. If the first week is the current solving stage, history data is randomly selected from built-in artificial files [4]. History data for each stage must be produced by solvers before processing to the next stage and it should include the following information for each individual roster:

- the last assignment of previous week.
- consecutive assignments of the same type as last day.
- total number of worked shifts.
- total number of worked weekends.

In the above list of constraints, H indicates hard constraints that must be satisfied by a solution to be considered feasible and S indicates soft constraints that incur a penalty if violated.

2.2 Problem Modification

Given that in each stage the future demand in this multi-stage NRP is considered as unknown, we apply the framework by Powell [2] which considers the exogenous information. The Markov Decision Process (MDP) notation is summarized as $\{S, A, W, Tr(S, A, W)\}$.

S is a **state** variable, split as pre-decision state and post-decision state. The pre-decision state is the start point and the post-decision state is a termination for each stage. For each stage t in the NRP, the pre-decision state variable corresponds to the combination of weekly schedules from stages 1 to $t-1$, and S is the empty set for the first stage. The post-decision state is the combination of weekly schedules including the one for the current stage t.

A is an **action** variable which determines the policy selected in the current stage. In the NRP, A is a weekly roster where each nurse is assigned a combination of integer variables indicating the shift type for each day. The feasibility of a solution is controlled by the selection of decisions.

W is defined as **exogenous information** which is available only within each stage t. In the NRP, W represents the weekly requirements (local constraints) described above.

The **transition function** $Tr(S, A, W)$ transfers a pre-decision state to the post-decision state with the decision A and the exogenous information W. In the NRP considered here, the transition function performs two roles, one is to update the solution with weekly roster A and week data W and the other one is to update the nurse historical information based on the value of A and W.

3   Proposed Algorithm

The structure of the proposed algorithm is exhibited in Fig. 1 and consists of three parts. First, the pre-process phase sets up the search space. Then, the local phase is an enhancement of our previous work [6] for solving the weekly

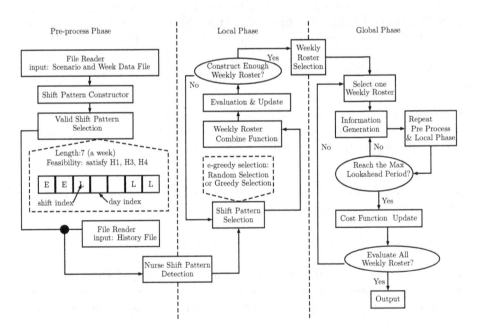

Fig. 1. Overview of the proposed algorithm applying ADP with combined policy functions

optimization problems. Finally, the global phase applies a lookahead policy for future demand evaluation. Each of these parts is explained below.

3.1   Pre-process Phase

If a shift pattern (SP) is defined as a weekly roster of a nurse, then a solution should be described as the combination of nurses' shift patterns. A solution is feasible if and only if each constructed SP satisfies all the hard constraints. Exploring infeasible solutions is not required in the principle-of-optimality approaches [2]. Instead, evaluating feasible-shift-pattern based solutions has the potential to make the search more efficient. With this purpose, the pre-process phase is designed to construct a reduced search space for the subsequent local and global phases.

The pseudocode of this pre-process phase is shown in Algorithm 1. Hard constraints selected to filter shift patterns belong to *global information* (Sect. 2.1) which each individual nurse roster is expected to obey. The set that contains all feasible shift patterns is defined as feasible set. Lines 2–6 are the selection steps, where *sp* indicates a single shift pattern and *vsp* represents the feasible set.

Once the feasible set is prepared, some shift patterns are not available to specific nurses with the consideration of nurse history data. For example, if the last assignment of a nurse in history data is a late shift, then any pattern starting with an early shift in the feasible set becomes infeasible for this nurse (2.1 **H3**). Lines 7–13 represent the specific shift pattern selection procedure of each nurse with the consideration of related history data.

Algorithm 1. Pre-process Phase

1: $vsp \leftarrow null$;
2: **repeat**
3:     $sp \leftarrow ShiftPatternConstructor()$;
4:     **if** sp satisfy hard constraints **then**
5:         add sp to vsp;
6: **until** no more action from constructor
7: **for** Each Nurse n **do**
8:     $ivsp \leftarrow null$;
9:     Collect the last assigned shift type x_{last};
10:     **for** each $sp \in vsp$ **do**
11:         Select x_1 from sp;
12:         **if** $\{x_{last}, x_1\}$ satisfy hard constraint **then**
13:             add sp to $ivsp$

3.2   Local Phase - Value Function Approximation

Given the output of the pre-process phase, the weekly nurse rostering optimization problem can be seen as selecting a proper shift pattern for each nurse, so as to satisfy constraint **H2** and minimize the soft constraints violation cost. The input to this phase are the *ivsp* for each nurse. Since the future demand is assumed not known in each particular week, the local optimal weekly roster is not guaranteed to be the one incorporated into the overall solution. Therefore, the output of this local phase is a selection of weekly rosters as depicted in Fig. 1.

$$Q(S, a) = r(S, a) + \gamma max_{a'} Q(\delta(S, a), a') \tag{1}$$

The Q-learning function, presented in Eq. (1), is applied to tackle the local phase problem. The aim is to update the value of S when changes are made by the selected a. In this multi-stage nurse rostering problem, S is a weekly roster and a is a list of selected shift patterns for nurses. Shift patterns are selected based on two methods. *Random Selection* is applied if S is not fully constructed or sample size of S is small. Shift patterns of unassigned nurses in the roster will be randomly selected. This selection is replaced by *Greedy Selection* after constructing a number of S. For a fully constructed S, shift pattern of one or a list of nurses is updated by the one with minimum cost, or equally described as highest reward, from previous steps. $r(S, a)$ is the reward function and calculated from two aspects, the overall constraint violation update and times of the selected a. The pseudocode of this local phase is shown in Algorithm 2.

Algorithm 2. Local Phase - Value Function Approximation

1: Initial value of max_iter, ϵ
2: $i \leftarrow 0$, $M \leftarrow Empty$ $S_{List} \leftarrow Empty$;
3: **while** $i < max\_iter$ **do**
4:     $Sol \leftarrow Empty$
5:     **for** Each Nurse n **do**
6:         $rnd \leftarrow RandomNumberGenerator()$
7:         **if** $rnd < \epsilon$ **then**
8:             $sp \leftarrow RandomSelection(ivsp)$
9:         **else**
10:            $sp \leftarrow GreedySelection(ivsp)$
11:        $Insert(Sol, sp)$
12:        $c = CostFunction(sp)$
13:        $UpdateValue(V(Sol), c)$
14:     $Add(S_{List}, Sol)$
15:     $e = ExpectedFunction(S_{List})$
16:     $\gamma = Parameter(V(Sol), e)$
17:     $UpdateValue(V(Sol), \gamma \times e)$
18:     $Update(\epsilon)$
19:     $i \leftarrow i + 1$
20: $WeeklyRosterSelection(S_{List}, M)$

A sample here is a weekly roster which is constructed by selecting shift patterns from each nurse. The shift pattern selection function in lines 6–10 uses

RandomSelection or *GreedySelection* which selects the shift pattern with minimum cost. This is known as ϵ-greedy selection function [2]. This shift pattern selection function ensures that the local phase constructs a weekly rosters set with a degree of variety and not only concentrating on the local optimum. The selected shift pattern *sp* is added to *Sol* in line 11. In line 12, *CostFunction* calculates the shift pattern cost (c) according to the violation of soft constraints **S1-S5** and then the value of this weekly roster is updated in line 13.

In line 14, the fully constructed weekly roster *Sol* is stored in the sample list S_{List}. Lines 15–18 correspond to the *Evaluation & Update* in Fig. 1. The purpose of the expected function is to indicate the average value of the constructed weekly roster while γ is an importance factor and its value is adjusted in the opposite direction to the value of the constructed weekly roster. For instance, if the cost value of a particular weekly roster is larger than the expected value, the value of γ is set to a smaller value, and vice verse.

The end of this local phase in line 20 results in the output set M which is a subset of S_{List}, i.e. a set of weekly rosters some with small constraint violation cost (due to the greedy selection) and others with possibly large cost (due to the random selection). This set M is the input to the global phase described in the following subsection.

3.3   Global Phase - Lookahead Policy

In the local phase, the weekly rosters are evaluated for the weekly constraints only, i.e. from **H1** to **H4** and from **S1** to **S5**. However, since in each week the future demand is unknown, the global constraints **S6** and **S7** are not considered. Then, this global phase evaluates the weekly rosters with artificial future demand through a lookahead period. The lookahead policy seeks to construct a potential solution within a lookahead period based on the weekly roster and artificial future demand in order to evaluate the solution for the global constraints. The input to this global phase is the set of weekly rosters M from the local phase. The output is one weekly roster only as the final solution to the weekly optimization problem. The pseudocode of the global phase is shown in Algorithm 3 which is applied to each weekly roster in M. The method *Information Generation* will be explained in Sect. 4, here we assume all the artificial future demand is obtained in advance.

$LK(S)$ is the lookahead value for each weekly roster S and calculated using Eq. 2. n is the nurse index. *stage* is the week index. T is the lookahead period. sp_n is a single shift pattern of nurse n in the weekly roster S. x_{nt} is a shift pattern at the lookahead stage t of nurse n. x_{nt} belongs to the valid shift pattern set VSP_{nt}.

$$LK(S) = \sum_{n=1}^{N} \sum_{t=stage}^{T+stage} min_n V(sp_n, x_{nt}) \tag{2}$$

This global phase incorporates the pre-process and local phases described above. For each nurse n, the valid shift pattern set VSP_{nt} is constructed in line 7 based on the current shift pattern sp_n and the artificial weekly demand at

---

Algorithm 3. Global Phase
_____

 1: Initial value of $LK(S)$;
 2: **for** Each Nurse n **do**
 3:     select sp_n from S
 4:     Initial $ideal\_sol = Insert(sp_n, \phi)$
 5:     $V(ideal\_sol) = V(sp_n)$
 6:     **for** $t \leftarrow stage \quad to \quad T + stage$ **do**
 7:         $VSP_{nt} \leftarrow Pre - processPhase$
 8:         $x_{nt} \leftarrow GreedySelection(VSP)$
 9:         $c \leftarrow CostFunction(x_{nt})$
10:         $UpdateValue(V(ideal\_sol), c)$
11:         $Insert(ideal\_sol, x_{nt})$

12:     $c \leftarrow CostFunction(ideal\_sol)$
13:     $UpdateValue(V(ideal\_sol), c)$
14:     $UpdateValue(LK(S), V(ideal\_sol))$
_____

lookahead period t. We select the shift pattern x_{nt} in VSP_{nt} with the lowest cost and build up an ideal individual assignment with the combination of sp_n and x_{nt} in lines 8 and 11. The initial value of this ideal solution is the same value of sp_n and is updated with the constraint violation cost of x_{nt} in line 10. Lines 7–11 are repeated until reaching the last lookahead stage $T + stage$. The value of $ideal\_sol$ is then added the constraint violation of **S6** and **S7**. This is the evaluation of a single shift pattern sp_n and this value is added to $LK(S)$ for each nurse n.

Once all the weekly rosters are evaluated through the Algorithm 3, the one with lowest $LK(S)$ will be selected as the final weekly solution and the nurse historical information is updated for the following week.

4   Experimental Design and Results Analysis

The problem instances for evaluating the proposed approach are selected from the Second International Nurse Rostering Competition (INRC-II) [4]. The are three sets of instances, all available at [8]. One is a test set with small number (up to 21) of nurses. Another is the competition set released to the competitors. The last set is a hidden set that was made available at the end of the competition. For the experiments here we use the first two data set only.

The proposed algorithm described in Sect. 3 was implemented in Java (JDK 1.7) and all computations were performed on an Intel (R) Core (TM) i7 CPU with 3.2 GHz and RAM 6 GB.

4.1   Experimental Settings

For a problem that considers 3 working shifts and 1 day off per day of the week, the total number of possible shift patterns is 16384 (4^7). The pre-process phase reduces this number to 1607 making possible to apply the proposed approach to solve large NRP instances. There are three different representations in the value function approximation, lookup table, parametric model and non-parametric model. As the search space is considerably small after the pre-process phase, we implement a lookup table in the local phase procedure.

The initial value of ϵ is set to 0.9 and is updated based on the Generalized Harmonic Step Size Function [2]. Through preliminary experimentation we tuned the size of the simulation sample $S_{List} = 100$ and the output set $M = 30$ in the local phase. Also through preliminary experiments and results analysis, we decided to select elements from S_{List} for M following the 1-6-3 rule. That is, 10% is selected from S_List with the lowest $V(S)$, the 90% of S_List is split into two subgroups, good and bad, based on the constraint violation cost. Then 60% is randomly selected from the good subgroup and 30% is randomly selected from the bad subgroup.

The cost value for both single shift pattern sp and weekly roster S is calculated using Eq. (3) where c_s is the soft constraint violation cost and V_{sc} is the number of violation for each constraint. The calculation of the constraint violation is fully described in [4].

$$c = \sum_{eachconstraint} c_s \times V_{sc} \tag{3}$$

The artificial future demand is generated by randomly selecting a week data file per week in the lookahead period. Back to the algorithm described in the Sect. 3, only one future path is evaluated for each weekly roster. Less evaluations of lookahead policy is not ideal but more evaluations consume much computation time and memory. By preliminary experiments we found that 1000 evaluations is the minimum to achieve the level of performance in our results while still using considerably short computation time. The value of $LK(S)$ is updated based on Eq. (4). All experimental results presented in the rest of this section correspond to 20 runs for each problem instance.

$$LK(S) = \frac{1}{k} \sum_{i=1}^{k} LK_i(S) \tag{4}$$

4.2   Lookahead Period Comparison

We tested various lookahead periods for each planning horizon. The lookahead period T for scenarios with 4 weeks is set as 1, 2 and 3 and as 3, 5 and 7 for scenarios with 8 weeks. All the scenarios from the test set were used for these experiments comparing the different values of T and results are presented in Table 1.

In the table, Obj is the average objective value and $Std.$ is the standard deviation. The performance of using longer lookahead period is not much better than when using a shorter one for the smallest problem instance (n005w4). But for the larger problem, either with longer planning horizon or larger number of nurses, the average objective value when using that largest T is the best,

Table 1. The average objective value and standard deviation obtained with various lookahead periods for each instance. Best values are indicated in bold.

	T = 1		T = 2		T = 3	
Instance	Obj.	Std.	Obj.	Std.	Obj.	Std.
n005w4_1	456	55.724	452	49.67	**451.5**	23.573
n005w4_2	436.5	35.6735	**430.5**	31.578	**430.5**	14.568
n005w4_3	541	67.456	530.5	54.674	**530**	33.584
n021w4_1	2176	435.754	2056.5	343.563	**1815**	185.683
n021w4_2	3059.5	563.743	2375.5	484.626	**2150**	254.673
n021w4_3	3415	447.784	2767.5	306.639	**2035**	186.460
	T = 3		T = 5		T = 7	
Instance	Obj.	Std.	Obj.	Std.	Obj.	Std.
n012w8_1	1527.5	435.375	1375.5	368.466	**1237.5**	235.256
n012w8_2	1747	373.692	1623.5	275.573	**1544**	205.574
n012w8_3	1928.5	563.681	1736.5	503.684	**1515.5**	385.678

as much as 20% improvement is achieved in instance $n021w4$. The standard deviation value is smaller as the value of T increases indicating that the algorithm performance is more robust with longer lookahead period.

4.3 Algorithm Validation and Comparison

Based on the observations from the experiments with the test set, the lookahead period was set to $T = 3$ for 4-week instances and to $T = 7$ for 8-week instances on experiments with the competition data set. Results are presented in Table 2.

A value of *99999* in the table indicates that the approach ran out of memory. The performance of the proposed ADP-CP is evaluated through two aspects for each instance. In the left part of the Table 2 we compare it with each individual policy. The solution constructed by individual simulation approach is a combination of optimal weekly rosters. The global constraints are considered only when solving the weekly optimization problem in the last stage. On the other hand, the individual lookahead policy focuses on the solution evaluation of global constraints but each weekly solution is solved with random selection approaches. Looking further has the benefit on the overall solution by comparing the value in columns 2 and 4. Local optimum is only concentrated on the assignment patterns, such as the consecutive working patterns and the consecutive days off. We select the instance $n030w4\_1$ as an example. The number of working shifts for each nurse is set as 4 to avoid local constraint violations. The total working days for each nurse is 16 in the final solution. However in some contract, the minimum total working days is 20. A significant large global constraint violation cost is added to the final objective value. A good weekly roster also improves the optimality of lookahead policy with the comparison of columns 3 and 4.

Table 2. Experimental results of the proposed ADP with combined policy (ADP-CP), individual simulation approach, individual lookahead policy, best and the worst results from the competition. Best values are indicated in bold.

Instance	Simulation	Lookahead	ADP-CP	Best	Worst
n030w4_1	1925	2725	**1780**	1745	9850
n030w4_2	2650	2710	**1610**	1935	10605
n030w8_1	5350	6645	**4830**	2295	21185
n030w8_2	6310	5820	**4855**	1900	21145
n040w4_1	8120	3945	**3270**	1765	14680
n040w4_2	6895	4260	**3735**	1910	14460
n040w8_1	14720	10125	**9305**	3105	35010
n040w8_2	19255	10165	**8975**	2975	33000
n050w4_1	5900	4070	**3535**	1525	17745
n050w4_2	6210	4070	**3030**	1480	15380
n050w8_1	19525	10045	**8965**	5560	43040
n050w8_2	13905	9725	**8420**	5475	42765
n060w4_1	18480	16977	**12282**	2830	19230
n060w4_2	20945	17794	**15019**	2950	20400
n060w8_1	20215	**9590**	9720	2840	44130
n060w8_2	17545	11000	**10160**	3200	44430
n080w4_1	23195	21870	**18350**	3474	26935
n080w4_2	26305	21435	**16885**	3535	27210
n080w8_1	48505	44880	**35975**	4845	64915
n080w8_2	47355	44065	**38800**	5105	66515
n100w4_1	19625	19295	**16045**	1445	33740
n100w4_2	20530	20270	**17885**	2070	33465
n100w8_1	53155	39550	**35690**	3095	85260
n100w8_2	50340	40755	**35440**	3135	87445
n120w4_1	99999	24075	**22960**	2470	36235
n120w4_2	99999	22680	**22065**	2530	36320
n120w8_1	99999	43215	**39170**	3555	83590
n120w8_2	99999	**40840**	41350	3435	82145

The right part of Table 2 seeks to validate our ADP-CP approach by comparing the quality of the solutions obtained to the *Best* and *Worst* reported for the competition. The performance of ADP-CP is close to the best in the instance *n030w4*. It also achieved a good gap from the best in instances *n040w4* and *n050w4*. However, the performance is not so close to the best solutions for larger problem instances. Nevertheless, the quality of the solutions produced with the proposed ADP-CP is in the middle among all the competition results which were produced by several different algorithms. We believe that this work has accomplished good progress in making possible the application of dynamic programming, with the approximation policies, for solving this complicated multi-stage nurse rostering problem. This is an important step towards making dynamic programming practical in its application for solving difficult combinatorial optimization problems when a multi-stage solving approach can be followed.

5   Conclusion

This paper proposed a three-phase approximate dynamic programming (ADP) algorithm to solve the multi-stage nurse rostering problem. This is a problem where a roster is constructed for each week with the future demand assumed not known and the history information for the previous week needs to be considered. The first phase of the proposed approach is a pre-process that generates a set of valid shift patterns. The second phase is a local phase that applies the value function approximation, to solve the weekly optimization problem and generate a set of weekly rosters. The third phase is a global phase that implements a lookahead policy to evaluate the effect of the future uncertainty within a lookahead period. The proposed ADP then combines value function approximation and lookahead policy. The instances from the Second Nurse Rostering Competition (INRC-II) are used in the experiments to validate the performance of this proposed algorithm. Experimental results show that the combined policy approach in the proposed algorithm produces better performance than the individual policies. Besides, the results obtained with the proposed algorithm on some of the INRC-II problem instances are close to the best solutions reported for the competition. Future works should be focused on improving the solution quality and reducing the computational time. These improvements could be achieved by applying different methods to evaluate the lookahead samples. Furthermore, improving the quality of weekly rosters could also benefit the lookahead policy as arguably better weekly rosters could help to achieve better results with shorter lookahead periods and also reduce the computation time.

References

1. Puterman, M.L.: Markov decision processes: discrete stochastic dynamic programming. Wiley, New York (2014)
2. Powell, W.B.: Approximate Dynamic Programming: Solving the curses of dimensionality, vol. 703. Wiley, Hoboken (2007)
3. Burke, E.K., de Causmaecker, P., Berghe, G.V., van Landeghem, H.: The state of the art of nurse rostering. J. Sched. **7**(6), 441–499 (2004)
4. Ceschia, S., Dang, N.T.T., de Causmaecker, P., Haspeslagh, S., Schaerf, A.: Second international nurse rostering competition (INRC-II)—problem description and rules—. arXiv preprint arXiv:1501.04177 (2015)

5. Tesauro, G.: Practical issues in temporal difference learning. In: Sutton, R.S. (ed.) Reinforcement Learning. The Springer International Series in Engineering and Computer Science (Knowledge Representation, Learning and Expert Systems), vol. 173, pp. 33–53. Springer, Boston (1992). https://doi.org/10.1007/978-1-4615-3618-5_3

6. Shi, P., Landa-Silva, D.: Dynamic programming with approximation function for nurse scheduling. In: Pardalos, P.M., Conca, P., Giuffrida, G., Nicosia, G. (eds.) MOD 2016. LNCS, vol. 10122, pp. 269–280. Springer, Cham (2016). https://doi.org/10.1007/978-3-319-51469-7_23

7. Dang, N.T.T., Ceschia, S., Schaerf, A., de Causmaecker, P., Haspeslagh, S.: Solving the multi-stage nurse rostering problem. In: Proceedings of the 11th International Conference of the Practice and Theory of Automated Timetabling, pp. 473–475 (2016)

8. INRC-II the second nurse rostering competition. http://mobiz.vives.be/inrc2/. Accessed 23 May 2016

A Data Mining Tool for Water Uses Classification Based on Multiple Classifier Systems

Iván Darío López[⊠] ⓘ, Cristian Heidelberg Valencia ⓘ,
and Juan Carlos Corrales ⓘ

Grupo de Ingeniería Telemática (GIT),
Universidad del Cauca, Popayán, Colombia
{navis, chpayan, jcorral}@unicauca.edu.co

Abstract. Water is not only vital for ecosystems, wildlife, and human consumption, but also for activities such as agriculture, agro-industry, and fishing, among others. However, in the same way as their water use has increased, it has also been detected an accelerated deterioration of its quality. In this sense, to have predictive knowledge about water quality conditions, can provide a significant relevance to many socio-economic sectors. In this paper, we present an approach to predict the water quality for different uses (aquaculture, irrigation, and human consumption) discovering knowledge from several datasets of American and Andean Watersheds. This proposal is based on Multiple Classifier Systems (MCS), including Bagging, Stacking, and Random Forest. Models as Naïve Bayes, KNN, C4.5, and Multilayer Perceptron are combined to increase the accuracy of the classification task. The experimental results obtained show that Random Forest and Stacking expose acceptable precision on different water-use datasets. However, Bagging with C4.5 was the most appropriate architecture for the problem addressed. These results indicate that MCS techniques can be used for improving accuracy and generalization capacity of the prediction tools used by stakeholder involvement in the water quality process.

Keywords: Classification · Machine learning · Multiple classifier systems
Water quality

1 Introduction

Water Quality (WQ) can be defined as the set of physical, chemical, biological, and radiological characteristics of surface and underground waterbodies. However, experts in water resources management, show that there is no single definition of water quality because it strictly depends on its use [1]. Thus, for example, water that cannot be used for human consumption may be used for other activities such as irrigation or aquaculture, among others, as it has specific characteristics that make it suitable for such use [2]. Some of this water is used by infrastructure installed by human and most of the extracted water is then returned to the environment after it has been used.

By implementing plans for water management, Watershed Management Authorities (WMA) collect samples from different water body points to determine the current status

© Springer International Publishing AG 2018
G. Nicosia et al. (Eds.): MOD 2017, LNCS 10710, pp. 362–375, 2018.
https://doi.org/10.1007/978-3-319-72926-8_30

of water quality. Traditionally, this task involves time, both sampling and construction of WQ data series, and in most cases, the management response is given too late. This, occasionally could represent an economic and hugely environmental lost, which affect territorial sustainability on a regional or global scale.

In Integrated Water Resources Management (IWRM), there is a dynamic relationship between governmental institutions and basin stakeholders, which must work together to ensure the viability of their decisions in order to achieve the sustainable development objectives. In this sense, determine the potential use of water in a watershed, allows formulating management alternatives associated to the land use for agricultural or agro-industrial activities developed there. Additionally, the planning processes and IWRM can be facilitated. Thus, measures or decisions to reduce the risk of contamination of water's surface and groundwater, can be supported.

In view of the above, it is necessary to have models or mechanisms to anticipate the materialization of pollution risk in a short-term, preventing negative effects that impact water resources quality. In this way, several researches in Machine Learning (ML), and specifically Multiple Classifier Systems (MCS) [3] has introduced algorithms and predictive techniques, which have the ability to classify different conditions of water, based on the analysis of collected historical data. In [4], a classification into quality classes based on either bio-indicator or physical and chemical indicator data is developed. This process requires multiple input data previously prepared at a special facility; in this sense, in-situ classification can be complex or even unfeasible. Bassiliades [5] presents an intelligent system for monitoring and predicting water quality parameters of Northern Greece such as temperature, pH, among others. This system does not perform water use classification although it does in-situ predictions of water quality parameters.

On the other hand, Muhammad [6] proposes a classification model for water quality. The performance of various classification models and algorithms, was analyzed and compared in order to identify the significant features that contributed in classifying water quality of Kinta River, Perak Malaysia. Finally, Partalas [7] studies a large ensemble (200 models) of Artificial Neural Networks (ANN) and Support Vector Machines (SVM) for predicting WQ. However, it does not perform any water use classification, focusing only on selecting the best model to predict WQ parameters measured by the system implemented in [5].

According to the previous literature review, it is important to develop models that combine different classifiers, primarily to cover the disadvantages of some individual techniques and improve them with the addition of other algorithms. In all reviewed studies, classification uses data which had not been immediately measured. One of the objectives of this proposal is to develop a classifier that can determine the possible water use base on in-situ parameter measurements, focusing the laboratory test only in those needed to corroborate the use of water classification. Moreover, no studies were found, which employ MCS to determine possible uses of water, and it has not been applied to water samples from Andean watersheds. This proposal is important because it would allow government authorities to take decisions about different productive sectors, which use water like an input for all their processes. In this way, these organisms would have a greater ability to forecast, in an integral and systemic capacity associated with the reality of ecosystems, land use, and human activities which makes it more flexible and useful.

This proposal presents various architectures for combining classifiers, with the aim of establishing the most accurate and improving the prediction accuracy to determine which type of use, a water sample belongs. This paper is organized as follows: the first part outlined the related studies to the main topics addressed around water quality prediction using MCS; subsequently, data sources for testing are described, MCS architectures are presented, explained and validated by experimental evaluation results. And the final part, presents the conclusions and future work of this research.

2   Data and Study Area

In the present study, different WQ databases were pre-processed to determine the possible use of water in a watershed. This approach was conducted using several classifiers and MCS architectures for combining classification models. In order to implement a predictive technique that offers accurate results using different WQ datasets; it is important to have several data sources which represent different uses of water. Three free access datasets from two data sources were employed as training data (Alviso estuary, Don Pedro Lake, and Juanchito station) and two datasets as test data (Piedras river watershed and Illinois river). Data sources are as follows:

2.1   United States Geological Survey (USGS)

USGS is a research organization of the United States federal government. This agency provides reliable water quality data in the United States for public access [8]. In this study, data with label (A) was selected. These datasets belong to two sampling sites in the state of California: the first one comprising the territory of Alviso; Guadalupe River and Coyote Creek end up in Alviso wetland through an estuary that flows into the San Francisco bay. One of the main activities is fish farming, the second one is Don Pedro Lake, located in Mariposa County, which covers an area of 32.56 km^2 where one of the main uses is the irrigation.

The fish farming dataset is composed of 14 variables as follows: *agency_cd, site_no, datetime, tz_cd, 01_Temperature, 01_Temperature_cd, 02_Spec-conduc, 02_Spec-conduc_cd, 04_Turbidity, 04_Turbidity_cd, 07_Susp-sed-conc, 07_Susp-sed-conc_cd, 08_Dissol-oxygen and 08_Dissol-oxygen_cd*. Each variable consists of 138200 instances; however, some variables contain more than 50% of missing values. Variables like *agency_cd, site_no, datetime, tz_cd* and all variables with the "_cd" suffix, were not used in this study; some of them for its alpha-numeric values, and the others because they have an alphabetic value according to its level of process: data approved for publication (A), which have been processed and thoroughly reviewed by the USGS staff; and provisional data subject to revision (P), which have no approval of the review staff. Additionally, during the variables inspection, *Turbidity* had multiple alpha-numeric values, for this reason this variable was not used. Finally, all missing values in this dataset were deleted.

On the other hand, the irrigation dataset is formed by 13 variables: *agency_cd, site_no, datetime, tz_cd, 01_Temperatura, 01_Temperatura_cd, 02_Spec-conduc, 02_Spec-conduc_cd, 03_Dissolved-oxygen(%sat), 03_Dissol-oxygen(%sat)_cd, 04_Dissol-oxygen(mgpl), 04_Dissol-oxygen(mgpl)_cd, 05_pH and 05_pH_cd*. Each variable has 1161 instances, in this case, the amount of missing values was less than 30% in all variables. In a similar way as the previous dataset, *agency_cd, site_no, datetime, tz_cd* and all the remain variables with the "_cd" suffix, were not used in this study based on the same reasons. In this case, only two variables had missing values and these were deleted.

2.2   Cauca River Modeling Project Phase II (PMC II)

PMC II is a project to monitor the Río Cauca water quality [9]. The study area was the stretch "Hormiguero-Mediacanoa", specifically "Puente Juanchito" monitoring station near Santiago de Cali city, Colombia. The Cauca River Basin is the second largest waterway of Colombia and crosses around 183 municipalities, representing approximately 41% of the Colombian population. Water quality variables in this dataset are as follows: *Date, pH, Temperature, Dissolved Oxygen,* and *Conductance*. Date variable consists on the corresponding date and time for each instance, *Dissolved Oxygen* has the saturation percentage, *Conductance,* the specific conductance in *microsiemens/cm,* and *Temperature* values are given in *Celsius degrees*. After inspect every instance in this dataset, it was determined that they satisfy the regulations in resolution 2115 of June 22, 2007 by the Ministries of Social Protection and Environment for physical characteristics of water quality for human consumption.

2.3   Río Piedras Watershed (Test Dataset 1)

Río Piedras watershed is located on the western slope of the Cordillera Central mountain range, west of Popayan, Colombia. This watershed covers an area of 58 km^2 and a variable altitude between 1900 m and 3800 m. In this zone, most of the population is formed by indigenous families belonging to Nasa and Coconucos ethnic groups and rural families. The representative economic activities include agriculture, livestock, fish farming, human consumption, among others [10]. Water quality variables in this dataset are the same as the PMCII data source, adding the *Dissolved Oxigen* variable in mg/l.

2.4   Illinois River (Test Dataset 2)

This river is a principal tributary of the Mississippi River, approximately 439 km long in the state of Illinois, U.S. [11]. It drains a large section of central Illinois, with a drainage basin of 74,479 km^2. Habitat loss from heavy siltation, and water pollution have eliminated most commercial fishing; therefore, irrigation is one of their mainly uses. WQ Variables considered in this dataset were: *temperature, specific conductance, dissolved oxygen in mg/l, dissolved oxygen in percent of saturation, pH,* and *turbidity*.

3 Models Description

Knowledge Discovery in Databases (KDD) is the process of discovering useful knowledge from a collection of data [12]. A database which stores data obtained by monitoring the water quality, is an important source of information. It may contain measurement data about the physical and chemical properties of the water at different measurement sites. This section presents the algorithms used in the present study (base classifiers). Additionally, three of the most popular methods for combining these algorithms are mentioned (MCS methods).

3.1 Base Classifiers

The goal in supervised learning, is to predict the value of a target feature on unlabeled instances and the learned model is also called a predictor [13]. For example, to determine the degree of pollution on a water body, three categories can be labeled as "high", "medium", and "low". The predictor should be able to forecast the label of an instance for which the label information is unknown, e.g., (dissolved oxygen = 21%, turbidity = 1.2 NTU, pollution degree = unknown). If the label is categorical, the task is also called classification and the learner is also called classifier. Classifiers used in this study are described below.

- **C4.5 Decision Tree (C4.5).** C4.5 is an algorithm extension of ID3 [14]; it generates a set of decision trees which can be used for classification tasks. In addition, it is one of the most popular algorithms in the Top 10 Algorithms in Data Mining [15].
- **K Nearest Neighbors (KNN).** KNN [16] is a classification method in supervised learning, used to estimate a density function $F(x|Cj)$, which determines the class membership for an instance. The input consists of the k closest training examples in the feature space.
- **Multilayer Perceptron (MLP).** MLP [17, 18] is a feedforward ANN model which maps sets of input data onto a set of appropriate outputs (in this case for classification). It consists of multiple layers of nodes in a directed graph, where each layer is fully connected to the next one.
- **Naïve Bayes Classifier.** This classifier belongs to the family of simple probabilistic classifiers which are based on a common principle: applying Bayes' theorem with strong independence assumptions between the features [19].

3.2 MCS Models

These methods train and combine multiple classifiers to solve the same problem. In contrast to the single learning approaches which try to construct one classifier from training data, MCS selects a set of learners and combine them [3, 13]. Three of the most used MCS were applied in this study and these are described below.

- **Bagging.** It is a machine learning ensemble meta-algorithm which attempts to reduce variance and helps to avoid overfitting [20]. For generating different base learners, Bagging adopts the bootstrap distribution (bootstrap sampling or data subsets for training the base classifiers) [21]. Bagging adopts the most popular strategies for aggregating the outputs of the base learners, that is, voting for classification and averaging for regression [13].

Three bootstrap sampling were selected in this study, due to the results were better than the other number of bootstraps. For each data subset in the MCS architecture, the same type of classifier was trained (C4.5, MLP, NB, or KNN). Finally, all classifications were combined in order to select the best-predicted class, in this case, the use of water.

- **Stacking.** This MCS method is a way of combining multiple models, that introduces the concept of a meta-classifier [22]. The procedure of stacking initially split the training set into two disjoint sets; train several base classifiers on the first part, and test the base learners on the second part. Using the predictions from base classifiers as the inputs, and the correct responses as the outputs, train a higher-level learner (meta-classifier). Three single classifiers were selected at the first level and the remaining learner was defined as meta-classifier; in this sense, all possible combinations in the architecture were used.

- **Random Forest (RF).** This ensemble learning method is a substantial variation of Bagging, combining this approach with a random selection of features. RF implements a large collection of non-correlated trees outputting the class that is the mode of the classes (classification) or mean prediction (regression) of the individual trees. The number of decision trees used in RF method for this study was 100, which is the default parameter in Weka tool.

4 Data Preprocessing

Initially, three datasets were obtained from separate files. These datasets represent three uses of water: fish farming, irrigation, and human consumption; the third is the only one that no needs a previous processing and it could be used directly, the other datasets were preprocessed to find erroneous and missing values in the attributes. Fish farming dataset had a large amount of missing values, for this reason, is not recommendable to apply imputation methods based on centrality. In contrast, other imputation methods were used, like Bootstrap and Predictive Mean Matching (PMM) in R software tool [23]. However, r-squared (R2) values of imputed instances were lower than 0.4; that means the imputed values were not acceptable. In view of the above, the same process was applied using MICE package with PMM, obtaining acceptable values for the imputed instances (similar distribution of data). In the same way, this process of imputation was used for irrigation dataset.

In a second phase, the three datasets were merged to generate a consolidated dataset, which was used in the evaluation tests applied to single classifiers and MCS. To increment the number of features, common WQ variables in at least two datasets were added to consolidated dataset, and then, missing values were assigned to each of them (*pH* and *dissolved oxygen* were added to fish farming instances, and *percent saturation of dissolved oxygen* to human consumption dataset). The class to predict in the consolidated dataset was *Type_of_Use*, which refers to the previously classified use of water for each instance.

Due to unbalanced classes presented in fish farming dataset, in a third phase, the minimum representative percentage of training samples for this dataset was evaluated. Experimental tests were conducted using MCS architectures with 3%, 5%, 10%, 15%, 30% y 100% of the total instances to determine if it is enough to use the entire fish farming dataset or a low percentage of this. Results are shown in Table 1.

Table 1. Percentage of correctly classified instances by C4.5, KNN, MLP and NB using different test percentages.

Test percentage (%)	Correctly classified instances (%)						
	C4.5	KNN	MLP	NB	Bagging-C4.5	RF	Stacking-C4.5
3	99.81	35.55	100	100	99.83	99.99	64.26
5	99.9	25.11	100	100	99.88	100	74.99
10	99.95	14.88	100	100	99.94	100	85.71
15	99.98	10.33	100	100	99.97	100	90
30	99.99	6.13	100	100	99.99	100	94.73
100	100	2.67	100	100	100	100	98.36

As can be seen from Table 1, the results are acceptable from 3% of the fish farming dataset, and there was a progressive increase to 100% using the total of instances, except the classifications performed by KNN with 1 neighbor, here the correctly classified instances decreased. It can be explained by the increment of the number of fish farming instances with missing values. However, 3% and 5% could be consider acceptable in order to obtain a high performance in classifiers, taking into account the high percent of correctly classify instances. As a result, the response and training times were lower, without affecting the classifier efficacy.

In order to support a better approach for the instances selection process, K-Means algorithm can be used to group samples in the dataset, and subsequently select instances of each group, ensuring an adequate and representative sample [24]. Self-Organizing Maps (SOM) [25] was used for this purpose on the 61000 fish farming instances. This technique is a type of Artificial Neural Network (ANN), which generates clusters automatically with the aim of infer the optimal number of groups based on the lowest sum of the squares of standard deviation for each group. With a large number of clusters, the sum of the squares decreases; nevertheless, the optimal number of groups was 6, due to the generated groups over this value, only labeled 1 neuron. SOM with 6 clusters is shown in Fig. 1, which shows different colors (blue, red, orange, green, brown, and purple) for optimal number of labeled clusters. Each neuron was assigned to a color, and there were no groups formed by a single neuron. Finally, using Weka software tool, K-Means was applied to fish farming data, and results show an acceptable clusters distribution in all cases.

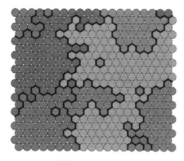

Fig. 1. SOM for finding the optimal number of clusters to select representative fish farming instances (Color figure online).

5   Results

C4.5, KNN, Multilayer Perceptron, and Naïve Bayes classifiers in Table 2, were used in the first evaluation test for the consolidated dataset. 3% of the total data in the fish farming dataset was used and 100% in the other two datasets. To apply this test, WEKA tool was used and all classifiers were configured using default parameters and the cross-validation process was used.

Table 2. Percentage of instances correctly classified using C4.5, KNN, MLP and Naïve Bayes.

C4.5	KNN	MLP	NB
99.85%	33.68%	100%	100%

Table 2 shows that Naïve Bayes and Multilayer Perceptron has the best results with 100% of instances correctly classified, C4.5 also has good results with more than 99.8%, but KNN classifier got only 33%. This could be due to the instances with missing values include in the dataset to maintain the same number of variables between the 3 datasets used in this study.

The second evaluation test was conducted by Multiple Classifier Systems with the aim of select the best architecture for combining classifiers in order to improve the classification task. In this sense, Random Forest, Bagging and Stacking methods with different configurations were used. Once again, all classifiers were configured using default parameters. In Bagging method, 4 configurations were used: (i) Bagging with Naïve Bayes, (ii) Bagging with KNN, (iii) Bagging with C4.5, and (iv) Bagging with Multilayer Perceptron. In Stacking method 4 configurations were used: (i) Stacking with C4.5 as meta-classifier and Naïve Bayes, Multilayer Perceptron and KNN as first level classifiers (Stacking-C4.5_NB-MLP-KNN), (ii) Stacking with Multilayer Perceptron as meta-classifier and Naïve Bayes, C4.5 y KNN as first level classifiers (Stacking-MLP_NB-C4.5-KNN), (iii) Stacking with Naïve Bayes as meta-classifier and Multilayer Perceptron, C4.5 and KNN as first level classifiers (Stacking-NB_MLP-C4.5-KNN), and finally (iv) Stacking with KNN as meta-classifier and Multilayer

Perceptron, C4.5 and Naïve Bayes as first level classifiers (Stacking-KNN_MLP-C4.5-NB). Results are showed in Table 3.

Table 3. Percentage of instances correctly classified using different architectures of MCS.

Random forest	Bagging				Stacking			
	NB	KNN	C4.5	MLP	C4.5[a] NB, MLP, KNN	MLP[a] NB, C4.5, KNN	NB[a] MLP, C4.5, KNN	KNN[a] MLP, C4.5, NB
100%	100%	35.6%	99.8%	100%	100%	100%	99.9%	100%

[a] These classifiers were used as meta-classifier in the MCS.

In the same way as the first evaluation test, Table 3 shows the results of the second test. Bagging with KNN architecture had the lowest results for classifying instances, although it improves the results obtained using only KNN. These results may be caused by the missing values added to the dataset, in this sense, only 1 neighbor were selected for this case. The other results were very accurate, with 100% of correctly classified instances.

Experimental results in both test shows that the classes in the consolidated dataset are highly separable. However, these experimental tests were performed using cross validation process (the training data are used in validation), which does not always guarantee more realistic situations to evaluate the classifier performance. Therefore, a new test was conducted using the same classifiers and architectures, but in this case, instead of the cross-validation method, two new test datasets were used. The first one corresponds to Rio Piedras, and the second one, to Illinois River. Table 4 shows the results, included the respective Confusion Matrix [26], obtained for the models that classified correctly some of the instances for human consumption.

As can be seen from the previous table, the percentage of correctly classified instances for test scenario 1, was around 63%. However, in some cases, the confusion matrix for the irrigation class had several instances classified as human consumption. For this reason, results may not be appropriate and to know this percentage is not enough to determine a possible use of water. The implications this could have, is that irrigation and fish farming comprises less rigorous quality parameters to meet, and a bad classification for human consumption would imply a risk for public health.

Taking into account the above, the aim of this study is to allow the decision-maker to reduce the sampling time and the number of tests necessaries to determine the type of use of a water sample using at least one of these classifiers or MCS directly on the in-situ collected data. In this sense, the best results were obtained by *Naïve Bayes* from the classic methods, *Bagging with Naïve Bayes*, and *Stacking-C4.5_NB-MLP-KNN*. Bagging correctly classified all the human consumption type, but could not differentiate between the other two water types. Stacking classified correctly 210 instances of human consumption and was able to make a little differentiation for the other two classes.

Table 4. Percentage of correctly classified instances using a test dataset.

Classifier/MCS	Test scenario 1					Test scenario 2				
	Percentage of correctly classified instances	Confusion Matrix[a]				Percentage of correctly classified instances	Confusion matrix			
Naïve bayes	65.4%	CA[b]	a	b	c	94.2%	CA	a	b	c
		a	185	0	159		a	344	0	0
		b	0	4533	0		b	0	4533	0
		c	0	2498	306		c	34	255	822
Random forest	61.5%	CA	a	b	c	78.3%	CA	a	b	C
		a	118	226	0		a	99	245	0
		b	0	4533	0		b	0	4533	0
		c	2553	174	77		c	127	838	146
Bagging with Naïve bayes	63.4%	CA	a	b	c	94.5%	CA	a	b	c
		a	344	0	0		a	344	0	0
		b	0	4533	0		b	0	4533	0
		c	0	2804	0		c	128	825	158
Bagging with C4.5	63.2%	CA	a	b	c	84.2%	CA	A	b	C
		a	329	15	0		a	332	12	0
		b	0	4533	0		b	0	4533	0
		c	2804	0	0		c		837	274
Stacking-C4.5_NB-MLP-KNN	65.7%	CA	a	b	c	93.8%	CA	A	b	C
		a	210	0	134		a	344	0	0
		b	0	4533	0		b	0	4533	0
		c	0	2498	306		c	25	288	798
Stacking-MLP_NB-C4.5-KNN	67.1%	CA	a	b	c	94.2%	CA	A	b	C
		a	182	0	162		a	344	0	0
		b	0	4533	0		b	0	4533	0
		c	2334	178	292		c	52	239	820

[a] Confusion Matrix, also known as an error matrix, is a table where each column represents the instances in a predicted class while each row represents the instances in an actual class
[b] CA: "Classified as". Classes, a: human consumption, b: fish farming, and c: irrigation

Based on results obtained for the realistic scenario 1, a single technique (a classifier or MCS) cannot be applied in all cases to determine the possible uses of a water sample. It can be seen that for human consumption data, the most appropriate option is *Bagging with Naïve Bayes* because it completely allows to separate this class from the other two classes. However, this MCS method cannot distinguish between irrigation and fish farming instances. In this sense, in WQ prediction for irrigation, both *Naïve Bayes* and *Stacking-C4.5_NB-MLP-KNN* can be used. Nevertheless, the last allows to differentiate better the human consumption samples; for irrigation, both were able to classify 306 of the 2804 instances of this class. Finally, to separate the water for fish farming, *Bagging with C4.5* can be used, although this technique only could be applied on fish farming samples, like *Bagging with NB* for human consumption; in contrast with selected classifiers for irrigation which had an acceptable performance on instances of human consumption and irrigation.

For test scenario 2, the irrigation data previously used were replaced by River Illinois dataset. In this case, there was an increase in the number of correctly classified instances belonging to the human consumption class, except for the Random Forest and Bagging with Naïve Bayes models. In addition, the percentage of correctly classified instances increased in all cases for the same class. This percentage increased too for irrigation class, while for the fish farming class it was maintained.

In the same way, Bagging with Naive Bayes, which failed to separate irrigation classes with the fish farming class with test dataset 1, using the test dataset 2, could differentiate most instances of this class. Nevertheless, 128 instances were classified as human consumption. This indicates that there was greater general separability of the 3 classes, but the effectiveness of the algorithm was slightly reduced to separate the human consumption class from the other two. In the case of Random Forest, the number of instances of irrigation classified incorrectly as human consumption decreased and the instances of this class correctly classified, most were classified as instances of the class fish farming. In the case of Naive Bayes, the number of correctly classified instances for irrigation increased considerably. However, 34 instances were classified as human consumption. Again, the general separability of the classes improved, but the effectiveness of separating the human consumption class from the other classes decreased. Similarly, with the other meta-classifiers, there was an increase in the separability of the 3 classes.

Furthermore, the Welch's t-test [27] in R software tool was made to see if the improve results in test scenario 2 are statistically different from test scenario 1. Table 5 shows the results, for this test the classified instances in each test scenario with the 6 selected classifiers/MSC were used.

Table 5. Welch's t-test p-value results.

Classifier/MSC p-value					
NB	Random forest	Bagging with Naïve Bayes	Bagging with C4.5	Stacking C4.5_NBMLP-KNN	Stacking MLP_NBC4.5-KNN
1.85e-08	<2.2e-16	0.1082	<2.2e-16	1.744e-10	<2.2e-16

As can be seem from the previous table, all the p-value or *observed significance level* [28] were small except for *Bagging with Naïve Bayes* value. This means that the differences shown between this MSC in the two test scenarios has no statistical significance. Statistical significance was found in all the remaining cases proving that the test scenario 2 are statistically different from the test scenario 1.

6   Conclusions

This paper has explained the importance of knowing the possible uses of water in a watershed. An adequate predictive tool would allow guiding the decision-making processes and could form the basis of decision-support systems for watershed management authority. These organizations or government entities may establish integrated control actions on water resources. Based on the above, this study compares different

MCS architectures for combining classifiers in order to determine which is the most accurate, taking into account particular conditions of real world. Experimentally, MCS techniques like *Bagging with Naïve Bayes,* and *Stacking with C4.5* as meta-classifier and *Naïve Bayes, Multilayer Perceptron, and KNN* as first level classifiers, performed best on instances classification.

In the last test, the results for Rio Piedras dataset were not as good as the first two, but it has a more realistic approach to the real-world situations. MCS improved the response over the classical methods not in the percentage of correctly classified instance, but in the correct differentiation between them; in general, between the human consumption class and the other two classes in the dataset. As a first hypothesis, we assume that the results could be better with more data to correctly differentiate between the classes. In the test dataset 1, the irrigation class did not have the *Dissolved oxygen* in *mg/l (milligrams per liter)*, so this had to be added and adapted to missing values. This could be the reason for the poor behavior in this test, compared to the other test where no test dataset was use.

The experimental results with the test dataset 2, prove that the hypothesis raised above is correct. By including the *dissolved oxygen variable* in *mg/l* (included as missing values in the initial tests) it improved the separability of the classes by increasing the percentages of correctly classified instances and the confusion matrices in all cases. However, an expert classifier was no longer obtained to separate the instances of human consumption like in test dataset 1. On the other hand, Stacking with C4.5 as meta-classifier and Naïve Bayes, Multilayer Perceptron, and KNN as first level classifiers could be used in this way considering the number of instances miss classified as human consumption and the separability between the other two classes. With this database addition, it is demonstrated that it is possible to use a single classifier to obtain an adequate response to the three classes considered in this study.

Finally, intelligent systems for use of water classification can be developed based on WQ parameters measured on site. However, in some cases, variations in these parameters are not significant, and only few features can determine the use of water in a watershed. In these cases, is recommendable to develop different ensemble methods or MCS in order to improve the classes differentiation, especially when the features present high levels of similarity. As a result, the sampling and lab-test time could be dramatically decrease using a WQ in-situ classification and it will have direct implications in the resources cost for these tasks making it less expensive and faster. The laboratory technician could focus the lab tests into the ones needed for the specific type of use, due to the in situ-classification system previously defined the use of water. The success of a prediction model depends on whether it can meet the requirements of the user. Different users might have different expectations, and it is difficult to know the "right expectation" of each one. A common strategy is to evaluate and estimate the performance of the models, and then let the user to decide whether a model is acceptable, or choose the best available model from a set of candidates.

Acknowledgements. The authors are grateful to the Telematics Engineering Group (GIT) and Environmental Studies Group (GEA) of the University of Cauca, Institute CINARA of the University of Valle, RICCLISA Program and AgroCloud project for supporting this research, and Colciencias (Colombia) for PhD scholarship granted to MsC. Iván Darío López.

References

1. Chapman, D.: Water Quality Assessments: A Guide to the Use of Biota, Sediments and Water in Environmental Monitoring. Taylor & Francis, Abingdon (1992)
2. Carbó, C.: Genética, patología, higiene y residuos animales (Genetics, Pathology, Hygiene and Animal Waste), vol. 4. Mundi-Prensa Libros (1995)
3. Dietterich, T.G.: Ensemble methods in machine learning. In: Kittler, J., Roli, F. (eds.) MCS 2000. LNCS, vol. 1857, pp. 1–15. Springer, Heidelberg (2000). https://doi.org/10.1007/3-540-45014-9_1
4. Grbović, J., Džeroski, S.: Knowledge discovery in a water quality database. In: Proceedings 1st International Conference on Knowledge Discovery and Data Mining (KDD 1995), Menlo Park, CA, pp. 81–86 (1995)
5. Bassiliades, N., Hatzikos, E., Koutitas, C., Antoniades, I., Vlahavas, I.: An intelligent system for monitoring and predicting water quality. In: Proceedings of the European Conference Towards eENVIRONMENT, Prague, Czech Republic, pp. 534–542 (2009)
6. Muhammad, S., Makhtar, M., Rozaimee, A., Aziz, A., Jamal, A.: Classification model for water quality using machine learning techniques. Int. J. Softw. Eng. Appl. 9(6), 45–52 (2015)
7. Partalas, I., Hatzikos, E., Tsoumakas, G., Vlahavas, I.: Ensemble selection for water quality prediction. In: Proceedings of the 10th International Conference on Engineering Applications of Neural Networks, Thessaloniki, p. 8 (2007)
8. U. S. Geological Survey, http://www.usgs.gov/. Accessed 14 April 2015
9. CVC - Segunda campaña de muestreo con propositos de calibracion del modelo de calidad del agua del rio cauca (Second sampling campaign with model calibration purposes of water quality in the Cauca river). Corporación Autónoma Regional del Valle del Cauca, Convenio Interadministrativo 0168, 27 noviembre 2002, vol. 15 (2005)
10. Borsdorf, A., Marchant, C., Mergili, M.: Agricultura Ecológica y Estrategias de Adaptación al Cambio Climático en la Cuenca del Río Piedras, pp. 7–73 (2013)
11. EPA - Watershed Characterization Report, https://ofmpub.epa.gov/waters10/watershed_characterization.control?pComID=3601974. Accessed 15 May 2017
12. Frawley, W., Piatetsky-Shapiro, G., Matheus, C.: Knowledge discovery in databases: an overview. AI Mag. 13(3), 57–70 (1992)
13. Zhou, Z.-H.: Ensemble Methods: Foundations and Algorithms. CRC Press, Boca Raton (2012)
14. Quinlan, J.: C4.5: Programs for Machine Learning. Morgan Kaufmann Publishers Inc., San Francisco (1993)
15. Wu, X.: Top 10 algorithms in data mining. Knowl. Inf. Syst. 14(1), 1–37 (2007)
16. Silverman, B., Jones, M.: An important contribution to nonparametric discriminant analysis and density estimation. Int. Stat. Rev. 57(3), 233 (1989)
17. Gurney, K.: An Introduction to Neural Networks. Taylor & Francis, Abington (2003)
18. Malsburg, C.: Frank Rosenblatt: Principles of Neurodynamics: Perceptrons and the Theory of Brain Mechanisms. In: Palm, D.G., Aertsen, D.A. (eds.) Brain Theory, pp. 245–248. Springer, Heidelberg (1986). https://doi.org/10.1007/978-3-642-70911-1_20
19. Russell, S., Norvig, P., Canny, J., Malik, J., Edwards, D.: Artificial Intelligence: A Modern Approach, vol. 2. Prentice Hall, Upper Saddle River (1994)
20. Breiman, L.: Stacked regressions. Mach. Learn. 24(1), 49–64 (1996)
21. Efron, B., Tibshirani, R.: An Introduction to the Bootstrap. CRC Press, New York (1994)
22. Wolpert, D.: Original contribution: stacked generalization. Neural Netw. 5(2), 241–259 (1992)

23. Vink, G., Frank, L., Pannekoek, J., van Buuren, S.: Predictive mean matching imputation of semicontinuous variables. Stat. Neerl. **68**(1), 61–90 (2014)
24. Corrales, D.C., Corrales, J.C., Sanchiz, A., Ledezma, A.: Sequential classifiers for network intrusion detection based on data selection process. In: Proceedings of the 2016 IEEE International Conference on Systems, Man, and Cybernetics (SMC 2016), Budapest, Hungary, p. 1 (2016)
25. Kohonen, T.: Self-organized formation of topologically correct feature maps. Biol. Cybern. **43**(1), 59–69 (1982)
26. Stehman, S.: Selecting and interpreting measures of thematic classification accuracy. Remote Sens. Environ. **62**(1), 77–89 (1997)
27. Welch, B.L.: The generalization of "Student's" problem when several different population variances are involved. Biometrika **34**(2), 28–35 (1947)
28. The Little Handbook of Statistical Practice. http://www.jerrydallal.com/LHSP/LHSP.htm. Accessed 06 July 2017

Parallelized Preconditioned Model Building Algorithm for Matrix Factorization

Kamer Kaya[⊠], Ş. İlker Birbil, M. Kaan Öztürk, and Amir Gohari

Faculty of Engineering and Natural Sciences, Sabancı University, Istanbul, Turkey
{kaya,sibirbil,mkozturk,amir}@sabanciuniv.edu

Abstract. Matrix factorization is a common task underlying several machine learning applications such as recommender systems, topic modeling, or compressed sensing. Given a large and possibly sparse matrix A, we seek two smaller matrices W and H such that their product is as close to A as possible. The objective is minimizing the sum of square errors in the approximation. Typically such problems involve hundreds of thousands of unknowns, so an optimizer must be exceptionally efficient. In this study, a new algorithm, Preconditioned Model Building is adapted to factorize matrices composed of movie ratings in the Movie-Lens data sets with 1, 10, and 20 million entries. We present experiments that compare the sequential MATLAB implementation of the PMB algorithm with other algorithms in the minFunc package. We also employ a lock-free sparse matrix factorization algorithm and provide a scalable shared-memory parallel implementation. We show that (a) the optimization performance of the PMB algorithm is comparable to the best algorithms in common use, and (b) the computational performance can be significantly increased with parallelization.

Keywords: Preconditioned model building · Matrix factorization
Multicore parallelism

1 Introduction

We investigate the performance of a novel optimization algorithm on the matrix factorization problem. The classic matrix factorization problem involves approximating a given matrix A as the product of two unknown matrices W and H:

$$A \approx WH, \tag{1}$$

where $A \in \mathbb{R}^{m \times n}$, $W \in \mathbb{R}^{m \times r}$, $H \in \mathbb{R}^{r \times n}$, with r a given integer (the *rank* of the factorization). In typical applications, A is sparse, and r is much smaller than either m or n. However, the resulting factor matrices W and H can be dense. The associated optimization problem is the minimization of the sum of squares of errors in the approximation

$$\min_{W,H} \sum_{i,j \in S} \left(A_{ij} - \sum_{k=1}^{r} W_{ik} H_{kj} \right)^2, \tag{2}$$

© Springer International Publishing AG 2018
G. Nicosia et al. (Eds.): MOD 2017, LNCS 10710, pp. 376–388, 2018.
https://doi.org/10.1007/978-3-319-72926-8_31

where the outer sum is over the set S of (i, j) pairs where A_{ij} is known (nonzero). If a good approximation with a small r can be found, the factors can be used to represent the original data in a more compressed form, with less redundancies.

One application of matrix factorization is in the field of recommendation systems, particularly content-based filtering. As a concrete example, suppose that each row of A corresponds to a particular user, each column to a particular movie, and the matrix element A_{ij} is a numeric value representing the rating given by user i to movie j. This matrix is very sparse, because most of the users have rated only a small fraction of all available movies. Furthermore, the data have redundancies, because the ratings given by users with common tastes and interests are likely to be correlated. After the original matrix A is factorized into factor matrices with relatively small rank r, we can multiply them back to obtain a full matrix A^*. The entries in A^* will then be estimates for the missing values in A. In other words, we can estimate whether a user would give a high ranking to a given movie, and display it as a recommendation to the user.

Intuitively, matrix factorization can be seen as discovering some hidden variables in the data. For example, the hidden dimensions can be movie genres, movies with a strong female character, movies that appeal to an adolescent audience, etc. [6]. If the input matrix comprises e-mails and the words in them, such as the now-public Enron e-mail data set, the hidden dimensions turn out to be topics like professional football, California blackout, and Enron downfall [1].

The power of matrix factorization as a recommender system is demonstrated in the Netflix Prize challenge. In this challenge, many different algorithms were compared with each other to see which one would improve the recommendation accuracy by more than 10%. The first algorithm that crossed this mark was based on matrix factorization [6].

Although other methods such as Principal Component Analysis or Latent Semantic Analysis can also be applied to that end, matrix factorization has the advantage that it does not regard empty matrix entries as zero values. The optimization problem considers only the sum of squares over existing values. This property reduces the error of the approximation [5].

Another application of matrix factorization is data compression, or representing the data in a low-dimensional subspace. Assume again that each row of A represents ratings of users. Then, from $A \approx WH$ it follows that the i-th row of A can be written as a linear combination of the rows of H, with coefficients taken from the i-th row of W:

$$A_{i,:} = \sum_{k=1}^{r} W_{ik} H_{k,:} \tag{3}$$

where the notation $A_{i,:}$ indicates the i-th row of matrix A. We can then interpret W_{ik} as a measure of user i's interest in movies that have property k. Similarly, we can interpret H_{kj} as a measure of how much of k is carried by the movie j.

Due to its ability to compress information, matrix factorization can also be used for unsupervised classification problems. To this end, the preferred variety is *nonnegative matrix factorization*, where both the data matrix A and the factor matrices W, H are constrained to have only nonnegative entries. With non-negativity, the linear combination (3) gives a recipe for constructing $A_{i,:}$ by adding ingredients $H_{k,:}$ in amounts of W_{ik}. Because no subtraction is involved, we can interpret the results in a more intuitive way [7]. In this study, we only consider unconstrained optimization, therefore nonnegative matrix factorization is beyond our scope at the moment. We solve only the *classic* matrix factorization problem, where entries can be negative real numbers.

For the experiments, we factorize user-movie rating matrices, provided by the MovieLens database. To minimize the objective function (2), we use the Preconditioned Model Building (PMB) method that we describe in Sect. 2.

We have developed a MATLAB implementation of the algorithm with no parallelization. We first factorize the MovieLens 1M rating matrix with PMB, as well as with other established optimization methods in the `minFunc` package. In Sect. 2.1 we show that the performance of PMB on this problem is comparable to the best ones that are in widespread use, and better than some others.

2   Preconditioned Model Building

In our recent work, we have proposed a new method that could be used as an alternative to line search procedure in unconstrained optimization algorithms [9,10]. From this perspective, the proposed method is another globalization mechanism that aids algorithms to converge from remote points to a local minimizer. The main idea of the proposed method is to build a series of quadratic model functions using trial points around the current iterate. With each trial point, the simpler quadratic model function is minimized and the next trial point is set to the location of the attained minimum. If this minimum point provides a sufficient decrease in the original objective function according to the Armijo condition, then it is accepted as the new step to move to the next iteration. Otherwise a new model is built around the incumbent trial point. As we construct a new quadratic model at each trial point, we aptly refer to this approach as model building (MB) algorithm in this paper.

At iteration k, MB takes an initial vector and uses it as the first trial point, s_k. To guarantee the convergence of the algorithm, this initial vector should be gradient related providing a sufficient descent. Let us formalize this discussion. Consider the unconstrained optimization problem of the form

$$\min_{x \in \mathbb{R}^n} f(x),$$

where $f : \mathbb{R}^n \mapsto \mathbb{R}$ is the objective function. Let x_k denote the point at iteration k. To obtain the next iterate x_{k+1}, the MB algorithm requires the initial vector s_k to satisfy the following two conditions:

$$m_0 \|\nabla f(x_k)\| \leq \|s_k\| \leq M_0 \|\nabla f(x_k)\|,$$
$$-\mu_0 \|\nabla f(x_k)\|^2 \leq s_k^\top \nabla f(x_k) \leq 0$$

$$(4)$$

for some $m_0, M_0, \mu_0 \in (0, \infty)$. If we simply set $s_k = -\nabla f(x_k)$, then both conditions are satisfied with $m_0 = M_0 = \mu_0 = 1$. This choice of s_k is, in fact, used when MB is first introduced by Öztoprak and Birbil [10].

Another way of setting the initial vector for the MB algorithm is to use a positive definite matrix. That is, we can set $s_k = -H_k \nabla f(x_k)$ and use this direction as an input to the MB algorithm. Since H_k is positive definite, the conditions in (4) are satisfied by taking m_0 as the minimum eigenvalue of H_k, and M_0 along with μ_0 as the maximum eigenvalue of H_k. Algorithm 1 shows explicitly the steps of our implementation, where the first trial step is determined after a preconditioner is computed (line 4). The model building steps are given between line 14 and line 21. The original algorithm in [10] takes $\eta \in (0, 1)$ as an input of the algorithm. In our implementation, we have observed that adjusting this parameter dynamically as shown in line 14 improves the performance.

Algorithm 1. Preconditioned Model Building

---

1 **Input:** x_0; $\rho = 10^{-4}$; $k = 0$

2 $f_k = f(x_k)$; $g_k = \nabla f(x_k)$;

3 **while** x_k *is not a stationary point* **do**

4 $\quad$ Compute the preconditioner H_k;

5 $\quad$ $s_k = -H_k g_k$;

6 $\quad$ **for** $t = 0, 1, 2, \cdots$ **do**

7 $\quad\quad$ $x_k^t = x_k + s_k$; $f_k^t = f(x_k^t)$; $g_k^t = \nabla f(x_k^t)$;

8 $\quad\quad$ $v_6 = s_k^\top g_k$; $\Delta f = f_k - f_k^t$;

9 $\quad\quad$ **if** $\Delta f \geq -\rho v_6$ **then**

10 $\quad\quad\quad$ $x_{k+1} = x_k^t$, $f_{k+1} = f_k^t$, $g_{k+1} = g_k^t$;

11 $\quad\quad\quad$ $k = k + 1$;

12 $\quad\quad\quad$ **break**;

13 $\quad\quad$ **end**

14 $\quad\quad$ $v_0 = s_k^\top g_k^t$; $\eta_1 = \frac{|\Delta f|}{v_6}$; $\eta_2 = \frac{|\Delta f|}{v_0}$; $\eta = \frac{\min(\eta_1, \eta_2)}{\eta_1 + \eta_2}$;

15 $\quad\quad$ $y = g_k^t - g_k$; $v_1 = y^\top s_k$; $v_2 = s_k^\top s_k$;

16 $\quad\quad$ $v_3 = y^\top y$; $v_4 = y^\top g_k$; $v_5 = g_k^\top g_k$;

17 $\quad\quad$ $\sigma = \frac{1}{2}(\sqrt{v_2}(\sqrt{v_3} + \frac{1}{\eta}\sqrt{v_5}) - v_1)$;

18 $\quad\quad$ $\theta = (v_1 + 2\sigma)^2 - v_2 v_3$;

19 $\quad\quad$ $c_g = -v_2/(2\sigma)$; $c_s = \frac{c_g}{\theta}(-(v_1 + 2\sigma)v_4 + v_3 v_6)$;

20 $\quad\quad$ $c_y = \frac{c_g}{\theta}(-(v_1 + 2\sigma)v_6 + v_2 v_4)$;

21 $\quad\quad$ $s_k = c_g g_k + c_s s_k + c_y y$;

22 $\quad$ **end**

23 **end**

The introduction of such a positive definite matrix H_k is also known as preconditioning. The advantage of preconditioning in the optimization context is to

Table 1. Metadata of MovieLens data sets.

Dataset	#users	#movies	#ratings	density
1M	6,040	3,952	1,000,209	0.042
10M	71,567	10,681	10,000,054	0.013
20M	138,493	27,278	20,000,263	0.005

incorporate second order information into the step evaluation [2]. Quasi-Newton methods obtain this information by making use of the gradient information collected in the previous iterations. The most famous one among the quasi-Newton methods is the limited BFGS method (L-BFGS) method [8]. In this current work, we have also used L-BFGS update mechanism for estimating our preconditioning matrices. Thus, we refer to the resulting procedure as Preconditioned Model Building (PMB) algorithm.

2.1 A First Comparison with Other Optimizers

The MovieLens data [4]. This is a public data set containing a large number of ratings of movies by individuals. The data is collected from the `movielens.org` web site, maintained by the GroupLens research group at the Univ. of Minnesota. Although the full data set keeps growing in time, stable data sets are available for benchmarking purposes. These are referred to as 1M, 10M, and 20M datasets. The names refer to the number of ratings contained in each data set.

Table 1 lists the number of users, number of movies and number of ratings in each data set. Each set contains users who have rated at least 20 movies. Ratings are integers between 1 and 5.

Numerical comparison. We solve the optimization problem (2) with PMB, as well as several other optimization functions commonly used in literature. We see that PMB does not have a significant handicap when compared against the other accepted methods.

We factorize the matrix of MovieLens 1M data set with several popular optimization algorithms, along with PMB. All of these factorizations are performed using MATLAB R2015b. For all methods except PMB, we have used `minFunc` package [11]. The codes for generating these results are available in the accompanying GitHub repository[1] for those who wish to replicate our results.[2]

Each algorithm is initialized with random matrix entries. Each entry is sampled from the uniform distribution $U(1,5)/\sqrt{r}$, where r is the rank of the factorization (set to 50), so that the resulting matrix product has entries mostly between 1 and 5. Algorithms are stopped after 500 iterations. The maximum number of function calls and the maximum number of iterations are both set to 1000.

---

[1] https://github.com/sibirbil/PMBSolve.

[2] The PMB results in this section are obtained with the MATLAB implementation, which is not parallelized and thus different from the results given in Sect. 4.

Table 2. Comparison of PMB with other optimizers for the 1M dataset with factorization rank 50, averaged over 50 runs.

Method	Mean final RMSE	95% confidence interval
Barzilai and Borwein	0.6436	(0.6096, 0.6855)
Cyclic Steepest Descent	0.5894	(0.5871, 0.5919)
Hessian-Free Newton	0.5561	(0.5544, 0.5581)
Conjugate Gradient (CG)	0.5558	(0.5548, 0.5568)
Scaled CG	0.5391	(0.5385, 0.5398)
PMB	0.5148	(0.5138, 0.5160)
Preconditioned CG	0.5020	(0.5002, 0.5038)
Limited memory BFGS	0.4954	(0.4944, 0.4965)

Every algorithm run is repeated 50 times with randomized initial points, and 95% confidence intervals for the mean values are estimated using bootstrapping.

The resulting RMS error values and gradient norm for each algorithm is shown on Table 2. We see that PMB is one of the most successful methods to solve this large matrix factorization problem.

3   Parallelization of PMB-Based Matrix Factorization

The PMB engine is implemented by using templates in C++11. Various optimization problems, e.g., matrix factorization as in this study, can be solved with the engine once the appropriate function/gradient computation source code is integrated. Moreover, this integration does not need a modification on the engine and a separate source file is sufficient.

The only time consuming part of the engine is the preconditioning; however, for the matrix factorization problem, preconditioning is only responsible for the 5% of the execution time. The rest is spent to the function and gradient computations for sparse factorization. Hence, in this work, we mainly focus on the function and gradient computations since they form the main bottleneck. The computations in the engine, mostly dot products, are also parallelized in a straightforward manner whlie optimizing the data reuse and memory accesses as much as possible. The execution time of the PMB for the matrix factorization problem dissected into three parts is given in Fig. 1. As the figure shows, the factorization-specific functions is responsible for most of the execution time.

3.1   Computational Tasks for Sparse Matrix Factorization

Given a sparse matrix A with τ entries, there are three tasks at each iteration:

1. Computing the error Δ_{ij} for known A_{ij} entries, i.e.,

$$\Delta_{ij} = (W_{i,:} \cdot H_{:,j}) - A_{ij} \text{ for all } (i.j) \in S \tag{5}$$

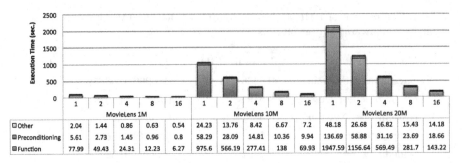

	MovieLens 1M					MovieLens 10M					MovieLens 20M				
	1	2	4	8	16	1	2	4	8	16	1	2	4	8	16
□ Other	2.04	1.44	0.86	0.63	0.54	24.23	13.76	8.42	6.67	7.2	48.18	26.68	16.82	15.43	14.18
▣ Preconditioning	5.61	2.73	1.45	0.96	0.8	58.29	28.09	14.81	10.36	9.94	136.69	58.88	31.16	23.69	18.66
▣ Function	77.99	49.43	24.31	12.23	6.27	975.6	566.19	277.41	138	69.93	1947.59	1156.64	569.49	281.7	143.22

Fig. 1. The execution time of PMB dissected into three parts for bottleneck detection: The most time consuming part, *Function*, computes the error for each A_{ij} and updates the factor matrices accordingly. The next part handles the *Preconditioning* stage. The *Other* parts of PMB, i.e., memory allocations, transfers etc., are considered as a third part for completeness.

where $\cdot$ is the dot product operator and $W_{i,:}$ and $H_{:,j}$ are row and column vectors corresponding to the i-th row of W and j-th column of H, respectively. This task simultaneously computes the overall function value $\sum_{(i,j)\in S} \Delta_{ij}^2$.

2. Computing the gradient entries for W; let $Z1$ be the matrix containing these entries. Then

$$Z1_{i,:} = \sum_{(i,j)\in S} \Delta_{ij} H_{:,j}^T. \tag{6}$$

3. Computing the gradient entries for H; let $Z2$ be the matrix containing these entries. Then

$$Z2_{:,j} = \sum_{(i,j)\in S} \Delta_{ij} W_{i,:}^T. \tag{7}$$

For all the tasks, the time complexity is $\mathcal{O}(r \times \tau)$ where r is the factorization rank.

3.2 Storing the Sparse Matrix and Auxiliary Data in Memory

We start by mentioning the common data structures for the implementation of the algorithms. For matrix factorization, the pattern and the numerical values of a sparse matrix is stored in both the compressed row storage (CRS) or compressed column storage (CCS) formats. These are well known storage formats for sparse matrices (see, e.g., Sect. 2.7 of Duff et al. [3]). Consider an $m \times n$ sparse matrix A with τ nonzeros. In CRS, the pattern of A is stored in three arrays:

- `colids[1,...,τ]` stores the column index of each entry.
- `vals[1,...,τ]` stores the corresponding numerical value of each entry. The column ids and values in a row are stored consecutively; and
- `ptrs[1,...,m + 1]` stores the location of the first entry of each row in array `colids` where `ptrs[m + 1] = τ + 1`. In particular, the column indices of the entries in row i are stored in `colids[ptrs[i],...,ptrs[i + 1] − 1]`. Similarly the values in the i-th row are stored in `values[ptrs[i],...,ptrs[i + 1] − 1]`

The CCS of a matrix A is the CRS of its transpose and vice-versa. In CCS, there are two pattern arrays (ccs)_rowids and (ccs)_ptrs, with functions similar to the first and the third arrays just described above. However, we will not need a (ccs)_vals array separately for the CCS format. Both formats are necessary for our parallel implementation.

The auxiliary sparse matrix Δ has τ entries and the same sparsity pattern of A. Hence, the same CRS/CCS pattern arrays can be used. We also use an extra delta array of size τ to store the Δ_{ij} values in CRS format. When a column-wise access to Δ is required (that will be necessary to avoid race conditions), we will utilize a static, precomputed ccs_trans array that translates the CCS-location to a CRS-location. The usage of this array will be described in more detail later.

The matrices W, $Z1$, H and $Z2$ are all dense and the first two and the last two contain $m \times r$ and $n \times r$ entries, respectively. To optimize the spatial locality of reference for the accesses to these matrices, we use the row-major layout for W and $Z1$ and the column-major layout for H and $Z2$.

3.3 Efficient and Lock-Free Parallel Implementation of the Tasks

The memory accesses for the sparse factorization problem can deteriorate the performance if they are not handled carefully. As mentioned above, our first task computes the Δ matrix which is used by the later tasks; if the implementation uses barriers in between the tasks, there will be (at least) τ memory accesses to the delta array by the second task. This overhead can be avoided when the first two tasks are integrated; in this version, each Δ_{ij} computed by (5) is immediately used by (6). This Δ_{ij} value is then stored in the corresponding entry of the delta array to be used later by the third task to compute (7).

The first task can be parallelized in two different ways: in the fine-grain approach, each Δ_{ij} computation can be assigned to a different thread, and in the coarse-grain approach, the values in $\Delta_{i,:}$ are assigned to the same thread. Although the former increases the degree of concurrency and eases load balancing, the latter is more appropriate for the integration of the first two tasks. As it can be seen by (6), a gradient entry $Z1_{ij}$ is modified for each entry in $\Delta_{i,:}$. Hence, when the fine-grain approach is taken and two threads independently compute and use $\Delta_{i,j'}$ and $\Delta_{i,j''}$, the entry $Z1_{ij}$ needs to be updated by both of these threads. To avoid such race conditions, expensive synchronization mechanisms are required. However, a lock-free implementation is possible when each row $\Delta_{i,:}$ (and hence $Z1_{i,:}$) is assigned to only a single thread. Since we access the elements of A and Δ in a row-wise manner, the CRS pattern and value arrays are used for this implementation.

A similar analysis of (7) implies that a lock-free parallel implementation of the third task is possible if the updates on each column of $Z2$ are solely assigned to a single thread. However, this requires an efficient access to the columns of Δ. Since the array delta is organized via CRS, the entries in a column of Δ are not consecutively stored in memory. On the other hand, with a CCS-to-CRS translator, one can access to these non-consecutive locations one after another. In our implementation, we use a helper array ccs_trans of size τ to convert the

```
void Task3(SparseMatrix* A, prec_t* W, prec_t* Z2) {
#pragma omp parallel for schedule(runtime)
  for (coord_t j = 0; j < A->n; j++) {
    prec_t *myZ2 = Z2 + (j * r);
    memset (myZ2, 0, sizeof(prec_t) * r);

    point_t start = A->ccs_ptrs[j];
    point_t end = A->ccs_ptrs[j + 1];
    for (point_t p = start; p < end; p++) {

      const prec_t *myW = W + (A->ccs_rowids[p] * r);
      prec_t dt = delta[A->ccs_trans[p]];

      for (int k = 0; k < r; k++) {
        myZ2[k] += myW[k] * dt;
      }
    }
  }
}
```
(a) Lock-free parallelization

```
    for (p = start; p < end - 3; p += 4) {
      const prec_t *W_1 = W + (A->ccs_rowids[p] * r);
      const prec_t *W_2 = W + (A->ccs_rowids[p + 1] * r);
      const prec_t *W_3 = W + (A->ccs_rowids[p + 2] * r);
      const prec_t *W_4 = W + (A->ccs_rowids[p + 3] * r);

      const prec_t dt_1 = delta[A->ccs_trans[p]];
      const prec_t dt_2 = delta[A->ccs_trans[p + 1]];
      const prec_t dt_3 = delta[A->ccs_trans[p + 2]];
      const prec_t dt_4 = delta[A->ccs_trans[p + 3]];

      for (int k = 0; k < LDIM; k++) {
        myZ2[k] += (W_1[k] * dt_1 + W_2[k] * dt_2 +
                    W_3[k] * dt_3 + W_4[k] * dt_4);
      }
    }
}
```
(b) With loop unrolling

Fig. 2. The lock-free parallelization of the third task is given on the left. The unrolled form of its middle loop which performs four iterations at once is given on the right. For simplicity, only the first part of the loop is given and the part that completes the remaining $|\Delta_{:,j}|$ mod 4 iterations is omitted.

CCS-locations to CRS-locations and access the correct Δ_{ij} values in the same column. The lock-free implementation of this task is given in Fig. 2a.

With pinpoint analysis, we identified the main bottleneck of the lock-free code in Fig. 2a as the memory updates in the innermost loop, which is expected since one needs to perform two data loads (from myW and myZ2) and a store (to myZ2) for each update. To reduce the accesses to/from myZ2, we unroll the middle loop and process multiple Δ and W values in the same line. In this way, we reduce the number of accesses to myZ2 by at most 4×. The loop-unrolled version of the third task is given in Fig. 2b. A similar loop-unrolling mechanism is applied to the integrated implementation of the first and second tasks, but a detailed explanation is omitted in the paper due to space limitations.

4   Experimental Results

All the simulation experiments in this section are performed on a single machine running on 64 bit CentOS 6.5 equipped with 384 GB RAM and a dual-socket Intel Xeon E7-4870 v2 clocked at 2.30 GHz, where each socket has 15 cores (30 in total). Each core has a 32 kB L1 and a 256 kB L2 cache, and each socket has a 30 MB L3 cache. All the codes are compiled with gcc 4.9.2 with the -O3 optimization flag enabled. For parallelization, we used OpenMP with (dynamic, 16) scheduling policy. For each datapoint in the figures and tables, we perform five experiments and presented the average.

We first investigate the impact of loop-unrolling. From now on, the integrated Δ and $Z1$ computation will be denoted as dtZ1. Similarly, we will use Z2 to denote the third task of Sect. 3.1. Figures 3 and 4 show the execution times of the lock-free implementation and its loop-optimized version for dtZ1 and Z2,

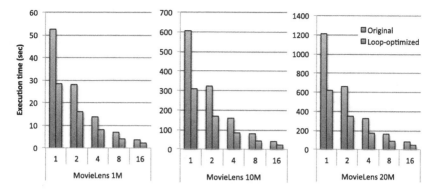

Fig. 3. The impact of loop-unrolling on the integrated `dtZ1` computation for all three datasets and $1, 2, 4, 8$ and 16 threads.

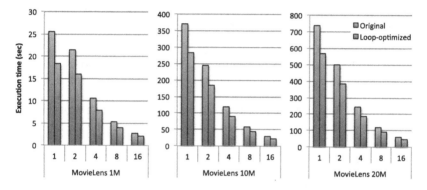

Fig. 4. The impact of loop-unrolling on the `Z2` computation for all three datasets and $1, 2, 4, 8$ and 16 threads.

respectively. As the figures show, unrolling the loop and perform four iterations at once significantly improves the performance of both tasks.[3]

The individual speedups of `dtZ1`, `Z2`, as well the overall speedup of the whole matrix factorization process, are given in Fig. 5 for MovieLens 10M and 20M datasets. As the figures show, with 16 threads, the speedup for the combined `dtZ1`, `Z2` (`Func`) is around $13\times$ whereas the overall speedup is around $11\times$. The overall speedup is smaller since except `dtZ1` and `Z2`, PMB performs only vector dot products which is a memory-bounded task infamous about its bad scalability.

We also experimented with a single-precision PMB implementation to see its impact on the performance. As expected, the performance is significantly improved; the performance is $1.36\times$, $1.52\times$, and $1.90\times$ better for `dtZ1`, `Z2`, and preconditioning, respectively, compared to the double-precision variant.

---

[3] We repeated this experiment by performing eight iterations at once but no further improvement is observed.

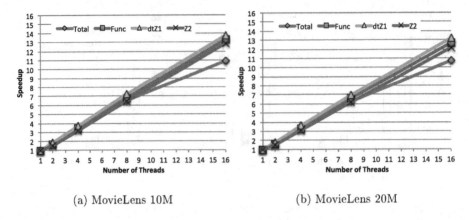

(a) MovieLens 10M   (b) MovieLens 20M

Fig. 5. Individual speedups with MovieLens 10M (left) and 20M (right) datasets and $1, 2, 4, 8$ and 16 threads for `dtZ1` and `Z2`. The charts also show their combined speedup (`Func`), and the overall speedup of whole execution.

Table 3. The execution times (in secs.) for `dtZ1`, `Z2`, and the Preconditioning phase of the loop-unrolled version when double and single precision arithmetic and data representation is used for single and 16-thread version.

Dataset	#threads	Double precision			Single precision			Improvement		
		dtZ1	Z2	Pre.	dtZ1	Z2	Pre.	dtZ1	Z2	Pre.
1M	1	28.3	18.3	5.6	22.7	12.5	2.9	1.25	1.46	1.93
	16	2.2	2.1	0.8	1.6	1.4	0.4	1.38	1.50	2.00
10M	1	308.2	283.3	57.7	230.3	174.0	33.1	1.34	1.63	1.74
	16	22.5	22.1	9.9	16.1	15.1	4.8	1.40	1.46	2.06
20M	1	618.0	567.1	132.1	462.0	359.4	70	1.34	1.58	1.89
	16	46.9	46.7	19.3	32.4	31.2	10.6	1.45	1.50	1.82
Average improvement								1.36	1.52	1.90

Although this improvement comes with a possible reduction on the accuracy, this is not the case for the datasets as the following experiment shows (Table 3).

Using this implementation, we factorize each of the 1M, 10M, and 20M rating matrices with factorization ranks 20 and 100. Each factorization is repeated 50 times with randomized initial conditions. The final RMSE values are then found by averaging, and confidence intervals are determined by bootstrap resampling. The algorithm stops when the number of iterations reaches 500 or when the absolute value of the largest element of the function gradient drops below 10^{-5}. Table 4 displays the results for this experiment.

We see that using single-precision version of PMB does not make a significant difference in the final RMSE value for matrix factorization, compared to the double-precision version. However, in every case, the single-precision version runs

Table 4. Final RMSE results from the factorization of MovieLens matrices with factorization rank 20/100 and single/double precision.

Dataset	Rank	Precision	Mean RMSE	95% confidence interval
1M	20	single	0.681926	(0.681265, 0.682619)
		double	0.682071	(0.681296, 0.682815)
	100	single	0.335744	(0.332367, 0.339220)
		double	0.338259	(0.334817, 0.341658)
10M	20	single	0.682580	(0.682190, 0.683038)
		double	0.683356	(0.682781, 0.683952)
	100	single	0.529536	(0.469352, 0.592387)
		double	0.528092	(0.468423, 0.590662)
20M	20	single	0.670611	(0.670027, 0.671292)
		double	0.679019	(0.669434, 0.697753)
	100	single	0.673502	(0.593685, 0.751497)
		double	0.667708	(0.589608, 0.745404)

faster by a factor of $1.5\times$–$2.0\times$. Therefore, in this particular problem, single-precision arithmetic can be preferred.

5   Conclusions

Preconditioned Model Building algorithm is a powerful optimizer that combines local model-building iterations with second-order information. Our results show that for the matrix factorization problem, the performance of the PMB algorithm is comparable to the best algorithms in general use.

Since it is cheap, the algorithm spends most of the execution time for the evaluation of the error and in the update of factor matrices. These computations are similar to the traditional sparse-matrix computations, therefore we can go around this bottleneck with appropriate parallelization techniques. Indeed, the PMB algorithm can be parallelized very well. The experiments show that there is little overhead thanks to the lock-free parallelization, and the speedup with 16 threads is about 11. Hence, the algorithm can be promising for large-scale optimization problems.

References

1. Berry, M.W., Browne, M., Langville, A.N., Pauca, V.P., Plemmons, R.J.: Algorithms and applications for approximate nonnegative matrix factorization. Comput. Stat. Data Anal. **52**(1), 155–173 (2007)
2. Bertsekas, D.P.: Nonlinear Programming. Athena Scientific, Belmont (1999)
3. Duff, I.S., Erisman, A.M., Reid, J.K.: Direct Methods for Sparse Matrices. Oxford University Press Inc., New York (1986)

4. Harper, F.M., Konstan, J.A.: The movielens datasets: history and context. ACM Trans. Interact. Intell. Syst. **5**(4), 19:1–19:19 (2015)
5. Hernando, A., Bobadilla, J., Ortega, F.: A non negative matrix factorization for collaborative filtering recommender systems based on a bayesian probabilistic model. Knowl.-Based Syst. **97**, 188–202 (2016)
6. Koren, Y., Bell, R., Volinsky, C.: Matrix factorization techniques for recommender systems. Computer **42**(8), 30–37 (2009)
7. Lee, D.D., Seung, H.S.: Learning the parts of objects by non-negative matrix factorization. Nature **401**(6755), 788–791 (1999)
8. Liu, D.C., Nocedal, J.: On the limited-memory BFGS method for large scale optimization. Math. Program. **45**, 503–528 (1989)
9. Öztoprak, F.: Parallel Algorithms for Nonlinear Optimization. Ph.D. thesis. Sabancı University (2011)
10. Öztoprak, F., Birbil, S.I.: An alternative globalization strategy for unconstrained optimization. arXiv preprint, arXiv:1705.05158 (2017) (To appear in Optimization)
11. Schmidt, M.: minFunc: unconstrained differentiable multivariate optimization in matlab, http://www.cs.ubc.ca/~schmidtm/Software/minFunc.html. Accessed 22 March 2017

A Quantitative Analysis on Required Network Bandwidth for Large-Scale Parallel Machine Learning

Mingxi Li[1], Yusuke Tanimura[1,2], and Hidemoto Nakada[1,2(✉)]

[1] University of Tsukuba, Tsukuba, Japan
`rei-meigi@aist.go.jp`
[2] National Institute of Advanced Industrial Science and Technology, Tokyo, Japan
{`yusuke.tanimura,hide-nakada`}`@aist.go.jp`

Abstract. Parallelization is essential for machine learning systems that deals with large-scale dataset. Data parallel machine leaning systems that are composed of multiple machine learning modules, exchange the parameter to synchronize the models in the modules through network. We investigate the network bandwidth requirements for various parameter exchange method using a cluster simulator called SimGrid. We have confirmed that (1) direct exchange methods are substantially more efficient than parameter server based methods, and (2) with proper exchange methods, the bisection-bandwidth of network does not affect the efficiency, which implies smaller investment on network facility will be sufficient.

1 Introduction

For modern machine learning systems, including deep learning systems, parallelization is inevitable since they have to process massive amount of training data. Data parallel machine learning systems train multiple machine learning models simultaneously on different subsets of training dataset, and synchronize the machine learning models periodically. There are mainly two methods to achieve the synchronization of models, i.e., the exchanging of parameters which constitute the models, or gradients of parameters. In a method, we use central server called **parameter server** to synchronize the parameter, and in the other method, we make the worker nodes directly communicate each other to exchange information and synchronize.

Synchronization requires network communication. This means synchronization efficiency is affected by system's network performance. This paper focus on the communication time for synchronization. Note that communication time is completely independent of algorithm, computation time, and aggregation time. The discussion could be applied to any algorithm involves repetitive parameter update. In order to know the relationship between the parameter synchronization method and the network structure of system, we use a distributed computing environment simulator, called SimGrid [4], to virtually build a large-scale environment, and performed a quantitative evaluation. We investigate the network

© Springer International Publishing AG 2018
G. Nicosia et al. (Eds.): MOD 2017, LNCS 10710, pp. 389–400, 2018.
https://doi.org/10.1007/978-3-319-72926-8_32

bandwidth requirement for several parameter exchange methods, focusing on the bisection bandwidth of the network. The bisection bandwidth is an important measure of network performance which is known to affect significantly on a specific class of applications' execution time. Detailed definition will be given in Sect. 3.

The contribution of the paper are the following; (1) We quantitatively evaluate the required bisection bandwidth for each parameter exchange method and confirmed that the bisection bandwidth affects significantly on some of the methods, but not on all the methods, (2) We show that by choosing exchange method properly, we can reduce the investment on the network facilities without sacrificing the performance.

The next section of this paper gives the overview of distributed machine learning systems focusing on the parameter exchange methods and introduce the distributed computing environment simulator SimGrid, which we use in this paper for the quantitative evaluation. Section 3 presents the cluster and network configuration we assume. Section 4 describes experimental setup and the result of the experiments. Section 5 gives detailed discussion on the result. Section 6 gives summary of the paper and the future work.

2   Background

2.1   Parameter Exchange Methods for Large Scale Machine Learning Systems

To parallelize machine learning systems, there are two methods; **Data Parallel** and **Model Parallel**. While data parallel method simultaneously trains multiple machine learning models synchronizing each other, model parallel parallelize inside a single machine learning model. These two methods could be used complementarily. In this paper, we focus only on the data parallel.

Data parallel machine learning methods could be categorized from two aspects; synchronicity and parameter exchange methods. For synchronicity, there are synchronous methods and asynchronous methods. Synchronous method means that all the machine learning models are forced to be exactly the same periodically, while asynchronous methods allow slight difference among the models. This paper deals with synchronous methods only. For the parameter exchange methods, we have two options; the parameter server based method and the direct communication method.

Parameter Server Based Method. This method utilizes centralized server, called parameter server, to exchange parameters [1,5,13]. The left diagram in Fig. 1 shows the parameter server based parameter exchange. The workers (machine learning modules) send parameters (or gradients) to the parameter server, the parameter server aggregates the parameter, and send back them to the workers. Often, multiple servers are used to form one parameter server to

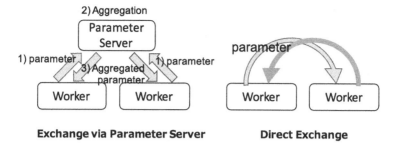

Exchange via Parameter Server   **Direct Exchange**

Fig. 1. Gradient exchange methods.

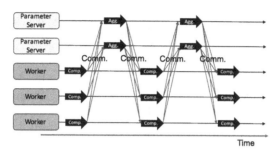

Fig. 2. Bulk synchronous parallel communication pattern with parameter servers.

shard the parameters and avoid overload each server. In this setting each parameter server take care of a certain subset of parameters.

Figure 2 shows the communication pattern among the parameter servers and the workers. Each worker sends/receives messages to/from all the parameter servers, periodically and repeatedly. This periodic communication pattern is called BSP Bulk Synchronous Parallel [17]. Note that there are no communications among workers nor among parameter servers. In other words, the communication between nodes forms bi-graph.

Direct Exchange Method. It is possible to synchronize the models without using central server. by repeating peer-to-peer exchange of parameters [18]. The left diagram in Fig. 3 shows the communication with 8 workers. Communication pattern likes this is known as **butterfly** communication, which is widely used, for example, by the allreduce in MPI [16]. In this communication, in each step i, each node N_m will exchange message with node N_l, where l is obtained by flipping ith bit of m. This means that in the earlier steps, the communication tends to take place locally, while in the later steps the communication spread out globally. It can exchange information with all the nodes within $Log_2 N$ steps of communication where N is the number of workers.

Cluster Aware Direct Exchange Method. We can further optimize the butterfly assuming the sub-cluster structure. To reduce the inter sub-cluster

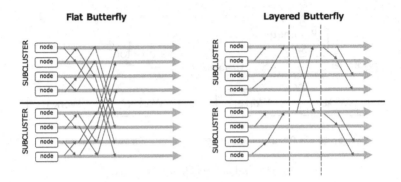

Fig. 3. Butterfly communication diagram. The left shows the 'flat' butterfly, while the right shows the '2-layered' butterfly communication.

communication, this method once gather the information inside the sub-clusters to the head nodes of sub-clusters, then perform butterfly among the head nodes of the sub-clusers, and then distribute the exchanged information in each cluster. We call this method **layered butterfly**. The right diagram in Fig. 3 shows the layered butterfly method.

This communication pattern requires $log_2 n + log_2 m + log_2 n$ steps where n is the number of nodes per sub-cluster and m is number of sub-clusters. Note that the flat butterfly shown above takes $log_2 N = log_2 nm = log_2 n + log_2 m$ steps; therefore the layered method requires $log_2 n$ more steps.

2.2   SimGrid: A Distributed Environment Simulator

SimGrid [2,4] is a simulation framework for distributed parallel applications. SimGrid is based on a discrete event simulation; it does not perform any real computation/communication. It just estimates times to perform computation/communication based on given parameters and records events like 'start/end of computation/communication'. The advantage of this type of simulator is that the simulation cost is relatively small. Even with single node computer, SimGrid can handle several thousands of communicating nodes.

To simulate a distributed system in SimGrid, users have to describe platform description and deployment description in XML, and the simulation code in C or C++.

3   Network Model

This section gives the network model assumed in this paper. We assume that a cluster is composed of smaller sub-clusters with 2-layered fat tree network [12] which uses switches with same number of port. Fat tree network is a tree network with *fatter* links near the root switch to avoid network congestion near the root switch. In this case we simulate fatter link by employing multiple root switch.

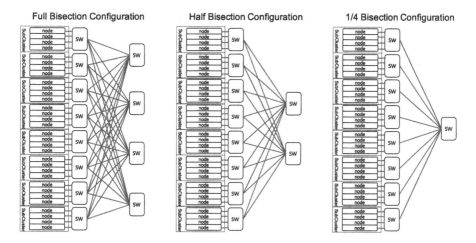

Fig. 4. Network connections with 8 port switches.

Here, we focus on **bisection bandwidth**, that is defined as the following; if the network is bisected into two partitions, the bisection bandwidth is the bandwidth available between the two partitions [7]. If the bisection bandwidth of a network equals to the total bandwidh of one half of the nodes, we call the network with '**full-bisection**' bandwidth.

Number of ports of the switches determines the maximum number of sub-clusters we can have. With n-node switch, n sub-clusters are the maximum, since all the higher layer switches have to be connected with the all the lower layer switches. Figure 4 shows the configuration with 8-ports switches. The left diagram shows the 'full-bisection' configuration with 8 port switches and 32 nodes in total. You can examine this by counting the number of connections between upper half and lower half of the cluster, 16. it is the same as the number of nodes on each half of the cluster. We can configure networks with less bisection bandwidth by reducing the number of upper layer switches, as shown in Fig. 4.

4   Experiments

We performed experiments with SimGrid using simulation with the network structure and the parameter exchange methods described below. We measured the time to perform one parameter exchange, no computation is took into account to focus on the communication cost only. Time for aggregating the parameters are also omitted in the simulation, since the time are relatively small and could be ignored.

4.1   Structure of the Cluster and the Network

We have setup clusters using 4, 8, 16, 32, and 64 port switches, with 2048, 512, 128, 32, 8 nodes in total respectively, with several bisection-bandwidth ratio. Table 1 shows the setups and the required number of switches for each setup.

Table 1. Number of switches required to construct the network.

#Ports/Switch	#Clusters	#Nodes/Cluster	#Nodes in Total	Bisection Ratio					
				1/32	1/16	1/8	1/4	1/2	1
64	64	32	2048	64 + 1	64 + 2	64 + 4	64 + 8	64 + 16	64 + 32
32	32	16	512	-	32 + 1	32 + 2	32 + 4	32 + 8	32 + 16
16	16	8	128	-	-	16 + 1	16 + 2	16 + 4	16 + 8
8	8	4	32	-	-	-	8 + 1	8 + 2	8 + 4
4	4	2	8	-	-	-	-	4 + 1	4 + 2

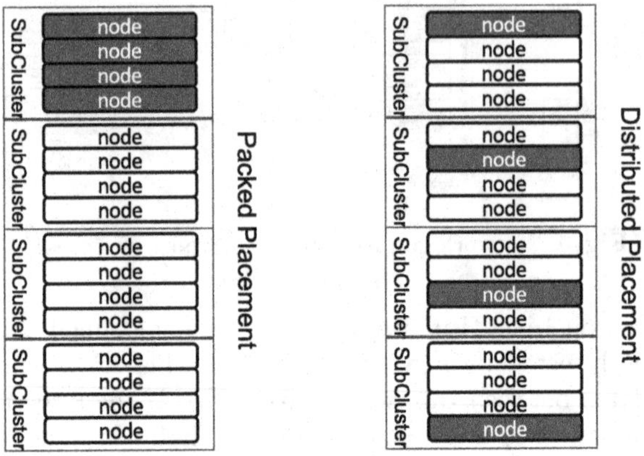

Fig. 5. Parameter server placements. The left diagram shows the 'packed' placement where one sub-cluster is dedicated for parameter servers, while the right diagram shows the 'distributed' placement where parameter servers are evenly distributed to all the sub-clusters. Note that the used node in each sub-cluster is determined in round-robin fashion to evenly distribute load to the upper layer switches.

The switch can handle $p/2$ connections where p is the number of ports. We set the bandwidth of links as 1 GB/s, assuming 10 G Ethernet with TCP overhead.

4.2 Parameter Exchange Methods

We test one parameter server based method and two butterfly based methods. For the parameter server based method, we assume $1/p$ of the nodes in the cluster are used for parameter servers while the others are used for workers. Note that the number of parameter servers equals to the number of nodes in each sub-cluster. We tested two placement strategy for parameter server based method. One is 'packed' and the other is 'distributed', shown in Fig. 5.

For the butterfly based methods, all the nodes are used for workers.[1] We test the simple flat-butterfly method with the layered-butterfly method.

---

[1] This means that the number of worker nodes is different between butterfly network based method and parameter server based method. (Number of worker nodes of parameter server based method is always n nodes fewer, where n is the number of nodes per sub-clusters.) However, even if the butterfly network based method reduces n nodes, the execution time is expected to be the same.

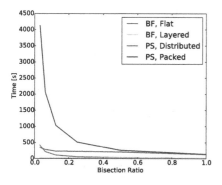

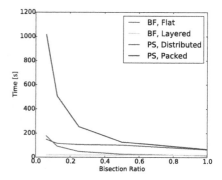

Fig. 6. The result with 64 subclusters. (Color figure online)

Fig. 7. The result with 32 subclusters. (Color figure online)

In summary, we test four settings; namely, parameter server with packed placement (**PS, packed**) and distributed placement (**PS, distributed**), flat butterfly (**BF, flat**), and layered butterfly (**BS, layered**).

4.3 Results of Experiments

We give the whole results in Table 2. Figures 6 and 7 show the results for 64 ports and 32 ports switches, respectively. The x-axis shows the bisection-ratio, where 1.0 means full-bisection bandwidth while 0.5 means half-bisection bandwidth. The y-axis shows the execution time to perform one parameter exchange. We can see that these two graphs are quite similar. We will discuss on the result in the next section.

5 Discussion

5.1 Discussion on the Results

From this result, it can be seen that the method using the parameter server is inferior to the butterfly network based method in basic performance. This is because the connections to the parameter servers becomes the bottleneck.

Parameter server method with packed placement setting exhibit significant performance drops as the bisection ratio decreases. On the other hand, with distributed placement setting, the parameter server method is hardly affected by the reduction of the bisection bandwidth. This is because of the network traffic is smoothed throughout the cluster by distributing the parameter server nodes.

Butterfly based methods are faster than parameter server based methods, in general. The flat butterfly method tends to be affected by the reduced bisection bandwidth, since it performs inter cluster communication heavily. In contrast, the layered butterfly method is not affected by the bisection bandwidth at all.

Table 2. Simulation results.

#Clusters	#Nodes	Exchange Method	Bisection Ratio					
			1/32	1/16	1/8	1/4	1/2	1
64	2048	BF, Flat	421.08	213.25	109.33	57.37	31.39	18.40
		BF, Layered	23.30	23.30	23.30	23.30	23.30	23.30
		PS, Distributed	347.30	281.40	229.91	224.79	203.63	129.08
		PS, Packed	4123.45	2061.72	1030.86	515.43	257.71	135.30
32	512	BF, Flat	-	177.53	90.93	47.63	25.98	15.15
		BF, Layered	-	19.07	19.07	19.07	19.07	19.07
		PS, Distributed	-	149.08	115.11	104.58	101.04	63.61
		PS, Packed	-	1014.50	507.25	253.62	126.81	66.58
16	128	BF, Flat	-	-	72.53	37.89	20.57	11.91
		BF, Layered	-	-	14.85	14.85	14.85	14.85
		PS, Distributed	-	-	58.33	49.41	45.77	31.16
		PS, Packed	-	-	247.42	123.71	61.86	32.47
8	32	BF, Flat	-	-	-	28.14	15.15	8.66
		BF, Layered	-	-	-	10.62	10.62	10.62
		PS, Distributed	-	-	-	24.14	21.56	14.65
		PS, Packed	-	-	-	57.73	28.87	15.15
4	8	BF, Flat	-	-	-	-	9.74	5.41
		BF, Layered	-	-	-	-	6.39	6.39
		PS, Distributed	-	-	-	-	8.35	6.39
		PS, Packed	-	-	-	-	12.37	6.49

Figure 8 shows a close up of Fig. 6. As shown in the figure, the flat butterfly is slightly faster than layered butterfly with full-bisection bandwidth. This is because the flat butterfly requires fewer steps than the layered one, as discussed in Sect. 2.1.

5.2   Data Size, Data Representation, Link Bandwidth

In the experiment above, we assumed the total data size of the parameter is 1 GB, which means 256 million parameters in 32 bit single precision floating point. Actual data size depends on the number of parameters and the data representation. Table 3 summarizes important networks and their number of parameters, with the published year. While the number of parameters get larger than 1 billion once in 2012, it decreases again because of the rise of deep and narrow networks.

Another important issue is the representation of the each datum(gradient or parameter). If we can represent the gradient with small sized representation, it will reduce the burden for communication. With 16 floating point, we can reduce the data size by half. Gradient quantization, which uses 8-bit integer to represent the gradient, can reduce the data to one-quarter. An extreme case is known as 1-bit SGD [14], where the gradient is represented as just one bit sign, reducing the size of data to 1/32.

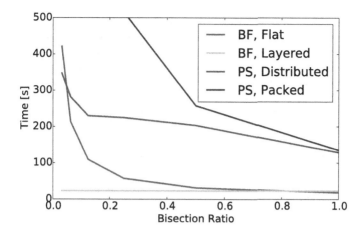

Fig. 8. The result with 64 subclusters, enlarged. (Color figure online)

Table 3. Number of parameters and data size for important neural networks.

Name	#Parameters	Data size			
		32 bit float	16 bit float	8 bit quantize	1 bit
Google Cat(2012) [11]	1 billion	4 GB	2 GB	1 GB	128 MB
AlexNet(2012) [10]	62 million	248 MB	124 MB	62 MB	8 MB
GoogLeNet(2015) [15]	6.8 million	27.2 MB	13.2 MB	6.8 MB	0.85 MB
ResNet 152(2016) [6]	2.5 million	10 MB	5 MB	2.5 MB	0.31 MB

Another important parameter is the network bandwidth of each link. In this experiment, we assumed link bandwidth as 1 GB/s, assuming 10 Gbit Eather network with protocol overheads. Link bandwidth is quite important for this setting, since it linearly speed up the communication. If we use a network with twice bandwidth, the communication time will become one half.

5.3   Computation Time and Parallelization Efficiency

In Bulk Synchronous Parallel computation, the ratio of communication time and computation time is quite important. When we assume that we exchange the gradient after one mini-batch execution, the computation time equals to the mini-batch computation time. To process a mini-batch with 256 images using AlexNetwork on single NVIDIA P-100, it requires around 1.4 s. To gain enough speed up with parallelization, the communication time have to be small enough compared to this computation time.

Figure 9 shows the expected parallelization efficiency for several settings, assuming the layered butterfly, 64 million parameters, and one second computation time. X-axis represents number of nodes and Y-axis represents the parallelization efficiency, where 1.0 is the ideal case. We changed the data representation (32 bit float and 1-bit) and link bandwidth (10 G and 40 G). In general,

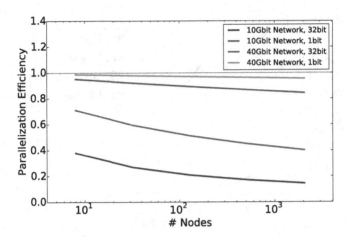

Fig. 9. Expected parallelization efficiency for layered butterfly, 64 million parameters, and 1 s of computation time. (Color figure online)

with more nodes, it get more difficult to get good efficiency. From this figure, we can conclude that both of the gradient data representation and link bandwidth are quite important to get better parallelization efficiency.

6   Conclusion

We have quantitatively evaluated the performance of several parameter exchange method for several possible network configuration to know the required bandwidth for each method. We have revealed that; (1) Parameter server based methods are substantially slower than the direct exchange methods, (2) Cluster aware direct exchange method (layered butterfly) outperforms naive exchange method (flat butterfly), except for the case with full-bisection bandwidth. (3) Cluster aware direct exchange method shows the constant good performance regardless of the bisection bandwidth ratio.

The implications are; (1) we should use direct exchange method over parameter server based method if possible, although they are much more difficult to implement, (2) if we end up using parameter server based method, we should distribute parameter servers with careful placement strategy to mitigate the network congestion, (3) we should invest in network link speed, since it is quite important for the performance, (4) we should not invest in *fatness* of the network, since it does not affect a lot, if we pick proper parameter exchange method.

Our future work include the followings:

- We will pursue parameter server based method with asynchronous settings [3,8]. While this work shows that the parameter server based methods are not efficient in term of performance, the methods have certain good characteristics such as easiness of implementation and suitability for fault tolerance.

By relaxing the strictness of parameter synchronization, there might be good chance to obtain better performance even with parameter based methods.

– We also would like to investigate asynchronous direct exchange methods. For example, in [9], the authors proposes gossip protocol based method, where each worker sends computed gradient to very limited number of peers only. This kind of methods significantly reduce the requirement for network bandwidth, in exchange for undesirable effects on the convergence.
– To pursue the asynchronous setting, we need to clarify the effect of 'gradient staleness' on convergence. Further experiments are required.

Acknowledgement. This paper is based on results obtained from a project commissioned by the New Energy and Industrial Technology Development Organization (NEDO). This work was supported by JSPS KAKENHI Grant Number JP16K00116.

References

1. Parameter server: http://parameterserver.org/. Accessed 20 June 2015
2. Simgrid: Versatile simulation of distributed systems. http://simgrid.gforge.inria.fr/index.php. Accessed 11 July 2016
3. Agarwal, A., Chapelle, O., Dudík, M., Langford, J.: A reliable effective terascale linear learning system. J. Mach. Learn. Res. **15**(1), 1111–1133 (2014). http://dl.acm.org/citation.cfm?id=2627435.2638571
4. Casanova, H., Giersch, A., Legrand, A., Quinson, M., Suter, F.: Versatile, scalable, and accurate simulation of distributed applications and platforms. J. Parallel Distrib. Comput. **74**(10), 2899–2917 (2014). http://hal.inria.fr/hal-01017319
5. Dean, J., Corrado, G.S., Monga, R., Chen, K., Devin, M., Le, Q.V., Mao, M.Z., Ranzato, M., Senior, A., Tucker, P., Yang, K., Ng, A.Y.: Large scale distributed deep networks. In: NIPS 2012: Neural Information Processing Systems (2012)
6. He, K., Zhang, X., Ren, S., Sun, J.: Deep residual learning for image recognition. In: Computer Vision and Pattern Recognition (CVPR) (2016)
7. Hennessy, J.L., Patterson, D.A.: Computer Architecture: A Quantitative Approach, 3rd edn. Morgan Kaufmann Publishers, Inc., San Francisco (2003)
8. Ho, Q., Cipar, J., Cui, H., Kim, J.K., Lee, S., Gibbons, P.B., Gibson, G.A., Ganger, G.R., Xing, E.P.: More effective distributed ml via a stale synchronous parallel parameter server. In: Proceedings of the 26th International Conference on Neural Information Processing Systems, NIPS 2013, pp. 1223–1231. Curran Associates Inc., USA (2013). http://dl.acm.org/citation.cfm?id=2999611.2999748
9. Jin, P., Yuan, Q., Jin, P., Keutzer, K.: How to scale distributed deep learning? In: ML Systems Workshop, NIPS 2016 (2016)
10. Krizhevsky, A., Sutskever, I., Hinton, G.E.: Imagenet classification with deep convolutional neural networks. In: Pereira, F., Burges, C.J.C., Bottou, L., Weinberger, K.Q. (eds.) Advances in Neural Information Processing Systems 25, pp. 1097–1105. Curran Associates Inc. (2012)
11. Le, Q., Ranzato, M., Monga, R., Devin, M., Chen, K., Corrado, G., Dean, J., Ng, A.: Building high-level features using large scale unsupervised learning. In: International Conference in Machine Learning (2012)
12. Leiserson, C.E.: Fat-trees: universal networks for hardware-efficient supercomputing. IEEE Trans. Comput. **34**(10), 892–901 (1985)

13. Li, M., Andersen, D.G., Park, J.W., Smola, A.J., Ahmed, A., Josifovski, V., Long, J., Shekita, E.J., Su, B.Y.: Scaling distributed machine learning with the parameter server. In: 11th USENIX Symposium on Operating Systems Design and Implementation (OSDI 2014), pp. 583–598. USENIX Association, Broomfield, October 2014. https://www.usenix.org/conference/osdi14/technical-sessions/presentation/li_mu

14. Seide, F., Fu, H., Droppo, J., Li, G., Yu, D.: 1-bit stochastic gradient descent and application to data-parallel distributed training of speech dnns. In: Interspeech 2014, September 2014

15. Szegedy, C., Liu, W., Jia, Y., Sermanet, P., Reed, S., Anguelov, D., Erhan, D., Vanhoucke, V., Rabinovich, A.: Going deeper with convolutions. In: Computer Vision and Pattern Recognition (CVPR) (2015). http://arxiv.org/abs/1409.4842

16. Thakur, R., Gropp, W.D.: Improving the performance of MPI collective communication on switched networks. Technical report, ANL/MCS-P1007-1102, Argonne National Laboratory, November 2002

17. Valiant, L.G.: A bridging model for parallel computation. Commun. ACM **33**(8), 103–111 (1990). http://doi.acm.org/10.1145/79173.79181

18. Zhao, H., Canny, J.: Butterfly mixing: accelerating incremental-update algorithms on clusters. In: Proceedings of the 2013 SIAM International Conference on Data Mining (2013)

Can Differential Evolution Be an Efficient Engine to Optimize Neural Networks?

Marco Baioletti, Gabriele Di Bari, Valentina Poggioni$^{(\boxtimes)}$, and Mirco Tracolli

University of Perugia, Perugia, Italy
{marco.baioletti,valentina.poggioni}@unipg.it, dbgabri@gmail.com,
mirco.theone@gmail.com

Abstract. In this paper we present an algorithm that optimizes artificial neural networks using Differential Evolution. The evolutionary algorithm is applied according the conventional neuroevolution approach, i.e. to evolve the network weights instead of backpropagation or other optimization methods based on backpropagation. A batch system, similar to that one used in stochastic gradient descent, is adopted to reduce the computation time. Preliminary experimental results are very encouraging because we obtained good performance also in real classification dataset like MNIST, that are usually considered prohibitive for this kind of approach.

1 Introduction

The raise of Deep Learning allowed to Neural Networks (NN) to come back on the crest of a wave since very complex problems have been solved with new architectures and optimization techniques [2,5,15,18]. Moreover this raise has been motivated also by the birth of new computational models using NNs, like Neural Turing Machines [11], Neural Programmer-Interpreters [26] or hybrid models [12].

According to these new trends also neuroevolution has been renewed [4,9, 13,14,23,32]. Several approaches have been proposed both to train the topology and the weigths of the networks. Compared to other neural network learning methods, neuroevolution is highly general, allowing learning without explicit targets, with non differentiable activation functions, and with recurrent networks [10,22]. An interesting analysis on the motivations why backpropagation (and its developments) is still the most used technique to train neural networks and evolutionary approaches are not sufficiently studied is presented in [23]. In that work a simple and efficient method to divide the training set in batches in order to train neural networks with a particular version of Differential Evolution (DE) is presented. Despite the performance are very interesting in terms of accuracy of the predictions the authors do not present experiments with large problems.

The advantages of replacing backpropagation, or other similar methods, with an evolutionary algorithm is clear: the fitness function to be optimized is not required to be differentiable or even continuous.

© Springer International Publishing AG 2018
G. Nicosia et al. (Eds.): MOD 2017, LNCS 10710, pp. 401–413, 2018.
https://doi.org/10.1007/978-3-319-72926-8_33

DE it is a well known evolutionary technique that demonstrated very good performance in several problems [31], has a quick convergence and is robust [25].

The main purpose of this paper is hence to show that DE can be effectively used to train neural networks also in case of quite large problems. We propose an algorithm that uses DE with population elements which encode the weights of the neural network. The system applies a batching system, with restart and elitism for different DE variants and mutation operators.

In order to prove that this mechanism is feasible, we tested it using state of art classification datasets chosen with different numbers of features and different number of records. Although the results are preliminary, they are very encouraging because in all the experiments the system reaches a very good accuracy, always comparable or even better than BPG, also when the network is larger (for instance in MNIST) than the ones presented in literature. Considering also that there is a room for further improvements, we are confident that this idea could be applicable also to larger networks.

The paper is organized as follows. Background knowledge about DE algorithm and neuroevolution are summarized in Sect. 2, related works are presented in Sect. 3, the system is described in Sect. 4 and experimental results are shown in Sect. 5. Conclusions are drawn in Sect. 6 where some ideas for future works are also depicted.

2   Background

2.1   Differential Evolution

Differential evolution (DE) is a metaheuristics that solves an optimization problem by iteratively improving a population of N candidate solutions with respect to a fitness function f. Usually, DE is used to solve continuous optimization problems, where the candidate solutions are numerical vectors of dimension D, but there exist many adaptions to solve combinatorial optimization problems, where the solutions are discrete objects [27]. The population evolution proceeds for a certain number of generations or terminates after a given criterion is met. The initial population can be generated with some strategies, the most used approach is to randomly generate each vector. In each generation, for every population element, a new vector is generated by means of a mutation and a crossover operators. Then, a selection operator is used to choose the vectors in the population for the next generation.

The most important operator used in DE is the *differential mutation*. For each vector x_i in the current generation, called *target vector*, a vector $\bar{y}_i$, called *donor vector*, is obtained as linear combination of some vectors in the population selected according to a given strategy. There exist many variants of the mutation operator (see for instance [7,8]). In our paper we have used DE/rand/1, where

$$\bar{y}_i = x_a + F(x_b - x_c)$$

and DE/current_to_best, where

$$\bar{y}_i = x_i + F(x_{best} - x_i) + F(x_a - x_b)$$

In these formulae, a, b, c are unique random indices different from i, *best* is the index of the best vector in the population and F is a real parameter.

Furthermore we have chosen to implement also DE with Global and Local Neighborhoods (DEGL) [6], indeed it works pretty well in neural networks learning, as explained in [24].

DEGL generates the mutant vector through the combination of two contributors. The first contributor is computed as:

$$L_i = x_i + \alpha(x_{i-best} - x_i) + \beta(x_a - x_b)$$

where x_{i-best} is the individual with best fitness in the neighborhood of target x_i and α, β are two constants with same role of F. The neighborhood of the element x_i contains a fixed number of other population elements, chosen at random.

The second contributor is computed as:

$$G_i = x_i + \alpha(x_{best} - x_i) + \beta(x_a - x_b)$$

where x_{best} is the individual with best fitness in the population. The two contributors are then combined as follow:

$$\bar{y}_i = wG_i + (1 - w)L_i$$

where $w \in [0, 1]$ is the interpolation factor between L_i and G_i.

The crossover operator creates a new vector y_i, called *trial vector*, by recombining the donor with the corresponding target vector by means of a given procedure. The crossover operator used in this paper is the binomial crossover regulated by a real parameter CR.

Finally, the usual selection operator compares each trial vector y_i with the corresponding target vector x_i and keeps the better of them in the population of the next generation.

3   Related Works

The first works applying DE to NN date back to the late '90s and the early 2000s [17,20] where the first applications of DE to train feed-forward NN are presented and analyzed. More recently, several other applications of evolutionary algorithms have been presented in the area of neuroevolution but they are different either for the evolutionary approach used or for the object of evolution [7,8].

In the first case the dominating approach used is the genetic one [10,28,33]. The approach is used also to optimize weights but it is very limited by being a discrete approach so it needs an encoding phase. Several authors proposed a direct representation of the real weights in genes either as a string of real values or as a string of characters, which are then interpreted as real values with a given precision using for example Gray-coded numbers. More adaptive approach has been suggested, for example in [29] or in the more recent [21]. In the first work the authors proposed a dynamic encoding which depends on the exploration and exploitation phases of the search. In the second one the authors propose a selfadaptive encoding, where the characters of the string are interpreted as a system of particles whose center of mass determines the encoded value. Other approaches have also used a direct encoding that exploits the particular structure of the problem. The methods are not general and cannot be extended to be applicable also to general cases [10]. In [14,16] a floating-point representation of the synaptic weights is used. In these cases the authors use the evolution strategy called CMA-ES for reinforcement learning applied to the pole balancing problem.

Among DE applications to neuroevolution it is worth to cite [9,19,24,32]. These works are different from our approach because they apply the evolution in a different way. In [19] the DE algorithm with a modified best mutation operation is used to enhance the search exploration of a PSO; this PSO is then used to train the NN and the global best value obtained is used as a seed by the BPG. In [9] three different methods (GA, DE and EDA) to evolve neural networks architectures are compared. In particular, the evolutionary methods are implemented to train the architecture of a network with one hidden layer, the learning factor and the seed for the weights initialization. In [24] the author studied the stagnation problem of DE approaches when used to train NN. He proposed to merge the DE with Global and Local neighborhood-based mutation operators algorithm with the Trigonometric mutation operator. In [32] the authors use the Adaptive DE (ADE) algorithm to choose the initial weights and the thresholds of networks. Also in this case the networks are trained by BPG. The authors proved that the system is effective to solve time series forecasting problems.

The paper which have the strongest connection with ours is undoubtely [23], where a Limited Evaluation Evolutionary Algorithm (LEEA) is applied to optimize the weigths of the network. The differences between the two papers are several. First of all, we use DE as evolutionary algorithm, while they employ an ad hoc evolutionary algorithm, similar in some aspects to a genetic algorithm. DE and the other enhancement methods allow our algorithm to train networks much larger than those used in [23]: while we are able to train a feed-forward neural network for MNIST (which has more than 7000 weights), the maximum size handled in [23] is less than 1500 weights. Another difference is the batching system: they use mini-batches which are changed at every generation, while we use larger batches which are changed after a certain number of generations. On the other hand, we use the validation set to compare the networks when the batch is changed, while they use a form of fitness inheritance.

4   The Algorithm

In this section we present our idea of applying Differential Evolution to optimize the weights of the connections in a feed-forward neural network.

Let $\mathcal{P}$ a population of np neural networks with a given fixed topology and fixed activation functions.

Since the DE works with continuous values, we can use a straightforward representation based on a one-to-one mapping between the weights of the neural network and individuals in DE population.

In details, suppose we have a feed-forward neural network with k levels, numbered from 0 to $k - 1$. Each network level l is defined by a real valued matrix $\mathbf{W}^{(l)}$ representing the connection weights and by the bias vector $\mathbf{b}^{(l)}$.

Then, each population element x_i is described by a sequence

$$\langle (\hat{\mathbf{W}}^{(i,0)}, \mathbf{b}^{(i,0)}), \ldots, (\hat{\mathbf{W}}^{(i,k-1)}, \mathbf{b}^{(i,k-1)}) \rangle,$$

where $\hat{\mathbf{W}}^{(i,l)}$ is the vector obtained by linearization of the matrix $\mathbf{W}^{(i,l)}$, for $l = 0, \ldots, k - 1$. For a given population element x_i, we denote by $x_i^{(h)}$ its h–th component, for $h = 0, \ldots, 2k - 1$, i.e. $x_i^{(h)} = \hat{\mathbf{W}}^{(i,h/2)}$, if h is even, while $x_i^{(h)} = \mathbf{b}^{(i,(h-1)/2)}$ if h is odd. Note that each component $x_i^{(h)}$ of a solution x_i is a vector whose size depends on the number of neurons of the associated levels.

The population elements are evolved by applying mutation and crossover operators in a componentwise way. For instance, the mutation rand/1 for the element x_i is applied in the following way: three indices a, b, c are randomly chosen in the set $\{1, \ldots, np\} \setminus \{i\}$ without repetition; then, the h–th component $\bar{y}_i^{(h)}$ of the donor element $\bar{y}_i$ is obtained as the linear combination

$$\bar{y}_i^{(h)} = x_a^{(h)} + F(x_b^{(h)} - x_c^{(h)})$$

for $h = 0, \ldots, 2L - 1$.

The evaluation of a population element in the selection operator is performed by computing the cross–entropy of the corresponding neural network. The optimization problem is then to find the neural network with the minimum cross-entropy value.

Anyway, this computation is the most time consuming operation in the overall algorithm and it will lead to unacceptable computation time if the cross–entropy considers the whole dataset. For this motivation we have decided to follow a batching method similar to the one proposed in [23].

The dataset D is split in three different sets: a training set TS used for the training phase, a validation set VS used for a uniform evaluation of the individuals selected at the end of each training phase, and a test set ES used to evaluate the performance of the best neural network.

Then, the training set TS is randomly partitioned in K batches of size B. This phase is very important because the records in each batch should follow more or less the same distribution as in TS. Otherwise, the risk is to train specialized networks without generalization ability.

At each generation the population is evaluated against only a limited number of training examples given by the size of the current batch, instead of evaluating the population against the whole training set. This allows to reduce the computational load, particularly on large training sets.

The fitness function used is then

$$H_{z'}(z) = -\sum_{i=1}^{B}\sum_{j=1}^{C} z'_{ij}\log(z_{ij})$$

where z'_{ij} and z_{ij} are respectively the predicted value and the true value for the classification of i-th record in the batch with respect to the j-th class and C is the number of classes.

The batch is changed after s generations (called *epoch*), so that the evolution has enough time to learn from the batch. If the algorithm is required to continue for more than K epochs, the batches are reused in a cyclic way, i.e. after the last batch, the first batch will be used again, and so on.

Since the fitness function depends also on the batch and we need a fixed way to compare the elements, at the end of every epoch the best neural network $best\_net_e$ of the epoch e is selected as the neural network in $\mathcal{P}$ which reaches the highest accuracy in the validation set VS. The best neural network $best\_net_{global}$ found so far is then eventually updated.

At the beginning of each epoch, the fitness of every element in $\mathcal{P}$ is re-evaluated by computing the cross-entropy on the new batch.

To avoid a premature convergence of the algorithm, a reset method is applied, i.e. discard all the current population, except the best element, and continue with

---

Algorithm 1. The algorithm DENN

Initialize the population;
Extract the $K = TS/B$ batches $batch_0, \ldots, batch_{K-1}$;
$h \leftarrow 0$;
for $e \leftarrow 1$ **to** tot_gen/s **do**
  Set the current batch as $batch_{e\ modK}$;
  Re-evaluate all the elements $(x_1, \ldots, x_{np})$;
  for $g \leftarrow 1$ **to** s **do**
    for $i \leftarrow 1$ **to** np **do**
      $y_i \leftarrow$ generate_offspring(x_i)
    for $i \leftarrow 1$ **to** np **do**
      if y_i *is better than* x_i *in terms of* H **then**
        $x_i \leftarrow y_i$
  $best\_net_e, best\_score_e \leftarrow$ best_score$(x_1, \ldots, x_{np})$;
  Update $best\_net_{global}, best\_score_{global}$;
  if $best\_net_{global}$ *is not changed* **then**
    if $h > counter$ **then**
      Reset the population;
      $h \leftarrow 0$;
    else
      $h \leftarrow h + 1$
return $best\_net_{global}$;

---

a new randomly generated population. The reset is performed at the end of each epoch e, if the the score of $best\_net_{global}$ has remained unchanged for a certain number *counter* of epochs.

The DE parameters F and CR have a great impact on the evolution and their values are not easy to be chosen. Therefore, we have decided to adopt the auto–adaptive scheme jDE [3]. This method evolves the values of these two parameters by a process which is strictly related to the selection operator of DE. In this way, the algorithm is able to dynamically select the best values of F and CR for the problem. The complete algorithm, called DENN, is depicted in Algorithm 1.

In the algorithm DENN, the function *generate_offspring* computes the mutation and the crossover operator in order to produce the trial element, while the function *best_score* returns the best network and its score among all the elements in the population.

5   Experimental Results

The main objective of these experiments is to assess the effectiveness of DE algorithm as an alternative to backpropagation, and other similar methods, for neural network optimization also in the case of quite large problems. The size of NNs handled in this paper are larger than those used in the previous works presented in literature. We run two kinds of experiments in order to (i) evaluate which combination of DE variant and mutation operator performs better and (ii) study which setting of algorithm parameters can provide the best results, also considering the computational effort. DENN has been implemented both as a TensorFlow plugin written in C++ and Python and as a stand-alone C++ program[1].

5.1   Datasets

We decided to test the system on recent classification datasets downloaded by the UCI repository[2], and on the well known MNIST[3] dataset. MAGIC, QSAR and GAS are datasets for classification problems that have been chosen because they differ for the number of features and records and therefore are well suited to assess the scalability of the system. Finally, we decided to test the system on the MNIST dataset because it is a classical challenge with well known results obtained by NN classification systems. Moreover, it is considered an interested challenge also in [23].

- MAGIC Gamma telescope: dataset with 10 features, 2 classes and 19020 records.
- QSAR biodegradation: dataset with 41 features, 2 classes and 1055 records.

---

[1] Source code available at https://github.com/Gabriele91/DENN.
[2] https://archive.ics.uci.edu/ml/datasets.
[3] http://yann.lecun.com/exdb/mnist/.

– GAS Sensor Array Drift: dataset with 128 features, 6 classes and 13910 records.
– MNIST: dataset with 784 features, 10 classes and 70000 records.

MAGIC, QSAR and GAS have been split in this way: the training set is composed of 80% of the records, the validation set is composed of another 10% of the records and the remaining 10% records are in the test set.

The MNIST dataset is already provided as a pair with separated training and test sets (TS, ES). Then we extracted the validation set VS from the training set by a uniform random sampling that preserves the distribution over the classes. Since a too small validation set could have a negative impact on the performance in term of accuracy, we chose, as in the other datsets, $|VS| = |TS| \cdot 10\%$.

5.2   Results

First of all we have analyzed the data in order to understand which are the parameters yielding to the best performance. The system depends from several parameters: some deriving from the use of DE (np, F, CR, the DE-variant and the mutation and crossover operators), other depending from the batching system (B, s) or from the application of the reset mechanism ($counter$).

A systematic battery of test has been run in order to study all the parameters. Due to the space limits, in this work just some data and graphics can be discussed. Other preliminary experiments are discussed in [1,30].

During this experimental phase, we have noted that the most important choice is selecting the DE variant (classical or jDE) and the mutation operator ($rand/1$, $current$-to-$best$, $DEGL$). Increasing other parameters, like np or B, can have a positive impact on the algorithm performance only when these values are below a certain threshold. When this threshold is overcome, either the computational time becomes too high or the results do not improve. This fact also agrees with (i) other traditional results on DE that in general suggest large (but not too large) populations and (ii) our initial idea about the batch size, according to which a trade-off between batch size and computational effort is necessary.

The batch size has been chosen to be proportional to the number of classes, in order to have, on average, a given number of examples for each class. Moreover we found that B influnces also the number of population elements np, and the steps s spent on the same batch to train the network. This is unsurprising because with a larger number of records in a batch, the population size (and the number of steps) should be larger as well.

In Fig. 1, for all the datasets analyzed in this work, the accuracy values of a NN without hidden layer trained with different settings of the algorithm are plotted. We compare the six different combinations of the DE variants and mutation operators ($DE+rand/1$; $DE+current$_to_$best$; $DE+degl$; $jDE+rand/1$; $jDE+current$_to_$best$; $jDE+DEGL$). Moreover, also the accuracy values obtained by the same NN trained with BPG (GD) are reported. The accuracy values plotted in the graphics are computed: (i) training the neural network for a given

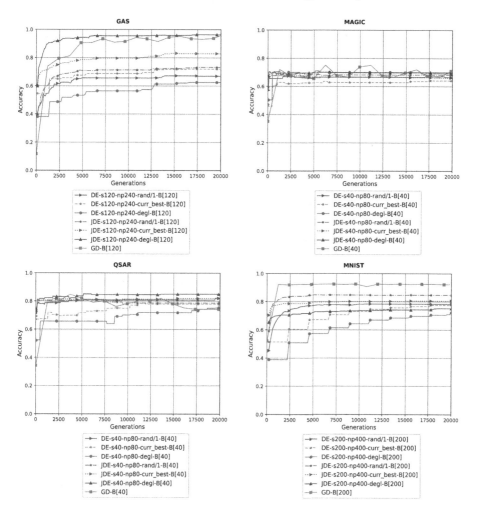

Fig. 1. Comparisons in term of accuracy among the different combinations DEvariant-mutationType on GAS, MAGIC, MNIST and QSAR datasets

number k of generations (on the x-axis from 1 to 20000) and (ii) running, on the test set, the NN that obtained the best evaluation on the validation set.

The values of the other parameters are: batch size $B = 20C$, where C is the number of classes in the dataset, population size $np = 2B$, number of training steps in the same batch $s = B$ and $counter = 10$. The values of $F = 0.5$ and $CR = 0.9$ are used only when the classical DE is applied. These data have been chosen after an extended experimental phase, partially showed in [1,30]. From these experiments we noted, for example, that fixing B and setting $s = B$ and $np = 2B$ we can obtain, on average, good results.

From the data plotted in the graphics in Fig. 1 we can conclude that in the most cases jDE performs better than standard DE. Both in the largest datasets

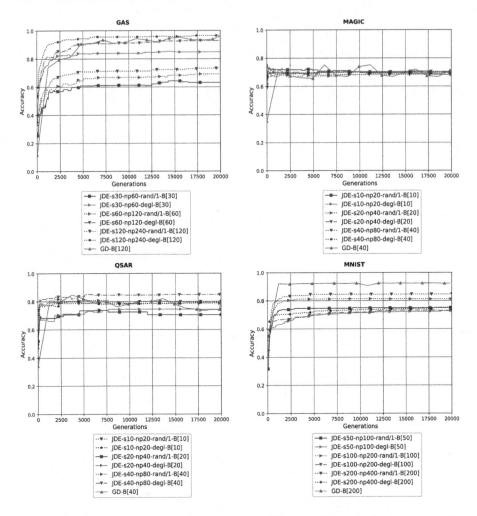

Fig. 2. Comparisons in term of accuracy among different setting of s (and consequently of B and np) on GAS, MAGIC, MNIST and QSAR datasets

(GAS and MNIST) and QSAR the differences are sharp, while in the case of MAGIC dataset the differences are so narrow that it is impossible to distinguish the best algorithm.

While the "winner" variant is undisputable (at least for these experiments), the same is not for the mutation operator: in the GAS dataset the difference between the combination $DE+DEGL$ and the second performing $jDE+current\_to\_best$ is clear, but in the MNIST dataset the best performing combination is $jDE+rand/1$, with $DE+DEGL$ only in 4th/5th position.

In Fig. 2 data on accuracy for different settings of B, s and np are plotted. In these graphics we can compare different values for the batch size B, set respectively to $B = 5C$, $B = 10C$, $B = 20C$, where C is the number of the classes.

The right setting of B is determinant for our algorithm because a too small value for B does not allow to reach good performance, while a too high value can increase too much the execution time and moreover can cause overfitting.

From the plots we can see that, excluding the cases where the differences are not significant, the best values are obtained with the highest values of B, both for the *rand/1* and *degl* mutation.

6   Conclusions and Future Works

In this paper we presented an algorithm based on Differential Evolution to train the weights of a neural network. This algorithm can be an effective alternative to the backpropagation method because of its intrinsic advantages deriving from the use of an evolutionary algorithm. The experiments presented show how the system is able to solve classification problems also in case of large image datasets, like MNIST, reaching satisfying accuracy very close to the state of the art. These results are very encouraging considering that the algorithm and the implementation could be improved and other enhancements are already under investigation.

The proposed approach allows also to handle computational models based on neural networks which do not need to be fully differentiable and this can lead to simpler models.

Future works include: the implementation of other DE variants and mutation/crossover operators; the application of the system both to other kind of problems like numerical estimation and to larger problems; the application to other computational models based on neural networks like Neural Turing Machines.

References

1. Bari, G.D.: Denn: Differential evolution for neural networks. Master thesis (2017)
2. Bengio, Y., Goodfellow, I.J., Courville, A.: Deep learning. Nature **521**, 436–444 (2015)
3. Brest, J., Boskovic, B., Mernik, M., Zumer, V.: Self-adapting control parameters in differential evolution: a comparative study on numerical benchmark problems. IEEE Trans. Evol. Comput. **10**(6), 646–657 (2006)
4. Cardamone, L., Loiacono, D., Lanzi, P.L.: Evolving competitive car controllers for racing games with neuroevolution. In: Proceedings of the 11th Annual Conference on Genetic and Evolutionary Computation, pp. 1179–1186. ACM (2009)
5. Collobert, R., Weston, J.: A unified architecture for natural language processing: deep neural networks with multitask learning. In: Proceedings of the 25th International Conference on Machine Learning, ICML 2008, pp. 160–167. ACM, New York (2008)
6. Das, S., Abraham, A., Chakraborty, U.K., Konar, A.: Differential evolution using a neighborhood-based mutation operator. IEEE Trans. Evol. Comput. **13**(3), 526–553 (2009)

7. Das, S., Mullick, S.S., Suganthan, P.: Recent advances in differential evolution an updated survey. Swarm Evol. Comput. **27**, 1–30 (2016)
8. Das, S., Suganthan, P.N.: Differential evolution: a survey of the state-of-the-art. IEEE Trans. Evol. Comput. **15**(1), 4–31 (2011)
9. Donate, J.P., Li, X., Sánchez, G.G., de Miguel, A.S.: Time series forecasting by evolving artificial neural networks with genetic algorithms, differential evolution and estimation of distribution algorithm. Neural Comput. Appl. **22**(1), 11–20 (2013)
10. Floreano, D., Dürr, P., Mattiussi, C.: Neuroevolution: from architectures to learning. Evol. Intell. **1**(1), 47–62 (2008)
11. Graves, A., Wayne, G., Danihelka, I.: Neural turing machines. arXiv preprint arXiv:1410.5401 (2014)
12. Graves, A., Wayne, G., Reynolds, M., Harley, T., Danihelka, I., Grabska-Barwińska, A., Colmenarejo, S.G., Grefenstette, E., Ramalho, T., Agapiou, J., et al.: Hybrid computing using a neural network with dynamic external memory. Nature **538**(7626), 471–476 (2016)
13. Hausknecht, M., Lehman, J., Miikkulainen, R., Stone, P.: A neuroevolution approach to general atari game playing. IEEE Trans. Comput. Intell. AI Games **6**(4), 355–366 (2014)
14. Heidrich-Meisner, V., Igel, C.: Neuroevolution strategies for episodic reinforcement learning. J. Algorithms **64**(4), 152–168 (2009)
15. Hinton, G., Deng, L., Yu, D., Dahl, G.E., Mohamed, A.R., Jaitly, N., Senior, A., Vanhoucke, V., Nguyen, P., Sainath, T.N., Kingsbury, B.: Deep neural networks for acoustic modeling in speech recognition: the shared views of four research groups. IEEE Signal Process. Mag. **29**(6), 82–97 (2012)
16. Igel, C.: Neuroevolution for reinforcement learning using evolution strategies. In: The 2003 Congress on Evolutionary Computation, 2003. CEC 2003, vol. 4, pp. 2588–2595 (2003)
17. Ilonen, J., Kamarainen, J.K., Lampinen, J.: Differential evolution training algorithm for feed-forward neural networks. Neural Process. Lett. **17**(1), 93–105 (2003)
18. Krizhevsky, A., Sutskever, I., Hinton, G.E.: Imagenet classification with deep convolutional neural networks. In: Pereira, F., Burges, C.J.C., Bottou, L., Weinberger, K.Q. (eds.) Advances in Neural Information Processing Systems 25, pp. 1097–1105. Curran Associates Inc. (2012)
19. Leema, N., Nehemiah, H.K., Kannan, A.: Neural network classifier optimization using differential evolution with global information and back propagation algorithm for clinical datasets. Appl. Soft Comput. **49**, 834–844 (2016). http://www.sciencedirect.com/science/article/pii/S1568494616303866
20. Masters, T., Land, W.: A new training algorithm for the general regression neural network. In: 1997 IEEE International Conference on Systems, Man, and Cybernetics, vol. 3, pp. 1990–1994 (1997)
21. Mattiussi, C., Dürr, P., Floreano, D.: Center of mass encoding: a self-adaptive representation with adjustable redundancy for real-valued parameters. In: Proceedings of the 9th Annual Conference on Genetic and Evolutionary Computation, GECCO 2007, pp. 1304–1311. ACM, New York (2007)
22. Miikkulainen, R.: Neuroevolution, pp. 716–720. Springer, Boston (2010). https://doi.org/10.1007/978-0-387-30164-8_589
23. Morse, G., Stanley, K.O.: Simple evolutionary optimization can rival stochastic gradient descent in neural networks. In: Proceedings of the Genetic and Evolutionary Computation Conference (GECCO) 2016, pp. 477–484. ACM, New York (2016)

24. Piotrowski, A.P.: Differential evolution algorithms applied to neural network training suffer from stagnation. Appl. Soft Comput. **21**, 382–406 (2014)
25. Price, K., Storn, R.M., Lampinen, J.A.: Differential Evolution: A Practical Approach to Global Optimization. Springer, Heidelberg (2006). https://doi.org/10.1007/3-540-31306-0
26. Reed, S., de Freitas, N.: Neural programmer-interpreters. Technical report, arXiv:1511.06279 (2015). http://arxiv.org/abs/1511.06279
27. Santucci, V., Baioletti, M., Milani, A.: Algebraic differential evolution algorithm for the permutation flowshop scheduling problem with total flowtime criterion. IEEE Trans. Evol. Comput. **20**(5), 682–694 (2016)
28. Schaffer, J.D., Whitley, D., Eshelman, L.J.: Combinations of genetic algorithms and neural networks: a survey of the state of the art. In: Proceedings of COGANN 1992: International Workshop on Combinations of Genetic Algorithms and Neural Networks, pp. 1–37 (1992)
29. Schraudolph, N.N., Belew, R.K.: Dynamic parameter encoding for genetic algorithms. Mach. Learn. **9**(1), 9–21 (1992)
30. Tracolli, M.: Enhancing denn with adaboost and self adaptation. Master thesis (2017)
31. Vesterstrom, J., Thomsen, R.: A comparative study of differential evolution, particle swarm optimization, and evolutionary algorithms on numerical benchmark problems. In: Proceedings of the 2004 Congress on Evolutionary Computation (IEEE Cat. No.04TH8753), vol. 2, pp. 1980–1987 (2004)
32. Wang, L., Zeng, Y., Chen, T.: Back propagation neural network with adaptive differential evolution algorithm for time series forecasting. Expert Syst. Appl. **42**(2), 855–863 (2015)
33. Yao, X.: Evolving artificial neural networks. Proc. IEEE **87**(9), 1423–1447 (1999)

BRKGA-VNS for Parallel-Batching Scheduling on a Single Machine with Step-Deteriorating Jobs and Release Times

Chunfeng Ma[1], Min Kong[1], Jun Pei[1,2(✉)], and Panos M. Pardalos[2]

[1] School of Management, Hefei University of Technology, Hefei 230009, China
feiyijun.ufl@gmail.com
[2] Department of Industrial and Systems Engineering,
Center for Applied Optimization, University of Florida, Gainesville 32611, USA

Abstract. This paper investigates the problem of scheduling step-deteriorating jobs with release times on a single parallel-batching machine. The processing time of each job can be represented as a simple non-linear step function of its starting time. The machine can process up to c jobs simultaneously as a batch. The objective is to minimize the makespan, and we show that the problem is strongly NP-hard. Then, a hybrid meta-heuristic algorithm BRKGA-VNS combining biased random-key genetic algorithm (BRKGA) and variable neighborhood search (VNS) is proposed to solve this problem. A heuristic algorithm H is developed based on the structural properties of the problem, and it is applied in the decoding procedure of the proposed algorithm. A series of computational experiments are conducted and the results show that the proposed hybrid algorithm can yield better solutions compared with BRKGA, PSO (Particle Swarm Optimization), and VNS.

Keywords: Parallel-batching · Step-deteriorating · Release times · Makespan

1 Introduction

In the past two decades, the parallel-batching scheduling problem has become an important research direction in the field of scheduling. Meanwhile, job release times and step-deteriorating jobs cannot be ignored in the practical situations. Some certain cases are indicated in many production scenarios, such as the surface treatment of the steel products. Consider the production of custom industrial steel products, such as engine case, vault doors or boiler covers, whereby iron ingots are first converted into different steel products in the mold workshop, and then surface treatment of the steel products will be carried out on a machine. The steel products have to reach a threshold temperature before it can be processed in batch by the machine into end products. If the starting time of batches is later than deterioration time, the batches need to be reheated or reprocessed before the machine can work on it. Consequently, extended time is required to produce each batch from steel products that has waited longer than a certain time interval. For this parallel-batching scheduling problem, the processing time of

© Springer International Publishing AG 2018
G. Nicosia et al. (Eds.): MOD 2017, LNCS 10710, pp. 414–425, 2018.
https://doi.org/10.1007/978-3-319-72926-8_34

batch is the largest processing time in the batch. The release time of batch is the largest release times among all the jobs in the batch.

The classical parallel-batching models relevant to release times have been addressed by many researchers. Li et al. provided a polynomial-time algorithm for the bounded problem $1|p - batch, p_j = b_j t, r_j, b < n|C_{max}$, and showed that the problem was binary NP-hard [1]. Ng et al. showed that $1|r_j, p - batch, p_j, c = 2|C_{max}$, was computationally intractable by performing a reduction from the following strongly NP-complete product partition problem [2].

Several papers dealing with deteriorating jobs in the context of batch scheduling have been published. Barketau et al. proposed a reprocessing model and allowed the defective items to continue deteriorating [3]. Ji and Cheng considered batch scheduling with linear deterioration in different situations. A fully polynomial time approximation scheme and a dynamic programming algorithm running in pseudo-polynomial time were proposed [4]. Recently, the serial-batching scheduling problems with step-deteriorating jobs were also studied, while many significant differences can be found in them. Mosheiov first studied the minimization of flowtime with step-deteriorating jobs, and proved that the problem was NP-hard, and also gave several heuristic procedures of some general models [5]. However, these studies focus on the linear deterioration in the context of batch scheduling problems and serial-batching scheduling problems with step-deteriorating jobs.

In this paper we address the problem of minimizing makespan on a single parallel-batching machine considering step-deteriorating jobs and non-identical release times. This problem has rarely been investigated in the scheduling literature, although parallel-batching problems with deteriorating or release times have been studied extensively. In many realistic manufacturing environments, this problem is worthwhile for us to consider three significant features together. The main contributions of this paper can be summarized as follows.

(1) The bounded parallel-batching scheduling problem with step-deteriorating jobs and release times is studied in this paper. To the best of our knowledge, this type of problem has been rarely discussed.

(2) Several structural properties on jobs batching and batching sequencing are addressed. Based on these properties, a heuristic algorithm H is proposed. Meanwhile, the heuristic algorithm is used in the decoding procedure of the meta-heuristic algorithm.

(3) Due to its NP-hard nature, a hybrid meta-heuristic algorithm BRKGA-VNS is presented. In this algorithm, a variant of VNS is applied to improve the effectiveness of iterative process for the biased random-key genetic algorithm.

The rest of this paper is organized as follows. Section 2 describes the problem. Section 3 introduces some basic lemmas. Section 4 proposes heuristic methods based on the lemmas. The computational experiments are conducted to evaluate the proposed methods in Sect. 5. Finally, In Sect. 6 we conclude the paper.

2 Notations and Problem Description

The notations used in this paper are described as Table 1.

Table 1. Parameters and description

Parameters	Description
j, x	Job index
i, k	Batch index
b	Identical extended time
d	Identical deterioration time
n	Total number of the jobs
m	Total number of the batches
c	Capacity of the parallel-batching machine
r_j	Job release time of $J_j, j = 1, 2, \cdots, n$
r_{max}	Maximum release time of all jobs
R^i	Batch release time of $B_i, i = 1, 2, \cdots, m$
a_j	Basic processing time of job $J_j, j = 1, 2, \cdots, n$
p_j	Actual processing time of job $J_j, j = 1, 2, \cdots, n$
p^i	Actual processing time of batch $B_i, i = 1, 2, \cdots, m$
C_i	Starting time of batch $B_i, i = 1, 2, \cdots, m$
C_{max}	The makespan of the schedule

We consider that a manufacturer which has a single parallel-batching machine can process at most c jobs simultaneously. There are $n(n \geq 1)$ jobs to be processed, and the job set is $J = \{J_1, J_2, \ldots, J_n\}$. The processing time of each job is modeled as a step function dependent upon its starting time. An identical extended time b is penalized when the starting time of job is later than an identical deterioration time d. The actual processing time of job J_j is $p_j = \begin{cases} a_j & s_j \leq d \\ a_j + b & otherwise \end{cases}$.

We need to make decisions on the job batching and batch sequencing simultaneously to minimize the makespan. The studied scheduling problem is described in Fig. 1. We investigate the actual situations in the surface treatment of the steel products. In most cases, the maximum release time of all jobs is no more than the deterioration time. Based on this situation, we just consider $r_{max} \leq d$.

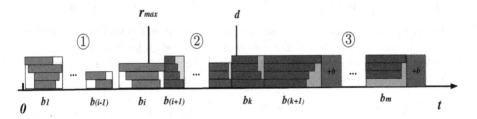

Fig. 1. The process of the studied scheduling problems

Using the conventional notation of Graham et al. [6], the type of scheduling problem is represented as $1|r_j, p-batch, p_j = a_j \text{ or } a_j+b, c|C_{max}$. When a batch is being processed, no job can be added into the batch and no job can be removed from the batch.

3   The Problem $1|r_j, p-batch, p_j = a_j \text{ or } a_j+b, c|C_{max}$

In this section, we first discuss two properties of general problem, and then discuss these properties of the special cases, which will be used in the problem with the objective of minimizing the makespan.

3.1   The Properties of General Problem

We study a single parallel-batching machine scheduling problem with the objective of minimizing the makespan. For the general problem, two properties on the jobs sequencing and batching argument are proposed as follows.

Lemma 1. Consider a schedule $\rho = (\rho_1, B_i, B_{i+1}, \rho_2)$, if $B_{i+1} \cap \{J_j\} = \{J_j\}$, $(j = 1, 2, \cdots, n)$, and there exists the situation that $r_j \leq R^i$, $p_j \leq p^i$, and $|b_i + 1| \leq c$, then putting job J_j in batch B_i would improve the solution.

Proof: It is obvious that putting job J_j in batch B_i would not affect the release times and processing time of batch B_i, but the release times and processing time of batch B_{i+1} may be reduced, this lemma is proved.   □

Lemma 2. Let schedule $\sigma = (\sigma_1, B_i, B_{i+1}, \sigma_2)$ represent an optimal schedule. A batch pair $(i, i+1)$ in schedule σ has the situations that $R^i \geq R^{i+1}$ and $p^i \geq p^{i+1}$, then swapping batch B_i and B_{i+1} would improve the solution.
Proof: It is easy to understand, we omit it.   □

3.2   The Properties of Special Cases

In this subsection, the properties of special cases are discussed. Firstly, we show that the problem $1|r_j, p-batch, p_j = a_j \text{ or } a_j+b, c|C_{max}$ is strongly NP-hard when batch's starting time is earlier than r_{max}, and we apply ERT-LPT (Earliest Release Time-Longest Processing Time) rule for the first procedure [7]. Secondly, we apply FBSPT (Full Batch Shortest Processing Time) rule when batch's starting time is between r_{max} and d. Thirdly, we show an optimal solution to solve the case when batch's starting time is more than d.

Lemma 3. The problem $1|r_j, p-batch, p_j = a_j \text{ or } a_j+b, c|C_{max}$ is strongly NP-hard.
Brucker et al. showed the problem $1|r_j, p-batch, p_j, b<n|C_{max}$ was strongly NP-hard [8]. In our problem, considering that $b = 0$ and $d = \infty$, then, the problem is reduced to $1|r_j, p-batch, p_j, b<n|C_{max}$. Hence, the problem $1|r_j, p-batch, p_j = a_j \text{ or } a_j+b, c|C_{max}$ is also strongly NP-hard.

Lemma 4. There is a satisfactory solution that batches are sequenced in the FBSPT rule when batch's starting time is between r_{max} and d.

From Lemma 4, we can get a satisfactory solution when batch's starting time is between r_{max} and d, at this point, all jobs are available and don't have the deteriorative effect, we should process more jobs before d. Therefore, it can be solved satisfactorily by the full batch shortest processing time rule, where all batches are full except for the one batch that makes next batch having deteriorative effect.

Lemma 5. When batch's starting time is later than d, there exists an optimal schedule that jobs are scheduled in the FBLPT (Full Batch Longest Processing Time) rule [9].

Based on the Lemma 5, we can obtain the optimal schedule when batch's starting time is later than d. Note that the remaining jobs are batched as full as possible such that all the remaining batches are full except the one with the highest index.

4   Heuristic

Once the job sequence is determined, a heuristic algorithm H is developed to calculate the makespan of this job set. The process of the proposed heuristic is implemented through three phases: (1) $C_i < r_{max}$, (2) $r_{max} \leq C_i \leq d$, and (3) $d < C_i$. The parameters and description of heuristic algorithm H are given in Table 2.

Table 2. Parameters and description

J	Set of all jobs, $J = \{J_1, J_2, \ldots, J_n\}$
$A(t)$	Set of jobs that are available at time t
$U(t)$	Set of jobs that have not been scheduled yet at time t
S	Schedule of the generated batches
C_{s-max}	The completion time of last batch in the schedule S
P, Q	Index of the last batch generated in the phase

For simplicity, a detailed description of the proposed heuristic algorithm is given in Table 3.

5   BRKGA-VNS Algorithm

5.1   Key Steps of BRKGA-VNS Algorithm

In this subsection, a BRKGA-VNS algorithm combining biased random-key genetic algorithm (BRKGA) and variable neighborhood search (VNS) algorithm is proposed to solve the studied problem. Goncalves and Resende first proposed biased random-key genetic algorithm (BRKGA), and gave a detailed analysis. Comparing BRKGA heuristic with other standard GAs, the results showed that the BRKGA heuristic were indeed competitive [10]. Resende and Ribeiro introduced an Application Programming Interface (API) for quick implementations of BRKGA heuristics, and apply the framework to a number of hard combinatorial optimization problems [11].

Table 3. The description of heuristic algorithm H

Heuristic algorithm H	
Step 1	Set $i = 1$, $B_1 = J_1$, $p^i = a_1$, $t = r_1$, $U(t) = J \setminus \{J_1\}$
Step 2	If $r_{max} \leq t \leq d$, go to step 6, and if $d < t$, go to step 10, Otherwise, go to step 3
Step 3	$A(t) = \{J_j \mid r_j \leq t\}$. Place the first job of $A(t)$ into S, $A(t) = A(t) \setminus \{J_1\}$, index jobs in schedule S as rule ERT-LPT. Update $U(t)$ and $A(t)$
Step 4	Judge whether each job in $A(t)$, if there exist a job J_x satisfying that $r_x \leq R^i, p_x \leq p^i$, then place job J_x into B_i
Step 5	Process schedule S. Set $t = C_{s-max}$, if $U(t) = \emptyset$, then stop, output $C_{max} = C_{s-max}$, otherwise, go to step 2
Step 6	Update $U(t)$ and $A(t)$, set $P = i$
Step 7	If $d < t$, go to step 10, Otherwise, go to step 8
Step 8	Place the first job of $A(t)$ into S, $A(t) = A(t) \setminus \{J_1\}$, index new jobs in schedule S as rule FBSPT. Update $U(t)$ and $A(t)$
Step 9	Process schedule S, Set $t = C_{s-max}$, if $U(t) = \emptyset$, then stop, output $C_{max} = C_{s-max}$, otherwise, go to step 6
Step 10	Update $U(t)$ and $A(t)$, set $Q = i$
Step 11	Place the first job of $A(t)$ into S, $A(t) = A(t) \setminus \{J_1\}$, index new jobs in schedule S as rule FBLPT. Update $U(t)$ and $A(t)$
Step 12	Process schedule S, Set $t = C_{s-max}$, if $U(t) = \emptyset$, then stop, output $C_{max} = C_{s-max}$, otherwise, go to step 10

As we know, VNS has received extensive attentions from researchers in recent years, and is widely used in various combinatorial optimization problems since it was improved by Hansen and Mladenović [12]. In order to speed up the convergence of the algorithm, we adopt the variable neighborhood search (VNS) algorithm to replace the iteration procedure in traditional BRKGA. The VNS algorithm is introduced to change the iterative process of BRKGA, and improve the process from the elite random-key vectors to the partition labeled TOP in the next population (The partition labeled TOP is a set of elite partition).

For the procedure of encoding and decoding, we can know that each vector r_i is composed of n randomly generated numbers (random keys) in the real interval (0,1]. The vector $r_i = r_{i1}, \cdots, r_{id}, \cdots, r_{in}$ is decoded in the following steps:

Table 4. The procedure of encoding and decoding

Step 1	For a vector $a = \{1, 2, \cdots, n\}$, sort all the elements in vector a by the non-decreasing order of λ_{id} to obtain a job list;
Step 2	Execute the Algorithm H to obtain the C_{max} of the job list;
Step 3	Output the C_{max} as the fitness of λ_i.

The algorithm framework of BRKGA-VNS is described in Table 4, and the part of VNS is reflected in the 6th line in the pseudocode (Table 5).

Table 5. The Pseudocode of BRKGA-VNS

Pseudocode of BRKGA-VNS

1.　　Initialize the parameters of algorithm, such as the maximum iterations L_{max}, the size of population N, the number of TOP solutions NB, and the number of BOT solutions NW. Set $L = 1$.

2.　　Randomly　initialize　the　population,　$P^{(L)} = \{P_1^{(L)}, \cdots, P_g^{(L)}, \cdots, P_N^{(L)}\}$,　where $P_g^{(L)} = \{P_{g1}^{(L)}, \cdots, P_{gj}^{(L)}, \cdots, P_{gn}^{(L)}\}$ is the gth individual in the population, $g = 1, \cdots, N, j = 1, \cdots, n$.

3.　　**While** $L < L_{max}$ **do**

4.　　Calculate the fitness of each individual in $P^{(L)}$ and sort the individuals by the non-increasing order of its fitness.

5.　　Divide $P^{(L)}$ into $P^{(L)e}$ and $P^{(L)\bar{e}}$ /* $P^{(L)e}$ is the set of NB elite individuals and $P^{(L)\bar{e}}$ is the remaining non-elite individuals

6.　　$P^{(L+1)e} \leftarrow RVNS(P^{(L)e})$ /*Execute the **RVNS** algorithm for each individual in $P^{(L)e}$ to obtain the elite individuals $P^{(L+1)e}$ in next generation.

7.　　$P^{(L+1)} \leftarrow P^{(L+1)} \cup P^{(L+1)e}$

8.　　**For** $g \leftarrow NW + 1$ to $N - NB$ **do**

9.　　Select parent $P_a^{(L)}$ from $P^{(L)e}$ randomly

10.　Select parent $P_b^{(L)}$ from $P^{(L)\bar{e}}$ randomly

11.　**For** $j \leftarrow 1$ to n **do**

12.　　**If** $rand > 0.5$ **then** /* $rand$ is a random number $\in (0, 1)$

13.　　　$P_{cj}^{(L)} \leftarrow P_{aj}^{(L)}$ /* $P_c^{(L)}$ is the offspring of parent $P_a^{(L)}$ and $P_b^{(L)}$

14.　　**Else**

15.　　　$P_{cj}^{(L)} \leftarrow P_{bj}^{(L)}$

16.　　**End if**

17.　**End for**

18.　$P^{(L+1)} \leftarrow P^{(L+1)} \cup \{P_c^{(L)}\}$

19.　$g ++$

20.　**End for**

21.　Generate NW mutants $P^{(L+1)m}$ each having n random-keys $\in (0, 1]$

22.　$P^{(L+1)} \leftarrow replace(P^{(L+1)m})$ /*Replace the NW worst solutions in $P^{(L+1)}$ with $P^{(L+1)m}$

23.　$L ++$

24.　**End while**

25.　**Output** $f* \leftarrow argmin\{f(P_g^{(L_{max})}) \mid P_g^{(L_{max})} \in P^{(L_{max})e}\}$

We also give the framework of the proposed algorithm as Fig. 2.

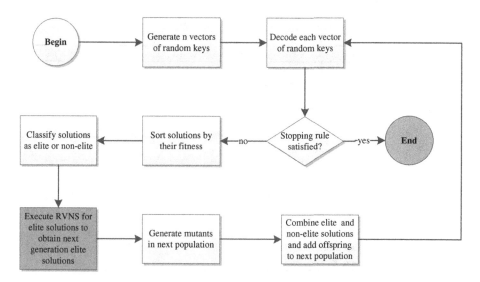

Fig. 2. The flowchart of the proposed hybrid BRKGA-VNS

5.2   RVNS

İn order to enhance solution quality of the elite individuals, we apply a variant of VNS named RVNS for them in each iterations [13]. Based on the characteristic of random-key encoding, a simple neighborhood structure is applied in this paper, $N_h(X)$ is used to denote the hth neighborhood of individual X, and the $N_h(X)$ is defined as follows (Table 6).

Table 6. Neighborhood structure

Neighborhood structure	
Step 1.	Set $u = 1$.
Step 2.	Randomly select a gene of individual X to perform mutation operation.
Step 3.	If $u \leq h$, then go to step 2. Otherwise, output solution X.

The difference between VNS and RVNS is that the RVNS remove the local search procedure from the basic VNS. The pseudocode of RVNS is given as below (Table 7).

5.3   Computational Experiments and Comparison

Based on the situation that we just consider $r_{max} \leq d$, the paper introduces the features that the release times do not exceed the deterioration time. In this subsection, a serial of computational experiments are conducted to test the performance of our proposed algorithm BRKGA-VNS, compared with BRKGA [10], VNS [12], and PSO [14]. The parameters of the test problems were randomly generated as Table 8.

Table 7. The pseudocode of RVNS

The pseudocode of RVNS	
1.	Give an initial solution $X = \{x_1, \cdots, x_n\}$, and set u_{max} ($u_{max} \leq n$)
2.	**While** the termination condition is not met, **do**
3.	$u \leftarrow 1$
4.	**While** $u < u_{max}$ **do**
5.	$X' \leftarrow Shake(X), X' \in N_u(X)$ /* Randomly generate a solution X' from $N_u(X)$ */
6.	**If** X is better than X, **then**
7.	$X \leftarrow X', u \leftarrow 1$
8.	**else**
9.	$u \leftarrow u + 1$
10.	**End if**
11.	**End while**
12.	**End while**
13	**Output** X

Table 8. Parameters setting

Notation	Definition	Value
n	The number of jobs	40,50,60,70.100,150,200,250
d	Identical deterioration time	U[180,300] for small scale, U[360,750] for large scale
b	Identical extended time	U[20,30]
c	Capacity of the parallel-batching machine	U[3, 8]
a_j	Basic processing time of job J_j j $= 1, 2, \cdots,$ n	U[10,30]
r_j	Release time of job J_j j $= 1, 2, \cdots,$ n	U[0,100]
NB	The number of TOP solutions	5
NW	The number of BOT solutions	5

Table 9. The results of the average objective value (Ave.Obj) and the minimum objective value (Max.Obj) for each algorithm

		BRKGA-VNS		BRKGA			PSO			VNS		
No.	n	Ave.Obj	Max.Obj	Ave.Obj	Max.Obj	Impr.	Ave.Obj	Max.Obj	Impr.	Ave.Obj	Max.Obj	Impr.
1	40	254.45	259.78	264.15	281.42	3.67	257.57	274.54	1.13	271.28	423.26	6.20
2	50	307.66	329.94	325.22	338.43	5.40	308.50	337.88	0.25	320.33	458.42	3.96
3	60	358.88	406.24	376.70	398.07	4.73	373.98	390.69	3.86	374.20	559.03	4.09
4	70	429.98	487.82	459.02	503.70	6.33	449.46	485.46	4.01	454.93	644.09	5.49
5	100	607.92	662.32	627.71	689.70	3.15	616.72	663.73	1.33	625.04	803.76	2.74
6	150	891.86	958.24	913.73	973.43	2.39	905.24	971.87	1.38	928.35	1156.86	3.93
7	200	956.11	1042.44	985.60	1070.25	2.99	976.54	1058.39	1.93	981.81	1118.24	2.62
8	250	1142.25	1186.47	1165.63	1219.24	2.01	1151.48	1193.65	0.77	1159.54	1342.51	1.49

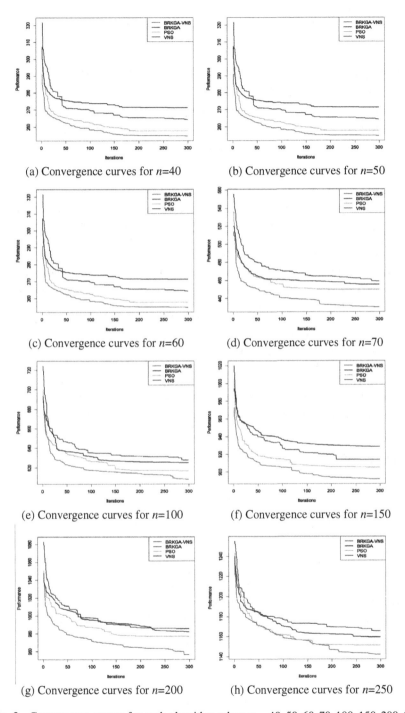

(a) Convergence curves for n=40

(b) Convergence curves for n=50

(c) Convergence curves for n=60

(d) Convergence curves for n=70

(e) Convergence curves for n=100

(f) Convergence curves for n=150

(g) Convergence curves for n=200

(h) Convergence curves for n=250

Fig. 3. Convergence curves for each algorithm when $n = 40, 50, 60, 70, 100, 150, 200, 250$

In order to evaluate the performance of BRKGA-VNS, the result of the average objective value (Ave.Obj) and the maximum objective value (Max.Obj) for the problem is listed in Table 9, the result of the improve objective value (Impr.(%)) obtain by $\frac{Ave.Obj_{BRKGA-VNS}-Ave.Obj_{other}}{Ave.Obj_{BRKGA-VNS}} \times 100$. We compare the effect of BRKGA-VNS with BRKGA, PSO and VNS when number of jobs is different, if other variables are same, all experimental results show that BRKGA-VNS is better, and the convergence curves of each algorithm are shown in Fig. 3.

All cases run 10 times to avoid the contingency of the experiment. To ensure the fairness of the experiment, the population size of BRKGA-VNS, BRKGA and PSO is set as 20. All the algorithms were implemented in C++ and run on a Lenovo computer running Window10 with a dual-core CPU Intel i3-3240@3.40 GHz and 4 GB RAM. All of algorithms perform 300 iterations in reasonable time, for example, we recorded that the program runs in 9.8 s at a time when n = 60, it is shown that the running time of BRKGA-VNS will not exceed 1 s. As can be observed from Table 9, we can conclude that proposed algorithm has better performance than the other algorithms with respect to solution quality, since the results obtained by BRKGA-VNS are better than those obtained by other algorithms among all the cases. Figure 3 shows the convergence curves of BRKGA-VNS, BRKGA, PSO, and VNS for each category so as to demonstrate the convergence performance of BRKGA-VNS more clearly. The figure shows the average of the best solution at each generation, we compare the problem size's effect on the average value when n = 40, 50, 60, 70, 100, 150, 200, 250, compared with BRKGA, PSO, and VNS, the hybrid BRKGA-VNS has both faster convergence speed and better results when solving the problems. Based on the computational results, it can be obtained that the BRKGA-VNS has better performance rate than those of other algorithms in solving the presented problem.

6 Conclusion

We study a single parallel-batching machine scheduling problem, it is motivated by the surface treatment of the steel products. We show that the problem is strongly NP-hard, and some structural properties are presented for both general problem and special cases. Based on these properties, a hybrid meta-heuristic BRKGA-VNS is proposed to solve the problem. And the results obtained by BRKGA-VNS are better than those obtained by BRKGA, PSO, and VNS among all the cases.

In future research work, we may devote to developing more effective meta-heuristic to solve some related parallel-batching scheduling problems, considering other objective functions. Moreover, the general batch scheduling models considering multiple unrelated parallel machines will be investigated, and more research problems from the real industry will be refined.

Acknowledgments. This work is supported by the National Natural Science Foundation of China (Nos. 71231004, 71601065, 71690235, 71501058, 71690230, 71601060), and Innovative Research Groups of the National Natural Science Foundation of China (71521001), the Humanities and Social Sciences Foundation of the Chinese Ministry of Education

(No. 15YJC630097), Anhui Province Natural Science Foundation (No. 1608085QG167). Panos M. Pardalos is partially supported by the project of Distinguished International Professor by the Chinese Ministry of Education (MS2014HFGY026).

References

1. Li, S., Ng, C.T., Cheng, T.C.E., et al.: Parallel-batch scheduling of deteriorating jobs with release dates to minimize the makespan. Eur. J. Oper. Res. **210**(3), 482–488 (2011)
2. Ng, C.T., Barketau, M.S., Cheng, T.C.E., et al.: "Product Partition" and related problems of scheduling and systems reliability: computational complexity and approximation. Eur. J. Oper. Res. **207**(2), 601–604 (2010)
3. Barketau, M.S., Cheng, T.C.E., Kovalyov, M.Y.: Batch scheduling of deteriorating reworkables. Eur. J. Oper. Res. **189**(3), 1317–1326 (2008)
4. Ji, M., Cheng, T.C.E.: Batch scheduling of simple linear deteriorating jobs on a single machine to minimize makespan. Eur. J. Oper. Res. **202**(1), 90–98 (2010)
5. Mor, B., Mosheiov, G.: Batch scheduling with step-deteriorating processing times to minimize flowtime. Naval Res. Logist. (NRL) **59**(8), 587–600 (2012)
6. Graham, R.L., Lawler, E.L., Lenstra, J.K., et al.: Optimization and approximation in deterministic sequencing and scheduling: a survey. Ann. Discret. Math. **5**, 287–326 (1979)
7. Damodaran, P., Velez-Gallego, M.C.: Heuristics for makespan minimization on parallel batch processing machines with unequal job ready times. Int. J. Adv. Manuf. Technol. **49**(9), 1119–1128 (2010)
8. Brucker, P., Gladky, A., Hoogeveen, H., et al.: Scheduling a batching machine. J. Sched. **1**(1), 31–54 (1998)
9. Lee, C.Y.: Minimizing makespan on a single batch processing machine with dynamic job arrivals. Int. J. Prod. Res. **37**(1), 219–236 (1999)
10. Gonçalves, J.F., Resende, M.G.C.: Biased random-key genetic algorithms for combinatorial optimization. J. Heuristics **17**(5), 487–525 (2011)
11. Resende, M.G.C., Ribeiro, C.C.: Biased ranom-key genetic algorithms: an advanced tutorial. In: Proceedings of the 2016 on Genetic and Evolutionary Computation Conference Companion. Association for Computing Machinery, pp. 483–514 (2016)
12. Hansen, P., Mladenović, N.: Variable Neighborhood Search. Search Methodologies, pp. 313–337. Springer, Ner York (2014). https://doi.org/10.1007/978-1-4614-6940-7_12
13. Hansen, P., Mladenović, N., Perez-Britos, D.: Variable neighborhood decomposition search. J. Heuristics **7**(4), 335–350 (2001)
14. Kennedy, J.: Particle Swarm Optimization. Encyclopedia of Machine Learning, pp. 760–766. Springer, New York (2011). https://doi.org/10.1007/978-0-387-30164-8_630

Petersen Graph is Uniformly Most-Reliable

Guillermo Rela[✉], Franco Robledo, and Pablo Romero

Facultad de Ingeniería, Instituto de Matemática y Estadística,
IMERL, Universidad de la República,
Montevideo, Uruguay
{grela,frobledo,promero}@fing.edu.uy

Abstract. A celebrated problem in network optimization is the all-terminal reliability maximization. We want to communicate a fixed number n of terminals, but we have a fixed budget constraint m. The goal is to build m links such that the all-terminal reliability is maximized in the resulting graph. In such case, the result is a uniformly most-reliable graph. The discovery of these graphs is a challenging problem that launched an interplay between extremal graph theory and computational optimization.

In this paper, we mathematically prove that Petersen graph is uniformly most-reliable. The paper is closed with a conjecture on the existence of other uniformly most-reliable graphs.

Keywords: Network reliability analysis
Uniformly most-reliable graphs · Petersen graph

1 Motivation

Historically, extremal graph theory is inspirational for network design [7]. In the second book ever written in graph theory, Berge challenges the readers to find the graph with maximum connectivity among all graphs with a fixed number of nodes and links. Frank Harary provided not only a full answer, but also found connected graphs with minimum and maximum diameter [12]. Gustav Kirchhoff solved linear time-invariant resistive circuits, and as corollary he introduced the Matrix-Tree theorem, where he counts the number of spanning trees of a connected graph (i.e., the tree-number) using the determinant of a matrix [14]. This breakthrough in electrical systems launched the theory of trees, which represent building blocks in communication design. However, the corresponding extremal problem is not well understood: find the graph with a fixed number of nodes and links that maximizes the tree-number.

All the previous problems are connectivity-based, and deterministic in nature. Network reliability analysis deals with probabilistic-based models, where the goal is to determine the probability of correct operation of a system [3,11]. In its most elementary setting, we are given a simple graph G with perfect nodes

© Springer International Publishing AG 2018
G. Nicosia et al. (Eds.): MOD 2017, LNCS 10710, pp. 426–435, 2018.
https://doi.org/10.1007/978-3-319-72926-8_35

but random link failures with identical and independent probability ρ. The all-terminal reliability is the probability that the resulting random graph remains connected.

Even though network reliability is probabilistic in nature, there is a strong interplay with the previous deterministic problems. The motivation of this paper is to have a better understanding of the interplay between network reliability analysis and inspirational problems from network connectivity, which are completely deterministic in nature.

This paper is organized as follows. Section 2 presents a formal definition of uniformly most-reliable graphs and reliability polynomials. Section 3 covers the body of related works on uniformly most-reliable graphs. The main contribution is presented in Sect. 4, where it is formally proved that Petersen graph is uniformly most-reliable. Section 5 presents concluding remarks and trends for future work.

2   Uniformly Most-Reliable Graphs

We are given a simple graph $G = (V, E)$, with perfect nodes and unreliable links that fail independently with identical failure probability ρ. The all-terminal reliability $R_G(\rho)$ measures the probability that the resulting random graph remains connected, and it is a polynomial in $\rho \in [0, 1]$. For convenience, in this paper we work with the unreliability polynomial $U_G(\rho) = 1 - R_G(\rho)$. Let us denote $p = |V|$ and $q = |E|$ the respective order and size of the graph G. Further, denote by $m_k(G)$, or simply m_k, the number of link-disconnecting sets with cardinality k, this is, the number of subsets $E' \subseteq E$ such that $|E'| = k$ and $G' = G - E'$ is disconnected. By sum-rule, the unreliability polynomial can be expressed as follows:

$$U_G(\rho) = \sum_{k=0}^{q} m_k \rho^k (1 - \rho)^{q-k}. \tag{1}$$

Let us denote (p, q)-graph to the family of graphs with p nodes and q links. Clearly, if we consider a fixed $\rho \in [0, 1]$, there is at least one graph H that attains the minimum unreliability, i.e., $U_H(\rho) \leq U_G(\rho)$ for all (p, q) graph G. Further, if the previous condition holds for all $\rho \in [0, 1]$ and all (p, q)-graphs G, the graph H is uniformly most-reliable.

3   Related Work

From inspection of Expression (1), we can see that if there exists some (p, q)-graph H such that $m_k(H) \leq m_k(G)$ for all k and all (p, q)-graph G, then H is uniformly most-reliable. Curiously enough, this sufficient criterion is not known to be necessary. However, to the best of our knowledge, the search of uniformly most-reliable graphs rests on the minimization of all the coefficients m_k. This approach is promoted by the following result, which can be proved using elementary calculus [1]:

Proposition 1.

(i) *If there exists some integer k such that $m_i(H) = m_i(G)$ for all $i < k$ but $m_k(H) < m_k(G)$, then there exists $\rho_0 > 0$ such that $U_H(\rho) < U_G(\rho)$ for all $\rho \in (0, \rho_0)$.*

(ii) *If there exists some integer k such that $m_i(H) = m_i(G)$ for all $i > k$ but $m_k(H) < m_k(G)$, then there exists $\rho_1 < 1$ such that $U_H(\rho) < U_G(\rho)$ for all $\rho \in (\rho_1, 1)$.*

By definition, there are no disconnecting sets with lower cardinality than the link connectivity λ. Therefore, $m_i(G) = 0$ for all $i < \lambda$, and by Proposition 1-(i) uniformly most-reliably graphs must have the maximum link-connectivity λ. Furthermore, the number of disconnecting sets m_λ must be minimized. On the other hand, $m_i(G) = \binom{q}{i}$ for all $i > q-p+1$, since trees are minimally connected with $q = p-1$ links. The number of connected sets with $q-p+1$ links is precisely the tree-number $\tau(G)$, so $m_{q-p+1}(G) = \binom{q}{q-p+1} - \tau(G)$. Using Proposition 1-(ii), the tree-number should be maximized. Prior observations directly connect this network design problem with extremal graph theory:

Corollary 1. *A uniformly most-reliable (p, q)-graph H must have the maximum tree-number $\tau(H)$, maximum connectivity $\lambda(H)$, and the minimum number of disconnecting sets $m_\lambda(H)$ among all (p, q)-graphs with maximum connectivity.*

For convenience we say that a (p, q)-graph, H, is t-optimal if $\tau(H) \geq \tau(G)$ for every (p, q) graph G. Briefly, Corollary 1 claims that uniformly most-reliable graphs must be t-optimal and max-λ min-m_λ, where λ denotes the edge connectivity.

Frank Harary found the maximum connectivity of a (p, q) graph. By Handshaking, the average-degree of every (p, q)-graph is $\frac{2q}{p}$. If $\delta(G)$ denotes the minimum degree, we immediately get that $\lambda(G) \leq \delta(G) \leq \lfloor \frac{2q}{p} \rfloor$. The candidate connectivity is $\lambda_{max} = \lfloor \frac{2q}{p} \rfloor$. It suffices to find a (p, q)-graph with connectivity λ_{max} whenever $p \geq q - 1$ (otherwise, the graph is not connected). The evidence is the following family of graphs [12]:

Definition 1 (Harary Graphs $H_{(n,k)}$). *Let n and k be positive integers. Harary graph $H_{(n,k)}$ consists of n nodes $\{v_0, \ldots, v_{n-1}\}$ equally spaced around a circle, and the following links:*

- *If k is even, each vertex is adjacent to the $k/2$ nearest nodes in each direction.*
- *If k is odd and n is even, $H_{(n,k)}$ is $H_{(n,k-1)}$ with additional links $\{v_i, v_{i+\frac{n}{2}}\}$ for each $i = 0, \ldots, \frac{n}{2}$.*
- *If k and n are both odd, $H_{(n,k)}$ is $H_{(n,k-1)}$ with additional links $\{v_i, v_{i+\frac{n-1}{2}}\}$ for each $i = 0, \ldots, n - 1$.*

We immediately check that Harary graphs have maximum connectivity $\lambda_{max} = \lfloor \frac{2q}{p} \rfloor$, so, they are max-$\lambda$. The number of disconnecting sets should be minimized as well; max-λ graphs that minimize the disconnecting sets with λ nodes are called max-λ min-m_λ graphs. Prior works from Bauer et al. fully characterize max-λ min-m_λ graphs [2]. A key idea is to observe that in a max-λ graph, the number of disconnecting sets m_λ is at least the number of nodes with degree λ. If this bound is achieved, a max-λ min-m_λ graph is retrieved. For that purpose, they define generalized Harary graphs, which are just an augmentation of the original Harary graphs with random matchings (this is, edges with non-adjacent nodes). In that way, the number of nodes with degree λ is minimized, and the authors show that no other disconnecting sets with that size exists.

By Corollary 1, Bauer et al. provide a family of graphs that contain all uniformly most-reliable graphs. Later works try to find uniformly $(p, p+i)$-most-reliable graphs for i small, by a simultaneous minimization of all the coefficients m_k. The cases $i = -1$ and $i = 0$ are trivial. Indeed, when $q = p - 1$ all the trees have the same reliability polynomial ρ^q, so they are uniformly most-reliable (the reliability is zero if the graph is not connected). When $i = 0$ we have $q = p$, and the elementary cycle C_p is t-optimal. All the other graphs with $p = q$ are not 2-connected, and by direct inspection we can see that C_p achieves the minimum coefficients m_k.

Perhaps the first non-trivial uniformly most-reliable graphs were found by Boesch et al. in 1991 [5]. A new reading of Bauer et al. construction lead them to find that Monma graphs are $(n, n+1)$ uniformly most-reliable graphs, whenever the number of nodes in each path differ by at most one. Interestingly enough, Clyde Monma et al. used these graphs for the design of minimum cost two-node connected metric networks [17]. Figure 1 depicts Monma graphs. The reader is invited to find a combinatorial proof of Monma's t-optimality when the length of the paths differ by at most one in [9].

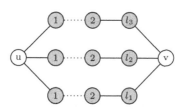

Fig. 1. Monma graph $M_{(l_1+1, l_2+1, l_3+1)}$.

A more challenging problem is to find $(n, n + 2)$ uniformly most-reliable graphs. Boesch et al. minimize the four effective terms m_0, m_1, m_2 and m_3 from Expression (1). An $(n, n + 2)$ max-λ min-m_λ graph already minimizes the first three terms. If in addition the tree-number is minimized all the coefficients are simultaneously minimized, and the result must be a uniformly most-reliable graph. The merit of the paper [5] is to adequately select the feasible graphs from Bauer et al. that minimizes the tree-number. Observe that K_4 can be partitioned into three perfect matchings, PM_1, PM_2 and PM_3.

The result is that we should insert $n - 4$ points in the six links of K_4 in such a way that:

(i) the number of inserted nodes in all the links differ by at most one, and
(ii) if we insert the same number of nodes in two different matchings $PM_i \neq PM_j$, then the number of nodes in the four links from $PM_i \cup PM_j$ are identical.

The resulting $(n, n + 2)$-graph defines, for every $n \geq 4$, a single graph up to isomorphism. The authors formally prove that the resulting graph is uniformly most-reliable $(n, n+2)$-graph. Furthermore, inspired by a previous research on t-optimality in multipartite graphs authored by Cheng [10], they conjecture that all uniformly most-reliable $(n, n + 3)$-graphs with more than 6 nodes are elementary subdivisions of $K_{(3,3)}$. This conjecture is correct, and it was proved by Wang [21].

In a recent paper, Romero formally proved that Wagner graph M_4 is uniformly most-reliable $(n, n + 4)$-graph [20]. To the best of our knowledge, the analysis of $(n, n + i)$-graphs for $i \geq 5$ is not available in the related literature.

Definition 2. *For every even natural n, Möbius graph M_n is constructed from the cycle C_{2n} adding n new links joining every pair of opposite nodes.*

Curiously enough, $M_2 = K_4$, $M_3 = K_{(3,3)}$ and M_4 are Möbius graphs M_n, and they are all uniformly most-reliable graphs. Furthermore, the discovery of $(n, n + 2)$ and $(n, n + 3)$ credited by Boesch et al. [5] and Wang [21] consider a partition of $K_{(3,3)}$ (resp. K_4) into three disjoint perfect matchings. The reader can check that all Möbius graphs can be partitioned into three such perfect matchings as well. In this sense, Möbius graphs apparently generalize the particular result for $K_{(3,3)}$ and K_4. This promotes the following:

Conjecture 1. All uniformly most-reliable $(n, n + i)$ graphs with $i < n$ are elementary subdivisions of Möbius graph M_n.

Conjecture 1 is not true. Indeed, we formally prove in Sect. 4 that Petersen graph serves as a counterexample for the case of $(n, n + 5)$-graphs. However, another (not so optimistic) conjecture is still open:

Conjecture 2. All uniformly most-reliable $(n, n + 4)$ graphs with $n \geq 8$ are elementary subdivisions of Wagner graph.

It is worth to remark that there are (p, q)-pairs where a uniformly most-reliable graphs does not exist [18]. The reader can consult [6] for a valuable survey on uniformly most-reliable graphs.

A full determination of t-optimal graphs for every (p, q)-pair is a related open problem. Indeed, a historical conjecture by Leggett and Bedrosian asserts that t-optimal graphs must be almost regular, that is, the degrees differ at most by one [16]. Even though closed formulas are available for the tree-number of specific graphs, the progress on t-optimality is effective on special regularity conditions [10], almost-complete graphs or other special graphs with few links [19] (Fig. 2).

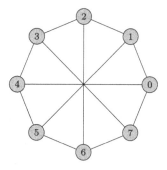

Fig. 2. Wagner Graph M_4

4   Petersen Graph

Petersen graph is the complement of the line-graph of K_5 (the reader can find alternative definitions in the book [13]). By Sachs theorem [4], its eigenvalues are 3 (simple), 1 (multiplicity 5) and -2 (multiplicity 4). Therefore, its tree-number is $\tau(P) = \frac{1}{10} \times (3 - 1)^5 \times (3 - (-2))^4 = 2000$.

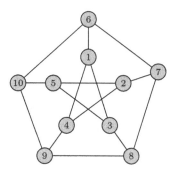

Fig. 3. Petersen graph

Figure 3 depicts Petersen graph. It is known that Petersen has the maximum tree-number among all cubic $(10, 15)$ graphs. Therefore, Petersen graph is the only candidate to be uniformly most-reliable $(10, 15)$-graph.

It is clearly super-λ with connectivity $\lambda = 3$, so $m_3 = 10$ is minimum among $(10, 15)$-graphs. From inspection we find that all the disconnecting sets with 4 links are either incident to a fixed node or fixed link, so $m_4 = 10 \times \binom{3}{3}\binom{12}{1} + 15 = 135$. Furthermore, all cubic $(10, 15)$-graphs possess the previous disconnecting sets. In order to count m_5 we observe that such disconnecting sets isolate nodes, links, 2-paths or 5-cycles, so, $m_5 = (10 \times \binom{12}{2} - 15) + 15 \times 10 + (\binom{10}{2} - 15) + 6 = 831$. Counting the complement, we know that $m_6 = \binom{15}{6} - \tau(P) = 3005$. For convenience, we say that a disconnecting set that isolates some node is trivial.

From now on, we will assume that the ground graph-set is always $(10, 15)$-graphs. The following two lemmas are preparatory for the main result, and their proofs only use combinatorial arguments:

Lemma 1. *The coefficient m_4 is minimized in Petersen graph.*

Proof. The result is trivial for cubic graphs. Consider an arbitrary $(10, 15)$-graph H, and denote $m_4 = 135$ the number of disconnecting sets in Petersen graph. If H has a bridge, then $m_4(H) \geq \binom{13}{3} \geq m_4$. It suffices to prove the result when $\delta(H) = 2$. If there exists non-adjacent nodes v_1 and v_2 such that $deg(v_1) = deg(v_2) = 2$, then $m_4(H) \geq 2 \times \binom{13}{2} - 1 > m_4$. If v_1 and v_2 are adjacent nodes then $m_4(H) \geq \binom{3}{2}\binom{12}{2} > m_4$. Finally, if there is a single node v such that $deg(v) = 2$, by Handshaking Lemma the degree-sequence must be $(4, 3, 3, 3, 3, 3, 3, 3, 3, 2)$. In this case, counting trivial disconnecting sets we know that $m_4(H) \geq \binom{4}{4} + 8 \times \binom{3}{3}\binom{12}{1} + \binom{3}{2}\binom{13}{2} > m_4$.

All cubic graphs with not more than 14 nodes were generated in [8]. There are only 19 cubic $(10, 15)$-graphs. The following result can be obtained by a computational test.

Lemma 2. *The coefficient m_5 is minimized in Petersen graph among all cubic $(10, 15)$-graphs.*

The following result is analogous to Lemma 1:

Lemma 3. *The coefficient m_5 is minimized in Petersen graph.*

Proof. By Lemma 2 we know that the result holds for in cubic graphs. We know that $m_5 = 831$ in Petersen graph. In the following, we remark that only trivial disconnecting sets are considered for counting. If H has a bridge, then $m_5(H) \geq \binom{14}{4} \geq m_5$. It suffices to prove the result when $\delta(H) = 2$. If $deg(v_1) = deg(v_2) = deg(v_3) = 2$ for three different nodes, we consider three disjoint and exhaustive cases:

(i) Non-adjacent nodes: $m_5(H) \geq 3 \times \binom{13}{3} - 3 \times 11 = 828$;

(ii) There are two adjacent nodes: $m_5(H) \geq \binom{13}{3} + \binom{3}{2}\binom{12}{3} + \binom{3}{3}\binom{12}{2} - 31 \geq m_5$;

(iii) There are two links among them: $m_5(H) \geq \binom{4}{2}\binom{11}{3} + \binom{4}{3}\binom{11}{2} + \binom{4}{4}\binom{11}{1} \gg m_5$

Assume that there are precisely two different nodes $v_1 \neq v_2$ such that $deg(v_1) = deg(v_2) = 2$. We know that $deg(v_i) = 3 + \delta_i$ for $i = 3, \ldots, 10$. By Handshaking Lemma we know that $30 = \sum_i deg(v_i)$, so $\sum_{i=3}^{10} \delta_i = 2$. Therefore, the only graphic degree-sequences with two degree-two nodes are $D_1 = (4, 4, 3, 3, 3, 3, 3, 3, 2, 2)$ and $D_2 = (5, 3, 3, 3, 3, 3, 3, 3, 2, 2)$. We consider four cases: D_1 or D_2 with adjacent or non-adjacent $(A$-$NA)$ nodes v_1 and v_2:

(i) D_1 and A: $m_5(H) \geq 2 \times \binom{4}{4}\binom{11}{1} + 6 \times \binom{3}{3}\binom{12}{2} + \binom{3}{2}\binom{12}{3} + \binom{3}{3}\binom{12}{2} - 6 \times 3 > m_5$;

(ii) D_1 and NA: $m_5(H) \geq 2 \times \binom{4}{4}\binom{11}{1} + 6 \times \binom{3}{3}\binom{12}{2} + 2 \times \binom{2}{2}\binom{13}{3} - 2 \times 5 - 11 > m_5$;

(iii) D_2 and A: $m_5(H) \geq \binom{5}{5} + 7 \times \binom{3}{3}\binom{12}{2} + \binom{3}{2}\binom{12}{3} + \binom{3}{3}\binom{12}{2} - 7 \times 3 > m_5$;

(iv) D_2 and NA: $m_5(H) \geq \binom{5}{5} + 7 \times \binom{3}{3}\binom{12}{2} + 2 \times \binom{2}{2}\binom{13}{3} - 2 \times 7 - 11 > m_5$.

Finally, we consider the case where there exists only one degree-2 node. In this case, the degree-sequence must be $(4, 3, 3, 3, 3, 3, 3, 3, 3, 2)$. Counting trivial disconnecting sets we get that $m_5(H) \geq \binom{4}{4}\binom{11}{1} + 8 \times \binom{3}{3}\binom{12}{2} + \binom{2}{2}\binom{13}{3} - 8 = 825$. However, it does not suffice to close the proof. We must find at least 6 non-trivial disconnecting sets. Observe that there exists at least 11 links whose extremes are nodes with degree 2 or 3, so $m_5(H) \geq 825 + 11 = 836 > m_5$.

Lemma 4. *Petersen graph is t-optimal.*

In order to prove that m_6 is minimized in Petersen graph we can distinguish several cases:

The first case is when $\delta(H) = 1$.

(i) If we have two nodes v_1 and v_2 such that $deg(v_1) = deg(v_2) = 1$
$m_6(H) \geq 2 \times \binom{14}{5} - \binom{13}{4} > m_6$.

(ii) If we have v_1, v_2, v_3, v_4 such that $deg(v_1) = 1$, $deg(v_2) = deg(v_3) = 2$, and $deg(v_4) = 3$, $m_6(H) \geq \binom{14}{5} + 2 \times \binom{13}{4} + \binom{12}{3} - 2 \times \binom{12}{3} - \binom{11}{2} - 2 \times \binom{10}{1} + \binom{10}{1} + \binom{9}{0} > m_6$.

(iii) nodes v_1, v_2, v_3, v_4, v_5, such that $deg(v_1) = 1$, $deg(v_2) = 2$, $deg(v_3) = deg(v_4) = deg(v_5) = 3$
$m_6(H) \geq \binom{14}{5} + \binom{13}{4} + 4 \times \binom{12}{3} - \binom{12}{3} + 4 \times \binom{11}{2} + 4 \times \binom{9}{0} > m_6$.

(iv) nodes $v_1, v_2, v_3, v_4, v_5, v_6, v_7, v_8$ such that $deg(v_1) = 1$, $deg(v_2) = deg(v_3) = deg(v_4) = deg(v_5) = deg(v_6) = deg(v_7) = deg(v_8) = 3$
$m_6(H) \geq \binom{14}{5} + 7 \times \binom{12}{3} - 7 \times \binom{11}{2} - \binom{7}{2} \times \binom{9}{0} > m_6$.

All the other cases with one node of degree one, are reduced to one of the above, or have more disconnecting sets than these graphs because they have one o more nodes with degree ≥ 4.

When $\delta(H) = 2$, we have these cases:

(i) nodes v_1, v_2, v_3, v_4, v_5 such that $deg(v_1) = deg(v_2) = deg(v_3) = deg(v_4) = deg(v_5) = 2$
$m_6(H) \geq 5 \times \binom{13}{4} - \binom{5}{2} \times \binom{11}{2} + \binom{5}{3} \times \binom{9}{0} > m_6$.

(ii) nodes $v_1, v_2, v_3, v_4, v_5, v_6, v_7, v_8$ such that $deg(v_1) = deg(v_2) = deg(v_3) = deg(v_4) = 2$, $deg(v_5) = deg(v_6) = deg(v_7) = deg(v_8) = 3$
$m_6(H) \geq 4 \times \binom{13}{4} - 4 \times 4 \times \binom{10}{1} + 4 \times \binom{12}{3} - 6 \times \binom{11}{2} - 6 \times \binom{9}{0} > m_6$.

(iii) nodes $v_1, v_2, v_3, v_4, v_5, v_6, v_7, v_8$ such that $deg(v_1) = deg(v_2) = 2$, $deg(v_3) = deg(v_4) = deg(v_5) = deg(v_6) = deg(v_7) = deg(v_8) = 3$. There are two cases, $(5, 3, 3, 3, 3, 3, 3, 3, 2, 2)$ and $(4, 4, 3, 3, 3, 3, 3, 3, 2, 2)$, using trivial and several non trivial disconnector sets $m_6(H) > m_6$ in both cases.

The last case is when $\delta(H) = 3$. There are 19 cubic graphs including Petersen, which have the lower m_6.

Cases $(..., 4, 2, 1), (...4, 3, 1), (...4, 1), (..., 4, 2, 2), (..., 4, 2)$ do not exist.

Theorem 1. *Petersen is uniformly most-reliable.*

Proof. Recall that Petersen is super-λ, so m_3 is minimized. Clearly, $m_i = 0$ for $i \in \{0, 1, 2\}$, and $m_i = \binom{15}{i}$ for all $(10, 15)$-graphs, when $i \geq 7$. Petersen is t-optimal, thus, it minimizes the coefficient m_6. Combining Lemmas 1, 3 and 4, we know that Petersen graph minimizes simultaneously all the coefficients m_i. Therefore, Petersen simultaneously minimizes all the coefficients m_i among $(10, 15)$-graphs, and thus it is uniformly most-reliable.

5   Conclusions and Trends for Future Work

Uniformly most-reliable graphs represent a synthesis in network reliability analysis. Finding them is a hard task not well understood. Prior works in the field try to globally minimize the coefficients of disconnecting sets. This methodology provides uniformly most-reliable $(n, n + i)$ graphs for $i \in \{-1, 0, 1, 2, 3, 4\}$. In this paper, we formally proved that Petersen graph is uniformly most-reliable $(n, n + 5)$ graph. This means that Conjecture 1 is false. This work reinforces Donald Knuth's statement that Petersen graph serves as a counterexample to several optimistic predictions in graph theory [15].

There are several trends for future work. A powerful methodology to find uniformly most-reliable graphs is not known. A full characterization of t-optimal graphs is an open problem. Conjecture 2 could be studied with a similar reasoning as in Boesch [5] and Wang [21].

Acknowledgements. This work is partially supported by Project 395 CSIC I+D *Sistemas Binarios Estocásticos Dinámicos.* We wish to thank Dr. Louis Petingi for his valuable comments on t-optimality throughout the writing of this manuscript.

References

1. Bauer, D., Boesch, F., Suffel, C., Van Slyke, R.: On the validity of a reduction of reliable network design to a graph extremal problem. IEEE Trans. Circuits Syst. **34**(12), 1579–1581 (1987)
2. Bauer, D., Boesch, F., Suffel, C., Tindell, R.: Combinatorial optimization problems in the analysis and design of probabilistic networks. Networks **15**(2), 257–271 (1985)
3. Beineke, L.W., Wilson, R.J., Oellermann, O.R.: Topics in Structural Graph Theory. Encyclopedia of Mathematics and its Applications. Cambridge University Press, Cambridge (2012)
4. Biggs, N.: Algebraic Graph Theory. Cambridge Mathematical Library. Cambridge University Press, Cambridge (1993)
5. Boesch, F.T., Li, X., Suffel, C.: On the existence of uniformly optimally reliable networks. Networks **21**(2), 181–194 (1991)
6. Boesch, F.T., Satyanarayana, A., Suffel, C.L.: A survey of some network reliability analysis and synthesis results. Networks **54**(2), 99–107 (2009)
7. Bollobás, B.: Extremal Graph Theory. Dover Books on Mathematics. Dover Publications, New York (2004)
8. Bussemake, F.C., Cobeljic, S., Cvetkovic, D.M., Seidel, J.J.: Cubic graphs on ≤ 14 vertices. J. Comb. Theor. B **23**(2), 234–235 (1977)

9. Canale, E., Piccini, J., Robledo, F., Romero, P.: Diameter-constrained reliability: complexity, factorization and exact computation in weak graphs. In: Proceedings of the Latin America Networking Conference on LANC 2014, pp. 1–7. ACM, New York (2014)

10. Cheng, C.-S.: Maximizing the total number of spanning trees in a graph: two related problems in graph theory and optimum design theory. J. Comb. Theor. B **31**(2), 240–248 (1981)

11. Colbourn, C.J.: Reliability issues in telecommunications network planning. In: Telecommunications Network Planning, chap. 9, pp. 135–146. Kluwer Academic Publishers (1999)

12. Harary, F.: The maximum connectivity of a graph. Proc. Natl. Acad. Sci. U.S.A. **48**(7), 1142–1146 (1962)

13. Holton, D.A., Sheehan, J.: The Petersen Graph. Australian Mathematical Society Lecture Series. Cambridge University Press, Cambridge (1993)

14. Kirchoff, G.: Über die auflösung der gleichungen, auf welche man bei der untersuchung der linearen verteilung galvanischer ströme geführt wird. Ann. Phys. Chem. **72**, 497–508 (1847)

15. Knuth, D.E.: The Art of Computer Programming: Introduction to Combinatiorial Algorithms and Boolean Functions. Addison-Wesley Series in Computer Science and Information Proceedings. Addison-Wesley, Reading (2008)

16. Leggett, J.D., Bedrosian, S.D.: On networks with the maximum numbers of trees. In: Proceedings of Eighth Midwest Symposium on Circuit Theory, pp. 1–8, June 1965

17. Monma, C., Munson, B.S., Pulleyblank, W.R.: Minimum-weight two-connected spanning networks. Math. Program. **46**(1–3), 153–171 (1990)

18. Myrvold, W., Cheung, K.H., Page, L.B., Perry, J.E.: Uniformly-most reliable networks do not always exist. Networks **21**(4), 417–419 (1991)

19. Petingi, L., Boesch, F., Suffel, C.: On the characterization of graphs with maximum number of spanning trees. Discrete Math. **179**(1), 155–166 (1998)

20. Romero, P.: Building uniformly most-reliable networks by iterative augmentation. In: Proceedings of the 9th International Workshop on Resilient Networks Design and Modeling (RNDM 2017), Alghero, Sardinia, Italy, September 2017 (to appear)

21. Wang, G.: A proof of Boesch's conjecture. Networks **24**(5), 277–284 (1994)

GRASP Heuristics for a Generalized Capacitated Ring Tree Problem

Gabriel Bayá[✉], Antonio Mauttone, Franco Robledo, and Pablo Romero

Departamento de Investigación Operativa, Facultad de Ingeniería,
Universidad de la República, Montevideo, Uruguay
{gbaya,mauttone,frobledo,promero}@fing.edu.uy

Abstract. This paper introduces a new mathematical optimization problem, inspired in the evolution of fiber optics communication. Real-life implementations must address a cost-robustness tradeoff. Typically, real topologies are hierarchically organized in backbone and access networks. The backbone is two-node-connected, while the access network usually considers either leaf nodes or elementary paths, directly connected to the backbone. We define the Capacitated Two-Node Survivable Tree Problem (CTNSTP for short). The backbone consists of at most m two-node-connected structures with a perfect depot as a common node. The access network consists of trees directly connected to the backbone. The CTNSTP belongs to the NP-Complete computational class. A GRASP heuristic enriched with a Variable Neighborhood Descent (VND) is provided. Certain neighborhoods of our VND include exact models based on Integer Linear Programming formulations. The comparison among recent works in the field confirm remarkable savings with the novel proposal.

Keywords: Network survivability · CTNSTP · GRASP · VND

1 Motivation

In fiber optics communication systems, robustness is essential, so, two-node-connected topologies are considered. A natural approach to accomplish two-node-connectivity (i.e. a node is connected to another one by two independent paths) is to connect all terminals in a ring or cycle in an economic way. This problem is called Traveling Salesman Problem or TSP, and it is widely studied in the scientific literature [12].

A cornerstone in the field of structural network design is authored by Monma et al. [13]. They study the Minimum-weight Two-Connected Spanning Problem (MW2CSP), briefly, how to connect terminals in the cheapest way, with a resulting two-node connected topology. They prove that the corresponding decision version for the MW2CSP belongs to the set of $\mathcal{NP}$-Complete decision problems. Furthermore, the cheapest Hamiltonian Tour (i.e., a ring that meets all the nodes) is not necessarily a global optimum. Specifically, the cost of the cheapest ring is upper-bounded by $4/3 \times opt$, being opt the cost of the best two-node-connected structure.

© Springer International Publishing AG 2018
G. Nicosia et al. (Eds.): MOD 2017, LNCS 10710, pp. 436–448, 2018.
https://doi.org/10.1007/978-3-319-72926-8_36

Inspired by fiber optics design, Labbé et al. introduce the Ring Star Problem, or RSP for short [11]. In that work the core is a ring, and the remaining terminals are linked to the ring as leaf-nodes. The objective is to find the minimum-cost topology meeting the previous constraints, given costs in the ring-connections and leaf-links. A further generalization, the Capacitated Ring Star Problem (CmRSP) is introduced by Baldacci et al. pressed by realistic solutions, where customers are geographically distributed [1]. The authors consider m rings with the depot as the only common node. The main difference with the RSP is the presence of m rings instead of one. Both optimization problems belong to the $\mathcal{N}\mathcal{P}$-Hard class, since they generalize the Hamiltonian Tour problem [9]. Therefore, the CmRSP has been heuristically addressed in several opportunities [8,17]. A trade-off between cost and robustness is proposed by Hill et al. [7], where the core is a ring again, but there are nodes from a secondary class that are connected to the ring by trees. The result is the Capacitated Ring-Tree Problem, or CRTP for short.

Recent works in structural network design replace rings by arbitrary two-connected components, inspired by savings predicted by Monma et al. For instance, Bayá et al. introduce the Capacitated m Two-Node Survivable Star Problem, or CmTNSSP [3] where the m rings of the CmRSP are replaced by two-connected components. Analogously, Recoba et al. introduce the Two-Node Connected Star Problem (TNCSP), which is precisely the RSP but with a two-node-connected core that replaces the ring [14].

In this paper, a extension for both the CRTP and CmRSP is introduced, where m two-connected structures are considered instead of cycles for each of the above problems, and the secondary nodes from the access network includes trees connected to these two-connected structures. The goal is to achieve flexibility and savings simultaneously. The main contributions of this paper are the following:

- The Capacitated Two-Node Survivable Tree Problem (CTNSTP) is introduced.
- Given its intractability, a heuristic resolution is developed. We adopted a GRASP approach enriched with a Variable Neighborhood Descent, or GRASP-VND.
- A fair comparison with prior works in the field is presented in order to highlight the benefits of this new proposal.

This article is organized in the following manner. The formal definition of the CTNSTP is presented in Sect. 2. A Greedy Randomized Adaptive Search Procedure (GRASP) is developed for its resolution in Sect. 3. The experimental analysis is conducted in Sect. 4. Concluding remarks and trends for future work are discussed in Sect. 5.

2 Capacitated Two-Node Survivable Tree Problem

The cost-robustness trade-off is a major engineering challenge to develop physical communication systems. Ideally, the underlying topology should be flexible enough to produce savings, but resilient to simple node/link failures in the backbone. Here we describe the closest works from a topological point of view. In fact, we present a topological extension of the CRTP, gaining on both flexibility and savings.

Voß and Hill recently introduced the CRTP [7]. They consider a simple and completed undirected graph $G = (V, E)$, a positive integer m and a partition $V = \{s\} \cup V_{T_1} \cup V_{T_2} \cup V_S$, being s the depot or source node, V_{T_1} the type-1 terminal nodes, V_{T_2} the type-2 terminal nodes and V_S the optional or Steiner nodes. The depot s has a capacity q_s, and there is a cost-matrix $C = (c_{i,j})$, $v_i, v_j \in V$. The goal is to choose a minimum cost spanning subgraph $H = \cup_{i=1}^{k} R_{l_i}$, where R_{l_i} are ring-trees. A ring-tree is a connected graph with at most n edges, where n is the number of nodes of such graph. The ring-trees R_{l_i} only meet on the depot $s \in R_{l_i}$ and have a length l_i, where l_i is the number of terminal nodes (type-1 or type-2) that belong to R_{l_i}. Every node from the set V_{T_1} belongs to precisely one ring-tree, while nodes from V_{T_2} belong to exactly one ring. Steiner nodes may belong to one ring-tree when their inclusion improves the total cost of solution. The capacity constraint implies that $l_i \leq q_s$ for all $i \in \{1, \ldots, m_p\}$, with $m_p \leq m$, being m the maximum number of ring-trees allowed.

Here, we consider a relaxation of CRTP, where rings are replaced by arbitrary 2-node-connected components. We obtain the Capacitated Two-Node Survivable Tree Problem, or CTNSTP (see Fig. 1). The CTNSTP also belongs to the $\mathcal{NP}$-Hard class, since the design of a single component ($m = 1$, $q_s = |V|$, $V_{T_1} = V_S = \emptyset$) is the minimum-cost 2-node-connected spanning network problem (MW2NCSN), which is $\mathcal{NP}$-Hard [13].

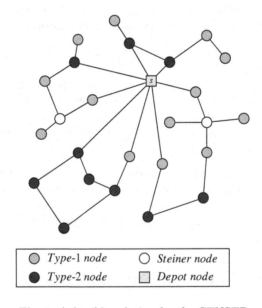

| ○ Type-1 node | ○ Steiner node |
| ● Type-2 node | □ Depot node |

Fig. 1. A feasible solution for the CTNSTP.

3 GRASP Resolution

Greedy Randomized Adaptive Search Procedure (GRASP) is a powerful multi-start or iterative process, with great success in telecommunications [16]. In GRASP, feasible solutions are produced in a first phase, while neighbor solutions are explored in a second phase. The best overall solution is returned as the result. There is a trade-off between greediness (intensification) and randomization (diversification), by means of a restricted candidate list. We invite the reader to consult [15] for a comprehensive study of this metaheuristic. Here, we sketch the main ingredients of our particular GRASP design, namely, Construction Phase and Local Search Phase.

3.1 Construction Phase

During the Construction Phase, components will be iteratively built. The goal is to produce a feasible solution that includes type-2 terminal nodes in 2-node-connected structures, and type-1 terminal nodes in both 2-node-connected structures and trees. Let us consider an arbitrary instance for the CTNSTP, a positive integer k and a maximum number of iterations $MaxIter$. In order to define our construction phase, the following four functions will be used:

(1) $Pick(m, G, C, MaxIter)$: returns m terminal nodes $v_1, \ldots, v_m$ (with $v_i \in V_{T_1} \cup V_{T_2}$) which maximize their total distance to the depot, out of $MaxIter$ trials. $Random(v_1, \ldots, v_m)$ selects randomly one node of the v_i $i = 1 \cdots m$.

(2) $Connect(G, C, s, node, k, non\_connected)$: returns a set of k node-disjoint paths between the depot s and $node$.

(3) $ChooseTwo(\mathscr{C}, node)$: chooses 2 paths out of k uniformly at random, for $node$. In this way a cycle between s and $node$ was built.

(4) $Insert(non\_connected, G, C)$: inserts type-2 nodes in the backbone and type-1 nodes in the backbone or in a tree, taken from the set $non\_connected$.

Algorithm 1. Construction Phase

1: **input** G, C, k, m, $MaxIter$
2: $G_{Sol} \leftarrow \emptyset$
3: $component\_nodes \leftarrow \emptyset$
4: $non\_connected \leftarrow V_{T_1} \cup V_{T_2}$
5: $\{v_1, \ldots, v_m\} \leftarrow Pick(m, G, R, MaxIter)$
6: **for** i=1 **to** m **do**
7:     $node = Random(v_1, \ldots, v_m)$
8:     $\mathscr{C} \leftarrow Connect(G, C, s, node, k, non\_connected)$
9:     $\mathscr{C}_i \leftarrow ChooseTwo(\mathscr{C}, node)$
10:    $G_{Sol} \leftarrow G_{Sol} \cup \mathscr{C}_i$
11:    $component\_nodes[i] \leftarrow component\_nodes[i] \cup \mathscr{C}_i$
12:    $non\_connected \leftarrow non\_connected - \mathscr{C}_i$
13: **end for**
14: $G_{Sol} \leftarrow G_{Sol} \cup Insert(non\_connected, G, C)$
15: **return** G_{Sol}

The previous functions will be called sequentially. *Pick* runs *MaxIter* independent random sets of m terminal nodes. It returns the set with minimum global cost between all the pairs of the set. Once the set $v_1, \ldots, v_m$ is obtained, *Connect* is called for each node v_i. It applies Bhandari's algorithm [4] in order to find the cheapest set of k node-disjoint paths between the depot and terminal v_i (type-1 or type-2). Function *ChooseTwo* just chooses uniformly at random two disjoint paths out of k. Finally, in *Insert*, non-connected type-1 and type-2 nodes are randomly chosen and iteratively added to the smallest component, meeting feasibility. In this way, the capacity constraint is met during the construction phase. Consider an isolated node v and a component $\mathscr{C}$ (see Fig. 2). All links that belong to other components will be deleted, and the costs of all links from $\mathscr{C}$ are set to 0. An artificial node v' is connected to every node from $\mathscr{C}$. Bhandari's algorithm is applied in order to find k (or possibly less) node-disjoint paths between v and v' in the resulting network. Only two disjoint paths between v and v' will be chosen. Finally, the resulting links that connect v with $\mathscr{C}$ are added to the solution. Type-1 nodes can either be inserted into an existing tree, or a new tree can be built for that specific purpose. We pre-check feasibility, inspecting the values of m, q, and the number of terminal nodes $|V_{T_1} \cup V_{T_2}|$ and was checked at the end of the construction phase. Unfeasible solutions are discarded, there is no re-feasibility solution process.

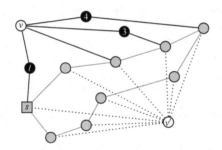

Fig. 2. Including node v into component $\mathscr{C}$.

3.2 Local Search Phase

The following functions determine different neighborhood structures. They are applied in order, whenever they produce savings, using Variable Neighborhood Descent. First, we present the key idea of each function:

1. **Swap-Nodes:** picks a random terminal and swaps it with its closest terminal.
2. **Move-Node:** removes a node, reconnects their neighbors and inserts the node into a tree or 2-node-connected structure.
3. **Crossing-Components:** finds close terminal nodes from different 2-node-connected structures, deletes adjacent links and reconnects the components.

4. **Add-Links:** random links are first added and then adequately deleted from some 2-node-connected structure.
5. **Tree-Convert:** removes a random number of type-1 nodes in the 2-node-connected structure of a component, and re-inserts them in a tree.
6. **Move-Steiner:** first removes and then inserts Steiner nodes.
7. **Best Component:** replacement of cycles by the best 2-node connected component, using an exact ILP based algorithm.

Also, in order not to stuck in local optima, a perturbation process takes place. Function *Shake* randomly disconnects a percentage p of terminal nodes and reconnects them in another way. Type-2 nodes are inserted into a 2-node-connected component, while type-1 nodes are inserted into existing trees or new trees are created, in a cost minimization manner. The procedure tries to keep a balance between the number of 2-connected structures and trees in the solution. *Shake* is called after the VND scheme, where the solution is a local optima for every neighborhood structure.

In the following paragraphs, the seven functions will be explained in full detail.

Swap-Nodes. This local search selects two nodes and makes an exchange (swapping) between them. This process starts with a random selection of a type-1 or type-2 terminal node and tests all possible ways to swap this node with another *close* node belonging to a 2-node-connected component (the same or other) or belonging to a tree. To clarify the concept *close* we define a neighborhood related to the considered node. Let i be a terminal node and a neighborhood N defined as follows:

$$N(i) = \left\{ j \in V_{T_1} \cup V_{T_2} : j \begin{array}{c} \text{are the } k \text{ nodes } closer \text{ to node } i \\ \text{taking into account costs } c_{ij} \\ \text{defined in original graph } G \end{array} \right\} \tag{1}$$

It should be noted that to apply the movement, both nodes must belong to the 2-node-connected structure of the component, or they must belong to different trees. The algorithm picks a random node i and proceeds as follows. Consider its closest node j. If j belongs to a tree (i belongs to a different tree to allow move) we exchange the nodes between trees removing each of them and inserting in the other tree using a Minimum Spanning Tree algorithm [10]. If j is a node that belongs to a 2-node-connected structure, this function connects adjacent nodes of j to node i and adjacent nodes of i to node j. Each time a swapping movement leads to improvement and keeps the feasibility, the current solution is updated.

Move-Node. This local search performs the extraction of all terminal nodes in a random order from their current positions in the solution, and relocates them to another positions either in the 2-node-connected structure of component or in a tree, improving the overall cost without losing feasibility. We extract a terminal node and we reconnect the adjacents to the extracted node. To make the insertion of the extracted node we consider the neighborhood $N(i)$ defined above. For each terminal node i we consider all possible insertions between k closest nodes, and select the movement that produces the lowest total cost. The algorithm repeats the same procedure for all $i \in V_{T_1} \cup V_{T_2}$ not even considered, by examining $N(i)$ until finally selecting the movement that produces the lowest total cost.

Crossing-Components. This local search takes two *close* nodes, each one in different 2-node-connected structure of a component, eliminates one of their adjacent edges (for each node) and connects each pair of nodes (in different component) by the edge that generates the best cost.

Add-Links. This local search inserts k edges in a 2-node-connected structure of a selected component. Afterwards the function considers all nodes of degree 3 or greater of the component, and removes one incident edge until leaving the node degree in 2, without losing feasibility. This process is performed several times in each component.

Tree-Convert. In this local search, k type-1 terminal nodes belonging to a 2-node-connected structure of a component are removed, then they are reinserted in the best positioned tree (if there are any) or a new tree is generated with the removed node and the best positioned node of the component.

Move-Steiner. This local search works by adequate deletion/insertion of Steiner nodes. The first stage of this local search considers all Steiner nodes belonging to the solution, and tries to remove them if the cost is reduced. Then, greedy insertions are tried in order to reduce the cost as well.

Best-Component. This local search is based on Integer Linear Programming. This is an exact local search and it always returns the best two-connected component, which may be a cycle. Further information about the model used in this local search can be found in [2]. Given a feasible solution to the problem, we identify all cycles that exist in each component. For each cycle we apply the best replacement by a 2-node-connected topology. As stated in Sect. 1, the best 2-node-connected solution covering a certain set of nodes is not necessarily a cycle, so this local search may include such topologies in our solution. In order to model this local search we used a particular case of **GSP** (*Generalized Steiner Problem*) [5], with 2-connectivity requirement between every pair of nodes.

4   Experimental Analysis

As far as we know, the closest work is the Capacitated Ring-Tree Problem or CRTP. In fact, the CTNSTP is a topological relaxation of the CRTP, and every

Table 1. Values found for instances with 26 nodes.

P	r_1	$|V_{T_2}|$	$|V_{T_1}|$	$|V_S|$	m	q	l_b	u_b	u_{b0}	u_{b1}	Δ	$t(s)$
Q-1	1	0	12	13	3	5	157	**157**	**157**	157	0,000	600
	0.75	3	9				210	**210**	215	**211**	−1,860	600
	0.5	6	6				227	**227**	**227**	227	0,000	600
	0.25	9	3				236	**236**	**236**	236	0,000	600
	0	12	0				242	**242**	**242**	242	0,000	600
Q-2	1	0	12	13	4	4	163	**163**	164	166	1,220	600
	0.75	3	9				207	**207**	**207**	207	0,000	600
	0.5	6	6				240	**240**	**240**	240	0,000	600
	0.25	9	3				249	**249**	**249**	249	0,000	600
	0	12	0				251	**251**	**251**	251	0,000	600
Q-3	1	0	12	13	5	3	170	**170**	173	175	1,156	600
	0.75	3	9				242	**242**	244	244	0,000	600
	0.5	6	6				251	**251**	**251**	253	0,797	600
	0.25	9	3				279	**279**	**279**	279	0,000	600
	0	12	0				279	**279**	**279**	279	0,000	600
Q-4	1	0	18	7	3	7	207	**207**	**207**	208	0,483	600
	0.75	4	14				256	**256**	**256**	256	0,000	600
	0.5	9	9				274	**274**	**274**	274	0,000	600
	0.25	13	5				292	**292**	**292**	292	0,000	600
	0	18	0				301	**301**	305	**301**	−1,311	600
Q-5	1	0	18	7	4	5	217	**217**	220	223	1,364	600
	0.75	4	14				285	**285**	**285**	288	1,053	600
	0.5	9	9				313	**313**	318	320	0,629	600
	0.25	13	5				334	**334**	**334**	334	0,000	600
	0	18	0				339	**339**	**339**	339	0,000	600
Q-6	1	0	18	7	5	4	227	**227**	231	232	0,433	600
	0.75	4	14				278	**278**	**278**	280	0,719	600
	0.5	9	9				336	**336**	**336**	336	0,000	600
	0.25	13	5				361	**361**	**361**	361	0,000	600
	0	18	0				375	**375**	**375**	375	0,000	600
Q-7	1	0	25	0	3	10	245	**245**	248	248	0,000	600
	0.75	6	19				294	**294**	**294**	296	0,680	600
	0.5	13	12				313	**313**	**313**	313	0,000	600
	0.25	18	7				327	**327**	**327**	327	0,000	600
	0	25	0				328	**328**	**328**	328	0,000	600
Q-8	1	0	25	0	4	7	252	**252**	267	268	0,375	600
	0.75	6	19				311	**311**	315	319	1,270	600
	0.5	13	12				345	**345**	**345**	347	0,580	600
	0.25	18	7				357	**357**	**357**	357	0,000	600
	0	25	0				362	**362**	**362**	362	0,000	600
Q-9	1	0	25	0	5	6	254	**254**	262	268	2,290	600
	0.75	6	19				319	**319**	322	326	1,242	600
	0.5	13	12				369	**369**	372	372	0,000	600
	0.25	18	7				378	**378**	379	**378**	−0,264	600
	0	25	0				396	**396**	397	**396**	−0,252	600

Table 2. Values found for instances with 51 nodes.

P	r_1	$|V_{T_2}|$	$|V_{T_1}|$	$|V_S|$	m	q	l_b	u_b	u_{b0}	u_{b1}	Δ	$t(s)$
Q-10	1	0	12	38	3	5	156	**156**	**156**	156	0,000	3600
	0.75	3	9				190	**190**	196	196	0,000	3600
	0.5	6	6				213	**213**	215	217	0,922	3600
	0.25	9	3				222	**222**	**222**	222	0,000	3600
	0	12	0				242	**242**	**242**	242	0,000	3600
Q-11	1	0	12	38	4	4	159	**159**	163	166	1,807	3600
	0.75	3	9				209	**209**	**209**	209	0,000	3600
	0.5	6	6				230	**230**	**230**	230	0,000	3600
	0.25	9	3				238	**238**	**238**	238	0,000	3600
	0	12	0				251	**251**	**251**	251	0,000	3600
Q-12	1	0	12	38	5	3	170	**170**	172	173	0,578	3600
	0.75	3	9				203	**203**	**203**	203	0,000	3600
	0.5	6	6				251	**251**	**251**	253	0,791	3600
	0.25	9	3				278	**278**	**278**	278	0,000	3600
	0	12	0				279	**279**	**279**	279	0,000	3600
Q-13	1	0	25	25	3	10	245	**245**	248	248	0,000	3600
	0.75	6	19				293	302	305	306	0,327	3600
	0.5	12	13				311	**311**	312	312	0,000	3600
	0.25	18	7				322	**322**	**322**	322	0,000	3600
	0	25	0				328	**328**	**328**	328	0,000	3600
Q-14	1	0	25	25	4	7	252	**252**	267	269	0,743	3600
	0.75	6	19				304	**304**	321	321	0,000	3600
	0.5	12	13				341	352	352	355	0,845	3600
	0.25	18	7				357	**357**	**357**	357	0,000	3600
	0	25	0				362	**362**	**362**	362	0,000	3600
Q-15	1	0	25	25	5	6	254	**254**	262	267	1,873	3600
	0.75	6	19				331	335	339	337	−0,593	3600
	0.5	12	13				359	370	372	372	0,000	3600
	0.25	18	7				372	387	387	**385**	−0,519	3600
	0	25	0				390	**390**	397	**392**	−1,276	3600
Q-16	1	0	37	13	3	14	304	**304**	**304**	304	0,000	3600
	0.75	9	28				350	375	375	377	0,531	3600
	0.5	18	19				364	376	378	**376**	−0,532	3600
	0.25	27	10				379	**379**	380	**379**	−0,264	3600
	0	37	0				380	**380**	381	**380**	−0,263	3600
Q-17	1	0	37	13	4	11	308	**308**	309	310	0,323	3600
	0.75	9	28				363	**363**	369	376	1,862	3600
	0.5	18	19				384	399	399	403	0,993	3600
	0.25	27	10				396	404	404	404	0,000	3600
	0	37	0				410	**410**	418	**412**	−1,456	3600
Q-18	1	0	37	13	5	9	314	**314**	**314**	314	0,000	3600
	0.75	9	28				374	408	408	412	0,971	3600
	0.5	18	19				401	431	431	435	0,920	3600
	0.25	27	10				417	436	436	**433**	−0,693	3600
	0	37	0				446	**446**	452	**446**	−1,345	3600
Q-19	1	0	50	0	3	19	376	**376**	377	380	0,789	3600
	0.75	12	38				418	427	436	438	0,457	3600
	0.5	25	25				435	445	447	450	0,667	3600
	0.25	37	13				451	**451**	454	454	0,000	3600
	0	50	0				462	**462**	473	**465**	−1,720	3600
Q-20	1	0	50	0	4	14	384	**384**	386	392	1,531	3600
	0.75	12	38				423	458	458	**456**	−0,439	3600
	0.5	25	25				448	493	493	496	0,605	3600
	0.25	37	13				471	502	502	**496**	−1,210	3600
	0	50	0				493	**493**	513	**499**	−2,806	3600
Q-21	1	0	50	0	5	12	390	**390**	392	396	1,010	3600
	0.75	12	38				447	491	501	506	0,988	3600
	0.5	25	25				478	526	526	531	0,942	3600
	0.25	37	13				497	525	525	**523**	−0,382	3600
	0	50	0				522	526	541	**526**	−2,852	3600

Table 3. Values found for instances with 76 nodes.

P	r_1	$|V_{T_2}|$	$|V_{T_1}|$	$|V_S|$	m	q	l_b	u_b	u_{b0}	u_{b1}	Δ	$t(s)$
Q-22	1	0	18	57	3	7	213	**213**	214	216	0,935	3600
	0.75	4	14				272	**272**	272	276	1,471	3600
	0.5	9	9				288	318	318	318	0,000	3600
	0.25	13	5				303	318	318	318	0,000	3600
	0	18	0				331	**331**	332	331	−0,301	3600
Q-23	1	0	18	57	4	5	232	**232**	235	236	0,426	3600
	0.75	4	14				302	309	312	314	0,641	3600
	0.5	9	9				336	**336**	336	336	0,000	3600
	0.25	13	5				359	369	369	**367**	−0,542	3600
	0	18	0				386	**386**	390	386	−1,026	3600
Q-24	1	0	18	57	5	4	257	**257**	259	265	2,317	3600
	0.75	4	14				325	**325**	325	326	0,308	3600
	0.5	9	9				368	379	379	379	0,000	3600
	0.25	13	5				397	**397**	**397**	397	0,000	3600
	0	18	0				448	**448**	451	**448**	−0,665	3600
Q-25	1	0	37	38	3	14	320	**320**	**320**	320	0,000	3600
	0.75	9	28				363	390	390	396	1,538	3600
	0.5	18	19				372	402	402	405	0,746	3600
	0.25	27	10				390	403	403	406	0,744	3600
	0	37	0				409	**409**	413	**409**	−0,969	3600
Q-26	1	0	37	38	4	11	326	**326**	336	339	0,893	3600
	0.75	9	28				382	402	402	408	1,493	3600
	0.5	18	19				410	455	455	459	0,879	3600
	0.25	27	10				418	460	460	**458**	−0,435	3600
	0	37	0				446	458	458	**454**	−0,873	3600
Q-27	1	0	37	38	5	9	340	**340**	343	350	2,041	3600
	0.75	9	28				407	446	446	**442**	−0,897	3600
	0.5	18	19				426	473	473	474	0,211	3600
	0.25	27	10				443	497	497	**485**	−2,414	3600
	0	37	0				477	506	506	**502**	−0,791	3600
Q-28	1	0	56	19	3	21	383	**383**	395	398	0,759	3600
	0.75	14	42				427	462	462	469	1,515	3600
	0.5	28	28				438	477	477	480	0,629	3600
	0.25	42	14				461	465	472	474	0,424	3600
	0	56	0				476	**476**	495	480	−3,030	3600
Q-29	1	0	56	19	4	16	389	**389**	402	406	0,995	3600
	0.75	14	42				441	488	488	489	0,205	3600
	0.5	28	28				466	520	520	525	0,962	3600
	0.25	42	14				492	532	532	**530**	−0,376	3600
	0	56	0				514	535	543	**536**	−1,289	3600
Q-30	1	0	56	19	5	13	399	**399**	414	420	1,449	3600
	0.75	14	42				469	533	533	536	0,563	3600
	0.5	28	28				493	554	554	554	0,000	3600
	0.25	42	14				512	558	558	549	−1,613	3600
	0	56	0				546	557	561	**554**	−1,248	3600
Q-31	1	0	75	0	3	28	473	**473**	478	483	1,046	3600
	0.75	18	57				516	551	551	566	2,722	3600
	0.5	37	38				537	564	564	566	0,355	3600
	0.25	56	19				554	564	573	**568**	−0,873	3600
	0	75	0				572	**572**	584	**575**	−1,541	3600
Q-32	1	0	75	0	4	21	482	**482**	494	500	1,215	3600
	0.75	18	57				531	573	573	575	0,349	3600
	0.5	37	38				552	612	612	614	0,327	3600
	0.25	56	19				586	618	618	**616**	−0,324	3600
	0	75	0				603	626	626	**620**	−0,958	3600
Q-33	1	0	75	0	5	17	488	**488**	495	501	1,212	3600
	0.75	18	57				552	623	623	630	1,124	3600
	0.5	37	38				585	623	623	625	0,321	3600
	0.25	56	19				608	656	656	**650**	−0,915	3600
	0	75	0				641	674	674	**667**	−1,039	3600

Table 4. Values found for instances with 101 nodes.

P	r_1	$|V_{T_2}|$	$|V_{T_1}|$	$|V_S|$	m	q	l_b	u_b	u_{b0}	u_{b1}	Δ	$t(s)$
Q-34	1	0	25	75	3	10	274	**274**	282	282	0,000	7200
	0.75	6	19				314	**314**	327	**325**	−0,612	7200
	0.5	12	13				337	353	353	**350**	−0,850	7200
	0.25	18	7				356	363	363	363	0,000	7200
	0	25	0				366	**366**	**366**	366	0,000	7200
Q-35	1	0	25	75	4	7	289	**289**	293	293	0,000	7200
	0.75	19	6				344	367	367	367	0,000	7200
	0.5	12	13				367	405	405	**404**	−0,247	7200
	0.25	18	7				385	416	416	418	0,481	7200
	0	25	0				409	425	425	**423**	−0,471	7200
Q-36	1	0	25	75	5	6	299	**299**	**299**	299	0,000	7200
	0.75	19	6				361	393	393	**390**	−0,763	7200
	0.5	12	13				378	403	403	**401**	−0,496	7200
	0.25	18	7				407	429	429	432	0,699	7200
	0	25	0				440	452	452	**450**	−0,442	7200
Q-37	1	0	50	50	3	19	411	**411**	**411**	411	0,000	7200
	0.75	12	38				457	492	492	**490**	−0,407	7200
	0.5	25	25				473	499	499	**496**	−0,601	7200
	0.25	37	13				483	503	503	**499**	−0,795	7200
	0	50	0				493	508	523	**516**	−1,338	7200
Q-38	1	0	50	50	4	14	415	**415**	420	423	0,714	7200
	0.75	12	38				460	480	480	481	0,208	7200
	0.5	25	25				484	517	517	**512**	−0,967	7200
	0.25	37	13				501	531	531	**528**	−0,565	7200
	0	50	0				525	537	537	**532**	−0,931	7200
Q-39	1	0	50	50	5	12	426	**426**	443	445	0,451	7200
	0.75	12	38				481	505	505	505	0,000	7200
	0.5	25	25				495	527	527	**524**	−0,569	7200
	0.25	37	13				523	564	564	**556**	−1,418	7200
	0	50	0				553	574	574	**570**	−0,697	7200
Q-40	1	0	75	25	3	28	511	**511**	516	517	0,194	7200
	0.75	18	57				555	594	594	**588**	−1,010	7200
	0.5	37	38				570	592	592	596	0,676	7200
	0.25	56	19				588	612	612	**610**	−0,327	7200
	0	75	0				606	**606**	622	**613**	−1,447	7200
Q-41	1	0	75	25	4	21	516	**516**	519	521	0,385	7200
	0.75	18	57				559	595	595	597	0,336	7200
	0.5	37	38				582	607	607	**603**	−0,659	7200
	0.25	56	19				603	619	619	**612**	−1,131	7200
	0	75	0				624	639	642	**632**	−1,558	7200
Q-42	1	0	75	25	5	17	522	**522**	529	531	0,378	7200
	0.75	18	57				584	653	653	654	0,153	7200
	0.5	37	38				598	645	645	**644**	−0,155	7200
	0.25	56	19				622	670	670	**662**	−1,194	7200
	0	75	0				649	689	689	**686**	−0,435	7200
Q-43	1	0	100	0	3	38	555	**555**	**555**	556	0,180	7200
	0.75	25	75				611	652	652	654	0,307	7200
	0.5	50	50				624	657	660	660	0,000	7200
	0.25	75	25				644	648	656	**652**	−0,610	7200
	0	100	0				663	**663**	683	**677**	−0,878	7200
Q-44	1	0	100	0	4	28	564	**564**	568	568	0,000	7200
	0.75	25	75				624	663	663	666	0,452	7200
	0.5	50	50				644	690	690	**682**	−1,159	7200
	0.25	75	25				665	683	691	**684**	−1,013	7200
	0	100	0				684	700	700	**692**	−1,143	7200
Q-45	1	0	100	0	5	23	570	**570**	576	580	0,694	7200
	0.75	25	75				629	695	695	698	0,432	7200
	0.5	50	50				674	717	717	722	0,697	7200
	0.25	75	25				689	730	730	**714**	−2,192	7200
	0	100	0				709	743	743	**733**	−1,346	7200

feasible solution of the latter is feasible in the former. We refer to the work on the CRTP in [7]. In that paper, a considerable number of problem instances used are solved to optimality and those that are unsolved have lower bounds that will guide us to measure the results generated by our application. We use the test instances provided by Hill and reported in [6, 7]. These instances are originated in the Class A instances of CmRSP in [1]. For each Class A instance, a partition of terminal nodes in type-1 and type-2 was made with different distribution of such kind of nodes, summarizing 5 instances for $[0, 0.25, 0.5, 0.75, 1]$ percentage of type-1 nodes. We performed 10 executions for each instance and we presented the best value.

Tables 1, 2, 3 and 4 present a contrast between the optimal solution (when it was reached, otherwise the lower bound) achieved for the CRTP in [7] and the optimal solution for the CTNSTP found in our metaheuristic. The acronyms are the following: P is the identifier of the instance, r_1 is the percentage of type-1 nodes, $|V_{T_2}|$, $|V_{T_1}|$ and $|V_S|$ are the number of type-2, type-1 and Steiner nodes of the instance respectively, m is the maximum number of components and q the capacity of each component. Acronyms l_b and u_b are the lower and upper bound in the exact resolution method [7], u_{b_0} is the cost of solution using the approximate method in [6] and u_{b_1} the optimum produced by our metaheuristic. The parameter Δ is a measure of our GRASP-VND effectiveness, we compare the results produced by our metaheuristic with the results reported in [6]. The relative reduction is $\Delta = \frac{u_{b_1} - u_{b_0}}{u_{b_0}}$. This means that our proposal outperforms the previous solution whenever $\Delta < 0$. Column u_b is in bold face when $u_b = l_b$ and therefore u_b is the global optimum. Column u_{b_0} is in bold face when the value in [6] reaches global optimum [7]. Column u_{b_1} (our value) is in bold face when it outperforms u_{b_0} [6]. Finally, the value $t(s)$ is a time limit (in seconds) imposed to each run of our algorithm; this means that it stops either when it performs the specified number of GRASP iterations or when it reaches the time limit. We can note that from 225 instances, we obtained the global optimum in 55 of them, better results in 73 instances, and the average gap was 0.099.

5   Conclusions and Trends for Future Work

The Capacitated Two-Node Survivable Tree Problem (CTNSTP) has been introduced. As far as we know, it has not been studied in prior literature. The need for redundancy and cheaper costs in network deployment is remarkable. Inspired by theoretical results and the related problem CmRSP, we propose an alternative problem where rings are replaced by arbitrary two-node connected components. Both problems are computationally intractable. Therefore, heuristics are suitable for large case scenarios. The CTNSTP has been heuristically addressed using a GRASP metaheuristic enriched with a Variable Neighborhood Descent (VND) and one exact local search. Results from the literature concerning CRTP were taken as reference for comparison. The reader can appreciate from the results that the resulting topologies in our CTNSTP produce relative savings with respect to the CRTP. However, the components obtained in the solutions were cycles instead of other two-connected topologies. Further research is needed in order to understand the nature of problem instances which influence these results. In general terms, we can say that our proposed heuristic algorithm improves results of other existing heuristics conceived for a more particular problem. Moreover, our algorithm is able to potentially solve a more general problem.

The problem introduced in this work can be extended in a suitable way to model delay-sensitive applications. To meet this goal, diameter constraints should be introduced to ensure connectivity between terminal nodes by a limited number of hops.

References

1. Baldacci, R., Dell'Amico, M., González, J.J.S.: The capacitated m-ring-star problem. Oper. Res. **55**(6), 1147–1162 (2007)
2. Bayá, G., Mauttone, A., Robledo, F.: The capacitated m two node survivable star problem. Yugoslav J. Oper. Res. **27**(3), 341–366 (2017)
3. Bayá, G., Mauttone, A., Robledo, F., Romero, P.: Capacitated m two-node survivable star problem. Electron. Notes Discrete Math. **52**, 253–260 (2016). INOC 2015 7th International Network Optimization Conference
4. Bhandari, R.: Optimal physical diversity algorithms and survivable networks. In: Proceedings of Second IEEE Symposium on Computers and Communications, 1997, pp. 433–441. IEEE (1997)
5. Dreyfus, S.E., Wagner, R.A.: The steiner problem in graphs. Networks **1**(3), 195–207 (1971)
6. Hill, A.: Multi-exchange neighborhoods for the capacitated ring tree problem. In: Dimov, I., Fidanova, S., Lirkov, I. (eds.) NMA 2014. LNCS, vol. 8962, pp. 85–94. Springer, Cham (2015). https://doi.org/10.1007/978-3-319-15585-2_10
7. Hill, A., Voß, S.: Optimal capacitated ring trees. EURO J. Comput. Optim. **4**(2), 137–166 (2016)
8. Hoshino, E.A., de Souza, C.C.: A branch-and-cut-and-price approach for the capacitated m-ring-star problem. Discrete Appl. Math. **160**(18), 2728–2741 (2012)
9. Karp, R.M.: Reducibility among combinatorial problems. In: Miller, R.E., Thatcher, J.W. (eds.) Complexity of Computer Computations, pp. 85–103. Plenum Press, New York (1972)
10. Kruskal, J.B.: On the shortest spanning subtree of a graph and the traveling salesman problem. Proc. Am. Math. Soc. **7**(1), 48–50 (1956)
11. Labbé, M., Laporte, G., Martín, I.R., González, J.J.S.: The ring star problem: polyhedral analysis and exact algorithm. Networks **43**(3), 177–189 (2004)
12. Laporte, G.: The traveling salesman problem: an overview of exact and approximate algorithms. Eur. J. Oper. Res. **59**(2), 231–247 (1992)
13. Monma, C., Munson, B.S., Pulleyblank, W.R.: Minimum-weight two-connected spanning networks. Math. Prog. **46**(1–3), 153–171 (1990)
14. Recoba, R., Robledo, F., Romero, P., Viera, O.: Two-node-connected star problem. Int. Trans. Oper. Res. **25**(2), 523–543 (2017)
15. Resende, M.G.C., Ribeiro, C.C.: GRASP: Greedy Randomized Adaptive Search Procedures. In: Burke, E.K., Kendall, G. (eds.) Search Methodologies, pp. 287–312. Springer, New York (2014). https://doi.org/10.1007/978-1-4614-6940-7_11
16. Robledo, F.: GRASP heuristics for Wide Area Network design. Ph.D. thesis, INRIA/IRISA, Université de Rennes I, Rennes, France (2005)
17. Zhang, Z., Qin, H., Lim, A.: A memetic algorithm for the capacitated m-ringstar problem. Appl. Intell. **40**(2), 305–321 (2014)

Data-Driven Job Dispatching in HPC Systems

Cristian Galleguillos[1,2(✉)], Alina Sîrbu[3], Zeynep Kiziltan[1], Ozalp Babaoglu[1], Andrea Borghesi[1], and Thomas Bridi[1]

[1] Department of Computer Science and Engineering,
University of Bologna, Bologna, Italy
{zeynep.kiziltan,ozalp.babaoglu,andrea.borghesi3,thomas.bridi}@unibo.it
[2] Escuela de Ing. Informática, Pontificia Universidad Católica de Valparaíso,
Valparaíso, Chile
cristian.galleguillos.m@mail.pucv.cl
[3] Department of Computer Science, University of Pisa, Pisa, Italy
alina.sirbu@unipi.it

Abstract. As High Performance Computing (HPC) systems get closer to exascale performance, job dispatching strategies become critical for keeping system utilization high while keeping waiting times low for jobs competing for HPC system resources. In this paper, we take a data-driven approach and investigate whether better dispatching decisions can be made by transforming the log data produced by an HPC system into useful knowledge about its workload. In particular, we focus on job duration, develop a data-driven approach to job duration prediction, and analyze the effect of different prediction approaches in making dispatching decisions using a real workload dataset collected from Eurora, a hybrid HPC system. Experiments on various dispatching methods show promising results.

1 Introduction

High Performance Computing (HPC) systems have become fundamental "instruments" for doing science much like microscopes and telescopes were during the previous century. The race towards exascale (10^{18} operations per sec.) HPC systems is in full swing with several efforts underway. Pushing current HPC systems to exascale performance requires a 50-fold increase in their speed and an order of magnitude increase in their energy efficiency [6]. While we can expect progress in hardware design to be a major contributor towards these goals, rest of the increase has to come from software techniques and from massive parallelism employing millions of processor cores. At these scales, job *dispatching* strategies become critical for keeping system utilization high while keeping waiting times low for jobs that are competing for HPC system resources.

HPC systems produce large amounts of data in the form of logs tracing resource consumption, errors and various other events during their operation. Data science can transform this raw data into knowledge through models built from historical data capable of anticipating unseen or future events. We believe

G. Nicosia et al. (Eds.): MOD 2017, LNCS 10710, pp. 449–461, 2018.
https://doi.org/10.1007/978-3-319-72926-8_37

that predictive computational models obtained through data-science tools will be indispensable for the operation and control of future HPC systems.

In an HPC system, a *dispatcher* decides which jobs to run next among those waiting in the queue (*scheduling*) and on which resources to run them (*allocation*). Ideally, dispatching decisions should complete all jobs in the shortest amount of time possible while keeping the system utilization high. This goal is achievable only with complete *a priori* knowledge of the workload which is rarely available at dispatching time. We therefore rely on predictive models to obtain useful knowledge about the workload from the log data of an HPC system, with the purpose of making better *dispatching decisions*. In particular, we focus on job duration, investigate an approach to job duration prediction based on the data available at job submission, and subsequently study the power of different approaches to prediction in making *dispatching decisions*. Our data-driven approach is based on a real workload dataset collected from Eurora [2], a hybrid HPC system equipped with CPU, GPU and MIC technologies to deliver high power efficiency. Experimental results on various dispatching methods show that job duration prediction can significantly benefit dispatching decisions in general, and specifically our simple data-driven approach can offer a valid alternative.

The contributions of this paper are twofold: (i) development of a simple yet effective data-driven approach for job duration prediction, (ii) analysis of the effect of different prediction approaches on various state-of-the-art dispatching methods.

The rest of the paper is organized as follows. In Sects. 2, 3 and 4, we describe the dataset used, the prediction approaches and the dispatching methods in consideration, respectively. In Sect. 5, we evaluate the reliability of the predictions and then their impact on dispatching decisions. We discuss the related work in Sect. 6 and conclude in Sect. 7 with indications for future work.

2   Data Description

The workload dataset used throughout this paper comes from the Eurora system which is hosted at CINECA[1], the largest datacenter in Italy, and was ranked first on the Green500 list in July 2013. Eurora has a modular architecture based on nodes (blades), each one having 2 octa-core CPUs and 2 expansion cards that can be configured to host an accelerator module. Of the 64 nodes, half of them host 2 powerful NVidia GPUs, meanwhile the other half are equipped with 2 Intel MIC accelerators. Each node has 16 GB of RAM memory. These 64 nodes are dedicated exclusively to computation, with the user interface being managed by a separate node. Eurora has been used by scientists across Italy to perform simulation studies from different fields, hence the workload is heterogeneous.

The workload data includes logs for over 400,000 jobs submitted between March 2014 and August 2015. For each job, we have information on the submission, start and end times, queue, wall-time, user and job name, together with

---

[1] The Italian Inter University Consortium for High Performance Computing (http://www.cineca.it).

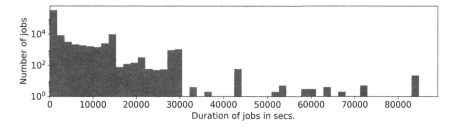

Fig. 1. Distribution of job durations on Eurora.

resources used and their allocation on the various nodes. The data has been collected through a dedicated monitoring system [6]. For our study, we selected the 10 busiest months, resulting in a total of 372,321 jobs. Figure 1 shows the distribution of job durations of the selected workload. The maximum job duration is 24 h. The figure demonstrates the existence of many short jobs and fewer longer jobs, with a long tailed distribution of job duration. As observed earlier, this is typical to HPC [21] and cloud systems [9], hence results on this system should apply to large scale computational infrastructures in general.

To evaluate the effects of prediction on different job types, we divided the jobs into classes: short jobs with duration of under 1 h, medium jobs with duration between 1 and 12 h, long jobs with duration over 12 h. In terms of frequency, 93.15% of jobs fall into the short class (the vast majority), 6.82% into the medium class and only 0.03% into the long class. We also computed the CPU time used by jobs in each class. It is the medium class that uses most resources, with 87.63% of the total while short and long jobs use only 10.77% and 1.6%, respectively.

3   Job Duration Prediction

Duration of jobs is an important consideration in dispatching decisions and knowing them at job submission time clearly facilitates better algorithms. Dispatching algorithms are often developed with the assumption that job durations are known [4,17]. Even if this is not practical, in some cases it may be possible to rely on user-provided estimates of job duration [7,17]. Many HPC systems allow users to define a wall-time value, and use a default value when users fail to provide one. This wall-time, which in the case of Eurora is set on a per-queue basis, can be considered a crude prediction of job duration.

It has been shown that in general user estimations are not reliable [17], while predefined wall-times are inflexible to account for all user needs. In these conditions, prediction of job duration through other means may prove to be an important resource. Here, we describe a simple data-driven heuristic algorithm that relies on user histories to predict job duration. The data-driven approach is particularly useful when user data can be stored for longer periods of time, which is increasingly feasible through modern Big Data tools and techniques.

Our heuristic constructs job profiles from the available workload data. The profile includes job name, queue name, user-declared wall-time, and the number of resources of each type (CPU, GPU, MIC, nodes) requested. Each user is analyzed separately. Prediction is based on the observation that jobs with the same or similar profiles have the same duration for long periods of time—there is a temporal locality of job durations. Then, at some point, the duration changes to a new set of values, which are again stable in time. This could be due for instance to changes in user behavior: a user first tests the code with short runs, then decides to run the real simulation which may last longer, then may decide to test again after having made changes, and so on. Another explanation could be switching between input datasets: the user performs repeated runs on one set of data, then moves to another. Hence, for each new job, our heuristic searches for the last job with a similar profile, and uses the duration of that job to predict the duration of the new one. We analyze users separately. The similar profile is identified using a set of consecutive rules. First, a full profile match is searched for, then if this does not exist in the user history, a profile where the job name has the same prefix is looked up. This follows from the observation that users often name jobs with similar durations with the same job name followed by a number (e.g. "job1", "job2"). If this is unsuccessful, we allow for resources used to differ, as long as the full job name, queue and wall-time are the same. If also this search fails, we look for the same match but with the name prefix rather than the exact name. If none of these rules give a match, we look for the last job with the same name, or, as a last resort, the same name prefix. If all rules fail, then we take the wall-time as the predicted duration. In all cases, the prediction is capped by the wall-time.

We have also used machine learning to predict job duration. However, results were not satisfactory (not shown for space reasons), with our simple heuristic providing much better performance. We believe this is due to the temporal locality observed in the data, and also due to the fact that jobs with the same profile may have several different durations depending on when they were submitted. This means that a regular regression model would try to fit a wide range of values with the same features, resulting in an averaging of the observed durations.

4   Job Dispatching Methods

Job dispatching in HPC systems is an optimization problem which has been studied extensively [11,15]. Since it is a hard problem [14], most of the proposed solutions are heuristic-based methods, which are fast but do not guarantee optimality. In this paper, we examine 5 of such methods reported in the literature. In the following, we give intuitions for the algorithms underlying these dispatching methods. We note that in the first three, we have adopted the *all-requested-computers-available* policy for resource allocation [24]. For each scheduled job, this policy searches sequentially the nodes in an attempt to find resources available for running the job, and if succeeds, it maps the job onto those nodes. The resource allocation policy of the remaining two are custom made and explained in their respective subsections. In all methods, the objective of the dispatcher is to minimize the total waiting time of the submitted jobs. The waiting time of a job is the time passed between its submission and its starting time.

Shortest job first, longest job first. Shortest Job First (SJF) and Longest Job First (LJF) use the estimated duration at scheduling time, sorting all jobs that have to be scheduled in ascending (or descending) order, and then mapping the shortest job (or the longest job) to a resource [23]. Both algorithms continue moving through the sorted list until no available resources remain for allocating to the current job. The aim of SJF is to reduce the waiting time of the short jobs, thus causing delays for the execution of the long jobs. Conversely, LJF reduces the waiting time of the long jobs, causing slowdown for short jobs.

EASY-Backfilling. A key element of many commercial dispatchers is the *back-filling* algorithm [24] which starts scheduling jobs stepping through a priority list such as SJF or LJF, or commonly (as also adopted here) using the jobs' submission order (first-in-first-out policy). If a job cannot be dispatched due to lack of available resources (blocked job), backfilling calculates the time in the future when enough resources will be released to run the blocked job, based on the estimated duration of running jobs. While the blocked job is waiting, the dispatcher maps other jobs in the queue over the available resources. If, however, the durations have been underestimated, the resources for the blocked job will not be available when needed, which can force termination of the running jobs. In such a case, EASY-Backfilling (EBF) [24] does not terminate the running jobs but instead delays the starting time of the blocked job. To keep all the jobs running until their termination, we have here adopted EBF.

Priority rule-based. As an extension of the first-in-first-out policy, many dispatchers sort the set of jobs to be scheduled by certain priority, running those with higher priorities first. This algorithm is referred to as Priority Rule-Based (PRB) [1,18] and is widely used in commercial HPC dispatchers[2,3]. In our work, the priority rules are based on [7] and sort the jobs to be scheduled in decreasing order of the jobs' urgency in leaving the queue. To determine if a job could wait in the queue, the ratio between the waiting time and the expected waiting time (assumed to be available for each queue the jobs are submitted) of the job is calculated. Then, jobs that are closer to surpass their expected waiting time have priority over the jobs that still could wait in the queue. As a tie breaker the "job demand" is used, which is the job's resource requirements multiplied by the estimated job duration. Hence, among the high priority jobs, those that have requested less resources and have shorter durations have further priority. The allocation process tries to assign each job to the nodes containing resources available for running the job. The nodes are also sorted by their current load (nodes with fewer free resources are preferred), thus trying to fit as many jobs as possible on the same node, to decrease the fragmentation of the system.

---

[2] Altair PBS Works (http://www.pbsworks.com/).
[3] SLURM Workload Manager (https://slurm.schedmd.com/).

Hybrid constraint programming method. One of the drawbacks of the heuristic methods is the limited exploration of the solution space. Recently, new approaches have been proposed to improve the performance of traditional scheduling, without violating the real-time requirements. For example, Bartolini et al. [5] propose a HPC job dispatcher based on Constraint Programming (CP) that is able to outperform traditional PRB methods. To increase scalability, Borghesi et al. introduce a hybrid approach combining CP and a heuristic algorithm [7] (CPH). We adopted this last method in this paper.

CPH is composed of two stages. The first corresponds to scheduling the jobs using CP with the objective of minimizing the total waiting time. The schedule is generated using a relaxed model of the problem which considers each resource type as one unique resource, i.e., CPU availability corresponds to the sum of the available CPUs of all the computing nodes, memory availability corresponds to the sum of the memory availability of all the computing nodes, and so on. The model is solved with a custom search strategy guided by a branching heuristic using the scheduling policy of PRB. Due to the problem complexity, we do not insist on finding optimal solutions but impose a time limit to bound the search; the best solution found within the limit is the scheduling decision. The preliminary schedule generated in the first stage may contain some inconsistencies because of considering the available resources as a whole. During the second stage, which corresponds to the resource allocation, any inconsistencies are removed. If a job can be mapped to a node then it will be dispatched, otherwise it will be postponed. The second stage uses the allocation policy of PRB.

5   Experimental Results

We have implemented a discrete event simulator for job submission and job dispatching, named AccaSim[4], and used it to simulate the Eurora system with the workload trace described in Sect. 2. AccaSim is a freely available Python library. At every time point, it checks if there are jobs to be dispatched. If so, it calls a dispatching method to generate a dispatching decision, and then simulates the running of the jobs on the system. AccaSim library already includes the implementations of the SJF, LJF and EBF dispatching methods. The PRB and CPH implementations are available for download in the AccaSim website. The experiments were ran on a CentOS machine equipped with Intel Xeon CPU E5-2640 Processor and 15 GB of RAM.

The simulation study considered the five dispatching methods described in Sect. 4 together with three estimations of job duration: prediction based on walltime (W), data-driven prediction presented in Sect. 3 (D) and real duration (R). The real duration was included to provide a baseline to which the other two predictions are compared. Therefore, for each of the five dispatching methods, there are three estimations of job duration, resulting in 15 combinations (e.g., for the SJF method we have SJF-W, SJF-D and SJF-R corresponding to wall-time prediction, data-driven prediction and real duration, respectively).

---

[4] https://sites.google.com/view/accasim.

To compare the quality of the dispatching decisions of the 15 combinations, we have selected two criteria. The first is *job slowdown*, a common metric for evaluating job scheduling algorithms [16], which quantifies the effect of each method on the jobs themselves and is directly perceived also by the HPC users. Slowdown of a job j is a normalized waiting time and is defined as $slowdown_j = (T_{w,j} + T_{r,j})/T_{r,j}$ where $T_{w,j}$ is the waiting time and $T_{r,j}$ is the duration of job j. A job waiting more than its duration has a higher slowdown than a job waiting less than its duration. The second criterion is the *number of queued jobs* at a given time. This metric is a measure of the effects of dispatching on the computing system itself, being directly related to system throughput: the lower the number of waiting jobs, the higher the throughput.

Next, we present the performance of our data-driven prediction of job duration and then continue with the evaluation of the various dispatching methods.

5.1 Prediction Performance

To evaluate the performance of our data-driven prediction of job duration, we compute the absolute error of the prediction and compare with that of the wall-time prediction. Over all the jobs in the system, our algorithm obtains a mean absolute error of 38.9 min. Using the wall-time, on the other hand, results in a mean absolute error of 225.11 min. Figure 2 shows the distribution of the absolute errors in the two cases, showing that in the data-driven case, these are concentrated towards small values, while in the case of the wall-time the distribution peaks at errors over 1 h. The plot shows clearly that our data-driven prediction produces much better results compared to wall-time prediction.

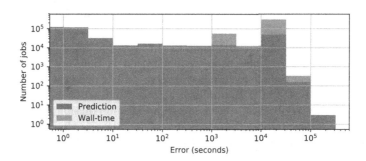

Fig. 2. Absolute data-driven prediction error, compared to wall-time prediction.

5.2 Dispatching Performance Using Prediction

To analyze the effects of prediction on job dispatching, we plot the distribution of our evaluation criteria for all 15 combinations of the dispatching methods and duration predictions. For easy visualization of distributions, we use boxplots that show the minimum and maximum values (top and bottom horizontal

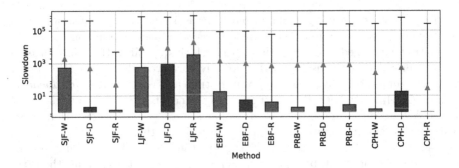

Fig. 3. Distribution of job slowdown for each method.

lines), the range between the 1st and 3rd quartiles (the colored box), the median (horizontal line within the box) and the mean (the triangles). Note that with the logarithmic scale on the vertical axis, some of these elements may be missing from the plots, meaning their value is zero.

Effects of Prediction on Jobs. The first analysis looks at job slowdown for all 372,321 jobs dispatched. Figure 3 shows the distribution of slowdown achieved by each dispatching method with each prediction type. For better visualization, we plot only the jobs where slowdown is different from 1 in at least one method-prediction combination. The removed jobs are those that are dispatched immediately as they arrive in the system, so are not relevant for our comparison. As the figure shows, the dispatching methods displaying best performance when the most basic and least effective prediction is used (wall-time) are PRB and CPH, while the methods performing worst are LJF and SJF. This is understandable since the latter methods are quite simple while the former employ more sophisticated reasoning.

An interesting effect when using real duration is that not all dispatching methods show a clear benefit. While we observe a clear decrease in slowdown in SJF, EBF and CPH, for LJF a significant increase in slowdown is present, while for PRB no change is observed. We understand that prediction does not always help the dispatching methods. One possible explanation is that the incomplete nature of the dispatching methods tends to lead to suboptimal decisions which can sometimes be compensated by underestimation of job durations, which will not be possible anymore with a (perfect) prediction.

When using our data-driven prediction in the dispatching methods, we expect the performance to stay between the wall-time prediction and the real job duration. Figure 3 shows that this is true for most methods. In the cases of SJF and EBF, the real job duration improves the results, so does our prediction, albeit less effectively. In the case of LJF, real job duration worsens the results, so does our prediction, but less severely. PRB, which already does not benefit from real job duration, does not benefit from our prediction either. The only dispatching method where the performance improves with perfect prediction but decreases

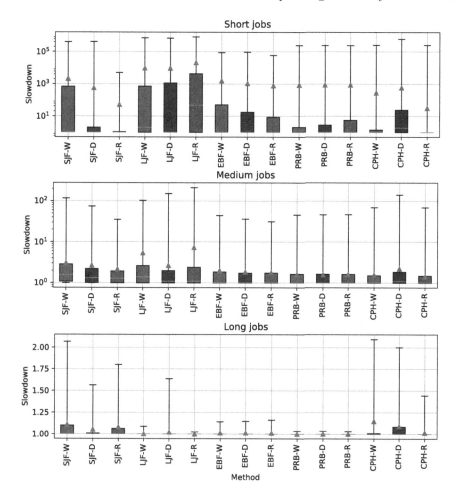

Fig. 4. Distribution of job slowdown for short, medium and long jobs for each method.

with our prediction is CPH. We believe this is because our data-driven prediction may sometimes underestimate job duration, which is never the case for wall-time and the real duration. CPH is not resilient to job duration underestimation, hence an imperfect prediction can actually be detrimental.

Even if PRB and CPH provide the best overall results, we observe that SJF comes in very close, with comparable slowdown, when adding prediction. However, the first two methods are more sophisticated and incur an overhead when building the dispatching decisions, while SJF is a very simple strategy. Hence, in the presence of predictions, one may prefer to use a simple method such as SJF over the heavier methods such as PRB and CPH.

To better understand the effects of prediction, we also look at the different job classes. Figure 4 shows box-plots of slowdown distributions for short, medium and long jobs. When prediction is beneficial, we see that the jobs that benefit

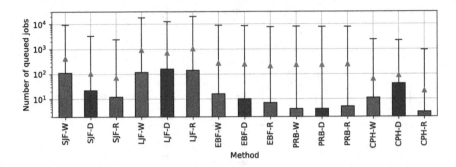

Fig. 5. Distribution of number of jobs waiting at every second, for each method.

most are the short ones. This is good news, given that a large number of our jobs are short, as we saw in Sect. 2. Some smaller differences are also visible on medium jobs, while on long jobs the methods seem to be quite comparable, with slightly larger slowdown values in CPH and SJF compared to the rest.

Effects of Prediction on the System. Besides effects on individual jobs, it is important to understand prediction's role in improving system-level behavior. For this we look at the size of the waiting queue. Figure 5 shows the distribution of the number of jobs in the queue at every second. We removed from the plot those time points where there were no jobs in the queue for any of the 15 combinations, because these corresponded to low system utilizations and have no value for our comparison. The figure shows that the effect on the system is similar to the performance measured by the slowdown. In particular, SJF and EBF are improved by prediction (both data-driven and real durations). PRB shows no difference, however queue size is already the shortest among all dispatching methods. LJF does not benefit from prediction, while CPH seems to be improved only by perfect and not by our data-driven prediction.

We also looked at the distribution of resource utilization (amount of CPUs, GPUs, MICs and memory used at each second), but we observed minor differences in the total use of resources between methods, so results are not shown.

6   Related Work

A number of previous efforts have developed techniques for predicting interesting aspects of workloads such as power consumption and job duration [10,20]. Borghesi et al. [8] propose a machine learning approach to forecast the mean power consumption of HPC applications using only information available at scheduling time, such as the resources requested, the maximum duration, the user, etc. Sirbu et al. [22] present a support vector machine model to predict the power consumption of jobs, taking also into account their variability.

Predicting the durations of HPC jobs have also been considered in previous research works, especially in relation to job dispatching [3,19]. Tsafrir et al. [12]

propose a model that uses the run times of the last two jobs to predict the duration of the next job. This prediction is then used for scheduling purposes. Their approach is lightweight and efficient, however, the prediction accuracy can be improved using more complex techniques like the ones proposed in this paper. Gaussier et al. [13] show the importance of estimating the duration of HPC jobs with backfilling schedulers. Their results clearly suggest that a backfilling policy benefits from accurate duration predictions; the only limitation is that their work focuses exclusively on a particular scheduling algorithm.

7 Conclusions

We have presented an analysis of the effect of job duration prediction on HPC job dispatching decisions, based on a real workload dataset collected from Eurora, a hybrid HPC system. We implemented five state-of-the-art dispatching methods and studied their performance in the presence of predictions based on a data-driven heuristic and on estimates based on wall-time. These two approaches to prediction were compared among themselves and also against a baseline: perfect prediction using the real job duration from the data.

Our conclusions are severalfold. First, our data-driven approach results in more effective predictions than the estimates based on wall-time. Second, even a perfect prediction does not necessarily benefit dispatching methods. One possible explanation is that the incomplete nature of the dispatching methods tend to lead to suboptimal decisions which can sometimes be compensated by under-estimation of job durations. Third, prediction is nevertheless beneficial in the majority of the methods we have considered, and in the presence of our data-driven prediction, a simple dispatching method can become a valid alternative to the sophisticated state-of-the-art methods. Finally, when using prediction is advantageous, the main beneficiaries are the short jobs. Given the prominent presence of short jobs in typical HPC [21] and cloud system [9] workloads, our conclusions should apply to large-scale computational infrastructures in general.

The dispatching methods presented here exploit prediction in their scheduling component. In future work, we will also develop allocation heuristics that can exploit prediction, especially in the case of hybrid HPC systems. Additionally, we plan to extend our job duration prediction heuristic to include resources shared with other jobs, similar to [22], to improve our prediction power. Finally, we plan to integrate power predictions [22] into the dispatchers, to optimize not only the system response, but also energy consumption, which is mandatory for building exascale systems.

Acknowledgments. We thank Dr. A. Bartolini, Prof. L. Benini, Prof. M. Milano and Dr. M. Lombardi for fruitful discussions on the work presented here and for providing access to the Eurora data, together with the SCAI group in Cineca. We acknowledge the Cineca PM-HPC award allowing access to HPC resources. C. Galleguillos has been supported by Postgraduate Grant PUCV 2017. A. Sîrbu has been partially funded by the E.U. project SoBigData Research Infrastructure—Big Data and Social Mining Ecosystem (grant agreement 654024).

References

1. Buddhakulsomsiri, J., Kim, D.S.: Priority rule-based heuristic for multi-mode resource-constrained project scheduling problems with resource vacations and activity splitting. Eur. J. Oper. Res. **178**(2), 374–390 (2007)
2. Cavazzoni, C.: EURORA: a european architecture toward exascale. In: FutureHPC@ICS, pp. 1:1–1:4. ACM (2012)
3. Chen, X., et al.: Predicting job completion times using system logs in supercomputing clusters. In: DSN Workshops, IEEE Computer Society (2013)
4. Chandio, A.A., et al.: A comparative study of job scheduling strategies in large-scale parallel computational systems. In: TrustCom/ISPA/IUCC, pp. 949–957. IEEE Computer Society (2013)
5. Bartolini, A., Borghesi, A., Bridi, T., Lombardi, M., Milano, M.: Proactive workload dispatching on the EURORA supercomputer. In: O'Sullivan, B. (ed.) CP 2014. LNCS, vol. 8656, pp. 765–780. Springer, Cham (2014). https://doi.org/10.1007/978-3-319-10428-7_55
6. Bartolini, A., et al.: Unveiling eurora - thermal and power characterization of the most energy-efficient supercomputer in the world. In: DATE, pp. 1–6. European Design and Automation Association (2014)
7. Borghesi, A., Collina, F., Lombardi, M., Milano, M., Benini, L.: Power capping in high performance computing systems. In: Pesant, G. (ed.) CP 2015. LNCS, vol. 9255, pp. 524–540. Springer, Cham (2015). https://doi.org/10.1007/978-3-319-23219-5_37
8. Borghesi, A., Bartolini, A., Lombardi, M., Milano, M., Benini, L.: Predictive modeling for job power consumption in HPC systems. In: Kunkel, J.M., Balaji, P., Dongarra, J. (eds.) ISC High Performance 2016. LNCS, vol. 9697, pp. 181–199. Springer, Cham (2016). https://doi.org/10.1007/978-3-319-41321-1_10
9. Reiss, C., et al.: Heterogeneity and dynamicity of clouds at scale: Google trace analysis. In: SoCC, p. 7. ACM (2012)
10. Storlie, C., et al.: Modeling and predicting power consumption of high performance computing jobs. arXiv:1412.5247 (2014, preprint)
11. Feitelson, D.G., Rudolph, L., Schwiegelshohn, U., Sevcik, K.C., Wong, P.: Theory and practice in parallel job scheduling. In: Feitelson, D.G., Rudolph, L. (eds.) JSSPP 1997. LNCS, vol. 1291, pp. 1–34. Springer, Heidelberg (1997). https://doi.org/10.1007/3-540-63574-2_14
12. Tsafrir, D., et al.: Backfilling using system-generated predictions rather than user runtime estimates. IEEE Trans. Parallel Distrib. Syst. **18**(6), 789–803 (2007)
13. Gaussier, É., et al.: Improving backfilling by using machine learning to predict running times. In: SC, pp. 64:1–64:10. ACM (2015)
14. Blazewicz, J., et al.: Scheduling subject to resource constraints: classification and complexity. Discret. Appl. Math. **5**(1), 11–24 (1983)
15. Cao, J., et al.: A taxonomy of application scheduling tools for high performance cluster computing. Clust. Comput. **9**(3), 355–371 (2006)
16. Feitelson, D.G.: Metrics for parallel job scheduling and their convergence. In: Feitelson, D.G., Rudolph, L. (eds.) JSSPP 2001. LNCS, vol. 2221, pp. 188–205. Springer, Heidelberg (2001). https://doi.org/10.1007/3-540-45540-X_11
17. Feitelson, D.G., Weil, A.M.: Utilization and predictability in scheduling the IBM SP2 with backfilling. In: IPPS/SPDP, pp. 542–546 (1998)
18. Haupt, R.: A survey of priority rule-based scheduling. Oper. Res. Spektrum **11**(1), 3–16 (1989)

19. Matsunaga, A.M., Fortes, J.A.B.: On the use of machine learning to predict the time and resources consumed by applications. In: CCGRID, pp. 495–504. IEEE Computer Society (2010)

20. Shoukourian, H., Wilde, T., et al.: Predicting the energy and power consumption of strong and weak scaling HPC applications. Supercomput. Front. Innov. $\mathbf{1}(2)$, 20–41 (2014)

21. Sîrbu, A., Babaoglu, O.: A holistic approach to log data analysis in high-performance computing systems: the case of IBM blue gene/q. In: Hunold, S., Costan, A., Giménez, D., Iosup, A., Ricci, L., Gómez Requena, M.E., Scarano, V., Varbanescu, A.L., Scott, S.L., Lankes, S., Weidendorfer, J., Alexander, M. (eds.) Euro-Par 2015. LNCS, vol. 9523, pp. 631–643. Springer, Cham (2015). https://doi.org/10.1007/978-3-319-27308-2_51

22. Sîrbu, A., Babaoglu, O.: Power consumption modeling and prediction in a hybrid CPU-GPU-MIC supercomputer. In: Dutot, P.-F., Trystram, D. (eds.) Euro-Par 2016. LNCS, vol. 9833, pp. 117–130. Springer, Cham (2016). https://doi.org/10.1007/978-3-319-43659-3_9

23. Streit, A.: Enhancements to the decision process of the self-tuning dynP scheduler. In: Feitelson, D.G., Rudolph, L., Schwiegelshohn, U. (eds.) JSSPP 2004. LNCS, vol. 3277, pp. 63–80. Springer, Heidelberg (2005). https://doi.org/10.1007/11407522_4

24. Wong, A.K.L., Goscinski, A.M.: Evaluating the easy-backfill job scheduling of static workloads on clusters. In: CLUSTER, IEEE Computer Society (2007)

AbstractNet: A Generative Model for High Density Inputs

Boris Musarais$^{(\boxtimes)}$ (iD)

ESGI, 242 Rue du Faubourg Saint-Antoine, 75012 Paris, France
boris.musarais@esgi.com

Abstract. This paper introduces AbstractNet, a generative model for high density inputs. The model suggests a method that uses unsupervised learning to generate feature maps. The model drastically improves the performances of raw audio generation by reducing the required amount of input data and computing power necessary to achieve a similar result when compared to the state of the art.

Keywords: Unsupervised learning · Generative model · Audio · Auto-Encoder
LSTM · RNN · Deep neural networks · Data compression

1 Introduction

Among the vast amount of fields that are concerned by the dimensionality problem, audio generation is a very intuitive example. Anyone can understand why feeding such dense data to a neural network would prove itself to be a challenge. A known approach to generating raw audio is to simply give the raw data to a recurrent neural network such as LSTMs [1] and let the model try to guess what the next frame will be as experimented in GRUV [2].The output would be added to the input and fed back to the network. Wavenet [3] suggested another approach, using dilated convolutions [4, 5]:

$$p(x) = \prod_{t=1}^{T} p(x_t | x_1, \dots, x_{t-1}) \qquad (1)$$

Formula (1) shows that the output has to be fed back to the input making it an autoregressive [6] model. Given the dimensionality of raw audio (up to 44100 samples per second), this approach requires a massive amount of time and computing power to produce raw audio. The larger the receptive field is, the longer it will take to compute (Fig. 1).

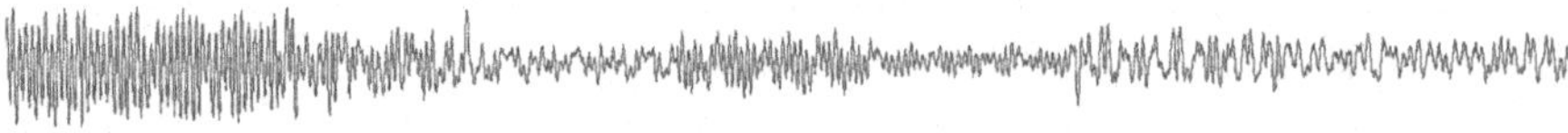

Fig. 1. Real-time generated waveform (16000 Hz).

Because it is so hard [7, 8] to train deep neural networks since they struggle to get an abstract overview of things, and need a massive amount of input data to do so, we will try and take a different approach; maybe there is a way to reduce the dimensionality of generative models.

© Springer International Publishing AG 2018
G. Nicosia et al. (Eds.): MOD 2017, LNCS 10710, pp. 462–469, 2018.
https://doi.org/10.1007/978-3-319-72926-8_38

2   AbstractNet

2.1   The Model

Instead of letting the model figure everything out from scratch, we could divide the learning process into two (or more) steps, guiding it so it can get an abstract view faster. As suggested in "The Sparsely-Gated Mixture-of-Experts Layer" [9] approach, sub-networks offer a solid approach to reducing the computing dimensionality of a dataset. Other works [10] also suggest using auto-encoders [11] as an effective optimization solution.

The ideal approach would be to generalize our model, making no assumptions about the input shape (other than the fact that is it defined by recurrent patterns).

Here is how we will do this:

- Let the network find low level features in a sample of our dataset that represents the entire population using unsupervised learning (more specifically auto-encoders)
- Compress the entire dataset using those features.
- Generate a high level feature map using a recurrent neural network that defines how lower features are to be distributed
- Generate a dense, low level input from the feature map using a specialized conditional generative model.

Considering some raw input data (an audio sample or anything else), this is what the model looks like (Fig. 2):

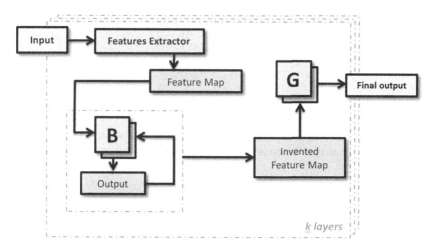

Fig. 2.   Architecture of the AbstractNet generation model

G: Generative model, used to rebuild low level signals from feature maps.
B: Recurrent generative model, used on a feature map. In the tests I used LSTMS [1].

This model can be stacked to go deeper; each layer is then conditioned on the previous one, allowing the system to understand more abstract concepts without the need of large datasets or great computing power!

Features Extraction. One of the key elements of AbstractNet is feature definition, this can be done programmatically (ex: amplitude and frequency are simple features that can be extracted manually), but the ideal setup is to have an unsupervised neural network extract the features from the low level raw input. The idea behind this approach is that a few samples of the large scale input data are enough to describe the low level features of a dataset. This means one should provide to the feature extractor at least one example for each feature present in the entire dataset (Obviously It is recommended to provide more than one).

The examples are then converted to a feature map by the auto-encoder (Fig. 3).

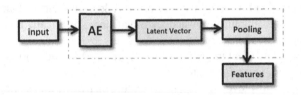

Fig. 3. Architecture of the features extractor model

AE: Auto-encoder

A feature map is represented as a set of features for each frame in the raw waveform:

$$x = \left\{ \{k_0, \dots, k_u\}_{[0:N]} \right\} \tag{2}$$

N: Feature map length
k: a feature
u: the amount of features (hidden units) defined in the auto-encoder

The feature map is then fed to a neural network whose task is solely to train on multi-dimensional feature maps. To control the output, the network is conditioned by the feature channel:

$$\hat{x}(c) = \{x, c\} \tag{3}$$

c: desired feature channel

As a matter of fact, we don't have to use the decoding part of the auto-encoder. Using a conditional [12] network can drastically improve the quality of the generated low level output; the de-noising aspect of the auto encoder is not really required (Fig. 4).

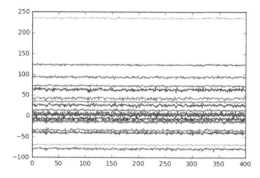

Fig. 4. A 20 channels generated feature map, in practice it is normalized.

Entropy pooling. To improve the receptive field and increase the coherence and flexibility of the generator, an entropy [13] map is injected in the input to condition the high level generator. This allows the network to have a higher level (thus more abstract) view (Fig. 5).

$$E(k) = \sum_{n=1}^{N} \log x^{k+n} \tag{4}$$

E: entropy of a given sample
k: sample start index
n: sample length
x: input data

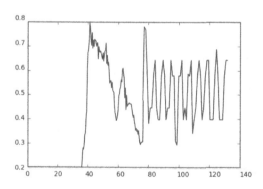

Fig. 5. An input sample with its entropy map one the left side, and the raw audio sample on the right side.

Quantization. In this generative model, previous outputs are fed back to the new input, this means that the error rate is subject to an exponential increase:

$$e(n) = \sum_{k=1}^{n} \varepsilon_k \tag{5}$$

e: error rate
n: generation iteration
ε: measured error

To improve the model's flexibility and its tolerance to unknown patterns, the inputs have been quantized [14], encoding each value to only 256 values:

$$x_n = \frac{\lfloor x_n * u \rfloor}{u} \tag{6}$$

x: input data
u: quantization (in our case 256)

2.2   Conditional AbstractNets

Just like the higher levels in the AbstractNet architecture control the lower levels output, one can manually inject his own conditions to an AbstractNet model to generate an output with custom characteristics:

$$p(x|c) = \prod_{n=1}^{T} p(x_n|x_1, \ldots, x_{n-1}, c) \tag{7}$$

c: a conditional input, similar to formula (3).

This is typically done for TTS [15, 16]. One could use this technique to define the words he wants the model to generate.

3   Experiments

To figure out if AbstractNet really is efficient at generating coherent high density signals, the best way is to test it and evaluate the quality of the generated output.

3.1   Piano Dataset

The generative model was trained on piano samples taken from DJ Oakawari's songs. The encoder was trained on small samples extracted from the dataset; the training phase took 3 h on an NVidia GTX860 M GPU.

For this test, a single AbstractNet layer was used. The model was implemented with Tensorflow. To improve the compression performances, I used Fast Fourier Transforms on the input signals; this allowed the encoder to ignore the time dimensionality. The decoder has been conditioned on previous waveforms to regenerate an audio sample when combined with a latent vector. The latent feature map for this test was 40 times smaller than the raw audio input; the output had a background noise due to the generator not being able to properly rebuild the audio sequence, even when conditioned on the previous samples. Several techniques [17] exist for noise removal, but it wasn't the point of this experiment.

Regardless, the model was able to generate high frequency audio samples (at 16000 Hz) in real time, a very promising result.

3.2 Complex Samples

More complex audio samples do not get compressed as effectively, because one sample (for example with dance music) contains more features than piano. This is where a deeper architecture could prove itself more efficient.

Activation Function. Because the experiments are not defined as a classification problem, but rather a regression problem; since we want to find the closest approximation of a signal, no specific activation function was used, only the identity function.

Error function. The cross entropy [18] function was used for error minimization during the experiments:

$$H(p, q) = - \sum_x p(x) \log q(x) \tag{8}$$

While I do not have the kind of computing power needed to train on video samples or to generate real time TTS, I believe this is a problem AbstractNet is able to approach with decent hardware. More tests have yet to be performed to explore the real potential of this technique; my tests were done on a laptop. The Auto-Encoder approach could also be tested with other techniques, such as Generative Adverserial Networks (GANs) [19], Variational-Auto-Encoders (VAE) [20] or even combined VAE + GAN [21]. I am eager to see the results of other people using the AbstractNet architecture on other data-sets with powerful machines and great ideas.

4 Conclusion

This paper presented AbstractNet, a generative model designed for dense inputs that takes a higher abstract view on the data, while drastically improving performances. By combining multiple layers of the AbstractNet architecture, the model could go further in its understanding of high level data structures, making the generated output more coherent. Because lower layers are conditional networks, it is possible to use them independently to generate low level features. This interesting behavior allows us to "talk" to the AbstractNet model and ask it to explain why it has behaved in a certain way. This means one can follow and understand the reasoning of an AbstractNet model to make it behave as expected.

Acknowledgements. I want to thank Alain Lioret from Université Paris 8, Aurélien Schlossman from Ariane Group, Nicolas Vidal, Martin Tricaud and everyone at Ecole Superieure De Génie Informatique (ESGI). I would also like to thank all the people who believed in this project.

References

1. Hochreiter, S., Schmidhuber, J.: Long short-term memory. Neural Comput. **9**(8), 1735–1780 (1997)
2. Nayebi, A., Vitelli, M.: GRUV: Algorithmic Music Generation using Recurrent Neural Networks (2015)
3. van den Oord, A., Dieleman, S., Zen, H., Simonyan, K., Vinyals, O., Graves, A., Kavukcuoglu, K.: Wavenet: A generative model for raw audio. CoRR abs/1609.03499 (2016)
4. Dutilleux, P.: An implementation of the "algorithme à trous" to compute the wavelet transform. In: Combes, J.M., Grossmann, A., Tchamitchian, P. (eds.) Wavelets. Inverse Problems and Theoretical Imaging, pp. 298–304. Springer, Heidelberg (1989). http://doi.org/10.1007/978-3-642-97177-8_29
5. Holschneider, M., Kronland-Martinet, R., Morlet, J., Tchamitchian, P.: A real-time algorithm for signal analysis with the help of the wavelet transform. In: Combes, J.M., Grossmann, A., Tchamitchian, P. (eds.) Wavelets. Inverse Problems and Theoretical Imaging, pp. 286–297. Springer, Heidelberg (1990). http://doi.org/10.1007/978-3-642-75988-8_28
6. Akaike, H.: Fitting autoregressive models for prediction. Ann. Inst. Stat. Math. **21**(1), 243–247 (1969)
7. Glorot, X., Bengio, Y.: Understanding the difficulty of training deep feedforward neural networks. In: Aistats, vol. 9, pp. 249–256, May 2010
8. Montana, D.J., Davis, L.: Training feedforward neural networks using genetic algorithms. In: IJCAI, vol. 89, pp. 762–767, August 1989
9. Shazeer, N., Mirhoseini, A., Maziarz, K., Davis, A., Le, Q., Hinton, G., Dean, J.: Outrageously large neural networks: The sparsely-gated mixture-of-experts layer. arXiv preprint arXiv:1701.06538 (2017)
10. Hinton, G.E., Salakhutdinov, R.R.: Reducing the dimensionality of data with neural networks. Science **313**(5786), 504–507 (2006)
11. Rumelhart, D.E., Hinton, G.E., Williams, R.J.: Learning Representations by Back-Propagating Errors (1988)
12. Adali, T., Liu, X., Sonmez, M.K.: Conditional distribution learning with neural networks and its application to channel equalization. IEEE Trans. Sig. Process. **45**(4), 1051–1064 (1997)
13. Cox, G.: On the relationship between entropy and meaning in music: an exploration with recurrent neural networks. In: Proceedings of the Annual Meeting of the Cognitive Science Society (2010)
14. Ahalt, S.C., Krishnamurthy, A.K., Chen, P., Melton, D.E.: Competitive learning algorithms for vector quantization. Neural Netw. **3**(3), 277–290 (1990)
15. Taylor, P.: Text-To-Speech Synthesis. Cambridge university press, Cambridge (2009)
16. Ze, H., Senior, A., Schuster, M.: Statistical parametric speech synthesis using deep neural networks. In: IEEE International Conference on Acoustics, Speech and Signal Processing (ICASSP), 2013, pp. 7962–7966. IEEE, May 2013
17. Rudin, L.I., Osher, S., Fatemi, E.: Nonlinear total variation based noise removal algorithms. Phys. D Nonlinear Phenom. **60**(1–4), 259–268 (1992)
18. Deng, L.Y.: The Cross-Entropy Method: A Unified Approach to Combinatorial Optimization, Monte-Carlo Simulation, and Machine Learning (2006)

19. Goodfellow, I., Pouget-Abadie, J., Mirza, M., Xu, B., Warde-Farley, D., Ozair, S., Bengio, Y.: Generative adversarial nets. In: Advances in Neural Information Processing Systems, pp. 2672–2680 (2014)
20. Kingma, D.P., Welling, M.: Auto-encoding variational bayes. arXiv preprint arXiv: 1312.6114 (2013)
21. Larsen, A.B.L., Sønderby, S.K., Larochelle, H., Winther, O.: Autoencoding beyond pixels using a learned similarity metric. arXiv preprint arXiv:1512.09300 (2015)

A Parallel Framework for Multi-Population Cultural Algorithm and Its Applications in TSP

Olgierd Unold$^{(\boxtimes)}$ and Radosław Tarnawski

Department of Computer Engineering, Faculty of Electronics,
Wroclaw University of Science and Technology,
Wyb. Wyspianskiego 27, 50-370 Wroclaw, Poland
olgierd.unold@pwr.edu.pl

Abstract. This paper presents a novel parallel framework based on the Multi-Population Cultural Algorithm (MPCA) scheme for optimization problems. Contrary to the existing variants of Cultural Algorithm (CA), the proposed parallel framework for MPCA (PFMPCA) allows the use of any implemented metaheuristic both in a belief, and in a population space. Furthermore, the proposed approach permits CA to evolve simultaneously multiple population and belief sub-spaces, leveraging the dual inheritance mechanism and utilizing multi-population approach. Moreover, each sub-population (in population or belief space) is able to communicate between each other. PFMPCA has been implemented on Graphics Processing Units (GPUs) using CUDA programming model. The performance of the developed framework was evaluated using asymmetric Travelling Salesman Problem (ATSP). The MPCA for TSP implemented by means of the parallel framework proves to have an extensible architecture designed to accommodate changes and good performances.

Keywords: Cultural Algorithm · Multi-Population
GPU computing · CUDA architecture · Travelling Salesman Problem
Ant Colony Optimization · Genetic Algorithm

1 Introduction

Cultural Algorithm is a branch of Evolutionary Algorithms proved to be an efficient approach for solving many optimization problems (eg. [2]), mainly due to so-called its *dual inheritance mechanism*. CA works using two evolution spaces, i.e. population (PS) and belief (BS), communicating through $accept - influence$ protocol. This double evolutionary mechanism allows often converging faster with a better solution in comparison with single-population based approaches [16].

The multi-population Cultural Algorithm, firstly introduced in [3], is an extended version of the basic CA. In general, in MPCA both population and belief spaces can contain more than one sub-spaces evolving independently. CA with multiple population and belief spaces enables evolutionary knowledge to

© Springer International Publishing AG 2018
G. Nicosia et al. (Eds.): MOD 2017, LNCS 10710, pp. 470–482, 2018.
https://doi.org/10.1007/978-3-319-72926-8_39

be exchanged among populations according to certain rules. It has been proved that MPCA can improve the speed of convergence and overcome premature convergence [7]. At present, rich work have been done about how to multiply populations in MPCAs. Some examples of MPCA include parallel co-operating CA [3], MPCA adopting knowledge migration [7], MP particle swarm CA [8], MPCA with differential evolution [20], transfer-agent MPCA [9], heterogeneous MPCA [12], GPU-accelerated CA [4], MPCA for community detection in social networks [22], and heritage dynamic CA [10]. Although it is not a typical MPCA architecture, it is worth mentioning here CA extended by sub-cultures [1].

Worth noting is the universal architecture of MPCA proposed by Guo et al. [7]. The so-called MPCA-KM consists of n sub-populations, where each sub-population is a basic CA containing belief and populations spaces. Each sub-population evolves independently, and at the constant intervals the evolved knowledge in the sub-populations is migrated to each other.

In this paper, we propose a novel parallel framework for MPCA. The introduced parallel framework PFMPCA enables to communicate between any belief or population space, not only placed in the same sub-population. This approach extends substantially mentioned above the MPCA-KM architecture. Moreover, the proposed framework has been implemented on GPUs using Compute Unified Device Architecture (CUDA) [13]. To our best knowledge, CUDA-based (MP)CA was presented only in [4], not counting our own implementation of parallel CA [18].

PFMPCA was experimentally compared with a sequential and a parallel Ant Colony Optimization (ACO), as well as a parallel Cultural Ant Colony Optimization (pCACO) [18] on a speed-up and solution quality over asymmetric Travelling Salesman Problems taken from TSPLIB library. In pCACO the parallel ACO works in population space, whereas parallel Genetic Algorithm evolves the belief space population. TSP is a typical $\mathcal{NP}$-hard optimization problem, where a travelling salesman wants to travel all cities but each city is supposed to be visited only once. Note that many others $\mathcal{NP}$-problems can be attributed to TSP, such a postman problem, product assembly line, clustering of data arrays, or even DNA sequencing [11].

The remainder of this paper is organised as follows. In Sect. 2, we briefly review the concepts of (sequential) Cultural Algorithm, (sequential) Ant Colony Optimization, Genetic Algorithm, General-Purpose Computing on GPUs, and Travelling Salesman Problem. The proposed framework is discussed in Sect. 3. Empirical experiments of evaluating the PFMPCA are conducted in Sect. 4. At last, Sect. 5 gives some concluding remarks and points out the further research directions.

2 Preliminaries

Cultural Algorithm [16] depicts cultural evolution as a process of dual inheritance from both a micro-evolutionary level (population space) and a macro-evolutionary level (belief space) (see Fig. 1). From the perspective of evolution, any computational framework according to the requirement of CA can be used to represent or describe both of the spaces. The population space is composed of the individuals representing the search space of possible solutions, whereas the belief space consists of the experienced knowledge acquired during evolution process. These two spaces interact with each other through operations *accept* and *influence*. The PS conducts the evolution and periodically contributes to the BS using *accept* operation to update BS. BS performs evolution as well and calls *influence* operation to direct the evolution process in PS. When *accept* operation is invoked, the solution is updated by the local best (i.e. the shortest path in TSP case) from PS, and when *influence* is fired, the global best solution in BS is transferred to PS. The fulfilled termination condition ends the algorithm.

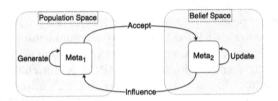

Fig. 1. Generic architecture of a Cultural Algorithm. Meta$_i$ describes any metaheuristic algorithm.

Ant Colony Optimization [5] has been inspired by real ant colony's foraging behaviour, where ants can often find the shortest path between a food source and their nest. For obvious reasons, ACO has been first applied to Travelling Salesman Problem [5]. The transition probability from city i to city j for the k-th ant is defined as follows $P_k(i,j) = [\tau(i,j)]^\alpha \cdot [\eta(i,j)]^\beta / \sum_{u \in J_k^i} [\tau(i,u)]^\alpha \cdot [\eta(i,u)]^\beta$ if $j \in J_k^i$, 0 otherwise, where J_k^i is the set of cities allowed to be visited by ant k from the city i, $\tau(i,j)$ is the amount of pheromone trail on the edge i to j, $\eta(i,j)$ denotes inverse of path length from i to j, α and β are parameters that control the relative importance of trail versus visibility.

After the ants end all their tours, the pheromone trails $\tau(i,j)$ are updated according to the formula $\tau(i,j) \leftarrow (1-\rho) \cdot \tau(i,j) + \sum_{k=1}^m \Delta\tau_k(i,j)$ where m is the number of ants, $(1-\rho)$ is the evaporation rate such that $(0 \le \rho \le 1)$, $\Delta\tau_k(i,j)$ is the amount of pheromones remaining on the path at current iteration for ant k. This amount is calculated as $\Delta\tau_k(i,j) = Q/L_k$ if $(i,j) \in k$-th ant tour, 0 otherwise, where Q is a constant, and L_k is the total length of k's ant tour.

$\mathcal{MAX}$-$\mathcal{MIN}$ Ant System ($\mathcal{MMAS}$) is one of the most efficient ACO-based algorithm [17], in which the amount of pheromone over the edge between two vertices i and j is restricted $\tau_{min} \le \tau(i,j) \le \tau_{max}$ to avoid search stagnation.

Genetic Algorithm (GA) [6] inspired by nature and proposed by Holland in 1975, encodes the parameters of a solution into the chromosome consisted of genes. GA randomly generates a set of chromosomes as the initial population.

Then, it randomly selects two chromosomes from the population to perform some genetic operators, like the crossover and the mutation operations, repeatedly until the result satisfies the termination condition. After performing the genetic operators, the system calculates the fitness value of each chromosome. Chromosomes with higher fitness values are to be selected into the gene pool for reproduction in the next generation. The application of GA as a robust population-based metaheuristic for TSP is already intensively studied since the 1980s, and many special dedicated representations and genetic operators were proposed.

General-Purpose Computing on GPUs. In the CUDA programming model a program to be executed is divided into CPU and GPU part (Heterogeneous Programming). In most cases the GPU (device) works as a coprocessor of CPU (host). The host and the device maintain their own separate memory spaces. CUDA program is based on functions executed in parallel by a given number of CUDA threads called kernels. Threads are grouped together into blocks, which are executed independently to each other. Threads within a block can communicate by sharing data through the shared memory, which is fast in access time but relatively small in size. In a single block, it is also possible to synchronise threads execution to coordinate memory accesses. To avoid data hazard, synchronisation and atomic (also related to global memory) functions are available. Threads in different blocks are independent. Blocks of threads are, in turn, grouped into grids. When a kernel is invoked it needs to know the number of blocks in a grid and the number of threads in a block. During kernel execution, blocks of the grid are independently scheduled among the GPUs Streaming Multiprocessors (SM). It is also possible to run multiple kernels concurrently and overlap data transfer using different CUDA streams. Each SM is composed of CUDA cores. SM executes on successive clock cycles a single warp of threads (32 related threads). Flow control instructions may force threads of the same warp to serialise their execution paths. Each thread has an access to local memory too, which can be registers or a specific region of the global memory. Registers are the fastest available GPUs memories, but the data cannot be stored in directly there. In the GPU, tens of thousands threads can be executed concurrently, which can significantly improve execution time.

Travelling Salesman Problem is a well-known combinatorial optimization task, proved to be $\mathcal{NP}$-hard. In TSP a set of cities to be visited and distances between them are given. TSP is to find the shortest way of visiting all the cities and returning to the starting point. Regarding a set of cities as vertices connected in pairs by weighted edges, the goal of TSP is to find a Hamiltonian cycle with the least weight in a complete weighted graph. There are many types of TSP, including symmetric euclidean and non-euclidean, asymmetric, dynamic, and special cases, like multiple TSP [21]. In this paper, the asymmetric TSP (ATSP) is considered, which is the more general version, where the distances between the cities are dependent on the direction of travelling of the edges.

3 Implementation of the Proposed Parallel Framework on GPUs

Proposed parallel framework for MPCA provides a set of functions in form of MPCA class methods. These methods were inspired by the GPU lock-free synchronization algorithm [19]. The main task of the framework class object is to create specific data structures in the device (GPU) global memory. A single framework class object is created for each of chosen CA metaheuristics. Any metaheuristic algorithm can be used in both of CA spaces. To connect the selected PS with a chosen BS metaheuristic in MPCA, the setup method is launched on the PS's object with a BS's object as an argument. This kind of connection allows the PS metaheuristic to make use of the BS metaheuristic provided data (CA *influence* operation). The other way, to connect the BS with the PS metaheuristic, BS metaheuristic's object runs setup method with the PS class object as an argument (CA *accept* operation is provided). It is possible to connect to any number (or none) of metaheuristics, even in the same CA space.

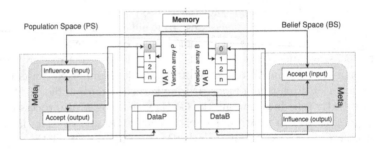

Fig. 2. Flow chart of a proposed framework for MPCA. Meta$_{i(j)}$ describes any metaheuristic algorithm

Every MPCA framework class object is indexed based on a number of running metaheuristics in each of CA space. The main purpose of index is to determine which version array (VA) element is taken into consideration during *influence* and *accept* operations (see *ID* in Algorithms 1 and 2). The example of created version arrays and the data matrices for each CA metaheuristics are shown in Fig. 2. As can be seen from the PS point of view *influence* operation is associated with (data) input, and the PS *accept* operation is associated with (data) output. From the BS point of view, the other way, *accept* operation is associated with (data) input, and the *influence* operation with (data) output.

The arrows (Fig. 2) indicate the data flow. Each specific data structure (including each VA element) can be written by only one metaheuristic. In our example (see Algorithm 1), the PS *influence* (input) operation starts by comparing BS's VA (*sbVerionArray*) element 0 with an element associated with the specific PS metaheuristic (in the same array). If the first element's value (index 0) is larger than PS metaheuristic's specific one (based on *ID*), then the

PS metaheuristic is allowed to (*inputAllowed*) read the data (DataB) provided by BS. After the indirect performance of the *influence* operation the PS meta-heuristic's element value in BS VA is set to the array first element value (see Algorithm 2). Otherwise, the value remains the same (data not read).

Algorithm 1. The PFMPCA method used before *input* operation execution (Fig. 3)	**Algorithm 2.** The PFMPCA method used after *input* operation execution (Fig. 3)
1 **if** *threadIndex == 0* **then** 2 inputAllowed = false; 3 **if** *sbVersionArray[ID] ¦ sbVersionArray[0]* **then** 4 | inputAllowed = true; 5 **end** 6 __threadfence(); 7 **end** 8 __syncthreads();	1 **if** *threadIndex == 0 and inputAllowed != false* **then** 2 sbVersionArray[ID] = sbVersionArray[0]; 3 inputAllowed = false; 4 __threadfence(); 5 **end** 6 __syncthreads();

Further, the *accept* (output) operation in the PS begins with comparison of every PS algorithm VA indices with the array index 0 (see Algorithm 3). If the sum of the same values occurrence (*count*) is equal to the number of connected BS algorithms (*numberOfConnections*), PS algorithm is allowed to process the output operation (*outputAllowed*). The chosen data is written to PS algorithm data matrix (DataP) and the PS algorithm array index 0 content is incremented or increased by some value (see Algorithm 4). Otherwise, the array index 0 value remains the same (data not written).

Algorithm 3. The PFMPCA method used before *output* operation execution (Fig. 3)	**Algorithm 4.** The PFMPCA method used after *output* operation execution (Fig. 3)
1 **if** *threadIndex == 0* **then** 2 count = 0; 3 outputAllowed = false; 4 **while** *i ¦ numberOfConnections* **do** 5 **if** *myVersionArray[0] ==* *myVersionArray[i]* **then** 6 | count += 1; 7 **else** 8 | break; 9 **end** 10 **end** 11 **if** *count == numberOfConnections* **then** 12 | outputAllowed = true; 13 **end** 14 __threadfence(); 15 **end** 16 __syncthreads();	1 **if** *threadIndex == 0 and outputAllowed != false* **then** 2 myVersionArray[0] += 1; 3 outputAllowed = false; 4 __threadfence(); 5 **end** 6 __syncthreads();

Generally speaking, VA element values provide information about which data array release is available to read (both *versionArray's* element 0) or which data array release was read by the other metaheuristics (both *versionArray's* other elements). Updating own data matrix is only possible if all connected meta-heuristics have read or signal to read provided data (*myVerionArray* element values the same). Reading data matrix is only possible if the new data is provided (difference between *sbVerionArray* element 0 and element based on metaheuris-tic *ID*). A similar flow takes place in case of Belief Space with the difference in *accept* (input) operation usage.

Population Space (PS). PS can make use of potentially any metaheuris-tic (ACO algorithm in our approach). The algorithm behind PS starts on a host (CPU) side with device's (GPU's) memory allocation. In our case it is a pheromone and a distance matrix. All algorithm specific parameters are set too and sent in form of a single structure. The other argument is connected directly with the proposed framework class object which contains all methods and data required to process CA operations. After the kernel launch the algorithm flow is redirected to the device.

At first specific instance problem values are calculated. They are used during pheromone amount update computation. First one (*reference_value*) is a product of the longest instance edge and the instance size. The second one (*pheromone_multiplier*) is a quotient of the instance size and the ants number parameter. Both are saved in a shared memory due to a broadcast mechanism usage. The path creation step is made independently by each CUDA thread which is associated with an individual ant. A single thread block is used.

The placement of CA operations provided by PFMPCA's methods is very flexible. In our case if the solution created each turn by the ant is the best (the shortest) one provided by that ant so far, it is aimed to be saved at an ant-related data matrix column (structure of arrays pattern is used). Considered PS data matrix is located in the global memory. Before solution save, VA elements are compared by framework method (see Algorithm 3). In case of permission, solution data is safely provided and the PS algorithm VA element (index 0) is updated (see Algorithm 4). Described operation can be interpreted as the first stage of the *accept* operation (output operation from the PS algorithm point of view).

At the end of the algorithm turn, right after pheromone trails evaporation process, PS algorithm is trying to run indirectly the *influence* operation with some probability (5% in our case). Once again framework operations are used. Firstly, to determine if associated BS algorithm data is ready to be read and probably contains some new data (see Algorithm 1). Next, to update related VA element (see Algorithm 2) if the data was used or we want to skip the current version of provided solutions. *Influence* operation takes only the best provided solution and updates pheromone matrix by adding base pheromone value to edges on its basis, but still limited by MAX-MIN rule. From the PS algorithm point of view it is the input operation, the second stage of the *influence* operation.

Algorithm ends with the last iteration. With the end of the last working PS algorithm ends the whole MPCA. Right before returning to host PS algorithm signals its end by setting chosen value (e.g. different than 0) to provided by the connected BS algorithm variable.

Belief Space (BS). On a basic level BS specific algorithm works very similar to any PS algorithm. First, on the host side, global memory and algorithm specific data is allocated. In our case a simple genetic algorithm is used and for this reason parent's and offspring's matrices are created (structure of arrays pattern is used). The parent's matrix is also preliminary initialized with a random set of TSP solutions (initial population). The algorithm is parameterized by a number of generations, a population size and a number of taken best PS generated solutions in CA *accept* operation. Once again MPCA framework class object is defined and sent to kernel function. The algorithm related kernel is launched just after the start-up of the other PS algorithm kernels. The BS algorithm work is redirected to the device.

Each thread in a thread block is related to a specific parent and an offspring. A single thread block is used. The offspring is created on the basis of the related parent and the randomly chosen one. If the new offspring is better (denotes a shorter path) than the predecessor (related parent), it takes his place in the memory. Otherwise, the parent remains unchanged. Optional offspring mutation is considered too. After crossover operation all individuals are sorted based on their fitness function values (bitonic mergesort algorithm is used). At this point, after creating potentially better solutions, we place the first stage of the *influence* operation (output operation from the BS algorithm point of view). If possible (see Algorithm 3), the solutions are sent to BS algorithm related data matrix. In the simplest case only one solution is provided. Every first attempt to update the algorithm related (CA) data matrix is successful due to fact that the algorithm VA is initialized with zeros (each element has the same value).

Before each new generation BS algorithm is trying to run indirectly the second stage of the *accept* operation with some probability (5% in our case). From the BS algorithm point of view it is the input operation. Algorithm tries to swap some number of the worst individuals (parents) for the new ones based on the solutions provided by related PS algorithm. The number of worst individuals to be swapped is set to the number of ants used in PS algorithm and for this reason the number of ants in related PS algorithm is generally smaller than the number of BS algorithm individuals. To control the process, MPCA framework operations are used once more (see Algorithms 1 and 2).

The BS algorithm ends with the end of the related PS algorithm. It is worth noting that the BS algorithm data matrix is available in the device's global memory until the end of the whole MPCA algorithm.

4 Experimental Results

In this study, we used a PC with one Intel Core i5-4670K (4 cores, 3.4 GHz) processor, a single Asus GeForce GTX770 2048 MB 256 bit DirectCU II OC, and Kingston 8192 MB 1600 MHz HyperX Blu Red CL10. The OS was Windows 7 Professional 64 bit. For CUDA program compilation, Microsoft Visual Studio 2012 and CUDA Toolkit 8.0 (V8.0.60) were used.

The instances on which we tested our algorithm were taken from the TSPLIB benchmark library [15]. In the experiments, we used 8 instances which were classified as asymmetric Travelling Salesman Problem. The testbed reflects the full range of TSPLIB asymmetric instances. 50 runs were performed for each instance.

The parameters settings are as follows, for ACO: $\tau_{baz} = 1000$, $\tau_{min} = 1$, $\tau_{max} = 5 \cdot \tau_{baz}$, $\alpha = 1$, $\beta = 3$, $Q = reference\_value$, $\rho = 0.1$, $m = 64$, $t_{end} = 100$; for pACO: $m = threads = 64$; for pCACO: $population\_size = 256$,

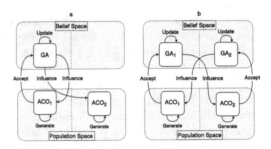

Fig. 3. MPCA structures used in experiments. (a) pMPCA$_{GA+2ACO}$, and (b) pMPCA$_{2GA+2ACO}$

$no\_of\_generation$ = 1000, $probability\_of\_inversion\_mutation$ = 0.02, $probability\_of\_crossover$ = 1.0; and for both pMPCA: probability of executing CA operation: 5%, ACO$_1$ $\rho = 0.1$, ACO$_2$ $\rho = 0.15$. The values were chosen accordingly to the previous work [18].

In proposed pMPCA$_{GA+2ACO}$ structure a single PS algorithm (ACO$_2$) is added to existing standard pCACO architecture (see Fig. 3a). It is not fully utilized because of no usage of solutions generated by ACO$_2$. None of CA metaheuristics is connected with ACO$_2$ by *accept* operation. However, the ACO$_2$ still creates solutions and is influenced by BS. Additionally, its ρ parameter is increased to 0.15 value. The purpose is to search the solution space in a different way, i.e. to check solutions closer to some specific solution provided by BS, since higher ρ value causes worse (rarely visited) paths to be chosen with much less probability over time. Furthermore, additional algorithm working in any of CA spaces gives us an extra opportunity to find the better solution in almost the same time at the expense of computing power and some higher memory usage.

Table 1 provides experimental results obtained with proposed parallel framework MPCA in two variants (see Fig. 3), compared to results gained by sequential Ant Colony Optimization (ACO), parallel Ant Colony Optimization (pACO), parallel Cultural Ant Colony Optimization (pCACO), pMPCA$_{GA+2ACO}$, and pMPCA$_{2GA+2ACO}$. Friedman post-hoc test with Holm p-value adjustment method was used to check if the differences were statistically significant at the significance level $\alpha = 0.01$. The symbol ↓ indicates that the corresponding method significantly degrades the result obtained with the best method.

The results show that pMPCA in both variations is able to find significantly better average solution than any other used algorithms. The architecture of MPCA with complete sub-populations, i.e. pMPCA$_{2GA+2ACO}$, outperforms in most cases the variant with the lack of BS metaheuristic in one of sub-populations pMPCA$_{GA+2ACO}$, when comparing two implementations of the proposed parallel MPCA framework. It is mostly due to fact that more active heuristics give us obviously higher probability of getting the better solution in total. Additionally, changed ρ value in one of ACO algorithms used in PS diversifies generated solutions. One ACO algorithm is searching much wider solution space, while the other is focused on creating solutions closer to some best ones provided so far. The pMPCA$_{2GA+2ACO}$ structure provided the best average solutions mainly because of an extra BS algorithm and a cross-connection between CA spaces, which makes an additional use of the ACO$_2$ data. The last

Table 1. Performance comparison of sequential Ant Colony Optimization (ACO) with parallel Ant Colony Optimization (pACO), parallel Cultural Ant Colony Optimization (pCACO), pMPCA$_{GA+2ACO}$ and pMPCA$_{2GA+2ACO}$, applied to eight asymmetric TSP instances, available in TSPLIB (best known solution in parentheses). For each case respectively from the top given are the best solution obtained, the average solution with standard deviation in parentheses, the average time in seconds to find the best solution in a run. Averages are taken over 50 trials, 64 ants in all ACO implementations were used. The best average solution is in bold. The symbol ↓ indicates that the corresponding method statistically significantly degrades the result obtained with the best method.

Instance	ACO	pACO	pCACO	pMPCA$_{GA+2ACO}$	pMPCA$_{2GA+2ACO}$
ftv44 (1613)	1624	1650	1623	1613	1613
	1683.92↓	1684.68↓	1666.72↓	1630.42↓	**1624.94**
	(13.87)	(9.88)	(15.42)	(2.26)	(2.27)
	4.3463	2.3485	2.3554	2.2601	2.2697
ftv55 (1608)	1635	1674	1635	1612	1612
	1696.84↓	1691.08↓	1676.50↓	1649.50↓	**1639.42**
	(17.57)	(17.84)	(18.80)	(17.98)	(8.42)
	7.1225	3.7326	3.7357	3.5631	3.5734
ftv64 (1839)	1905	1902	1879	1861	1856
	1941.32↓	1930.50↓	1911.72↓	1892.32↓	**1886.46**
	(11.80)	(14.80)	(13.16)	(13.69)	(10.86)
	9.4212	5.1313	5.1454	4.8429	4.8531
ftv70 (1950)	2093	2068	2003	1970	1970
	2148.74↓	2143.90↓	2102.38↓	1996.70	**1990.5**
	(18.87)	(24.31)	(33.97)	(18.58)	(16.53)
	10.7068	6.1712	6.1960	5.7289	5.7296
kro124p (36230)	38682	39397	38450	37406	37401
	39937.98↓	39968.78↓	39362.12↓	38105.98↓	**37876.88**
	(417.49)	(324.13)	(377.94)	(341.63)	(242.46)
	21.0575	12.9537	12.9688	11.9853	12.0963
ftv170(2755)	3130	3150	3083	3013	2983
	3269.88↓	3274.62↓	3268.30↓	3118.36↓	**3059.62**
	(53.41)	(51.29)	(63.11)	(51.94)	(29.04)
	59.1272	38.7157	39.4822	35.8026	36.4662
rbg323 (1326)	1466	1467	1468	1460	1466
	1485.82↓	1486.40↓	1483.76↓	1482.14	**1481.30**
	(6.03)	(7.25)	(7.49)	(7.27)	(6.64)
	211.2016	140.3757	147.4379	134.0678	135.9170
rbg443 (2720)	3271	3273	3212	3262	3259
	3304.88↓	3306.24↓	3262.18↓	3288.68↓	**3284.36**
	(13.26)	(12.90)	(17.10)	(11.79)	(11.79)
	390.3430	268.2758	283.8038	258.7033	261.3430

approach has also often got smaller standard deviation compared to the others, what suggest being more stable in finding better solutions. Both pMPCA algorithms have comparable average time with other approaches despite the fact that more algorithms is used at the same time. It is because population space, as well as belief space, use different streams, and therefore both spaces are invisible to themselves.

In Figs. 4 and 5 we present the results of comparing the average tour and convergence speed between ACO, parallel ACO, parallel Cultural ACO, pMPCA$_{GA+2ACO}$ and pMPCA$_{2GA+2ACO}$ for two exemplary datasets ftv64 and ftv170. Both figures show that the proposed MPCA framework achieves average shorter tour than pACO and pCACO with similar or better speed of convergence.

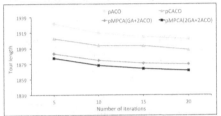
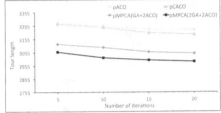

Fig. 4. Evolution of average tour length of selected ftv64 dataset for pACO, pCACO, pMPCA$_{GA+2ACO}$ and pMPCA$_{2GA+2ACO}$

Fig. 5. Evolution of average tour length of selected ftv170 dataset for pACO, pCACO, pMPCA$_{GA+2ACO}$ and pMPCA$_{2GA+2ACO}$

5   Conclusion

In this paper, we propose a novel parallel framework based on the Multi-population Cultural Algorithm for solving asymmetric Travelling Salesman Problem on a GPU. The results show that the proposed PFMPCA finds significantly better average solution than compared algorithms, i.e. sequential and parallel Ant Colony Optimization, and parallel Cultural Ant Colony Optimization. What is interesting, both presented parallel MPCA variants have comparable or even better average time with other parallel methods despite the fact, that more heuristics are used at the same time. It is possible thanks to PFMPCA communication methods and an effective use of GPU lock-free synchronization algorithm.

Future work is to research other heuristics for evolving belief space (like very promising African Buffalo Optimization [14]), to test another architectures of the framework and to improve some weaknesses of the proposed PFMPCA. One of them is that only best solution coming from evolved sub-population can be exchanged with the other one. Sometimes it can lead to premature convergence.

Acknowledgments. The work was supported by statutory grant of the Wroclaw University of Science and Technology, Poland.

Author Contribution. OU initiated and designed the study, supervised the work, made statistical tests. RT implemented the framework, performed the experiments. Both authors wrote and approved the final manuscript.

References

1. Ali, M.Z.: Using cultural algorithms to solve optimization problems with a social fabric approach. Ph.D. thesis, Wayne State University (2008)
2. Ali, M.Z., Awad, N.H., Suganthan, P.N., Reynolds, R.G.: A modified cultural algorithm with a balanced performance for the differential evolution frameworks. Knowl. Based Syst. **111**, 73–86 (2016)
3. Digalakis, J.G., Margaritis, K.G.: A multipopulation cultural algorithm for the electrical generator scheduling problem. Mathe. Comput. Simul. **60**(3), 293–301 (2002)
4. Dong, J., Yuan, B.: GPU-accelerated standard and multi-population cultural algorithms. In: 2013 International Conference on Service Sciences (ICSS), pp. 129–133. IEEE (2013)
5. Dorigo, M.: Optimization, learning and natural algorithms. Ph.D. thesis, Politecnico di Milano, Italy (1992)
6. Goldberg, D.E., Holland, J.H.: Genetic algorithms and machine learning. Mach. Learn. **3**(2), 95–99 (1988)
7. Guo, Y.N., Cheng, J., Cao, Y.Y., Lin, Y.: A novel multi-population cultural algorithm adopting knowledge migration. Soft Comput. **15**(5), 897–905 (2011)
8. Guo, Y.N., Liu, D.: Multi-population cooperative particle swarm cultural algorithms. In: 2011 Seventh International Conference on Natural Computation (ICNC), vol. 3, pp. 1351–1355. IEEE (2011)
9. Hlynka, A.W., Kobti, Z.: Knowledge sharing through agent migration with multi-population cultural algorithm. In: FLAIRS Conference (2013)
10. Hlynka, A.W., Kobti, Z.: Heritage-dynamic cultural algorithm for multi-population solutions. In: 2016 IEEE Congress on Evolutionary Computation (CEC), pp. 4398–4404. IEEE (2016)
11. Hoffman, K.L., Padberg, M., Rinaldi, G.: Traveling salesman problem. In: Gass, S.I., Fu, M.C. (eds.) Encyclopedia of Operations Research and Management Science, pp. 1573–1578. Springer, New York (2013). https://doi.org/10.1007/978-1-4419-1153-7_1068
12. Kobti, Z., et al.: Heterogeneous multi-population cultural algorithm. In: 2013 IEEE Congress on Evolutionary Computation (CEC), pp. 292–299. IEEE (2013)
13. Nvidia: Nvidia CUDA (2017). http://nvidia.com/cuda
14. Odili, J.B., Mohmad Kahar, M.N.: Solving the traveling salesman's problem using the african buffalo optimization. Comput. Intell. Neurosci. **2016**, 3 (2016)
15. Reinhelt, G.: TSPLIB: a library of sample instances for the TSP (and related problems) from various sources and of various types (2017). http://comopt.ifi.uniheidelberg.de/software/TSPLIB95
16. Reynolds, R.G.: An introduction to cultural algorithms. In: Proceedings of the Third Annual Conference on Evolutionary Programming, Singapore, pp. 131–139 (1994)

17. Stützle, T., Hoos, H.: MAX-MIN ant system and local search for the traveling salesman problem. In: IEEE International Conference on Evolutionary Computation, pp. 309–314. IEEE (1997)
18. Unold, O., Tarnawski, R.: Cultural Ant Colony Optimization on GPUs for Travelling Salesman Problem. In: Pardalos, P.M., Conca, P., Giuffrida, G., Nicosia, G. (eds.) MOD 2016. LNCS, vol. 10122, pp. 317–329. Springer, Cham (2016). https://doi.org/10.1007/978-3-319-51469-7_27
19. Xiao, S., Feng, W.: Inter-block GPU communication via fast barrier synchronization. In: 2010 IEEE International Symposium on Parallel & Distributed Processing (IPDPS), pp. 1–12. IEEE (2010)
20. Xu, W., Wang, R., Zhang, L., Gu, X.: A multi-population cultural algorithm with adaptive diversity preservation and its application in ammonia synthesis process. Neural Comput. Appl. 21(6), 1129–1140 (2012)
21. Yuan, S., Skinner, B., Huang, S., Liu, D.: A new crossover approach for solving the multiple travelling salesmen problem using genetic algorithms. Eur. J. Oper. Res. 228(1), 72–82 (2013)
22. Zadeh, P.M., Kobti, Z.: A multi-population cultural algorithm for community detection in social networks. Procedia Comput. Sci. 52, 342–349 (2015)

Honey Yield Forecast Using Radial Basis Functions

Humberto Rocha[1,2(✉)] and Joana Dias[1,2]

[1] Faculdade de Economia, CeBER, Universidade de Coimbra,
3004-512 Coimbra, Portugal
`hrocha@mat.uc.pt, joana@fe.uc.pt`
[2] INESC-Coimbra, 3030-290 Coimbra, Portugal

Abstract. Honey yields are difficult to predict and have been usually associated with weather conditions. Although some specific meteorological variables have been associated with honey yields, the reported relationships concern a specific geographical region of the globe for a given time frame and cannot be used for different regions, where climate may behave differently. In this study, Radial Basis Function (RBF) interpolation models were used to explore the relationships between weather variables and honey yields. RBF interpolation models can produce excellent interpolants, even for poorly distributed data points, capable of mimicking well unknown responses providing reliable surrogates that can be used either for prediction or to extract relationships between variables. The selection of the predictors is of the utmost importance and an automated forward-backward variable screening procedure was tailored for selecting variables with good predicting ability. Honey forecasts for Andalusia, the first Spanish autonomous community in honey production, were obtained using RBF models considering subsets of variables calculated by the variable screening procedure.

Keywords: Honey yield · Weather · Radial basis functions
Variable screening

1 Introduction

Honey has been used by humans for at least 8000 years [2]. Its production and economic interest have grown to the present day. However, annual production has large fluctuations mainly associated with weather conditions [5,18]. While some studies claim that temperatures in May, June and July are particularly important predictors of honey yields [6,7,9], other claim that variation in honey yields could be more related to March temperatures and rainfall, sunshine and temperature from April to July [4,5]. In fact, the precise relationships between weather conditions and honey yields are not well established yet. Furthermore, the relationships between weather conditions and honey yields already reported concern a specific geographical region of the globe for a given time window and cannot be used for different regions, where climate may behave differently.

© Springer International Publishing AG 2018
G. Nicosia et al. (Eds.): MOD 2017, LNCS 10710, pp. 483–495, 2018.
https://doi.org/10.1007/978-3-319-72926-8_40

In this study, the relationships between honey yield and a large number of weather variables were explored aiming to forecast honey production in Andalusia, a Spanish autonomous community. Radial basis functions (RBF) were used to interpolate the data and provide the predictive models. RBF regression has been successfully applied in different contexts, including aeronautics [13,14] or radiotherapy [16,17]. RBF models proved to mimic well unknown responses providing reliable surrogates that can be used either for prediction or to extract relationships between variables [15]. The selection of the predictors is of the utmost importance and a variable screening approach is presented. The remainder of the paper is organized as follows. Honey production and weather data in Andalusia are presented in the next Sect. 2. In Sect. 3 we briefly describe RBF interpolation. Section 4 presents the variable screening strategy proposed. Results are presented in Sect. 5 followed by the conclusion's Sect. 6.

2   Honey Production and Weather Data in Andalusia

Spain is the largest EU honey producer which is the second world producer after China [3]. Andalusia is the first Spanish autonomous community regarding honey production (6887 tonnes) and honey bee hives (562503 units), according to the latest statistical data released by the Spanish Ministry of Agriculture, Food and Environment [1]. There are two different types of honey bee hives in Andalusia: fixed comb hives – traditional hive types that require permanent damage of the comb for harvesting – and movable comb hives – modern hive types that include top-bar hives, horizontal frame hives or vertical stackable frame hives. In Andalusia, about 97% of the hives are modern hive types and we will only consider these type of hives for our forecast.

Andalusia is in the south of Spain, east of Portugal and the Atlantic Ocean and north of the Mediterranean Sea and Africa. It is the second largest in area of the Spanish autonomous communities with $87268 \, km^2$. Andalusia is divided into eight provinces – Almeria, Cádiz, Córdoba, Granada, Huelva, Jaén, Málaga and Seville – with distinct weather conditions. It is covered by a set of automated agroclimatic stations that can perform various meteorological measurements [8]. Since the honey yields of the different provinces are also available, instead of averaging different weather conditions causing a larger weather bias, forecast was made for each region considering the corresponding weather data. Only the five largest honey producer provinces (Córdoba, Granada, Huelva, Málaga and Seville) were considered. Historical data of honey yields and number of hives for the time frame in study (2001–2015) for each province is presented in Table 1. Historical data of the weather variables considered – rainfall (mm), evapotranspiration (mm), minimum temperature (°C), maximum temperature (°C), mean temperature (°C) and relative humidity (%) – are available in Appendix.

3   Radial Basis Functions Models

RBF interpolation models can produce response surfaces capable of to exploring the nonlinear relationships between different input or explanatory variables

Table 1. Tonnes of honey yields and number of hives for the five largest honey producer provinces of Andalusia.

Year	Córdoba		Granada		Huelva		Málaga		Seville	
	Honey yield	# hives	Honey yield	# hives	Honey yield	# hives	Honey yield	# hives	Honey yield	# hives
2001	268	44739	504	33600	990	66000	761	63398	1131	76644
2002	297	49529	586	39083	1323	66150	872	62263	1369	80989
2003	193	53700	595	39690	944	72580	966	64404	1499	90847
2004	149	41461	656	43723	514	51360	765	63757	894	71238
2005	126	41901	671	44757	433	61832	438	67417	827	82692
2006	473	42987	680	45357	779	64954	697	69669	879	82694
2007	756	45800	565	43466	992	66156	1138	71145	808	79515
2008	435	43531	522	40155	941	67200	987	70488	950	93880
2009	616	43990	546	42020	987	65784	900	66651	1083	97500
2010	600	44804	568	43665	891	66015	915	67783	1066	97463
2011	749	46825	594	45671	882	67813	1000	67450	964	94315
2012	242	48385	613	47127	674	67425	761	76069	954	94173
2013	675	45000	633	48705	972	67041	937	78093	1270	97314
2014	458	45825	660	50791	1105	69060	939	78254	1471	101463
2015	442	58935	700	53856	1066	71056	876	97316	1479	106494

and output or response variable(s). Moreover, RBFs can be used to predict unknown responses given the values of the explanatory variables. It was shown that stochastic models coincide with the corresponding RBF models [21]. For a set of data points in a high dimensional space, even if scarce or poorly distributed, a RBF interpolation model (surface) can always be calculated. However, the RBF model behavior between data points, is highly dependent on the basis function considered. For a given data set, some RBFs can provide desirable trends while other may exhibit undesirable trends. Thus, instead of a typical *a priori* choice based either on the literature or on authors' preferences, it is advisable to select the most adequate RBF for the data set at hand considering numerical metrics [15]. A brief description of RBF interpolation is provided next.

3.1 RBF Interpolation

Let $y(\mathbf{x})$ denote the response for a given data point $\mathbf{x}$ of n components (variables) such that the value of y is only known at a finite set of N input data points $\mathbf{x}^1, \ldots, \mathbf{x}^N$, i.e., only $y(\mathbf{x}^k)$ $(k = 1, \ldots, N)$ are known. A RBF interpolation model $h(\mathbf{x})$ can be generically represented as

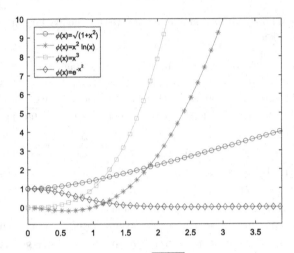

Fig. 1. Graphs of multiquadric, $\phi(x) = \sqrt{1 + x^2}$, thin plate spline, $\phi(x) = x^2 \ln(x)$, cubic spline, $\phi(x) = x^3$, and Gaussian, $\phi(x) = \exp(-x^2)$ RBFs.

$$h(\mathbf{x}) = \sum_{j=1}^{N} \alpha_j \phi(\|\mathbf{x} - \mathbf{x}^j\|), \tag{1}$$

where $\phi(x)$ is the selected RBF, α_j are the coefficients determined by the interpolation equations $h(\mathbf{x}^k) = y(\mathbf{x}^k)$ $(k = 1, \ldots, N)$, $\|\mathbf{x} - \mathbf{x}^j\|$ corresponds to the parameterized distance between $\mathbf{x}$ and $\mathbf{x}^j$,

$$\|\mathbf{x} - \mathbf{x}^j\| = \sqrt{\sum_{i=1}^{n} |\theta_i| (x_i - x_i^j)^2},$$

and $\theta_1, \ldots, \theta_n$ are scalars [15]. Coefficients $\alpha_1, \ldots, \alpha_N$ in Eq. (1) are computed for fixed parameters θ_i using the interpolation equations of the following linear system:

$$\sum_{j=1}^{N} \alpha_j \phi(\|\mathbf{x}^k - \mathbf{x}^j\|) = y(\mathbf{x}^k), \quad \text{for } k = 1, \ldots, N. \tag{2}$$

Multiquadric, $\phi(x) = \sqrt{1 + x^2}$, thin plate spline, $\phi(x) = x^2 \ln x$, cubic spline, $\phi(x) = x^3$, and Gaussian, $\phi(x) = \exp(-x^2)$, are examples of RBFs that are commonly used to model linear, almost quadratic and cubic growth rates, as well as exponential decay of the response, respectively [12] – see Fig. 1.

3.2 Cross-Validation

Calculation of the RBF model $h(\mathbf{x})$ in Eq. (1) requires the selection of a RBF $\phi(x)$ and the choice of model parameters $\theta_1, \ldots, \theta_n$. While selection of the most appropriate RBF for the given data set can be done iteratively by testing the different possible choices of $\phi(x)$, there is an infinite number of possible choices for $\theta_1,\ldots,\theta_n$. For different fixed sets of model parameters $\theta_1,\ldots,\theta_n$, distinct models with different behaviors between data points are calculated for a given selection of $\phi(x)$. Cross-validation (CV) can be used for model parameter tuning leading to models with enhanced prediction capability [19]. Furthermore, the most appropriate basis function $\phi(x)$ can be numerically computed using prediction accuracy (CV error) as main criterion. The leave-one-out CV procedure can be used in model parameter tuning for RBF interpolation [15]:

Algorithm 1. Leave-one-out cross-validation for RBF interpolation

Input:

- $\mathbf{x}^1,\ldots,\mathbf{x}^N$, N input data points with n components.
- $y(\mathbf{x}^1),\ldots,y(\mathbf{x}^N)$, response of the N input data points.

Iteration:

1. Fix a set of model parameters $\theta_1,\ldots,\theta_n$.
2. For $j=1,\ldots,N$, construct the RBF model $h_{-j}(\mathbf{x})$ of the data points $(\mathbf{x}^k, y(\mathbf{x}^k))$ for $1 \leq k \leq N, k \neq j$.
3. Set prediction error as the following CV root mean square error:

$$E^{CV}(\theta_1,\ldots,\theta_n) = \sqrt{\frac{1}{N}\sum_{j=1}^{N}(h_{-j}(\mathbf{x}^j) - y(\mathbf{x}^j))^2}. \tag{3}$$

The goal of model parameter tuning by CV is to find $\theta_1,\ldots,\theta_n$ that minimize the CV error, $E^{CV}(\theta_1,\ldots,\theta_n)$, so that the interpolation model has the highest prediction accuracy when CV error is the measure. Using different θ_i allows the model parameter tuning to scale each variable x_i based on its significance in modeling the variance in the response, thus, has the benefit of implicit variable screening built in the model parameter tuning.

4   Variable Screening

A regression model with too many input variables may have several disadvantages including an increasing difficulty on model parameter optimization or data overfitting. A standard variable screening procedure aims to identify a subset of the input variables that have significant impact on the response $y(\mathbf{x})$. In other words, if the change of $y(\mathbf{x})$ with respect to a given variable is negligible, then the subset of the input variables should not include such variable.

Variable screening methods that require the response values for specific input vectors, such as ANOVA, cannot be used in this study. Other existing variable screening techniques require specific conditions. E.g., the main effects estimate (MEE) method, proposed by Tu and Jones [20], generally requires a uniform distribution of the existing input vectors in a rectangular domain of the input space which is not the case.

Forward or backward variable screening methods are typically used to determine the explanatory power of input variables of polynomial models (linear regression) that are independent of the data distribution. Here, we assume that forward and backward variable screening methods are valid for variable selection in nonlinear models. In general, under this assumption, the forward and backward variable screening methods can be formally applied for variable selection if the data is fitted by a regression model that is independent of data distribution. We propose a generalization of a combined forward-backward variable screening procedure, described in Algorithm 2, that is based in the predicting ability instead of the typically used coefficient of determination (R^2). In the first iteration of this procedure, input vectors with a single variable at a time are fitted using RBF models (1) and the CV error (3). The best model and corresponding variable correspond to the smallest CV error which is a proxy for the prediction error. In the second iteration, input vectors with two variables, fixing the one found in the first iteration, are fitted using RBF models (1) and the CV error (3). The second variable that, along with the fixed first variable, forms the best prediction pair of variables is fixed for the third iteration. This procedure continues until the prediction error (CV error) fails to improve. Note that, at successful iteration k, we may not find the best subset of k predicting variables, i.e. the set of k variables that corresponds to the smallest CV error. E.g., at iteration two we only tested $n - 1$ possibilities – the pairs constituted by the first fixed variable and each of the remaining $n - 1$ variables – instead of all possibilities – $\binom{n}{2} = \frac{n!}{2!(n-2)!}$. Thus, at the end of the forward procedure we proceed with a backward procedure aiming to further improve the CV error. The rational behind this procedure is identical except that instead of being added, a variable is removed at each iteration.

5   Computational Results

Our tests were performed on a 2.60 Ghz Intel Core i7-6700HQ PC with 16 GB RAM and we used MATLAB (R2016a) [10]. Optimal RBF model parameters $\theta_1, \ldots, \theta_n$ of (3) were computed by minimizing the CV error using a MATLAB implementation (*fminsearch*) of a derivative-free optimization algorithm called Nelder-Mead [11]. The optimal CV error obtained for the different basis functions tested was used as proxy of their prediction ability [15]. Thin plate spline RBF was selected as basis function since the corresponding RBF models presented the lowest CV errors.

The strategy sketched to forecast the honey yield in Andalusia for each of the years in study, 2001–2015, was the following:

– Remove the data concerning the year to forecast for each of the five provinces of Andalusia – Córdoba, Granada, Huelva, Málaga and Seville – guaranteeing that no bias is introduced in the results;

Algorithm 2. Forward-backward variable screening

Input:

- $\mathbf{x}^1, \ldots, \mathbf{x}^N$, N input data points with n components (variables) each $- x_1, \ldots, x_n$
- $y(\mathbf{x}^1), \ldots, y(\mathbf{x}^N)$, response of the N input data points
- $\hat{\mathbf{x}}^1, \ldots, \hat{\mathbf{x}}^N$, N empty input data points with 0 components (variables) each

Forward screening:

$CV_{best} \leftarrow +\infty$
Improve $\leftarrow 1$
While Improve
    For $i = 1$ to n
        If x_i **is not** a variable of input data $\hat{\mathbf{x}}^1, \ldots, \hat{\mathbf{x}}^N$
            $\breve{\mathbf{x}}^1, \ldots, \breve{\mathbf{x}}^N \leftarrow \hat{\mathbf{x}}^1, \ldots, \hat{\mathbf{x}}^N \oplus x_i$, where operation $\oplus$ adds variable x_i to the set
                of input vectors $\hat{\mathbf{x}}^1, \ldots, \hat{\mathbf{x}}^N$
            Construct the RBF model $h_i(\breve{\mathbf{x}})$ of the data points $(\breve{\mathbf{x}}^k, y(\mathbf{x}^k))$ for $1 \le k \le N$
            and compute CV_i using (3) to measure the prediction error
        Else
            $CV_i \leftarrow +\infty$
        End If
    End For
    If $\operatorname{argmin}_{1 \le i \le n} \mathrm{CV_i} < \mathrm{CV_{best}}$
        $CV_{best} \leftarrow \operatorname{argmin}_{1 \le i \le n} \mathrm{CV_i}$
        $\hat{\mathbf{x}}^1, \ldots, \hat{\mathbf{x}}^N \leftarrow \hat{\mathbf{x}}^1, \ldots, \hat{\mathbf{x}}^N \oplus x_i$
    Else
        Improve $\leftarrow 0$
    End If
End While

Backward screening:

Improve $\leftarrow 1$
While Improve
    For $i = 1$ to n
        If x_i **is** a variable of input data $\hat{\mathbf{x}}^1, \ldots, \hat{\mathbf{x}}^N$
            $\breve{\mathbf{x}}^1, \ldots, \breve{\mathbf{x}}^N \leftarrow \hat{\mathbf{x}}^1, \ldots, \hat{\mathbf{x}}^N \ominus x_i$, where operation $\ominus$ removes variable x_i
                from the set of input vectors $\hat{\mathbf{x}}^1, \ldots, \hat{\mathbf{x}}^N$
            Construct the RBF model $h_i(\breve{\mathbf{x}})$ of the data points $(\breve{\mathbf{x}}^k, y(\mathbf{x}^k))$ for $1 \le k \le N$
            and compute CV_i using (3) to measure the prediction error
        Else
            $CV_i \leftarrow +\infty$
        End If
    End For
    If $\operatorname{argmin}_{1 \le i \le n} \mathrm{CV_i} < \mathrm{CV_{best}}$
        $CV_{best} \leftarrow \operatorname{argmin}_{1 \le i \le n} \mathrm{CV_i}$
        $\hat{\mathbf{x}}^1, \ldots, \hat{\mathbf{x}}^N \leftarrow \hat{\mathbf{x}}^1, \ldots, \hat{\mathbf{x}}^N \ominus x_i$
    Else
        Improve $\leftarrow 0$
    End If
End While

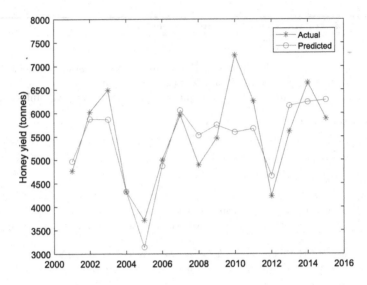

Fig. 2. Actual honey yields in Andalusia compared with RBF forecast.

– Consider the remaining data from the five provinces to:
  • find a subset of variables using Algorithm 2;
  • fit the Thin plate RBF models using the subset of variables found;
  • estimate the honey yield for that year for each province;
– Considering the average contribution of each province to the overall honey yield of Andalusia, calculate five different honey yield estimates for that year for Andalusia;
– Consider the median of the five previous predictions as the final estimate of honey yield for Andalusia in that year.

Forecast results following this strategy are displayed in Fig. 2. The mean prediction error was 7.9% which is quite good for such an irregular series. Apart from one year (2010), forecast for all the remaining years are very close to actual honey yield. Furthermore, honey yield trend is well captured. We have to highlight the importance of variable screening. Selecting a subset of variables with good predicting ability enables a better forecast. To calculate the production forecast for each year, that year is eliminated from the data for all provinces. This means that the variable screening procedure do not consider any data from the year to forecast. This leads to different subsets of variables being considered for the forecast of different years. Thus, it is not useful to enumerate the different subsets of predicting variables as they depend on the year (and the geographical region). Nevertheless, some variables appear more often in the different subsets including the minimum temperature in April, the maximum temperature in June and evapotranspiration in September. It is interesting to report as well that the number of hives was often absent of the subset of best predicting variables. Although more hives could be expected to lead to higher honey productions, figures show otherwise. If we plot the number of hives and corresponding total

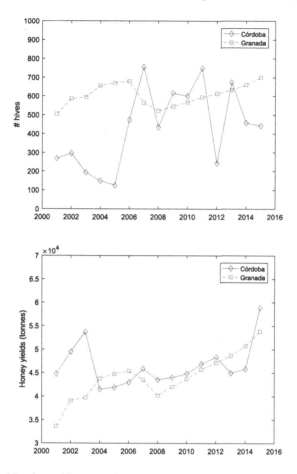

Fig. 3. Number of hives and honey yields for Córdoba and Granada.

honey production for Córdoba and Granada (see Fig. 3), it is straightforward to see that larger number of hives do not correspond to an increased production. Furthermore, for the same province, increase in the number of hives randomly reflects an increased production. Note that, by simple inspection of Table 2 it is possible to verify that weather conditions are quite different for these two provinces which might solely explain the differences in honey yield.

6   Conclusions

Honey yields are difficult to predict and have been usually associated with weather conditions. Although some particular meteorological variables have been associated with honey yields, extrapolating the reported relationships to different regions of the globe or even for different temporal periods is not straightforward.

Thus, the selection of weather variables should be performed using data of the specific regions to be studied and considering adequate time frames.

In this study, we propose an automated forward-backward variable screening procedure that lead to subsets of variables with good predicting ability. RBFs models were used to fit the data and guide the variable screening algorithm. RBF interpolation models can provide excellent interpolants even for poorly distributed data points. Instead of an a priori choice of a RBF basis, the numerical choice of the most adequate RBF is advised. We used the CV error as proxy of the prediction error to decide which RBF basis should be used.

For the subsets of variables obtained using the variable screening procedure, RBF models obtained high quality honey yield predictions. A set of forecasts for Andalusia, obtained from the extrapolation of forecasts for the different provinces considered, allowed a better final annual forecast obtained by excluding extreme values. The variables considered for the RBF models change for different years. Therefore, unlike other studies where specific variables are identified as the most relevant, the only conclusion that can be safely drawn is that meteorological variables are good predictors of honey production but they depend on the geographic region and the time frame considered. The reverse problem of using honey yields to acknowledge climate changes should be as interesting and challenging as the problem addressed here.

Acknowledgements. This work has been supported by the Fundação para a Ciência e a Tecnologia (FCT) under project grant UID/MULTI/00308/2013.

Appendix

Table 2. The weather variables, 2001–2015, means and standard deviations (SD).

	Córdoba		Granada		Huelva		Málaga		Seville	
	Mean	SD	Mean	SD	Mean	SD	Mean	SD	Mean	SD
Rainfall (mm)										
January	53,2	37,8	31,8	23,1	54,4	36,1	39,3	39,7	54,5	35,1
February	74,5	50,4	37	31,8	68,9	67,9	60	61,5	69	59,2
March	80,9	74,1	38,5	23,2	83,8	58,1	69,6	50,5	61,6	59,2
April	57,3	42,9	40,7	21,9	61,9	39,5	35,3	21,7	39,5	26,3
May	36,7	31,1	34	19,1	30,4	30,4	18,3	17,1	31,3	32,4
June	7,7	14,4	17,5	12,4	7,1	13,4	1	1,6	7,6	9
July	0,6	2	3	6,5	1	2,8	1	1,1	2,3	5,3
August	7,2	13,8	9,8	12,5	8,7	17,9	0,7	1,3	3,9	12,2
September	33,9	31,4	23,6	15,9	27,8	30,4	31,4	32,2	28,5	22,2

(Continued)

Table 2. *(Continued)*

	Córdoba		Granada		Huelva		Málaga		Seville	
	Mean	SD	Mean	SD	Mean	SD	Mean	SD	Mean	SD
October	82,4	50,3	30,8	26	105,6	51,4	55	46	73,2	38,4
November	73,5	57,6	38,4	23,3	80,1	67,8	79,6	70,5	72,8	52,7
December	85	99,4	39,1	42,4	78	79,1	67,9	71,6	84,4	99,5
Evapotranspiration (mm)										
January	31,1	5,5	46,1	6,4	35,9	3,5	44	5,1	32,4	2,9
February	44,4	5,7	54,1	6,9	50,4	6	55,9	5,8	45,5	5,6
March	78,2	8,3	86,7	9,9	83,8	10,4	86,7	12,5	80,4	8,5
April	108,1	9,2	107,8	14	114,4	10,9	114,7	11,3	112,4	9,2
May	146,2	13	143,3	18,6	158,6	15,1	151,9	12,5	154,2	14,5
June	180,5	9,7	178,2	12,1	189,7	15,3	180,5	9,4	187,1	10,8
July	203,1	9,4	210,5	8,8	214,5	10,9	196,1	8,5	209,3	7,7
August	181,3	10	184,4	15,3	186,3	8,7	174,5	7,2	187,1	7,3
September	120,1	9	122,4	9,2	124,5	9,9	121,7	10,8	125,6	8,6
October	74,2	6,2	85,3	9,9	78,1	6,5	79,2	8,7	77,8	6,7
November	39,8	5	48,9	8,3	41,6	10,4	49,8	6,5	40,8	4,2
December	28,5	3,7	39,3	5,2	31,5	3,5	39	3,6	29,3	3,3
Minimum temperature (°C)										
January	2,7	2,1	0,6	1	6,2	1,3	6,2	1,3	3,8	1,7
February	3,4	2,4	0,6	1,5	6,3	1,7	6,8	1,7	4,2	2,3
March	6,4	1,6	3	0,9	8,4	0,8	8,7	0,8	7	1,2
April	8,9	1,2	5,1	1,1	10,3	1	10,7	0,9	9,4	1,1
May	11,5	1,1	8,4	1,3	12,9	1,1	13,4	0,8	12,3	1,2
June	15,2	0,9	13,1	1,2	16,4	1	17,2	0,9	16,2	1
July	17	0,8	16,1	1	18	0,7	19,5	0,8	18,1	0,7
August	17,7	0,9	16	0,6	18,5	1	20,3	0,9	18,8	1
September	15,4	1	12,4	0,8	16,6	0,6	17,7	0,4	16,3	0,8
October	12	1,1	9	0,9	14,1	0,9	14,2	0,8	12,8	1,5
November	6,3	2,2	3,7	1,3	9,3	1,2	9,7	1,5	7,2	1,8
December	3,5	1,4	1,2	1,1	6,9	1	7,2	0,9	4,7	1,4
Maximum temperature (°C)										
January	14,7	1	10,5	1,6	15,3	1	17,1	1	14,6	1
February	16,2	1,4	10,8	1,9	16,2	1	17,4	1,1	15,9	1,3
March	19,9	1,4	14,3	1,4	19,1	1,3	19,7	1,2	19,3	1,3
April	23,2	1,4	17	1,9	21,7	1,2	22,1	1	22,4	1,4
May	27,9	2,2	21,2	2,4	25,8	2	25,4	1,2	27	2,2

(Continued)

Table 2. *(Continued)*

	Córdoba		Granada		Huelva		Málaga		Seville	
	Mean	SD	Mean	SD	Mean	SD	Mean	SD	Mean	SD
June	33,7	1,4	27,5	1,6	30,2	1,3	29,8	1,2	32,4	1,5
July	37,1	1,3	31,6	1,1	33,2	1,3	32,5	1,1	35,8	1,2
August	36,8	1,1	30,9	1,1	32,9	1,1	32,6	1	35,6	1
September	31,3	1,2	25	1,2	28,6	1,4	29	0,7	30,4	1,1
October	25,6	1,7	20,3	1,7	24,3	1,6	25	1,2	25,2	1,6
November	18,6	1,4	13,6	1,8	18,8	1,2	20	1,3	18,6	1,3
December	15,2	1,3	11	1,6	15,8	0,9	17,4	1	15,3	1,3
Mean temperature (°C)										
January	8	1,3	5	1,1	10,2	0,9	11,4	0,9	8,7	1,1
February	9,2	1,5	5,3	1,6	10,8	1,1	12	1,2	9,7	1,4
March	12,7	0,7	8,3	1,1	13,4	0,8	14,1	0,8	12,9	0,8
April	15,8	1,1	10,8	1,4	15,7	1	16,4	0,8	15,7	1,1
May	19,7	1,5	14,7	1,8	19,2	1,5	19,5	0,9	19,7	1,6
June	24,8	1,1	20,2	1,4	23,2	1,1	23,7	1	24,4	1,2
July	27,7	1	23,7	1	25,5	0,9	26,3	0,9	27,3	1
August	27,5	0,8	23,1	0,9	25,4	1	26,5	0,7	27,3	1
September	23,1	0,9	18,2	0,9	22,1	0,8	23,2	0,4	23,2	0,8
October	18,1	0,9	14,2	1,2	18,7	1,1	19,3	0,7	18,7	1,1
November	11,7	1,4	8,2	1,5	13,6	1	14,6	1,1	12,5	1,2
December	8,5	1	5,5	1,3	10,9	0,9	12,1	0,7	9,5	1,1
Relative humidity (%)										
January	81,9	5,8	62,4	8,6	78,8	5,1	70,7	4,4	81	6,1
February	77,4	7,5	62,5	7,9	74	7,9	68,2	4,9	76	9,2
March	71,8	6,9	59,9	7,3	71,4	6,8	67,1	7	70,8	7,6
April	67,1	5,9	60,2	6,9	68,4	6,3	63	7,1	65,4	6
May	57,7	7,2	54,3	8,4	59,2	5,2	56,7	5,6	55,4	7,3
June	47,1	4,4	44,3	5,5	52,3	4,8	51,5	3,8	47,6	4,9
July	38	4,5	34,5	4,4	47,1	4,5	51,2	4	40,3	5
August	39,8	4,2	38,5	4,6	51,1	3	55,3	3,6	42,3	4,4
September	53,8	6,4	51,9	6,1	63	7,1	61,9	4,3	55,7	7,3
October	69,2	6,4	57,6	6,7	71,7	4,9	69,6	3,7	67,4	5,3
November	78	7,5	63,2	10,3	73,6	7,7	70,5	6	74,4	8,3
December	81,6	5,3	64,3	5,4	78,7	5,4	73,2	3,7	80,2	4,9

References

1. Anuário de Estadistica del Ministerio de Agricultura, Alimentación y Medio Ambiente. http://www.mapama.gob.es/
2. Crane, E.: The Archaeology of Beekeeping. Cornell University Press, Ithaca (1983)
3. FAOSTAT. http://www.fao.org/faostat/
4. Holmes, W.: Weather and honey yields. Scott. Beekeep. **75**, 190–192 (1988)
5. Holmes, W.: The influence of weather on annual yields of honey. J. Agric. Sci. **139**, 95–102 (2002)
6. Hurst, G.W.: Honey production and summer temperatures. Meteorol. Mag. **96**, 116–120 (1967)
7. Hurst, G.W.: Temperatures inhigh summer, and honey production. Meteorol. Mag. **99**, 75–82 (1970)
8. Instituto de investigación y formación agraria y pesquera. www.juntadeandalucia.es/agriculturaypesca/ifapa/ria/
9. Krishnamurti, B.: A brief analysis of eleven years (1928–1938) records of scale hives at the Rothamsted Bee Laboratory. Bee World **20**, 121–123 (1939)
10. MATLAB 2016a: Natick. The MathWorks Inc., Massachusetts (2016)
11. Nelder, J., Mead, R.: A simplex method for function minimization. Comput. J. **7**, 308–313 (1965)
12. Powell, M.: Radial basis function methods for interpolation to functions of many variables. HERMIS Int. J. Comput. Math. Appl. **3**, 1–23 (2002)
13. Rocha, H., Li, W., Hahn, A.: Principal component regression for fitting wing weight data of subsonic transports. J. Aircr. **43**, 1925–1936 (2006)
14. Rocha, H.: Model parameter tuning by cross validation and global optimization: application to the wing weight fitting problem. Struct. Multi. Optim. **37**, 197–202 (2008)
15. Rocha, H.: On the selection of the most adequate radial basis function. Appl. Math. Model. **33**, 1573–1583 (2009)
16. Rocha, H., Dias, J.M., Ferreira, B.C., Lopes, M.C.: Selection of intensity modulated radiation therapy treatment beam directions using radial basis functions within a pattern search methods framework. J. Global Optim. **57**, 1065–1089 (2013)
17. Rocha, H., Dias, J.M., Ferreira, B.C., Lopes, M.C.: Beam angle optimization for intensity-modulated radiation therapy using a guided pattern search method. Phys. Med. Biol. **58**, 2939 (2013)
18. Switanek, M., Crailsheim, K., Truhetz, H., Brodschneider, R.: Modelling seasonal effects of temperature and precipitation on honey bee winter mortality in a temperate climate. Sci. Total Environ. **579**, 1581–1587 (2017)
19. Tu, J.: Cross-validated multivariate metamodeling methods for physics-based computer simulations. In: Proceedings of the IMAC-XXI (2003)
20. Tu, J., Jones, D.R.: Variable screening in metamodel design by cross-validated moving least squares method. In: Proceedings of the 44th AIAA (2003)
21. Zilinskas, A.: On similarities between two models of global optimization: statistical models and radial basis functions. J. Glob. Optim. **48**, 173–182 (2010)

Graph Fragmentation Problem for Natural Disaster Management

Natalia Castro$^{(\boxtimes)}$, Graciela Ferreira, Franco Robledo, and Pablo Romero

Facultad de Ingeniería, Instituto de Matemática y Estadística IMERL,
Universidad de la República, Montevideo, Uruguay
{ncastro,gferreira,frobledo,promero}@fing.edu.uy

Abstract. Natural disasters represent a threaten for the existence of human beings. Given its remarkable importance, operational researchers should contribute to provide rationale decisions.

In this paper we study a purely combinatorial problem that models management disasters, called Graph Fragmentation Problem, or GFP for short. The problem belongs to the $\mathcal{NP}$-Hard class. As corollary, finding the optimal protection scheme is prohibitive for large populations. First, we review the problem and its properties. Then, we introduce a mathematical programming formulation and exact resolution for small instances. Finally, we discuss feasible model extensions and trends for future work.

1 Motivation

History reveals painful memories full of pandemics, lighting shocks and fires. The Spanish flu from 1918 was deadlier than any war in history, and half the population of the world has been exposed to the virus [14]. An infernal fire in October 1871 ravaged part of Chicago, leaving more than 90.000 homeless and 300 deaths [10].

We encourage operational researchers to be engaged with society, and provide means to cope with natural disasters. In this paper, we follow the research line introduced in [12]. There, a single individual of a population is exposed to a natural disaster, and the disaster is immediately propagated through neighbors. Our task is to determine a sub-population that is protected beforehand, subject to a budget constraint. Clearly, the notion of protection depends on the specific application (location of fire-stations, isolation in electric systems, vaccination against a pandemics).

This paper is organized as follows. Section 2 presents the background of the problem under study, and its origin from epidemic modelling. Section 3 presents a formal definition of the GFP. Theoretical results for the GFP are presented in Sect. 4, together with the main approaches to address the problem. The main contributions are offered in Sects. 5, 6 and 7. Specifically, a mathematical programming formulation for the GFP is introduced in Sect. 5. Lower and upper bounds are obtained in Sect. 6, inspired by relaxations. An experimental analysis is carried out in Sect. 7, where we test the performance of our exact solution

© Springer International Publishing AG 2018
G. Nicosia et al. (Eds.): MOD 2017, LNCS 10710, pp. 496–505, 2018.
https://doi.org/10.1007/978-3-319-72926-8_41

for the GFP under different test cases coming from real-life applications. It is worth to notice that the literature in the exact analysis of the GFP is scarce, and here we provide the first steps towards the development of optimal protection schemes under this fundamental model. Section 8 presents feasible model extensions, concluding remarks and trends for future work.

2   Background

A cornerstone in epidemic model is classical SIR (Susceptible - Infected - Removed). In SIR it is assumed a fully-mixed infinite population with random contacts. More realistic models are available from authoritative literature in the field [2,9]. They consider a graph and epidemic spread governed by probabilistic rules. The authors claim that node-protection (choosing which nodes to remove, so that the epidemic cannot propagate through them) is a presumably hard task, but they do not provide hints nor mathematical proofs.

The Graph Fragmentation Problem, or GFP, represents a worst case analysis of an abstract epidemic modelling. In [12], a realistic SIR-based model is provided, and the Graph Fragmentation Problem (GFP) is introduced as an extremal analysis of highly virulent scenarios. Incidentally, it models other catastrophic events, such as fire-fighting and electric shocks (the formal model is presented in Sect. 3). There, only Greedy-based heuristics are presented, and there is no complexity analysis. A GRASP heuristic enriched with a path-relinking post-optimization stage is developed in [13].

The first result on computational complexity is offered for the GFP in [11]. The authors prove that the GFP belongs to the class of $\mathcal{NP}$-Hard problems. This theoretical result confirms the intuition from epidemiologists that finding an optimal node-protection mechanism is a hard task.

Curiously enough, in a more recent paper, the optimal protection scheme is found in all acyclic graphs, elementary cycles and some bipartite graphs [1]. In contrast, GFP presents a strong inapproximability result for general graphs. More specifically, there is no approximation algorithm with factor lower than $3/2$, unless $\mathcal{P} = \mathcal{NP}$.

3   Graph Fragmentation Problem

We are given a population represented by a graph $G = (V, E)$, and a budget constraint B, which is a natural number B such that $0 \leq B \leq |V|$. We can choose B nodes and protect them: we delete the nodes from G obtaining a subgraph G', so that the chosen nodes cannot be affected by the disaster. The nature picks a node v uniformly at random from G'. The disaster kills all the members of the same connected component as v.

The goal is to minimize the expected number of deaths. Mathematically, if the subgraph G' has $V' = n$ nodes and k connected components with orders $n_1, \ldots, n_k$, the probability to choose component i is n_i/n. Therefore, the expected number of deaths is $E(G') = \sum_{i=1}^{k} n_i p_i$, with $p_i = n_i/n$. The goal of the Graph Fragmentation Problem (GFP) is to choose the protected set in order to minimize the expected number of deaths:

$$\min_{U \subseteq V} \sum_{i=1}^{k} \frac{n_i^2}{n}$$

$$s.t. |U| \leq B.$$

Observe that the denominator n is constant for a fixed instance (G, B) in the GFP. Therefore, our problem is to minimize the Euclidean norm of the vector $n = (n_1, \ldots, n_k)$, or Constrained Euclidean Norm Minimization (CENM):

$$\min_{U \subseteq V} \|n_{G-U}\|^2$$

$$s.t. |U| \leq B,$$

where $n_{G-U} = (n_1, \ldots, n_k)$ is the vector with the orders of the connected components from $G' = G - U$. Observe that the objective function $\|n_{G-U}\|^2$ is minimized when the resulting graph $G' = G - U$ has isolated nodes. The reader is invited to consult [1,4,6] for a discussion of related vulnerability metrics.

4   Analysis

In this section we highlight the main ideas on the analysis of the GFP for a better understanding of the problem. The following problem will be used to characterize the computational complexity of the GFP.

Definition 1 (Minimum Cardinality Vertex Cover)
Instance: simple graph $G = (V, E)$ and positive integer k.
Does there exist a node-set U such that $|U| \leq k$ and every link is incident to some node from U?

Recall that Minimum Cardinality Vertex Cover belongs to Karp list of 21 $\mathcal{NP}$-Complete decision problems [8].

Theorem 1. *The GFP belongs to the class of $\mathcal{NP}$-Hard problems.*

Proof. The graph $G' = G - U$ has isolated nodes if and only if U is a vertex cover, where $|U| \leq B$. Thus, the GFP is at least as hard as Minimum Cardinality Vertex Cover.

The following problem will be considered in order to prove a stronger inapproximability result for the GFP:

Definition 2 (Multiway k-cut)
Instance: simple graph $G = (V, E)$, terminal set $K \subseteq V$ with $|K| = k$, positive integer B.
Does there exist a separator set $U \subseteq V - K$ with $|U| \leq B$ such that each terminal node belongs to different components in $G - U$?

We know that Multiway 2-cut is in $\mathscr{P}$. A polynomial time algorithm is provided by Ford and Fulkerson [5]. However, Multiway k-cut is $\mathscr{NP}$-Complete for every fixed $k \geq 3$ [3].

Theorem 2. *It is $\mathscr{NP}$-hard to approximate GFP within $\frac{5}{3} - \varepsilon$, for any $\varepsilon > 0$.*

Proof. Consider an instance of Multiway 3-cut with ground graph $G = (V, E)$, distinguished nodes $\{v_1, v_2, v_3\}$ and positive integer B. Replace those nodes by large cliques $\{K_N, K_N, K_N\}$, where $N >> |V|$. The order of the new graph G^* is roughly $3N$. If the instance accepts a 3-cut, the cost in the GFP with instance (G^*, B) is roughly N. Otherwise, the expected number of dead nodes is never lower than $\frac{(2N)^2 + N^2}{3N} = \frac{5N}{3}$. Therefore, an approximation algorithm with factor $5/3$ would decide if G with distinguished nodes $\{v_1, v_2, v_3\}$ accepts a 3-cut using B nodes. The existence of such algorithm implies the solution of 3-cut. □

Even though the GFP does not accept an optimal solution in polynomial time (unless $\mathscr{P} = \mathscr{NP}$), there exists a dynamic programming-based polynomial time method to find the optimal solution in acyclic graphs:

Theorem 3. *If G is acyclic, there exists a polynomial time algorithm to find the best protection scheme with B nodes.*

Proof. First, consider arbitrary graphs G_1 and G_2 that accept a polynomial time algorithm for any B, then we can solve the problem for $G = G_1 \cup G_2$, using all partitions $B = B_1 + B_2$. This reasoning holds for disjoint branches of a rooted tree (the root is arbitrary in this context). We can consider leaf nodes and their parents, and proceed with disjoint branches as before. The number of stages in a dynamic programming algorithm is not more than the height of the tree (which is not greater than the order of the graph). Finally, the result hold for acyclic graphs. Just connect all the trees by a fixed auxiliary node and consider the previous algorithm for the resulting tree. The reader is invited to consult [1] for technical details. □

Theorem 4. *The size of the connected components in $G' = G - U$ must be as even as possible.*

Proof. Let $n_1 \geq n_2 \geq \ldots \geq n_k$ be the orders of the connected components in G'. If $|n_1 - n_k| \geq 2$, a straight calculation shows that $\|(n_1, \ldots, n_k)\|^2 \geq \|(n_1 - 1, \ldots, n_k + 1)\|^2$. This means that the cost in the GFP is reduced whenever the size of the components in G' are as even as possible. □

There is no general result for cyclic graphs in general. However, the following result holds for the elementary cycle:

Theorem 5. *The best protection scheme is known for the cycle C_n.*

Proof. Delete an arbitrary node, and obtain an elementary path. Then, protect $B - 1$ nodes in such a way that the resulting sub-paths are as even as possible. By Theorem 4, the resulting graph provides the minimum-cost protection scheme.

Let us further analyze the GFP for bipartite graphs. Consider $G = (V_1 \cup V_2, E)$ where $E \subseteq V_1 \times V_2$. Recall that König theorem asserts that the minimum cardinality of a vertex cover in bipartite graphs is precisely the size of the maximum matching L. This number can be found by Ford and Fulkerson algorithm: connect all nodes from V_1 to a source s, all the nodes from V_2 to a sink t, and find the max-flow with unit capacities in the links. If $B \geq L$, all nodes from a vertex cover can be protected, and they can be found in polynomial time. We obtain the following:

Theorem 6. *The optimality for the GFP can be found in polynomial time for all bipartite graphs whenever B is not lower than the maximum matching.*

The computational complexity for the GFP remains open for bipartite graphs in general.

5   Mathematical Programming Formulation

An integer quadratic programming model (IQP) for the GFP is developed. In the model we consider a directed graph $G^d = (V, E')$, where every link from G is replaced by two one-way links. Consider the following model variables:

- n_k: size of connected component k;
- $U_i \in \{0,1\}$, $i \in V$: node $i \in U$ (or not);
- $x_{ij}^k \in \{0,1\}$, $(i,j) \in E$: link (i,j) belongs to component k in G;
- $N_i^k \in \{0,1\}$, $i \in V$: node i belongs to the component k;
- $y_{ij}^{u,v} \in \{0,1\}$, $(i,j) \in E$, $u,v \in V$: there is some u-v-path that includes (i,j) in the way $i \rightarrow j$.

The mathematical programming model is the following:

$$\text{min.} \sum_{i=1...K} n_i^2/(n-B) \tag{1}$$

$$\text{s.t.} \sum_{j\in V} U_j \leq B, \tag{2}$$

$$\sum_{j\in V} N_j^k = n_k, \forall k = 1 \dots K \tag{3}$$

$$\sum_{k=1...K} N_j^k = 1 - U_j, \forall j \in V \tag{4}$$

$$N_i^k + N_j^s \leq 1, \forall i,j \in V, i \neq j, (i,j) \in E, \forall k, s \in K, s \neq k, \tag{5}$$

$$\sum_{k=1...K} x_{ij}^k \leq (1 - U_i), \forall (i,j) \in E, i,j \in V \tag{6}$$

$$\sum_{k=1...K} x_{ij}^k \leq (1 - U_j), \forall (i,j) \in E, i,j \in V \tag{7}$$

$$N_i^k + N_j^k \leq 1 + x_{ij}^k, \forall k \in 1 \dots K, \forall (i,j) \in E, i,j \in V \tag{8}$$

$$y_{i,j}^{u,v} + y_{j,i}^{u,v} \leq \sum_{k=1...K} x_{ij}^k, \forall u,v \in V, \forall (i,j) \in E, i,j \in V \qquad (9)$$

$$\sum_{(u,j)\in E'} y_{u,j}^{u,v} \geq N_u^k + N_v^k - 1, \forall u,v \in V, \forall k \in 1...K \qquad (10)$$

$$\sum_{(u,j)\in E'} y_{u,j}^{u,v} = \sum_{(i,v)\in E'} y_{i,v}^{u,v}, \forall u,v \in V \qquad (11)$$

$$\sum_{(r,j)\in E'} y_{r,j}^{u,v} = \sum_{(i,r)\in E'} y_{i,r}^{u,v}, \forall r,u,v \in V, r \neq u, r \neq v \qquad (12)$$

The objective function captures de cost of the GFP (1). Inequality (2) represents the budget constraint. The size of each connected component is found using Constraint (3). Constraint (4) set $N_j^k = 0$ for every k whenever j is picked for protection. Furthermore, if j is not picked for protection, exactly one member of the variable-set $\{N_j^k\}_{k=1...n}$ must be set to 1. Constraint (5) avoid the existence of a path between different connected components. In Constraints (6) and (7), the variable $x_{i,j}^k$ is set to 0 when at least one of i or j are protected. Constraint (8) respects the definition of the binary variable $x_{i,j}^k$. Constraints (9)–(12) represent Kirchhoff equations, that ensure connectivity in each component. The binary variables $y_{i,j}^{u,v}$ represent the u-v flow that is carried in the link (i,j). Constraint (9) avoids two-way flows. Constraints (10)–(12) model this flow.

This is an IQP formulation or more general, a mixed integer quadratic problem (MIQP). It is well known that it is NP-hard. However, it is important to remark that, differently from MILP or ILP, the source of complexity of IQP is not restricted to the integrality requirement on its variables.

6   Bounds for the GFP

A lower bound is found by a natural relaxation of the problem, where the variables n_k, x_{ij}^k, N_i^k and y_{ij}^{uv} assume real values. Although this problem is also MIQP, only U_i variables remain binary. In order to find an upper-bound, the objective function is modified, and as a result we obtain an integer linear program. Observe that all the constraints are linear. Since we preserve all constraints, a feasible solution for the GFP is produced. The new objective function is to minimize the size of the largest component.

The upper bound of GFP is an ILP, also NP-hard. It is modeled as follows:

$$\min Z \text{ with:}$$
$$s.t. \, n_k \leq Z, \forall k$$
$$Eqs. \, (2) - (12)$$

7   Proof of Concept

This section presents the exact analysis that is product of our mathematical programming model under selected real-life networks. The model was implemented in CPLEX 12.6.3.0, MIP solver, and the executions were performed on an eight-core Intel i7 processor at 3.07 GHz, 16 GB RAM. As a proof-of-concept, four graphs coming from real-life applications were considered:

- The electrical optical network EON considered by Gouveia et al. [7]. See Fig. 1(b).
- The National Science Foundation Network form the USA, also considered in the previous study [7]. See Fig. 1(c).
- The Uruguayan Academic Network, RAU2, depicted in Fig. 1(d).
- ARPANET (Advanced Research Projects Agency Network), depicted in Fig. 1(e).

We also considered a toy example in which the analysis is straight (see graph N_1 from Fig. 1(a)). Table 1 summarizes the main characteristics of the graphs considered in the experimental analysis. Columns LB, UB and Opt stand for lower-bound, upper-bound and optimal value, respectively. The optimal value was calculated solving the exact model developed in Sect. 5.

Table 1. Results

Graph	$|V|$	$|E|$	B	$|V| - B$	LB	UB	Opt
N1	9	8	1	8	1.03	4.00	3.50
RAU2	10	17	2	8	1.06	2.75	2.75
NFSNET	14	52	5	9	1.13	4.56	4.56
EON	19	36	6	13	1.06	3.46	3.00
ARPANET	20	25	5	15	1.04	2.87	2.60

The gap between the upper bound (UB) and the optimal value in the GFP (opt) is small under all instances. This highlights the fact that the size of the connected components should be as even as possible, in a strict correspondence with Theorem 4.

Curiously enough, if we consider ARPANET with budget 6 instead, the optimal solution could not be found in a reasonable time (less than 48 h). However, the bounds are efficiently found in that case, where either some variables assume real values or a the objective is replaced by a linear one.

Note that a trivial lower bound for GFP is 1. In effect, when we protect and remove B nodes in any graph, there are $|V| - B$ remaining nodes and in the best case, these result all disconnected. Then, there would be $|V| - B$ connected components with size 1 and the value of objective function for lower bound is 1. As shown in the Table 1, all values found for proposed lower bound, are very near to the trivial lower bound. It would be desirable to improve these values in future work or to research if this gap has some theoretical basis.

The complexity of the quadratic objective function promotes further research in the analysis of exact and approach algorithms for the GFP.

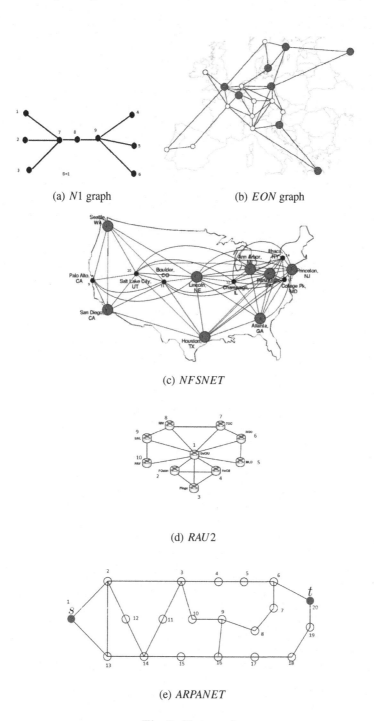

(a) *N*1 graph

(b) *EON* graph

(c) *NFSNET*

(d) *RAU*2

(e) *ARPANET*

Fig. 1. Test graphs

8   Conclusions and Trends for Future Work

We strongly believe that operational researchers should be engaged with the society in providing means to cope with risk analysis and natural disasters. A purely combinatorial problem is studied in this paper, called Graph Fragmentation Problem or GFP. The GFP belongs to the class of $\mathcal{NP}$-Hard problems, and there is no hope to find efficient algorithms to solve it optimally, unless $\mathcal{P} = \mathcal{NP}$. However, it is possible to solve cases where the population is configured with no cycles or elementary cycles.

A new mathematical programming formulation for the GFP is introduced in this paper, together with bounds. Exact resolutions in CPLEX confirm the fact that optimal solutions can be obtained for the GFP under small populations. Furthermore, there computational efficiency of an integer linear programming relaxation is notorious, and provides feasible solutions with small gaps for the optimal GFP.

Further research includes the development of heuristic methods, extended models, and the interplay between related relaxations. Observe that in the GFP it is assumed that a singleton is picked uniformly at random. This selection law could be modified, and the attacker could select relevant individuals from the system first. As future work, we would like to understand this generalization of the GFP with weighted nodes and adaptive protection schemes. Game theory provides a means to find optimal answers to different attacking systems.

Acknowledgements. This work is partially supported by Project 395 CSIC I+D *Sistemas Binarios Estocásticos Dinámicos.*

References

1. Aprile, M., Castro, N., Robledo, F., Romero, P.G.: Analysis of node-resilience strategies under natural disasters. In: International Conference on Design of Reliable Communication Networks 2017 (DRCN 2017), Munich, Germany, pp. 93–100, March 2017
2. Ball, F., Sirl, D.: Acquaintance vaccination in an epidemic on a random graph with specified degree distribution. J. Appl. Probab. **50**(4), 1147–1168 (2013)
3. Dahlhaus, E., Johnson, D.S., Papadimitriou, C.H., Seymour, P.D., Yannakakis, M.: The complexity of multiway cuts (extended abstract). In: Proceedings of the Twenty-Fourth Annual ACM Symposium on Theory of Computing, STOC 1992, New York, NY, USA, pp. 241–251. ACM (1992)
4. Dinh, T.N., Xuan, Y., Thai, M.T., Pardalos, P.M., Znati, T.: On new approaches of assessing network vulnerability: hardness and approximation. IEEE/ACM Trans. Netw. **20**(2), 609–619 (2012)
5. Ford, L.R., Fulkerson, D.R.: Maximal flow through a network. Can. J. Mathe. **8**, 399–404

6. Gomes, T., Tapolcai, J., Esposito, C., Hutchison, D., Kuipers, F., Rak, J., de Sousa, A., Iossifides, A., Travanca, R., Andr, J., Jorge, L., Martins, L., Ugalde, P.O., Pai, A., Pezaros, D., Jouet, S., Secci, S., Tornatore, M.: A survey of strategies for communication networks to protect against large-scale natural disasters. In: 2016 8th International Workshop on Resilient Networks Design and Modeling (RNDM), pp. 11–22, September 2016

7. Gouveia, L., Simonetti, L., Uchoa, E.: Modeling hop-constrained and diameter-constrained minimum spanning tree problems as steiner tree problems over layered graphs. Math. Program. **128**(1), 123–148 (2011)

8. Karp, R.M.: Reducibility among combinatorial problems. In: Miller, R.E., Thatcher, J.W. (eds.) Complexity of Computer Computations, pp. 85–103. Plenum Press (1972)

9. Newman, M.E.J.: The structure and function of complex networks. SIAM Rev. **45**(2), 167–256 (2003)

10. Pauly, J.J.: The great Chicago fire as a national event. Am. Q. **36**(5), 668–683 (1984)

11. Piccini, J., Robledo, F., Romero, P.: Analysis and complexity of pandemics. In: 2016 8th International Workshop on Resilient Networks Design and Modeling (RNDM), pp. 224–230, September 2016

12. Piccini, J., Robledo, F., Romero, P.: Node-immunization strategies in a stochastic epidemic model. In: Pardalos, P., Pavone, M., Farinella, G.M., Cutello, V. (eds.) MOD 2015. LNCS, vol. 9432, pp. 222–232. Springer, Cham (2015). https://doi.org/10.1007/978-3-319-27926-8_19

13. Piccini, J., Robledo, F., Romero, P.: Graph fragmentation problem. In: Proceedings of 5th the International Conference on Operations Research and Enterprise Systems, pp. 137–144 (2016)

14. Taubenberger, J.K., Reid, A.H., Lourens, R.M., Wang, R., Jin, G., Fanning, T.G.: Molecular virology: was the 1918 pandemic caused by a bird flu? Was the 1918 flu avian in origin? (Reply). Nature **440**, 9–10 (2006)

Job Sequencing with One Common and Multiple Secondary Resources: A Problem Motivated from Particle Therapy for Cancer Treatment

Matthias Horn[1]([✉]), Günther Raidl[1], and Christian Blum[2]

[1] Institute of Computer Graphics and Algorithms, TU Wien, Vienna, Austria
{horn,raidl}@ac.tuwien.ac.at
[2] Artificial Intelligence Research Institute (IIIA-CSIC), Campus of the UAB,
Bellaterra, Spain
christian.blum@iiia.csic.es

Abstract. We consider in this work the problem of scheduling a set of jobs without preemption, where each job requires two resources: (1) a common resource, shared by all jobs, is required during a part of the job's processing period, while (2) a secondary resource, which is shared with only a subset of the other jobs, is required during the job's whole processing period. This problem models, for example, the scheduling of patients during one day in a particle therapy facility for cancer treatment. First, we show that the tackled problem is NP-hard. We then present a construction heuristic and a novel A* algorithm, both on the basis of an effective lower bound calculation. For comparison, we also model the problem as a mixed-integer linear program (MILP). An extensive experimental evaluation on three types of problem instances shows that A* typically works extremely well, even in the context of large instances with up to 1000 jobs. When our A* does not terminate with proven optimality, which might happen due to excessive memory requirements, it still returns an approximate solution with a usually small optimality gap. In contrast, solving the MILP model with the MILP solver CPLEX is not competitive except for very small problem instances.

1 Introduction

This work considers the following combinatorial optimization problem. A finite set of jobs must be processed without preemption. Each job requires two resources: (1) a common resource, shared by all jobs, is required during a certain part of the job's processing period, while (2) a secondary resource, which is shared with only a subset of the other jobs, is required during the job's whole processing period. This is the case, for example, in the context of the production of certain products where some raw material is put into specific fixtures or

We gratefully acknowledge the financial support of the Doctoral Program "Vienna Graduate School on Computational Optimization" funded by Austrian Science Foundation under Project No W1260-N35.

G. Nicosia et al. (Eds.): MOD 2017, LNCS 10710, pp. 506–518, 2018.
https://doi.org/10.1007/978-3-319-72926-8_42

molds (the secondary resources), which are then sequentially processed on a single machine (the common resource). Finally, some further postprocessing (e.g., cooling) might be required before the fixtures/molds are available for further usage again. In order to perform this process as efficiently as possible, the aim is to minimizing the makespan, i.e., the total time required to finish the processing of all jobs. In the following we refer to this problem as *Job Sequencing with One Common and Multiple Secondary Resources* (JSOCMSR).

The technical definition of the problem, which is provided later on, was inspired by a more specific application scenario: the scheduling of patients in radiotherapy for cancer treatment [2,7] and particle therapy for cancer treatment [8]. In modern particle therapy, carbon or proton particles are accelerated in cyclotrons or synchrotrons to almost the speed of light and from there directed into a treatment room where a patient is radiated. A number of differently equipped treatment rooms is available (typically two to four) and the particle beam can only be directed into one of these rooms at a time. For each patient it is known in advance in which room she or he has to be treated in dependence on her/his specific needs. Moreover, each patient requires a certain preparation (such as positioning, fixation, possibly sedation) in the room before the actual irradiation can start. Upon finishing the irradiation of a patient, some further time is usually needed for medical inspections before the patient can actually leave the room and the treatment of a next patient can start. Note that the available rooms correspond to the secondary resources mentioned above, while the particle beam is the common resource. The scheduling of a set of patients at, e.g., one day in such a facility is considered.

For further information on particle therapy patient scheduling, in which JSOCMSR appears as sub-problem, the interested reader is referred to [8]. The whole practical scenario has to consider a time horizon of several weeks, additional resources, their availability time windows, and a combination of more advanced objectives.

The JSOCMSR is rather easy to solve when (1) only the common resource usage is the bottleneck and enough secondary resources are available or (2) the pre- and postprocessing times in which only the secondary resources are required are negligible in comparison to the jobs' total processing times. In such cases the jobs can, essentially, be performed in almost an arbitrary ordering. The problem, however, becomes challenging when pre- and postprocessing times are substantial and many jobs require the same secondary resources. In this work we consider such difficult scenarios.

1.1   Contribution of This Work

In addition to formally proving that the JSOCMSR is NP-hard, we provide a lower bound on the makespan objective, which is then exploited both in the context of a constructive heuristic and a novel A* algorithm. The latter works on a special graph structure that allows to efficiently exploit symmetries and features a diving mechanism in order to obtain also heuristic solutions in regular intervals. In addition, we present a mixed-integer linear programming (MILP) model for the JSOCMSR. Our experiments show that the A* algorithm performs excellently. Even many large problem instances with up to 1000 jobs can be solved to proven optimality. There are, however, also difficult problem instances for which A* terminates early due to excessive memory requirements. In these cases, heuristic solutions together with lower bounds and typically small optimality gaps are returned. In comparison, solving the MILP model by the general purpose MILP solver CPLEX[1] cannot compete with A*, as only solutions to rather small problem instances can be obtained in reasonable time.

2   Related Work

In the literature there are only few publications dealing with scenarios similar to JSOCMSR. Veen et al. [10] studied a related problem in which the common resource corresponds to a machine on which the jobs are processed and secondary resources needed in a pre- and postprocessing are called templates. An important restriction in their problem is that the postprocessing times are assumed to be negligible compared to the total processing times of the jobs. This implies that the starting time of each job only depends on its immediate predecessor. More specifically, a job j requiring a different resource than its predecessor j' can always be started after a setup time only depending on job j, while a job requiring the same resource can always be started after a postprocessing time only depending on job j'. Due to these characteristics, this problem can be interpreted as a traveling salesman problem (TSP) with a special cost structure. It is shown that this problem can be solved efficiently in time $O(n \log n)$.

Somewhat related is the no-wait flowshop problem; see [1] for a survey on this problem and related ones. Here, each job needs to be processed on each of m machines in the same order and the processing of the job on a successive machine always has to take place immediately after its processing has finished on the preceding machine. This problem can be solved in time $O(n \log n)$ for two machines via a transformation to a specially structured TSP [4]. In contrast, for three and more machines the problem is NP-hard, although it can still be transformed into a specially structured TSP. Röck [9] proved that the problem is strongly NP-hard for three machines by a reduction from the 3D-matching problem.

A more general problem as which our JSOCMSR can be modeled is the Resource-Constrained Project Scheduling Problem (RCPSP) with maximal time lags. We obtain a corresponding RCPSP instance from a JSOCMSR instance by splitting each job into three activities which are the preprocessing, the main part also requiring the common resource, and the postprocessing. These activities must be performed for each job in this order with maximal time lags of zero, and all resource requirements must be respected. For a survey on RCP-SPs with various extensions and respective solution methods see Hartmann and Briskorn [6]. For practically solving the JSOCMSR, however, such a mapping does not seem to be effective due to the increased number of required activities and since specificities of the problem are not exploited.

---

[1] https://www-01.ibm.com/software/commerce/optimization/cplex-optimizer.

3   Problem Definition and Complexity

An instance of JSOCMSR consists of a set of n jobs $J = \{1, \ldots, n\}$, the common resource 0, and a set of m secondary resources $R = \{1, \ldots, m\}$. By $R_0 = \{0\} \cup R$ we denote the set of all resources. Each job $j \in J$ has a total processing time $p_j > 0$ during which it fully requires a secondary resource $q_j \in R$. Furthermore, each job j requires the common resource 0 from a time $p_j^{\text{pre}} \geq 0$ on, counted from the job's start, for a duration p_j^0 with $0 < p_j^0 \leq p_j - p_j^{\text{pre}}$. A solution to the problem is described by the jobs' starting times $s = (s_j)_{j \in J}$ with $s_j \geq 0$. Such a solution s is feasible if no two jobs require a resource at the same time.

The objective is to find a feasible schedule that minimizes the finishing time of the job that finishes latest. This optimization criterion is known as the *makespan*, and it can be calculated for a solution s by

$$\text{MS}(s) = \max\{s_j + p_j \mid j \in J\}. \tag{1}$$

As each job requires the common resource 0, and only one job can use this resource at a time, a solution implies a total ordering of the jobs. Vice versa, any ordering—i.e., permutation—$\pi = (\pi_i)_{i=1,\ldots,n}$, of the jobs in J can be decoded into a feasible solution in the straight-forward greedy way by scheduling each job in the given order at the earliest feasible time. We call a schedule in which, for a certain job permutation π, each job is scheduled at its earliest time, a *normalized schedule*. Obviously, any optimal solution is either a normalized schedule or there exists a corresponding normalized schedule with the same objective value. We therefore also use the notation $\text{MS}(\pi)$ for the makespan of the normalized solution induced by the job permutation π.

For convenience we further define the duration of the postprocessing time by $p_j^{\text{post}} = p_j - p_j^{\text{pre}} - p_j^0$, $\forall j \in J$ and denote by $J_r = \{j \in J \mid q_r = r\}$ the subset of jobs requiring resource $r \in R$ as secondary resource. Note that $J = \bigcup_{r \in R} J_r$. The minimal makespan over all feasible solutions, i.e., the optimal solution value, is denoted by MS^*.

3.1   Computational Complexity

Let the decision variant of JSOCMSR be the problem in which it has to be determined if there exists a feasible solution with a makespan corresponding to a given constant MS^*.

Theorem 1. *The decision variant of JSOCMSR is NP-complete for $m \geq 2$.*

Proof. Our problem is in class NP since a solution can be checked in polynomial time. We show that JSOCMSR is NP-complete by a polynomial reduction from the well-known NP-complete *Partition Problem* (PP) [3], which is stated as follows: Given a finite set of positive integers $A \subset \mathbb{N}$, partition it into two disjoint subsets A_1 and A_2 such that $\sum_{a \in A_1} a = \sum_{a \in A_2} a$.

We transform an instance of the PP into an instance of the JSOCMSRC as follows. Let $m = 2$ and J consist of the following jobs:

– For each $a \in A$ there is a corresponding job $j \in \{1,\ldots,|A|\} \subset J$ with processing time $p_j = a$ requiring resource $q_j = 1$ and the common resource 0 the whole time, i.e., $p_j^0 = p_j$ and $p_j^{\mathrm{pre}} = 0$.
– Furthermore, there are two jobs $j \in \{|A|+1, |A|+2\} \subset J$ with processing times $p_j = \frac{1}{2}\sum_{a \in A} a + 1$ requiring resource $q_j = 2$ the whole time but the common resource 0 just at the first time slot, i.e., $p^0 = 1$ and $p_j^{\mathrm{pre}} = 0$.

Let $\mathrm{MS}^* = p_{|A|+1} + p_{|A|+2} = \sum_{a \in A} a + 2$. A feasible solution to JSOCMSR with makespan MS^* must have the jobs $|A| + 1$ and $|A| + 2$ scheduled sequentially without any gap and all other jobs in parallel to those two. A corresponding solution to the PP can immediately be derived by considering the integers associated with the jobs scheduled in parallel to job $|A+1|$ as A_1 and those scheduled in parallel to job $|A+2|$ as A_2. The obtained solution to the PP must be feasible since $\sum_{a \in A_1} a = \sum_{a \in A_2} a = \frac{1}{2}\sum_{a \in A} a$ holds as the jobs corresponding to the integers do not overlap and there is exactly $\frac{1}{2}\sum_{a \in A} a$ time left at the common resource 0 when processing jobs $|A| + 1$ and $|A| + 2$, respectively. It also follows that if there is no JSOCMSR solution with makespan MS^*, then there cannot exist a feasible solution to the PP.

Clearly, the described transformation of a PP instance into a JSOCMSR instance as well as the derivation of the PP solution from the obtained schedule can both be done in time $O(|A|)$, i.e., polynomial time.

Consequently, the decision variant of the JSOCMSR is NP-complete.   □

Corollary 1. *The makespan minimization variant of JSOCMSR is NP-hard.*

3.2 Lower and Upper Bounds

For an instance of JSOCMSR a lower bound for the makespan can be calculated on the basis of each resource $r \in R$ by taking the total time $\sum_{j \in J_r} p_j$. Similarly, one more lower bound can also be obtained from the total time resource 0 is required, i.e., $\sum_{j \in J} p_j^0$. The latter can further be improved by adding the minimal time for preprocessing and postprocessing for the first and last scheduled jobs, respectively. Taking the maximum of these $m + 1$ individual lower bounds yields

$$\mathrm{MS}^{\mathrm{LB}} = \max\left(\min_{j,j' \in J \,|\, j \neq j' \vee |J|=1}(p_j^{\mathrm{pre}} + p_{j'}^{\mathrm{post}}) + \sum_{j \in J} p_j^0, \ \max_{r \in R} \sum_{j \in J_r} p_j\right). \quad (2)$$

Figure 1 illustrates these relationships.

A trivial upper bound is obtained when scheduling all jobs strictly sequentially, yielding $\mathrm{MS}^{\mathrm{UB}} = \sum_{j \in J} p_j$. It follows that taking any normalized solution has an approximation factor of no more than m, since $\mathrm{MS}^{\mathrm{UB}} \leq m \cdot \mathrm{MS}^{\mathrm{LB}}$.

4 Least Lower Bound Heuristic

We construct a heuristic solution by iteratively selecting a not yet scheduled job and always appending it at the end of the current partial schedule at the

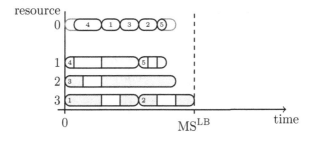

Fig. 1. Resource-specific individual lower bounds and the overall lower bound MS^{LB} for an example instance with $n = 5$ jobs and $m = 3$ secondary resources.

earliest possible time. The crucial aspect is the greedy selection of the job to be scheduled next, which is based on the lower bound calculation from Sect. 3.2. Therefore we call this heuristic *Least Lower Bound Heuristic* (LLBH).

Let π^{P} be the current partial job permutation representing the current normalized schedule and $J' \subseteq J$ be the set of remaining unscheduled jobs. Given π^{P}, the *earliest availability time* for each resource—that is, the time from which on the resource might be used by a next yet unscheduled job—can be calculated from the respective finishing time of the last job using this resource:

$$t_0 = \begin{cases} \max_{j \in J \setminus J'} s_j + p_j^{\text{pre}} + p_j^0 & \text{for } J' \neq J \\ 0 & \text{else} \end{cases} \tag{3}$$

$$t_r = \begin{cases} \max_{j \in J_r \setminus J'} s_j + p_j & \text{for } J_r \setminus J' \neq \emptyset \\ 0 & \text{else} \end{cases} \quad \forall r \in R \tag{4}$$

These times, however, can possibly be further increased (*trimmed*) as the earliest usage time of resource $r \in R$ also depends on the remaining unscheduled jobs and the earliest usage time of the common resource 0. We therefore apply the rule

$$t_r \leftarrow \max(t_r, t_0 - \max_{j \in J_r \cap J'} p_j^{\text{pre}}) \quad \forall r \in R \mid J_r \cap J' \neq \emptyset. \tag{5}$$

Moreover, also t_0 might be increased as its earliest usage time also depends on the remaining unscheduled jobs and the earliest usage times of their secondary resources. These relations are considered by applying the rule

$$t_0 \leftarrow \max\left(t_0, \min_{j \in J'}(t_{q_j} + p_j^{\text{pre}})\right) = \max\left(t_0, \min_{r \in R \mid J_r \cap J' \neq \emptyset}(t_r + \min_{j \in J_r \cap J'} p_j^{\text{pre}})\right). \tag{6}$$

Further note that after a successful increase of t_0 by rule (6), some resource $r \in R$ might become available for a further increase of its t_r by the respective rule (5). We therefore apply all these trimming rules repeatedly until no further increase can be achieved.

Following our general lower bound calculation for the makespan in (2), it is now possible to derive a more specific lower bound for a given partial permutation π^{P} considering any possible extension to a complete solution on the basis of each resource $r \in R \mid J_r \cap J' \neq \emptyset$ by

$$\mathrm{MS}_r^{\mathrm{LB}}(\pi^{\mathrm{P}}) = \begin{cases} t_r + \sum_{j \in J_r \cap J'} p_j & \text{for } J_r \cap J' \neq \emptyset \\ 0 & \text{else} \end{cases} \qquad \forall r \in R. \qquad (7)$$

Note that we define $\mathrm{MS}_r^{\mathrm{LB}}(\pi^{\mathrm{P}}) = 0$ for any resource r that is not required by any remaining job in J' since these bounds should not be relevant for our further considerations.

A lower bound w.r.t. the common resource 0 can be calculated similarly by

$$\mathrm{MS}_0^{\mathrm{LB}}(\pi^{\mathrm{P}}) = \max \left(t_0 + \min_{j \in J'} p_j^{\mathrm{post}}, \min_{j,j' \in J' \mid j \neq j' \vee |J'| = 1} (t_{q_j} + p_j^{\mathrm{pre}} + p_{j'}^{\mathrm{post}}) \right) + \sum_{j \in J'} p_j^0. \qquad (8)$$

Clearly, an overall lower bound for the partial solution π^{P} is obtained from the maximum of the individual bounds

$$\mathrm{MS}_{\max}^{\mathrm{LB}}(\pi^{\mathrm{P}}) = \max_{r \in R_0} \mathrm{MS}_r^{\mathrm{LB}}(\pi^{\mathrm{P}}). \qquad (9)$$

For selecting the next job in LLBH to be appended to π^{P}, we always consider the impact of each job $j \in J'$ on each individual bound $\mathrm{MS}_r^{\mathrm{LB}}$, $r \in R_0$, as this gives a more fine-grained discrimination than just considering the impact on the overall bound $\mathrm{MS}_{\max}^{\mathrm{LB}}(\pi^{\mathrm{P}})$, which would often lead to ties.

More specifically, let $\boldsymbol{f}(\pi^{\mathrm{P}}) = (f_0(\pi^{\mathrm{P}}), \ldots, f_m(\pi^{\mathrm{P}}))$ be the vector of the bounds $\mathrm{MS}_r^{\mathrm{LB}}(\pi^{\mathrm{P}})$ for $r \in R_0$ sorted in *non-increasing value order*, i.e., $f_0(\pi^{\mathrm{P}}) = \mathrm{MS}_{\max}^{\mathrm{LB}}(\pi^{\mathrm{P}}) \geq f_1(\pi^{\mathrm{P}}) \geq \ldots \geq f_m(\pi^{\mathrm{P}})$ holds.

Let $\pi^{\mathrm{P}} \oplus j$ denote the partial solution obtained by appending job $j \in J'$ to π^{P}. We consider $\pi^{\mathrm{P}} \oplus j$ *better than* $\pi^{\mathrm{P}} \oplus j'$ for $j, j' \in J'$ iff

$$\exists i \in \{0, \ldots, m\} \mid f_i(\pi^{\mathrm{P}} \oplus j) < f_i(\pi^{\mathrm{P}} \oplus j') \wedge \forall i' < i : f_{i'}(\pi^{\mathrm{P}} \oplus j) = f_{i'}(\pi^{\mathrm{P}} \oplus j'). \qquad (10)$$

In other words, the sorted vectors $\boldsymbol{f}(\pi^{\mathrm{P}} \oplus j)$ and $\boldsymbol{f}(\pi^{\mathrm{P}} \oplus j')$ are compared in a lexicographic order.

LLBH always selects in each iteration a job $j \in J'$ yielding a (locally) best extension. In the case when multiple extensions have equal $\boldsymbol{f}$-vectors, one of them is chosen at random.

5   Mixed Integer Linear Programming Formulation

The position-based mixed integer linear program (MILP) described in the following models solutions to the JSOCMSR in terms of permutations of all jobs. Index $i \in \{1, \ldots, n\}$ refers hereby to position i in a permutation. Variables $x_{j,i} \in \{0, 1\}$, for all $j \in J$ and $i \in \{1, \ldots, n\}$, are set to one iff job j is assigned to position i in the permutation. Variables $s_i \geq 0$ represent the starting time of the jobs scheduled at each position $i = 1, \ldots, n$ in the permutation. Finally, $MS \geq 0$ is the makespan variable to be minimized.

$$\min MS \tag{11}$$

$$\sum_{j \in J} x_{j,i} = 1 \qquad\qquad i = 1, \ldots, n \tag{12}$$

$$\sum_{i=1}^{n} x_{j,i} = 1 \qquad\qquad j \in J \tag{13}$$

$$s_i + \sum_{j \in J} x_{j,i} \cdot p_j \leq MS \qquad\qquad i = 1, \ldots, n \tag{14}$$

$$s_1 = 0 \tag{15}$$

$$s_i + \sum_{j \in J} x_{j,i} \cdot p_j^{\text{pre}} \geq s_{i-1} + \sum_{j \in J} x_{j,i-1} \cdot (p_j^{\text{pre}} + p_j^0) \qquad i = 2, \ldots, n \tag{16}$$

$$s_{i'} - s_i + \sum_{j \in J_r} x_{j,i'}(M + p_j) + \sum_{j \in J_r} x_{j,i} M \leq 2M$$
$$i = 2, \ldots, n, \ i' = 1, \ldots, i-1, \ r \in R \tag{17}$$

$$x_{j,i} \in \{0, 1\} \qquad\qquad j \in J, \ i = 1, \ldots, n \tag{18}$$

$$s_i \geq 0 \qquad\qquad i = 1, \ldots, n \tag{19}$$

$$MS \geq 0 \tag{20}$$

Hereby, Eq. (12) ensure that exactly one job is assigned to the i-th position of the permutation and (13) ensure that each job is assigned to exactly one position. The makespan is determined by inequalities (14). Equation (15) sets the starting time of the first job in the permutation to zero, and the remaining two sets of inequalities make sure that no resource is used by more than one job at a time. Hereby, inequalities (16) take care of the common resource 0, while (17) consider the secondary resources. The Big-M constant in these latter inequalities is set to the makespan obtained by LLBH.

6   A* Algorithm

Based on the solution construction principle of LLBH it is also possible to perform a more systematic search for a proven optimal solution following the concept of A* search [5]. Our A* algorithm searches in a graph whose nodes correspond to partial solutions and whose arcs represent the extensions of partial solutions by appending not yet scheduled nodes. More precisely, each node in this graph maintains the following information:

1. The *unordered* set $\hat{J} \subset J$ of already scheduled jobs, implemented by a bit-vector.
2. A set of *Non-Dominated Times* (NDT) records, where each NDT record corresponds to an individual, more specific partial solution with an indirectly given ordering for the scheduled jobs by storing:

- the vector $t = (t_r)_{r \in R_0}$ of the trimmed earliest usage times t_r for all resources as defined by (3)–(6);
- the last scheduled job $j^{\text{last}} \in \hat{J}$ after which t was obtained;
- an evaluation vector f' similar to f that will be defined below.

Thus, each node aggregates all partial solutions π^P having the same jobs $\hat{J}$ scheduled, and each NDT record provides more specific information for each (non-dominated) partial solution. For a given node/NDT record, the corresponding ordering of the scheduled jobs $\hat{J}$ can be derived in a reverse iterative manner by considering the fitting preceding node/NDT records, always continuing with the node $\hat{J} \setminus \{j^{\text{last}}\}$ and an NDT record with times t_r allowing to schedule job j^{last} without exceeding the t_r values of the last node/NDT record.

Initially a starting node/NDT record corresponding to the empty schedule is generated with $\hat{J} = \emptyset$, $t = \mathbf{0}$, $j^{\text{last}} = \text{none}$, and $f' = (\text{MS}^{\text{LB}}, \ldots, \text{MS}^{\text{LB}})$. The goal node is a node with $\hat{J} = J$, corresponding to all complete solutions.

The set of all so far considered nodes is implemented by a hash-table with $\hat{J}$ as key. Furthermore, the A* algorithm maintains a priority queue containing references to all *open* node/NDT record pairs, i.e., the non-dominated partial solutions that have not yet been expanded. The order criterion in this priority queue extends the *is-better* relation (10) from the LLBH heuristic by considering the number of remaining unscheduled jobs $|J \setminus \hat{J}|(\pi^P)$ as secondary criterion after $\text{MS}^{\text{LB}}_{\text{max}}(\pi^P)$, i.e., vectors

$$f' := (\text{MS}^{\text{LB}}_{\text{max}}(\pi^P) = f_0(\pi^P), |J \setminus \hat{J}|(\pi^P), f_1(\pi^P), \ldots, f_m(\pi^P)) \tag{21}$$

are lexicographically compared. This enhanced relation implies that partial solutions with more scheduled jobs are preferred over partial solutions with the same $\text{MS}^{\text{LB}}_{\text{max}}$ but fewer scheduled jobs, and thus the search adopts depth-first search characteristics when $\text{MS}^{\text{LB}}_{\text{max}}$ does not change. In this way, complete solutions are obtained earlier.

Algorithm 1 sketches our A* algorithm. In each major iteration, a best node/NDT record pair is taken from the priority queue and expanded by considering the addition of each job $j \in J \setminus \hat{J}$. Hereby, the corresponding node is looked up or created when it does not yet exist and a respective NDT record is determined by calculating the earliest usage times t and the evaluation vector f. The possibly multiple NDT records in the node are checked for dominance: Only non-identical and non-dominated entries are kept. An NDT record with time vector t *dominates* (symbol $\lhd$) another NDT record with time vector t' iff $\forall r \in R_0 \ (t_r \leq t'_r) \wedge \exists r \in R_0 \ (t_r < t'_r)$. The A* algorithm stops with a proven optimal solution when the goal node representing a complete solution is selected for expansion.

Diving: The A* algorithm described above aims at finding a proven optimal solution as quickly as possible. It usually does not yield intermediate complete solutions significantly earlier than when terminating with the proven optimum.

To also obtain intermediate heuristic solutions we extended our A* algorithm by *diving* for a complete solution at regular intervals: At the very beginning

---

Algorithm 1. A* Algorithm for JSOCMSR

1: Initialize priority queue Q with $(\emptyset, (\mathbf{0}, \text{none}, (\text{MS}^{\text{LB}}, \ldots, \text{MS}^{\text{LB}})))$
2: $iter \leftarrow 0$
3: **repeat**
4:    **if** $iter \bmod \delta = 0$ **then**
5:       $\pi \leftarrow$ perform diving to obtain complete solution
6:       $\pi^{\text{best}} \leftarrow \pi$ **if** new best complete solution
7:    **end if**
8:    $(\hat{J}, (\boldsymbol{t}, j^{\text{last}}, \boldsymbol{f}')) \leftarrow Q.\text{pop}()$
9:    **if** $|\hat{J}| = n$ **then**
10:       **return** proven optimal solution π^{best}
11:    **end if**
12:    **for all** $j \in J \setminus \hat{J}$ **do**
13:       find or create node N with $\hat{J}(N) = \hat{J} \cup \{j\}$
14:       calculate new NDT record $(\boldsymbol{t}_{\text{new}}, j, \boldsymbol{f}'_{\text{new}})$ from $\boldsymbol{t}$
15:       **if** $\not\exists (\boldsymbol{t}_{\text{dom}}, j^{\text{last}}_{\text{dom}}, \boldsymbol{f}'_{\text{dom}}) \in \text{NDTs}(N) \mid \boldsymbol{t}_{\text{dom}} \trianglelefteq \boldsymbol{t}_{\text{new}}$ **then**
16:          Remove every $(\boldsymbol{t}_{\text{d}}, j'_{\text{d}}, \boldsymbol{f}'_{\text{d}}) \in \text{NDTs}(N) \mid \boldsymbol{t}_{\text{new}} \triangleleft \boldsymbol{t}_{\text{d}}$
17:          Add $(\boldsymbol{t}_{\text{new}}, j, \boldsymbol{f}'_{\text{new}})$ to $\text{NDTs}(N)$
18:          $Q.\text{push}(\hat{J}(N), (\boldsymbol{t}_{\text{new}}, j, \boldsymbol{f}'_{\text{new}}))$
19:          **if** $|\hat{J}(N)| = n$ **then**
20:             $\pi \leftarrow$ derive complete solution from $(\hat{J}(N), (\boldsymbol{t}_{\text{new}}, j, \boldsymbol{f}'_{\text{new}}))$
21:             $\pi^{\text{best}} \leftarrow \pi$ **if** new best complete solution
22:          **end if**
23:       **end if**
24:    **end for**
25:    $iter \leftarrow iter + 1$
26: **until** time- or memory-limit reached
27: **return** heuristic solution π^{best} and lower bound f_0

---

and after each δ regular iterations, the algorithm switches from its classical best-first strategy temporarily to a greedy completion strategy which follows in essence LLBH. The currently selected node is expanded by considering all feasible extensions, and each extension is evaluated by calculating the respective evaluation vector $\boldsymbol{f}'$. From all these extensions, only those that are new and non-dominated—i.e., no corresponding node/NDT entry exists yet—are kept. Should no extension remain in this way, diving terminates unsuccessfully. Otherwise, a best extension is selected from this set according to the lexicographic comparison of the $\boldsymbol{f}'$ vectors, and the diving continues by expanding this node/NDT record pair next. This methodology guarantees that always not yet expanded nodes are further expanded and the diving, if successful, always yields a different solution.

7 Computational Results

To test our algorithms we created two non-trivial sets of random instances. Set B exhibits a *balanced* (B) workload over all resources R, whereas set S has a *skewed* (S) workload. Each set consists of 50 instances for each combination of

$n \in \{10, 20, 50, 100, 200, 500, 1000\}$ jobs and $m \in \{2, 3, 5\}$ secondary resources. The required resource q_j for each job $j \in J$ was randomly sampled from the discrete uniform distribution $\mathcal{U}\{1, m\}$ for the balanced set B but in a skewed way for set S: There, resource m is chosen with twice the probability of each of the resources 1 to $m - 1$. The preprocessing times p_j^{pre} and postprocessing times p_j^{post} were sampled from $\mathcal{U}\{0, 1000\}$ for both instance sets, while times p_j^0 were sampled from $\mathcal{U}\{1, 1000\}$ in case of set B and $\mathcal{U}\{1, 2500\}$ in case of set S.

A third set of instances was derived from the work on patient scheduling for particle therapy in [8]. This set, called P, comprises 699 instances that are expected in practical day-scenarios of this application. We partitioned the whole set into groups with up to 10, 11 to 20, 21 to 50, and 51 to 100 jobs with 51, 39, 207 and 402 instances, respectively. All these instances use $m = 3$ secondary resources. All three instance sets are available from https://www.ac.tuwien.ac. at/research/problem-instances#JSOCMSR.

The algorithms were implemented using G++ 5.4.1. All tests were done on a single core of an Intel Xeon E5649 with 2.53 GHz with a CPU-time limit of 900 s and a memory limit of 15 GB RAM. The MILP from Sect. 5 was solved with CPLEX 12.7. In A* diving was performed every $\delta = 1000$-th iteration.

Table 1 lists aggregated results for each combination of instance type and the different numbers of jobs and secondary resources. Columns opt state the percentage of instances that could be solved to proven optimality. Columns %-gap list average optimality gaps of final solutions π, which are calculated by $100 \cdot (\text{MS}(\pi) - \text{LB})/\text{LB}$, where LB is the lower bound returned from A* in case of LLBH and A* and the lower bound returned from CPLEX in case of CPLEX. Columns $\sigma_{\text{\%-gap}}$ provide corresponding standard deviations. Columns t show the median running times in seconds. In case of MILP, optimality gaps are list only if solutions for all 50 instances could be obtained.

These results give a rather clear picture: While A* performs very well on essentially all instance sets and sizes—its largest average optimality gaps are <5%—CPLEX applied to our MILP model cannot compete at all. CPLEX is not even able to solve all instances with 10 jobs to optimality, and generally does not yield any solution for instances with 200 and more jobs. With only few exceptions, instances of set B are generally rather easy to solve for A* to either optimality or with a small remaining gap of less than 0.2%. Median running times are here fractions of a second for $n \leq 500$ and under three seconds for $n = 1000$. Here we could observe that the general lower bound MS^{LB} is usually very tight and especially for $m = 2$ often already corresponds to the optimal solution value. Skewed instances of type S but also most instances of type P are more difficult to solve. Especially for set S and $m \in \{2, 3\}$, A* was only able to solve instances up to size 20 consistently to optimality. The reason when A* did not terminate with proven optimality was always that the memory limit had been reached. However, thanks to A*'s diving, heuristic solutions with small remaining optimality gaps could still be found. The LLBH is—as expected—always very fast, nevertheless providing excellent solutions, although without specific performance guarantees.

Table 1. Average results of LLBH, A*, and CPLEX for instances of sets B, S, and P.

type	n	m	LLBH opt[%]	%-gap	$\sigma_{\%\text{-gap}}$	t[s]	A* Search opt[%]	%-gap	$\sigma_{\%\text{-gap}}$	t[s]	MILP/CPLEX opt[%]	%-gap	$\sigma_{\%\text{-gap}}$	t[s]
B	10	2	90	0.197	0.87	<0.1	100	0.000	0.00	<0.1	40	0.007	0.01	22.6
B	20	2	96	0.074	0.37	<0.1	100	0.000	0.00	<0.1	-	-	-	900.1
B	50	2	100	0.000	0.00	<0.1	100	0.000	0.00	<0.1	-	-	-	900.0
B	100	2	100	0.000	0.00	<0.1	100	0.000	0.00	<0.1	-	-	-	900.0
B	200	2	100	0.000	0.00	<0.1	100	0.000	0.00	<0.1	-	-	-	900.0
B	500	2	100	0.000	0.00	0.5	100	0.000	0.00	0.4	-	-	-	900.0
B	1000	2	100	0.000	0.00	3.8	100	0.000	0.00	2.6	-	-	-	900.0
B	10	3	74	1.133	2.48	<0.1	100	0.000	0.00	<0.1	48	0.007	0.01	19.2
B	20	3	76	0.767	1.65	<0.1	100	0.000	0.00	<0.1	2	-	-	900.1
B	50	3	74	0.752	1.40	<0.1	92	0.078	0.30	<0.1	-	-	-	900.0
B	100	3	68	0.632	1.16	<0.1	82	0.168	0.39	<0.1	-	-	-	900.0
B	200	3	68	0.405	0.81	<0.1	82	0.172	0.42	<0.1	-	-	-	900.0
B	500	3	64	0.294	0.46	0.5	68	0.117	0.21	0.4	-	-	-	900.0
B	1000	3	68	0.127	0.25	3.8	76	0.062	0.16	2.7	-	-	-	900.0
B	10	5	50	2.320	3.27	<0.1	100	0.000	0.00	<0.1	74	0.004	0.01	2.2
B	20	5	42	1.634	2.31	<0.1	100	0.000	0.00	<0.1	44	-	-	900.0
B	50	5	52	0.475	0.78	<0.1	94	0.016	0.07	<0.1	34	-	-	900.0
B	100	5	52	0.247	0.45	<0.1	88	0.016	0.06	<0.1	-	-	-	900.0
B	200	5	74	0.076	0.17	<0.1	96	0.002	0.01	<0.1	-	-	-	900.0
B	500	5	80	0.014	0.04	0.5	96	0.001	0.01	0.4	-	-	-	900.0
B	1000	5	76	0.006	0.01	3.8	98	0.000	0.00	2.6	-	-	-	900.0
S	10	2	40	1.387	1.84	<0.1	100	0.000	0.00	<0.1	60	0.004	0.01	2.8
S	20	2	14	1.675	1.41	<0.1	100	0.000	0.00	19.3	2	11.986	10.09	900.1
S	50	2	0	4.739	2.58	<0.1	0	3.374	2.32	154.2	-	-	-	900.1
S	100	2	0	4.122	1.70	<0.1	0	3.271	1.57	153.1	-	-	-	900.0
S	200	2	0	3.678	1.01	<0.1	0	3.163	0.98	166.1	-	-	-	900.0
S	500	2	0	3.662	0.75	0.5	0	3.360	0.70	201.5	-	-	-	900.0
S	1000	2	0	3.626	0.50	3.8	0	3.453	0.48	241.1	-	-	-	900.0
S	10	3	44	1.343	1.73	<0.1	100	0.000	0.00	<0.1	50	0.006	0.01	4.2
S	20	3	20	2.323	1.86	<0.1	100	0.000	0.00	15.2	28	-	-	900.0
S	50	3	18	4.170	2.96	<0.1	20	2.807	2.34	163.3	8	-	-	900.0
S	100	3	18	4.506	3.11	<0.1	20	3.593	2.64	181.4	-	-	-	900.0
S	200	3	10	4.545	2.91	<0.1	10	4.011	2.70	194.1	-	-	-	900.0
S	500	3	0	4.960	1.94	0.5	0	4.672	1.92	236.5	-	-	-	900.0
S	1000	3	0	5.018	1.46	3.8	0	4.852	1.41	246.3	-	-	-	900.0
S	10	5	46	1.496	1.87	<0.1	100	0.000	0.00	<0.1	66	0.004	0.01	0.2
S	20	5	64	0.890	1.80	<0.1	100	0.000	0.00	<0.1	82	0.616	2.42	0.8
S	50	5	74	0.275	0.85	<0.1	88	0.097	0.49	<0.1	84	-	-	16.6
S	100	5	88	0.044	0.17	<0.1	98	0.014	0.10	<0.1	46	-	-	890.5
S	200	5	86	0.010	0.03	<0.1	100	0.000	0.00	<0.1	-	-	-	900.0
S	500	5	96	0.002	0.01	0.5	100	0.000	0.00	0.4	-	-	-	900.0
S	1000	5	96	0.001	0.01	3.8	100	0.000	0.00	2.6	-	-	-	900.0
P	≤10	3	82	0.366	0.93	<0.1	100	0.000	0.00	<0.1	63	0.611	0.84	<0.1
P	≤20	3	64	0.374	0.75	<0.1	100	0.000	0.00	<0.1	59	7.512	17.35	63.5
P	≤50	3	62	0.554	0.96	<0.1	80	0.219	0.55	<0.1	27	-	-	900.0
P	≤100	3	65	0.497	1.01	<0.1	77	0.247	0.58	<0.1	4	-	-	900.0

8   Conclusions

In this work we introduced the problem of scheduling a set of jobs, where each job requires two resources: a common resource shared by all jobs for part of their processing, and a secondary resource for the whole processing time. Despite that we could show this problem to be NP-hard, we came up with an excellent lower bound for the makespan, which we exploited in the fast constructive heuristic LLBH and the complete A* search. The A* algorithm features in particular a special graph structure in which each node corresponds to an unordered set of already scheduled jobs in combination with a set of NDT records representing individual non-dominated partial solutions. Hereby it is possible to exploit symmetries and reduce the memory consumption. A diving mechanism is further used to obtain heuristic solutions in regular intervals. It turns out that A* works mostly extremely well. However, some instances especially with skewed resource workloads and competing resources are occasionally hard to solve. The focus of further research will be to better understand difficult instances, to consider extended variants of this problem and to develop advanced heuristic methods.

References

1. Allahverdi, A.: A survey of scheduling problems with no-wait in process. Eur. J. Oper. Res. **255**(3), 665–686 (2016)
2. Conforti, D., Guerriero, F., Guido, R.: Optimization models for radiotherapy patient scheduling. 4OR **6**(3), 263–278 (2008)
3. Garey, M.R., Johnson, D.S.: Computers and Intractability: A Guide to the Theory of NP-Completeness. W. H. Freeman and Co., New York (1979)
4. Gilmore, P.C., Gomory, R.E.: Sequencing a one-state variable machine: a solvable case of the traveling salesman problem. Oper. Res. **12**(5), 655–679 (1964)
5. Hart, P., Nilsson, N., Raphael, B.: A formal basis for the heuristic determination of minimum cost paths. IEEE Trans. Syst. Sci. Cybern. **4**(2), 100–107 (1968)
6. Hartmann, S., Briskorn, D.: A survey of variants and extensions of the resource-constrained project scheduling problem. Eur. J. Oper. Res. **207**(1), 1–14 (2010)
7. Kapamara, T., Sheibani, K., Haas, O., Petrovic, D., Reeves, C.: A review of scheduling problems in radiotherapy. In: Proceedings of the International Control Systems Engineering Conference, pp. 207–211. Coventry University Publishing, Coventry (2006)
8. Maschler, J., Riedler, M., Stock, M., Raidl, G.R.: Particle therapy patient scheduling: first heuristic approaches. In: PATAT 2016: Proceedings of the 11th International Conference of the Practice and Theory of Automated Timetabling, Udine, Italy, pp. 223–244 (2016)
9. Röck, H.: The three-machine no-wait flow shop is NP-complete. J. ACM **31**(2), 336–345 (1984)
10. Van der Veen, J.A.A., Wöginger, G.J., Zhang, S.: Sequencing jobs that require common resources on a single machine: a solvable case of the TSP. Math. Program. **82**(1–2), 235–254 (1998)

Robust Reinforcement Learning
with a Stochastic Value Function

Reiji Hatsugai$^{(\boxtimes)}$ and Mary Inaba

The University of Tokyo, Tokyo, Japan
reiji@g.ecc.u-tokyo.ac.jp

Abstract. The field of reinforcement learning has been significantly advanced by the application of deep learning. The Deep Deterministic Policy Gradient(DDPG), an actor-critic method for continuous control, can derive satisfactory policies by use of a deep neural network. However, in common with other deep neural networks, the DDPG requires a large number of training samples and careful hyperparameter tuning.

In this paper, we propose a Stochastic Value Function (SVF) that treats a value function such as the Q function as a stochastic variable that can be sampled from $N(\mu_Q, \sigma_Q)$. To learn the appropriate value functions, we use Bayesian regression with KL divergence in place of simple regression with squared errors. We demonstrate that the technique used in Trust Region Policy Optimization (TRPO) can provide efficient learning. We implemented DDPG with SVF (DDPG-SVF) and confirmed (1) that DDPG-SVF converged well, with high sampling efficiency, (2) that DDPG-SVF obtained good results while requiring less hyperparameter tuning, and (3) that the TRPO technique offers an effective way of addressing the hyperparameter tuning problem.

1   Introduction

As deep learning is able to process complex sensory inputs and raw pixels, it can be applied to a wide range of tasks, and reinforcement learning with deep learning (Deep RL) has made great progress in addressing challenging problems. However, most existing deep reinforcement learning methods have low sampling efficiency, due to overfitting that arises from the use of deep neural networks to approximate the value functions. This limits the application of Deep RL to real physical tasks, such as robot control.

To address this limitation, we propose the use of a Stochastic Value Function (SVF). SVF applies value functions such as the Q function (action state value function) and V function (state value function) to reinforcement learning and treats them as stochastic variables that are sampled from $N(\mu_Q, \sigma_Q)$.

Many existing value-based reinforcement learning methods train their value functions by using regression based on a Bellman equation [1,2]. For training the SVF, we use Bayesian regression with Kullback-Leibler (KL) divergence rather than simple regression with L2 divergence. Furthermore, we show theoretically

© Springer International Publishing AG 2018
G. Nicosia et al. (Eds.): MOD 2017, LNCS 10710, pp. 519–526, 2018.
https://doi.org/10.1007/978-3-319-72926-8_43

that the natural gradient [3] with the Hessian-free optimization in Trust Region Policy Optimization (TRPO) can be applied.

We implemented the Deep Deterministic Policy Gradient with SVF (DDPG-SVF), had it learn to solve a pendulum problem and a lunar lander problem, and compared the performance of DDPG and DDPG-SVF with and without the trust region method. Our experimental results demonstrated that DDPG-SVF required fewer samples and was more robust, since delicate, precise tuning of the hyperparameters of the deep neural network was unnecessary. We also confirmed the effectiveness of the natural gradient with the Hessian-free optimization.

2 Background

The goal of reinforcement learning is to identify the policy that maximizes the expected discounted cumulative rewards received by the agent. At time step t and state s_t, the agent chooses action a_t sampled from $\pi(a_t|s_t)$, belonging to state space $\mathcal{S}$ and action space $\mathcal{A}$, respectively. π maps states to a probabilistic distribution over the action space $\pi \colon \mathcal{S} \to \mathcal{P}(\mathcal{A})$. When the environment transitions to a new state s_{t+1} following the model dynamic $p(s_{t+1}|s_t, a_t)$, the agent receives the reward $r(s_t, a_t)$. γ is a discount factor. In this context, a state action value function Q^π can be defined. This represents the discounted expected cumulative rewards from taking action a_t in state s_t and then adopting policy π and model dynamics $p(s_{t+1}|s_t, a_t)$.

$$Q^\pi(s, a) = E_{\pi, p}\Big[\sum_{k=0}^{\infty} \gamma^k r(s_{t+k})\Big| s_t = s, a_t = a\Big]. \tag{1}$$

We can derive recursive equation, Bellman Equation.

$$Q^\pi(s_t, a_t) = E_p\big[r(s_t, a_t) + \gamma E_\pi\big[Q^\pi(s_{t+1}, a_{t+1})\big]\big]. \tag{2}$$

2.1 DDPG

In deterministic policy setting [2], we can avoid calculating the expected Q for π. Instead, π is given by the function $\mu \colon \mathcal{S} \to \mathcal{A}$.

$$Q^\mu(s_t, a_t) = E_p\left[r(s_t, a_t) + \gamma Q^\mu(s_{t+1}, \mu(s_{t+1}))\right]. \tag{3}$$

We can then evaluate the right-hand side of Eq. 3 without sampling from the policy. This type of learning is known as *off-policy* reinforcement learning [1, 2, 4, 5] and makes use of trajectories sampled from policies that are different from the current policy, improving the efficiency of sampling.

In many value-based reinforcement learning methods, Q is parameterized by θ_Q and updated by minimizing the square of the Bellman residual with respect to θ_Q.

$$Loss(\theta_Q) = E_p\left[(Q^\mu(s_t, a_t|\theta_Q) - y_t)^2\right]. \tag{4}$$

Here, $y_t = r(s_t) + \gamma Q^\mu(s_{t+1}, \mu(s_{t+1})|\theta'_Q)$, where y_t is the target value. For stable learning, the target parameters θ'_Q can be used to evaluate the target value [1,2]. The target parameters are updated by $\theta_I \leftarrow (1-\tau)\theta_I + \tau\theta$. In DDPG, policy is deterministic and is parameterized by θ_μ and described by $\mu(s|\theta_\mu)$. μ is updated by gradient with respect to θ_μ of J, which is the expected return from the state distribution, where $J = E_\rho[r(s, \mu(s))]$ and ρ is the state distribution. We can then derive the gradient of J with respect to θ_μ as follows.

$$\nabla_{\theta_\mu} J = E_p \left[\nabla_a Q^\mu(s, a|\theta_Q)\big|_{s=s_t, a=\mu(s_t)} \nabla_{\theta_\mu} \mu(s|\theta_\mu)\big|_{s=s_t} \right]. \tag{5}$$

It has been proven that this gradient is equal to the policy gradient [6] at the limit of policy's variance to zero [7].

However, it is known that the simple regression from Eq. 4 can cause overfitting, slowing the fitting of new samples. To make matters worse, the biased gradient estimate derived from Q in Eq. 5, makes the learning policy prone to divergence.

2.2 TRPO

In TRPO, a surrogate objective is derived that gives the lower bound of the true objective. This surrogate objective is in a KL penalized form.

$$\max_\theta \sum_n \frac{\pi_\theta(a_n|s_n)}{\pi_{\theta_{old}}(a_n|s_n)} A^{\pi_{\theta_{old}}} - CD_{KL}\left[\pi_{\theta_{old}}||\pi_\theta\right], \tag{6}$$

where D_{KL} is the KL divergence.

KL penalized maximazation is then treated as a KL constrained maximization problem.

$$\max_\theta L_{\pi_{\theta_{old}}}(\pi_\theta) = \sum_n \frac{\pi_\theta(a_n|s_n)}{\pi_{\theta_{old}}(a_n|s_n)} A^{\pi_{\theta_{old}}}, \tag{7}$$

$$s.t.\, D_{KL}\left[\pi_{\theta_{old}}||\pi_\theta\right] \le \delta. \tag{8}$$

The right direction for this maximization problem can be computed using a linear approximation of L and a quadratic approximation of KL. Equation 6 then becomes a quadratic optimization problem.

$$\max_\theta g(\theta - \theta_{old}) - \frac{C}{2}(\theta - \theta_{old})^T F(\theta - \theta_{old}), \tag{9}$$

where $g = \frac{\partial}{\partial\theta} L_{\pi_{\theta_{old}}}(\pi_\theta)$ and $F = \frac{\partial^2}{\partial\theta^2} D_{KL}\left[\pi_{\theta_{old}}||\pi_\theta\right]$. The right direction can then be computed as $F^{-1}g$, using conjugate gradient descent [8]. Many automatic differentiation libraries, such as TensorFlow [9], can be used to compute $F^{-1}g$ with a computation time $\mathcal{O}(n)$ and without requiring $\mathcal{O}(n^2)$ memory usage (where n is number of parameters). Line search is then performed with KL constrained by Eq. 8. In many reinforcement learning tasks, TRPO offers the best performance and the most stable learning [10]. However, as TRPO is an *on-policy* learning method, it requires a large batch of trajectories for updating, reducing sampling efficiency.

3   Proposal

As noted in Sect. 2, DDPG is sampling efficient but not stable, whereas TRPO is stable but less sampling efficient. We therefore propose the use of a SVF and specifically, a DDPG with SVF (DDPG-SVF) to increase stability and sampling efficiency. Like DDPG, DDPG-SVF uses a deep neural network for reinforcement learning but treats the Q function as a stochastic variable sampled from a Gasussian distribution $N(\mu_Q, \sigma_Q)$. The network architecture of DDPG-SVF is shown in Fig. 1. To train the SVF, we use Bayesian regression rather than the simple regression of Eq. 4. In Bayesian regression, a loss function is defined by KL divergence between an approximated posterior distribution $q(\mu_Q, \sigma_Q|\theta_Q)$ and the true posterior distribution $p(\mu_Q, \sigma_Q|D)$. Here, D is the sample set.

$$Loss(\theta_Q) = KL\left[q(\mu_Q, \sigma_Q|\theta_Q)||p(\mu_Q, \sigma_Q|D)\right] \tag{10}$$
$$= KL[q(\mu_Q, \sigma_Q|\theta_Q)||p(\mu_Q, \sigma_Q)] - E_q[log(p(D|\mu_Q, \sigma_Q))] + const. \tag{11}$$

By applying a reparameterization trick [11] and analytically computing the KL divergence between the approximated posterior $q(\mu_Q, \sigma_Q|\theta_Q)$ and prior distribution $p(\mu_Q, \sigma_Q)$ (both of which are Gaussian), Eq. 11 can be differentiated with respect to θ_Q. We then update θ_Q to minimize Eq. 11 using gradient-based methods. As the distribution of the samples changes continuously in reinforcement learning, the distribution obtained one time step earlier is used as the prior distribution.

Equation 11 can be regarded as yielding the *negative log-liklihood* and KL constraint.

$$\min_{\theta_Q} -E_{q(\mu_Q, \sigma_Q|\theta_Q)}[log(p(D|\mu_Q, \sigma_Q))] + KL[q(\mu_Q, \sigma_Q|\theta_Q)||p(\mu_Q, \sigma_Q)]. \tag{12}$$

$$\min_{\theta_Q} -E_{q(\mu_Q, \sigma_Q\theta_Q)}[log(p(D|\mu_Q, \sigma_Q))]$$
$$s.t.\ KL[q(\mu_Q, \sigma_Q|\theta_Q)||p(\mu_Q, \sigma_Q)] \leq \delta. \tag{13}$$

Equation 12 can be approximated by Eq. 13, and the SVF can be trained by the natural gradient with the Hessian-free algorithm of TRPO [12]. While μ can be trained in the same way as DDPG. Algorithm 1 is the full DDPG-SVF algorithm with and without the trust region.

We can reduce the biases of the Q estimate and gradient of Q with respect to actions when updating μ, because the biases arising from overfitting are reduced by replacing simple regression with Bayesian regression. We apply a natural gradient, which can exit a saddle point more quickly than a first-order gradient with fewer computational resources.

4   Experiments

We designed a Pendulum test and a LunarLander test in OpenAI Gym [13] and used them to compare the performance of three models: (a) DDPG-SVF with

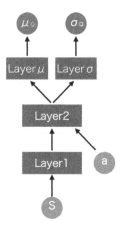

Fig. 1. The network architecture of
DDPG-SVF

Fig. 2. Pendulum

Fig. 3. LunarLander

trust region, (b) DDPG-SVF without trust region, and (c) conventional DDPG. Our main focus was on (A) sampling efficiency and (B) robustness.

4.1 Experimental Methods

The Pendulum test is a continuous control task, the main purpose of which is to stabilize the pendulum in an inverted position. An agent can send a signal discribing the torque moment as an action. Figure 2 shows the task. The LunarLander test is a continuous control task, the main purpose of which is to land between flags on the moon softly. An agent can send signals as power of engines. Figure 3 shows the task. The network architecture of DDPG-SVF is shown in Fig. 1. This architecture differed from that of DDPG only in the final layer and, the network sizes were the same, giving an output size of 400 for Layer 1 and 300 for Layer 2. The same initialization method was used. This network architecture was used in all experiments. The other hyperparameters are given in Table 1 and it was based on original DDPG paper [2]. We multiple a scale reward and a reward for scaling reward.

Algorithm 1. DDPG-SVF

Initialize μ_Q, σ_Q with weights θ_Q, μ with weights θ_μ
Initialize target networks μ'_Q, σ'_Q with $\theta_Q \to \theta'_Q$, μ with $\theta_\mu \to \theta'_\mu$
Initialize replay buffer R
for *episode* $= 1$ to M **do**
   Initialize random process N
   Get initial state s_1 from environment
   for $t = 1$ to T **do**
      Select action $a_t = \mu(s_t|\theta_\mu) + N_t$
      Execute action a_t and get reward r_t, next state s_{t+1} and terminal signal T_t
      Store transition $(s_t, a_t, r_t, s_{t+1}, T_t)$
      Sample M transitions from R
      Compute $y_i = r_i + \gamma \mu_Q(s_{i+1}, \mu'(s_{i+1}))$, $L = \frac{1}{M} \sum -log(p(y_i|\mu_Q, \sigma_Q))$, D_{KL}
      if use trust region **then**
         Compute $F^{-1}g$ by CG
         Line search with constraint $D_{KL} \leq \delta$
      else
         Minimize $Loss = L + D_{KL}$
      end if
      Update target networks $\theta_Q \to \theta_{Q'}$, $\theta_\mu \to \theta_{\mu'}$
   end for
end for

Table 1. Hyperparameter settings

	Algo	Batch size	π lr	Q lr	τ	γ	Scale reward	Game
SQ-tr	DDPG	32	1E−04	−	1E−03	0.99	1.0	Pendulum
SQ-no-tr	DDPG	32	1E−04	1E−03	1E−03	0.99	1.0	Pendulum
dQ	DDPG	32	1E−04	1E−03	1E−03	0.99	1.0	Pendulum

4.2 Sampling Efficiency

Due to overfitting, the DDPG had the lowest sampling efficiency. The only variation in the hyperparameter settings was the Q learning rate for the stochastic Q function when applying the trust region method, as this method does not require the learning rate. Figure 4 shows that whereas the conventional DDPG model required approximately 300 learning episodes, the DDPG-SVF without trust region required 180, and the DDPG-SVF with trust-region only required 120 in the Pendulum test. Figure 5 shows that SVF with and without trust region learn more quickly than conventional DDPG and SVF with trust region is the most stable learning in the LunarLander test. This demonstrated the effectiveness of both the Bayesian regression with KL divergence and the trust region method.

4.3 Robustness

In the conventional DDPG method, if the estimation of the Q function is biased, the policy gradient also becomes biased, and significant hyperparameter tuning is required. As the DDPG-SVF suffers less from overfitting, we anticipated that biased estimation would be improved. To investigate the robustness of the model, we applied 135 hyperparameter settings to each method. These parameter settings were the Cartesian products of Table 2. We ran 500 episodes for each setting and sorted by the obtained score. The plots are shown in Fig. 6, in which the horizontal axis is the sorted order ratio and the vertical axis is the scores. It can be seen that the conventional DDPG model performed well only as far as the upper 40%, after which the scores decreased significantly. In contrast, the DDPG-SVF without trust region had a threshold at approximately 55%. When using the DDPG-SVF with trust region, more than 90% of the hyperparameters had a score higher than those of the other two methods.

Table 2. Hyperparameters

π lr	1e−4, 5e−4, 1e−3, 5e−3, 1e−2
τ	1e−4, 1e−3, 1e−2
γ	0.8, 0.9, 0.99
Scale reward	0.1, 0.5, 1.0

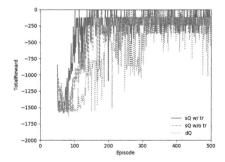

Fig. 4. Pendulum learning curve   **Fig. 5.** LunarLander learning curve

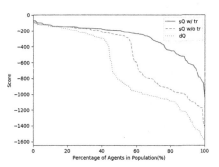

Fig. 6. Pendulum robustness

5   Concluding Remarks

We proposed a DDPG with SVF (DDPG-SVF), in which the Q function is treated as a stochastic variable sampled from the Gaussian distribution $N(\mu_Q, \sigma_Q)$ and the learning is based on KL divergence rather than the squared errors. Our experimental results demonstrated that the DDPG-SVF was able to improve robustness while reducing the number of trials required. Our SVF approach is completely modular and can be applied to a range of other value-based reinforcement learning methods, including DQN [1]. An uncertainty signal of the value function, such as σ_Q, should be applied in reinforcement learning algorithms.

References

1. Mnih, V., Kavukcuoglu, K., Silver, D., Rusu, A.A., Veness, J., Bellemare, M.G., Graves, A., Riedmiller, M., Fidjeland, A.K., Ostrovski, G., Petersen, S., Beattie, C., Sadik, A., Antonoglou, I., King, H., Kumaran, D., Wierstra, D., Legg, S., Hassabis, D.: Human-level control through deep reinforcement learning. Nature **518**(7540), 529–533 (2015)
2. Lillicrap, T.P., Hunt, J.J., Pritzel, A., Heess, N., Erez, T., Tassa, Y., Silver, D., Wierstra, D.: Continuous control with deep reinforcement learning. arXiv.org, September 2015
3. Amari, S.-I.: Natural gradient works efficiently in learning. Neural Comput. **10**(2), 251–276 (1998)
4. Precup, D.: Eligibility traces for off-policy policy evaluation. Computer Science Department Faculty Publication Series, p. 80 (2000)
5. Watkins, C.J.C.H., Dayan, P.: Q-learning. Mach. Learn. **8**(3), 279–292 (1992)
6. Sutton, R.S., McAllester, D.A., Singh, S.P., Mansour, Y.: Policy gradient methods for reinforcement learning with function approximation. In: Advances in Neural Information Processing Systems, pp. 1057–1063 (2000)
7. Silver, D., Lever, G., Heess, N., Degris, T., Wierstra, D., Riedmiller, M.: Deterministic policy gradient algorithms. In: Proceedings of the 31st International Conference on Machine Learning (ICML-2014), pp. 387–395 (2014)
8. Martens, J., Sutskever, I.: Training deep and recurrent networks with hessian-free optimization. In: Montavon, G., Orr, G.B., Müller, K.-R. (eds.) Neural Networks: Tricks of the Trade. LNCS, vol. 7700, pp. 479–535. Springer, Heidelberg (2012). https://doi.org/10.1007/978-3-642-35289-8_27
9. Abadi, M., Agarwal, A., Barham, P., Brevdo, E., Chen, Z., Citro, C., Corrado, G.S., Davis, A., Dean, J., Devin, M., et al.: Tensorflow: large-scale machine learning on heterogeneous distributed systems. arXiv preprint arXiv:1603.04467 (2016)
10. Duan, Y., Chen, X., Houthooft, R., Schulman, J., Abbeel, P.: Benchmarking deep reinforcement learning for continuous control. In: International Conference on Machine Learning, pp. 1329–1338 (2016)
11. Kingma, D.P., Welling, M.: Auto-encoding variational bayes. arXiv.org, December 2013
12. Schulman, J., Levine, S., Moritz, P., Jordan, M.I., Abbeel, P.: Trust region policy optimization. arXiv.org, February 2015
13. Brockman, G., Cheung, V., Pettersson, L., Schneider, J., Schulman, J., Tang, J., Zaremba, W.: OpenAI Gym. arXiv preprint arXiv:1606.01540 (2016)

Finding Smooth Graphs with Small Independence Numbers

Benedikt Klocker[✉], Herbert Fleischner, and Günther R. Raidl

Institute of Computer Graphics and Algorithms, TU Wien,
Favoritenstraße 9–11/186-1, 1040 Vienna, Austria
{klocker,fleischner,raidl}@ac.tuwien.ac.at

Abstract. In this paper we formulate an algorithm for finding smooth graphs with small independence numbers. To this end we formalize a family of satisfaction problems and propose a branch-and-bound-based approach for solving them. Strong bounds are obtained by exploiting graph-theoretic aspects including new results obtained in cooperation with leading graph theorists. Based on a partial solution we derive a lower bound by computing an independent set on a partial graph and finding a lower bound on the size of possible extensions.

The algorithm is used to test conjectured lower bounds on the independence numbers of smooth graphs and some subclasses of smooth graphs. In particular for the whole class of smooth graphs we test the lower bound of $2n/7$ for all smooth graphs with at least $n \geq 12$ vertices and can proof the correctness for all $12 \leq n \leq 24$. Furthermore, we apply the algorithm on different subclasses, such as all triangle free smooth graphs.

Keywords: Branch and bound · Smooth graphs
Combinatorial optimization

1 Introduction

In graph theory independent sets are well studied objects and the independence number of a graph is a central characteristic which is strongly related to many important properties. One natural research subject is to find lower and upper bounds for the independence number for general graphs, see for example [6], or for specific subclasses of graphs, see for example [12].

In this paper we focus on the independence number of smooth graphs, a subclass of 4-regular Hamiltonian graphs. For a complete definition of smooth graphs see Sect. 2. This work is motivated by the works of Fleischner, Sabidussi and Sarvanov [2,3], three renowned graph theorists, who already studied smooth graphs and their independence number in depth from a graph-theoretic perspective.

This work is supported by the Austrian Science Fund (FWF) under grant P27615 and the Vienna Graduate School on Computational Optimization, grant W1260.

© Springer International Publishing AG 2018
G. Nicosia et al. (Eds.): MOD 2017, LNCS 10710, pp. 527–539, 2018.
https://doi.org/10.1007/978-3-319-72926-8_44

We are interested in lower bounds on the independence number of smooth graphs. Sarvanov conjectured that every smooth graph G with $n > 11$ vertices has independence number $\alpha(G) \geq \frac{2}{7}n$ [11]. The main goal of this work is to design an algorithm which can check lower bounds on the independence number for smooth graphs and either prove them for all graphs with a given number of vertices or disprove them by finding a graph with a smaller independence number.

By using Brooks' Theorem [1] we get a lower bound on the independence number for all 4-regular graphs. It states that every 4-regular graph with n vertices that is not the K_5 can be colored with 4 colors which implies that it has an independent set of size at least $n/4$. This property, together with the fact that we only consider graphs containing a Hamiltonian cycle and therefore having an independence number of at most $n/2$, give us an interval of interesting possible lower bounds.

We will describe a branch-and-bound algorithm which heavily depends on the graph-theoretic results and bounds to search through the space of possible graphs in an efficient way [10]. The main idea is to use a heuristic to compute a large independent set together with the graph-theoretic bounds to detect infeasible subproblems as early as possible. For complete solutions we use an integer linear programming (ILP) model to compute their independence number and to check if they are feasible.

In the next section we will formally define smooth graphs and state the problem framework, and in Sect. 3 we present a branch-and-bound approach that solves the problem. In Sect. 4 we will infer some useful bounds and properties using already existing graph-theoretic results, and in Sect. 5 we will describe how to use those bounds and properties to compute a usually very tight bound on the independence number of a partial solution in order to detect infeasibility as early as possible. In Sect. 6 we will present some computational results for four different problem variants. Finally, we will conclude with Sect. 7 and propose promising further work.

2   Problem Formulation

In the context of this paper we only consider loopless undirected graphs, which may contain multiple edges, and just write graph for this type of graphs. A graph is called r-*regular* if every vertex has degree r. We are interested in 4-regular Hamiltonian graphs $G = (V, E)$, in which a Hamiltonian cycle $H \subseteq E$ exists. If we consider the graph $G \setminus H$ after removing the cycle H we get a 2-regular graph which consists of a set of cycles. We call the cycles of $G \setminus H$ the *inner cycles* of G. Such a graph is called *smooth* if the inner cycles are "non-selfcrossing" in the sense that the cyclic order of its vertices agrees with their cyclic order of H. An example for a smooth graph is given in Fig. 1.

The *independence number* of a graph is the size of its largest independent set. Based on Sarvanov's conjecture [11] we formulate the following problem. Given $n \in \mathbb{N}$ as input, does there exist a smooth graph with n vertices and independence

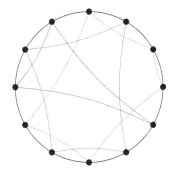

Fig. 1. Smooth graph with twelve vertices and three inner cycles in different colors (Color figure online)

number smaller than $\frac{2}{7}n$? This problem can be generalized to the following family of problems. Given $n \in \mathbb{N}$ as input, does there exist a smooth graph with n vertices that satisfies some properties $\mathcal{P}$ and has independence number smaller than qn for some given factor $q \in \left(\frac{1}{4}, \frac{1}{2}\right]$? We call this problem *Existence of Smooth Graphs with Small Independence Number* or short $\mathrm{ESSI}(q, \mathcal{P})$.

3   Algorithmic Approach

In this section we present a branch-and-bound approach that solves $\mathrm{ESSI}(q, \mathcal{P})$, i.e. it checks for a given $n \in \mathbb{N}$ if there exists a smooth graph with n vertices and independence number smaller than qn that satisfies the conditions $\mathcal{P}$. The conditions of $\mathcal{P}$ can get added to the branch-and-bound approach in a problem-specific manner.

3.1   Solution Representation

If we assume that the Hamiltonian cycle and therefore the order of the vertices in the Hamiltonian cycle is given, every inner cycle of a smooth graph is already uniquely determined if we only know the set of its vertices. W.l.o.g. we assume the vertex set $V = \{1, \ldots, n\}$ to be ordered so that the Hamiltonian cycle $\{\{1, 2\}, \{2, 3\}, \ldots, \{1, n\}\}$ is fixed. Therefore, we only have to partition the vertex set $\{1, \ldots, n\}$ into sets of size at least three and the result represents a smooth graph. For the rest of the algorithmic description section we will use a partitioning of the ordered vertex set $\{1, \ldots, n\}$ into sets of size at least three as a solution representation.

3.2   Core Algorithm

The core algorithm is based on the branch-and-bound principle. The branching is done by assigning the next not yet assigned vertex in the order of the Hamiltonian cycle to an already existing partition or to a new partition. The start solution is the solution where no vertex is assigned. After assigning a vertex to a partition we check if the resulting partial solution satisfies all bounds and if there is a theoretical possibility to complete it to a solution that satisfies the wanted conditions. We call a partial solution that fails this check an infeasible partial solution. If the current partial solution is infeasible, we can cut off this branch and continue with the next partial solution. The infeasibility check of partial solutions is described in more detail in Sect. 5.

Whenever the branching reaches a complete solution, where all vertices are assigned to partitions, we compute its independence number and check the conditions $\mathcal{P}$. Note that computing the independence number is NP-hard for the class of smooth graphs [2]. We compute it by solving the integer linear program

$$\max \left\{ \sum_{v \in V} x_v \,\middle|\, x_v \in \{0,1\} \ \forall v \in V \wedge x_v + x_w \leq 1 \ \forall \{v, w\} \in E \right\}.$$

As search strategy we use depth first search. Although for searching through the whole tree in order to obtain all feasible graphs, the search strategy is irrelevant since we are not reusing information of found solutions, it may be relevant for finding a feasible solution as fast as possible.

4   Bounds and Other Useful Properties

To reduce the search space for our problem we first derive some bounds and other properties for smooth graphs that may have an independence number smaller than qn. We will mainly use the results of Fleischner, Sabidussi and Sarvanov to infer bounds and other properties [2,3]. Those will then be useful for checking infeasibility and recognizing infeasible partial solutions as early as possible.

We consider the problem $\mathrm{ESSI}(q, \mathcal{P})$ and we assume that the satisfaction properties $\mathcal{P}$ and the factor q are fixed. For the rest of this section we will assume that G^* is a smooth graph with n vertices that satisfies the properties $\mathcal{P}$ and has independence number $\alpha(G^*) < qn$, i.e. G^* is a solution to the problem $\mathrm{ESSI}(q, \mathcal{P})$. Let r^* be the number of inner cycles of G^*.

Fleischner and Sarvanov proved in [3] the following theorem.

Theorem 1. *Let G be a smooth graph with n vertices and r the number of inner cycles. Then the following holds.*

$$\alpha(G) \geq \frac{n - r}{3} \tag{1}$$

We use this theorem to compute a lower bound of r^*.

Corollary 1. *For G^* and r^* the following holds.*

$$r^* \geq n - 3\lceil qn \rceil + 3 \tag{2}$$

Proof. Since the independence number $\alpha(G^*)$ is integral we get from (1) that $\alpha(G^*) \geq \left\lceil \frac{n-r^*}{3} \right\rceil$.

$$\alpha(G^*) < qn \Rightarrow \left\lceil \frac{n-r^*}{3} \right\rceil < qn \Leftrightarrow \left\lceil \frac{n-r^*}{3} \right\rceil \leq \lceil qn \rceil - 1$$

$$\Leftrightarrow \frac{n-r^*}{3} \leq \lceil qn \rceil - 1 \Leftrightarrow r^* \geq n - 3\lceil qn \rceil + 3$$

Inequality (1) can be strengthened if we exclude one special graph, which we call $G^{(2)}$. $G^{(2)}$ is defined for even n and is the unique simple smooth graph with only two inner cycles. $G^{(2)}$ is unique since the only possibility to being simple and having only two inner cycles is if all even vertices are in one inner cycle and all odd vertices are in another inner cycle. By excluding $G^{(2)}$ Fleischner and Sarvanov [3] proved the following stronger inequality.

Theorem 2. *Let G be a smooth graph with n vertices that is not isomorphic to $G^{(2)}$ and let r be the number of inner cycles. Then the following holds.*

$$\alpha \geq \frac{n-r+1}{3} \tag{3}$$

Fleischner and Sarvanov stated this theorem with another equivalent condition. They proved Theorem 2 first for multigraphs and then showed that it also holds for simple graphs that have three consecutive vertices in different inner cycles. Putting this two conditions together we get that two consecutive vertices lie in different cycles, since the graph must be simple. Therefore, if three consecutive vertices never lie in three different inner cycles it must hold that vertex k and vertex $k + 2$ always lie in the same inner cycle. This further implies that all even vertices form one inner cycle and so do all odd vertices. Therefore, the only graph that does not satisfy both conditions is $G^{(2)}$.

As before we can use this theorem to compute a stronger lower bound for r^*.

Corollary 2. *If G^* is not isomorphic to $G^{(2)}$ the following holds.*

$$r^* \geq n - 3\lceil qn \rceil + 4 \tag{4}$$

Proof. The proof is analogous to the proof of Corollary 1 by replacing (1) with (3).

Another useful theorem is the following from [4].

Theorem 3 (Cycle-Plus-Triangles Theorem). *Let G be a smooth graph where all inner cycles are triangles, i.e. have length three. Then G is 3-colorable.*

In [3] the following corollary of the cycle-plus-triangle theorem is stated.

Corollary 3. *Let G be a smooth graph with n vertices where all inner cycles have length smaller than or equal to four. Let r be the number of inner cycles and r_3 be the number of inner cycles of length three. Then the following holds.*

$$\alpha(G) \geq \frac{n - (r - r_3)}{3} \tag{5}$$

Let for the following corollary r_3^* be the number of inner cycles of length three of G^*.

Corollary 4. *G^* has either an inner cycle with length greater than four or the following holds.*

$$r^* \geq n - 3\lceil qn \rceil + 3 + r_3^* \tag{6}$$

Proof. The proof is analogue to the proof of Corollary 1 by replacing (1) with (5).

Until now, we only provided lower bounds for r^*, but by using Theorem 3 we can also compute the following upper bound.

Corollary 5. *Let G^* and r^* be as described at the beginning of the section. Then $r^* < qn$ holds.*

Proof. We remove vertices for each inner cycle with length greater than three until every inner cycle has length three. For each removed vertex we connect the two neighbors in the inner cycle and the two neighbors in the Hamiltonian cycle. The result is a smooth graph G' with $n' = 3r^*$ where all inner cycles are triangles. Removing vertices and adding edges can only decrease the independence number since every independent set in the transformed graph is also an independent set in the original graph. Therefore, we know $\alpha(G') \leq \alpha(G^*)$ and we can conclude the proof using Theorem 3 as follows.

$$qn > \alpha(G) \geq \alpha(G') = \frac{n'}{3} = r^*$$

5 Checking Infeasibility

To check if a given partial solution is infeasible, we use the bounds and properties from Sect. 4, and compute an as tight lower bound for the independence number of any completion of the partial solution as possible. Let S be a partial solution, i.e. S is a partitioning of a subset of the vertices of G.

To be able to use the lower bound from Corollary 2 for r, we need to exclude the graph $G^{(2)}$. To do this we check the conditions $\mathcal{P}$ for the unique graph $G^{(2)}$ and compute the independence number of it before we execute the branch and bound algorithm. Let r^{LB} be the lower bound for the number of inner cycles r which we get from (4). Furthermore, let $r^{\mathrm{UB}} = \lfloor qn \rfloor$ be the upper bound for the number of inner cycles r which we get from Corollary 5.

If $|S| > r^{UB}$ the given partial solution is infeasible. Let $k = \sum_{P \in S} |P|$ be the number of fixed vertices in S and

$$\ell := \sum_{P \in S: |P| < 3} 3 - |P|$$

the number of vertices that are at least needed to complete all partitions of S. Furthermore, let $R_i := |\{P \in S : |P| \geq i\}|$ be the number of partitions in S with at least i vertices. Now we can show the following theorem.

Theorem 4. *Let S be a partial solution and r^{UB}, r^{LB}, k, ℓ and $(R_i)_{i \geq 3}$ be as described above. With that we can define the following value.*

$$m := \max \left[0, \min \left(5 - \max \left(3, \max_{P \in S} |P| \right), n - 3\lceil qn \rceil + 3 - R_4 \right) \right].$$

If there exists a feasible completion of S the following holds.

$$k + \ell + m + 3\max(0, r^{LB} - |S|) \leq n \qquad (7)$$

Proof. First of all every completion of S must complete all partitions $P \in S$ with $|P| < 3$, which implies that at least ℓ vertices must be added to the k existing ones. If $|S| < r^{LB}$ we know that a completion of S with the desired properties must have at least r^{LB} different partitions and therefore $3(r^{LB} - |S|)$ additional vertices must be added.

By Corollary 4 either the completion must contain a partition of size at least five or (6) must hold. To get a partition of size five we can add $\max(0, 5 - \max(3, \max_{P \in S} |P|))$ additional vertices to the largest partition. Otherwise, to satisfy (6) we need to have $n - 3\lceil qn \rceil + 3$ many partitions of size at least four. We have at the moment R_4 many inner cycles with length at least four and therefore we need $\max(0, n - 3\lceil qn \rceil + 3 - R_4)$ many additional vertices to get enough inner cycles of length four.

Plugging everything together and considering that in total we have n vertices we get (7). □

If (7) is violated we know that S is infeasible.

We covered now the cases where we can determine that S is infeasible without even computing an independent set. Now we compute an independent set on the partial graph of S, which is the graph induced by all fixed vertices $V_S = \bigcup_{P \in S} P$. By the branching rules we know that $V_S = \{1, \dots, k\}$ for some $k \leq n$.

The partial graph $G_S = (V_S, E_S)$ consists of the fixed vertices and all possible edges between those vertices. Since we do not know if a partition $P \in S$ with $|P| \geq 3$ is already complete or not, we also do not know if the vertices $\min(P)$ and $\max(P)$ are connected or not. We want that every independent set in G_S is also an independent set in G and therefore we have to add those edges to E_S.

$$E_S := \{\{a, b\} \in E_G : a, b \in V_S\} \cup \{\{\min(P), \max(P)\} : P \in S, |P| \geq 3\}$$

To compute an independent set on G_S we use the minimum-degree greedy algorithm [8]. In each iteration this algorithm adds a vertex with the minimum degree to the independent set and removes the vertex and all its neighbors from the graph. Besides good approximation ratios the greedy algorithm is also fast, it can be implemented in $\mathcal{O}(n)$ time.

Let I be the independent set found by the minimum-degree greedy on the graph G_S. Our goal is now to find a good lower bound on how many additional vertices can be added to I in each completion of S.

Theorem 5. *Let S be a partial solution and I an independent set on the graph G_S. Furthermore, let k, ℓ, m and r^{UB} be as described above and let*

$$V_I^{\max} := |I \cap \{1, k\}| + |I \cap \{\min P : P \in S\}| + |I \cap \{\max P : P \in S\}|.$$

Then there exists for every completion G of S an independent set I_G with

$$|I_G| \geq |I| + \frac{\left[n - k - V_I^{\max} - \min\left(r^{UB} - |S|, \frac{n-k-\ell-m}{3}\right)\right]}{3}. \tag{8}$$

Proof. Let G be an arbitrary completion of S. First of all we upper bound the number of inner cycles r of G. Clearly we know $r \leq r^{UB}$. Furthermore, by using the same reduction as in the proof of Theorem 4 we get

$$k + \ell + m + 3\max(0, r - |S|) \leq n \Rightarrow r \leq \frac{n - k - \ell - m}{3} + |S|. \tag{9}$$

Now we can compute a lower bound on the independence number of G. Let $V_I \subseteq V_G \setminus V_S$ be the set of all vertices in G that are not in V_S and are adjacent to one of the vertices in I. The vertices of V_I are either connected to I via the Hamiltonian cycle, which is only possible if the vertex 1 or the vertex k is in I, or via an inner cycle, which is only possible for the end vertices $\min P$ and $\max P$ of an inner cycle $P \in S$. Therefore we can bound the size of V_I by

$$|V_I| \leq |I \cap \{1, k\}| + |I \cap (\{\min P : P \in S\} + |I \cap \{\max P : P \in S\})| = V_I^{\max}.$$

We consider now the residual graph G^{rem} after removing the vertices V_S and V_I from G, which is a graph with $n - k - |V_I|$ vertices. We complete the independent set I by an algorithm that is similar to the minimum-degree greedy algorithm. Instead of always taking a vertex with the minimum degree we take the minimum remaining vertex, i.e. the first vertex in the order of the Hamiltonian cycle that is not adjacent to any vertex in the independent set so far.

Let $I_0 = I$ be the start set and I_i the set after iteration i and let v_i be the vertex added in iteration i. Furthermore, let P_i be the partition in G of the vertex v_i and G_i be the remaining graph in iteration i, $G_0 = G^{\mathrm{rem}}$. We distinguish two cases, the case if $v_i = \min(P_i)$ is the first vertex in P_i or not. Since we selected v_i as the first vertex in the order of the Hamiltonian cycle which is still in G_{i-1} we know that the preceding neighbor of v_i in the Hamiltonian cycle is not in G_{i-1} and therefore we obtain that the degree $d_{G_{i-1}}(v_i)$ of v_i in G_{i-1} is smaller than or equal to three. If $v_i \neq \min(P_i)$ we also know that one neighbor in the inner cycle containing v_i is a predecessor of v_i in the Hamiltonian cycle and therefore it is also not in G_{i-1}, which gives us $d_{G_{i-1}}(v_i) \leq 2$. Summing up over all iterations we get

$$n - k - |V_I| = \sum_{i=1}^{x} d_{G_{i-1}}(v_i) + 1 \leq x + 3(r - |S|) + 2(x - r + |S|)$$

$$\Rightarrow x \geq \frac{n - k - |V_I| - r + |S|}{3} \geq \frac{n - k - |V_I| - \min\left(r^{\mathrm{UB}} - |S|, \frac{n-k-\ell-m}{3}\right)}{3}.$$

In total, we constructed a new independent set I_G with $|I| + x$ elements and therefore (8) holds.

If $\mathcal{P}$ is not empty we can calculate problem specific bounds for those constraints and check them. To summarize this section Algorithm 1 describes the whole procedure for checking infeasibility.

Algorithm 1. Checking Infeasibility

INPUT: n, q, $\mathcal{P}$ and a partial solution S
Compute r^{LB}, r^{UB}, k, ℓ, m
if $|S| > r^{\mathrm{UB}}$ **then**
    return *infeasible*
end if
if (7) is not satisfied **then**
    return *infeasible*
end if
Construct G_S and apply minimum-degree greedy to get independent set I
Compute $V_I^{\max}$
if $|I| + \frac{\left[n - k - V_I^{\max} - \min\left(r^{\mathrm{UB}} - |S|, \frac{n-k-\ell-m}{3}\right)\right]}{3} \geq qn$ **then**
    return *infeasible*
end if
if Problem specific bound check for $\mathcal{P}$ fails **then**
    return *infeasible*
end if
return *possibly feasible*

5.1 Symmetry Breaking

Until now the branch and bound procedure will consider many isomorphic graphs, such as all rotations alongside the Hamiltonian cycle and their reversals. In this section we will describe how we break those symmetries.

To this end we define the gap sequence of a complete solution. Let S be a complete solution, i.e., a partitioning of the vertex set $V = \{1, \ldots, n\}$. Let $P_i \in S$ be the partition of vertex i and let g_i be the gap between vertex i and its successor j in the partition P_i, i.e., let $j = \min\{j \in P_i : j > i\}$ if this set is not empty or $j = \min\{j \in P_i : j < i\}$ otherwise and $g_i = j - i$ if $j > i$ or $j + n - i$ otherwise. We call the sequence $(g_i)_{i=1}^n$ the gap sequence of S.

If two S have the same gap sequence they are not only isomorphic but also exactly the same according to the vertex labeling. We break those symmetries by ensuring that the gap sequence is minimal according to the lexicographical order under all rotations alongside the Hamiltonian cycle and their reversals. Be aware that rotating alongside the Hamiltonian cycle simply means shifting the gap sequence, but reversing the Hamiltonian cycle is a non-trivial change in the gap sequence.

We can compute the gap sequence not only for complete solutions but also for partial solutions. In some cases the next gap is not yet known and instead of calculating a gap we can calculate a lower bound and an upper bound for the gap. With the lower and upper bounds we can check if there is a rotation that always leads to a smaller gap sequence. We can also compute lower and upper bounds for the reversed gap sequence and also check if reversing leads to a smaller gap sequence.

If we found a rotation or a reversed rotation that always leads to a smaller gap sequence, we can fathom the current branch and continue with the next one. The motivations behind the choices of $\mathcal{P}$ are explained subsequently.

6   Computational Results

In this section we will present computational results for instances to four different problems. Our algorithm is implemented in C++ and compiled with g++ 4.8.4. To solve the ILP model for finding a maximum independent set we used Gurobi 7.0.1 [7]. All tests were performed on a single core of an Intel Xeon E5540 processor with 2.53 GHz and 2 GB RAM.

We consider four different variants of the problem. The first and original variant is with $q_1 = \frac{2}{7}$ and with an empty constraint set $\mathcal{P}_1 = \emptyset$. The second problem is also with $q_2 = \frac{2}{7}$ but with the additional constraint that all inner cycles have length at most four, i.e. $\mathcal{P}_2 = \{(R_5 = 0)\}$. The third problem is with $q_3 = \frac{5}{16}$ and $\mathcal{P}_3 = \{(\text{all inner cycles have length } 4)\}$. The fourth problem is with $q_4 = 0.334$ and $\mathcal{P}_4 = \{(\text{G contains no triangles})\}$.

6.1   Problem 1

We tested the implementation for $n \in \{6, \ldots, 29\}$. The algorithm found for $n = 8$ one feasible solution and $n = 11$ two feasible solutions. For all larger n it could not find any feasible solutions. Furthermore, the algorithm was able to finish the branch-and-bound search for all $n \leq 24$, which proves that for $n = 8$ and $n = 11$ the found feasible solutions are the only ones and for all other $n \leq 24$ there does not exist any feasible solution. For $n > 24$ it could not finish the search within 5,000,000 s.

The interesting values of n are the ones where $2n/7$ is only a little bit larger than $\lfloor 2n/7 \rfloor$, since then it may be easier to find a graph with independence number $\lfloor 2n/7 \rfloor$. Therefore, we are especially interested in the values $n \equiv 1 \pmod 7$

and $n \equiv 4 \pmod 7$. Table 1 summarizes the results and running times for those values and compares them with the results of Problem 2. Column $t[s]$ shows the run time in seconds and column *Candidates* the number of complete solutions that got checked by the ILP solver.

Table 1. Results for selected values of n for Problem 1 and Problem 2

	Problem 1		Problem 2	
n^{**}	$t[s]$	Candidates	$t[s]$	Candidates
8	<1	1	<1	1
11	<1	3	<1	1
15	<1	5	<1	<1
18	94	2,298	33	259
22	25,443	5,795	5,047	145
25	>5,000,000	>330,000	4,868,324	160,556
29	>5,000,000	>1,463	>5,000,000	>60,713

6.2 Problem 2

Problem 2 is a more restricted variant of Problem 1 and was tested to check if the restriction helps speeding up the search. Especially the bound corresponding to the value m can be improved through this restriction. We tested again all inputs $n \in \{6, \ldots, 29\}$. For $n = 8$ and $n = 11$ the algorithms found one solution, the second solution of $n = 11$ contains an inner cycle of length five. For all larger n it also could not find any feasible solution.

Through the speedup compared to Problem 1 the algorithm was able to finish the search for all $n \leq 28$ and therefore proves for all $11 < n \leq 28$ that there does not exist a feasible solution. For $n = 29$ it could not finish the search within 5,000,000 s. Table 1 summarizes the results and running times and compares them with Problem 1.

6.3 Problem 3

Fleischner conjectured that smooth graphs only containing inner cycles of length four with at least 12 vertices have independence number at least $5n/16$ [5]. This was the motivation to consider this problem with $q_3 = \frac{5}{16}$. Our algorithm was able to disprove the conjecture by finding 36 smooth graphs with 20 vertices and independence number $6 < qn = 20 \cdot 5/16$ containing only inner cycles of length four. Furthermore, it could find feasible graphs with 24 vertices and independence number $7 < qn = 24 \cdot 5/16$.

Clearly we only have to consider values for n with $n \equiv 0 \pmod 4$. For $n = 8$ we found the same graph as in Problem 1 and 2, for $n = 12$ and $n = 16$ the algorithm could prove that there are no feasible graphs. For $n = 20$ it could finish the search and prove that the found 36 feasible graphs are the only ones but for $n = 24$ the search did not finish in under 5,000,000 s.

The run time for $n = 20$ was 11 min and for $n = 24$ it was 11 h. For $n = 28$ the algorithm could not finish in reasonable time and also did not find a feasible solution in the first 5,000,000 s run time.

6.4   Problem 4

For triangle-free smooth graphs it is proven that $4n/13$ is a valid lower bound for the independence number [9]. This raises the question if it is possible to reach this lower bound or if there exists a stronger lower bound. We use $q = 0.334$ since we want to check if there exist triangle-free smooth graphs with independence number smaller than or equal to $n/3$ and therefore we could use for q any value $1/3 + \varepsilon$ with a small $\varepsilon > 0$. The algorithm was not able to find a graph with independence number smaller than $n/3$ but it was able to find graphs with independence number $n/3$. It could solve the instances up to $n = 26$ in under 5,000,000 s.

7   Conclusion and Further Work

In this paper we formalized a family of problems for finding smooth graphs with small independence numbers. We proposed an algorithm for solving problems of this family which is based on branch and bound. To increase the efficiency of the algorithm by computing good bounds, we used graph-theoretic results to obtain properties and bounds for the number of inner cycles and their sizes. Using those results we proposed a procedure for computing a strong lower bound on the independence number of partial solutions to detect infeasibility as early as possible. We applied our algorithm to four different problems and reported the results and the running times for different graph sizes. Doing this we could disprove one conjecture and find more support for other conjectures for small graphs.

Further work may be to compare different heuristics for computing independent sets for partial solutions. Furthermore, one idea could be to search for a minimal feasible graph, which may enable some reduction properties and therefore some stronger bounds. Additionally, it would be interesting to use a metaheuristic to solve our problems, which would allow to search larger smooth graphs with small independence numbers heuristically.

References

1. Brooks, R.L.: On colouring the nodes of a network. Mathematical Proceedings of the Cambridge Philosophical Society **37**, 194–197 (1941)
2. Fleischner, H., Sabidussi, G., Sarvanov, V.I.: Maximum independent sets in 3-and 4-regular Hamiltonian graphs. Discrete Math. **310**(20), 2742–2749 (2010)

3. Fleischner, H., Sarvanov, V.I.: Small maximum independent sets in Hamiltonian four-regular graphs. Rep. Nat. Acad. Sci. Belarus **57**(1), 10 (2013)
4. Fleischner, H., Stiebitz, M.: A solution to a colouring problem of P. Erdős. Discrete Math. **101**(1–3), 39–48 (1992)
5. Fleischner, H.: Institute of Computer Graphics and Algorithms. TU Wien, Personal communication (2016)
6. Griggs, J.R.: Lower bounds on the independence number in terms of the degrees. J. Comb. Theor. Ser. B **34**(1), 22–39 (1983)
7. Gurobi Optimization Inc: Gurobi optimizer reference manual, version 7.0.1 (2016)
8. Halldórsson, M., Radhakrishnan, J.: Greed is good: approximating independent sets in sparse and bounded-degree graphs. Algorithmica **18**(1), 145–163 (1997)
9. Jones, K.F.: Independence in graphs with maximum degree four. J. Comb. Theor. Ser. B **37**(3), 254–269 (1984)
10. Lawler, E.L., Wood, D.E.: Branch-and-bound methods: a survey. Oper. Res. **14**(4), 699–719 (1966)
11. Sarvanov, V.I.: Institute of Mathematics at the National Academy of Sciences of Belarus. Personal communication (2016)
12. Shearer, J.B.: A note on the independence number of triangle-free graphs. Discrete Math. **46**(1), 83–87 (1983)

BioHIPI: Biomedical Hadoop Image Processing Interface

Francesco Calimeri[1], Mirco Caracciolo[1], Aldo Marzullo[1(✉)],
and Claudio Stamile[2]

[1] Department of Mathematics and Computer Science,
University of Calabria, Rende, Italy
{calimeri,marzullo}@mat.unical.it, mircocaracciolo@gmail.com
[2] Department of Electrical Engineering (ESAT), STADIUS,
Katholieke Universiteit Leuven, Leuven, Belgium
Claudio.Stamile@esat.kuleuven.be

Abstract. Nowadays, the importance of collecting large amounts of data is becoming increasingly crucial, along with the application of efficient and effective analysis techniques, in many areas. One of the most important field in which Big Data is becoming of fundamental importance is the biomedical domain, also due to the decreasing cost of acquiring and analyzing biomedical data. Furthermore, the emergence of more accessible technologies and the increasing speed-up of algorithms, also thanks to parallelization techniques, is helping at making the application of Big Data in healthcare a fast-growing field.

This paper presents a novel framework, Biomedical Hadoop Image Processing Interface (BioHIPI), capable of storing biomedical image collections in a Distributed File System (DFS) for exploiting the parallel processing of Big Data on a cluster of machines. The work is based on the *Apache Hadoop* technology and makes use of the *Hadoop Distributed File System* (HDFS) for storing images, the *MapReduce* libraries for parallel programming for processing, and *Yet Another Resource Negotiator* (YARN) to run processes on the cluster.

Keywords: Big Data · Hadoop · Image processing

1 Introduction

The generation of *Big Data* by biomedical institutes and research is significant, and still expected to grow. Large amounts of data in different formats are continuously generated and have to be stored, managed and analyzed in order to provide suitable information for many applications [4,5].

Traditional solutions can be useful for low-cost storage; however, performance is usually not satisfactory. There are several solutions for such problems related to Big-Data, in this respect; a relevant role in developing and disseminating such solutions is played by technologies like Hadoop [8], a framework that allows

G. Nicosia et al. (Eds.): MOD 2017, LNCS 10710, pp. 540–548, 2018.
https://doi.org/10.1007/978-3-319-72926-8_45

the distributed processing of large data sets across clusters of computers using simple programming models.

In this paper we present a novel framework, namely *BioHIPI*[1], based on the use of Apache Hadoop technology, relying on the Hadoop Distributed File System (HDFS) for storing images, MapReduce for parallel programming for processing, and Yet Another Resource Negotiator (YARN) to run processes on the cluster. BioHIPI is built on the existing Hadoop Image Processing Interface (HIPI) [6], an image library designed to be used with Apache Hadoop. Useful features for importing and analyzing biomedical images have been implemented, thanks to a simple and intuitive representation. Currently, biomedical image formats supported by BioHIPI include *NIfTI* and *DICOM*; it also provides support for processing two-dimensional images such as JPEGs and PNGs. BioHIPI was born from the need for bringing this new parallel programming approach, based on MapReduce, also to biomedical image processing. It allows to import different image formats in a single bundle (BioHIB) and process them all together in one single run. For this purpose, the infrastructure related to the addition of new input formats has been improved with respect to HIPI, making the framework more flexible and scalable. BioHIPI is implemented in Java and can be used as a library to define, among other things, more sophisticated algorithms for biomedical image processing, and/or, new user interfaces to simplify the data collection process.

The remainder of the paper is structured as follows. In Sect. 2 we provide an overview of the Apache Hadoop framework. In Sect. 3, we illustrate the related literature. In Sect. 4, a detailed description of our framework is provided. In Sect. 5 we present our experimental evaluation. Eventually, in Sect. 6 we draw our conclusions.

2   Apache Hadoop

Hadoop is a software framework originally created in 2004 by Doug Cutting, and became a top-level Apache Software Foundation project in January 2008. It can be installed on a Linux cluster to permit large scale distributed data analysis. Hadoop is currently used by many researchers, both in academic and industrial contexts (Yahoo is one of the largest among such contributors), and the community of users has been growing rapidly [1,7].

2.1   Hadoop Distributed File System (HDFS)

Among the main components, Hadoop features a robust filesystem inspired by Google's file system [2]. The Hadoop Distributed File System (HDFS) is written in Java, and is designed in order to run on commodity hardware, where stored data are partitioned and replicated on clusters' nodes; it is fault-tolerant, and has been developed to work on large clusters and be distributed on low-cost

---

[1] Source code is available at https://github.com/memoclaudio/BioHipi.

machines. HDFS relies on a master/slave architecture (in which the master can be a bottleneck) and has been tested with good results on clusters with thousands of nodes. The filesystem provides a high data access bandwidth, and is designed for applications that need to access entire datasets or large parts of them. It is a special, specialized filesystem, whose primary purpose is to contain input and output data for MapReduce.

2.2 MapReduce

MapReduce is a programming model for processing large amounts of data, which allows high parallelization of calculations [9]. MapReduce works by complying to the *Divide et Impera* paradigm: a problem can be decomposed into a sequence of smaller problems to deal with individually. The approach used in MapReduce is based on two separate *map* and *reduce* operations, often applied in pairs: first the map function, and then the reduce function. This allows one to transform and aggregate a list of elements $L[x_1, x_2, ..., x_n]$:

- The *map* operation takes as argument a function $f(x)$ and applies it to all the elements of the list L:

$$map(L[x_1, x_2, ..., x_n], f(x)) \rightarrow L[f(x_1), f(x_2), ..., f(x_n)]$$

- The *reduce* operation takes as argument a list of elements L, a starting value v and a function g. In detail, g is applied to the first element of the list L, and the result is stored in a temporary variable. This temporary value and the second element of L are the arguments for the next application of the function g. The process is repeated until all the elements in L are processed.

$$reduce(L[x_1, x_2, ..., x_n], v, g) \rightarrow g(g(g(g(v, x_1)), x_2)..., x_{n-1}), x_n)$$

Intuitively, the *map* is parallelizable, as the calls are independent from each other; on the other hand, the same does not hold for the *reduce* function (even if, in many applications, list elements can be grouped into parts on which the reduce function can be applied independently).

The MapReduce paradigm operates on sets of key-value pairs. Program execution is divided into a Map and a Reduce stage, separated by data transfer between nodes in the cluster. More precisely, the workflow can be decomposed in three different stages: *map, shuffle, reduce*. In this scenario, the *map* phase works directly on the input and the map function is invoked on every set of key-value pairs, producing a multi-set of elements. The *shuffle* phase is managed by the system, and groups, per each key, every pairs generated in the previous step. The *reduce* step is invoked on each multi-set generated and returns a multi-set of pairs. All pairs created in the *reduce* phase constitute the actual output of the computation.

2.3 Yet Another Resource Negotiator (YARN)

Apache Hadoop YARN is a sub-project of Hadoop and it is the main characteristic of the second version, which is used to develop our framework [10]. It allows multiple access engines (either open-source or proprietary) to use Hadoop as the common standard for batch, interactive and real-time engines that can simultaneously access the same data set. YARN combines a central resource manager that reconciles the way applications use Hadoop system resources with node manager agents that monitor the processing operations of individual cluster nodes. Running on commodity hardware clusters, Hadoop has attracted particular interest as a staging area and data store for large volumes of structured and unstructured data intended for use in analytics applications. Separating HDFS from MapReduce with YARN makes the Hadoop environment more suitable for practical applications where waiting for the whole time for batch jobs to finish might be an issue.

3 Hadoop Image Processing Interface (HIPI)

Standard Hadoop MapReduce programs struggle in representing image input and output data in a useful format. For example, with current methods, in order to distribute a set of images over Map nodes, the user needs to pass the images as a String. Each image then needs to be decoded in each map task in order to get access to the pixel information. This technique is not only computationally inefficient, but also not convenient for the programmer. Thus, it involves significant overhead to obtain standard floating-point image representation with such an approach. Hadoop Image Processing Interface (HIPI) [6] is an image library designed to be used with Apache Hadoop: it provides a solution for storing a large collection of images on the HDFS, and make them available for efficient distributed processing with MapReduce style parallel programs.

In this work an Hadoop framework for image analysis is proposed; the project consists of an extension of HIPI[2] for biomedical images, and currently supports, among others, NIfTI, RDA (for SIEMENS spectroscopy) and DICOM formats.

4 Biomedical Hadoop Image Processing Interface (BioHIPI)

We introduce next BioHIPI, a library for parallel image processing based on the HIPI framework [6]. Specifically, HIPI provides useful tools for distributing a large collection of two-dimensional images on Hadoop's HDFS. Each image collection is represented within the HDFS as a single entity called HipiImageBundle (HIB).

To make the HIPI code more suited to our needs, we reorganized the framework classes that are considered unnecessary, and the addition of new formats has

---

[2] http://hipi.cs.virginia.edu/.

been made faster and more intuitive. This is achieved by implementing a better management of the CodecManager class, which provides the codecs needed for decoding and encoding specific formats. Images are represented by two classes, one for the header, which is called *BioHipiImageHeader*, and one for the image data, which is called *BioHipiImage*. These two classes constitute respectively the key and the value for each pair used as input for the Hadoop MapReduce computations. A typical BioHIPI workflow is illustrated in Fig. 1.

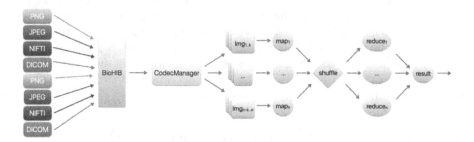

Fig. 1. Representation of a typical BioHIPI process

4.1   BioHIPIImageBundle

The first step needed to analyze biomedical images is the creation of a BioHIB, where the dataset is copied within the HDFS as a single file, termed *datafile*. There is also an index file in HDFS containing information useful for control functions that is updated in parallel with the data file. This setup allows us to easily access images across the entire bundle without having to read in every image [6].

4.2   BioHIPIImageHeader

BioHIPIImageHeader is the class used by the framework to encapsulate all the information useful for image recognition, such as DICOM or NIfTI image format specifications, before it is analyzed. The information is characterized by: a Java enumerative called *BioHipiImageFormat*, which indicates the image format (JPEG, PNG, NIfTI, DICOM) and a Java hash table, called *metatada*, consisting of a list of elements represented as a pair (key/value) of strings. This allows easier access to information and more efficient storage within a BioHIB.

4.3   BioHIPIImage

BioHIPIImage is an abstract class that can be extended in order to implement specific analysis capabilities for each type of image. Extensions currently available support raster, NIfTI, and DICOM image formats. The classes to represent these images are *RasterImage*, *NiftiImage* and *DicomImage*, respectively.

RasterImage. *RasterImage* is a concrete class that extends *BioHipiImage* to represent bidimensional images in the form of a floating-point number array containing RGB pixels. Storage is carried out using the raster-scan technique where color values are stored in three different steps: first the red channel, then the green channel and finally the blue channel. This type of storage requires the number of bands to be three.

NiftiImage. *NiftiImage* is a concrete class that extends *BioHipiImage* to represent biomedical images in NIfTI format as four-dimensional arrays of double-precision numbers. The array is the voxel values.

DicomImage. *DicomImage* is a concrete class that extends *BioHipiImage* to represent biomedical images in DICOM 2D format. This class allows one to get all the information contained in the file header, in the form of a Java Object. DICOM 3D images are built using a *DicomImage* sequence, and the voxel values are manipulated by using the *ImageJ* library [3].

4.4   CodecManager

CodecManager is the class that provides the necessary codecs for decoding and encoding supported images, such as JPEG, PNG, NIfTI, and DICOM. Codecs consist of an implementation of two interfaces:

- ImageDecoder: a Java interface that allows to deploy *decodeHeader* and *decodeImage* functions. *DecodeHeader*, by means of a Java input stream, is implied in the process of reading and decoding the image header, generating a *BioHipiImageHeader*. Specifically, the latter is filled with metadata suitable for the image format. DecodeImage is used in the process of reading and decoding the image, creating the concrete, full-feature class extending *BioHipiImage*.
- ImageEncoder: a Java interface that enables the implementation of the *encodeImage* function. The purpose is to encode and write the image into a Java output stream.

Concrete implementation of *ImageEncoder* and *ImageDecoder* are relative to each format: Raster, DICOM and NIfTI. These implementation make uses of different Java libraries to encode and decode the image.

5   Experimental Analysis

The intuitive operation of the BioHIPI framework has been implemented in practical examples, and tested in order to asses how simple algorithms can handle large amounts of data. Tests were performed on a reference dataset composed by 9 NIfTI images (604 MB), 59 DICOM images (12.4 MB), 201 JPEG images (12.7 MB), and 196 PNG images (120.2 MB). Execution times resulting from experiments are reported in Table 1; it is worth noting that we tested the framework on the following four different tasks:

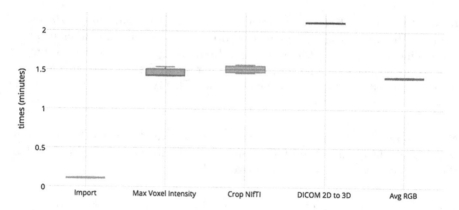

Fig. 2. Execution times for each experiment (averaged over three runs)

- **Maximum voxel intensity search in a set of NIfTI images**: This case shows how to obtain the maximum value of the voxel assigned to each coordinate. A collection of images with different formats are processed by the map function. For each image, a set of key/value pairs is returned, indicating respectively the coordinates and the value of the voxel. Outputs are produced based on the constant value of the key, in order to automatically create, in the *shuffle* phase, a list of voxel values that characterize a specific coordinate position (x, y, z, t) of different images. In the *reduce* function, the analysis of the lists is performed to get the maximum voxel value for each x, y, z, t coordinate.
- **Crop of a set of NIfTI images**: The goal of this use case is to extract specific local regions in a set of NIfTI images. For each image, a set of key/value pairs is returned, each indicating the metadata of the *BioHipiImageHeader* (the key) and the NIfTI image containing the specific area (the value).
- **Converting a set of DICOM 2D image in a DICOM 3D image**: In this test, we created DICOM 3D images starting from a sequence of DICOM 2D images. A collection of images in different formats is passed to the map function that only examines DICOM images, excluding images with different file format. For each image, a set of key/value pairs is returned, each indicating the patient's name (the key) and DICOM image (the value) with the data to be analyzed. Outputs produced in the *map* phase are selected based on the constant value of the key, so that the *shuffle* step automatically creates a sequence of DICOM 2D images that are useful for creating a DICOM 3D image. In the Reduce function, the DICOM 3D image is generated getting voxel intensities by means of the ImageJ library.
- **Compute the average of the RGB values of a collection of raster images**: This case shows how to obtain the image representing the average of RGB colors from a collection of raster images. A collection of images with different formats is passed to the map function that only examines Raster images, excluding different format ones. For each image, a set of key/value

pairs is returned, indicating respectively a standard reference value and Raster images containing the average RGB color.

Table 1 shows execution times for each experiment of our experimental campaign. Graphical representation is reported in Fig. 2.

Table 1. Execution times for each test case of the experimental campaign, including time required to import the dataset. Results are averaged on three runs for each experiment (standard deviation in parenthesis)

	Average time
Import	0 m 11 s ($\pm 0,01$)
Max voxel intensity	1 m 46 s ($\pm 0,06$)
Crop NIfTI	0 m 51 s ($\pm 0,05$)
DICOM 2D to 3D	2 m 11 s ($\pm 0,01$)
Avg RGB	1 m 41 s ($\pm 0,01$)

6   Conclusion

In this work we introduced BioHIPI, a new framework for managing Big Data in the domain of biomedical images. More in detail, we took advantage from Hadoop's MapReduce paradigm in order to facilitate the use and organization of large collections of biomedical images. To achieve the goal, an existing framework, HIPI, has been properly extended to support different and complex standard file formats for the representation of biomedical images, namely the NIfTI and DICOM formats.

Our preliminary experimental activities showed that the use of BioHIPI within projects that use Big Data in the context of image analysis can potentially significantly simplify the development process, and speed up the analysis. The framework features significant room for future interesting improvements, especially the addition of other formats for biomedical images, like RDA file for MRI spectroscopy.

Acknowledgments. Claudio Stamile is funded by an EU MC ITN TRANSACT 2012 (316679) project. Francesco Calimeri has been partially supported by the Italian Ministry for Economic Development (MISE) under project "PIUCultura – Paradigmi Innovativi per l'Utilizzo della Cultura" (n. F/020016/01-02/X27), and by the EU under project "Smarter Solutions in the Big Data World (S2BDW)" (n. F/050389/01-03/X32) funded within the call "HORIZON2020" PON I&C 2014-2020.

References

1. Henschen, D.: Emerging Options: MapReduce, Hadoop: Young, But Impressive. Information Week (2010). 24
2. Ghemawat, S., Gobioff, H., Leung, S.-T.: The Google file system. In: Proceedings of the 19th ACM Symposium on Operating Systems Principles (SOSP 2003), pp. 29–43 (2003)
3. Schindelin, J., Rueden, C.T., Hiner, M.C., Eliceiri, K.W.: The ImageJ ecosystem: an open platform for biomedical image analysis. Mol. Reprod. Dev. **82**(7–8), 518–529 (2015)
4. Margolis, R., Derr, L., Dunn, M., Huerta, M., Larkin, J., Sheehan, J., Mark, G., Green, E.D.: The National Institutes of Health's Big Data to Knowledge (BD2K) initiative: capitalizing on biomedical big data. J. Am. Med. Inform. Assoc. **21**(6), 957–958 (2014)
5. Luo, J., Wu, M., Gopukumar, D., Zhao, Y.: Big data application in biomedical research and health care: a literature review. Biomed. Inf. Insights **8**, 1–10 (2016)
6. Sweeney, C., Liu, L., Arietta, S., Lawrence, J.: HIPI: a Hadoop image processing interface for image-based MapReduce tasks. University of Virginia (2011)
7. Taylor, R.C.: An overview of the Hadoop/MapReduce/HBase framework and its current applications in bioinformatics. BMC Bioinf. **11**(Suppl 12), S1 (2010)
8. White, T.: Hadoop: The Definitive Guide. O'Reilly Media Inc., Newton (2012)
9. Dean, J., Sanjay, G.: MapReduce: simplified data processing on large clusters. Commun. ACM **51**(1), 107–113 (2008)
10. Vavilapalli, V.K., Murthy, A.C., Douglas, C., Agarwal, S., Konar, M., Evans, R., Graves, T., Lowe, J., Shah, H., Seth, S., Saha, B., Curino, C., O'Malley, O., Radia, S., Reed, B., Baldeschwieler, E.: Apache Hadoop YARN: yet another resource negotiator. In: Proceedings of the 4th Annual Symposium on Cloud Computing (SOCC 2013), Article 5 (2013)

Evaluating the Dispatching Policies for a Regional Network of Emergency Departments Exploiting Health Care Big Data

Roberto Aringhieri[1]([⊠]), Davide Dell'Anna[2], Davide Duma[1], and Michele Sonnessa[3]

[1] Computer Science Department, Università degli Studi di Torino, Corso Svizzera 185, 10149 Torino, Italy
roberto.aringhieri@unito.it

[2] Department of Information and Computing Sciences, Utrecht University, Princetonplein 5, 3584 CC Utrecht, The Netherlands

[3] Department of Economics and Business Studies, Università degli Studi di Genova, Via Vivaldi 5, 16126 Genova, Italy

Abstract. The Emergency Department (ED) is responsible to provide medical and surgical care to patients arriving at the hospital in need of immediate care. At the regional level, the EDs system can be seen as a network of EDs cooperating to maximise the outputs (number of patients served, average waiting time, ...) and outcomes in terms of the provided care quality. In this paper we discuss how quantitative analysis based on health care big data can provide a tool to evaluate the dispatching policies for the network of emergency departments operating in Piedmont, Italy: the basic idea is to exploit clusters of EDs in such a way to fairly distribute the workload. Further, we discuss how big data can enable a novel methodological approach to the health system analysis.

Keywords: Emergency care pathway · Health systems · Big data

1 Introduction

The Emergency Department (ED) is responsible to provide medical and surgical care to patients arriving at the hospital in need of immediate care. At the regional level, the ED system can be seen as a network of EDs cooperating to maximise the outputs (number of patients served, average waiting time, ...) and outcomes in terms of the provided care quality. Many EDs, especially those serving a large amount of people, complain about the large number of non-urgent patients usually transported by the Emergency Medical Service (EMS) ambulances. Further, EMSs usually do (or can) not take into account the ED workload level when assigning and transporting a patient to an ED. When a peak of emergency demand arises, EDs suffer from increasing overcrowding [10]. The systematic review reported in [9] describes the causes, effects and solutions to the ED overcrowding. The causes described are non-urgent visits of

© Springer International Publishing AG 2018
G. Nicosia et al. (Eds.): MOD 2017, LNCS 10710, pp. 549–561, 2018.
https://doi.org/10.1007/978-3-319-72926-8_46

patients, influenza seasons, hospital closures, ambulance diversion, inadequate staffing, delay in diagnostics and hospital bed shortages. As described before, ED overcrowding leads to delayed patient care. This results in an increased risk of mortality, patients who left without being seen (LWBS) and also financial losses. Both academic literature [2] and the ED managers argue that the efficiency and the equity of the ED system can depend on the interplay between the EMS and the ED network.

The development of models for the analysis of a health system as a whole is one of the main challenges in the health care management field. The basic idea is to have a tool capable to validate management policies at health system level modelling the patient flow through the care pathway. As a matter of fact, the current trend in the analysis of health care systems is to shift the attention from single departments to the entire health care chain in such a way to increase patient's safety and satisfaction, and to optimise the use of the resources.

In order to apply such an approach to the analysis of a regional ED network, one of the main difficulties is the collection of all the information regarding the transportation of the patients from the emergency scene to the ED. Nevertheless this problem can be now overcome exploiting the immense amounts of data generated by health care systems. Health Care Big Data (HCBD) are a key enabling technology to support detailed health system analysis: exploiting the HCBD, one can replicate the behaviour of the health system modelling how each single patient flows within her/his care pathway.

In this paper we discuss how quantitative analysis based on the HCBD can provide a tool to evaluate dispatching policies for a regional network of emergency departments: the basic idea is to exploit clusters of EDs in such a way to fairly distribute the workload. We present a simulation model based on the case study of the Piedmont in Italy, and powered by the knowledge provided by the analysis of regional HCBD.

The paper is organised as follows. In Sect. 2, we introduce the general concept of clinical pathway and its application to the emergency care. Further, we discuss how big data can enable a novel methodological approach to the health system analysis. In Sect. 3, we first discuss the case study under consideration and then we report how we implemented the simulation model. In Sect. 4, we report a quantitative analysis of the results obtained running the simulation model. Conclusions and future works are discussed in Sect. 5.

2   Clinical Pathways and Health System Analysis

The current development of the health care systems is aimed to recognise the central role of the patient as opposed to the one of the health care providers. In this context, Clinical Pathways (CPs) shift the attention from a single health benefit to the health care chain that is involved in the illness episode treatment. A CP can be conceived as an algorithm based on a flow chart that details all decisions, treatments, and reports related to a patient with a given pathology, with a logic based on sequential stages [8]. A CP is therefore "the path" that a patient suffering from a disease traverses in the National Health System. For this reason, they can be considered an operational tool in the clinical treatment of diseases, from a patient-focused point of view [12]. Many papers show that, appropriately implemented, CPs have the potential to increase patient outcome, to reduce patient length of stay and to limit variability in care, thereby yielding cost savings [7,14]. On the contrary [1], limited attention has been dedicated to study how CP can optimise the use of resources (see, e.g., [4,11]).

The development of models for the analysis of a health system as a whole is one of the main challenges in the Health Care Management field [13]. The basic idea is to have a tool capable to validate management policies at health system level modelling the patient flow through the corresponding CP. Literature indicates that System Dynamics (SD) seems to be the most appropriate methodology. A first attempt has been made by Wolstenholme during his collaboration with the NHS. In [17], he applies SD to the development of national policy guidelines for the U.K. health service. The tested policies include the use of "intermediate care" facilities aimed at preventing patients needing hospital treatment. Intermediate care, and the consequent reductions in the overall length of stay of all patients in community care, is demonstrated here to have a much deeper effect on total patient wait times than more obvious solutions, such as increasing acute hospital bed capacity. More generally, as discussed in [18], the key message is that affordable and sustainable downstream capacity additions in patient pathways can be identified, which both alleviate upstream problems and reduce the effort for their management.

A SD model has been used as a central part of a whole-system review of emergency and on-demand health care in Nottingham, as reported in [6]: due to a growing emergency care demands, the hospital systems were unlikely to achieve some government performance and quality targets. Such a model discovered a range of undesirable outcomes associated with the growing demand and, at the same time, suggested policies capable to mitigate such impacts. In [16], the authors were interested in determining whether SD can be an appropriate methodology to model the patient flow in a hospital, and to analyse it from a strategic planning perspective. The SD model were developed in collaboration with the General Campus at The Ottawa Hospital with particular attention to the delays experienced by patients in the ED. The authors reported about the modelling techniques, validation and scenarios tested, accompanied by their comments regarding the appropriateness of SD for such a strategic analysis.

From a modelling point of view, SD is a simulation methodology whose main elements are stocks and flows: a stock is any entity that accumulates or depletes over time; a flow is the rate of change of a stock. For instance, in health care a stock can represent the waiting list for a surgery, that is a number of people requiring a surgery, while a flow can be the rate of a new insertion in the list. One of the main limitation of using SD for health system analysis is that patients are indistinguishable from each other within stocks and flows. On the contrary, health care services are generally characterised by a large variety of different patients suffering from the same diseases and flowing in the same care pathway.

From the above remarks, Discrete Event Simulation (DES) seems the more appropriate methodology to model such a large variety of patients flowing in their corresponding pathway because of DES has the capability of representing each single patient (or entity) within one of more pathways. Further, DES can easily enable the application of optimisation algorithms to take the best (or the most rational) decision regarding a single or a group of entities modelling a single or a group of patients.

It is worth noting that such a modelling approach requires a lot of detailed data, that is all the data needed to replicate the behaviour of each single patient flowing in its corresponding pathway. Moreover, in terms of health system analysis, such a model requires the availability of all the data for all patients flowing in all pathways of the same type in the health system under consideration. A defining characteristic of today's data-rich society is the collection, storage, processing and analysis of immense amounts of data. This characteristic is cross-sectoral and applies also to health care.

Therefore, we argue that the HCBD can power a detailed health system analysis using DES methodology: exploiting the HCBD, one can replicate the behaviour of the health system modelling how each single patient flows within her/his care pathway. The novelty of the paper is therefore the use of the DES methodology for the health system analysis exploiting the Big Data in order to better represent the variety of the patients accessing the health system.

3   A DES Model Powered by the HCBD

In this section we report about the development of a Discrete Event Simulation (DES) model powered by the Health Care Big Data (HCBD) for the analysis of the dispatching policies for a regional Emergency Department (ED) network. First we present the specific case study (Sect. 3.1) and then we discuss our two-phase DES model (Sect. 3.2).

3.1   The ED Network Operating in Piedmont Region

Piedmont (Italian: Piemonte) is one of the 20 regions of Italy. It has an area of 25,402 km^2 and a population of about 4.6 million. The capital of Piedmont is Turin. Piedmont is organised in 7 provinces. The province of Cuneo is the largest one while the province of Turin is the most populated one: actually, about 2.3 million of inhabitants are living in, and 1.4 million are living in the area of Turin. Figure 1 reports the number of inhabitants living in Piedmont and in the province of Turin, divided in different age classes.

According to the 2015's report of the "Programma Nazionale Esiti" by the Ministry of Health, the waiting time for a urgent and a non-urgent code could exceed respectively 60 min and 450 min, in the worst case. In other similar Italian regions, such waiting times are about 20% lower. We remark that in Italy, the Regions are in charge of providing the health services in accordance with the minimal level decided at the national level by the Ministry of Health.

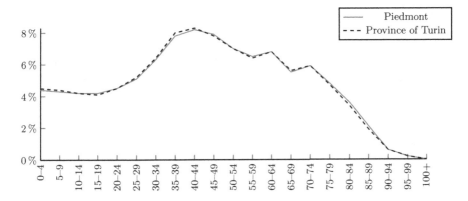

Fig. 1. Population of Piedmont and province of Turin: age distribution (ISTAT 2011).

This comparative analysis demonstrates the need of investigating the reasons of such differences and, eventually, to individuate some possible improvements.

From more than 10 years, the Piedmont region is collecting data about the regional health system, and released a regional law to unify the flows of data gathered from all the health care providers operating in Piedmont, that is, local health agencies, hospitals, and all the private structures in agreement. Such a regional law guarantees the quality of the data collected in accordance with the national standards: all the information must respect a standard format and their consistency is checked for financial reasons since health providers are reimbursed w.r.t. the number and the type of treatments.

Concerning the access to the network of EDs, the HCBD contains all the information regarding the access: encrypted patient ID, patient registered residence, times (arrival, discharge, ...), urgency code, ED, treatment(s), etc. Each year they collect all the information regarding about $1,800,000$ accesses to the regional network of EDs: for instance, in 2013, there were $1,768,800$ accesses; among them, the 90.53% were non-urgent. The network is composed of 49 EDs, mostly – about 20 – located in the province of Turin. The EMS usually transports patients to the closest ED, apart some particular – limited in number – cases.

3.2 A Two-Phase DES Model

We propose a quantitative model for the analysis of the network of EDs operating in Piedmont. The proposed model is organised in two phases, and it operates on a time horizon of one month. The first phase is devoted to data analysis concerning the time horizon taken into account in order to determine the appropriate value of the parameters of the DES model, which is the main part of the second phase. As a matter of fact, the emergency demand depends on the day of the week and the time of day [15]. Further, a not accurate forecasting can lead to managerial solutions that worsen the EMS performance, and by consequence the quality of the access to the ED, even if more resources are used, as discussed in [3].

Dynamic Estimation of the Parameters. In order to have a proper representation of the main parameters of the network of EDs, the first phase of our quantitative model concerns the analysis of the big data relative to the time horizon considered in the running experiments. Parameters and their corresponding distributions are empirically computed over adequate time intervals in such a way to fit the model on a given and fixed time horizon, and to replicate both the patients flow and their management by the EDs.

The main parameters dynamically evaluated are the emergency demands and their urgency code, the capacity and the service time of each ED, and how the patients are distributed to EDs with respect to their geographic origin. The general evaluation procedure consists in scrolling the data concerning each access in chronological order to keep track of the information needed to estimate the considered parameters and their corresponding empirical distribution, as we describe in the following.

Emergency demands. The emergency demand consists in the number of accesses to the whole regional ED network. Such a distribution is computed counting the average number of accesses in each time interval of 30 min over each day of the time horizon considered.

Urgency distribution. The urgency distribution measures the percentage of patients having a urgent or a non urgent code with respect to the origin of the patients.

Service time of each ED. The service time of each ED is estimated using the information regarding the time on which the patient has been take over by the ED, and the time on which the patient has been discharged. The service time has been estimated by the code of urgency.

Capacity of each ED. An ED usually has a formal capacity defined a priori. On the contrary, the real practice showed that the real capacity could be different. Further, we should take into account variations in the staffing. From these considerations, we estimate the capacity of the ED by counting the maximum number of patients that are in the ED at the same time. We compute such a value for each interval of three hours in a day, for each day in the time horizon. The capacity of each interval of three hours is finally set to the value corresponding to the 90-percentile of all the values measured in the same interval and in the same day of the week.

Patient geographical distribution. From the data of the patients, we estimate the number of the patients coming from a city identified by its postal code. We also estimate the number of patients that accessed an ED from a given city.

Although the percentage of patients transported by the EMS could be evaluated dynamically, our preliminary analysis showed that such a parameter currently ranges in $[13.3\%, 14.2\%]$. Therefore, we decided to set this parameter in such a way to study the interplay between the EMS and the ED network varying such a percentage in Sect. 4.

Figure 2 reports an example of the distribution of the daily arrival of the patients derived from the about $150,000$ accesses to the ED network in July 2001.

Note that the figure reports ticks of 1 h (instead of 30 min) only to improve its readability.

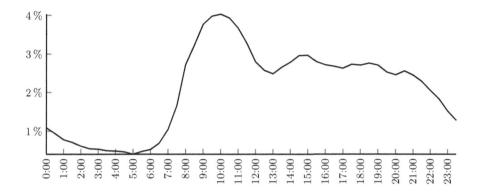

Fig. 2. Distribution of the patient arrivals during the day (July 2011).

The DES Model. We propose a DES model to represent the pathway of the patient entering in the ED network. Our DES model is based on a straightforward representation of the flowchart depicted in Fig. 3.

An emergency request of a patient is generated in accordance with the geographical distribution of the patients and the arrival distribution. At the moment of its generation, an ED is associated to the patient pursuant to the distribution of the patients accessing each ED, which usually corresponds to the closest one. Such an emergency request can be served or not by an EMS ambulance. When the request is not served by the EMS, we assume that the patient reaches – in some way – the ED previously associated. On the contrary, the transportation of the patient is in charge of the EMS. In our model, the ambulance transports the patient to the associated ED only if the urgency code is high (red or yellow in the Italian system), otherwise the EMS can decide where to transport the patient in accordance with some policies (dispatching decision for non urgent patients). After arriving at the ED, the patient will wait for the treatment, which usually lasts for a time distributed following the service time distribution dynamically estimated. When the patient will be discharged, he/she will exit from the model.

The considered dispatching policies are two. The first one, say P_0, dispatches a non urgent patient to the ED associated to the patient at the moment of its generation, without any change. The second one, say P_1, dispatches a non urgent patient following a service state policy, that is, at the moment t, the patient is dispatched to the ED h having minimal ratio r_h^t

$$r_h^t = \frac{w_h^t + s_h^t}{c_h},\tag{1}$$

in which the values w_h^t and s_h^t are respectively the number of patients waiting and receiving health care, and c_h is the estimated capacity of the ED. This policy

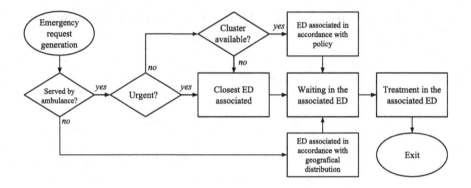

Fig. 3. The flowchart representing the emergency pathway.

is suggested by the fact that Piedmont region is building an ICT infrastructure to share the real-time information regarding the workload of the EDs.

The policy P_1 does not consider all the ED network but only those belonging to a cluster of EDs. A cluster of EDs is a subset of all the EDs operating in Piedmont that can be reached in no more than 30 min from a given origin. We identified 5 different clusters in Piedmont, denoted by C_i, $i = 1, \ldots 5$. The largest one C_1, composed of 20 EDs, is located in the Turin area. The clusters C_2 (province of Alessandria) and C_3 (province of Cuneo) are composed of 7 and 6 EDs, respectively. Finally, two smaller clusters, composed of 2 EDs each, are located in the area of "Valli di Lanzo" (C_4) and in the area of Alba and Bra (C_5).

The proposed DES model is quite flexible: as a matter of fact, the ED network operating during the time horizon considered can be obtained by simply activating the dispatching policy P_0. Note that this also provide a tool to evaluate the ED network as a whole system, instead of having simpler measures as those reported in the "Programma Nazionale Esiti".

Implementation details. The dynamic estimation of the parameters has been implemented in Python 2.7. A script evaluates data concerning the time horizon of interest from the input data-set and generates an Excel file with the parameters of the distributions described above.

Apart from the emergency demand, that has been evaluated at the regional level calculating, as mentioned before, the average number of accesses in each time interval of 30 min over each day of the time horizon considered, the rest of parameters takes also into account the origin of the patients and/or the related EDs. Urgency code distribution has been estimated by distinguishing for each ED four different codes (from 1 to 4). The accessing distribution has been estimated considering both the distribution of provenance of patients and the distribution of accesses of the EDs, mitigating in this way the possibility of not considering patients collected from an ED in a different location from their city of provenance. Finally the service time distribution has been estimated considering both the ED and the gravity of patients.

The DES model has been implemented using AnyLogic 7.2 [5]. At simulation start-up it takes in input the file before generated and uses it to initialise the parameters. Custom distributions have been used for the parameters above described, while specific objects (respectively Service and ResourcePool, Schedule and Agent) have been used for the definition of the EDs, their capacities (varying pursuant to the hour of the day) and for the patients. When a patient is generated it is assigned to him a provenance, the destination hospital (pursuant to the selected policy), an urgency code and the expected service time. The routing of patients has been implemented using two matrices, associating each patient provenance to one (in case of policy P_0) or more (policy P_1) possible EDs of destination.

4   Quantitative Analysis

In this section we report the quantitative analysis performed to test our two-phase DES model.

In our analysis, we considered four different months in 2011. Table 1 provides more details about the input of our model. For each month considered, the table reports the total number of accesses considered and their classification with respect to the urgency code (1 represents the more urgent code while 4 the less one) and their origin with respect to the property of belonging or not to a cluster.

We would remark that the total number of accesses considers only those accesses for which at least one between the origin of the patient or the ED of destination is correctly reported in the data. Finally, the last column of the table reports the percentage of the accesses to an ED belonging to one of the five clusters. This means that the majority of the patients can be served by an ED belonging to a cluster. Further, the cluster C_1, composed of 20 EDs over 49, treats more than the 50% of the accesses.

Table 1. Description of the data considered in our quantitative analysis.

	Total accesses	Requests by urgency			Requests by clusters					
		3–4	2	1	C_1	C_2	C_3	C_4	C_5	
Jan	126,698	107,773	17,688	1,237	69,773	11,480	12,467	3,701	4,201	80.21%
Feb	116,961	99,806	16,074	1,081	64,876	9,819	11,548	3,379	4,008	80.05%
Jun	132,654	113,734	17,562	1,358	70,292	11,672	12,632	3,568	4,451	77.36%
Jul	123,758	106,404	15,970	1,384	62,505	11,027	12,836	3,507	4,196	76.01%

Our quantitative analysis consists in using the two-phase DES model to solve the four instances arising from the four months in Table 1. For each instance, a test consists in solving the instance by varying the percentage of patients transported by the EMS, denoted by p_E, in the interval $[7\%, 27\%]$ with a step of 5%. The rationale is to study the interplay between the EMS and the ED network, as discussed in Sect. 3.2 and suggested in [2]. Finally, the results for each solution are the average values among those obtained by running the two-phase DES model 100 times, each time starting from a different initial conditions in such a way to have independent and identically distributed repetitions.

Table 2. P_1 vs. P_0: waiting time reduction Δ_w in minutes.

	p_E	7%	12%	17%	22%	27%	Avg. Δ_w
Jan	All	15.51	25.91	34.70	42.16	50.79	33.82
	EMS	17.75	21.63	24.73	27.74	34.37	25.24
	No EMS	15.43	26.67	37.01	46.62	57.39	36.62
Feb	All	6.81	13.10	19.75	26.29	31.87	19.56
	EMS	−4.31	1.62	6.68	11.33	15.70	6.21
	No EMS	7.67	14.70	22.48	30.60	37.99	22.69
Jun	All	19.70	39.12	64.51	75.73	80.82	55.98
	EMS	5.78	19.88	45.51	58.88	66.45	39.30
	No EMS	20.81	41.85	68.54	80.63	86.28	59.62
Jul	All	8.27	13.19	17.64	21.15	24.23	16.90
	EMS	−3.86	−2.62	0.22	3.89	7.55	1.04
	No EMS	9.20	15.35	21.20	26.03	30.43	20.44
Avg. Δ_w	All	12.57	22.83	34.15	41.33	46.93	
	EMS	3.84	10.13	19.28	25.46	31.02	
	No EMS	13.28	24.64	37.31	45.97	53.02	

Table 2 shows the results of our quantitative analysis reporting the waiting time reduction Δ_w considering the whole network of EDs. Such values are computed as follows: for a given dispatching policy $i = 0, 1$, we compute the average waiting time w_{ij} for each ED $j = 1, \ldots, 49$, and then we set W_{P_i} equals to the average of all the values w_{ij}; finally, $\Delta_w = W_{P_1} - W_{P_0}$. Note that P_1 is better than P_0 when $\Delta_w > 0$.

The results prove a general improvements of the waiting times, which improves further as soon as the percentage of the patients transported by the EMS increases. It is worth noting that the different results for each different instances depend on the different composition of the emergency demand reported in Table 1 (see, e.g., the last column reporting the percentage of the accesses to an ED belonging to a cluster).

Table 3 shows the results of our quantitative analysis reporting the waiting time reduction considering the cluster C_1, that is the bigger one in terms of both the number of EDs and the number of accesses. Although the general

Table 3. P_1 vs. P_0, cluster C_1: waiting time reduction Δ_w in minutes.

	p_E	7%	12%	17%	22%	27%	Avg. Δ_w
Jan	All	11.88	19.75	22.96	24.24	27.55	21.28
	EMS	30.81	29.65	24.95	22.54	24.52	26.49
	No EMS	2.63	7.74	9.03	9.02	12.11	8.10
Feb	All	−3.15	−1.11	1.66	4.57	6.64	1.72
	EMS	6.41	5.40	5.06	5.86	7.27	6.00
	No EMS	−9.57	−9.80	−8.32	−6.37	−4.85	−7.78
Jun	All	2.19	14.53	36.13	46.32	51.54	30.14
	EMS	9.89	15.69	29.30	36.75	42.02	26.73
	No EMS	−2.19	9.17	31.60	42.18	47.85	25.72
Jul	All	8.43	11.60	11.96	10.58	8.86	10.29
	EMS	−0.76	6.17	9.52	9.87	9.47	6.85
	No EMS	12.59	16.96	18.11	17.17	15.89	16.14
Avg. Δ_w	All	4.84	11.19	18.18	21.43	23.65	
	EMS	11.59	14.23	17.21	18.75	20.82	
	No EMS	0.87	6.02	12.60	15.50	17.75	

improvement is inferior than those for the whole network, such results confirm the comments done for the whole network.

5 Conclusions and Future Developments

We presented a two-phase DES model to evaluate the dispatching policies for the regional network of emergency departments powered by the knowledge provided by the analysis of regional health care big data. The model has been tested on the case study of the Piedmont in Italy showing that there is room to improve its efficiency. Further, we observed that such an improvement is more significant as soon as the percentage of the patients transported by the EMS increases. This remark has an evident managerial implication that would not have been possible without an analysis of the entire ED network.

More generally, the results showed the effectiveness of the proposed approach in terms of the capability of modelling a whole health care system through a discrete event simulation approach, which exploits the availability of the health care big data. As discussed in [17,18], there could be a significant difference between the formal description of the health system and the its real functioning. To overcome this modelling problem, our idea is to retrieve a picture of the system from the big data through the dynamic estimation of the parameters, which allow to fit the model on a given time horizon replicating both the patients flow and their management.

Future developments will be follow two main research lines. The first one is to improve the current model adding a more detailed representation of the transportation network and predictive dispatching policies. The second one is to validate such a methodological approach on a more complex health care network, such as those of the hospitals with their specialties.

References

1. Aringhieri, R., Addis, B., Tànfani, E., Testi, A.: Clinical pathways: insights from a multidisciplinary literature survey. In: Proceedings ORAHS 2012 (2012), ISBN: 978-90-365-3396-6
2. Aringhieri, R., Bruni, M., Khodaparasti, S., van Essen, J.: Emergency medical services and beyond: addressing new challenges through a wide literature review. Comput. Oper. Res. **78**, 349–368 (2017)
3. Aringhieri, R., Carello, G., Morale, D.: Supporting decision making to improve the performance of an Italian emergency medical service. Ann. Oper. Res. **236**, 131–148 (2016)
4. Aringhieri, R., Duma, D.: The optimization of a surgical clinical pathway. In: Obaidat, M.S., Ören, T., Kacprzyk, J., Filipe, J. (eds.) Simulation and Modeling Methodologies, Technologies and Applications. AISC, vol. 402, pp. 313–331. Springer, Cham (2015). https://doi.org/10.1007/978-3-319-26470-7_16
5. Borshchev, A.: The Big Book of Simulation Modeling. Multimethod Modeling with AnyLogic, vol. 6 (2013), ISBN: 978-0-9895731-7-7
6. Brailsford, S., Lattimer, V., Tarnaras, P., Turnbull, J.: Emergency and on-demand health care: modelling a large complex system. J. Oper. Res. Soc. **55**(1), 34–42 (2004)
7. Cardoen, B., Demeulemeester, E.: Capacity of clinical pathways - a strategic multi-level evaluation tool. J. Med. Syst. **32**(6), 443–452 (2008)
8. De Bleser, L., Depreitere, R., De Waele, K., Vanhaecht, K., Vlayen, J., Sermeus, W.: Defining pathways. J. Nurs. Manag. **14**, 553–563 (2006)
9. Hoot, N., Aronsky, D.: Systematic review of emergency department crowding: causes, effects, and solutions. Ann. Emerg. Med. **52**(2), 126–136 (2008)
10. Hwang, U., Concato, J.: Care in the emergency department: how crowded is over-crowded? Acad. Emerg. Med. **11**(10), 1097–1101 (2004)
11. Ozcan, Y., Tànfani, E., Testi, A.: Improving the performance of surgery-based clinical pathways: a simulation-optimization approach. Health Care Manag. Sci. **20**, 1–15 (2017)
12. Panella, M., Marchisio, S., Stanislao, F.: Reducing clinical variations with clinical pathways: Do pathways work? Int. J. Qual. Health Care **15**, 509–521 (2003)
13. Proudlove, N., Black, S., Fletcher, A.: Or and the challenge to improve the NHS: modelling for insight and improvement in in-patient flows. J. Oper. Res. Soc. **58**(2), 145–158 (2007)
14. Rotter, T., Kinsman, L., James, E., Machotta, A., Gothe, H., Willis, J., Snow, P., Kugler, J.: Clinical pathways: effects on professional practice, patient outcomes, length of stay and hospital costs (review). The Cochrane Library, vol. 7 (2010)
15. Setzler, H., Saydam, C., Park, S.: EMS call volume predictions: a comparative study. Comput. Oper. Res. **36**(6), 1843–1851 (2009)
16. Vanderby, S., Carter, M.: An evaluation of the applicability of system dynamics to patient flow modelling. J. Oper. Res. Soc. **61**(11), 1572–1581 (2010)

17. Wolstenholme, E.: A patient flow perspective of U.K. health services: exploring the case for new "intermediate care" initiatives. Syst. Dyn. Rev. **15**(3), 253–271 (1999)
18. Wolstenholme, E., Monk, D., McKelvie, D., Arnold, S.: Coping but not coping in health and social care: masking the reality of running organisations beyond safe design capacity. Syst. Dyn. Rev. **23**(4), 371–389 (2007)

Refining Partial Invalidations for Indexed Algebraic Dynamic Programming

Christopher Bacher$^{(\boxtimes)}$ and Günther R. Raidl

Algorithms and Complexity Group, TU Wien, Vienna, Austria
{bacher,raidl}@ac.tuwien.ac.at

Abstract. We consider dynamic programs modelled in a variant of the Algebraic Dynamic Programming (ADP) framework which allows us to develop general purpose solvers for Dynamic Programming problems. In such dynamic programs the information accumulated in memoization tables is usually lost if the input data of the problem instance changes. We analyze those changes and how they affect the information stored for subproblems of a dynamic program. We then present the theory for a new algorithm for partial invalidation and incremental evaluation of ADPs based on a previous simpler algorithm. The new algorithm should reduce the amount of discarded information in Dynamic Programming tables and to speed up the reevaluation of dynamic programs in the face of changing inputs. In future work we will integrate the algorithms into a framework currently under development to conduct thorough experiments on their practical efficieny.

Keywords: Algebraic Dynamic Programming · Formal grammars
Incremental evaluation · Logic programming

1 Introduction

The principles of Dynamic Programming are well known since the 1950s: separate a problem into smaller problems and recombine the optimal results to these to obtain the optimal result of the larger problem—the essence of Bellman's Principle of Optimality [1]—then store the results for later reuse. In practice however, the implementation of the principle is largely problem dependent.

This is one of the main disadvantages of Dynamic Programming (DP). The dynamic program is often formalized as a set of recurrent functions which exhibit the optimality principle. Based on this formalization, one starts to implement the according algorithm which at the end is often hardly recognizable to solve the original formulation. Each improvement, heuristic, bounding, and other tricks employed to increase the performance of the dynamic program is lost once a different DP problem needs to be solved.

Other general problem solving techniques like (Mixed) Integer Linear Programming, Quadratic Programming, Constraint Programming, SAT, or Answer Set Programming, have a large arsenal of efficient and easy to use solver software.

© Springer International Publishing AG 2018
G. Nicosia et al. (Eds.): MOD 2017, LNCS 10710, pp. 562–573, 2018.
https://doi.org/10.1007/978-3-319-72926-8_47

So far solvers or frameworks for DP problems are rather rare and often restricted to specific classes of problems. Since the early 2000s a new approach for modelling and solving dynamic programs called Algebraic Dynamic Programming (ADP) has been promoted [2].

ADP models dynamic programs as formal grammars and originally targeted problems on sequence data [3] which often occur in the field of bioinformatics. The area saw additional work starting from improved implementations [4,5] to other grammar types like multi-tape grammars, or multiple context-free grammars [6]. Other approaches expand the applicability of ADP to set-like data structures [7,8] which opens new problems to be tackled with ADP.

We use a variant of ADP which we developed to expand the reach of approach to other combinatorial optimization problems. Especially we consider the following scenario.

DP as technique is often employed to solve problems on large data sets, e.g., string alignment problems, or shortest path variants. As the data sets may be of substantial size even polynomial time algorithms may require long runtimes for solving. Should the data set now change, the dynamic program has to be reevaluated losing the information from the previous run.

The scenario which motivated our work on changing inputs initially, is the use of DP algorithms as solution decoders embedded in metaheuristics. An example for this are route-first-cluster-second algorithms for Vehicle Routing Problems [9], which represent solution candidates as permutations of customers to be visited. A DP algorithm is then used to partition the permutation into separate tours. If local search moves are applied to such an encoding the solution candidate changes in limited areas leaving large portions unmodified. Nevertheless, the DP algorithm has to redecode the entire solution, instead of solely the changed parts.

In this work we describe the theory for a new algorithm which determines based on the type of change of the input data, which parts of the dynamic program need to be recomputed. We base the algorithm for *partial invalidation* on a simpler variant which we develop in [10]. Experiments on the practical efficiency are out of the scope of this work and will be conducted on a per application basis with a solver framework currently under development.

Section 2 provides the formalism for our ADP variant and contains a simple example model to illustrate the concepts. Section 3 describes a simple algorithm for partial invalidation and incremental reevaluation from [10]. In Sect. 4 we present the new algorithm which is based on logic programming and is intended to improve upon the simpler variant. Section 5 concludes the article.

2 Indexed Algebraic Dynamic Programming

Before we explain partial invalidation and its extension, we provide a short description of the Indexed Algebraic Dynamic Programming (IADP) formalism. An in-depth treatment of IADP can be found in [10].

In ADP variants, dynamic programs (DPs) are described using formal grammars. IADP is a variant of ADP which uses explicit indices on the symbols of the grammar.

Although the index-freeness of ADP is one of main benefits mentioned in former work [2, 4] we reintroduce indices to allow for easier modelling of several problems like the Knapsack Problem, or the Resource-constrained Shortest Path Problem [10].

An IADP consists of three parts. First, the indexed grammar describes the search space of the DP. Second, evaluation algebras define how target values, e.g., costs and distances, are computed for a solution in the search space. Third, dominance criteria determine which target values are memoized in the DP tables.

Definition 1 (Indexed Grammar). *An indexed grammar consists of a set of indexed terminal symbols $\hat{\mathcal{A}}$, a set of indexed non-terminal symbols $\hat{\mathcal{N}}$, a set of productions $\hat{\mathcal{P}}$, and an axiom $S \in \hat{\mathcal{N}}$. An indexed symbol $A_{\mathcal{I}}$ consists of a name A and an index tuple $\mathcal{I} = (i_1, \ldots)$ where each index i_k has an associated domain $\mathcal{D}(i_k)$. We denote with $A_{\check{\mathcal{I}}}$ a concrete indexed symbol with an instantiated index tuple $\check{\mathcal{I}} \in \mathcal{D}(\mathcal{I})$.*

Definition 2 (Tables and Memoization). *Each non-terminal $X_{\mathcal{I}}$ is backed by a table $\bar{X}$ which is accessed by its index tuples $\bar{X}[\check{\mathcal{I}}]$ (or a sub-tuple thereof, see [10]). A table stores all non-dominated, i.e., optimal target values. Furthermore, for each target value a set of back-pointers is stored referring to the subproblems, i.e., symbols which were used to calculate the value.*

For the purpose of this work we distinguish between two different forms of productions: top-down and bottom-up. Top-down productions specify how a non-terminal $X_{\mathcal{I}}$ can be decomposed into smaller subproblems, i.e., non-terminals, and atomic parts of the input, i.e., terminals. In contrast, bottom-up productions describe how several subproblems can be aggregated to form larger subproblems.

Other ADP variants do not distinguish between top-down and bottom-up productions. Though they may use both forms of evaluation as in [11]. The reason for this is that other ADP approaches usually operate on restricted sets of input data, e.g., sequences, and can convert between both forms of productions automatically. Due to the more general nature of IADP an automatic conversion is only applicable in special cases.

Definition 3 (Top-Down Productions). *A top-down production for a non-terminal $N_{\mathcal{I}_0}$ is written as*

$$N_{\mathcal{I}_0} \to X_{\mathcal{V}_1}^1 \ldots X_{\mathcal{V}_l}^l \quad \forall \mathcal{I}_Z = (z_1, \ldots, z_h) \in \mathcal{S}$$

with $X^k \in \hat{\mathcal{A}} \cup \hat{\mathcal{N}}$, $1 \leqslant k \leqslant l$ where $\mathcal{V}_k$ are expressions for computing the index tuples $\check{\mathcal{I}}_k$ from $\mathcal{I}_0$, $\mathcal{I}_Z$ and the input instance $\mathcal{C}$. Top-down productions can be all-quantified over some set $\mathcal{S}$ of free index tuples $\mathcal{I}_Z$ to generate productions in dependence of the input $\mathcal{C}$. We use $\mathcal{P}(N_{\mathcal{I}_0})$ to refer to all productions of the given non-terminal.

The evaluation of the top-down productions starts at the axiomatic symbol S and repeatedly expands all applicable productions until either all remaining productions are cut off due to feasibility constraints (see below), or reach terminal symbols. If a symbol is encountered that has been reached before it is not reexpanded but once optimal target values for the subproblem are known they are reused.

Definition 4 (Bottom-Up Productions). *A bottom-up production for a non-terminal $N_{\mathcal{I}_0}$ is written as*

$$\left\{ X_{\mathcal{I}_1}^1 \ldots X_{\mathcal{I}_l}^l \right\} \dashrightarrow N_{\mathcal{I}_0} \quad \langle p_1 \rangle \ldots \langle p_m \rangle$$

with $X^k \in \hat{\mathcal{A}} \cup \hat{\mathcal{N}}$, $1 \leqslant k \leqslant l$. The relationships between the index tuples $\mathcal{I}_k$ are given in the form of propagators p_h, $1 \leqslant h \leqslant m$. A propagator p_h defined for an index $i_{kj} \in \mathcal{I}_k$ is able to compute index values for other index tuples $\mathcal{I}_{k'}$.

Bottom-up evaluation starts with the set of all terminal symbols that are known to exist in solution candidates. They form the initial *trigger set*. Each production containing an element of the trigger set is activated and uses the propagators to determine values for the other symbols. Once target values are known for all symbols of a specific index assignment of the left-hand side a target value for the right-hand side is computed. If the computed target value is not dominated (see below) then the left-hand side symbol enters the trigger set.

Definition 5 (Conditions and Constraints). *The feasibility of a production (top-down and bottom-up) can be restricted using constraints written as $[c_i]$, $1 \leqslant i \leqslant m$ on the right-hand side of a production. Each c_i is a predicate over the index values occurring in the production and over the input instance $\mathcal{C}$. A production may only be applied if all constraints are satisfied.*

Definition 6 (Evaluation Algebras). *To compute target value we use evaluation algebras. An evaluation algebra σ recursively computes target values using combinator functions f_a and $f_{A^{(i)}}$ for each terminal a and production $A^{(i)}$ as follows.*

$$\sigma_{\mathcal{C}}(a_{\check{\mathcal{I}}}) = \left\{ f_a(\check{\mathcal{I}}, \mathcal{C}) \right\}, \qquad\qquad a_{\check{\mathcal{I}}} \in \hat{\mathcal{A}}$$

$$\sigma(A_{\check{\mathcal{I}}}) = \bigcup_{(A_{\check{\mathcal{I}}}^{(i)} \to \gamma_i) \in \mathcal{P}(A_{\check{\mathcal{I}}})} f_{A^{(i)}} \left[\bigtimes_{X_{\mathcal{V}_j}^{(j)} \in \gamma_i} \sigma\left(X_{\mathcal{V}_j(\check{\mathcal{I}}, \mathcal{C})}^{(j)} \right) \right], \qquad A \in \hat{\mathcal{N}}$$

Note that $\sigma(A_{\check{\mathcal{I}}})$ computes a set of target values using the cross-product of target value sets of each subproblem of a production. No filtering is applied by default. For formulating optimization problems we need a form of filtering of target values. This is done by using dominance criteria.

Definition 7 (Dominance Criteria). *Let A and B two arbitrary target value tuples each containing one entry per evaluation algebra. A dominance relation $A \succ B$ defines whether A dominates B. If this the case then B must not be considered further during the evaluation, i.e., it can be replaced in the DP tables.*

The computed non-dominated target values are then stored in the symbol's table. Using appropriate dominance criteria we can easily go beyond simple minimization or maximization problems and tackle for example pareto-optimization problems.

A final definition is given to provide a semantic structure for our IADP models.

Definition 8 (Yield-Normalform). *A grammar and a set of evaluation algebras are in yield-normalform if for a given yield, i.e., the word of terminals derived for a solution candidate, the computed target values are the same, regardless of the path taken through the productions.*

In general we expect all IADPs to be in yield-normalform as this property assigns a clear meaning to the derived terminal symbols. A solution is then uniquely represented by its yield.

2.1   Modelling a Simple Example

To illustrate the concepts described in Sect. 2, we present a model for the well-known 0–1 knapsack problem [10,11]. In the 0–1 knapsack problem we want to find a subset of a set of items $\mathcal{I}$ with maximal value while respecting the capacity Q of the knapsack. Each item i is assigned a weight $w(i)$ and a price $p(i)$.

The model uses the terminal π_i to represent an item $i \in \mathcal{I}$ and the non-terminal $B_{i,q}$ to model a knapsack with capacity q and for which we considered all items up to π_i (exclusive) for packing. The productions (4) and (5) then either pack another item and ask for the optimal packing of a smaller knapsack, or stop packing by producing the special terminal ε. The evaluation algebra and the dominance criterion (6)–(8) then turn the grammar into a maximization model.

$$\text{nonterm. } B(\text{i: int}, \text{q: real}) \tag{1}$$

$$\text{term. } \pi(\text{i: int}) \tag{2}$$

$$\text{axiom } B_{0,Q} \tag{3}$$

$$B_{i,q} \to \pi_j \, B_{j+1,q-w(j)} \qquad [0 \leqslant q - w(j)] \quad \forall j \in \mathcal{I} : j \geqslant i \tag{4}$$

$$\mid \varepsilon \tag{5}$$

$$\sigma(B) := \sigma(\pi^{(1)}) + \sigma(B^{(1)}) \mid 0 \tag{6}$$

$$\sigma(\pi_i) := p(i) \tag{7}$$

$$X \succ Y \Leftrightarrow \sigma(X) > \sigma(Y) \tag{8}$$

3   Partial Invalidation

Once the undominated target values of an IADP have been obtained and the solutions have been reconstructed, the intermediate results of the non-terminals are usually no longer needed. Freeing up the memory allocated for the symbol tables is the typical action to take.

In some situations, however, it might be beneficial to retain the information stored in the symbol tables. For example consider the case when several DP instances have to be solved which share parts of the input, or if the input changes over time, e.g., when using a DP for candidate decoding inside a metaheuristic. In such situations parts of the tabled information may be reused to solve the modified DP instance without reevaluating the full IADP.

In [10] we present a first algorithm for identifying substructures in an IADP which are not affected by a specific change made to the input. We will summarize the algorithm as it forms the basis of our improvements discussed in Sect. 4.

The main question is how a change in the input may reflect itself on the structure of the IADP instance, i.e., its yield language, derivation DAGs, and undominated target values. Based on the formal definition of an indexed grammar, the input $\mathcal{C}$ can be accessed by the index expression of the symbols and by conditions/constraints. Further when looking closer on the definition of an evaluation algebra, we see that only the terminals may use information from the instance for the calculation of target values. However, we exclude changes to the structure of the evaluation algebras themselves.

Our procedure is called *partial invalidation* as it implicitly determines a set of table entries of the IADP affected by the modification of the input. The algorithm then determines recursively a set of table entries which are no longer guaranteed to be valid.

For the algorithm to work properly we require the grammar and used evaluation algebras to be in yield-normalform. Further we need to introduce additional bookkeeping for determining which indexed symbols depend on each other or on which parts of the input. We do this by keeping track of *touches*.

An indexed symbol $X_{\bar{\imath}}$ is said to *touch* another symbol $Y_{\bar{\jmath}}$ if during runtime some production $X_{\bar{\imath}} \to \ldots Y_{\bar{\jmath}} \ldots$ is expanded. Similarly a symbol is said to touch a part of the input if the input part occurs either in a condition/constraint or in the index calculation of the symbol on the right-hand side of a production. Further in the case of all-quantified productions an input part is only considered touched if it occurs in the filter portion of the quantification, as if the filter would be expressed as—a less efficient—constraint. In constrast, if the input part may be generated by the quantification, the possibility of generation is not enough to touch it. Only if the input part is actually generated and bound by some element of the right-hand side of the created production, the part is touched. How a part of the input $\mathcal{C}$ is defined is a decision of the user, but usually each constant and matrix cell is considered an atomic part. Each symbol and input part stores a list of indexed symbols touching the element, denoted by $\Gamma(.)$.

We distinguish between the following changes to the input $\mathcal{C}$. First, we inspect the case where an input part is modified or removed. Let $\mathcal{C}_x$ be the modified/removed input part.

---

function INVALIDATE(X)
    for all $Y \in \Gamma(X)$ **do**
        if Y is not marked as **uncomputed then**
            $\bar{Y} = \emptyset$
            Mark Y as **uncomputed**
            INVALIDATE(Y)
        end if
    end for
    $\Gamma(X) = \emptyset$
end function

Procedure INVALIDATE($\mathcal{C}_x$) recursively steps back along the touched symbols and resets the data associated with the touching symbol. The only symbols which remain set are those for which it is known that they never could be effected by the modification. Note that the axiom of the grammar is always invalidated.

Second, we have to handle the case when an input part is added to the instance. This may only occur if an element is added to a set-like data structure of the input. Such an addition at least changes the set of productions generated by all-quantors and may invalidate computed solutions as conditions/constraints are modified by the set extension. To handle additions easily we require the IADP to be terminal-quantified, which is defined as follows.

Definition 9 (Terminal-quantifiedness). *An IADP is terminal-quantified iff each element e of a set-like input part $\mathcal{E} \in \mathcal{C}$ which is quantified over can be mapped by a function $f_{\hat{A}} \colon \mathcal{E} \to 2^{\hat{A}}$ to the set of all terminals which can only be generated if $e \in \mathcal{E}$.*

First we use the INVALIDATE(.) procedure to propagate changes of conditions/constraints due to the modification of the set-like input part $\mathcal{E}$. Afterwards we use the bottom-up evaluation approach to recompute the missing values. Restarting the bottom-up evaluation algorithm ensures that only the necessary parts of the IADP are recomputed. The procedure stops as soon as no more updates have been performed.

4   Refining Partial Invalidation

In many IADP models the two procedures described in Sect. 3 act in an extremely greedy way. Depending on the modified input part large portions of the computed derivation DAGs may be invalidated. We therefore look for special cases in IADPs and input modifications which allow us to restrict the set of invalidated table cells as well as determining which parts of the program actually need recomputation.

We try to characterise the effects of input modifications on the affected table cells. Let $X_{\check{\mathcal{I}}}$ be an indexed symbol and $\bar{X}[\check{\mathcal{I}}]$ its associated table cell. Next we define the following sets.

Definition 10 (Feasibility Set). *The feasibility set of an indexed symbol $X_{\check{\mathcal{I}}}$ are the generated productions $\hat{\mathcal{P}}(X_{\check{\mathcal{I}}})$ whose conditions/constraints are satisfied.*

Definition 11 (Optimality Set and Optimal Subproblems). *The optimality set of an indexed symbol $X_{\check{\mathcal{I}}}$ are those productions $\hat{\mathcal{P}}_{\mathrm{opt}}(X_{\check{\mathcal{I}}})$ whose target values are stored in $\bar{X}[\check{\mathcal{I}}]$. An optimal subproblem is any symbol $Y_{\check{\mathcal{J}}} \in \gamma$ with some $(X_{\check{\mathcal{I}}} \to \gamma) \in \hat{\mathcal{P}}_{\mathrm{opt}}(X_{\check{\mathcal{I}}})$. As a shorthand we write $Y_{\check{\mathcal{J}}} \in \hat{\mathcal{P}}_{\mathrm{opt}}(X_{\check{\mathcal{I}}})$.*

When modifying the input, we can supply additional change-relevant information in the form of *facts* for each indexed symbol $X_{\check{\mathcal{I}}}$, e.g., once the IADP has been solved we can add the fact *optimal subproblem* to all symbols satisfying the definition, in another case the user may supply explicit facts like stating that the occurring input change may only remove elements from the optimality sets. Based on those facts we express the improved invalidation procedures as logic programs, i.e., as inference rules in the style of Prolog or DLV. Inference rules may use the `affected(X)` and `touched(X,Y)` predicates which state that X is affected by the input modification and that X touched Y.

The rules are of the form as shown in (9). The **heads** are inferred iff all **bodies** are known hold and no **weak-negation** is known hold. As a special case procedural ACTIONS can be triggered in the same way.

$$\text{head}_1(X_1,\ldots),\ldots,\text{ACTION}_1(X_1,\ldots),\ldots :\text{-} \text{body}_1(X_1,\ldots), \ldots,$$
$$\textbf{not}\,\text{weak-negation}_1(Z_1,\ldots), \quad (9)$$
$$\ldots$$

We analyze the actions that should be taken to invalidate or update the information stored for an indexed symbol based on different categories of facts. First, the feasibility sets of symbols may either be known to stay the same, potentially be reduced, i.e., child symbols dropping from the set, or potentially be extended, i.e., child symbols entering the set. We use the predicates `feas-may-drop(X,Y)`, `feas-may-enter(X,Y)`, as well as their strong negations `feas-may-not-drop(X,Y)`, `feas-may-not-enter(X,Y)`. Similarly, the optimality sets of symbols may either stay the same, be potentially reduced, or extended. The predicates `opt-may-drop(X,Y)`, `opt-may-enter(X,Y)`, and their counterparts `opt-may-not-drop(X,Y)`, `opt-may-not-enter(X,Y)` indicate that a production containing Y on the right-hand side may (not) be dropped from or enter the optimality set. Third, `opt-sub(X,Y)` states that X has an optimal subproblem Y.

$$\text{MAYDROP}(X,Y) \text{:- opt-may-drop}(X,Y), \text{opt-sub}(X,Y),$$
$$\text{affected}(Y). \tag{10}$$
$$\text{opt-may-drop}(X,Y) \text{:- feas-may-drop}(X,Y), \text{opt-sub}(X,Y),$$
$$\textbf{not } \text{opt-may-not-drop}(X,Y). \tag{11}$$

If we labelled a symbol X with opt-may-drop and we know that the child Y causing the change is an optimal subproblem, then we only have to inspect the target values computed from the optimal subproblem. If the table cell is modified then the change needs to be propagated to the symbols touching X. For this behaviour the change to the feasibility set is not relevant as in the worst case the optimal subproblem may drop from the set which affects the table cell equivalently.

---

function MAYDROP(X, Y)
    Remove the back-pointers to Y from all target values in $\bar{X}$
    Remove target values $\mathcal{U}$ without back-pointers from $\bar{X}$
    if $\mathcal{U} \neq \emptyset$ **then**
        Mark affected(X).
        Add rule opt-may-drop(Z,X):-touched(Z,X),$\mathcal{U} \not\subseteq \bar{X}$.
    end if
end function

On the other hand if it cannot be inferred that Y is an optimal subproblem of X then no changes to the optimality set of X can occur. We indicate a case without action once in (12) though we refrain from stating such cases for the remainder of the work.

$$\text{NOACTION:- opt-may-drop}(X,Y), \text{affected}(Y), \textbf{not } \text{opt-sub}(X,Y). \tag{12}$$

The open cases where opt-sub(X,Y) is known, but opt-may-drop(X,Y) cannot be inferred are expressed by (13) and (14). Target values derived from Y may change in this case and need to be recomputed. Depending on the change to the feasibility set of X might be known to change or not.

$$\text{UPDATEOPT}(X,Y,\bot) \text{:- opt-sub}(X,Y), \text{affected}(Y),$$
$$\textbf{not } \text{opt-may-drop}(X,Y), \tag{13}$$
$$\textbf{not } \text{feas-may-drop}(X,Y).$$
$$\text{UPDATEOPT}(X,Y,\top) \text{:- opt-sub}(X,Y), \text{affected}(Y),$$
$$\text{feas-may-drop}(X,Y) \tag{14}$$
$$\textbf{not } \text{opt-may-drop}(X,Y).$$

Independent of the knowledge that Y is an optimal subproblem of X or not, other productions containing Y may enter the optimality set. Rule (15) captures this eventuality.

---

function UPDATEOPT(X, Y, *check*)

    Compute $\mathcal{U} = \left\{ \sigma(\gamma) \mid (X \to \gamma) \in \hat{\mathcal{P}}_{\mathrm{opt}}(X), Y \in \gamma, \text{if } \textit{check} \text{ then } \text{CHECKFEAS}(\gamma) \right\}$

    Update $\bar{X}$ with $\mathcal{U}$

    if $\exists u \in \mathcal{U} : u \in \bar{X}$ **then**

        Mark **affected**(X).

        Try to infer **opt-may-enter**(Z, X) for all **touched**(Z, X).

        Try to infer **opt-may-drop**(Z, X) for all **touched**(Z, X).

    else

        MAYDROP(X, Y)

    end if

end function

$$\text{MAYENTER}(X, Y) \text{:- } \textbf{opt-may-enter}(X, Y), \textbf{affected}(Y). \tag{15}$$

$$\textbf{opt-may-enter}(X, Y) \text{:- } \textbf{feas-may-enter}(X, Y), \tag{16}$$
$$\textbf{not opt-may-not-enter(X, Y)}.$$

---

function MAYENTER(X, Y)

    Compute $\mathcal{U} = \left\{ \sigma(\gamma) \mid (X \to \gamma) \notin \hat{\mathcal{P}}_{\mathrm{opt}}(X), Y \in \gamma, \text{CHECKFEAS}(\gamma) \right\}$

    Update $\bar{X}$ with $\mathcal{U}$

    if $\exists u \in \mathcal{U} : u \in \bar{X}$ **then**

        Mark **affected**(X).

        Try to infer **opt-may-enter**(Z, X) for all **touched**(Z, X).

        Try to infer **opt-may-drop**(Z, X) for all **touched**(Z, X).

    end if

end function

Once the refined invalidation procedure has finished, we have to check if symbols exist that have been labelled as **affected** but no action could be derived—including NOACTION. In such a case we fall back to the naive INVALIDATION procedure for those symbols to guarantee that no invalid information is retained in the DP tables. Afterwards, for each symbol which has been marked as **uncomputed** or **incomplete** we have to invoke either the top-down or bottom-up evaluation procedure to reevaluate the IADP.

Finally, we have to handle the case where new terminals can be generated. As no touch information is available for those symbols we have to resort to the bottom-up evaluation procedure as in the naive partial invalidation algorithm to handle those. Nevertheless, as potentially less information has been invalidated the bottom-up procedure is expected to complete the DP tables faster.

Besides the rules given in this section the user of an IADP framework may add additional inference rules to enhance the procedure further.

5 Conclusion

In this work we presented the theory for a new algorithm for invalidating Dynamic Programming tables of computed DPs in the face of changing inputs. The *partial invalidation* algorithm operates on dynamic programs modelled in the IADP framework which is also suitable for general purpose solver development. With the help of a logic programming framework the algorithm tries to infer special cases for the invalidation of table cells and tries to abort the procedure as soon as possible. The logic program powering the refined partial invalidation algorithm can be easily extended by a framework user by adding new rules and facts.

By using the algorithms we expect to make the embedding of DPs into other methods like metaheuristics more efficient. Using DPs as solution decoders in heuristics is a common approach to improve the quality of the solutions. Though applying modifications, e.g., local search moves, to the solutions requires a redecoding. By applying the described algorithms this often repeated effort might be reduced, thereby speeding up the overall optimization process.

In future work we will integrate both the simple and the refined algorithm variant into an IADP framework which is currently under development. Experiments over several problem types still need to be conducted to determine the practical effectiveness of the approach. Further, new generic inference rules are of interest which may be added if the IADP under consideration sastifies specific requirements, e.g., monotonic evaluation algebras, or monotonic constraints.

References

1. Bellman, R.: Dynamic Programming. Princeton University Press, Princeton (1957)
2. Giegerich, R., Meyer, C.: Algebraic Dynamic Programming. In: Kirchner, H., Ringeissen, C. (eds.) AMAST 2002. LNCS, vol. 2422, pp. 349–364. Springer, Heidelberg (2002). https://doi.org/10.1007/3-540-45719-4_24
3. Giegerich, R., Meyer, C., Steffen, P.: A discipline of dynamic programming over sequence data. Sci. Comput. Program. **51**(3), 215–263 (2004)
4. Sauthoff, G., Janssen, S., Giegerich, R.: Bellman's GAP: a declarative language for dynamic programming. In: Proceedings of the 13th International ACM SIGPLAN Symposium on Principles and Practices of Declarative Programming, pp. 29–40. ACM (2011)
5. Sauthoff, G., Möhl, M., Janssen, S., Giegerich, R.: Bellman's GAP–a language and compiler for dynamic programming in sequence analysis. Bioinformatics **29**(5), 551–560 (2013)
6. Algebraic dynamic programming for multiple context-free grammars: Riechert, M., Höner zu Siederdissen, C., Stadler, P.F. Theoret. Comput. Sci. **639**, 91–109 (2016)
7. Höner zu Siederdissen, C., Prohaska, S.J., Stadler, P.F.: Dynamic Programming for Set Data Types. In: Campos, S. (ed.) BSB 2014. LNCS, vol. 8826, pp. 57–64. Springer, Cham (2014). https://doi.org/10.1007/978-3-319-12418-6_8
8. Prohaska, S.J., Stadler, P.F.: Algebraic dynamic programming over general data structures. BMC Bioinform. **16**(19), 1–13 (2015)

9. Prins, C., Labadi, N., Reghioui, M.: Tour splitting algorithms for vehicle routing problems. Int. J. Prod. Res. **47**(2), 507–535 (2009)
10. Bacher, C., Raidl, G.R.: Extending algebraic dynamic programming for modelling and solving combinatorial optimization problems. Technical report, Algorithms and Complexity Group, TU Wien, Vienna, Austria (2017). in Preparation)
11. Sauthoff, G.: Bellman's GAP: A 2nd Generation Language and System for Algebraic Dynamic Programming. Ph.D. thesis, Bielefeld University (2010)

Subject Recognition Using Wrist-Worn Triaxial Accelerometer Data

Stefano Mauceri[1,2]([✉]), Louis Smith[1,2], James Sweeney[1,2],
and James McDermott[1,2]

[1] Natural Computing Research and Applications Group, School of Business,
University College Dublin, Dublin, Ireland
stefano.mauceri@ucdconnect.ie, {james.sweeney,james.mcdermott2}@ucd.ie
[2] ICON Plc, Dublin, Ireland
louis.smith@iconplc.com
http://ncra.ucd.ie, http://www.iconplc.com

Abstract. This study demonstrates how a subject can be identified by
the means of accelerometer data generated through wrist-worn devices in
the context of clinical trials where data integrity is of utmost importance.
A custom vector of features extracted from the daily accelerometer time
series is defined. Feature selection is adapted to take account of the
sequential structure in features. Several classifiers are compared within
three different learning frameworks: binary, multi-class and one-class.
A simple algorithm like logistic regression shows excellent performance
in the binary and multi-class frameworks.

Keywords: Accelerometer data · Anomaly detection · Classification
Clinical trials

1 Introduction

Clinical trials are an essential research tool to progress medical knowledge, drug
development and patient care. In this regard, data from wrist-worn accelerometers could play a key role in monitoring the efficacy of treatment options on
movement disorders or the impact of drugs on subjects' free living activity levels
[7]. However, fraud or other misconduct are a well known issue in clinical trials
[5]. For these reasons, an important task is to confirm that a given device is worn
only by the intended subject for the whole trial period.

An example is perhaps the best way to clarify the study aim. Researchers are
designing a clinical trial which involves the enrolment of a subject. The plan is
to provide the subject with a device to assess his/her physical activity without
requiring frequent visits to the research site. How can they be sure that the
subject doesn't give the device to someone else during the trial periods and so
doing invalidates the integrity of the collected data? The first step is to ask the
subject to spend some time at the research centre so that some accelerometer
data can be gathered. Afterwards, this reference data can be used to develop a
machine learning model to confirm subject's identity.

© Springer International Publishing AG 2018
G. Nicosia et al. (Eds.): MOD 2017, LNCS 10710, pp. 574–585, 2018.
https://doi.org/10.1007/978-3-319-72926-8_48

A possibility would be to make use of a *supervised binary classification method* to distinguish those days which belong to the intended subject from those which don't. Alternatively the same data could be employed to implement a *supervised multi-class classification method*. In this case, a group of subjects would be involved simultaneously and the aim would be to recognise each single subject. However, both binary and multi-class methods assume, in a sense, a fixed population of "others", well-represented in the data, which is not realistic. Therefore the best option may be a *semi-supervised one-class classification method* where learning is focused only on the intended subject.

The purpose of this work is to investigate the most appropriate learning framework, the most effective classification algorithm and the optimal set of features to perform subject recognition from accelerometer data. Feature selection is conceived to improve interpretability.

The paper proceeds as follows: Sect. 2 mentions some related studies; the dataset in use, the set of features and the performance metrics employed are described in Sect. 3. Section 4 describes the experimental design, the hyperparameter tuning and the feature selection strategies. Results are examined in Sect. 5 while study conclusions are summarised in Sect. 6.

2   Related Work

Wearable devices are seen as an inexpensive and unobtrusive diagnostic tool therefore they are gaining a primary role in the field of medical research especially in sleep medicine [1,12]. In recent years their employment in motion related studies, especially outside the laboratory environment, is increasing. Two central applications are the assessment of physical activity [14], and the study of movement disorders [13]. Wearable devices can embed a variety of sensors able to track human activity information. Of these, only triaxial acceleration is considered in the present study.

In biometric recognition a heterogeneous set of human characteristics such as, but not only: facial features [16], gait [15], and voice [10] are exploited to identify individuals. Of these, facial features and voice are of little relevance in the present study, but gait is of some relevance. In common with this study, one of the main approaches in the field involves the analysis of periodic signals gathered from dedicated accelerometers. Gait is periodic and in the case of human beings a gait cycle is said to be two steps and its frequency is around 1 Hz [3]. A gait cycle is the object of many gait recognition approaches.

The proposed approach takes into account only acceleration signals. Unlike the aforementioned biometric applications for gait recognition, here the focus is on a time window of 24 h and one data point per minute. To the best of the authors' knowledge this represents a novel application of accelerometer data.

Of the various approaches to the classification problem the binary and multi-class ones are well known. Some classification algorithms allow to use more than two classes however others don't. These can be turned into multi-class algorithms by constructing n binary classification problems, one for each class. In contrast, the one-class classification framework is less common. In this case learning is driven only by positive examples (accelerometer time-series from the intended subject in this application) while negative examples (accelerometer time-series from someone else) are either not present or not properly sampled [8]. Recently, one-class learning techniques are gaining increasing attention. Such approaches are valuable, for example, in detecting anomalies in security-related tasks achieving good performance in terms of accuracy and computational cost [2].

3   Wrist-Worn Triaxial Accelerometer Data

First this section introduces the dataset in use (Sect. 3.1), the data preparation steps (Sect. 3.2) and the feature set (Sect. 3.3). Continuing, the classification setup is described: training, validation and test sets (Sect. 3.4), data transformation (Sect. 3.5), class labels and performance metrics (Sect. 3.6).

3.1   Dataset Description

Data are collected through wrist-worn accelerometers. The type of devices in use are ActiGraph (Pensacola, Florida), model GT9X Link. Each unit is equipped with a triaxial accelerometer able to measure linear acceleration within a range of ± 16 g per axis and capable of data recording at up to 100 Hz[1].

The raw data consists of date-time tags and the acceleration recorded along each of the 3 axis. Using the square root of the sum of the squares of the acceleration along the 3 axis rounded to the closest integer is obtained a new variable named *magnitude*. This variable is employed to create univariate daily time series at the resolution of 1 min. Each daily time series X is transformed to $\log(1+X)$. A visual representation of log transformed magnitude recordings for one typical day is shown in Fig. 1.

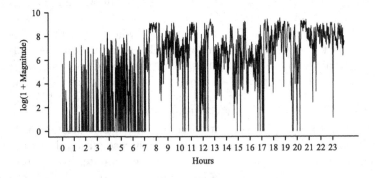

Fig. 1. Magnitude recordings for one day for one subject

---

[1] For further information see: http://www.actigraphcorp.com/.

A sample of 9 volunteers, 3 females and 6 males, wore the mentioned device on their wrist for a period of approximately 45 days. All the participants are adults working from Monday to Friday; they are required to wear the devices every day from Monday to Sunday in free living conditions. All the subjects are office workers based in the same location.

3.2 Data Preparation

Before the data is used certain preparations are needed. Exploratory analysis portray the fact that subjects' behaviour on Saturdays differs from Sundays and both differ from weekdays. Moreover, the number of weekend days (Saturdays and Sundays) available per subject is relatively small. For these reasons, weekend days are removed. Also, days where one or more hours of recordings are missing, are dropped, as this would be in contrast with the present approach.

Statistical analysis of the data show that it's common for a daily cumulative magnitude to assume a minimum value around 1 million but a few days are found in the range from 0 to 500,000. Deepening the analysis of these days is observed a prevalence of non-wear time over wear time. Practically speaking, non-wear time corresponds to a sequence of recordings all with the same magnitude equal to 0. This can happen because subjects take off their devices and leave them lying still somewhere. Therefore, days where the cumulative magnitude is lower than 500,000 are pruned.

Devices in use have an internal memory of 4 GB and this requires weekly transfer of data to a larger hard drive. This time is also used to re-charge devices' batteries. As a consequence, there are a few days per subject where can be found a series of missing values. This issue is addressed scanning through each daily time series and when a sequence of contiguous missing values of size N (with N ≤ 60) is found it's filled with a copy of the N values that precede.

3.3 Feature Set

A set of 25 features, listed below, is manually constructed by drawing on the statistical and time series analysis literature [4]. The feature set includes: mean, standard deviation, mode frequency, median, variance, kurtosis, skewness, minimum, maximum, number of outliers [6], number of observations above mean, number of observations below mean, number of observations greater than zero, average of the absolute value of consecutive changes in the series, autocorrelation with lags of 1, 2, and 3 min, the 5 coefficients of an ARIMA model with parameters (2, 1, 2), time reversal asymmetry statistic [4], histogram based estimation of entropy in samples of 6 min, number of peaks in segments of ±6 min. In the last two mentioned meta-features the interval length of 6 min is chosen simply because it allows to divide one hour in ten equally sized segments. Computation of entropy, autocorrelation and the parameters of the ARIMA model can result in a 'nan' for those sequences where nearly all the values are equal to zero. In these cases a zero is returned instead.

While almost all the mentioned features are well known, further clarification is required for the number of outliers and the time reversal asymmetry statistic: their formulae are shown below.

$$Outliers(X) = \sum_{i=1}^{n} \left[\frac{0.6745 * (x_i - Med(X))}{Med(|X - Med(X)|)} > 3.5 \right], \quad TRS(X) = \frac{\langle (x_i - x_{i-l})^3 \rangle}{\langle (x_i - x_{i-l})^2 \rangle}$$

Meta-features. The assembly of the feature vector is a fundamental part of this study. For each hour of the day all the features of the feature set are extracted. As a result, 25 vectors of 24 scalars are obtained. Each of these vectors is referred to as a *meta-feature*.

3.4   Training, Validation and Test Sets

In the present dataset the number of available days differs from subject to subject. In order to construct a balanced training and test set the subject with the least number of days (23 days) is taken as reference. From those subjects where more days are available only 23 are randomly selected.

For each subject selected days are randomly split using a ratio of approximately 80% to 20% in training and test instances: 18 of the 23 days are included in the training set, while the remaining 5 are included in the test set. Thus a training set of 162 instances (18 days × 9 subjects) and a test set of 45 instances (5 days × 9 subjects) are obtained.

For feature selection and hyper-parameter tuning cross validation is used with a 2:1 split of the training data.

3.5   Data Transformation

The 600 features have different scales thus column-wise standardisation is applied by removing the mean and scaling to unit variance. Standardisation parameters of a given column (μ and σ) of the training set are retained to apply the same transformation on the corresponding column of the test set. The same is done for each training and validation set generated through cross-validation.

This applies to the binary, multi-class and one-class experiments. In the one-class experiment, it would be common practice to standardise data for one subject at a time. However, it's noticed that classification performance is positively influenced when the entire dataset is standardised at once.

3.6   Class Labels and Performance Metrics

In the multi-class experiment each subject involved in the study is assigned a number from 1 to 9 corresponding to his/her class label. In this case all the classes are represented equally i.e. the test set is made of 5 instances per subject.

In the binary and one-class experiments there are only two class labels: 1 and 0. Subjects are tested one at the time. The intended subject is assigned the label 1 while the label 0 is assigned to all the others. Now, the number of instances in each of the two classes is quite different i.e. 5 true instances vs. 40 false ones.

Results are presented through 4 different metrics: the area under the receiver operating characteristic curve (AUROC) which is insensitive to class imbalance, precision, recall, and specificity.

In general the ROC curve shows how the TPR and the FPR change as the classification threshold changes. ROC curves can be used to compare classifier and make cost vs. benefit decisions. However, in this study the focus isn't on tweaking classifiers thresholds. In the binary and multi-class experiments decision thresholds are internally defined by the algorithms in use and what is observed is a vector of class labels as outcome of the classification task. In the one-class experiment test cases which show a score greater than the 5th percentile of the training scores are considered as not anomaly. The score is calculated as a density or distance measure according to the algorithm in use. The vector of class labels is employed to compute the AUROC using only 3 points $(0, 0)$, (TPR, FPR), $(1, 1)$, and the trapezoidal rule for integration. While the AUROC is regarded as a standard performance evaluation criterion in binary classification problems, its extension to the multi-class case requires precise definition. In the present study it's computed considering one class at a time (one vs. all approach). In other words, for each subject, the TPR and the FPR are estimated and used to work out his/her AUROC. The multi-class AUROC is then the mean over subjects.

4   Experiments

Three sets of experiments are carried out: (1) *binary classification*, (2) *multi-class classification*, and (3) *one-class classification*.

Each experiment is divided in 4 stages. First, classifiers (as listed in Sects. 4.1 and 4.2) with default hyper-parameters are tested exploiting all the available meta-features. Second, classifiers are tested after their hyper-parameters are tuned using a grid search approach and all the available meta-features are employed. Third, classifiers with their default hyper-parameters are tested after for each of them a subset of optimal meta-features is selected using a forward stepwise feature selection algorithm. Lastly, classifiers are tested after their hyper-parameters are tuned and only the optimal subset of meta-features found in the previous stage are employed.

4.1   Binary and Multi-class Classification Algorithms

A set of 6 classifiers is tested for both the binary and multi-class experiments. The algorithms appraised are: *Gaussian Naïve Bayes, K-Nearest Neighbours, Logistic Regression, Multi-Layer Perceptron, Random Forest* and *Support Vector Classification*. Furthermore, two baselines are used: (1) *most frequent*: which always predicts the most frequent label in the training set, (2) *uniform*: which predicts with uniform probability at random among the labels in the training set.

4.2   One-Class Classification Algorithms

In the one-class classification experiment a set of 5 classifiers is tested. The algorithms evaluated are: *Gaussian Mixture, Isolation Forest* [9], *Kernel Density, One-Class Support Vector Machine* and *Single Spherical Gaussian Density*[2]. While no baseline classifiers are directly employed in this experiment for comparisons it's possible to refer to the baseline in the binary case.

4.3   Feature Selection

For each classifier, a forward feature selection algorithm is employed. It differs from a standard one in two ways. First, each feature is repeated 24 times (for 24 h), giving a meta-feature. For interpretability of the final model, it's necessary to either select or remove each meta-feature en masse. Therefore, to select or remove one meta-feature it's necessary to select or remove all 24 associated columns in the training data. Second, because of the small size of the training data, there is high variability among training-validation splits. Therefore, it's necessary to take the average performance over many splits when carrying out feature selection. The feature selection algorithm runs in two stages:

Stage 1: a training-validation split is carried out, then a forward stepwise pass is run, adding at each step the meta-feature which gives the best validation performance. The selection order of the meta-features is recorded. This is repeated 100 times and for each meta-feature its mean selection order is calculated. The meta-features are sorted in increasing mean selection order.

Stage 2: the algorithm iterates over the list of meta-features, adding one at a time to the model and recording its mean validation performance over 100 training-validation splits. It's observed that stopping at the first decrease in validation performance doesn't lead to the optimal stopping-point, so instead the optimal number and selection of meta-features is found by taking the highest validation performance during the full iteration.

The number of features selected per classifier is shown below in Table 1.

Table 1. Feature selection details

Algorithm	Binary	Multi-class	Algorithm	One-class
Gaussian Naïve Bayes	9	10	Gaussian Mixture	1
K-Nearest Neighbours	16	15	Isolation Forest	13
Logistic Regression	22	10	Kernel Density	7
Multi-Layer Perceptron	19	11	One-Class SVM	1
Random Forest	4	4	SSGD	2
SVC	2	6		

[2] All classifiers use Scikit-Learn [11].

4.4   Hyper-parameter Tuning

For each algorithm hyper-parameters are tuned. The list of hyper-parameters searched for each algorithm is shown below. Two exceptions are: *Gaussian Naïve Bayes* and *Single Spherical Gaussian Density* where there are no hyper-parameters which can affect performance. Moreover, in 3 cases the hyper-parameter set found in the implementation in use is altered. In *Kernel Density* the bandwidth is set to 10^5 due to the high dimensional space, for the same reason in *One Class SVM* gamma is set to 10^{-5} ($gamma = \frac{1}{bandwidth}$); in *Gaussian Mixture* the number of components is set equal to 1 due to the limited number of training cases.

The grid search method employed generates all the possible combinations of hyper-parameters. Then, all the combinations are tested and the setting with the best performance is retained.

- *K-Nearest Neighbours*
  - k: {2, 3, 4}
  - leaf_size (leaf size of the k-d tree): {25, 30, 35}
- *Logistic Regression*
  - C (penalty parameter of the error term): {0.01, 0.1, 0.5, 1}
  - fit_intercept (whether to use intercept_scaling or not): {True, False}
  - intercept_scaling (intercept scaling factor): {0.1, 0.4, 0.8}
  - solver (optimiser): {lbfgs, newton-cg, sag}
- *Multi-Layer Perceptron*
  - activation: {identity, logistic, relu, tanh}
  - learning_rate: {adaptive, constant, invscaling}
  - max_iter (maximum number of iterations): {500, 1,000, 1,500}
  - solver (optimiser): {adam, lbfgs, sgd}
- *Random Forest*
  - n_estimators (number of trees in the forest): {50, 100, 150}
  - max_features (features to consider for the split): {auto, $\log_2(N)$, None}
  - min_samples_split (samples required to split a node): {4, 6, 8, 10, 12}
- *Support Vector Classification*
  - C (penalty parameter of the error term): {0.001, 0.01, 0.1, 0.9, 1.5}
  - gamma (inverse of the bandwidth): {0.001, 0.01, 0.05, 0.1}
- *Gaussian Mixture*
  - covariance_type: {diagonal, full, spherical, tied}
  - n_components (number of components): {1, 2, 4, 8}
- *Isolation Forest*
  - n estimators (number of base estimators): {5,000, 10,000, 15,000}
  - contamination (proportion of anomalies in the dataset): {0.05}
- *Kernel Density*
  - bandwidth: {10^{-1}, 10^0, 10^1, 10^2, 10^3, 10^4, 10^5, 10^6}
  - kernel: {cosine, epanechnikov, exponential, gaussian, linear, tophat}
  - metric: {canberra, euclidean, manhattan}
- *One-Class SVM*
  - gamma (inverse of the bandwidth): {10^{-2}, 10^{-4}, 10^{-6}}
  - kernel: {cosine, epanechnikov, exponential, gaussian, linear, tophat}
  - nu: {0.1, 0.5, 0.9}

5   Results and Discussion

Performance achieved in the binary and multi-class classification experiments along with the two baselines are shown in Table 2. For the one-class classification experiment results are summarised in Table 3. In both the tables for each classifier results are listed according to the 4 stages described in Sect. 4.

Table 2. Binary and multi-class classification experimental results – 1. Default hyper-parameters. 2. Grid search on hyper-parameters. 3. Feature selection. 4. Grid search after feature selection.

	Algorithm	Binary classification				Multi-class classification			
		AUROC	Precision	Recall	Specificity	AUROC	Precision	Recall	Specificity
	Most Frequent	50	0	0	100	6	1	11	90
	Uniform	45	9	42	48	55	21	20	90
1	Gaussian Naïve Bayes	75	37	71	78	74	60	53	94
2		–	–	–	–	–	–	–	–
3		74	29	82	66	84	74	71	96
4		–	–	–	–	–	–	–	–
1	K-Nearest Neighbours	77	80	56	99	74	63	64	96
2		82	91	64	99	73	61	62	95
3		77	75	56	98	80	68	64	96
4		81	84	64	98	80	74	64	96
1	Logistic Regression	84	52	78	91	86	77	76	97
2		83	57	73	93	86	77	76	97
3		85	54	78	92	86	77	76	97
4		83	65	71	95	84	71	71	96
1	Multi-Layer Perceptron	82	69	69	96	81	74	67	96
2		83	60	71	94	83	69	69	96
3		78	66	60	96	85	75	73	97
4		86	60	80	93	84	73	71	96
1	Random Forest	55	28	11	100	84	78	71	96
2		57	41	13	100	76	62	58	95
3		64	78	29	100	73	52	51	94
4		66	73	33	99	76	58	58	95
1	SVC	56	22	11	100	80	71	64	96
2		50	0	0	100	74	62	64	96
3		60	50	20	100	83	72	69	96
4		68	78	36	100	81	76	67	96

While all the algorithms demonstrate better performance than baselines, in respect to the learning framework, it cannot be said that one approach is clearly better than the others. The gap between the binary and the multi-class approach is limited for the majority of the classifiers except for *Random Forest* and *SVC* which work out better in the multi-class scenario. Overall, the one-class approach shows slightly lower performance.

Looking at the results per subject, not provided in this text, can be seen how for some subjects the AUROC score is close or equal to 100%.

Table 3. One-class classification experimental results – 1. Default hyper-parameters. 2. Grid search on hyper-parameters. 3. Feature selection. 4. Grid search after feature selection.

	Algorithm	One-class classification			
		AUROC	Precision	Recall	Specificity
1	Gaussian Mixture	50	0	0	100
2		68	34	64	71
3		50	13	7	93
4		62	20	76	49
1	Isolation Forest	58	14	89	27
2		53	12	87	20
3		61	15	93	29
4		61	15	87	34
1	Kernel Density	51	12	87	15
2		78	43	73	82
3		55	12	89	21
4		79	31	89	70
1	One-Class SVM	66	41	49	83
2		74	30	84	63
3		69	41	60	79
4		71	48	56	87
1	SSGD	51	12	87	15
2		–	–	–	–
3		59	15	87	32
4		–	–	–	–

The best results, in terms of AUROC, are gained by the *Logistic Regression* classifier, 85% in the binary experiment and 86% in the multi-class experiment.

The second best classifier is the *Multi-Layer Perceptron*, which peaks at 86% AUROC in the binary case and it's consistent in the multi-class case. While the *Logistic Regression* algorithm is deterministic ceteris paribus over different runs, *Multi-Layer Perceptron*'s results can fluctuate ceteris paribus over different runs.

K-Nearest Neighbours and *Gaussian Naïve Bayes* follow. In the binary experiment *Random Forest* and *SVC* have lower performance compared to the previously mentioned classifiers, however the gap decreases in the multi-class case.

In the one-class experiment only *Kernel Density* and *One-Class SVM* can achieve 79% and 74% AUROC respectively while all the other classifiers have poor performance.

The impact of hyper-parameter tuning is not always positive in the binary and multi-class cases, but sometimes the performance is improved. The choice of hyper-parameters improve performance greatly for the *Kernel Density* and the *One-Class SVM* classifiers in the one-class experiment.

Considering the feature selection strategy results fluctuate. Performance isn't boosted when the optimal sub-set of meta-features is selected. However, for most classifiers the number of meta-features can be remarkably decreased (as shown in Table 1) with a negligible impact on performance and this can allow higher model interpretability. Top 5 most frequently selected features are: autocorrelation (at different lags), mean, average of the absolute value of consecutive changes, median, and histogram based estimation of entropy.

6   Conclusions and Further Work

The present study describes a problem driven research on subject recognition using wrist-worn triaxial accelerometer data. It's shown how accelerometer data gathered over an entire day can outline a unique representation of a subject and therefore allow his/her identification.

Well-established classifiers are tested using a customised vector of meta-features within three different learning frameworks. Overall, the three different learning frameworks seem to achieve similar performance. The binary and multi-class frameworks show slightly better results and this raises the question of how to construct an appropriate training set for each subject. Nonetheless, it would be interesting to study how these results vary if one or more subjects are removed from the training set and employed only in the test set. In this scenario the one-class approach may perform better than the other two.

The hyper-parameters are tuned. The feature set in use and a suitable adaptation of the forward feature selection algorithm allow high model interpretability. Results are promising: a simple algorithm like logistic regression achieve the best score in the binary and multi-class scenarios (86% AUROC).

This work paves the way to the employment of accelerometer data as an unobtrusive and practical tool in medical research. Further investigations are needed to address the reliability of the proposed approach on larger datasets.

Acknowledgements. This research is funded by ICON plc.

References

1. Ancoli-Israel, S., Cole, R., Alessi, C., Chambers, M., Moorcroft, W., Pollak, C.P.: The role of actigraphy in the study of sleep and circadian rhythms. Sleep **26**(3), 342–392 (2003)
2. Cao, V.L., Nicolau, M., McDermott, J.: One-class classification for anomaly detection with kernel density estimation and genetic programming. In: Heywood, M.I., McDermott, J., Castelli, M., Costa, E., Sim, K. (eds.) EuroGP 2016. LNCS, vol. 9594, pp. 3–18. Springer, Cham (2016). https://doi.org/10.1007/978-3-319-30668-1_1

3. Fernandez-Lopez, P., Liu-Jimenez, J., Sanchez-Redondo, C., Sanchez-Reillo, R.: Gait recognition using smartphone. In: 2016 IEEE International Carnahan Conference on Security Technology (ICCST), pp. 1–7. IEEE (2016)

4. Fulcher, B.D., Jones, N.S.: Highly comparative feature-based time-series classification. IEEE Trans. Knowl. Data Eng. **26**(12), 3026–3037 (2014)

5. George, S.L.: Research misconduct and data fraud in clinical trials: prevalence and causal factors. Int. J. Clin. Oncol. **21**(1), 15–21 (2016)

6. Iglewicz, B., Hoaglin, D.C.: How to Detect and Handle Outliers, vol. 16. ASQ Press, Milwaukee (1993)

7. Kelly, L.A., McMillan, D.G., Anderson, A., Fippinger, M., Fillerup, G., Rider, J.: Validity of actigraphs uniaxial and triaxial accelerometers for assessment of physical activity in adults in laboratory conditions. BMC Med. Phys. **13**(1), 5 (2013)

8. Khan, S.S., Madden, M.G.: A survey of recent trends in one class classification. In: Coyle, L., Freyne, J. (eds.) AICS 2009. LNCS (LNAI), vol. 6206, pp. 188–197. Springer, Heidelberg (2010). https://doi.org/10.1007/978-3-642-17080-5_21

9. Liu, F.T., Ting, K.M., Zhou, Z.H.: Isolation forest. In: Eighth IEEE International Conference on Data Mining, ICDM 2008, pp. 413–422. IEEE (2008)

10. Loughran, R., Agapitos, A., Kattan, A., Brabazon, A., O'Neill, M.: Speaker verification on unbalanced data with genetic programming. In: Squillero, G., Burelli, P. (eds.) EvoApplications 2016. LNCS, vol. 9597, pp. 737–753. Springer, Cham (2016). https://doi.org/10.1007/978-3-319-31204-0_47

11. Pedregosa, F., Varoquaux, G., Gramfort, A., Michel, V., Thirion, B., Grisel, O., Blondel, M., Prettenhofer, P., Weiss, R., Dubourg, V., Vanderplas, J., Passos, A., Cournapeau, D., Brucher, M., Perrot, M., Duchesnay, E.: Scikit-learn: machine learning in Python. J. Mach. Learn. Res. **12**, 2825–2830 (2011)

12. Rajna, P., Szomszed, A.: Actigraphy: a valuable diagnostic tool or a luxury investigation? (Neuropsychiatric aspects). Ideggyogy. Sz. **62**(9–10), 308–316 (2009)

13. Teskey, W.J., Elhabiby, M., El-Sheimy, N.: Inertial sensing to determine movement disorder motion present before and after treatment. Sensors **12**(3), 3512–3527 (2012)

14. Trost, S.G., McIver, K.L., Pate, R.R.: Conducting accelerometer-based activity assessments in field-based research. Med. Sci. Sports Exerc. **37**(11), S531 (2005)

15. Wang, L., Ning, H., Tan, T., Hu, W.: Fusion of static and dynamic body biometrics for gait recognition. IEEE Trans. Circuits Syst. Video Technol. **14**(2), 149–158 (2004)

16. Zhao, W., Chellappa, R., Phillips, P.J., Rosenfeld, A.: Face recognition: a literature survey. ACM Comput. Surv. (CSUR) **35**(4), 399–458 (2003)

Detection of Age-Related Changes in Networks of B Cells by Multivariate Time-Series Analysis

Alberto Castellini$^{(\boxtimes)}$ and Giuditta Franco

Department of Computer Science, Verona University,
Strada Le Grazie 15, 37134 Verona, Italy
{alberto.castellini,giuditta.franco}@univr.it

Abstract. Immunosenescence concerns the gradual deterioration of the immune system due to aging. Recent advances in cellular phenotyping have enabled key improvements in this context during the last decades. In this work we present a novel extensions and integration of data-driven models for describing age-related changes in the network of relationships among cell quantities of eight peripheral B lymphocyte subpopulations. Our dataset contains about six thousands samples of patients having an age between one day and ninety-six years, where for each patient, cell quantities of eight peripheral B lymphocyte subpopulations were measured. By correlation-based multiple time series segmentation we generate four sets of age-related networks depending on the number of age segments. We first analyze a partition in 30 very short segments, then segmentations in 5, 3 and 2 segments. Moving from a fine to a large grain segmentation, different aspects of the dataset are highlighted and analyzed.

1 Introduction

In last decades main mechanisms of the human immune system, one of the most complex and adaptive systems known in nature, were investigated from different perspectives. In this context, aging gathered much attention as a complex process which negatively impacts this system [11,27]. In fact, it may be assumed that the immune system, as a network of interacting cells, evolves during human life in terms of different aspects. Presence/absence, strength and types of interactions, for instance, can change during the life of a person due to the exposure to multiple foreign challenges through childhood, via young and mature adulthood, to the decline of old age [23].

The immune network theory, formulated by Jerne [15] in 1974 and subsequently developed by Parelson [22], was a starting point to describe the dynamics of lymphocyte interactions from a quantitative and systemic point of view [21]. More recently, advances in cellular phenotyping have enabled to elucidate several functioning mechanisms of the immune system, such as those underling immunosenescence [7,13], having a notable social and economical impact in the design of new therapies and vaccines. Latest studies showed that the majority of

© Springer International Publishing AG 2018
G. Nicosia et al. (Eds.): MOD 2017, LNCS 10710, pp. 586–597, 2018.
https://doi.org/10.1007/978-3-319-72926-8_49

lymphocyte biological variability seems to be age-dependent [26] and immunosenescence seems to be characterized by a decrease in cell-mediated immune functions, where defects in T- and B-cell functions coexist [27].

In this work we proposes a pairwise correlation based analysis over a dataset of B-cell quantities, measured in about six thousands patients having an age between one day and ninety-six years. For each patient, cell quantities of eight peripheral B lymphocyte subpopulations were measured and related time-series, obtained by ordering patients according to their age, analyzed. Pearson correlation was computed to find out age-related relationships between different B-cell subpopulations. This kind of analysis allowed us to visualize our data as (correlation based) networks over the available types of B-cells. Differences in such age-related networks were investigated in terms of a time-series segmentation problem, where a partition (in intervals) of the time line (i.e., patient age) emerges as an optimal one to maximize a measure of dissimilarity between corresponding B-cell networks (see Fig. 1).

Fig. 1. Age-related changes in immune system network.

Segmentation of multiple time-series is a complex problem, since different data partitioning may show different aspects of underlying processes and these aspects could have non-synchronous evidence. Main methodologies in the field [16,24] take inspiration and extend methods of motif discovery in time-series [6,19] or are based on clustering [8,25]. The choice of the information measure of course has a strong influence on the identification of segments (i.e., time intervals/clusters of time points) and change points. A possible measure is represented by the parameters of the (multivariate) mathematical models fitting the data in each segment, since they represent some aspects of the information in the segment itself. Comparing these parameters between couples of adjacent segments and maximizing their differences is a way to identify good segmentations. If linear regression models are used, then predictor coefficients are compared. In [4] a constant (to all age intervals) network was provided by setting general assumptions, while an age-dependent one was found by restricting statistical thresholds to validate our multivariate linear models. A previous model proposed for this dataset [2] describes a possible sequence of (ex-vivo observed) B cell maturation steps in human body. It was based on Metabolic P systems [9,10,20], with linear regulation maps generated by regression techniques and genetic algorithms [3,5].

The main contribution of this work concerns the generation and analysis of four sets of age-related networks depending on the number of age segments. We first analyze a partition in 30 very short segments, then we analyze a segmentation in 5, 3 and 2 segments, where the age-intervals of corresponding segments increase while the number of segments decrease. Moving from a fine to a large grain segmentation, different aspects of the dataset are highlighted. We used a brute-force algorithm for testing every possible segmentation with specific number of segments, and evaluated these segmentations according to a segmentation performance measure based on the average correlation difference between segments. A preliminary comparison between these networks and those computed in [4] is provided.

The rest of the paper is organized as follows. The dataset and algorithm are described in Sect. 2. A discussion on initial results is presented in Sect. 3 and Sect. 4 reports some conclusions and proposals for future work.

2   Material and Methods

This section describes the immunological dataset and the segmentation method used to generate age-related networks of B cell subpopulations.

2.1   Dataset

Data were collected at the University Hospital of Verona (Italy) from 2001 to 2012 as measures of amount of B cells exhibiting the combinations of receptor clusters *CD27*, *CD23* and *CD5* in 5,954 patients. There were 2,910 males and 3,045 females (male/female ratio: 0.95) and the median age of the patients was 37 years (range: 0–95 years). More details on the dataset and the clinical method used to collect it may be found in [2,4,26]. The names of population size variables corresponding to each cell phenotype are displayed in Table 1. In other terms, B cell phenotype of 8 subpopulations (indicated by presence and absence of three receptor clusters), may be abstractly described by random variables accounting for quantities of corresponding cell in each patient.

The dataset is a matrix of 5,954 rows and eight columns, in which rows (i.e., patients) can be sorted by age, obtaining a kind of multivariate time-series where patient age represents time. Given the definition of our problem (see previous section), this is a particular case where it is reasonable to reduce cross-sectional data into multivariate time-series. In fact, if we sort the data according to the age of patients, we have a screenshot of the human immune system (or, more specifically, of the B-cell network) along the lifetime of a metapatient, who may

Table 1. Dataset variables.

X_1 = CD5+ CD23+ CD27-	X_5 = CD5- CD23- CD27+
X_2 = CD5- CD23+ CD27-	X_6 = CD5+ CD23- CD27+
X_3 = CD5- CD23- CD27-	X_7 = CD5+ CD23+ CD27+
X_4 = CD5+ CD23- CD27-	X_8 = CD5- CD23+ CD27+

be assumed to have a basic functioning system (the number of patients is high enough, to be able to neglect possible known or unknown diseases, defeacts on the system).

2.2   Algorithm

The segmentation algorithm here used is a brute-force partitioning method, based on the maximization of differences in correlation between adjacent segments. The multiple time-series of patients (sorted by age) is initially split in 30 *primitive segments*, each containing 200 patients who result in being of very similar age. We assume that the time-series is stationary in each primitive segment because the age effect is irrelevant in such small intervals.

Time-series segmentations are generated by partitioning the vector of primitive segments $(1, \ldots, 30)$ in n segments $s_1, \ldots, s_n$, such that $2 \leq n \leq 30$. *Segment* $s_i = (k_i, k_{i+1})$, $1 \leq k_i < k_{i+1} \leq 30$, $i = 1, \ldots, n$ contains all patients in primitive segments between k_i and k_{i+1}. For instance, segment $(2, 4)$ contains all primitive segments from 2 to 4 (namely, patients from 200 to 799 in the age-sorted list of patients). A *segmentation* having n segments is then represented by an n-uple $S^n = (k_1, \ldots, k_n)$, where elements $1 \leq k_1 < \ldots < k_n = 30$ are the indexes of the primitive segments that delimit the end of each segment in the segmentation. For instance, segmentation $(3, 7, 20, 30)$ contains four segments, namely $(1, 3)$, $(4, 7)$, $(8, 20)$ and $(21, 30)$, hence patients are segmented as $(1, 599)$, $(600, 1399)$, $(1400, 1999)$ and $(2000, 5954)$.

Given a segment s_i, we indicate by $c_i = (c_{i,1}, \ldots, c_{i,28})$ the vector containing the correlation of each couple of different cells in the age interval of segment s_i. The number of elements in c_i is 28 because there are 28 possible pairs of different cell types in our dataset. We compute the vector of absolute differences of correlation between two adjacent segments s_i and s_{i+1} as $d_{i,i+1} = (d_{i,i+1}^1, \ldots, d_{i,i+1}^{28}) = |c_{i+1} - c_i|$. Let us analyze an example from Fig. 2 which shows the (unique) segmentation in 30 primitive segments. The first element of c_1 is the correlation between X_1 and X_7 in the first age segment (i.e., 0.0–1.0 years), which has a value of 0.49 (see first column of matrix *(a)*). The first element of c_2 is the correlation between X_1 and X_7 in the second age segment (i.e., 1.0–2.5 years), which has a value of 0.76 (see second column of matrix *(a)*). The absolute difference of these two values is the value of the first element of vector $d_{1,2}$, namely 0.27 (see first column of matrix *(b)*).

Given a specific number of segments n, our goal is to identify the segmentation $\hat{S}_n = (\hat{k}_1, \ldots, \hat{k}_n)$ which maximizes the overall absolute differences of correlation between adjacent segments. To this end we define the performance measure of a generic segmentation S_n with n segments as:

$$m(S_n) = \frac{\sum_{i=1}^{n-1}(\sum_{j=1}^{28}(d_{i,i+1}^j(S_n)))}{28 \cdot (n-1)} \tag{1}$$

which represents the average difference in correlation between all adjacent segments in S_n and depends on the specific segmentation points $k_1, \ldots, k_n$ in which the multiple time-series is partitioned. The algorithm described in Table 2 aims at identifying the segmentation S_n that maximizes $m(S_n)$.

Table 2. Brute-force algorithm for generating of the best segmentation with n segments according to the segmentation performance measure in Eq. (1).

Best_segmentation(n)

Input: n: number of segments

```
1.  # Initialization
2.  bestSegmentation=NULL;
3.  bestPerformance=-Inf;
4.  # Search for best segmentation
5.  for each segmentation S_n^h = (k_1,...,k_n) | 1 ≤ k_1 < ... < k_n = 30 {
6.      compute segmentation performance p = m(S_n^h);
7.      if(p>bestPerformance) {
8.          bestSegmentation=S_n^h;
9.          bestPerformance=p;
10.     }
11. }
12. return (bestSegmentation,bestPerformance);
```

The advantage of using this algorithm is that all possible partitioning of the 30 primitive segments in n parts is tested. On the other hand, the number of these partitions is the binomial coefficient $\binom{29}{n}$ which grows very quickly as n increases or decreases to $n/2$. Table 3 shows the number of possible segmentations depending on n and the time needed by the brute-force algorithm to find the best segmentation. The algorithm was implemented in R language and it run on a laptop with processor Intel® QuadCore™ $i7$-$3537U$ 2.00 GHz and 8 GB of RAM.

Table 3. Algorithm performance depending on the number of segments n.

n	# segmentations	Time
2	29	0.15 s
3	406	1.95 s
4	3654	16.83 s
5	23751	1.96 min
6	118755	11.52 min
7	475020	1.15 h
30	1	0.01 s

3   Results

We started to analyze the basic segmentation in 30 intervals (maximal granularity) having 200 patients each. Correlation analysis on this case are reported in Fig. 2, where we may see: the pairwise correlation 28×30 matrix (a), the absolute value of differences of correlations between two consecutive segments reported in a 28×29 matrix (b), and the sorted row-wise average vector (c). Two examples of correlation values (computed on the 30 segments) are graphed in (e): those between X_1 and X_7, and between X_2 and X_5, where the dotted horizontal line denotes the 0.5 threshold.

Only absolute values are considered, since we are interested in selecting couple of relatively high correlated variables (rather than in the verse they are correlated). Matrix in (a) is filtered across the threshold value 0.5, so to obtain the bicolor matrix in (d), which corresponds to a sequence of 30 networks over the variables. Four of these networks, corresponding to the segments 1, 2, 12, 29, are reported in (f), where we may notice a different presence of edges (i.e., different couples of highly correlated variables) for different age intervals, and a decrease of presence of edges with the age increasing (segment 12 corresponds to the range 25–29 years old and segment 29 to the range 73–78 years, see Fig. 2). The 30-partition has the advantage to group patients having very similar age, then detecting actual relationships between variables which cannot be seen macroscopically from the time-series, which are due to data-driven age-independent structural properties of the B-cell network. However, we aim at finding a partition with a minor number of segments, having better performance in terms of dissimilarity between corresponding networks, but still keeping the property to exhibit an age-independent network which fits actual relationships among variables over the intervals.

We run the brute force segmentation algorithm, to analyze performances of first n-partitions with $n = 2, \ldots, 7$, which results are reported in Figs. 3 and 4. Performance measure of all possible partitions with 2,3,...,7 segments was computed, and best values are respectively reported in Fig. 3(a), where the maximum (m = 0.31) is obtained for a tripartition. It generates one B cell network for patients up to one year old, another one for the interval 1–34.8 years, and the last

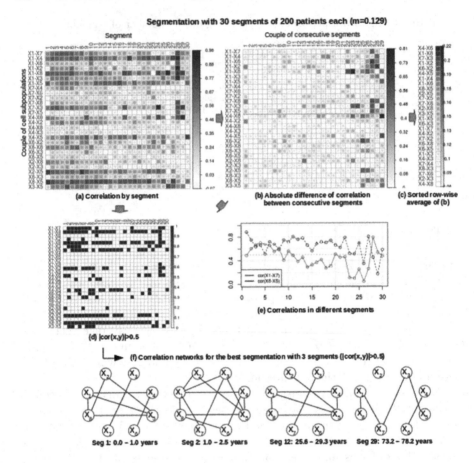

Fig. 2. Initial partition of 30 intervals, each with 200 patients. Correlation matrix (a), correlation differences between consecutive segments in absolute value (b), corresponding sorted row average vector (c) having components average 0.129. Pairwise correlations greater than 0.5 (in absolute value) in (d), reported in (f) as networks for a specific sample of segments: 1, 2, 12, 29. Absolute values of pairwise correlations of (X_1, X_7) and (X_2, X_5) (e).

one for ages over 34.8. One only correlation is kept for the whole life, between X_1 and X_4, two phenotypes which differ only for the expression of CD23, as it was also the case for linear models proposed in [4].

After one year of age, B cell network has many more edges (that is, it has new variable couples highly correlated), and there are several edges which appear only in the range of age [1–34.8] as it is evident by observing the central column of matrix in (c), which is almost all dark (high correlation) with row neighbours white (scarse correlation). However the network of the second segment is not an enrichment of the previous one, because a couple of correlations, between X_4, X_6 and between X_2, X_8, get lost (as they decrease under 0.5 in absolute value). Both

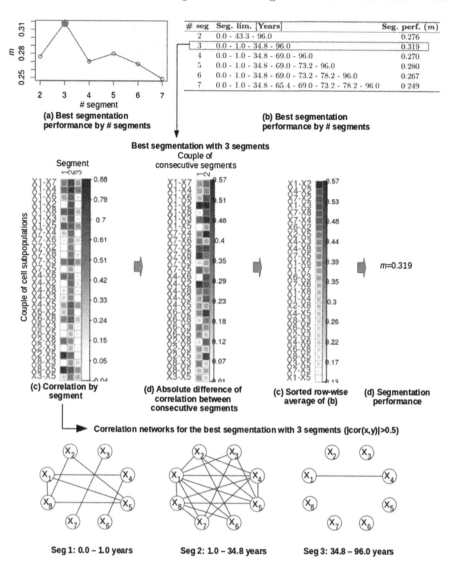

Fig. 3. Tripartition corresponding to maximal segmentation performance ($m = 0.31$). Best values for segmentation performance m over partitions with 2, 3, 4, 5, 6, 7 segments in (a), corresponding m values and ages delimiting the segments in (b). On the tripartition with m maximum: pairwise correlations (c) generating networks on the bottom with a filter of 0.5, correlations differences in abs value (d), and corresponding sorted row-wise average (c).

these edges connect two cell phenotypes which differ each other for only one receptor expression: CD27. This confirms observations by linear models in [4], where however relationship between X_2, X_8 was present until higher age (23 years).

We notice that tripartition is also a natural segmentation we observe if we look at the multivariate time-series (see Fig. 3 in [2]), having all a peak around one year, and a descendent tail after about 30 years. We presume this granularity keeps track of a sort of macro age-dependent correlation in the three segments, and should be further investigated in its stationarity component. Namely, ARIMA models could be considered to eliminate the non-stationarity [14]. A good point of the tripartition is to find a couple of change points which are maintained, among the others, in partitions with more segments (see table (b) in Fig. 3): 1.0 and 34.8 years. The decline of the immune system however happens inside the interval 34.8–96.0, which in the following we analyze within a finer grain. Due to these observations, we pass to analyze our data at a major granularity, related to the second best values for m, which are 0.27 and 0.28, corresponding to partitions with 2 and 5 segments respectively, visualized in Fig. 4.

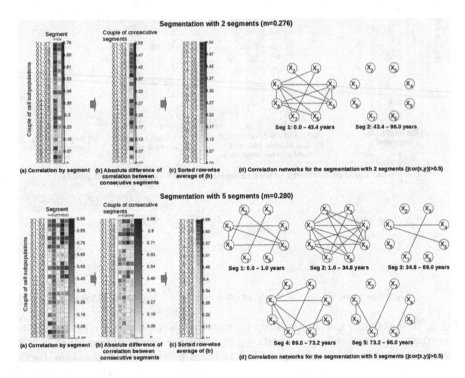

Fig. 4. Bipartition (top) and 5-partition (bottom), corresponding to second best segmentation performance (0.27 and 0.28 respectively). Correlation matrix (a), correlation differences between consecutive segments in absolute value (b), corresponding sorted row average vector (c) and networks with absolute value of pairwise correlations over 0.5 (d).

We may notice that the bipartition has an excessively low granularity, since the network corresponding to the second interval of ages (starting with 43.4 years) has no edges - that is, pairwise correlations have all an absolute value smaller than 0.5. On the other hand, the partition with 5 segments turns out to be quite informative and interesting, by finding new stable change points at 69 and 73.2 (see Table (b) of Fig. 3). Other less significant change points are found by next partitions in 6 and 7 segments, where respectively 78.2 was first added, and then 65.4. Hence, according to this model, the B cell network alterations, observed as the decline of defence in eldery people, should be investigated around specific age ranges, at 69 and 73 (and of course 78) years.

In bottom of Fig. 4, a good performant 5-partition (with performance measure 0.28) is described in terms of age-related B cell networks (d). It is the model we propose in this paper, with same two networks than in the tripartition until 34.8 years, and three last networks suggesting the dynamics of mature adulthood, to the decline of old age. In our data driven correlation networks, B cell networks change dramatically at 69 years with an increasing of all node degrees (correlations of all variables). Namely, in this age range, we notice an irreversible lost of correlation between X_1-X_5 (cell phenotypes with all three opposite receptors) and an irreversible recover of both connections X_4-X_6 and X_1-X_8, while pairwise correlation between X_3-X_7 (cell phenotyes having all or none of the receptors expressed) is the only one which keeps a value greater than 0.5 until the end. A biomedical validation of this model will be the next step for future work, in order to test and eventually improve it.

4   Conclusion and Ongoing Work

A recent broad interest is focused on the lifetime aging of immune system, in terms of changes of immune mechanisms of an individual during his/her infancy, growing/mature age and senescence. In particular, efficient and fast computational methods are proposed in the machine learning literature (for data clustering, and feature extraction) to infer new knowledge from given data. In this context we are currently considering different types of methodologies for multiple time-series analysis, such as, segmentation [16], change-point detection [1,17,18]. Namely, in this paper a simple algorithm allowed us to preliminary analyze some statistically validated partitions of ages where the B cell networks of immune system have change points of interest. In our correlation data driven model, years 69 and 73 seems to be critical for the decline studied in immunosenescence.

The model proposed in this paper may be naturally extended (by improving the partition algorithm, the selected thresholds, by investigating the intermediate zone, with 8–28 intervals, by heuristics from the literature) and improved, with more sophisticated statistical analysis of our specific dataset. For instance we are currently considering a recent approach, proposed in [12], where subsequence clustering of multivariate time-series is profitably used for discovering repeated patterns in temporal data. Once these patterns have been discovered, the initial dataset can be interpreted as a temporal sequence of only a small number of states (namely clusters or segments). Patterns are defined by Markov Random Field (MRF) characterizing the interactions between different variables in typical subsequences of specific clusters. Based on this graphical representation, a simultaneous segmentation of time-series data may be efficiently realized.

Acknowledgments. Authors would like to thank Antonio Vella (department of pathology and diagnostics, University Hospital of Verona) for providing the dataset used in this work and for interesting discussions on the role of B cells in the immune system.

References

1. Barnett, I., Onnela, J.-P.: Change point detection in correlation networks. Sci. Rep. **6**(18893), 1–11 (2016)
2. Castellini, A., Franco, G., Manca, V., Ortolani, R., Vella, A.: Towards an MP model for B lymphocytes maturation. In: Ibarra, O.H., Kari, L., Kopecki, S. (eds.) UCNC 2014. LNCS, vol. 8553, pp. 80–92. Springer, Cham (2014). https://doi.org/10.1007/978-3-319-08123-6_7
3. Castellini, A., Franco, G., Pagliarini, R.: Data analysis pipeline from laboratory to MP models. Nat. Comput. **10**(1), 55–76 (2011)
4. Castellini, A., Franco, G., Vella, A.: Age-related relationships among peripheral B lymphocyte subpopulations. In: 2017 IEEE Congress of Evolutionary Computation - CEC, pp. 1864–1871 (2017). Springer, Berlin, Germany
5. Castellini, A., Paltrinieri, D., Manca, V.: MP-GeneticSynth: inferring biological network regulations from time series. Bioinformatics **31**(5), 785–787 (2015)
6. Chiu, B., Keogh, E., Lonardi, S.: Probabilistic discovery of time series motifs. In: Proceedings of the 9th ACM SIGKDD International Conference on Knowledge Discovery and Data Mining, KDD 2003, pp. 493–498. ACM (2003)
7. Davey, F.R., Huntington, S.: Age-related variation in lymphocyte subpopulations. Gerontology **23**, 381–389 (1977)
8. Duchêne, F., Garbay, C., Rialle, V.: Learning recurrent behaviors from heterogeneous multivariate time-series. Artif. Intell. Med. **39**(1), 25–47 (2007)
9. Franco, G., Jonoska, N., Osborn, B., Plaas, A.: Knee joint injury and repair modeled by membrane systems. BioSystems **91**(3), 473–488 (2008)
10. Franco, G., Manca, V.: A membrane system for the leukocyte selective recruitment. In: Martín-Vide, C., Mauri, G., Păun, G., Rozenberg, G., Salomaa, A. (eds.) WMC 2003. LNCS, vol. 2933, pp. 181–190. Springer, Heidelberg (2004). https://doi.org/10.1007/978-3-540-24619-0_13
11. Gruver, A.L., Hudson, L.L., Sempowski, G.D.: Immunosenescence of ageing. J. Pathol. **211**(2), 144–156 (2007)
12. Hallac, D., Vare, S., Boyd, S., Leskovec, J.: Toeplitz inverse covariance-based clustering of multivariate time series data. In: Proceedings of the 23rd ACM SIGKDD International Conference on Knowledge Discovery and Data Mining, KDD 2017, pp. 215–223. ACM, New York (2017)
13. Hicks, M.J., Jones, J.F., Minnich, L.L., Wigle, K.A., Thies, A.C., Layton, J.M.: Age-related changes in T- and B-lymphocyte subpopulations in the peripheral blood. Arch. Pathol. Lab. Med. **107**(10), 518–523 (1983)
14. Hyndman, R.J., Athanasopoulos, G.: Forecasting: Principles and Practice, 2nd edn., O Texts (2014)

15. Jerne, N.K.: Towards a network theory of the immune system. Annales d'immunologie **125C**(1–2), 373–389 (1974)
16. Keogh, E., Chu, S., Hart, D., Pazzani, M.: Segmenting time series: a survey and novel approach. In: Data mining in Time Series Databases, pp. 1–22. World Scientific, Singapore (1993)
17. Lavielle, M.: Detection of multiple changes in a sequence of dependent variables. Stochast. Process Appl. **83**(1), 79–102 (1999)
18. Lavielle, M., Teyssière, G.: Detection of multiple change-points in multivariate time series. Lith. Math. J. **46**(3), 287–306 (2006)
19. Lin, J., Keogh, E., Lonardi, S., Patel, P.: Finding Motifs in time series. In: Proceedings of the Second Workshop on Temporal Data Mining, pp. 52–68. ACM (2002)
20. Manca, V., Castellini, A., Franco, G., Marchetti, L., Pagliarini, R.: Metabolic P systems: a discrete model for biological dynamics. Chin. J. Electron. **22**(4), 717–723 (2013)
21. Menshikov, I., Beduleva, L., Frolov, M., Abisheva, N., Khramova, T., Stolyarova, E., Fomina, K.: The idiotypic network in the regulation of autoimmunity: theoretical and experimental studies. J. Theor. Biol. **21**(375), 32–9 (2015)
22. Perelson, A.S.: Immune network theory. Immunol. Rev. **110**, 5–33 (1989)
23. Simon, A.K., Hollander, G.A., McMichael, A.: Evolution of the immune system in humans from infancy to old age. Proc. R. Soc. B Biol. Sci. **282**, 20143085 (2015)
24. Terzi, E., Tsaparas, P.: Efficient algorithms for sequence segmentation. In: Proceedings of the 2006 SIAM International Conference on Data Mining, pp. 316–327. SIAM (2006)
25. Vahdatpour, A., Amini, N., Sarrafzadeh, M.: Toward unsupervised activity discovery using multi-dimensional motif detection in time series. In: Proceedings of the 21st International Joint Conference on Artifical Intelligence, IJCAI 2009, pp. 1261–1266 (2009)
26. Veneri, D., Ortolani, R., Franchini, M., Tridente, G., Pizzolo, G., Vella, A.: Expression of CD27 and CD23 on peripheral blood B lymphocytes in humans of different ages. Blood Transfus. **7**, 29–34 (2009)
27. Weiskopf, D., Weinberger, B., Grubeck-Loebenstein, B.: The aging of the immune system. Transpl. Int. **22**, 1041–1050 (2009)

Author Index

Printed in the United States
By Bookmasters